T0318635

# Viral Proteases
# and Their Inhibitors

# Viral Proteases and Their Inhibitors

Edited by

**Satya P. Gupta**
Department of Applied Sciences,
National Institute of Technical Teachers' Training and Research
Bhopal, India

ACADEMIC PRESS
An imprint of Elsevier

Academic Press is an imprint of Elsevier
125 London Wall, London EC2Y 5AS, United Kingdom
525 B Street, Suite 1800, San Diego, CA 92101-4495, United States
50 Hampshire Street, 5th Floor, Cambridge, MA 02139, United States
The Boulevard, Langford Lane, Kidlington, Oxford OX5 1GB, United Kingdom

**Notices**
Knowledge and best practice in this field are constantly changing. As new research and experience
broaden our understanding, changes in research methods, professional practices, or medical treatment may become necessary.

Practitioners and researchers must always rely on their own experience and knowledge in evaluating
and using any information, methods, compounds, or experiments described herein. In using such
information or methods they should be mindful of their own safety and the safety of others, including parties for whom they have a professional responsibility.

To the fullest extent of the law, neither the Publisher nor the authors, contributors, or editors, assume
any liability for any injury and/or damage to persons or property as a matter of products liability,
negligence or otherwise, or from any use or operation of any methods, products, instructions, or
ideas contained in the material herein.

**Library of Congress Cataloging-in-Publication Data**
A catalog record for this book is available from the Library of Congress

**British Library Cataloguing-in-Publication Data**
A catalogue record for this book is available from the British Library

ISBN: 978-0-12-809712-0

For information on all Academic Press publications visit our website at
https://www.elsevier.com/books-and-journals

Working together
to grow libraries in
developing countries

www.elsevier.com • www.bookaid.org

*Publisher:* Sara Tenney
*Acquisition Editor:* Linda Versteeg-Buschman
*Editorial Project Manager:* Pat Gonzalez
*Production Project Manager:* Lucía Pérez
*Designer:* Mark Rogers

Typeset by Thomson Digital

# Contents

# Contributors

**Nilanjan Adhikari**, Jadavpur University, Kolkata, West Bengal, India

**Megha Aggarwal**, Indian Institute of Technology, Roorkee, Uttarakhand, India

**Sandip K. Baidya**, Jadavpur University, Kolkata, West Bengal, India

**Sudhanshu S. Behera**, DFARD, Bhubaneswar, Odisha, India

**Akalabya Bissoyi**, National Institute of Technology, Raipur, Chhattisgarh, India

**Arindam Bit**, National Institute of Technology, Raipur, Chhattisgarh, India

**Debasis Das**, Sree Chaitanya College, Habra, West Bengal, India

**Sonal Dubey**, Krupanidhi College of Pharmacy, Bengaluru, Karnataka, India

**Hanan Elhaes**, Ain Shams University, Cairo, Egypt

**Benazir Fatma**, Indian Institute of Technology, Roorkee, Uttarakhand, India

**Satya P. Gupta**, National Institute of Technical Teachers' Training and Research, Bhopal, Madhya Pradesh, India

**Dimitra Hadjipavlou-Litina**, Aristotle University of Thessaloniki, Thessaloniki, Greece

**Medhat Ibrahim**, National Research Center, Dokki, Giza, Egypt

**Tarun Jha**, Jadavpur University, Kolkata, West Bengal, India

**Rita Kakkar**, University of Delhi, Delhi, India

**Kriti Kashyap**, University of Delhi, Delhi, India

**Hongmin Li**, State University of New York, Albany, NY, United States

**Zhong Li**, Wadsworth Center, New York State Department of Health, Albany, NY, United States

**Rajat Mudgal**, Indian Institute of Technology, Roorkee, Uttarakhand, India

**Dolly A. Parasrampuria**, Global Clinical Pharmacology, Quantitative Sciences, Janssen Research & Development, Philadelphia, PA, United States

**Ashish Patel**, National Institute of Technology, Raipur, Chhattisgarh, India

**Vaishali M. Patil**, Kharvel Subharti College of Pharmacy, Swami Vivekanand Subharti University, Meerut, Uttar Pradesh, India

**Subrat K. Pattanayak**, National Institute of Technology, Raipur, Chhattisgarh, India

**Yenamandra S. Prabhakar**, Central Drug Research Institute, CSIR, Lucknow, Uttar Pradesh, India

**Achintya Saha**, University of Calcutta, Kolkata, West Bengal, India

**Noha A. Saleh**, Cairo University, Giza, Egypt

**Debabrata Satpathy**, National Institute of Technology, Raipur, Chhattisgarh, India

**Basheerulla Shaik**, National Institute of Technical Teachers' Training and Research, Bhopal, Madhya Pradesh, India

**Anjana Sharma**, Meerut Institute of Engineering and Technology, Meerut, Uttar Pradesh, India

**Abhishek K. Singh**, University of Allahabad, Allahabad, Uttar Pradesh, India

**Shailly Tomar**, Indian Institute of Technology, Roorkee, Uttarakhand, India

**Jing Zhang**, Wadsworth Center, New York State Department of Health, Albany, NY, United States

# Preface

Viruses are major pathogenic agents that can cause a variety of fatal diseases, such as AIDS, hepatitis, respiratory diseases, and many more in humans, plants, and animals. The most prominent of them have been adenoviruses, alphaviruses, flaviviruses, hepatitis C virus (HCV), herpesviruses, human immunodeficiency virus of type 1 (HIV-1), and picornaviruses. This book comprises chapters discussing in detail the structures and functions of these viruses and their proteases. Proteases in viruses play crucial roles in their replication, and thus represent very attractive targets for the design and development of potent antiviral agents. The book contains 14 chapters, each written by highly acclaimed authors. Chapter 1, entitled *Fundamentals of Viruses and Their Proteases*, written by Sharma and Gupta, presents an introductory remark on the structure and functions of various viruses and their proteases. Chapter 2, *Design and Development of Some Viral Protease Inhibitors by QSAR and Molecular Modeling Studies* by Saleh et al., presents some QSAR and molecular modeling studies on the inhibitors of several viral proteases. This chapter provides an idea as to how one can rationalize the drug discovery. Gupta et al. in Chapter 3, *Advances in Studies on Adenovirus Proteases and Their Inhibitors*, have presented in detail the structure and function of adenovirus proteases and the various studies on the design and development of their potent inhibitors. Alphaviruses, such as Chikungunya virus and others, have been widely known to cause fever, rash, and rheumatic diseases. These diseases are considered as neglected tropical diseases for which there are no current antiviral therapies or vaccines available. Therefore, in Chapter 4, Alphavirus *Nonstructural Proteases and Their Inhibitors*, Bissoyi et al. have described the techniques available to prevent *Alphavirus* infection and treat *Alphavirus*-associated malignancies. In addition, they have also discussed the recent outcomes in the fields of synthetic and natural medicinal chemistry research that were solely aimed toward fighting *Alphavirus* infection. For the design and development of drugs against any virus based on its proteases, it is essential that one should know the structure and function of that protease. Therefore Chapter 5, which has been written by Tomar and Aggarwal, entitled *Structure and Function of* Alphavirus *Protease*, describes the X-ray crystallographic structures of catalytically active capsid protease and nsP2 protease from different alphaviruses, and considers the proteolytic activity of both these proteases for the development of potent drugs against alphaviruses. In Chapter 6, entitled *Flavivirus Protease: An Antiviral Target*, Tomar et al. have vividly presented the role of the protease in *Flavivirus* and its inhibitors. Flaviviral-specific NS3 protease, in association with its cofactor NS2B, is the key enzyme responsible for the proteolytic processing of *Flavivirus* polyprotein, and thus a detail on structural and functional aspects of NS2B–NS3 *Flavivirus* protease is important for the development of antiflaviviral agents. Further, a detailed survey on the flaviviral protease; recent drug development studies targeting the NS3 active site, as well as studies targeting an NS2B/NS3 interaction site determined from flavivirus protease; and crystal structures have been presented in Chapter 7, entitled Flavivirus *NS2B/NS3 Protease: Structure, Function, and Inhibition*.

HCV infection is a major health burden worldwide, with an estimated 170–200 million chronically infected individuals. HCV NS5B, the RNA-dependent RNA polymerase of HCV, catalyzes the HCV RNA replication, and it is an attractive validated target for the development of antiviral therapy for HCV. Therefore, in Chapter 8 entitled *Design and Development of HCV NS5B Polymerase Inhibitors*, Das has presented details of design and development of HCV NS5B polymerase inhibitors. The major structural proteins and enzymes essential for replication of retroviruses, including HIV, are high–molecular weight polyproteins. HIV-1 encodes a protease that is essential for the production of the infectious virus. Thus Chapter 9 *Studies on HIV-1 Protease and its Inhibitors* written by Dubey presents a detailed account of the structure and functions of HIV-1 protease and the development of its inhibitors. Patil and Gupta have delved into picornaviral proteases and their inhibitors in Chapter 10, entitled *Studies on Picornaviral Proteases and Their Inhibitors*, to provide guidelines toward the development of potential irreversible and noncovalent inhibitors of 3C proteinases (3C$^{pro}$) as antiviral agents. 3C proteinases have a significant role in the replication of picornaviral proteases. Severe acute respiratory syndrome (SARS), caused by SARS *Coronavirus* (SARS-CoV), is a dreadful infection worldwide, having economic and medical importance and is a global threat for health. Therefore, Adhikari et al. have discussed the development of anti-SARS and anti-HRV drugs based on QSAR techniques in Chapter 11, entitled as *Structural Insight Into the Viral 3C-Like Protease Inhibitors: Comparative SAR/QSAR Approaches*. Chapter 12, written by Kashyap and Kakkar on *Herpesvirus Proteases: Structure, Function, and Inhibition*, describes the progress made on antiherpetic drugs targeting the herpesvirus proteases. Herpesviruses are involved in the development of atherosclerosis, restenosis after coronary angioplasty, accelerated atherosclerosis in recipients of heart transplants, and the induction of a prothrombotic phenotype in vascular endothelial cells. Therefore the study on drugs against herpesviruses is a great need of the time. Herpesvirus proteases have been found to be excellent targets for drug development and therefore the structure, function, and inhibition of these proteases have been presented in great details in this chapter. To develop any kind of drugs, the SAR and QSAR of drugs provide a great help. Therefore, in Chapter 13, Hadjipavlou-Litina and Gupta have presented a detailed account of antiherpetic drugs developed based on SAR and QSAR. In the last chapter, Chapter 14 entitled *Delicate Dance in Treating Hepatitis C Infections and Overcoming Resistance*, Parasrampuria has discussed about HCV, diseases related to it, and the status of its inhibition. The chapter covers the life cycle of HCV, structures of drug target proteins in it, and the development of drugs based on important targets. Thus, all the chapters of this book, which are written by the experts in the fields, are quite interesting and useful for further development of antiviral drugs. I have enjoyed reading all the chapters and hope the readers will do too. I thank all the authors for contributing these excellent chapters for this book.

**Satya P. Gupta**
National Institute of Technical Teachers' Training and Research,
Bhopal, Madhya Pradesh, India

Chapter 1

# Fundamentals of Viruses and Their Proteases

Anjana Sharma*, Satya P. Gupta**
*Meerut Institute of Engineering and Technology, Meerut, Uttar Pradesh, India; **National Institute of Technical Teachers' Training and Research, Bhopal, Madhya Pradesh, India

## 1  INTRODUCTION

Viruses are small obligate intracellular parasites that may contain either an RNA or DNA genome encapsulated by a protective virus-coded protein coat. Both RNA and DNA viruses infect host organisms by entering and replicating in host cells. A virus makes multiple copies of itself in one cell, releases these copies to infect new host cells, and makes more copies. This process ultimately results in disease progression. Virus particles containing the viral genome are packaged in a protein coat called the capsid. This protein coat is surrounded by a lipid bilayer consisting of viral proteins that assist the virus in binding to the host cell membranes. The viral envelope plays a vital role in viral infection, including virus attachment to specific receptors cells. Then the penetration of enveloped viruses through fusion of the viral envelope with the host cell membrane takes place, which may or may not involve receptor-mediated endocytosis, while nonenveloped viruses penetrate through receptor-mediated endocytosis, releasing viral content into the cells and packaging newly formed viral particles. This process involves bringing together newly formed genomic nucleic acid and structural proteins to form the nucleocapsid of the virus.

Viral proteases reveal many new strategies for proteolysis, in addition to cleaving a peptide bond. Proteases have been identified in a wide range of viruses, without any correlation to capsid complexity, lipid envelope, or the nature of their genomes (Kräusslich and Wimmer, 1998). They are found to be present in both enveloped and nonenveloped viruses. The proteases present in various viruses specifically belong to the family of serine proteases as in hepatitis C virus (HCV), *Flavivirus*, and herpesviruses; the family of cysteine proteases as in adenoviruses (AdVs); or the family of aspartyl protease as in human immunodeficiency virus of type 1 (HIV1) (Marcin and Marcin, 2013). The present chapter presents introductory remarks to all kinds of viral proteases that may act simply as an appetizer to the readers, discussing in detail about the different

Viral Proteases and Their Inhibitors. http://dx.doi.org/10.1016/B978-0-12-809712-0.00001-0

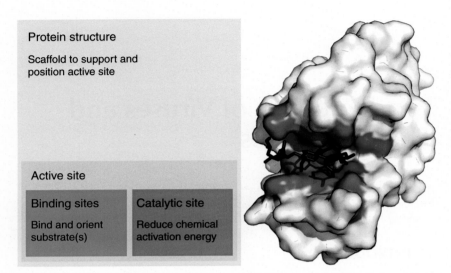

**FIGURE 1.1**  A representation of a protein structure showing its active site where the binding site is shown in *blue*, catalytic site in *red*, and the substrate (peptidoglycan) in *black*. *(Image by Shafee, T. Available from Wikimedia Commons: https://commons.wikimedia.org/w/index.php?curid=45801894)*

viruses, their structures, functions, and inhibition of their proteases. This chapter first presents the biochemistry of a protease and then introduces briefly about different viruses and their proteases.

## 2  BIOCHEMISTRY OF PROTEASES

A protease, also called a peptidase or proteinase, is the enzyme that performs proteolysis, that is, protein catabolism by hydrolysis of peptide bonds. Different classes of protease can perform the same reaction by completely different catalytic mechanisms. For catalytic action, proteases or enzymes possess an active site that consists of a binding site and a catalytic site as shown in Fig. 1.1, both constituting residues. Residues in the binding site form temporary bonds with the substrate and those in the catalytic site catalyze the reaction. The active site in the enzyme is usually a groove or pocket located in a deep tunnel within the enzyme. The residues of the catalytic site are typically very close to the binding site, and some residues can have dual roles, in both binding and catalysis. Once the substrate is bound and oriented in the active site, catalysis can begin.

Usually proteases possess a catalytic triad. As shown in Fig. 1.2, a catalytic triad refers to a group of three amino acid residues that function together at the center of the active site. An acid–base–nucleophile triad is a common motif for generating a nucleophilic residue (usually serine or cysteine amino acid, but occasionally threonine) for covalent catalysis. In three-dimensional structures of the enzyme, the residues of a catalytic triad can be far from each other along the

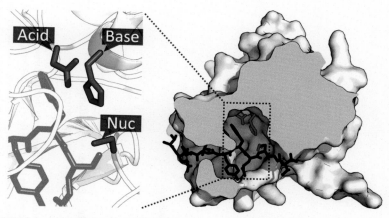

**FIGURE 1.2** **Example of a protease [tobacco etch virus (TEV) protease, TEV] with a catalytic triad consisting of an aspartate (acid), histidine (base), and serine (nucleophile).** The substrate *(black)* is bound by the binding site to orient it next to the triad. *(Image by Shafee, T. Available from Wikimedia Commons: https://commons.wikimedia.org/w/index.php?curid=28958186)*

amino acid sequence in the primary structure. Some examples of catalytic triads are Ser-His-Asp, Cys-His-Asp, Ser-His-His, Ser-Glu-Asp, and Ser-*cis*Ser-Lys. Enzymes that contain a catalytic triad use it for one of two reaction types: either to split a substrate (hydrolases) or to transfer one portion of a substrate to a second substrate (transferases). Catalytic triads perform covalent catalysis using a residue as a nucleophile.

Proteases can be found in all members of the Animalia, Plantae, Fungi, Bacteria, Archaea, and Virus kingdoms. All proteases can be grouped on the basis of their catalytic residue, and thus they are classified into seven groups (Oda, 2012), namely, serine proteases (serine alcohol), cysteine proteases (cysteine thiol), threonine proteases (threonine secondary alcohol), aspartic proteases (aspartic carboxylic acid), glutamic proteases (glutamic carboxylic acid), metalloproteases (usually zinc), and asparagine peptide lyases (asparagine), where terms within parentheses refer to their catalytic residue. In the enzymes, the residues of the catalytic site are typically very close to the binding site, and some residues can have dual roles in both binding and catalysis.

Proteases are usually involved in breaking long protein chains into shorter fragments by splitting the peptide bonds that link amino acid residues. Some of them that detach the terminal amino acids from the protein chain, such as aminopeptidases and carboxypeptidase A, are called exopeptidases, and some that attack internal peptide bonds of a protein, such as trypsin, chymotrypsin, pepsin, papain, and elastase, are called endopeptidases. The catalytic mechanism of these proteases involves either the activation of a water molecule, which acts as a nucleophile to hydrolyze the peptide bond (Fig. 1.3, one-step catalysis), or the use of a nucleophilic residue in a catalytic triad to perform

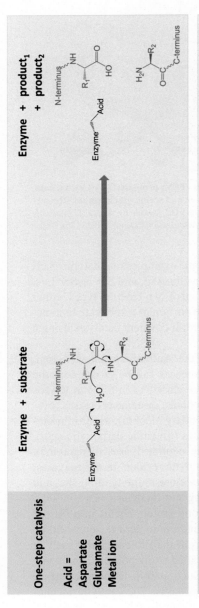
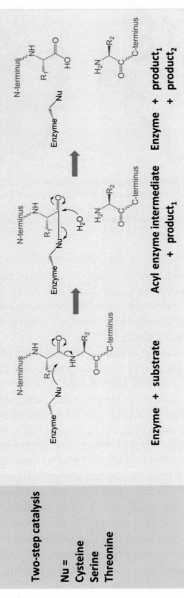

FIGURE 1.3  **The two catalytic mechanisms adopted by proteases: one-step catalysis, showing the activation of a water molecule that performs a nucleophilic attack on the peptide bond to hydrolyze it, and a two-step catalysis, showing the use of a nucleophilic residue in a catalytic triad to perform a nucleophilic attack that covalently links the protease to the substrate protein, releasing the first half of the product.** This covalent acyl enzyme intermediate is then hydrolyzed by activated water to complete the catalysis by releasing the second half of the product and regenerating the free enzyme. Enzyme is shown in *black*, substrate protein in *red*, and water in *blue*. *(Image by Shafee, T. Available from Wikimedia Commons: https://commons.wikimedia.org/w/index.php?curid=42551231)*

a nucleophilic attack that covalently links the protease to the substrate protein, releasing the first half of the product. This covalent acyl enzyme intermediate is then hydrolyzed by activated water to complete the catalysis through the release of the second half of the product and regeneration of the free enzyme (Fig. 1.3, two-step catalysis). Proteases that adopt the first mechanism are aspartic, glutamic, and metalloproteases and those that adopt second mechanism are serine, threonine, and cysteine proteases.

Regarding the specificity of proteases, it is said that proteolysis is highly promiscuous, such that a wide range of protein substrates are hydrolyzed. However, some proteases are highly specific and cleave only the substrate with a certain sequence, for example, tobacco etch virus (TEV) protease, which is specific for the sequence: ENLYFQ/S (where / = cleavage site).

## 3 VIRUSES

Viruses can be categorized into two major categories: enveloped and nonenveloped. Further classification is based upon the method of viral mRNA synthesis, that is, the Baltimore classification developed by Baltimore (1971), depending on their type of genome [DNA, RNA, single stranded (ss), or double stranded (ds), etc.] and their method of replication.

### 3.1 Enveloped/Nonenveloped Viruses

Enveloped viruses have a coat of lipids and proteins over which there are projection of spikes. They are able to induce antibody- and cell-mediated immunity. They are less virulent and are often released by budding and rarely cause cell lysis. They are highly sensitive to heat and light, while infectivity gets lost upon drying. Generally, they cannot survive inside the gastrointestinal tract. Their transmission is through blood, secretions, or organ transplantation. On the other hand, nonenveloped viruses have only a capsid protein coat and are more virulent. They have more lytic actions as compared to enveloped viruses, survive longer in host cells, and are more resistant to harsh environmental conditions, and thus survive in the gastrointestinal tract for a long time. Their mode of transmission is via fomites, dust, or intake of contaminated fecal or oral matter. They induce antibodies and can retain their infectivity even after drying. Some of the examples of nonenveloped viruses are *Rotavirus*, hepatitis A virus (HAV), adenovirus (AdV),, etc., and that of enveloped viruses are *Ebolavirus*, HIV, hepatitis B virus, Togavirus, *Flavivirus*, influenza virus, etc.

### 3.2 DNA and RNA Viruses

A DNA virus is a virus that has DNA as its genetic material and replicates using a DNA-dependent DNA polymerase. They can be either dsDNA or ssDNA viruses. ssDNA is usually expanded to double stranded in infected cells. These two classes of DNA viruses are also called as Group I and Group II viruses, respectively. Group I (dsDNA) viruses are highly dependent on the cell cycle of host cell. Before they are able to replicate, they must enter the host cell, where they forcefully induce the cell to undergo cell division. They require host cell polymerases to replicate the viral genome. Examples of Group I viruses are Herpesviridae, Adenoviridae, and Papoviridae. The examples of Group II (ss-DNA) viruses are Anelloviridae, Circoviridae, and Parvoviridae, which infect vertebrates; Geminiviridae and Nanoviridae, which infect plants; and the Microviridae, which infects prokaryotes. Most of Group II viruses contain circular genomes, except paroviruses.

RNA viruses are the viruses that have RNA as their genetic material. This nucleic acid is usually ssRNA, but may also be dsRNA. The example of RNA viruses are *Influenzavirus*, HCV, severe acute respiratory syndrome (SARS) *Coronavirus*, and polio and measles viruses. RNA viruses also are dsRNA or ssRNA viruses, which are put in Group III and Group IV or Group V, respectively, according to the Baltimore classification system. Viruses with RNA as their genetic material, but that include DNA intermediates in their replication cycle, are called retroviruses and comprise Group VI of the Baltimore classification. Notable human retroviruses include HIV1 and HIV2 that cause AIDS. Unlike DNA viruses, dsRNA viruses do not require host replication polymerases, rather they replicate in the core "capsid" that is in the cytoplasm. The ssRNA viruses are grouped into positive sense or negative sense according to sense of polarity of RNA. The replication of these viruses occurs in the cytoplasm and is not dependent on the cell cycle, such as that of DNA viruses. Well-known examples of positive-sense ssRNA viruses include picornaviruses [human rhinoviruses and foot-and-mouth disease virus (FMDV)], togaviruses (alphaviruses), and flaviviruses [HCV, West Nile virus (WNV), and yellow fever virus (YFV)]. The examples of negative-sense ssRNA viruses include viruses, such as measles virus, mumps virus, rabies virus, etc.

## 4 VIRUSES CONTAINING PROTEASES

The most important viruses that contain proteases and whose proteases have been exploited to design drugs of high therapeutic values have been mainly alphaviruses, HCV, HIV, picornaviruses, herpesviruses, AdVs, and flaviviruses. We are presenting here a brief discussion of these viruses and their proteases.

### 4.1 Adenoviruses

AdVs are common pathogens of humans, other mammals, and birds. They were first isolated in 1950s while studying the growth of poliovirus in the adenoidal tissue, hence their name (Rowe et al., 1953). It belongs to the family of Adenoviridae, which includes viruses affecting other mammals and birds, and is divided into five genera: *Mastadenovirus*, *Aviadenovirus*, *Atadenovirus*, *Siadenovirus*, and *Ichtoadenovirus* (Martin et al., 2007).

Besides adenoids, they can also infect other sites, such as the respiratory tract, eye, as well as gut, causing acute respiratory disease, pneumonia, gastroenteritis, and other disorders in humans. AdVs are icosahedral, nonenveloped viruses with a diameter of about 1000 Å, and a viral genome containing about 36,000 base pairs (Horwitz, 2001). The AdV genome is linear, nonsegmented dsDNA, which is approximately 30–38 kbp that allows the virus to carry 30–40 genes theoretically. In spite of its larger structure, in the Baltimore classification it is still a simple virus that largely relies on the host cell for survival and replication. The viral genome has an interesting feature that it has a terminal 55-kDa protein associated with each of the 5′ ends of the linear dsDNA, which act as primers in DNA replication and ensure that the ends of virus linear genome are adequately replicated. They are highly diversified. In humans there are 51 immunologically distinct human AdV serotypes in 6 species (A–F). Different serotypes are associated with various conditions, such as:

- Respiratory disease: mainly species human adenovirus (HAdV) B and C
- Conjunctivitis (HAdV B and D)
- Gastroenteritis (HAdV F serotypes 40 and 41)

Segerman et al. (2003) subdivided the species B into B1 and B2. There could be some correlation between the species and their tissue tropism and clinical properties. Thus the species B1, C, and E are responsible for causing respiratory disease, whereas species B, D, and E can induce ocular disease. Russell (2005) described that species F is the main cause for gastroenteritis, and B2 viruses infect the kidneys and urinary tract. Entry of AdVs into the host cell involves two types of interactions. One of the interactions for HAdV serotypes is via CD46 type of receptor, while another one is by coxsackievirus adenovirus receptor (CAR) serotypes. It was also found that the major histocampatibility complex (MHC) molecules and sialic acid residues also function in the same way. This is followed by secondary interactions, where the penton base protein interacts with an integrin molecule. This is the coreceptor that stimulates the entry of AdV-stimulated cell signaling, that is, entry of virions into the host cell (Wu and Nemerow, 2004).

### 4.1.1 Adenovirus Protease

AdV protease is a cysteine protease. Its structure consists of a central mixed five-stranded β-sheet that is surrounded by helices on both sides. Despite of a unique fold, the arrangement of the catalytic residues of AdV protease is same as that of papain, including the catalytic triad (Cys122-His54-Glu71), as well as the oxyanion hole (Gln115) of the protease. The overall organization of structure is almost similar to that of papain and Picornavirus leader protease, while some positions of structural features are different in the primary sequences of the proteases. The structural conservation of active site suggests that AdV protease is likely to have the same catalytic mechanism as papain (Ding et al., 1996).

## 4.2 Alphaviruses

Alphaviruses are mosquitoborne small, enveloped viruses belonging to the family of Togaviridae. Strauss and Strauss (1994) classified the virus depending upon the geographical origin: as New World alphaviruses or Old World alphaviruses. Old World alphaviruses, such as Chikungunya virus (CHIKV), O'Nyong–Nyong virus (ONNV), Ross River virus (RRV), Semliki Forest virus (SFV), and Sindbis virus (SINV), cause a fever, rash, and arthralgia syndrome, while New World alphaviruses, such as Eastern equine encephalitis virus (EEEV), Venezuelan equine encephalitis virus (VEEV), and Western equine encephalitis virus (WEEV), typically cause encephalitis in humans and other mammals (Ryman and Klimstra, 2008).

Alphaviruses are icosahedral in shape, small, enveloped, and about 70 nm in diameter (Fuller, 1987; Mancini et al., 2000; Morgan et al., 1961). It contains positive-sense ssRNA genome approximately 14 kb in length (Simmons and Strauss, 1972; Strauss et al., 1984). Its subgenomic 26S RNA encodes the structural proteins, while nonstructural proteins are translated from genomic RNA (Cancedda and Shatkin, 1979; Simmons and Strauss, 1972; Strauss et al., 1984). The two polyproteins are cleaved posttranslationally by viral cysteine and host proteases. The structural proteins (C, E3, E2, 6K, and E1) and their cleavage intermediates are involved in budding and viral encapsidation, while RNA replication takes place by nonstructural proteins (nsP1–4) and their cleaved intermediates also (Hardy and Strauss, 1989; Melançon and Garoff, 1987; Strauss et al., 1984). After entry, the *Alphavirus* particles undergo disassembly and release genomic RNA into the cytoplasm of infected cells. From the viral genome, translation of nonstructural (P1234) and structural polyproteins takes place from open reading frames (Glanville et al., 1976). There is a *cis*-cleavage between nsP3 and nsP4 to yield P123 and nsP4, which forms an unstable replication complex (RC) and synthesizes negative-strand RNA. The nsP1 and P23 are *trans*-cleaved products of polyprotein, so nsP1, P23, and nsP4 form an RC within virus-induced cytopathic vacuoles that are active in negative-strand synthesis, as well as genomic RNA synthesis, but not in subgenomic RNA synthesis (Dé et al., 1996; Froshauer et al., 1988; Kujala et al., 2001; LaStarza et al., 1994; Salonen et al., 2003). The negative-strand synthesis is inactivated after complete cleavage to nsP1, nsP2, nsP3, and nsP4, and the synthesis of positive-strand genomic RNA is switched on by the stable RC. The *Alphavirus* nsP2 protein is a multifunctional enzyme with the N-terminus, having RNA helicase, nucleoside triphosphatase (NTPase), and RNA-dependent 5′ triphosphatase (RTPase) activities (de Cedrón et al., 1999; Karpe et al., 2011; Pastorino et al., 2008), while its C-terminus contains the protease domain (Merits et al., 2001; Strauss et al., 1992). The nsP2 is considered to be an essential protein for viral replication and propagation due to its proteolytic activities (Leung et al., 2011). The antiviral gene expression is also inhibited after its translocation into the nucleus. The proteolytic activity of nsP2 has been characterized in alphaviruses to be similar to that of papain-like cysteine protease, with a cysteine–histidine catalytic dyad in the active site.

## 4.3 Flaviviruses

Flaviviruses, belonging to the family of Flaviviridae, are a group of nearly 70 enveloped RNA viruses, which cause life-threatening diseases in both humans and animals. Mostly they are arthropodborne viruses and are transmitted to vertebrates via mosquito or tick bites (Gubler et al., 2007). The various members of this genus are Dengue virus (DENV), YFV, WNV, Japanese encephalitis virus (JEV), and St. Louis encephalitis virus (SLEV), all highly pathogenic to humans. According to WHO, DENV (serotypes DENV1–4) causes Dengue fever, which may progress to life-threatening shock syndrome and Dengue hemorrhagic fever. More than 200,000 cases suffering from of YFV and 50,000 cases suffering from Japanese encephalitis virus are reported worldwide every year. Flaviviruses are enveloped viruses, having positive-sense ssRNA genome. They are introduced into the host cell by mosquito or tick during its blood meal. They enter into target cell via receptor-mediated endocytosis. The released RNA encodes a polyprotein precursor.

The *Flavivirus* genome is approximately 11 kb with a 5′ cap structure, but devoid of 3′ polyadenylation tail (Lindenbach et al., 2007). Both 5′ and 3′ ends play important roles in virus replication, viral protein translation, and virion assembly (Alvarez et al., 2006; Filomatori et al., 2006; Khromykh et al., 2003; Markoff, 2003; Polacek et al., 2009; Villordo and Gamarnik, 2008; Yu et al., 2008). The released RNA encodes a polyprotein precursor of approximately 3400 amino acids. This polyprotein is processed by host cell signalases and the viral NS2B–NS3 protease (NS2B–NS3pro) into three structural proteins, capsid (C), envelope (E), and membrane protein (M), and seven nonstructural proteins, NS1, NS2A, NS2B, NS3, NS4A NS4B, and NS5. The C-terminal region of NS3 contains an NTPase, a 5′ terminal RTPase, and an RNA helicase (Xu et al., 2005). NS5 possesses an RNA methyltransferase (Dong et al., 2012, 2014; Egloff et al., 2002; Ray et al., 2006) and RNA-dependent RNA polymerase activities (Malet et al., 2008; Yap et al., 2007). Together with the viral RNA, viral cofactors, and host cell cofactors, NS3 and NS5 form the virus RC assemble on the intracellular membrane to amplify the viral genome (Murray et al., 2008; Paul and Bartenschlager, 2013). Therefore, in principle, functional inhibition of the viral NS proteins and/or disruption of the RC underlie target-based anti-*Flavivirus* drug development (Bollati et al., 2010; Lim et al., 2013; Noble and Shi, 2012; Sampath and Padmanabhan, 2009). *Flavivirus* NS3 (69 kDa) is the second largest viral protein after NS5 in the *Flavivirus* genome, and plays several essential roles in the viral life cycle. NS3 has an N-terminal protease chymotrypsin-like domain that cleaves the viral polyprotein precursor to release individual NS proteins, and a C-terminal NTPase-dependent

RNA helicase [with an helicase superfamily 2 (SF2)–like fold] involved in genome replication and viral RNA synthesis (Lescar et al., 2008; Luo et al., 2010).

### 4.3.1    NS2B–NS3 Protease

The N-terminal domain of NS3 is a chymotrypsin-like serine protease that cleaves the viral polyprotein both in *cis*- and *trans*-forms (Chambers et al., 1990a,b; Li et al., 2005a,b). To function as an active enzyme, the NS3 protease requires the NS2B cofactor (Falgout et al., 1991; Jan et al., 1995; Yusof et al., 2000; Zhang et al., 1992). NS2B is an integral membrane protein of 14 kDa that contains three domains: two transmembrane segments located at the N- and C-termini and a central region of 47 amino acids (spanning amino acids 49–96) that acts as an essential protein cofactor of the NS3 protease (Clum et al., 1997). The *Flavivirus* NS3 protein is neither soluble nor catalytically active as a protease in vitro, suggesting that it does not fold properly without the NS2B protein, that must be either provided in *cis*- (Xu et al., 2005) or in *trans*-form (Kim et al., 2013; Phong et al., 2011; Wu et al., 2003).

### 4.3.2    NS3 NTPase/RNA Helicase (NS3hel)

The C-terminal domain of the NS3 protein (aa 180–618) belongs to the SF2 family (Fairman-Williams et al., 2010; Gorbalenya and Koonin, 1993). The overall structure can be broken up into three subdomains. Subdomains 1 and 2 adopt the RecA-like fold (Rao and Rossmann, 1973; Story and Steitz, 1992) and contain eight conserved motifs essential for RNA binding, ATP hydrolysis, and communication between both binding sites (Fairman-Williams et al., 2010; Gorbalenya and Koonin, 1993; Pyle, 2008). The third subdomain forms the ss-RNA binding tunnel. There is also an evidence suggesting that subdomain 3 mediates the interaction between NS3 and NS5, and hence disrupting this interaction could constitute a strategy for the design of antiviral compounds (Brooks et al., 2002; Fang et al., 2013; Tay et al., 2015). NS3 also has RTPase activity, which shares the same active site for ATP binding and hydrolysis (Wang et al., 2009). RNA 5' triphosphate hydrolysis is the first step for viral RNA capping (Decroly et al., 2012). Viruses carrying a defective or impaired NS3 helicase gene cannot replicate properly, indicating an essential role for NS3 helicase/RTPase activity in virus replication (Matusan et al., 2001).

## 4.4    Hepatitis C Virus

HCV belongs to Flaviviridae family, which also includes other viruses, for example, St. Louis encephalitis virus and WNV, responsible for severe human diseases, such as yellow and Dengue fevers. HCV is an enveloped virus having a 9.6-kb plus-strand RNA genome that encodes a long polyprotein precursor of ~3000 amino acids (Rosenberg, 2001), which is proteolytically processed by viral and cellular proteases to produce structural (nucleocapsid, E1, and

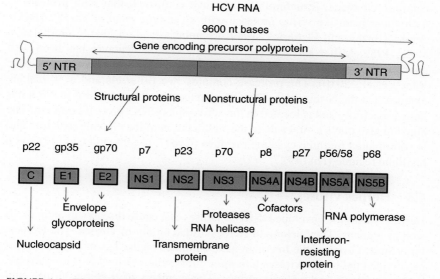

FIGURE 1.4 **Diagrammatic representations of structural and nonstructural proteins in a hepatitis C virus (HCV) RNA genome.** *NTR,* Nontranslated region. *(Image by Colm, G. Available from Wikimedia Commons: https://commons.wikimedia.org/wiki/File:HCV_genome.png)*

E2) (Santolini et al., 1994) and nonstructural proteins (NS1, NS2, NS3, NS4A, NS4B, NS5A, and NS5B) (Fig. 1.4). The structural glycosylated proteins, E1 and E2, act as the ligands for cellular receptors. For HCV entry, human CD81 receptors are required, which bind directly with E2 protein of HCV. This CD81 receptor is a widely distributed cell surface tetraspanin that participates in the formation of molecular complexes on various cell types, such as hepatocytes, natural killer cells, B lymphocytes, and T lymphocytes. Thus HCV not only exploits hepatocytes, but also modulates the host immune response. About 130 million people, estimating about 3% of the global population, are infected with HCV. Chronic HCV infection may lead to progressive liver injury, cirrhosis, and in some cases hepatocellular carcinoma (Sun et al., 2011).

### 4.4.1 HCV Protease

One of the HCV proteases that is responsible for viral replication in the cell cycle, is NS3-4A, a serine protease, which is a noncovalent heterodimer consisting of a catalytic subunit NS3 (the N-terminal one-third of NS3 protein) and an activating cofactor NS4A protein. NS3-4A has been considered as one of the most attractive targets for developing novel anti-HCV therapies. NS3 is also known as p70, a 70-kDa cleavage product of HCV polyprotein. A cleavage has to occur between NS2 and NS3 for the catalytic activity of NS3, and this is done by NS2-3 protease, an enzyme responsible for catalytic cleavage. Structural studies have confirmed that NS3 protease is a chymotrypsin-like serine protease with a catalytic triad of Ser139-His57-Asp81 (Bazan and Fletterick, 1989; Chambers et al., 1990a,b; Gorbalenya et al., 1989; Miller and Purcell, 1990). Its crystal structure has been well characterized (De Francesco and Steinkuhler, 2000; Morikawa et al., 2011; Raney et al., 2010). The protease contains two β-barrels and a 30-residue extension at the N-terminus. The C-terminus, which is two-third of NS3, has helicase activity and is located in the active site of NS3 protease domain. The zinc ion also plays a vital role in the stabilization of the enzyme structure and is coordinated by three cysteine residues (Cys97, Cys99, and Cys145) and a histidine residue (His149).

## 4.5 Herpes Viruses

Herpes viruses are enveloped dsDNA viruses that cause diseases in animals, as well as humans. They belong to Herpesviridae family of viruses. The Herpesvirus Study Group of the International Committee on the Taxonomy of Viruses has divided the herpes viruses into three subfamilies, termed α-Herpesvirinae, β-Herpesvirinae, and γ-Herpesvirinae. The members of α-Herpesvirinae subfamily are characterized by a variable host range, a short replicative cycle in the host, rapid growth and spread in cell culture, and the establishment of latent infections in sensory ganglia (Roizman et al., 1981). They are often referred to as neurotropic herpesviruses. Among the human herpesviruses (HHVs), herpes simplex virus 1 (HSV1), herpes simplex virus 2 (HSV2), and varicella zoster virus (VZV) belong to the α-Herpesvirinae subfamily. The β-Herpesvirinae members are characterized by a restricted host range with a long reproductive cycle in cell culture and in the infected host, which often results in the development of a carrier state. Latency is established in lymphocytes, secretory glands, and cells of the kidney, as well as other cell types (Kimberlin, 1998; Roizman et al., 1981). The members of this subfamily are HHV6, HHV7, and cytomegalovirus (CMV). In another subfamily, γ-Herpesvirinae, which is characterized by a restricted host range with replication and latency occurring in lymphoid tissues, some members have demonstrated lytic growth in epithelial, endothelial, and fibroblastic cells (Kieff and Liebowitz, 1990). In lymphoblastic cells, viral replication is restricted to either T or B cells. The γ-Herpesvirinae subfamily is further divided into two genera, *Lymphocryptovirus* and *Rhadinovirus*. Among the HHVs, Epstein–Barr virus (EBV) is a member of the *Lymphocryptovirus* genus, while the newly discovered HHV8 is a member of the *Rhadinovirus* genus.

The icosahedral nucleocapsids of herpesviruses have diameter of about 1250 Å. The genomes are between 130 and 250 kbp. They consist of four distinct components: the core, capsid, tegument, and envelope (Roizman and Sears, 1993). The core consists of 162 capsomeres and a dsDNA genome arranged in a torus shape that is located inside an icosadeltahedral capsid that is approximately 100 nm in size (Furlong et al., 1972). Tegument is an amorphous structure located between the capsid and the viral envelope that contains numerous proteins. Its structure is generally asymmetrical (Roffman et al., 1990). Presumably, the tegument is responsible for connecting the capsid to the envelope and acting as a reservoir for viral proteins that are required during the initial stages of viral infection (Batterson and Roizman, 1983; Pellett et al., 1985). The herpes virion's outermost structure is enveloped, which is derived from cell nuclear membranes and contains several viral glycoproteins (gps). The size of mature herpesviruses ranges from 120 to 300 nm, owing to differences in the size of the individual viral teguments. Cellular entry of infectious virus involves the attachment to a cell surface receptor followed by fusion of the outer viral envelope with a cell membrane. The genomic DNA is transported through the cell nuclear membrane into the nucleus, where transcription and replication occur.

### 4.5.1   Herpesvirus Protease

Herpesvirus protease is required for the life cycle of virus, as it carries out the maturational processing of the viral assembly protein. It was first identified in 1991 from studies on HSV1 and HCMV. The assembly protein is required to form a scaffold required for the formation of herpesvirus capsid. A temperature-sensitive mutant of HSV1 is not able to package the viral genome and produces empty capsids at the nonpermissive temperature (Gao et al., 1994; Preston et al., 1983), confirming the functional requirement of protease in completion of viral life cycle and spread of virus. The protease and assembly proteins are encoded by overlapping genes. The gene for the assembly protein uses the 3′ portion of the gene for the protease. The maturational processing of the assembly protein occurs at M-site adjacent to C-terminus of assembly protein. Additionally, the protease catalyzes cleavage at R-site, which releases an N-terminal fragment of about 250 residues from the full-length protease gene product. The N-fragment retains all catalytic activity of the protease protein, and is generally referred to as the herpesvirus protease. In HCMV protease, two additional cleavages are catalyzed by an enzyme within the protease itself at the residues 143 and 209. The cleavage at residue 143 produces a two-chain form of the protease that is still catalytically active (Holwerda et al., 1994; O'Boyle et al., 1995). In solution, herpesvirus protease exists in monomer–dimer equilibrium, where the dimer is the only active form. The crystal structures of all herpesvirus proteases have a novel polypeptide backbone of serine proteases. It contains a central, mostly antiparallel, seven-stranded β-barrel, which is surrounded by eight helices. The active site of the protease contains a novel Ser132-His63-His157 catalytic triad. The first two residues, Ser132 and His63, are determined from biochemical and mutagenesis studies, while the third one, His157, is not fully understood, but its removal can lead to a loss of catalytic activity by >20,000-fold.

## 4.6 Human Immunodeficiency Virus of Type 1

The HIV is an enveloped retrovirus with a positive-sense RNA genome. It belongs to the genus Lentivirdae (Horwitz, 2001). HIV is of two types, HIV1 and HIV2, where HIV1 was found to be the causative agent of AIDS (Gallo and Montagnier, 1988). These viruses store their genetic information as ribonucleic acid (RNA), while most viruses do so as deoxyribonucleic acid (DNA). Before viral replication can take place, the RNA must be converted to DNA by reverse transcriptase (RT) enzyme. HIV1 has an outer envelope consisting of a lipid bilayer with spikes of glycoproteins (gp), gp41 and gp120. These gps are attached in such a way that gp120 protrudes from the surface of the virus. Inside this envelope is a nucleocapsid (p17), which surrounds a central core of protein, p24. The core contains two copies of ssRNA (the virus genome). Gene expression is believed to be regulated by proteins, p7 and p9, that are bound to the RNA. Multiple molecules of the enzyme reverse transcriptase are found in the core. This enzyme is responsible for converting the viral RNA into proviral DNA (Abbas et al., 2000). HIV particles are spherical with a diameter of about 1000 Å. RNA genome of nucleocapsid is about 13 kbp in length. Upon entry into the host cell, the viral RNA is transcribed to DNA that integrates with host genome to produce the provirus. HIV1 infection leads to impairment of immune cells, mainly CD4$^+$ T cells, monocytes, and macrophages, that ultimately about after 10–15 years leads to AIDS, which is characterized by low CD4$^+$ cell counts. It is estimated that about 36 million people worldwide are affected by this virus. HIV1 transmission takes place by sexual contact, contaminated needles, vertical transmission from mother to child, blood transfusions, blood products, and organ/tissue transplants (Negishi, 1993). Various neurological complications affecting the nervous system have been seen in HIV-infected individuals. These include rare opportunistic infections and neoplastic diseases, such as cerebral toxoplasmosis, cryptococcal meningitis, and primary central nervous system lymphoma, as well as syndromes caused by or directly related to HIV1 infection, such as dementia and various neuromuscular complications (Krebs et al., 2000). As the immune system of HIV1-infected people becomes weak, opportunistic infections can also occur by organisms, such as bacteria, viruses, fungi, and parasites (Schneider, 2000).

### 4.6.1 HIV1 Protease

HIV1 protease plays a major in viral maturation, which is important for the production of infectious virus particles. It is an aspartic protease and the catalytic site has the characteristic Asp-Thr-Gly (Asp25, Asp26, and Asp27) sequence common to all aspartic proteases. It is a symmetric homodimer consisting of 99 amino acids per monomer (Fig. 1.5). The protease contains a β-sheet with the two aspartic acids Asp25 from the two monomers, forming the central active site. The active site cavity, the flexible flaps, and the dimer interface are three important regions of the protease structure. Protease inhibitors are competitive inhibitors that bind at the active site of the protease with the flaps

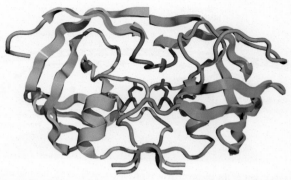

FIGURE 1.5   **The human immunodeficiency virus type 1 (HIV1) protease dimer** *(green and* *cyan)* **with active site Asp25** *(red)* **complexed with a polypeptide substrate** *(magenta).* *(Image* *available from Wikipedia: https://en.wikipedia.org/w/index.php?curid=18176790)*

folded into a closed conformation over the active site. The flaps were seen in closed and open conformations in the crystal structures of inhibitor-bound, as well as free protease (Liu et al., 2006; Perryman et al., 2004; Rose et al., 1998; Scott and Schiffer, 2000; Spinelli et al., 1991). This process of opening and clos-ing of the flaps enables the substrate to enter and leave the active site of protease. Inhibitor bound at the active site with the flaps in closed conformations keeps the enzyme in a locked-down state and prevents the processing of substrates. The protease active site cavity comprises the residues Arg8, Leu23, Asp25, Gly27, Ala28, Asp29, Asp30, Val32, Lys45, Ile47, Met46, Gly48, Gly49, Ile50, Phe53, Leu76, Thr80, Pro81, Val82, and Ile84. The residues forming the substrate-binding site are mainly hydrophobic in nature; the exceptions are the catalytic residues Asp25 and Asp29, which form hydrogen bonds with the peptide main chain groups, and Arg8, Asp30, and Lys45, which can interact with polar side chains or distal main chain groups in longer peptides (Tie et al., 2005).

## 4.7   Picornaviruses

Picornaviruses are small, nonenveloped viruses, having positive-sense ssRNA genome, and belong to the family of Picornaviridae (Racaniello, 2001). This family consists of 46 species having 26 genera, the best known are *Enterovirus* (poliovirus, rhinovirus, coxsackievirus, and echovirus), *Apthovirus* (FMDV), *Cardiovirus* (encephalomyocarditis virus, EMCV), Thieler's virus, and *Hepa-tovirus* (HAV) (Knowles et al., 2012). The diameter of picornaviruses is about 300 Å. Replication begins with the attachment of the virus to a specific cellular receptor that leads to virus internalization and destabilization of the capsid and release of the genome from the endosome into the cytoplasm (Bergelson, 2010). After binding to the receptor, the virus uses host and viral proteins to com-plete its replication cycle (Fig. 1.6). In fact, to complete a round of infection,

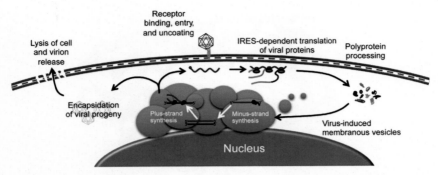

**FIGURE 1.6 Overview of the picornavirus replication cycle.** *IRES,* Internal ribosome entry site. *(From Cathcart, A.L., Baggs, E.L., Semler, B.L., 2014. Picornaviruses: Pathogenesis and Molecular Biology. Reference Module in Biomedical Sciences, third ed. Elsevier, UC, Irvine, (pp. 1–13).)*

a number of complex activities are performed and coordinated. These include replication of the plus-strand RNA via minus-strand intermediates, translation, proteolytic processing, inhibition of host cell transcription/translation, virion assembly, and cell lysis.

## 4.7.1 Picornavirus Proteases

The viral genome, having about 8000 bases, encodes a single polyprotein, which is processed cotranslationally by proteases including the 3C, 2A, and leader proteases (Kay and Dunn, 1990; Lawson and Semler, 1990; Racaniello, 2001; Ryan and Flint, 1997). Most of the processing is carried out by 3C, while 2A and leader proteases are present in only some of the picornaviruses. The 2A protease catalyzes the release of structural polyprotein at the N-terminus, while the leader protease releases itself from the polyprotein located at C-terminus (Strelbel and Beck, 1986). The 2A and leader proteases also inhibit host cell protein synthesis by cleaving host cell eukaryotic initiation factors 4G (eIF4G) during translation from 5′ capped mRNA of the host cells, whereas translation from viral mRNA is not affected (Racaniello, 2001). It has been found recently that cleaved eIF4G can still support translation from capped mRNA, but not as efficiently as from viral RNA (Ali et al., 2001). Additionally, other enzymes can also cleave eIF4G, including 3C protease of FMDV (Belsham et al., 2000), HIV protease (Ventoso et al., 2001), cellular caspases, and other proteases (Zamora et al., 2002).

The 3C and 2A proteases are cysteine proteases (Kräusslich and Wimmer, 1998), as they both contain a Gly-X-Cys-Gly motif, similar to that of the active site of chymotrysin-like serine proteases, which is Gly-Asp-Ser-Gly (Bazan and Fletterick, 1989; Choi et al., 1997; Zamora et al., 2002). Crystal structures of these proteases present in picornaviruses and other viruses, such as HAV and human rhinovirus 2 and 14, confirm that they are chymotrysin-like serine proteases (Allaire et al., 1994; Bergmann et al., 1997; Matthews et al., 1994; Mosimann et al., 1997; Peterson et al., 1999; Seipelt et al., 1999). The 3C protease of human rhinovirus 14 (HRV14) has 182 residues, while 2A protease of HRV2 has 142 residues (Peterson et al., 1999)

The active site of 3C protease of HRV14 contains a catalytic triad of Cys147-His40-Glu71, while in HAV 3C protease Glu71 is replaced by Tyr143. The catalytic triad of 2A proteases of HRV2 contains Cys106, His18, and Asp35. Crystal structures of the FMDV leader protease has confirmed that it is also a papain-like cysteine protease having 150 residues, with a catalytic triad of Cys51, His148, and Asp175 (Gorbalenya et al., 1991; Guarne et al., 1998, 2000; Piccone et al., 1995; Roberts and Belsham, 1995; Seipelt et al., 1999). This leader protease contributes to the inhibition of host protein synthesis by cleaving eIF4G at a different site than that of 2A.

## 5 CONCLUSIONS

All of the prominent viruses, namely AdVs, alphaviruses, flaviviruses, HCV, herpesviruses, HIV1, and picornaviruses, contain proteases that play crucial roles in their replication and thus are important targets for the discovery of potent antiviral drugs. The process of protein catabolism by hydrolysis of peptide bond is catalyzed by the enzyme called protease. It may also be known as peptidase or proteinase. Different classes of protease can perform the same reaction by completely different catalytic mechanisms. The catalytic mechanism of these proteases involves either the activation of a water molecule, which performs a nucleophilic attack on the peptide bond to hydrolyze it or uses a nucleophilic residue in a catalytic triad to perform a nucleophilic attack that covalently links the protease to the substrate protein, releasing the first half of the product. Regarding the specificity of proteases, it is said that proteolysis is highly promiscuous, such that a wide range of protein substrates are hydrolyzed. Proteases, being proteins themselves, are cleaved by other protease molecules, sometimes of the same variety. Some proteases, such as TEV protease, have high specificity and cleave only a very restricted set of substrate sequences. Virus proteases are now extensively studied so that they can be exploited for drug development.

## REFERENCES

Abbas, A.K., Lichtman, A.H., Pober, J.S., 2000. Cellular and Molecular Immunology. W. B. Saunders, Philadelphia, PA.

Ali, I.K., McKendrick, L., Morley, S.J., Jackson, R.J., 2001. Truncated initiation factor eIF4G lacking an eIF4E binding site can support capped mRNA translation. EMBO J. 20, 4233–4242.

Allaire, M., Chernaia, M.M., Malcolm, B.A., James, M.N., 1994. Picornaviral 3C cysteine proteinases have a fold similar to chymotrypsin-like serine proteinases. Nature 369, 72–76.

Alvarez, D.E., Lodeiro, M.F., Filomatori, C.V., Fucito, S., Mondotte, J.A., Gamarnik, A.V., 2006. Structural and functional analysis of dengue virus RNA. Novartis Found. Symp. 277, 120–132, (Discussion 132-125, 251-123).

Baltimore, D., 1971. Expression of animal virus genomes. Bacteriol. Rev. 35 (3), 235–241.

Batterson, W., Roizman, B., 1983. Characterization of the herpes simplex virion-associated factor responsible for the induction of alpha genes. J. Virol. 46, 371–377.

Bazan, J.F., Fletterick, R.J., 1989. Detection of a trypsin-like serine protease domain in flaviviruses and pestiviruses. Virology 171, 637–639.

Belsham, G.J., MacLnerney, G.M., Ross-Smith, N.J., 2000. Foot-and-mouth disease virus 3C protease induces cleavage of translation initiation factors eIF4A and eIF4G within infected cells. J. Virol. 74, 272–280.

Bergelson, J.M., 2010. Receptors. In: Ehrenfeld, E., Domingo, E., Roos, R.P. (Eds.), The Picornaviruses. ASM Press, Washington, DC, pp. 73–86.

Bergmann, E.M., Mosimann, S.C., Chernaia, M.M., Malcolm, B.A., James, M.N.G., 1997. The refined crystal structure of the 3C gene product from hepatitis A virus: specific proteinase activity and RNA recognition. J. Virol. 71, 2436–2448.

Bollati, M., Alvarez, K., Assenberg, R., Baronti, C., Canard, B., Cook, S., Coutard, B., Decroly, E., de Lamballerie, X., Gould, E.A., Grard, G., Grimes, J.M., Hilgenfeld, R., Jansson, A.M., Malet, H., Mancini, E.J., Mastrangelo, E., Mattevi, A., Milani, M., Moureau, G., Neyts, J., Owens, R.J., Ren, J., Selisko, B., Speroni, S., Steuber, H., Stuart, D.I., Unge, T., Bolognesi, M., 2010. Structure and functionality in *Flavivirus* NS-proteins: perspectives for drug design. Antiviral Res. 87, 125–148.

Brooks, A.J., Johansson, M., John, A.V., Xu, Y., Jans, D.A., Vasudevan, S.G., 2002. The interdomain region of dengue NS5 protein that binds to the viral helicase NS3 contains independently functional importin beta 1 and importin alpha/beta recognized nuclear localization signals. J. Biol. Chem. 277, 36399–36407.

Cancedda, R., Shatkin, A.J., 1979. Ribosome-protected fragments from Sindbis 42-S and 26-S RNAs. Eur. J. Biochem. 94, 41–50.

Chambers, T.J., McCourt, D.W., Rice, C.M., 1990a. Production of yellow fever virus proteins in infected cells: identification of discrete polyprotein species and analysis of cleavage kinetics using region-specific polyclonal antisera. Virology 177, 159–174.

Chambers, T.J., Weir, R.C., Grakoui, A., McCourt, D.W., Bazan, J.F., Fletterick, R.J., Rice, C.M., 1990b. Evidence that the N-terminal domain of nonstructural protein NS3 from yellow fever virus is a serine protease responsible for site-specific cleavages in the viral polyprotein. Proc. Natl. Acad. Sci. USA 87, 8898–8902.

Choi, H.K., Lu, G., Lee, S., Wengler, G., Rossmann, M.G., 1997. Structure of Semliki Forest virus core protein. Proteins Struct. Funct. Genet. 27, 345–359.

Clum, S., Ebner, K.E., Padmanabhan, R., 1997. Cotranslational membrane insertion of the serine proteinase precursor NS2B–NS3(Pro) of dengue virus type 2 is required for efficient in vitro processing and is mediated through the hydrophobic regions of NS2B. J. Biol. Chem. 272, 30715–30723.

de Cedrón, M.G., Ehsani, N., Mikkola, M.L., Kääriäinen, L., García, J.A., 1999. RNA helicase activity of Semliki Forest virus replicase protein NSP2. FEBS Lett. 448, 19–22.

Dé, I., Sawicki, S.G., Sawicki, D.L., 1996. Sindbis virus RNA-negative mutants that fail to convert from minus-strand to plus-strand synthesis: role of the nsP2 protein. J. Virol. 70, 2706–2719.

De Francesco, R., Steinkuhler, C., 2000. Structure and function of the hepatitis C virus NS3-NS4A serine proteinase. Curr. Top. Microbiol. Immunol. 242, 149–169.

Decroly, E., Ferron, F., Lescar, J., Canard, B., 2012. Conventional and unconventional mechanisms for capping viral mRNA. Nat. Rev. Microbiol. 10, 51–65.

Ding, J., Mcgrath, W.J., Sweet, R.M., Mangel, W.F., 1996. Crystal structure of the human adenovirus proteinase with its 11 amino acid cofactor. EMBO J. 15, 1778–1783.

Dong, H., Chang, D.C., Hua, M.H., Lim, S.P., Chionh, Y.H., Hia, F., Lee, Y.H., Kukkaro, P., Lok, S.M., Dedon, P.C., Shi, P.Y., 2012. 20-O methylation of internal adenosine by *Flavivirus* NS5 methyltransferase. PLoS Pathog. 8, e1002642.

Dong, H., Fink, K., Zust, R., Lim, S.P., Qin, C.F., Shi, P.Y., 2014. *Flavivirus* RNA methylation. J. Gen. Virol. 95, 763–778.

Egloff, M.P., Benarroch, D., Selisko, B., Romette, J.L., Canard, B., 2002. An RNA cap (nucleoside-20-O-)-methyltransferase in the *Flavivirus* RNA polymerase NS5: crystal structure and functional characterization. EMBO J. 21, 2757–2768.

Fairman-Williams, M.E., Guenther, U.P., Jankowsky, E., 2010. SF1 and SF2 helicases: family matters. Curr. Opin. Struct. Biol. 20, 313–324.

Falgout, B., Pethel, M., Zhang, Y.M., Lai, C.J., 1991. Both nonstructural proteins NS2B and NS3 are required for the proteolytic processing of dengue virus nonstructural proteins. J. Virol. 65, 2467–2475.

Fang, D.A., Huang, X.M., Zhang, Z.Q., Xu, D.P., Zhou, Y.F., Zhang, M.Y., Liu, K., Duan, J.R., Shi, W.G., 2013. Molecular cloning and expression analysis of chymotrypsin-like serine protease from the redclaw crayfish (*Cherax quadricarinatus*): a possible role in the junior and adult innate immune systems. Fish Shellfish Immunol. 34, 1546–1552.

Filomatori, C.V., Lodeiro, M.F., Alvarez, D.E., Samsa, M.M., Pietrasanta, L., Gamarnik, A.V., 2006. A 50 RNA element promotes dengue virus RNA synthesis on a circular genome. Genes Dev. 20, 2238–2249.

Froshauer, S.A., Kartenbeck, J., Helenius, A., 1988. Alphavirus RNA replicase is located on the cytoplasmic surface of endosomes and lysosomes. J. Cell. Biol. 107, 2075–2086.

Fuller, S.D., 1987. The T = 4 envelope of Sindbis virus is organized by interactions with a complementary T = 3 capsid. Cell 48, 923–934.

Furlong, D., Swift, H., Roizman, B., 1972. Arrangement of herpesvirus deoxyribonucleic acid in the core. J. Virol. 10, 1071–1074.

Gallo, R.C., Montagnier, L., 1988. AIDS in 1988. Sci. Am. 259, 41–48.

Gao, M., Matusick-Kumar, L., Hurlburt, W., DiTusa, S.F., Newcomb, W.W., Brown, J.C., McCann, III, P.J., Deckman, I., Colonno, R.J., 1994. The protease of herpes simplex virus type 1 is essential for functional capsid formation and viral growth. J. Virol. 68, 3702–3712.

Glanville, N.T., Ranki, M., Morser, J., 1976. Initiation of translocation directed by 42S and 26S RNAs from Semliki Forest virus in vitro. Proc. Natl. Acad. Sci. USA 73, 3059–3063.

Gorbalenya, A.E., Koonin, E.V., 1993. Helicases: amino acid sequence comparisons and structure–function relationships. Curr. Opin. Struct. Biol. 3, 419–429.

Gorbalenya, A.E., Donchenko, A.P., Koonin, E.V., Blinov, V.M., 1989. N-terminal domains of putative helicases of flavi- and pestiviruses may be serine proteases. Nucleic Acids Res. 17, 3889–3897.

Gorbalenya, A.E., Koonin, E.V., Lai, M.M., 1991. Putative papain-related thiol proteases of positive-strand RNA viruses. Identification of rubi- and aphthovirus proteases and delineation of a novel conserved domain associated with proteases of rubi-, alpha- and coronaviruses. FEBS Lett. 288, 201–205.

Guarne, A., Hampoelz, B., Glaser, W., Carpena, X., Tormo, J., Fita, I., Skern, T., 2000. Structural and biochemical features distinguish the foot-and-mouth disease virus leader proteinase from other papain-like enzymes. J. Mol. Biol. 302, 1227–1240.

Guarne, A., Tormo, J., Kirchweger, R., Pfistermueller, D., Fita, I., Skern, T., 1998. Structure of the foot-and-mouth disease virus leader protease: a papain-like fold adapted for self-processing and eIF4G recognition. EMBO J. 17, 7469–7479.

Gubler, D.J., Kuno, G., Markoff, L., 2007. Flaviviruses. In: Knipe, D.M., Howley, P.M. (Eds.), Fields Virology. fifth ed. Lippincott, Williams, and Wilkins, Philadelphia, PA, pp. 1153–1252.

Hardy, W.R., Strauss, J., 1989. Processing the nonstructural polyproteins of Sindbis virus: nonstructural proteinase is in the C-terminal half of nsP2 and functions both in *cis* and in *trans*. J. Virol. 63, 4653–4664.

Holwerda, B.C., Wittwer, A.J., Duffin, K.L., Smith, C., Toth, M.V., Cam, L.S., Wiegand, R.C., Bryant, M.L., 1994. Activity of two-chain recombinant human cytomegalovirus protease. J. Biol. Chem. 269, 25911–25915.

Horwitz, M., 2001. Fields, B.M., Knipe, D.M., Howley, P.M. (Eds.), Fields Virology, vol. 2, Lippincott William and Wilkins, Philadelphia, PA, (International Committee on taxonomy of viruses, 2002, Lentivirus National Institute of Health).

Jan, L.R., Yang, C.S., Trent, D.W., Falgout, B., Lai, C.J., 1995. Processing of Japanese encephalitis virus non-structural proteins: NS2B–NS3 complex and heterologous proteases. J. Gen. Virol. 76 (Pt. 3), 573–580.

Karpe, Y.A., Aher, P.P., Lole, K.S., 2011. NTPase and 5′-RNA triphosphatase activities of Chikungunya virus nsP2 protein. PLoS One 6, e22336.

Kay, J., Dunn, B.M., 1990. Viral proteinases: weakness in strength. Biochim. Biophys. Acta 1048, 1–18.

Khromykh, A.A., Kondratieva, N., Sgro, J.Y., Palmenberg, A., Westaway, E.G., 2003. Significance in replication of the terminal nucleotides of the *Flavivirus* genome. J. Virol. 77, 10623–10629.

Kieff, E., Liebowitz, D., 1990. Epstein-Barr virus and its replication. In: Fields, B.M., Knipe, D.M., Chanock, R.M., Hirsch, M.S., Melnick, J.L., Monath, T.P. et al., (Eds.), Fields Virology. second ed. Raven Press, New York, NY, pp. 1889–1920.

Kim, Y.M., Gayen, S., Kang, C., Joy, J., Huang, Q., Chen, A.S., Wee, J.L., Ang, M.J., Lim, H.A., Hung, A.W., Li, R., Noble, C.G., Lee le, T., Yip, A., Wang, Q.Y., Chia, C.S., Hill, J., Shi, P.Y., Keller, T.H., 2013. NMR analysis of a novel enzymatically active unlinked dengue NS2B–NS3 protease complex. J. Biol. Chem. 288, 12891–12900.

Kimberlin, D.W., 1998. Human herpesviruses 6 and 7: identification of newly recognized viral pathogens and their association with human disease. Pediatr. Infect. Dis. J. 17, 59–68.

Knowles, N.J., Hovi, T., Hyypiä, T., King, A.M.Q., Lindberg, A.M., Pallansch, M.A., Palmenberg, A.C., Simmonds, P., Skern, T., et al., 2012. Picornaviridae. In: King, A.M.Q., Adams, M.J., Carstens, E.B., Lefkowitz, E.J. (Eds.), Virus Taxonomy: Classification and Nomenclature of Viruses: Ninth Report of the International Committee on Taxonomy of Viruses. Elsevier, San Diego, CA, pp. 855–880.

Krebs, F.C., Ross, H., McAllister, J., Wigdahl, B., 2000. HIV-1-associated central nervous system dysfunction. Adv. Pharmacol. 49, 315–385.

Kräusslich, H.G., Wimmer, E., 1998. Viral proteinases. Annu. Rev. Biochem. 57, 701–754.

Kujala, P., Ikäheimonen, A., Ehsani, N., Vihinen, H., Kääriäinen, L., Auvinen, P., 2001. Biogenesis of the Semliki Forest virus RNA replication complex. J. Virol. 75, 3873–3884.

LaStarza, M.W., Lemm, J.A., Rice, C.M., 1994. Genetic analysis of the nsP3 region of Sindbis virus: evidence for roles in minus-strand and subgenomic RNA synthesis. J. Virol. 68, 5781–5791.

Lawson, M.A., Semler, B.L., 1990. Picornavirus protein processing—enzymes, substrates, and genetic regulation. Curr. Top. Microbiol. Immunol. 161, 49–87.

Lescar, J., Luo, D., Xu, T., Sampath, A., Lim, S.P., Canard, B., Vasudevan, S.G., 2008. Towards the design of antiviral inhibitors against flaviviruses: the case for the multifunctional NS3 protein from Dengue virus as a target. Antiviral Res. 80, 94–101.

Leung, J.Y., Ng, M.M., Chu, J.J., 2011. Replication of alphaviruses: a review on the entry process of alphaviruses into cells. Adv. Virol. 2011, (Article ID 249640).

Li, J., Lim, S.P., Beer, D., Patel, V., Wen, D., Tumanut, C., Tully, D.C., Williams, J.A., Jiricek, J., Priestle, J.P., Harris, J.L., Vasudevan, S.G., 2005a. Functional profiling of recombinant NS3 proteases from all four serotypes of Dengue virus using tetrapeptide and octapeptide substrate libraries. J. Biol. Chem. 280, 28766–28774.

Li, X.D., Sun, L., Seth, R.B., Pineda, G., Chen, Z.J., 2005b. Hepatitis C virus protease NS3/4A cleaves mitochondrial antiviral signaling protein off the mitochondria to evade innate immunity. Proc. Natl. Acad. Sci. USA 102, 17717–17722.

Lim, S.P., Wang, Q.Y., Noble, C.G., Chen, Y.L., Dong, H., Zou, B., Yokokawa, F., Nilar, S., Smith, P., Beer, D., Lescar, J., Shi, P.Y., 2013. Ten years of dengue drug discovery: progress and prospects. Antiviral Res. 100, 500–519.

Lindenbach, B.D., Thiel, H.J., Rice, C.M., 2007. Flaviviridae: The Viruses and Their Replication. Lippincott-Raven Publishers, Philadelphia, PA.

Liu, F., Kovalevsky, A.Y., Louis, J.M., Boross, P.I., Wang, Y.F., Harrison, R.W., Weber, I.T., 2006. Mechanism of drug resistance revealed by the crystal structure of the unliganded HIV-1 protease with F53L mutation. J. Mol. Biol. 358, 1191–1199.

Luo, D., Wei, N., Doan, D.N., Paradkar, P.N., Chong, Y., Davidson, A.D., Kotaka, M., Lescar, J., Vasudevan, S.G., 2010. Flexibility between the protease and helicase domains of the dengue virus NS3 protein conferred by the linker region and its functional implications. J. Biol. Chem. 285, 18817–18827.

Malet, H., Massé, N., Selisko, B., Romette, J.-L., Alvarez, K., Guillemot, J.-C., Tolou, H., Yap, T.L., Vasudevan, S.G., Lescar, J., Canard, B., 2008. The *Flavivirus* polymerase as a target for drug discovery. Antiviral Res. 80, 23–35.

Mancini, E.J., Clarke, M., Gowen, B., Rutten, T., Fuller, S.D., 2000. Cryo-electron microscopy reveals the functional organization of an enveloped virus, Semliki forest virus. Mol. Cell. 5, 255–266.

Marcin, S., Marcin, S., 2013. Viral Proteases as targets for drug design. Curr. Pharm. Design 19, 1126–1153.

Markoff, L., 2003. 5′- and 3′-noncoding regions in *Flavivirus* RNA. Adv. Virus Res. 59, 177–228.

Martin, M.A., Knipe, D.M., Fields, B.N., Howley, P.M., Griffin, D., Lamb, R., 2007. Fields' Virology. Wolters Kluwer Health/Lippincott Williams and Wilkins, Philadelphia, PA, (p. 2395).

Matthews, D.A., Smith, W.W., Ferre, R.A., Condon, B., Budahazi, G., McElroy, H.E., Gribskov, C.L., Worland, S., 1994. Structure of human rhinovirus 3C protease reveals a trypsin-like polypeptide fold, RNA-binding site, and means for cleaving precursor polyprotein. Cell 77, 761–771.

Matusan, A.E., Pryor, M.J., Davidson, A.D., Wright, P.J., 2001. Mutagenesis of the dengue virus type 2 NS3 protein within and outside helicase motifs: effects on enzyme activity and virus replication. J. Virol. 75, 9633–9643.

Melançon, P., Garoff, H., 1987. Processing of the Semliki forest virus structural polyprotein: role of the capsid F protease. J. Virol. 61, 1301–1309.

Merits, A., Vasiljeva, L., Ahola, T., Kaariainen, L., Auvinen, P., 2001. Proteolytic processing of Semliki forest virus-specific nonstructural polyprotein by nsP2 protease. J. Gen. Virol. 82, 765–773.

Miller, R.H., Purcell, R.H., 1990. Hepatitis C virus shares amino acid sequence similarity with pestiviruses and flaviviruses as well as members of two plant virus supergroups. Proc. Natl. Acad. Sci. USA 87, 2057.

Morgan, C., Howe, C., Rose, H.M., 1961. Structure and development of viruses as observed in the electron microscope. V. Western equine encephalomyelitis virus. J. Exp. Med. 113, 219–234.

Morikawa, K., Lange, C.M., Gouttenoire, J., Meylan, E., Brass, V., Penin, F., Moradpour, D., 2011. Non-structural protein 3-4A: the Swiss army knife of hepatitis C virus. J. Viral Hepat. 18, 305–315.

Mosimann, S.C., Chernaia, M.M., Sia, S., Plotch, S., James, M.N.G., 1997. Refined X-ray crystallographic structure of the poliovirus 3C gene product. J. Mol. Biol. 273, 1032–1047.

Murray, C.L., Jones, C.T., Rice, C.M., 2008. Architects of assembly: roles of Flaviviridae nonstructural proteins in virion morphogenesis. Nat. Rev. Microbiol. 6, 699–708.

Negishi, M., 1993. Preventive methods against HIV transmission. Nippon Rinsho 51, 509–511.

Noble, C.G., Shi, P.Y., 2012. Structural biology of dengue virus enzymes: towards rational design of therapeutics. Antiviral Res. 96, 115–126.

O'Boyle, II, D.R., Wager-Smith, K., Stevens, III, J.T., Wein Heimer, S.P., 1995. The effect of internal autocleavage on kinetic properties of the human cytomegalovirus protease catalytic domain. J. Biol. Chem. 270, 4753–4758.

Oda, K., 2012. New families of carboxyl peptidases: serine-carboxyl peptidases and glutamic peptidases. J. Biochem. 151, 13–25.

Pastorino, B.A., Peyrefitte, C.N., Almeras, L., Grandadam, M., Rolland, D., Tolou, H.J., Bessaud, M., 2008. Expression and biochemical characterization of nsP2 cysteine protease of Chikungunya virus. Virus Res. 131, 293–298.

Paul, D., Bartenschlager, R., 2013. Architecture and biogenesis of plus-strand RNA virus replication factories. World J. Virol. 2, 32–48.

Pellett, P.E., McKnight, J.L., Jenkins, F.J., Roizman, B., 1985. Nucleotide sequence and predicted amino acid sequence of a protein encoded in a small herpes simplex virus DNA fragment capable of trans-inducing alpha genes. Proc. Natl. Acad. Sci. USA 82, 5870–5874.

Perryman, A.L., Lin, J.H., McCammon, J.A., 2004. HIV-1 protease molecular dynamics of a wildtype and of the V82F/I84V mutant: possible contributions to drug resistance and a potential new target site for drugs. Protein Sci. 13, 1108–1123.

Peterson, J.F.W., Cherne, M.M., Liebig, H.D., Skern, T., Kuechler, E., James, M.N.G., 1999. The structure of the 2A proteinase from a common cold virus: a proteinase responsible for the shutoff of host-cell protein synthesis. EMBO J. 18, 5643.

Phong, W.Y., Moreland, N.J., Lim, S.P., Wen, D., Paradkar, P.N., Vasudevan, S.G., 2011. Dengue protease activity: the structural integrity and interaction of NS2B with NS3 protease and its potential as a drug target. Biosci. Rep. 31, 399–409.

Piccone, M.E., Zellner, M., Kumosinki, T.F., Mason, P.W., Grubman, M.J.J., 1995. Identification of the active-site residues of the L proteinase of foot-and-mouth disease virus. Virology 69, 4950–4956.

Polacek, C., Foley, J.E., Harris, E., 2009. Conformational changes in the solution structure of the dengue virus 5′ end in the presence and absence of the 3′ untranslated region. J. Virol. 83, 1161–1166.

Preston, V.G., Coates, J.A.V., Rixon, F.J., 1983. Identification and characterization of a herpes simplex virus gene product required for encapsidation of virus DNA. J. Virol. 45, 1056–1064.

Pyle, A.M., 2008. Translocation and unwinding mechanisms of RNA and DNA helicases. Annu. Rev. Biophys. 37, 317–336.

Racaniello, V.R., 2001. Picornaviridae: the viruses and their replication. Knipe, D.M., Howley, P.M. (Eds.), Fields Virology, vol. 1, Lippincott Williams and Wilkins, Philadelphia, PA.

Raney, K.D., Sharma, S.D., Moustafa, I.M., Cameron, C.E., 2010. Hepatitis C virus non-structural protein 3 (HCV NS3): a multifunctional antiviral target. J. Biol. Chem. 285, 22725–22731.

Rao, S.T., Rossmann, M.G., 1973. Comparison of super-secondary structures in proteins. J. Mol. Biol. 76, 241–256.

Ray, D., Shah, A., Tilgner, M., Guo, Y., Zhao, Y., Dong, H., Deas, T.S., Zhou, Y., Li, H., Shi, P.Y., 2006. West Nile virus 5′-cap structure is formed by sequential guanine N-7 and ribose 2′-O methylations by nonstructural protein 5. J. Virol. 80, 8362–8370.

Roberts, P.J., Belsham, G.J., 1995. Identification of critical amino acids within the foot and mouth disease virus leader protein, a cysteine protease. Virology 213, 140–146.

Roffman, E., Albert, J.P., Goff, J.P., Frenkel, N., 1990. Putative site for the acquisition of human herpesvirus 6 virion tegument. J. Virol. 64, 6308–6313.

Roizman, B., Sears, A.E., 1993. Herpes simplex viruses and their replication. In: Roizman, B., Whitley, R.J., Lopez, C. (Eds.), The Human Herpesviruses. first ed. Raven Press, New York, NY, pp. 11–68.

Roizman, B., Carmichael, L.E., Deinhardt, F., de-The, G., Nahmias, A.J., Plowright, W., Rap, F., Sheldrick, P., Takahashi, M., Wolf, K., 1981. Herpesviridae. Definition, provisional nomenclature, and taxonomy. The Herpesvirus Study Group, the International Committee on Taxonomy of Viruses. Intervirology 16, 201–217.

Rose, R.B., Craik, C.S., Stroud, R.M., 1998. Domain flexibility in retroviral proteases: structural implications for drug resistant mutations. Biochemistry 37, 2607–2621.

Rosenberg, S., 2001. Recent advances in the molecular biology of hepatitis C virus. J. Mol. Biol. 313, 451–464.

Rowe, W.P., Huebner, R.J., Gilmore, L.K., Parrott, R.H., Ward, T.G., 1953. Isolation of a cytopathogenic agent from human adenoids undergoing spontaneous degeneration in tissue culture. Proc. Soc. Exp. Biol. Med. 84 (3), 570–573.

Russell, W.C., 2005. Adenoviruses. In: Mahy, B.W.J., ter Meulen, V. (Eds.), Topley and Wilson's Microbiology and Microbial Infections. ninth ed. Hodder Arnold, London, pp. 439–447.

Ryan, M., Flint, M., 1997. Virus-encoded proteinases of the picornavirus super-group. J. Gen. Virol. 78, 699–723.

Ryman, K.D., Klimstra, W.B., 2008. Host responses to alphavirus infection. Immunol. Rev. 225, 27–45.

Salonen, A.H., Vasiljeva, L., Merits, A., Kääriäinen, L., Magden, J., Jokitalo, E., 2003. Properly folded nonstructural polyprotein directs the Semliki forest virus replication complex to the endosomal compartment. J. Virol. 77, 1691–1702.

Sampath, A., Padmanabhan, R., 2009. Molecular targets for *Flavivirus* drug discovery. Antiviral Res. 81, 6–15.

Santolini, E., Migliaccio, G., La Monica, N., 1994. Biosynthesis and biochemical properties of the hepatitis C virus core protein. J. Virol. 68, 3631–3641.

Schneider, M.M., 2000. Change in natural history of opportunistic infections in HIV-infected patients. Neth. J. Med. 57, 1–3.

Scott, W.R., Schiffer, C.A., 2000. Curling of flap tips in HIV-1 protease as a mechanism for substrate entry and tolerance of drug resistance. Structure 8, 1259–1265.

Segerman, A., Arnberg, N., Erikson, A., Lindman, K., Wadell, G., 2003. There are two different species B adenovirus receptors: sBAR, common to species B1 and B2 adenoviruses, and sB2AR, exclusively used by species B2 adenoviruses. J. Virol. 77, 1157–1162.

Seipelt, J., Guarne, A., Bergmann, E., James, M., Sommergruber, W., Fita, I., Skern, T., 1999. The structures of picornaviral proteinases. Virus Res. 62, 159–168.

Simmons, D.T., Strauss, J., 1972. Replication of Sindbis virus. I. Relative size and genetic content of 26 S and 49 S RNA. J. Mol. Biol. 71, 599–613.

Spinelli, S., Liu, Q.Z., Alzari, P.M., Hirel, P.H., Poljak, R.J., 1991. The three-dimensional structure of the aspartyl protease from the HIV-1 isolate BRU. Biochimie 73, 1391–1396.

Story, R.M., Steitz, T.A., 1992. Structure of the recA protein–ADP complex. Nature 355, 374–376.

Strauss, J.H., Strauss, E.G., 1994. The alphaviruses: gene expression, replication, and evolution. Microbiol. Rev. 58, 491–562.

Strauss, E.G., De Groot, R.J., Levinson, R., Strauss, J.H., 1992. Identification of the active site residues in the nsP2 proteinase of Sindbis virus. Virology 191, 932–940.

Strauss, E.G., Rice, C.M., Strauss, J., 1984. Complete nucleotide sequence of the genomic RNA of Sindbis virus. Virology 133, 92–110.

Strelbel, K., Beck, F., 1986. A second protease of foot-and-mouth disease virus. J. Virol. 58, 893–899.

Sun, J., Cao, N., Zhang, X.M., Yang, Y.S., Zhang, Y.B., Wang, X.M., Zhu, H.L., 2011. Oxadia-zole derivatives containing 1,4-benzodioxan as potential immunosuppressive agents against RAW264.7 cells. Bioorg. Med. Chem. 19, 4895–4902.

Tay, M.Y., Saw, W.G., Zhao, Y., Chan, K.W., Singh, D., Chong, Y., Forwood, J.K., Ooi, E.E., Gruber, G., Lescar, J., Luo, D., Vasudevan, S.G., 2015. The C-terminal 50 amino acid residues of Dengue NS3 protein are important for NS3-7NS5 interaction and viral replication. J. Biol. Chem. 290, 2379–2394.

Tie, Y., Boross, P.I., Wang, Y.F., Gaddis, L., Liu, F., Chen, X., Tozser, J., Harrison, R.W., Weber, I.T., 2005. Molecular basis for substrate recognition and drug resistance from 1.1 to 1. 6 angstroms resolution crystal structures of HIV-1 protease mutants with substrate analogs. FEBS J. 272, 5265–5277.

Ventoso, I., Blanco, R., Perales, C., Carrasco, L., 2001. HIV-1 protease cleaves eukaryotic initiation factor 4G and inhibits cap-dependent translation. Proc. Natl. Acad. Sci. USA 98, 12966–12971.

Villordo, S.M., Gamarnik, A.V., 2008. Genome cyclization as strategy for *Flavivirus* RNA replication. Virus Res. 139, 230–239.

Wang, C.C., Huang, Z.S., Chiang, P.L., Chen, C.T., Wu, H.N., 2009. Analysis of the nucleoside triphosphatase, RNA triphosphatase, and unwinding activities of the helicase domain of dengue virus NS3 protein. FEBS Lett. 583 (4), 691–696.

Wu, E., Nemerow, G.R., 2004. Virus yoga the role of flexibility in virus host cell recognition. Trends Microbiol. 12, 162–168.

Wu, C.F., Wang, S.H., Sun, C.M., Hu, S.T., Syu, W.J., 2003. Activation of dengue protease autocleavage at the NS2B–NS3 junction by recombinant NS3 and GSTNS2B fusion proteins. J. Virol. Methods 114, 45–54.

Xu, T., Sampath, A., Chao, A., Wen, D., Nanao, M., Chene, P., Vasudevan, S.G., Lescar, J., 2005. Structure of the Dengue virus helicase/nucleoside triphosphatase catalytic domain at a resolution of 2.4 Å. J. Virol. 79, 10278–10288.

Yap, T.L., Xu, T., Chen, Y.L., Malet, H., Egloff, M.-P., Canard, B., Vasudevan, S.G., Lescar, J., 2007. The crystal structure of the Dengue virus RNA-dependent RNA polymerase at 1.85 Å resolution. J. Virol. 81, 4753–4765.

Yu, I.M., Zhang, W., Holdaway, H.A., Li, L., Kostyuchenko, V.A., Chipman, P.R., Kuhn, R.J., Rossmann, M.G., Chen, J., 2008. Structure of the immature dengue virus at low pH primes proteolytic maturation. Science 319, 1834–1837.

Yusof, R., Clum, S., Wetzel, M., Murthy, H.M., Padmanabhan, R., 2000. Purified NS2B/ NS3 serine protease of dengue virus type 2 exhibits cofactor NS2B dependence for cleavage of substrates with dibasic amino acids in vitro. J. Biol. Chem. 275, 9963–9969.

Zamora, M., Marissen, W.E., Lloyd, R.E., 2002. Multiple eF4GI specific protease activities present in uninfected and polio-virus infected cells. J. Virol. 76, 165–177.

Zhang, L., Mohan, P.M., Padmanabhan, R., 1992. Processing and localization of Dengue virus type 2 polyprotein precursor NS3–NS4A–NS4B–NS5. J. Virol. 66, 7549–7554.

# FURTHER READING

Ralston, R., Thudium, K., Berger, K., Kuo, C., Gervase, B., Hall, J., Selby, M., Kuo, G., Hougton, M., Choo, Q.L., 1993. Characterization of hepatitis C virus envelope glycoprotein complexes expressed by recombinant vaccinia viruses. J. Virol. 67, 6753–6761.

Chapter 2

# Design and Development of Some Viral Protease Inhibitors by QSAR and Molecular Modeling Studies

**Noha A. Saleh\*, Hanan Elhaes\*\*, Medhat Ibrahim†**
*\*Cairo University, Giza, Egypt; \*\*Ain Shams University, Cairo, Egypt; †National Research Center, Dokki, Giza, Egypt*

## 1 INTRODUCTION

Molecular modeling is rapidly growing according to its applications in many areas of research. It is now widely used to study the molecular structure of large systems and in physics, chemistry, and biology. Molecular modeling is used to simulate the molecular behavior in chemical or biological systems (Leach, 1996). Accordingly, it is one of the leading techniques working with wide range of applications, such as drug design, biomaterials, emerging materials, and spectroscopy. Molecular modeling could be defined as a class of computerized work which applies the laws of physics supported with experimental data that can be used either for analyzing molecules including number and types of atoms, nature of the bonds, bond lengths, angles and dihedral angles, molecular energy, geometry optimization, enthalpy, and vibrational frequency of molecular systems. Molecular modeling can also describe nucleophilicity, electrophilicity, electrostatic potentials, and predict molecular and biological properties for understanding the structure–activity relationships to provide rationale to drug design) (Cohen, 1996). It calculates the energy of the simulated structure using basically the molecular mechanics (MM) based on classical potential functions (Foresman and Frisch, 1996) or quantum mechanical methods if electronic interactions are also to be taken. Spectroscopic features and the reaction paths of the given chemical systems can also be included in the calculations (Dorsett and White, 2000). Quantum mechanical methods use semiempirical methods, ab initio methods, and density functional theory. Both MM and quantum mechanical methods are applied in molecular modeling (Foresman, 1996; Foresman and Frisch, 1996).

Viral Proteases and Their Inhibitors. http://dx.doi.org/10.1016/B978-0-12-809712-0.00002-2

For certain arrangement of the atoms and molecules in certain system these models enable the energy to be calculated and hence, indicate the variation or the change of energy of the system with changing the position of the atoms and molecules. The second step in molecular modeling calculations is the type of calculation, such as calculation of energy minimization, molecular dynamics (MD), Monte Carlo simulation, or conformational search. Analyzing the calculations is very important step not only for obtaining certain properties but also to check if the calculations are performed in proper way. The ultimate goal of molecular modeling simulations is to help in selecting the best possible compound to synthesize, thus, saving time, efforts, and money. One of the main problems in molecular modeling is selecting a suitable theory for a given problem that may lead to reliable results. The choice of the theory depends on a compromise between the computational time, the availability of fast computers, and the intended degree of accuracy of the calculations.

## 2 QUANTITATIVE STRUCTURE–ACTIVITY RELATIONSHIP

Treating biological interactions is one of the main goals of molecular modeling, while pharmaceutical field requires certain care with some kind of special calculations. Consequently, rather than using MM and quantum mechanical methods, there are simpler approaches to deal with biological and pharmacological problems.

Experimental synthesis of thousands of compounds and their activity evaluation consumes lot of time and money. While computational work rationalizes the synthesis and thus drastically reduces the time and the cost of the drug discovery. The quantitative structure–activity relationship (QSAR) approach could solve the major problems of drug design. The QSAR is a computational tool, which quantifies the relationship between the physicochemical properties of the drug with its biological activity and thus this produces a mathematical model, which guides us as to how the structural or physicochemical properties of the molecules should be changed so that the activity of the compounds may be increased. While the structural properties may be related to the number of single/double bonds, number of hetero atoms, or number of hydrogen bond donors or acceptors in the molecules, etc., the physicochemical properties of the molecules are defined by their hydrophobicity ($\log P$), steric constants ($E_s$), molar refractivity (MR), and several electronic properties that can be theoretically evaluated by quantum mechanics. Using quantum mechanics, one can evaluate some crucial electronic parameters, useful in QSAR study, such as the highest occupied molecular orbital and the lowest unoccupied molecular orbital energies, total dipole moments, charge density at atoms, molecular polarizability, Mulliken's electronegativity, frontier electron densities, etc. (Cohen, 1996; Gupta, 2011). The first use of QSAR is attributed to Hansch and Leo (1995). He developed equations, which related biological activity to electronic characteristics and hydrophobicity of the molecules. There were hundreds of physical, structural, and chemical properties used in QSAR approach during the last decades (Atkin and de Paula, 2002). QSAR greatly helps to identify the specific groups and substituents that can increase the activity of compounds. The 3D molecular modeling studies greatly aid as to how one should modify the structure of a compound so that it may have the optimum interaction with the receptor. In molecular modeling, one can visualize how a molecule interacts with the active site of the receptor and if there is any further scope to have an even better interaction. Thus QSAR have been of great help to design and develop useful drugs (Garg and Bhhatarai, 2006).

Further, for drug design, it is important to study and know specifically the conformational properties of the molecules in solution and orientation within the receptor for the interaction. This is what we call molecular recognition, which is an interesting point of research in biological systems. The interactions require two molecules where they could interact with each other involving repulsion and attractions; one molecule could be called as ligand and the other as receptor. In such interactions, there occurs the change in degrees of freedom of both the partners, which has large impact on the free energy of binding (Bissantz et al., 2010).

Molecular modeling at different level of theories became very important tool not only for describing biological interactions but also for homology modeling of receptors, functional site location, characterization of ligand-binding sites in proteins, docking of small molecules into protein binding sites, protein–protein interaction studies, and molecular dynamic simulations. In some sense the computational results are utilized even to improve the experimental routes for further investigations (Choe and Chang, 2002). To study the interactions between large molecular entities, MM became a very handy tool where the computational method neglects explicit treatment of electrons. Newtonian mechanics has led to describe another method, useful in drug design, called MD. It is a computer simulation method for studying the physical movements of atoms and molecules, and is thus a type of N-body simulation. The atoms and molecules are allowed to interact for a fixed period of time, giving a view of the dynamical evolution of the system (van Gunsteren and Berendsen, 1977).

## 3 VIRAL PROTEASES

Proteases are the enzymes that perform protein catabolism by hydrolysis of peptide bonds in polypeptide chain. This hydrolytic function of proteases converts the polyproteins into functional proteins. The protease enzymes have essential roles in lifecycle or replication of any living cell (human, animals, plants, bacteria, and viruses). Due to their essential biological roles, they constitute good targets to inhibit many types of viral infection. This section of the chapter presents the various kinds of viruses, their proteases, and the development of their inhibitors. To know and study the behavior of substrates and inhibitors of proteases, it is important to know the residues nomenclature, which is devised by Schechter and Berger. Starting from the cleavage amide bond by the protease (substrate), the N-terminal side are numbered P1, P2, P3 etc., and the C-terminal P1′, P2′, P3′, etc.. The corresponding interaction sites in the enzyme are called S1, S2, S3, and S1′, S2′, S3′, etc. (Schechter and Berger, 1968).

## 3.1 Adenoviruses

Adenoviruses (AdVs) constitute one of the five classes of Adenoviridae family of viruses (Harrach et al., 2011). Adenovirus, a DNA virus, was first isolated in the 1950s in adenoid tissue–derived cell cultures, hence the name (Rowe et al., 1953). These primary cell cultures were often noted to spontaneously degenerate over time, and adenoviruses are now known to be a common cause of asymptomatic respiratory tract infection that produces in vitro cytolysis in these tissues. Adenoviruses cause a series of diseases in infected humans, such as gastrointestinal and ocular surface infections, respiratory, opportunistic infections in immune-deficient individuals, and obesity (Dhurandhar et al., 1992).

Adenoviruses are nonenveloped, icosahedral viruses that are very distinct when visualized by electron microscopy. The virion diameter ranges from 70 to 100 nm. The viral genome is double-stranded DNA molecule, linear, nonsegmented, and ranging in size from 26 to 48 kb. The capsid or coated protein comprises 252 capsomeres of which 240 are hexons (surrounded by 6 other capsomeres) or 12 pentons (surrounded by 5 capsomeres). The pentons form the angles of the icosahedron and each has a glycoprotein fiber extending from it. The length of four proteins and genome of double stranded DNA in the core of each virion is nearly 36 kb (Horwitz, 2001; Kozak et al., 2015). The adenovirus is divided into over 100 serotypes, 60 of which can cause human infection. These genome types have been identified and subsequently classified into seven species (A to G) on the basis of the variety of pathologies and tissue tropisms (Horwitz, 2001). Adenoviruses have been also isolated from a wide range of animal species, including lizards, fowl, dogs, bats, horses, monkeys (genus *Callicebus*), skunks, and sea lions (Marek et al., 2014; Kozak et al., 2015; Wellehan et al., 2004).

### 3.1.1 Adenovirus Protease AVP (Adenain) Structure and Function

The adenovirus genome is coded with at least 50 proteins. In adenovirus serotype 2, 6 of these proteins (pVI, pVII, pVIII, pTP, pIIIa, and L2 11 kDa) are synthesized as precursors, which are processed late in infection (Cortes-Hinojosa et al., 2015). In the temperature-sensitive mutant 1 of human adenovirus type 2, the location of protease gene on the viral genome is identified between nucleotides21778 and 22390 (Webster et al., 1993).

The protease enzyme of adenovirus has been referred to as 23K or L3-23K protein, being a product of the L3transcription region. Recently, it has been abbreviated as EP and AVP and named adenain (Webster et al.,1993;

Yeh-Kai et al., 1983). Based on the serotype of adenovirus, the adenain (AVP) is a single-chain protein with 201–214 amino acids without any disulfide bond or any known posttranslational modifications (Krausslich and Wimmer, 1988). Recombinant adenain (AVP) has little activity compared with that in disrupted virions. There are three cofactors, which promote the activity and function of adenain (AVP). The first cofactor is pVIc, which has 11 amino acid residues. It is from the C-terminus of the virion precursor protein pVI. These amino acids are Gly-Val-Gln-Ser-Leu-Lys-Arg-Arg-Arg-Cys-Phe in adenovirus 2, Ad2, which forms a disulfide bond with Cys104 of the enzyme. The second cofactor is the viral DNA (Cortes-Hinojosa et al., 2015; Tihanyi et al., 1993). The cofactors increase the catalytic rate constant for substrate hydrolysis: 300-fold with pVIc and 6000-fold with both pVIc and Ad2 DNA (Mangel et al., 1993). The third cofactor is a cytoplasmic cofactor, a nonviral actin (Mangel et al., 1996).

Roles of cofactors can be brief, as the apoenzyme is inactived before and during the virion assembly and thus it cannot cleave precursor proteins of the virion. When the virions are assembled, the proteinase encounters pVI and cleaves out pVIc, which binds to the proteinase. Then the complex of AVP-pVIc uses the viral DNA to cut the remaining precursor proteins. The viral DNA in this case is used as a guide wire and cofactor. The structure of adenain (AVP) complexed with its 11 residues cofactor was determined at 2.6 Å resolution (Brown et al., 2002). The AVP has catalytic triad residues. These amino acid residues are His54, Glu71, and Cys122. The Cys122 and His54 lie in the middle of the active site groove and considered as conserved amino acids among adenovirus serotypes. The pVIc residues are 32 Å away from the active-site nucleophile Cys122. The Val2′ amino acid of pVIc is the closest to the active site, its side chain is 14.5 Å from Cys122 (Brown et al., 2002). The adenain (AVP) is essential for uncoating virion and is released from the infected cell. It converts the immature noninfectious virions to mature infectious virions. The function of adenain (AVP) is the hydrolysis of precursor proteins at specific consensus sites and the cleavage of viral proteins at other sites (Ding et al., 1996). There are five AVP precursor proteins, PVI, PVII, PVIII, 87K, and 11K. The adenain (AVP) has substrate specificity, which is determined by the secondary structure around a Gly-Ala or Ala-Gly peptide bond (Ruzindana-Umunyana et al., 2002). Webster et al. (1993) identified two consensus sites for cleavage by adenain (AVP). These are (Met,Ile,Leu)-X-Gly-Gly---X or (Met,Ile,Leu)-X-Gly-X---Gly, where X is any amino acid (Gottlieb et al., 1983). The consensus cleavage type Gly-X---Gly is hydrolyzed three- to fourfold faster than Gly-Gly---X site (Weber, 1999).

### 3.1.2    Adenoviral Protease AVP (Adenain) Inhibitors

There are no vaccines or drugs available for specific treatment of enteric adenovirus infection, and symptoms are rarely sufficiently severe to require rehydration. Most compounds reported as adenovirus inhibitors are nucleoside or nucleotide analogs, such as cidofovir, 2′-nor-cyclic GMP, ganciclovir (GCV),

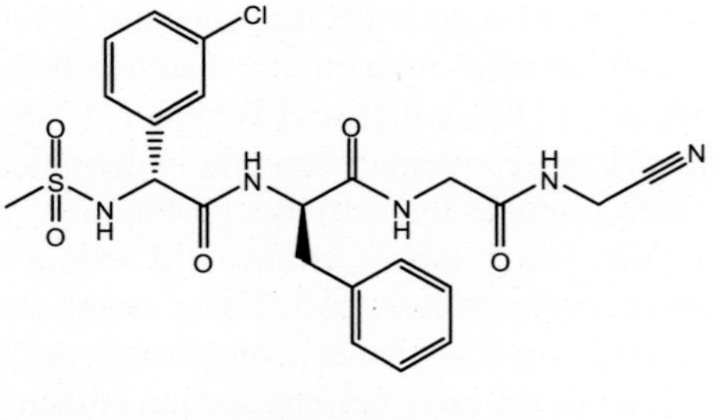

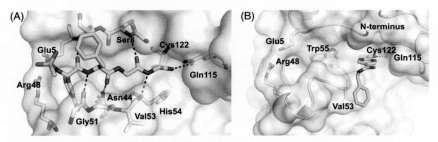

(1), Tetrapeptide nitrile 1 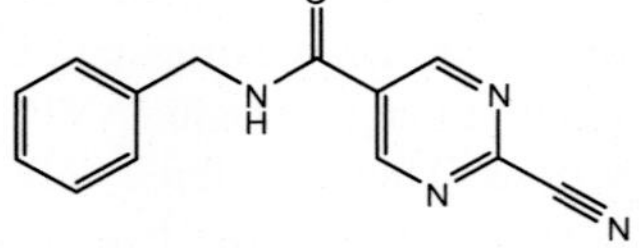            (2), Pyrimidine nitrile 2

FIGURE 2.1    The chemical structure of tetrapeptide nitrile 1 and pyrimidine nitrile 2.

and ribavirin (Diouri et al., 1996). Cidofovir and ribavirin are used in clinical studies with many outcomes. In addition, a number of compounds introduced have potency against adenoviruses in cell culture, but not yet used in patients (Lenaerts et al., 2008).

The adenovirus protease (adenain) is involved in several critical steps during virus propagation, including early and late stages of the replication cycle (Lenaerts and Naesens, 2006). Thus the specific inhibition of adenain may offer an efficacious treatment strategy for adenoviral infections. Beside several experimental studies, many theoretical studies have also been performed to investigate AVP inhibitors. The two-pronged fast track hit discovery approach was used to identify tetrapeptide nitrile 1 (**1**) and pyrimidine nitrile 2 (**2**) as antiadenain compounds (Fig. 2.1). In this study, the X-ray cocrystal structures of both hits in complex with adenain prior to the initiation of hit-to-lead activities were obtained. In the complex, both the inhibitors were found to be covalently bound to the catalytic Cys residue (Cys122) of adenain through their nitrile group (Fig. 2.2) (McGrath et al., 2001). However, tetrapeptide nitrile 1 proved to be a low nM inhibitor of adenain in vitro, but inactive in the replication of virus. Later, however, this compound was developed with increased permeability

FIGURE 2.2    **X-ray cocrystal structures**. Inhibitor tetrapeptide nitrile 1 bound in the active site of AVP2 (A, PDB code 4PIE) and inhibitor pyrimidine nitrile bound in the active site of AVP2 (B, PDB code 4PID). *(From Mac Sweeney, A., Grosche, P., Ellis, D., 2014. Discovery and structure-based optimization of adenain inhibitors. ACS Med. Chem. Lett. 5 (8), 937–941.)*

and, hence, with improved cellular potency by structure-guided design in which attempt was made to decrease the peptidic character of tetrapeptide nitrile 1. Guided by the X-ray cocrystal structure of adenain bound with **1**, a novel peptidomimetic scaffold was developed to form two inhibitor series. Based on the information provided by the X-ray cocrystal structures and molecular modeling studies, it was suggested that the two inhibitor groups have important implications for further medicinal chemistry optimization of inhibitors specific for adenain (Mac Sweeney et al., 2014). Generally, there is a growing need for specific and effective antiadenoviral therapies and development of inhibitors of adenain.

## 3.2 Hepatitis C Virus

Certain viral infections, different from hepatitis A and B, were found to be caused by some virus. This virus was initially called non-A non-B hepatitis virus (NANBH). It is now well identified and called hepatitis C virus (HCV) (Choo et al., 1989). HCV is a hepatotropic virus that infects 150–200 million of people, that is, ~3% of the world's population. Its infection percentage is very high in Asia and the North of Africa (Choo et al., 1989). HCV causes both acute and chronic infections. Acute infection is a nonlife threatening disease and ranges from being asymptomatic to causing a self-limited hepatitis. About 15%–45% of acutely infected patients spontaneously clear HCV within few months after infection. The remaining 55%–85% of patients develop chronic infection (Tibbs, 1997).

HCV is a small enveloped virus of 40–80 nm size, having a single-stranded RNA genome, which has ~9600 nucleotides, positive polarity and high conserved 59 untranslated region followed by a single open reading frame. The role of single open reading frame is encoded as polyprotein of 3010–3033 residues (Zhang and Windsor, 2013). Due to the error-prone RNA-dependent RNA polymerase in HCV, it produces highly variable progeny. It causes the formation of 7 genotypes and more than 80 subtypes of HCV (Catanese et al., 2013); through the viral and cellular proteases, the polyprotein is cleaved co- and posttranscriptionally into 10 polypeptides with a specific function. These functional proteins are the three structural proteins (core, E1, and E2) and seven nonstructural proteins (p7, NS2, NS3, NS4A, NS4B, NS5A, and NS5B) as seen in Fig. 2.3

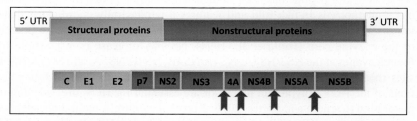

FIGURE 2.3    **The HCV genome.** The polyprotein encoded to three structural proteins followed by seven nonstructural proteins. The cleavage sites are marked as arrows for NS3-4A serine protease function.

(Jackowiak et al., 2014). E1 and E2 are the two envelope glycoproteins. These glycoproteins are important for cell attachment and entry. The core capsid protein forms the viral nucleocapsid P7, responsible for the formation of the ion-channel that plays a role in virus assembly and in the release of infectious virions (Jackowiak et al., 2014; Moradpour and Penin, 2013). NS2 is auto-protease, which affects the replication indirectly and participates in the virus assembly. NS3 protease plays a potent role in the nonstructural proteins processing, as well as some host cellular proteins. NS3 helicase plays a role in viral RNA unwinding and replication. NS4A is considered as a cofactor of NS3 that causes the high stability of NS3 and is responsible for the activity of protease and helicase. NS4B participates in remodeling of membrane and packaging of virus. NS5A is complexed with viral RNA genome and several cellular and viral proteins. Finally, NS5B is considered as the RNA dependent RNA polymerase that is important for replication and HCV assembly (Atoom et al., 2014). The structure and function of NS5B protease and its inhibitors have been widely studied and discussed (Kaushik-Basu et al., 2014; Musmuca et al., 2010; Patil et al., 2011, 2014; Sofia et al., 2012; Zhang et al., 2011). Therefore, we concentrate here on NS3 protease, its structure, function, and inhibitors.

### 3.2.1 NS3 Protease Structure and Function

A key enzyme of the HCV innate immune evasion strategy is the NS3-NS4A protease. This protease is formed from two subunits, NS3 and NS4A. NS3 contains the protease domain (N-terminal, serine protease) and RNA helicase domain (C-terminal). The NS3 protease domain has 181 residues, while the RNA helicase domain has 456 amino acid residues. Sometimes, the NS3 protease is described as chymotryps in-like and trypsin-like protease (Carnero and Fortes, 2016). NS4A contains 54 amphipathic amino acids, with a hydrophobic N-terminus and a hydrophilic C-terminus. NS4A is considered as NS3 protease cofactor and plays a role in assisting in the membrane-localization of NS3 and other viral replicase components (Saleh et al., 2015a). The 2.5 Å resolution X-ray crystal structure of the NS3 protease domain complexed with a synthetic NS4A activator peptide has been studied. This study reported that the protease has a chymotryps in-like fold and features a tetrahedrally coordinated metal ion distal to the active site. The NS4A peptide intercalates within a β sheet of the enzyme core (Failla et al., 1994). NS3/4A protease is responsible for four cleavages at the NS3/NS4A, NS4A/NS4B, NS4B/NS5A, and NS5A/NS5B sites (natural substrates) (Fig. 2.3). These proteolytic cleavages at four junctions of the HCV polyprotein precursor are essential for HCV lifecycle and its maturation (Carnero and Fortes, 2016; Kim et al., 1996). A natural decapeptide substrate for the HCV NS3-4A protease is described as $NH_2$-P6-P5-P4-P3-P2-P1-P1'-P2'-P3'-P4'-OH, with the scissile bond located between the P1 and P1' residues. It has six residues on the N-terminal domain and four residues on the C-terminal domain. The consensus site for natural decapeptide substrate sequence is (Asp/Glu)-X-X-X-X-(Cys/Thr)↓(Ser/Ala)-X-X-X, whereas X indicates variable residue (Manabe et al., 1994).

The three-dimensional structure of the NS3/NS4A complex consists of two structural domains, each having a twisted β sheet incorporating a "Greek key" motif. The C-terminal site has the conventional six-stranded β barrel, followed by a structurally conserved α helix. The core of this barrel is packed with hydrophobic residues that emanate from all six strands of the β barrel. The N-terminal site contains eight β strands rather than six, including the one contributed by NS4A (Failla et al., 1994). The active site of NS3 protease is shallow on the surface of the enzyme. The NS3 protease active site has conserved catalytic triad constituted of residues His57, Asp81 (in the N-terminal site of the NS3 protease) and Ser139 (in the C-terminal site). Substitution of any of the catalytic triad residues causes cleavage at NS3/NS4A, NS4A/NS4B, NS4B/NS5A, and NS5A/NS5B sites (Carnero and Fortes, 2016). The C-terminal domain contains a zinc ion tetrahedrally coordinated with four residues, Cys97, Cys99, Cys145, and His149. The zinc ion plays a significant role in the stability of NS3 protease structure. This metal ion is located at one end of the β barrel, at least 20 Å away from the catalytic Ser139 (Failla et al., 1994).

### 3.2.2 NS3 Protease Inhibitors

There is no vaccine for prevention of HCV infection. However, there are two challenges against the development of an effective vaccine. The first challenge is the little evidence of a neutralizing antibody response to infection. The second one is the extreme variability of the virus that causes potency restrictions of vaccines derived from only one viral strain (Ortqvist, 2010). The traditional therapy of HCV infection is a combination of pegylated interferon (IFN) alpha (α) with the antiviral nucleoside analog ribavirin (Jo et al., 2006). The mechanism of these treatments begins with the binding of interferons to specific receptors in cell surface. This causes an intracellular signaling cascade ultimately resulting in the proteins synthesis that mediates pleiotropic activities, including antiviral, antiproliferative, and immunomodulatory responses (McHutchison et al., 1998). Although the IFN-α causes the decrease of HCV viral load and the improvement of liver function that may reduce the risk of the development of hepatocellular carcinoma. This treatment has several disadvantages, which forces the discontinuation of this therapy. The disadvantages of IFNα-ribavirin therapy are its poor effect, dependence on HCV genotype, high cost, and many side effects, such as influenza-like symptoms, depressions, and hemolytic anemia (Carney and Gale, 2006).

Today, there are several drugs that have been approved for the treatment of HCV (Fig. 2.4), which targets the viral NS3 protease, such as boceprevir (**3**) and telaprevir (**4**), the RNA dependent RNA polymerase NS5B, such as sofosbuvir (**5**) or NS5A protease, such as ledipasvir (**6**), and dataclasvir (**7**)

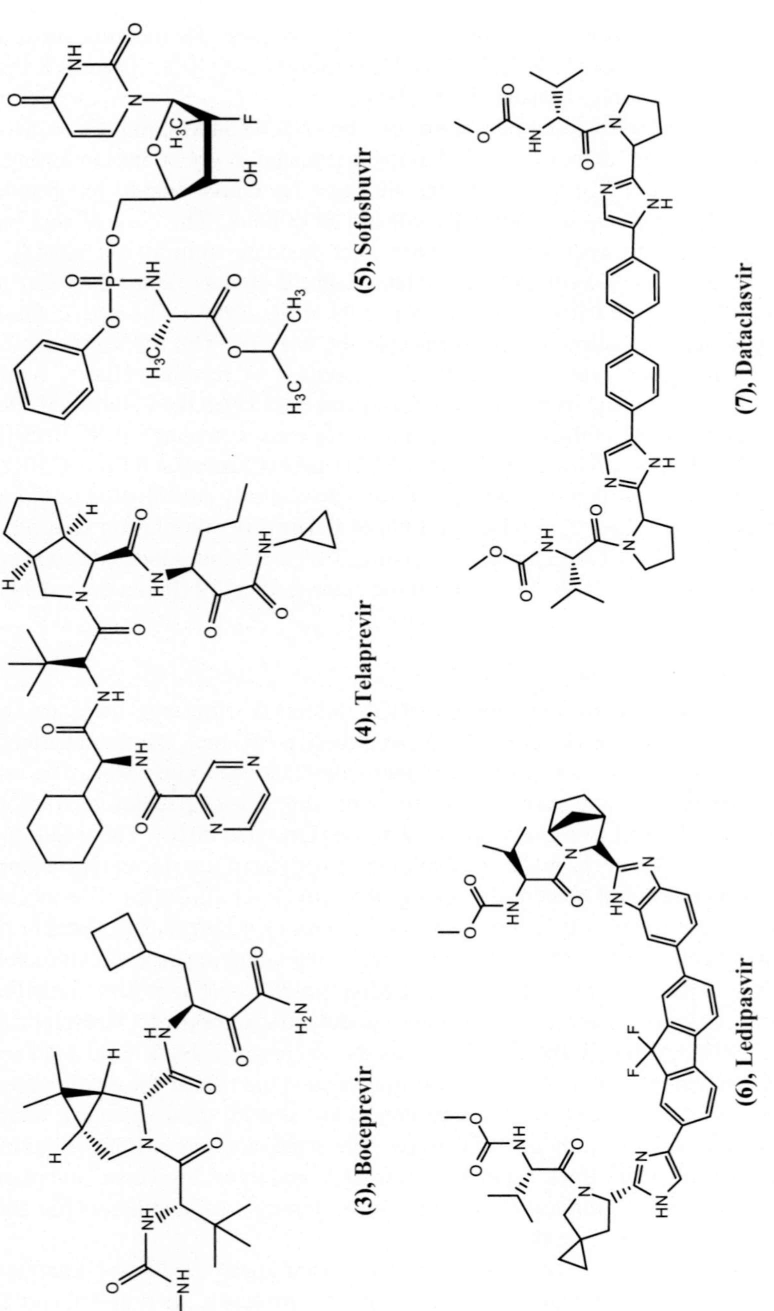

FIGURE 2.4 The structure of some approved HCV inhibitors. Boceprevir, telaprevir, sofosbuvir, ledipasvir, and dataclasvir.

(3), Boceprevir

(4), Telaprevir

(5), Sofosbuvir

(6), Ledipasvir

(7), Dataclasvir

(Poordad and Dieterich, 2012). However, still there is an urgent need for more effective antiviral therapeutics. The major obstacle in achieving that goal is the inability to cultivate HCV in vitro and the absence of a small animal model susceptible to HCV infection. This motivated more research efforts in this field. The only animal that can be infected by HCV is chimpanzee, but its use is limited by ethical reasons, its scarcity and high maintenance costs (Xie et al., 1998).

The computer aided drug design and molecular modeling overcome this obstacle and also save money. There are many molecular modeling and theoretical studies on anti-HCV research especially on HCV-NS3 protease inhibitors. Molecular modeling and three-dimensional quantitative structure–activity relationship (3D-QSAR) studies with comparative molecular field analysis (CoMFA) have provided guidelines to synthesize and modify new HCVNS3 peptidomimetic inhibitors. New tripeptide inhibitors, incorporating α- and β-amino acids at P1, P2, and P3 positions of a potent tripeptide scaffold, were studied to understand the structural implications (Nurbo et al., 2008). There were efforts to investigate nonpeptic HCVNS3 protease inhibitors by performing the molecular dynamics simulations combined with MM–PBSA (molecular mechanics and Poisson–Boltzmann surface area). Thus one could predict the binding mode of the polyphenol inhibitors, which included a glucopyranose ring and two or more galloyl residues substituted at the hydroxyl moiety in the binding pocket of the HCV NS3serine protease. This study suggested that the most favorable binding mode is that two galloyl residues at 3 and 4 positions of the glucopyranose ring of the inhibitors interact with Ser139, Gly137, Ala157, and Asp81 through hydrogen bond interactions and with Ala156 and Hie57 through hydrophobic interaction that are essential for the activities of the these inhibitors (Li et al., 2007). A series of indole derivatives were designed and their molecular modeling simulation study, including fitting to a 3D pharmacophore model in addition to their docking into the NS3 active site, was performed to see their action as HCV NS3 protease inhibitors. Then the compounds were synthesized and biologically evaluated in vitro using HCV protease binding assay. The experimental results were consistent with molecular modeling study (Ismail and Hattori, 2011). Novel two groups of peptidomimetic hexapeptide compounds were introduced as HCV NS3 protease inhibitors. The hexapeptide is an amino acid sequence of NS5A/NS5B substrate (Glu-Asp-Val-Val-Cys-Cys). This amino acid sequence is responsible for the Egyptian genotype 4a. The first group of hexapeptides were found to bind to a cellulose monomer at the positions 2, 3, or 6, while in the second group of these peptides were found to bind to a cellulose dimer at the positions 2, 3, 6, 2′, 3′, or 6′. Molecular modeling semiempirical PM3 calculations were used to calculate the electronic properties of the molecules to be compared with those of natural substrate. Computational results showed that the second group of hexapeptides were more stable, hydrophilic, and soluble in polar solutions (Ibrahim et al., 2012b, 2013).

Some modifications were made in the structures of the two NS3 protease inhibitors, telaprevir and boceprevir, to improve their biological activity. Modifications were done, by adding some functional groups at different positions of the compounds. When physicochemical and electronic properties of these modified structures were calculated, the compound with 1,3-dithiolane ring at position R2 had more favorable electronic and physicochemical properties than telaprevir and the fluorinated sulfonamide at position R1; 1,3-dithiolane ring at position R2; and cyclopropane at position R3 in boceprevir were found to increase the activity of boceprevir because of their electronic and physicochemical properties (Saleh et al., 2014, 2015b). The integration of structure-based drug design "molecular docking" and ligand-based drug design "QSAR" were used in order to evaluate the potency of some suggested HCVNS3 protease inhibitors. The design of these compounds was not only based on the N-terminal hexapeptide product of hydrolysis at NS5A/NS5B junction (Glu-Asp-Val-Val-Cys-Cys), but also based on the (Asp-Asp-Ile-Val-Pro-Cys) at HCV NS5A/5B junction to form three modified groups. The first group had linear peptides but the second and third groups had P1–P3 and P2–P4 macrocyclic structures, respectively. While the QSAR results exhibited the preference for macrocycles, the docking results showed that these macrocycles had the best docking scores. Its docking interaction with the NS3 protease is represented in Fig. 2.5 (Mostafa et al., 2014).

The charged compounds also were studied theoretically as HCV NS3 protease inhibitors. In this study, the QSAR descriptors were calculated for both neutral and charged compounds that were built from tetrapeptides, hexapeptides, or macrocyclic structures by linking the side chains of residues (P1 and P3). The peptide structure of these compounds was based on the sequences of NS3/NS4A, NS4A/NS4B, NS4B/NS5A, and NS5A/NS5B junctions. The P1–P3 macrocycle of NS4A/NS4B hexapeptide sequence (DEMEEC) either neutral or charged was the most stable and hydrophilic in nature. These best compounds were reoptimized at higher level of methods, B3LYP/6-31G (d, p) and HF/6-31G (d, p) methods. This study concluded that the macrocyclic compounds had low conformational flexibility and thus they bind more tightly to the active site than linear peptides and are not hydrolyzed by cellular proteases (Ezat et al., 2015). Molecular modeling and docking studies guided the design of novel nonpeptidomemitic inhibitors, that is, a series of azetidine macrocyclics. This approach predicted that the quinoline motif orients in close proximity to the catalytic triad and the stability of the HCV protease inhibitor is increased. Structure-based drug design showed that the azetidine group has excellent activity and selectivity in biochemical assays and also has lower sensitivity to mutants in the current families of anti-HCV protease compounds. But further biological profiling of this group led to disappointing metabolic stability and the series was terminated. Further study led to a new series that led to develop a clinical candidate IDX320 (**8**) (Parsy et al., 2015).

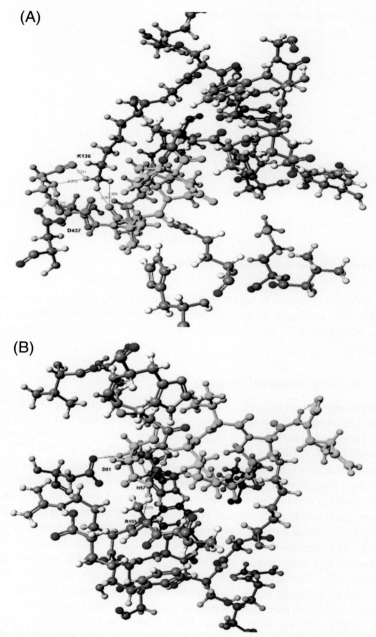

**FIGURE 2.5  The binding mode of some suggested macrocycles compounds with NS3 protease.** The compounds shown are (A) 5A/5B hexapeptide sequence and (B) 5A/5B azahexapeptide sequence. The compounds are shown in *yellow* and NS3 protease the active site residues are colored (carbon, oxygen, nitrogen, hydrogen, and sulfur are shown in *gray, red, blue, white*, and *yellow* colors, respectively). Hydrogen bonds are shown as *blue lines*. *(From Mostafa, H.I., El-bialy, N.S., Ezat, A.A., et al., 2014. QSAR analysis and molecular docking simulation of suggested peptidomimetic NS3 protease inhibitors. Curr. Comput. Aided Drug Des. 10 (1), 28–40.)*

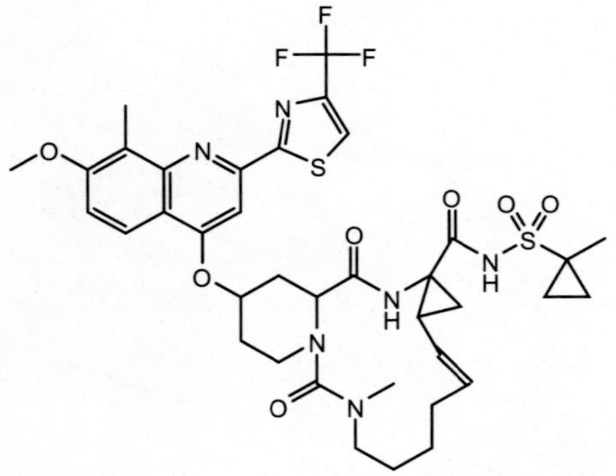

**(8), IDX320**

FIGURE 2.6    The chemical structure of IDX320.

The IDX320 structure is represented in Fig. 2.6. Finally, it is essential to develop and introduce more effective HCV inhibitors, due to many error-prone NS5B in HCV which led to the production of variable progenies and subtypes.

## 3.3    Human-Immunodeficiency Virus

The first information about human-immunodeficiency virus (HIV) was known in the US when many deaths occurred among patients that had pneumocyctis carinii pneumonia and Kaposi's sarcoma. However, after all trials for treatments of the different conditions, the patients still died (Gottlieb et al., 1981). In 1983, a new human retrovirus was isolated from such patients. Such patients suffered from long fatal illness, which was named as acquired immunodeficiency syndrome (AIDS) and its causative agent was recognized to be a retrovirus called human immunodeficiency virus of type 1 (HIV-1) (Barre-Sinoussi et al., 1983). The HIV infection pathogenesis resulted in decreasing the number of CD4+ T-cells. The decrease in the number of CD4+ T-cells leads inexorably to the appearance and development of AIDS that leads to death of the infected people (Fauci, 1993). The AIDS is thus considered the most advanced stage of HIV infection, an epidemic with significant health challenges worldwide. HIV-1 has spread to almost all parts of the world. In 2007, Joint United Nations Program on HIV/AIDS (AIDS Epidemic Update, 2007) reported that 33.2 millions of people were living with HIV worldwide and 2.1 millions of people were killed by AIDS. During the year 2007, 2.5 millions of people were newly infected with HIV (UNAIDS/WHO, 2007). According to the World Health Organization (WHO), a total of 37.20 millions of people are living with AIDS and 1.20 million died in 2014 alone (Islam and Pillay, 2016). Although the number of mortality has decreased, still the rate of people living with AIDS was increasing every year.

AIDS is considered a dreadful virulent disease still causing havoc worldwide. The shattering potential of this viral disease has not been fully realized because there is no curative treatment for this fatal disease. There are two types of HIV that cause AIDS, type 1 and type 2 (HIV-1 and HIV-2). It is the former, which is widely spread in the world, the latter one is mostly restricted to the western African regions (Guyader et al., 1987). The HIV-1 virion is spherical in shape with diameter of around 100 nm. A lipid bilayer membrane envelops the viral particle. The envelope is gp160 precursor that is cleaved by a furin (a ubiquitous subtilisin-like proprotein convertase) into two heterodimers (gp120 and gp41). In the gp120 subunit (SU), there are five variable loops (V1–V5), while the gp41 subunit has more conserved sequence. This is because it houses the fusion machinery. Through the gp41 (the transmembrane, TM protein), the glycoproteins are anchored to the virus and beneath p17 [the membrane A (MA) matrix shell protein] lines the inner surface of the particle. Both nucleocapsid proteins NC (p7) and the viral enzymes cause stabilization of the two copies of the viral RNA. The viral RNA is encapsulated by a cone-shaped structure consisting of capsid proteins CA (p24). The three viral enzymes are reverse transcriptase (RT) (p66, p51), integrase (IN) (p31), and protease (PR) (p11). The genetic information of HIV-1 carries three genes; gag, pol, and env. These genes, expressed as large unfunctional polyproteins, are processed into the functional proteins by either the viral protease (gag and pol) or host cellular enzymes (env). The gag encodes CA, MA, and NC, the pol encodes the viral enzymes PR, RT, and IN, and env processing results in the formation of functional SU and TM proteins (Fig. 2.7) (Ward and Wilson, 2015; Sluis-Cremer and Tachedjian, 2002). HIV-1 attacks immune system especially helper T-lymphocytes as host cell. The infection is initiated via the binding between gp120 virus and CD4 protein of host cell that acts as receptor for HIV. The replication of HIV may cause cell death with release of the more of infective and matured viral particles. This apoptosis plays important role in decreasing T cell counts during HIV infection.

### 3.3.1   HIV Protease Structure and Function

The HIV-1 protease (HIV-1 PR) is a virus-specific aspartic protease responsible for processing the polyproteins of gag and gag-pol during virion maturation and for the proliferation of the retrovirus (Fig. 2.7). The activity of HIV-1 PR

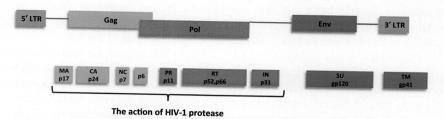

The action of HIV-1 protease

FIGURE 2.7    **The HIV-1 genome and sites at which the function of HIV-1 protease appeared.**

enzyme is essential for virus infectivity, and thus it is an important target for the development of anti-AIDS drugs. The HIV-1 PR can recognize Phe-Pro or Tyr-Pro sequence as the retrovirus-specific cleavage site. The HIV-1 PR was structurally characterized by X-ray crystallography for the first time in 1989 (Wlodawer and Erickson, 1993). The HIV-1 PR is a homodimer with C2 symmetry. The homodimeric protein consists of two identical 99 amino acid subunits, each of which has one catalytic aspartic acid Asp25, 25'. According to Ishima et al. (2001), amino acid sequences of mature HIV-1 protease are PQITL-WQRPLVTIKIGGQLKEALLDTGADDTVLEEMSLPGRWKPKMIGGIG-GFIKVRQYDQILIEICGHKAIGTVLVGPTPVNIIGRNLLTQIGCTLNF. The C2-symmetric active site is located at the dimer interface and each subunit contributes one catalytic aspartic acid in the tripeptide sequence, Asp25, 25' -Thr26, 26' -Gly27, 27'. The protease subunit fold contains a compact structure of β-strands and a short α-helix near the C-terminal. The antiparallel β-strands, constituted by residues 44–57 from both subunits, form a flexible "flap" region that is thought to fold down over the active site during catalysis in order to bind with substrate and exclude water (Sluis-Cremer and Tachedjian, 2002). The flap region regulates substrate entry into the active site by its conformation, which can be open, semiopen or closed (Hornak et al., 2006). The intermolecular interactions stabilize the two subunits. The four-stranded antiparallel β-sheets form the largest part of the interface which consists of residues 1–5 in N-terminal and residues 95–99 in C-terminal of the protease monomers (Sluis-Cremer and Tachedjian, 2002). The active site contains eight C2-symmetric subsites (S4, S3, S2, S1, S1', S2', S3', and S4'), which have the binding sites for the P4, P3, P2, P1, P1', P2', P3', and P4' residues of an octapeptide substrate numbered from the scissile bond. The scissile bond is defined as the bond at which the hydrolysis cleavage occurs (Andersson et al., 2003).

### 3.3.2 HIV Protease Inhibitors

HIV is one of the most fatal infectious diseases of the late 20th century with no effective treatment for this serious global threat and because of the resistance by the mutation-prone HIV there are no effective drugs till today. Therefore, there is an urgent need for developing new inhibitors with greater potential (Edward, 2008). One of the challenges in AIDS therapy is that the HIV virus develops drug-resistant variants rapidly by genetic mutation because of two reasons. The first reason is that HIV has a very high reproduction rate and the second reason is the large number of mutated forms, which occur during the transcription of viral RNA to DNA. These factors finally produce high error rate of the RT.

Anti-HIV drugs targeting the HIV-1 PR or RT may result into dramatic suppression of viral replication in infected individuals but fail to achieve complete viral suppression. HIV inhibitors can be divided into five classes according to their therapeutic target in the HIV-1 replication cycle. These classes are nucleoside reverse transcriptase inhibitors (NRTI), nonnucleoside RT inhibitors, protease inhibitors, fusion inhibitors, and cocktail therapy or highly

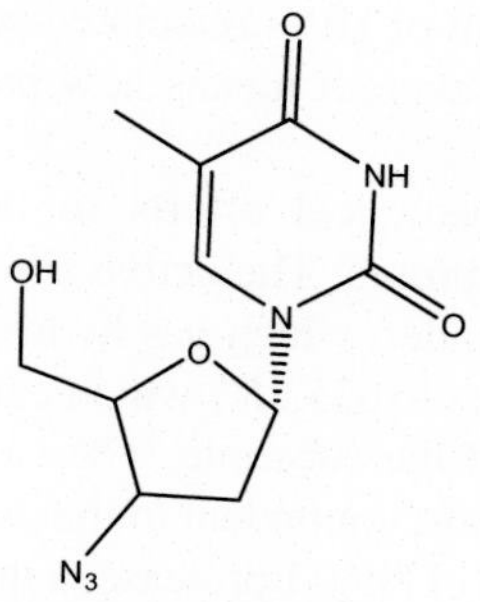

**(9), AZT (Azidothymidine)**

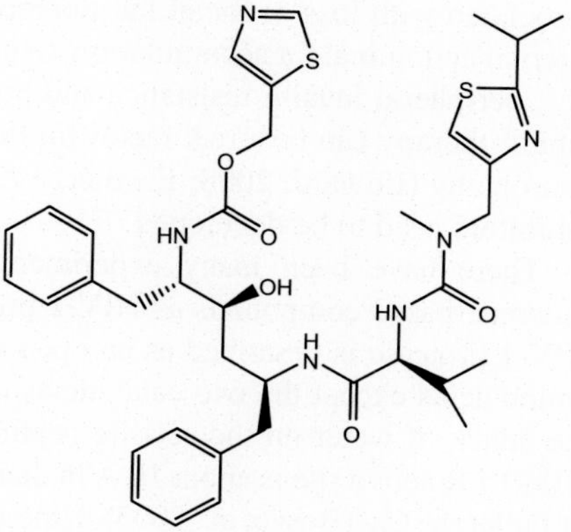

**(10), Ritonavir (Norvir)**

FIGURE 2.8    **The chemical geometries of AZT and Norvir.**

active antiretroviral therapy (Edward, 2008; Jenny, 2006). The first drug approved against HIV-1 is Zidovudine (ZDV) also called Azidothymidine (AZT) (**9**) (Fig. 2.8). It is a nucleoside analog that acts as an HIV-1 RT inhibitor. This is still widely used in combination with other antiretroviral drugs (Odriozola et al., 2003). Unfortunately, taking AZT may result in more severe side effects. AZT has common side effects that include nausea, vomiting, headache, dizziness, fatigue, weakness, muscle pain, and kidney disorders. It can also lead to severe bone marrow suppression with anemia, usually in the first few months of administration. These toxicities are more severe and more common in people with damaged immune systems (Herdewijn and De Clercq, 1990).

Generally, the significant problems in using anti-HIV drugs are the drug intolerance and drug toxicity. NRTI therapy causes liver toxicity with steatosis and lactic acidosis, abdominal pain, nausea, or vomiting, and with mortality rate near 50% (Arenas-Pinto et al., 2003). Moreover, mitochondrial toxicity occurs during NRTI therapy. A major problem with this toxicity is its time dependency and thus delayed effects. A major problem with this toxicity is its time dependency and therefore delayed onset. Due to the long-term therapy for several months with nucleoside analogs, the multiorgan side effects are observed. In some cases, reversal of symptoms was obtained after cessation of the drugs, and in some cases toxicity persisted despite drug discontinuation, occasionally with a fatal outcome (Brinkman et al., 1998). The therapy with HIV-1 PR inhibitors, especially with ritonavir (**10**) (Fig. 2.8), is associated with hepatic transaminase elevation (Sulkowski et al., 2000). The side effects of PR inhibitors therapy are gastrointestinal symptoms including nausea, vomiting, and diarrhea. Their use may be accompanied by a peculiar adipose tissue redistribution known as protease inhibitor-associated lipodystrophy, though this phenomenon may also occur in AIDS affected persons not taking protease inhibitors. This syndrome is associated with loss of facial fat, dorsocervical tissue accumulation, increased internal abdominal fat accumulation, hyperlipidemia (often exceeding 1000 mg/dL), peripheral insulin resistance and impaired glucose tolerance. Protease inhibitor therapy can be a risk factor for development of HIV-associated sensory neuropathy (Edward, 2008; Flexner, 1998). In essence, it means new protease inhibitors need to be developed.

There have been many experimental and theoretical efforts to develop fullerene-based compounds as HIV-1 protease inhibitors. The active site of the HIV-1 protease is described as an open-ended cylinder, which has hydrophobic amino acids except the two catalytic aspartic acids (Asp25,25′) which catalyzes the attack of water on the scissile peptide bond of the substrate. The cavity of HIV-1 PR active site is about 10 Å in diameter, almost equivalent to the diameter of Fullerene $C_{60}$ (Bosi et al., 2003). Kinetic analysis of HIV-1 protease in the presence of various water-soluble fullerene derivatives suggests a competitive mode of inhibition. This is due to the water-soluble fullerene derivatives that are able to form hydrogen bonds with the catalytic aspartic acids, as well as van der Waals contacts with the nonpolar HIV-1 protease residues. Various structural modifications have been tried to increase the hydrophobicity of the molecules in order to have the compounds with increased biological and pharmacological activity (Lee et al., 2007). Friedman et al. (1993) performed molecular modeling studies for two fullerene derivatives using the program DOCK3. They discovered that HIV-1 protease could be complexed and inhibited by van der Waals interactions by introducing a $C_{60}$ molecule into its hydrophobic catalytic site. They suggested that adding the functional groups at a specific position on fullerene surface might enable the compound to have electrostatic and hydrogen bond interactions with Asp25 and Asp25′ and increase the binding constant by several orders of magnitude (Lee et al., 2007). The same research group synthesized the first fullerene inhibitor of HIV-1 protease. This compound showed a good fitting in the active site of the HIV-1 protease with strong van der Waals interaction. MD simulations were performed for the HIV-1 protease complexed with lead fullerene-based inhibitor (diphenyl $C_{60}$ alcohol) in the three protonated states of enzyme: unprotonated, monoprotonated, and diprotonated states at Asp25 and Asp25′. This study was performed to investigate the behavior of these compounds as HIV protease inhibitors (Lee et al., 2007). Some theoretical studies were devoted toward the solution of the problems limiting the use of fullerenes as HIV protease inhibitors. This was carried out through theoretical investigation of novel fulleropyrrolidines derivatives using extensive molecular modeling. These novel fulleropyrrolidines derivatives were modified by the addition of chalcogene atoms (O, S, or Se) and hydroxymethylcarbonyl (HMC) group to form two groups of hydroxy-chalca-acetic acid-(4-pyrrolidin-1-yl-phenyl) ester[$C_{60}$-$C_2H_4N$-(4-XCOCH$_2$OH)$C_6H_4$] and hydroxy-chalcoacetic acid-[2-(2-hydroxy-acetylchalcanyl)-4-pyrrolidin-1-yl-phenyl] ester [$C_{60}$-$C_2H_4N$-(3,4-XCOCH$_2$OH)$C_6H_4$] where X is O, S, or Se atom. The first group has one function group of (HMC + O, S, or Se), while the second group has two functional groups of (HMC + O, S, or Se). These

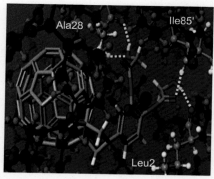

FIGURE 2.9    **The docking interaction between the ligands (with two oxygen atoms and HMC groups) and HIV-1 protease.** The cylinder molecule is considered the studied compound and the ball & cylinder molecule is considered the amino acid residues of protein. The *yellow dashed line* consider the hydrogen bonds between the ligands and HIV-1 protease. *(From Saleh, N., 2015. The QSAR and docking calculations of fullerene derivatives as HIV-1 protease inhibitors. Spectrochimica Acta 136, 1523–1529.)*

compounds provided a tight fit between $C_{60}$ and the cavity of the active site of HIV-1 protease (Zhang and Windsor, 2013; Jackowiak et al., 2014). This was due to the hydrophobic (van der Waals) interaction between the outer surface of $C_{60}$ and the hydrophobic residues of the active site of HIV-1 protease cavity. Also these compounds could form hydrogen bonding interactions with the two polar residues of the conserved aspartic acid through HMC group . A QSAR study indicated that the oxygen molecule of the second group, [$C_{60}$] fulleropyrrolidine-1-hydroxy-acetic acid 2-(2-hydroxy-acetoxy)-phenyl ester [$C_{60}$–$C_2H_4N$–(4-$OCOCH_2OH$) $C_6H_4$], could be the most promising HIV-1 protease inhibitor (Ibrahim et al., 2010a,b, 2012a). Thus the best compound was modified to increase its potency by repeating the function groups (HMC + O atom) from one to five function groups at different positions on phenyl ring. The docking was performed to investigate the mode of interaction between the compounds and HIV-1 PR active site. This docking showed that the compound with two oxygen atoms +HMC groups interacts with the active site of HIV-1 protease due to hydrophobic nature of HIV-1 protease active site residues except the two aspartic acids. The mode of this interaction is shown in Fig. 2.9. Thus the increase in the hydrophilicity and polarity of the compound would reduce this hydrophobic interaction between the compound and the protease (Saleh, 2015).

The inhibitory potencies of peptidomimetic protease inhibitors containing a novel dihydroxy ethylenediamine—Phe-Ψ[CHOH–CHOH]-Pro—core were predicted by molecular modeling approach (Frecer et al., 2008). A series of nonpeptide HIV-1 protease inhibitors were also studied by molecular docking to analyze the interaction mode of compounds with the enzyme and by QSAR to predict bioactivities of some new compounds (Deeb et al., 2012). The docking and QSAR analysis were also performed to evaluate the activity of new

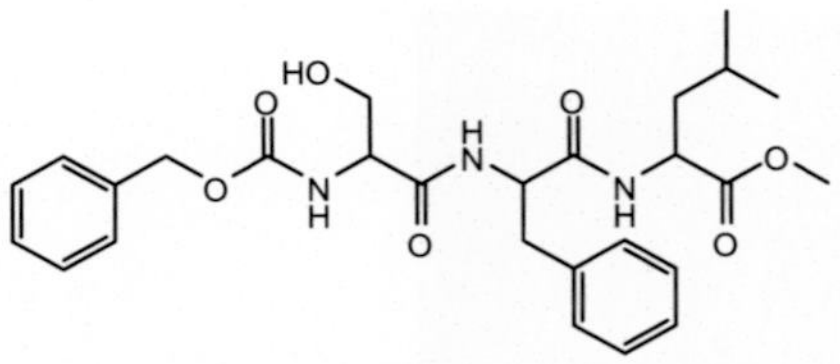

**(11), NSC111887**

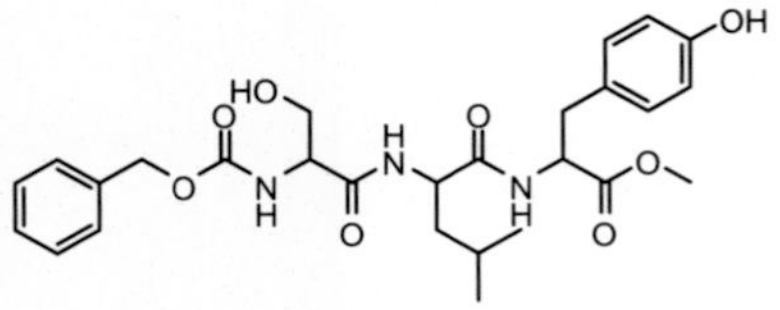

**(12), NSC121217**

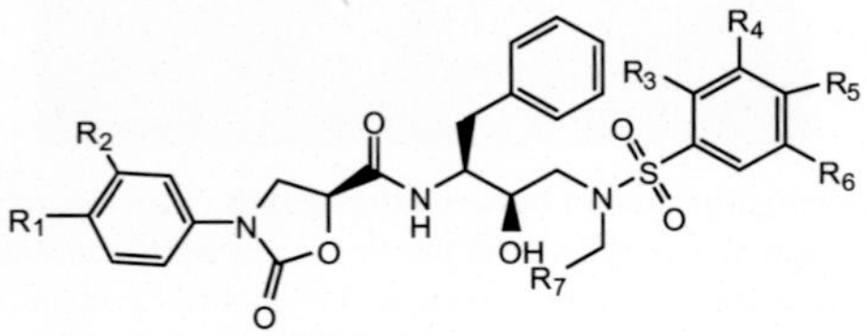

**(13), Oxazolidinone derivatives**

FIGURE 2.10   **The general chemical structure of NSC111887, NSC121217, and oxazolidinone derivatives.**

suggested HIV-1 protease inhibitors containing hydroxyethylamine core, different hydroxyprolinamide P2 ligands (Yan et al., 2012), some P1-substituted biaryl amprenavir derivatives, a series of tetrahydropyrimid-2-ones (Rao et al., 2013) and 71-membered library of bicyclic peptidomimetics based on the 6,8-dioxa-3-azabicyclo (3.2.1)-octane scaffold (Calugi et al., 2014). A new series of HIV-1 protease inhibitors related too phthalamide derivatives, that could act as the P2–P3 ligands in the active site of the enzyme at S2 and S3 subsites, were designed and synthesized. The variety of acyclic, cyclic, and heterocyclicamide derivatives were investigated. Also substituted prolines and oxazoles, which contain hydrogen bond donor and acceptor groups, were studied for specific interactions in the S3 subsite. A number of inhibitors exhibited excellent enzyme inhibitory and antiviral activity and formed a strong hydrogen bond with HIV-1 protease (Ghosh et al., 2015). The multistage virtual screening method might reduce the false-positive rate and improve the prediction accuracy. Some suggested HIV-1 protease inhibitors were investigated by multistage virtual screening method. The method sequentially applied support vector machine, shape similarity, pharmacophore modeling, and molecular docking. According to the theoretical study, six compounds were selected for further bioassay and two molecules, NSC111887 (**11**) and NSC121217 (**12**) (Fig. 2.10), showed inhibitory potency against HIV-1 protease (Wei et al., 2015). Oxazolidinone derivatives (**13**) (Fig. 2.10) were also studied as HIV-1 PR inhibitors. The 2D and 3D QSAR studies on oxazolidinone derivatives indicated that the significance of the presence of two aromatic and three hydrogen bond acceptor features (Ravichandran et al., 2016). Generally, compounds were evaluated theoretically and found promising HIV-1 protease inhibitors.

## 3.4 Picornavirus

Picornaviridae family is one of the largest viral families that is considered medically and economically most important class of human and animal viral pathogens (Ehrenfeld et al., 2010). The picornavirus is one of genera of Picornaviridae family. The name picornavirus is derived from the metric termpico, meaning very small RNA, referring to the type of nucleic acid in the viral genome. Thus *"pico-rna-virus"* literally means "very small RNA virus" (Melnick et al., 1963). Picornaviruses are nonenveloped, small, positive-sense single-stranded RNA genome packed in an icosahedric capsid with 60 subunits or protomers, arranged around axes of five-, three-, and twofold symmetry with 22–30 nm in diameter (Jubelt and Lipton, 2014). The picornavirus genome varies between 7 and 8.8 kb in length and shares a conserved organization throughout the family (Knowles et al., 2012). The single-stranded RNA consists of three regions. The first region is a 650–1300 nucleotide 5′-untranslated region. The 5′-end is covalently linked to the small viral protein of 22 amino acids (Lee et al., 1977). The second region consists of a single and long open reading frame encoding the viral polyprotein that is cleaved during translation by virally encoded proteases. Finally, the third region is a relatively short (~100 nm) 3′-untranslated region followed by a poly(A) tract of variable length (Toyoda et al., 2007).

Picornaviruses are masters of efficient traveling. They are adopted by a plethora of hosts, including plants, insects, and vertebrates. There are 285 different types of picornavirus, which form 29 species distributed among 14 genera. The picornavirus genera include *Aphthovirus, Avihepatovirus, Cardiovirus, Enterovirus, Erbovirus, Hepatovirus, Kobuvirus, Parechovirus, Sapelovirus, Senecavirus, Teschovirus, Tremovirus,* and the newly proposed genus *Cosavirus* (Fig. 2.11). Picornaviruses cause outbreaks and serious diseases, since picornavirus infections rapidly induce several "predatory" modifications of the cell that generally leads to cell death. Five of the picornavirus genera include pathogens infecting humans. They are *Enterovirus, Parechovirus, Hepatovirus, Cardiovirus, Kobuvirus,* and *Cosavirus* (Fig. 2.11) (Stanway et al., 2005).

FIGURE 2.11   **Various picornavirus genera.**

The *Enterovirus* genus is considered as the largest among the *Picornavirus* genera with 219 virus types in 10 species circulated worldwide. Seven species are known as enteroviruses that infect humans or other animals. Human enteroviruses were originally classified according to their pathogenicity in humans and animals. Human infections caused by viruses of the *Enterovirus* genus are often mild or asymptomatic with upper respiratory tract symptoms and/or exanthemas, although they may also be associated with more severe illnesses, such as agammaglobulinemia, chronic dilated cardiomyopathy, myocarditis, and chronic fatigue syndrome (Norder et al., 2011). The *Enterovirus* genus has three species of human rhinoviruses, HRV-A, -B, and -C, and leads to upper respiratory illness (Palmenberg et al., 2009).

The second class of picornavirus is constituted of human parechoviruses that infect mainly young children. They cause the gastrointestinal and respiratory tract infections, and may lead to acute flaccid paralysis, encephalitis, septicmeningitis, myocarditis, neonatal sepsis, and Reye syndrome (Wolthers et al., 2008). Since parechoviruses serotypes have been identified recently, the diseases related with them are not well studied (Norder el al., 2011).

The *Hepatovirus* genus has only one species, hepatitis A species. This species is responsible for hepatitis A (Martin and Lemon, 2006). The hepatitis A is mostly self-limiting. Only 10% of infected children are symptomatic, whereas 70% of infected adults have clinical hepatitis (Brundage and Fitzpatrick, 2006).

*Cardiovirus* has two species, encephalomyocarditis virus and theilovirus, which are mainly animal viruses. The encephalomyocarditis virus contains a single serotype. The encephalomyocarditis virus spreads worldwide with rodents being its natural reservoir. Its infection with other mammalian species is sometime fatal: it causes mortalities and is associated with myocarditis and encephalitis. Theiloviruses are classified into five types, two of which infects humans. These two types of viruses are vilyuisk human encephalomyelitis virus and saffold virus, which cause respiratory diseases, gastroenteritis, myocarditis, acute, and chronic encephalitis (Blinkova et al., 2009).

The *Kobuvirus* genus is classified into two species, *aichi virus* and bovine *Kobuvirus*. *Aichi virus* was discovered and isolated in 1989 as the likely cause of oyster associated gastroenteritis in Japan (Yamashita et al., 1991). Currently, *aichi virus* is subdivided into three genotypes (A, B, and C), all cause human infection (Ambert-Balay et al., 2008).

Members of the newly proposed *Cosavirus* genus were first identified and isolated in 2008 from both healthy children and children with acute flaccid paralysis in Pakistan (Kapoor et al., 2008). After that, the human *Cosavirus* was reported in an Australian child with acute diarrhea and in Chinese and Brazilian children with and without diarrhea (Dai et al., 2010). Finally, its infection was reported in an adult with diarrhea in Thailand (Khamrin et al., 2012). *Cosavirus* genus is classified into five genomic groups (from A to E) (Holtz et al., 2008).

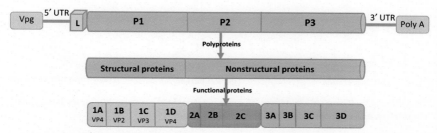

**FIGURE 2.12   The picornavirus genome.** Polyprotein and functional structure and nonstructural proteins. The 5′-untranslated region is covalently linked to the small viral protein and the 3′-untranslated region is linked to a poly A, the small leader (L) peptide bind to P1 region.

### 3.4.1   Picornavirus Protease Structure and Function

The open reading frame is translated into a single large polyprotein, of approximately 2200 amino-acid residues, which is subsequently cleaved by viral protease(s) into 11–12 individual viral mature proteins. The single open reading frame is divided into three regions: the first regions (P1), encoding the structural proteins, and the second and third regions (P2 and P3), encoding the nonstructural proteins (Fig. 2.12). The overall organization of picornavirus polyproteins starts with the structural proteins of the capsid, located at the amino-terminal portion of the polypeptide, and then is followed by the nonstructural proteins. In some genera, the P1 region is preceded by a small leader (L) peptide (Martinez-Salas and Ryan, 2010). The structural proteins are four capsid proteins (VP), which are VP2 (1A), VP2 (1B), VP3 (1C), and VP1 (1D). The number of VP is according to their migration in a polyacrylamide gel, and 1A, 1B, 1C, and 1D represent the order of the virus (capsid) proteins on the genome in P1. The nonstructural proteins are seven in number: 2A, 2B, 2C, 3A, 3B, 3C, and 3D (Fig. 2.12), in which 3B is the small basic protein, covalently linked to the 5′ end of the genome. The 2C is a helicase and 3D is an RNA-dependent RNA polymerase (Lee et al., 1977; Rueckert, 1985).

2A, 3B, and 3C are viral proteases. The first cleavage reaction in all enteroviral polyproteins is catalyzed by 2A, but the principal viral protease is 3C. Overall, the protein fold of 3C proteins is similar in picornavirus. The conserved catalytic triad residues in 3C are cysteine, histidine, and glutamic or aspartic acid residues (Cys…His…Glu/Asp) (Marcotte et al., 2007). According to the Schechter and Berger nomenclature, the cleavage of substrate protein occurs at Gln-Gly dipeptide with alanine in P4 and a proline residue in the P2′ position (Norder et al., 2011). The several host cellular proteins also are cleaved by the 3C in case of *Enterovirus* infections (Dougherty et al., 2010). The 3C protease consists of N-terminal α helix (14 amino-acid residues), and then it folds into two β barrels (amino-acid residues 15–79 and 99–173) formed by six antiparallel strands. The two barrels pack together, with a relative orientation of ~90, to form an extended groove for substrate binding. The active site of 3C protease is located in the cleft between the two β barrels. The active site cleft consists of

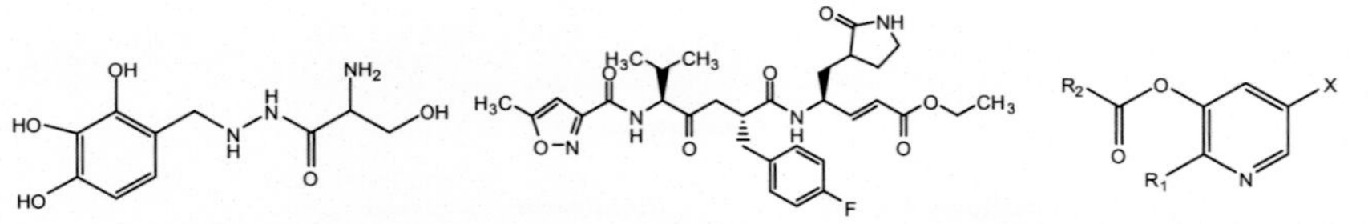

(14), Benserazide       (15), Rupintrivir       (16), Nonpeptidic inhibitors

FIGURE 2.13   **Chemical structures of some picornavirus protease inhibitors benserazide, rupintrivir, and nonpeptidic inhibitors.**

catalytic triad residues and an electrophilic oxyanion hole. The stabilization of the tetrahedral intermediate of the substrate covalently linked to the protease occurs by electrophilic oxyanion hole that consists of the amide groups of Gly145, Gln146, and Cys147. The nucleophilic Cys147 of catalytic triad is located in the second β barrel, while the His40 and Glu71 are located in the first β barrel (Norder et al., 2011; Sheng et al., 2011).

### 3.4.2 Picornavirus Protease Inhibitors

Picornaviruses are forked into many genera, species, and viruses that cause many different types of diseases, requiring specific antiviral inhibitors and the therapies. Proteolytical cleavage of the picornaviral polyproteins is essential for viral replication. Therefore, viral proteases are attractive targets for the development of antiviral therapy. In this section, some examples of molecular modeling and theoretical studies of antipicornavirus proteases are discussed.

Human coxsackievirus B3 is one of the members of the genus *Enterovirus* within the picornaviridae family. It can cause acute viral myocarditis in children and adolescents and induce serious diseases affecting pancreas or central nervous system. Acute myocarditis can lead to chronic myocarditis and dilated cardiomyopathy that may require cardiac transplantation (Chandirasegaran et al., 2014). The benserazide (**14**) was discovered as a novel inhibitor targeting coxsackievirus 3C protease (Fig. 2.13). Benserazide was characterized as a noncompetitive and allosteric inhibitor that has no electrophilic functional groups. Based on the molecular docking study of benserazide and its analogs, the putative allosteric binding site was recognized in the protease that could be the target of 2,3,4-trihydroxybenzyl moiety of benserazide (Bo-Kyoung et al., 2015). Further optimization was performed by the introduction of various aryl-alkyl substituted hydrazide moieties instead of the serine moiety. Among the optimized compounds, a 4-hydroxyphenylpentanehydrazide derivative was found to exhibit the most potent inhibitory activity. The results of the docking study indicated that this compound could partially bind to the P1 and P2 pockets of the active site. The 2,3,4-trihydroxybenzyl moiety formed hydrogen bonds strongly with Thr142 and Asn165 in the P1 pocket. The *para* hydroxyl group could bind to the P2 pocket through hydrogen bonding. The benzyl group of the compound showed an interaction with imidazole ring on the His40 by π—π bond (Bo-Kyoung et al., 2016). The peptidomimetic antiviral agents against 3C protease of *coxsackievirus* B3 were developed based on heterocyclic moieties, such as quinoline, benzimidazoline, and methyl imidazo-pyridine moities. In addition to the enzymatic and cell-based antiviral assays, the binding modes of the analogs in the 3C protease active site was determined by molecular docking studies, which indicated that the quinoline-substituted analogs could have several important hydrogen bond and hydrophobic interactions in the active site (Bo-Kyoung et al., 2012).

Human rhinovirus is the primary cause of upper respiratory infection or common cold and is linked to exacerbation of underlying respiratory diseases, such as asthma and chronic obstructive pulmonary disease (Denlinger et al., 2011). Rupintrivir (**15**) compound was developed as a peptidomimetic and a reversible inhibitor targeting human rhinovirus 3C protease (Fig. 2.13). It has α,β-unsaturated ester moiety that covalently interacts with a catalytic cysteine residue (Cys147) in the active site. Rupintrivir was evaluated in clinical trials but it failed in a study of naturally infected patients, because its ethyl ester moiety is rapidly hydrolyzed to yield an inactive acid form (Hsyu et al., 2002). Proline- and azetidine-based analogs of Rupintrivir was designed to act as an antihuman rhinovirus 3C protease inhibitors. These inhibitors targeted the P2 pocket of the binding site. The X-ray crystallography, quantum mechanical calculations, as well as QSAR studies were performed for potency optimization of rupintrivir analogs where the P2 residue was a cyclic amino acid. All calculations were performed at B3LYP/6-31G* and B3LYP/6-311G**. The hypothesis was that the replacement of the P2 substituent with a cyclic amino acid, such as L-proline, would stabilize the bound conformation of the compounds, leading to an increase in binding affinity. The computational study also suggested that changing ring size could potentially affect the rotational barrier of the phenyl ring and that the azetidine ring would offer the lowest rotational barrier. It was concluded that the compounds had the activity against a broad spectrum of human rhinovirus serotypes (Kawatkar et al., 2016). Some nonpeptidic inhibitors that were a series of novel hetero aromatic esters (**16**) were explored as human rhinovirus 3C protease inhibitors (Im et al., 2009). A compound with a 5-bromopyridinyl group was found to be the most potent, and the molecular docking study exhibited that a compound with 4-hydroxyquinolinone moiety had a new interaction with S1 pocket (Fig. 2.14) (Im et al., 2009).

# 4  CONCLUSIONS

The present QSAR models and molecular modeling approach provide interesting insights, and probable hints, for the development of inhibitors of various viral proteases. Although the approached is considered one of the important

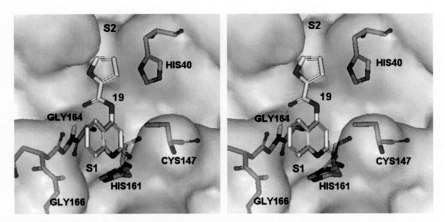

**FIGURE 2.14** **Stereo view of initial binding mode of compound with 4-hydroxyquinolinone moiety (19) and HRV 3Cpro.** The nitrogen, oxygen, and sulfur atoms are colored *blue, red,* and *orange,* respectively. Hydrogen bonds are displayed as *green dashed lines. (From Im, I., Lee, E.S., Choi, S.J., et al., 2009. Structure-activity relationships of heteroaromatic esters as human rhinovirus 3C protease inhibitors. Bioorg. Med. Chem. Lett. 19 (13), 3632–3636.)*

steps to understand such interactions and paves the way toward solution but additional theoretical and experimental approaches are still needed to be done in order to confirm or to improve the models presented in this chapter.

## REFERENCES

AIDS Epidemic Update, December 2007, Joint United Nations Programme on HIV/AIDS (UNAIDS/WHO).

Ambert-Balay, K., Lorrot, M., Bon, F., Giraudon, H., Kaplon, J., Wolfer, M., Lebon, P., Gendrel, D., Pothier, P., 2008. Prevalence and genetic diversity of *Aichi virus* strains in stool samples from community and hospitalized patients. J. Clin. Microbiol. 46, 1252–1258.

Andersson, H.O., Fridborg, K., Löwgren, S., Alterman, M., Mühlman, A., Björsne, M., Garg, N., Kvarnström, I., Schaal, W., Classon, B., Karlen, A., Danielsson, U.H., Ahlsen, G., Nillroth, U., Vrang, L., Öberg, B., Samuelsson, B., Hallberg, A., Unge, T., 2003. Optimization of P1-P3 groups in symmetric and asymmetric HIV-1 protease inhibitors. Eur. J. Biochem. 270, 1746–1758.

Arenas-Pinto, A., Grant, A.D., Edwards, S., Weller, I.V., 2003. Lactic acidosis in HIV infected patients: a systematic review of published cases. Sexually Trans. Infect. 79, 340–343.

Atkin, P.W., de Paula, J., 2002. Physical Chemistry. W.H. Freeman, New York, NY, Sections 21.2–21.4.

Atoom, A.M., Taylor, N.G., Russell, R.S., 2014. The elusive function of the hepatitis C virus p7 protein. Virology 462–463, 377–387.

Barre-Sinoussi, F., Chermann, J.C., Rey, F., Nugeyre, M.T., Chamaret, S., Gruest, J., Dauguet, C., Axler-Blin, C., Vezinet-Brun, F., Rouzioux, C., Rozenbaum, W., Montagnier, L., 1983. Isolation of a T-lymphotropic retrovirus from a patient at risk for acquired immune deficiency syndrome (AIDS). Science 220, 868–871.

Bissantz, C., Kuhn, B., Stahl, M., 2010. A medicinal chemist's guide to molecular interactions. J. Med Chem 53, 5061–5084.

Blinkova, O., Kapoor, A., Victoria, J., Jones, M., Wolfe, N., Naeem, A., Shaukat, S., Sharif, S., Alam, M.M., Angez, M., Zaidi, S., Delwart, E.L., 2009. Cardioviruses are genetically diverse and cause common enteric infections in South Asian children. J. Virol 83, 4631–4641.

Bo-Kyoung, K., Hyojin, K., Eun-Seok, J., Eun-Seon, J., Lak, S.J., Yong-Chul, K., 2016. 2, 3, 4-Trihydroxybenzyl-hydrazide analogues as novel potentcoxsackievirus B3 3C protease inhibitors. Eur. J. Med. Chem. 120, 202–216.

Bo-Kyoung, K., Jeong-Hyun, K., Na-Ri, K., Won-Gil, L., So-Deok, L., Soo-Hyeon, Y., Eun-Seok, J., Yong-Chul, K., 2012. Development of anti-*coxsackievirus* agents targeting 3C protease. Bioorg. Med. Chem. Lett. 22, 6952–6956.

Bo-Kyoung, K., Joong-Heui, C., Pyeonghwa, J., Youngjin, L., Jia, J.L., Kyoung, R.P., Soo, H.E., Yong-Chul, K., 2015. Benserazide, the first allosteric inhibitor of *coxsackievirus* B3 3C protease. FEBS Lett. 589, 1795–1801.

Bosi, S., Ros, T.D., Spalluto, G., Prato, M., 2003. Fullerene derivatives: an attractive tool for biological applications. Eur. J. Med. Chem. 38, 913–923.

Brinkman, K., ter Hofstede, H.J.M., Burger, D.M., Smeitink, J.A.M., Koopmans, P.P., 1998. Adverse effects of reverse transcriptase inhibitors: mitochondrial toxicity as common pathway. AIDS 12, 1735–1744.

Brown, M.T., McBride, K.M., Baniecki, M.L., Reich, N.C., Marriott, G., Mangel, W.F., 2002. Actin can act as a cofactor for a viral proteinase, in the cleavage of the cytoskeleton. J. Biol. Chem. 277, 46298–46303.

Brundage, S.C., Fitzpatrick, A.N., 2006. Hepatitis A. Am. Fam. Phys. 73, 2162–2168.

Calugi, C., Guarna, A., Trabocchi, A., 2014. Identification of constrained peptidomimetic chemotypes as HIV protease inhibitors. Eur. J. Med. Chem. 84, 444–453.

Carnero, E., Fortes, P., 2016. HCV infection, IFN response and the coding and non-coding host cell genome. Virus Res 212, 85–102.

Carney, D.S., Gale, Jr., M., 2006. HCV regulation of host defense, in: hepatitis C viruses: genomes and molecular biology. Horizon Stander Biosci., 375–398.

Catanese, M.T., Uryu, K., Kopp, M., Edwards, T.J., Andrus, L., Rice, W.J., Silvestry, M., Kuhn, R.J., Rice, C.M., 2013. Ultrastructural analysis of hepatitis C virus particles. Proc. Natl. Acad. Sci. USA 110, 9505–9510.

Chandirasegaran, M., Arunakumar, G., Jay, R., 2014. Intricacies of cardiac damage in *coxsackievirus* B3 infection: implications for therapy. Int. J. Cardiol. 177, 330–339.

Choe, J.I., Chang, S.K., 2002. Molecular modeling of complexation behavior of p-tert-butylcalix[5] arene derivative toward butylammonium ions. Bull. Korean Chem. Soc. 23, 48–52.

Choo, Q.L., Kuo, G., Weiner, A.J., Overby, L.R., Bradley, D.W., Houghton, M., 1989. Isolation of a cDNA clone derived from a blood-borne non-A, non-B viral hepatitis genome. Science 244, 359–362.

Cohen, N.C., 1996. Guidebook on Molecular Modeling in Drug Design, first ed. Academic press, Inc., San Diego, CA, USA.

Cortes-Hinojosa, G., Gulland, F.M.D., Goldstein, T., Venn-Watson, S., Rivera, R., Waltzed, T.B., Salemi, M., Wellehan, Jr., J.F.X., 2015. Phylogenomic characterization of California sea lion adenovirus-1. Infect. Genet. Evol. 31, 270–276.

Deeb, O., da Cunha, E.F.F., Cormanich, R.A., Ramalho, T.C., Freitas, M.P., 2012. Computer-assisted assessment of potentially useful non-peptide HIV-1 protease inhibitors. Chemometr. Intell. Lab. Syst. 116, 123–127.

Dai, X.Q., Hua, X.G., Shan, T.L., Delwart, E., Zhao, W., 2010. Human *Cosavirus* infections in children in China. J. Clin. Virol 48, 228–229.

Denlinger, L.C., Sorkness, R.L., Lee, W.M., Evans, M.D., Wolff, M.J., Mathur, S.K., Crisafi, G.M., Gaworski, K.L., Pappas, T.E., Vrtis, R.F., Kelly, E.A., Gern, J.E., Jarjour, N.N., 2011. A peripheral blood lymphopenia correlated with increased responsiveness. Am. J. Respir. Crit. Care Med. 184, 1007–1014.

Ding, J., McGrath, W.J., Sweet, R.M., Mangel, W.F., 1996. Crystal structure of the human adenovirus proteinase with its 11 amino acid cofactor. EMBO J. 15, 1778–1783.

Dhurandhar, N.V., Kulkarni, P., M. Ajinkya, S., Sherikar, A., 1992. Effect of adenovirus infection on adiposity in chicken. Vet. Microbiol. 31, 101–107.

Diouri, M., Keyvani-Amineh, H., Geoghegan, K.F., Weber, J.M., 1996. Cleavage efficiency by adenovirus protease is site-dependent. J. Biol. Chem. 271, 32511–32514.

Dorsett, H., White, A., 2000. Overview of Molecular Modelling and ab initio Molecular Orbital Methods Suitable for use With Energetic Materials. DSTO Aeronautical and Maritime Research Laboratory, Salisbury, South Australia.

Dougherty, J.D., Park, N., Gustin, K.E., Lloyd, R.E., 2010. Interference with cellular gene expression. In: Ehrenfeld, E., Domingo, E., Roos, R.P. (Eds.), The Picornaviruses. ASM Press, Washington, pp. 165–180.

Edward, K.C., 2008, Pathology of AIDS, Version 19, Utah state University Library. Available from: http://library.med.utah.edu/webpath/AIDS2008

Ehrenfeld, E., Domingo, E., Roos, R.P. (Eds.), 2010. The Picornaviruses. ASM Press, Washington, pp. 493.

Ezat, A.A., Mostafa, H.I., El-bialy, N.S., Saleh, N.A., Ibrahim, M., 2015. Computational approaches to study peptidomimetic and macrocyclic HCV NS3 protease inhibitors. J. Comput. Theor. Nanosci. 12, 52–59.

Failla, C., Tomei, L., De Francesco, R., 1994. Both NS3 and NS4A are required for proteolytic processing of hepatitis C virus nonstructural proteins. J. Virol. 68, 3753–3760.

Fauci, A.S., 1993. Multifactorial nature of human immunodeficiency virus disease: implications for therapy. Science 262, 1011–1018.

Flexner, C., 1998. HIV-Protease Inhibitors. New Eng. J. Med. 338, 1281–1292.

Foresman, J.B., 1996. In: Swift, M.L., Zielinski, T.J. (Eds.), Ab initio Techniques in Chemistry: Interpretation and Visualization. ACS BOOKS, Washington, DC, Chapter 14.

Foresman, J.B., Frisch, A., 1996. Exploring Chemistry With Electronic Structure Methods, second ed. Gaussian Inc, Pittsburgh, PA.

Frecer, V., Berti, F., Benedetti, F., Miertus, S., 2008. Design of peptidomimetic inhibitors of aspartic protease of HIV-1containing PheΨPro core and displaying favourable ADME-related properties. J. Mol. Graph. Model. 27, 376–387.

Friedman, S.H., Decamp, D.L., Sijbesma, R.P., Srdanov, G., Wudl, F., Kenyon, G.L., 1993. Inhibition of the HIV- 1 protease by fullerene derivatives: model building studies and experimental verification. Am. Chem. SOC. 115 (15), 6506–6509.

Garg, R., Bhhatarai, B., 2006. QSAR and molecular modeling studies of HIV protease inhibitors. Topic Heterocyclic Chem. 3, 181–271.

Ghosh, A.K., Takayama, J., Kassekert, L.A., Ella-Menye, J., Yashchuk, S., Agniswamy, J., Wangb, Y., Aoki, M., Amano, M., Weber, I.T., Mitsuya, H., 2015. Structure-based design, synthesis, X-ray studies, and biological evaluation of novel HIV-1 protease inhibitors containing isophthalamide-derived P2-ligands. Bioorg. Med. Chem. Lett. 25, 4903–4909.

Gottlieb, M.S., Groopman, J.E., Weinstein, W.M., Fahey, J.L., Detels, R., 1983. The acquired immunodeficiency syndrome. Ann. Int. Med. 99 (2), 208–220.

Gottlieb, M.S., Schroff, R., Schanker, H.M., Weisman, J.D., Fan, P.T., Wolf, R.A., Saxon, A., 1981. *Pneumocystis* carinii pneumonia and mucosal candidiasis in previously healthy homosexual men: evidence of a new acquired cellular immunodeficiency. New Eng. J. Med. 305, 1425–1431.

Gupta, S.P., 2011. QSAR and Molecular Modeling. Springer Basel/Anamaya, New Delhi.

Guyader, M., Emerman, M., Sonigo, P., Clavel, F., Montagnier, L., Alizon, M., 1987. Genome organization and transactivation of the human immunodeficiency virus type 2. Nature 326, 662–669.

Hansch, C., Leo, A., 1995. Exploring QSAR: Fundamentals and Applications in Chemistry and Biology. American Chemical Society, Washington, DC.

Harrach, B., Benkő, M., Both, G.W., Brown, M., Davison, A.J., Echavarría, M., Hess, M., Jones, M.S., Kajon, A., 2011. In: King, A.M.Q., Adams, M.J., Carstens, E.B., Lefkowitz, E.J. (Eds.), Virus Taxonomy: Classification and Nomenclature of Viruses: Ninth Report of the International Committee on Taxonomy of Viruses. Elsevier, SanDiego, pp. 125–141.

Herdewijn, P., De Clercq, E., 1990. Dideoxynucleoside analogues as inhibitors of HIV replication. In: De Clercq, E. (Ed.), Design of Anti-Aids Drugs. Elsevier Science, Amsterdam, p. 141.

Holtz, L.R., Finkbeiner, S.R., Kirkwood, C.D., Wang, D., 2008. Identification of a novel picornavirus related to cosaviruses in a child with acute diarrhea +. Virol. J. 5, 159.

Hornak, V., Oku, A., Rizzo, R.C., Simmerling, C., 2006. HIV-1 protease flaps spontaneously open and recluse in molecular dynamics simulations. Proc. Natl. Acad. Sci. USA 103, 915–920.

Horwitz, M., 2001. Adenoviruses. In: Knipe, D.M., Howley, P.M., Griffin, D.E., Lamb, R.A., Martin, M.A., Roizman, B., Straus, S.E. (Eds.), Fields Virology. fourth ed. Lippincott Williams & Wilkins, Philadelphia, P.A, pp. 2149–2171.

Hsyu, P.-H., Pithavala, Y.K., Gersten, M., Penning, C.A., Kerr, B.A., 2002. The antiviral compound enviroxime targets. Antimicrob. Agents Chemother. 46, 392–397.

Ibrahim, M., Saleh, N.A., Elshemey, W.M., Elsayed, A.A., 2010a. Computational notes on fullerene based system as HIV-1 protease inhibitors. J. Comput. Theor. Nanosci 7, 224–227.

Ibrahim, M., Saleh, N.A., Elshemey, W.M., Elsayed, A.A., 2012a. Fullerene derivative as anti-HIV protease inhibitor: molecular modeling and QSAR approaches. Mini Rev. Med. Chem. 12, 447–451.

Ibrahim, M., Saleh, N.A., Elshemey, W.M., Elsayed, A.A., 2012b. Hexapeptide functionality of cellulose as ns3 protease inhibitors. Med. Chem. 8, 826–830.

Ibrahim, M., Saleh, N.A., Elshemey, W.M., Elsayed, A.A., 2013. QSAR properties of novel peptidomimetic NS3 protease inhibitors. J. Comput. Theor. Nanosci. 10, 785–788.

Ibrahim, M., Saleh, N.A., Hameed, A.J., Elshemey, W.M., Elsayed, A.A., 2010b. Structural and electronic properties of new fullerene derivatives and their possible application as HIV-1 protease inhibitors. Spectrochimica Acta 75, 702–709.

Im, I., Lee, E.S., Choi, S.J., Lee, J., Kim, Y., 2009. Structure–activity relationships of heteroaromatic esters as human rhinovirus 3C protease inhibitors. Bioorg. Med. Chem. Lett 19, 3632–3636.

Ishima, R., Ghirlando, R., Tözsér, J., Gronenborn, A.M., Torchia, D.A., Louis, J.M., 2001. Folded monomer of HIV-1 protease. J. Biol. Chem. 276, 49110–49116.

Islam, Md. A., Pillay, T.S., 2016. Simplified molecular input line entry system-based descriptors in QSAR modeling for HIV-protease inhibitors. Chemometr. Intell. Lab. Syst. 153, 67–74.

Ismail, N.S., Hattori, M., 2011. Molecular modeling based approach, synthesis and in vitro assay to new in dole inhibitors of hepatitis C NS3/4A serine protease. Bioorg. Med. Chem. 19, 374–383.

Jackowiak, P., Kuls, K., Budzko, L., Mania, A., Figlerowicz, M., Figlerowicz, M., 2014. Phylogeny and molecular evolution of the hepatitis C virus. Infect. Genet. Evol. 21, 67–82.

Jenny, E., 2006. Design and synthesis of novel HIV-1 protease inhibitors comprising a tertiary alcohol in the transition-state mimic, Ph.D. dissertation, University of Uppsala, Sweden,

Jo, M., Nakamura, N., Kakiuchi, N., Komatsu, K., Shimotohno, K., Hattori, M., 2006. Inhibitory effect of Yunnan traditional medicines of hepatitis C viral polymerase. J. Nat. Med. 60, 217–224.

Jubelt, B., Lipton, H.L., 2014. Enterovirus/Picornavirus infections. In: Tselis, A.C., Booss, J. (Eds.), Handbook of Clinical Neurology, third series, Neurovirology vol. 123. Elsevier B.V..

Kapoor, A., Victoria, J., Simmonds, P., Slikas, E., Chieochansin, T., Naeem, A., Shaukat, S., Sharif, S., Alam, M.M., Angez, M., Wang, C., Shafer, R.W., Zaidi, S., Delwart, E., 2008. A highly prevalent and genetically diversified Picornaviridae genus in South Asian children. Proc. Natl. Acad. Sci. USA 105, 20482–20487.

Kaushik-Basu, N., Theresa, P.J., Manvar, D., Kondepudi, S., Battu, M.B., Sriram, D., Basu, A., Yogeeswari, P., 2014. Multiple pharmacophore modeling, 3D-QSAR and high-throughput virtual screening of hepatitis C virus NS5B polymerase inhibitors. J. Chem. Inform. Model. 54, 539–552.

Kawatkar, S.P., Gagnon, M., Hoesch, V., Tiong-Yip, C., Johnson, K., Ekb, M., Nilsson, E., Lister, T., Olsson, L., Patel, J., Yu, Q., 2016. Design and structure activity relationships of novel inhibitorsof human *rhinovirus* 3C protease. Bioorg. Med. Chem. Lett. 26, 3248–3252.

Khamrin, P., Chaimongkol, N., Malasao, R., Suantai, B., Saikhruang, W., Kongsricharoern, T., Ukarapol, N., Okitsu, S., Shimizu, H., Hayakawa, S., Ushijima, H., Maneekarn, N., 2012. Detection and molecular characterization of *Cosavirus* in adults with diarrhea, Thailand. Virus Genes 44, 244–246.

Kim, J.L., Morgenstern, K.A., Lin, C., Fox, T., Dwyer, M.D., Landro, J.A., Chambers, S.P., Markland, W., Lepre, C.A., O'Malley, E.T., Harbeson, S.L., Rice, C.M., Murcko, M.A., Caron, P.R., Thomson, J.A., 1996. Crystal structure of the hepatitis C virus NS3 protease domain complexed with a synthetic NS4A cofactor peptide. Cell 87, 343–355.

Knowles, N.J., Hovi, T., Hyypia, T., et al., 2012. Picornaviridae. In: King, A.M.Q., Adams, M.J., Carsten, E.B., Lefkowitz, E.J. (Eds.), Virus Taxonomy: Classification and Nomenclature of Viruses. Ninth Report of the International Committee on Taxonomy of Viruses. Elsevier, San Diego, pp. 855–880.

Kozak, R.A., Ackford, J.G., Slaine, P., Li, A., Carman, S., Campbell, D., Welch, M.K., Kropinski, A.M., Nagy, É., 2015. Characterization of a novel adenovirus isolated from a skunk. Virology 485, 16–24.

Krausslich, H.G., Wimmer, E., 1988. Viral proteinases. Annu. Rev. Biochem. 57, 701–754.

Leach, A.R., 1996. Molecular Modelling Principle and Applications. Addison Wesley Longman Limited, Harlow, England.

Lee, V.S., Nimmanpipug, P., Aruksakunwong, O., Promsri, S., Sompornpisut, P., Hannongbua, S., 2007. Structural analysis of lead fullerene-based inhibitor bound to human immunodeficiency virus type 1 protease in solution from molecular dynamics simulations. J. Mol. Graph. Model. 26, 558–570.

Lee, Y.F., Nomoto, A., Detjen, B.M., Wimmer, E., 1977. A protein covalently linked to poliovirus genome RNA. Proc. Natl. Acad. Sci. USA 74, 59–63.

Lenaerts, L., De Clercq, E., Naesens, L., 2008. Clinical features and treatment of adenovirus infections. Rev. Med. Virol. 18, 357–374.

Lenaerts, L., Naesens, L., 2006. Mini-review antiviral therapy for adenovirus infections. Antivir. Res. 71, 172–180.

Li, X., Zhang, W., Qiao, X., Xu, X., 2007. Prediction of binding for a kind of non-peptic HCV NS3 serine protease inhibitors from plants by molecular docking and MM-PBSA method. Bioorg. Med. Chem. 15, 220-206.

Mac Sweeney, A., Grosche, P., Ellis, D., Combrink, K., Erbel, P., Hughes, N., Sirockin, F., Melkko, S., Bernardi, A., Ramage, P., Jarousse, N., Altmann, E., 2014. Discovery and structure-based optimization of adenain inhibitors. ACS Med. Chem. Lett. 5, 937–941.

Manabe, S., Fuke, I., Tanishita, O., Kaji, C., Gomi, Y., Yoshida, S., Mori, C., Takamizawa, A., Yosida, I., Okayama, H., 1994. Production of nonstructural proteins of hepatitis C virus requires a putative viral protease encoded by NS3. Virology 198, 636–644.

Mangel, W.F., McGrath, W.J., Toledo, D.L., Anderson, C.W., 1993. Viral DNA and a viral peptide can act as cofactors of adenovirus virion proteinase activity. Nature 361, 274–275.

Mangel, W.F., Toledo, D.L., Brown, M.T., Martin, J.H., McGrath, W.J., 1996. Characterization of three components of human adenovirusproteinase activity in vitro. J. Biol. Chem. 271, 536–543.

Marcotte, L.L., Wass, A.B., Gohara, D.W., Pathak, H.B., Arnold, J.J., Filman, D.J., Cameron, C.E., Hogle, J.M., 2007. Crystal structure of poliovirus 3CD protein, virally encoded protease and precursor to the RNA-dependent RNA polymerase. J. Virol. 81, 3583–3596.

Marek, A., Ballmann, M.Z., Kosiol, C., Harrach, B., Schlotterer, C., Hess, M., 2014. Whole-genome sequences of two turkey adenovirus types reveal the existence of two unknown lineages that merit the establishment of novel species within the genus *Aviadenovirus*. J. Gen. Virol. 95, 156–170.

Martin, A., Lemon, S.M., 2006. Hepatitis A virus: from discovery to vaccines. Hepatology 43, S164–S172.

Martinez-Salas, E., Ryan, M.D., 2010. Translation and protein processing. In: Ehrenfeld, E., Domingo, E., Roos, R.P. (Eds.), The Picornaviruses. ASM Press, Washington, pp. 141–161.

McGrath, W.J., Baniecki, M.L., Peters, E., Green, D.T., Mangel, W.F., 2001. Roles of two conserved cysteine residues in the activation of human adenovirus proteinase. Biochemistry 40, 14468–14474.

McHutchison, J.G., Gordon, S.C., Schiff, E.R., Shiffman, M.L., Lee, W.M., Rustgi, V.K., Goodman, Z.D., Ling, M.H., Cort, S., Albrecht, J.K., 1998. Interferon Alfa-2B alone or in combination with ribavirin as initial treatment for chronic hepatitis C. Hepatitis Interventional Therapy Group. New Eng. J. Med. 339, 1485–1492.

Melnick, J.L., Cockburn, W.C., Dalldorf, G., 1963. Picornavirus group. Virology 19, 114–116.

Moradpour, D., Penin, F., 2013. Hepatitis C virus proteins: from structure to function. Curr. Top Microbiol Immunol 369, 113–142.

Mostafa, H.I., El-bialy, N.S., Ezat, A.A., Saleh, N.A., Ibrahim, M., 2014. QSAR analysis and molecular docking simulation of suggested peptidomimetic NS3 protease inhibitors. Curr. Comput. Aided Drug Des. 10, 28–40.

Musmuca, I., Caroli, A., Mai, A., Kaushik-Basu, N., Arora, P., Ragno, R., 2010. Combining 3-D quantitative structure-activity relationship with ligand based and structure based alignment procedures for in silico screening of new hepatitis C virus NS5B polymerase inhibitors. J. Chem. Inform. Model. 50, 662–676.

Norder, H., De Palma, A.M., Selisko, B., Costenaro, L., Papageorgiou, N., Arnan, C., Coutard, B., Lantez, V., De Lamballerie, X., Baronti, C., Solà, M., Tan, J., Neyts, J., Canard, B., Coll, M., Gorbalenya, A.E., Hilgenfeld, R., 2011. Picornavirus non-structural proteins as targets for new anti-virals with broad activity. Antiviral Res. 89, 204–218.

Nurbo, J., Peterson, S.D., Dahl, G., Danielson, U.H., Karlen, A., Sandstrom, A., 2008. Beta-amino acid substitutions and structure-based CoMFA modeling of hepatitis C virus NS3 protease inhibitors. Bioorg. Med. Chem. 16, 5590–5605.

Odriozola, L., Cruchaga, C., Andréola, M., Dollé, V., Nguyen, C.H., Tarrago-Litvak, L., Pérez-Mediavilla, A., Martinez-Irujo, J.J., 2003. Non-nucleoside inhibitors of HIV-1 reverse transcriptase inhibit phosphorolysis and resensitize the 3'-azido-3'-deoxythymidine (AZT)-resistant polymerase to AZT-5'-triphosphate. J. Biol. Chem. 43, 42710–42716.

Ortqvist, P., 2010. On the design and synthesis of hepatitis C virus NS3 protease inhibitors. Dissertations from the Faculty of Pharmacy, Uppsala, Sweden Tripeptides to achiral compounds. Digital Comprehensive Summaries of Uppsala, p. 85.

Palmenberg, A.C., Spiro, D., Kuzmickas, R., Wang, S., Djikeng, A., Rathe, J.A., Fraser-Liggett, C.M., Liggett, S.B., 2009. Sequencing and analyses of all known human rhinovirus genomes reveal structure and evolution. Science 324, 55–59.

Parsy, C.C., Alexandre, F., Bidau, V., Bonnaterre, F., Brandt, G., Caillet, C., Cappelle, S., Chaves, D., Convard, T., Derock, M., Gloux, D., Griffon, Y., Lallos, L.B., Leroy, F., Liuzzi, M., Loi, A., Moulat, L., Chiara, M., Rahali, H., Roques, V., Rosinovsky, E., Savin, S., Seifer, M., Standring, D., Surleraux, D., 2015. Discovery and structural diversity of the hepatitis C virus NS3/4Aserine protease inhibitor series leading to clinical candidate IDX320. Bioorg. Med. Chem. Lett. 25, 5427–5436.

Patil, V.M., Gupta, S.P., Masand, N., 2014. QSAR analysis of some heterocyclic compounds as HCV NS5B polymerase inhibitors, Journal of Engineering. Sci. Manag. Educ. 7, 215–219.

Patil, V.M., Gupta, S.P., Samanta, S., Masand, N., 2011. Current perspective of HCV NS5B inhibitors: a review. Curr. Med. Chem. 18, 5564–5597.

Poordad, F., Dieterich, D., 2012. Treating hepatitis C: current standard of care and emerging direct-acting antiviral agents. J. Viral Hepat. 19, 449–464.

Rao, S.N., Balaji, G.A., Balaji, V.N., 2013. Docking and 3-D QSAR studies on the binding of tetrahydropyrimid-2-oneHIV-1 protease inhibitors. J. Mol. Struct. 1042, 86–103.

Ravichandran, V., Venkateskumar, K., Shalini, S., Harish, R., 2016. Exploring the structure–activity relationship of oxazolidinones as HIV-1 protease inhibitors: QSAR and pharmacophore modelling studies. Chemometr Intell. Lab. Syst. 154, 52–61.

Rowe, W.P., Huebner, R.J., Gilmore, L.K., Parrott, R.H., Ward, T.G., 1953. Isolation of cytopathogenic agent from human adenoids undergoing spontaneous degeneration in tissue culture. Proc. Soc. Exp. Biol. Med. 84, 570–573.

Rueckert, R.R., 1985. Picornaviruses and their replication. In: Fields, B.N., Knipe, D.M., Charnock, R.M. et al., (Eds.), Virology. Raven Press, New York, pp. 705–738.

Ruzindana-Umunyana, A., Imbeault, L., Joseph, M.W., 2002. Substrate specificity of adenovirus protease. Virus Res. 89, 41–52.

Saleh, N.A., 2015. The QSAR and docking calculations of fullerene derivatives as HIV-1 protease inhibitors. Spectrochim. Acta 136, 1523–1529.

Saleh, N.A., Elfiky, A.A., Ezat, A.A., Elshemey, W.M., Ibrahim, M., 2014. The electronic and QSAR properties of modified telaprevir compounds as HCV NS3 protease inhibitors. J. Comput. Theor. Nanosci. 11, 544–548.

Saleh, N.A., Elshemey, W.M., Elsayed, A.A., Ibrahim, M., 2015a. NS3 serine protease as target for anti-hepatitis C virus. Rev. Theor. Sci. 3, 257–263.

Saleh, N.A., Ezat, A.A., Elfiky, A.A., Elshemey, W.M., Ibrahim, M., 2015b. Theoretical study on modified boceprevir compounds as ns3 protease inhibitors. J. Comput. Theor. Nanosci. 12, 371–375.

Schechter, I., Berger, A., 1968. On the active site of pro- teases. 3. Mapping the active site of papain; specific peptide inhibitors of papain. Biochem. Biophys. Res. Commun. 32, 898–902.

Sheng, C., Jing, W., Tingting, F., Bo, Q., Li, G., Xiaobo, L., Jianwei, W., Meitian, W., Qi, J., 2011. Crystal structure of human *Enterovirus* 71 3C protease. J. Mol. Biol. 408, 449–461.

Sluis-Cremer, N., Tachedjian, G., 2002. Modulation of the oligomeric structures of HIV-1 retroviral enzymes by synthetic peptides and small molecules. Eur. J. Biochem. 269, 5103–5111.

Sofia, M.J., Chang, W., Furman, P.A., Mosley, R.T., Ross, B.S., 2012. Nucleoside, nucleotide, and non-nucleoside inhibitors of hepatitis C virus NS5B RNA-dependent RNA-polymerase. J. Med. Chem. 55, 2481–2531.

Stanway, G., Brown, F., Christian, P., Hovi, T., Hyypia, T., King, A.M.Q., Knowles, N.J., Lemon, S.M., Minor, P.D., Pallansch, M.A., Palmenberg, A.C., Skern, T., 2005. Family Picornaviridae. In: Fauquet, C.M., Mayo, M.A., Maniloff, J., Desselberger, U., Ball, L.A. (Eds.), Virus Taxonomy. Eighth Report of the International Committee on Taxonomy of Viruses. Elsevier/Academic Press, London, pp. 757–778.

Sulkowski, M.S., Thomas, D.L., Chaisson, R.E., Moore, R.D., 2000. Hepatotoxicity associated with antiretroviral therapy in adults infected with human immunodeficiency virus and the role of hepatitis C or B virus infection. J. Am. Med. Assoc. 283, 74–80.

Tibbs, C.J., 1997. Tropical aspects of viral hepatitis. Hepatitis C. Trans. R. Soc. Trop. Med. Hyg. 91, 121–124.

Tihanyi, K., Bourbonniere, M., Houde, A., Rancourt, C., Weber, J.M., 1993. Isolation and properties of adenovirus type 2proteinase. J. Biol. Chem. 268, 1780–1785.

Toyoda, H., Franco, D., Fujita, K., Paul, A.V., Wimmer, E., 2007. Replication of poliovirus requires binding of the poly (rC) binding protein to the clover leaf as well as to the adjacent C-rich spacer sequence between the cloverleaf and the internal ribosomal entry site. J. Virol. 81, 10017–10028.

van Gunsteren, W.F., Berendsen, H.J.C., 1977. Algorithms for macromolecular dynamics and constraint dynamics. Mol. Phys. 34, 1311–1327.

Ward, A.B., Wilson, I.A., 2015. Insights into the trimeric HIV-1envelope glycoprotein structure. Trends Biochem. Sci. 40, 101–107.

Weber, J.M., 1999. Role of endopeptidase in adenovirusinfection. In: Seth, P. (Ed.), Adenoviruses: From Basic Research to Gene Therapy Application. Humana Press, Clifton, NJ.

Webster, A., Hay, R.T., Kemp, G., 1993. The adenovirus protease is activated by a virus-coded disulphide-linked Pepticle. Cell 72, 97–104.

Wei, Y., Li, J., Chen, Z., Wang, F., Huang, W., Hong, Z., Lin, J., 2015. Multistage virtual screening and identification of novel HIV-1 protease inhibitors by integrating SVM, shape, pharmacophore and docking methods. Eur. J. Med. Chem. 101, 409–418.

Wellehan, J.F., Johnson, A.J., Harrach, B., Benko, M., Pessier, A.P., Johnson, C.M., Garner, M.M., Childress, A., Jacobson, E.R., 2004. Detection and analysis of six lizard adenoviruses by consensus primer PCR provides further evidence of areptilian origin for the atadenoviruses. J. Virol. 78, 13366–13369.

Wlodawer, A., Erickson, J.W., 1993. Structure-based inhibitors of Hiv-1 protease. Annu. Rev. Biochem. 62, 543–585.

Wolthers, K.C., Benschop, K.S., Schinkel, J., Molenkamp, R., Bergevoet, R.M., Spijkerman, I.J., Kraakman, H.C., Pajkrt, D., 2008. Humanparechoviruses as an important viral cause of sepsis like illness and meningitis in young children. Clin. Infect. Dis. 47, 358–363.

Xie, Z.C., Riezu, J.I., Lasarte, Guillen, J., Su, J.H., Civeira, M.P., Prieto, J., 1998. Transmission of hepatitis C virus infection to tree shrews. Virology 244, 513–520.

Yamashita, T., Kobayashi, S., Sakae, K., Nakata, S., Chiba, S., Ishihara, Y., Isomura, S., 1991. Isolation of cytopathic small round viruses with BS-C-1 cells from patients with gastroenteritis. J. Infect. Dis. 164, 954–957.

Yan, J., Huang, N., Li, S., Yang, L., Xing, W., Zheng, Y., Hu, Y., 2012. Synthesis and biological evaluation of novel amprenavir-basedP1-substituted bi-aryl derivatives as ultra-potent HIV-1 protease inhibitors. Bioorg. Med. Chem. Lett. 22, 1976–1979.

Yeh-Kai, L., Akusjarvi, G., Alestrom, P., Pettersson, U., Tremblay, M., Weber, J., 1983. Genetic identification of an endoproteinase encoded by the adenovirus genome. J. Mol. Biol. 167, 217–222.

Zhang, H.X., Li, Y., Wang, X., Xiao, Z.T., Wang, Y.H., 2011. Insight into the structural requirements of benzothiadiazine scaffold-based derivatives as hepatitis C virus NS5B polymerase inhibitors using 3DQSAR, molecular docking and molecular dynamics. Curr. Med. Chem. 18, 4019–4028.

Zhang, R., Windsor, W.T., 2013. In vitro kinetic profiling of hepatitis C virus NS3 protease inhibitors by progress curve analysis. Methods Mol. Biol. 1030, 59–79.

## FURTHER READING

Grosche, P., Sirockin, F., Mac Sweeney, A., Ramage, P., Erbel, P., Melkko, S., Bernardi, A., Hughes, N., Ellis, D., Keith, D., Combrink, K., Jarousse, N., Altmann, E., 2015. Structure-based design and optimization of potent inhibitors of the adenoviral protease. Bioorg. Med. Chem. Lett. 25, 438–443.

Membreno, F.E., Lawitz, E.J., 2011. The HCV NS5B nucleoside and non-nucleoside inhibitors. Clin. Liver. Dis. 15, 611–626.

Tremblay, M.L., Déry, C.V., Talbot, B.G., Wber, J., 1983. In vitro cleavage specificity of the adenovirus type 2 proteinase. Biochimica et Biophysica Acta 743, 239–245.

Verloop, A., 1987. The STERIMOL Approach to Drug Design. Marcel Dekker, New York.

Zhang, X., 2016. Direct anti-HCV agents. Acta Pharmaceutica Sinica, 26–31.

Chapter 3

# Advances in Studies on Adenovirus Proteases and Their Inhibitors

**Satya P. Gupta\*, Basheerulla Shaik\*, Yenamandra S. Prabhakar\*\***
*\*National Institute of Technical Teachers' Training and Research, Bhopal, Madhya Pradesh, India; \*\*Central Drug Research Institute, CSIR, Lucknow, Uttar Pradesh, India*

## 1  INTRODUCTION

Adenoviruses (AdVs) belong to the family of Adenovirdae viruses. They are medium-sized, nonenveloped (without an outer lipid bilayer) viruses that have an icosahedral nucleocapsid containing a double-stranded DNA (dsDNA) genome. They are called adenoviruses (AdVs), as they were initially isolated from human adenoids (Rowe et al., 1953), which are masses of enlarged lymphatic tissue between the back of the nose and the throat, often hindering speaking and breathing in young children. AdVs cause many diseases, from common cold to life-threatening multiorgan diseases, weakening immune system. They not only infect human beings, but also many other vertebrates. In humans, more than 50 distinct serotypes of AdV have been found. In fact, it has been reported that there are 57 known AdV types in humans (Martin et al., 2007), which are classified into 7 different species (A–G) as follows:

A: 12, 18, 31
B: 3, 7, 11, 14, 16, 21, 34, 35, 50, 55
C: 1, 2, 5, 6, 57
D: 8, 9, 10, 13, 15, 17, 19, 20, 22, 23, 24, 25, 26, 27, 28, 29, 30, 32, 33, 36, 37, 38, 39, 42, 43, 44, 45, 46, 47, 48, 49, 51, 53, 54, 56
E: 4
F: 40, 41
G: 52

Viral Proteases and Their Inhibitors. http://dx.doi.org/10.1016/B978-0-12-809712-0.00003-4

**59**

Including all those present in other vertebrates, Adenoviridae can be divided into five genera: *Mastadenovirus*, *Aviadenovirus*, *Atadenovirus*, *Siadenovirus*, and *Ichtadenovirus*.

AdVs are nonenveloped viruses, whose virion has a unique "spike" or fiber associated with each penton base of the capsid. This spike aids in attachment of the virion to the host cell via the receptor on the surface of the host cell. The AdV genome is linear, nonsegmented dsDNA that is between 26 and 48 kbp. It has a terminal 55-kDa protein associated with each of the 5' ends of the linear dsDNA. AdVs are very simple viruses whose survival and replication heavily depend on the host cell.

## 2 ADENOVIRUS LIFE CYCLE

The entry of virus into the host cell starts from the initial attachment of virion particles to the host cell surface through binding of the fiber knob to the coxsackievirus B and adenovirus receptor (CAR), a type 1 transmembrane protein in the immunoglobulin superfamily. CAR is present in many human tissues, including heart, lung, liver, and brain (Howitt et al., 2003). Major histocompatibility complex (MHC) molecules and sialic acid residues have also been reported functioning in this capacity as well. Major histocompatibility complex is a set of cell surface proteins essential for the acquired immune system to recognize foreign molecules in vertebrates, which in turn determines histocompatibility.

After the virion particles attach to the cell surface, an exposed tripeptide Arg-Gly-Asp (RGD) motif on the penton base interacts with members of the αv integrin family, triggering virus internalization by clathrin-dependent, receptor-mediated endocytosis (Meier et al., 2002; Stewart et al., 1997). Integrins, a family of cell surface proteins, act as receptors for cell adhesion molecules. It is assumed that after virus internalization the acidic environment of the endosome induces escape of virions into the cytoplasm where dynein (a family of cytoskeletal motor proteins that moves along microtubules in cells) mediates their trafficking along microtubules toward the nucleus. In the nucleus, they are docked with the nuclear pore complex (NPC) (Kelkar et al., 2004; Trotman et al., 2001). Disassembly of the capsid at the NPC allows for import of the viral genome and commencement of the viral transcriptional program. In the nucleus, newly synthesized capsid and core proteins are imported from the cytosol, which assemble into new viral particles. The new genomes are packaged by the coordinated action of viral proteins IIIa, L1 52/55k, L4 33k, L4 22k, and IVa2 among themselves and with the viral DNA packaging sequence (Guimet and Hearing, 2013; Gustin and Imperiale, 1998; Ostapchuk et al., 2005, 2006, 2011; Perez-Romero et al., 2005; Wu et al., 2013; Zhang and Imperiale, 2000). The genome packaging produces the so-called young virions, which are finally transformed by proteolytic maturation to infectious AdV particle (Ishibashi and Maizel, 1974; Weber, 1999).

## 3  ADENOVIRUS MATURATION PROTEASE

The adenovirus protease (AVP) plays an important role during the maturation of the virus. It synthesizes AdV shell proteins IIIa, VI, and VIII, as well as core proteins VII, μ, and terminal protein as precursors and processes them during assembly (Blanche et al., 2001; Challberg and Kelly, 1981; Mangel et al., 1996; Weber, 1976). These six precursor proteins act as substrates for AVP. Another potential substrate, as it contains an AVP consensus sequence motif, is polypeptide L1 52/55k. As recently discussed by Mangel et al. (2014), late in AdV assembly, AVP becomes activated and cleaves multiple copies of three capsid and three core proteins (Fig. 3.1). Proteolytic maturation is an absolute requirement for rendering the viral particle infectious. It has been shown that the L1 52/55k protein, which is present in empty capsids, but not in mature virions, and is required for genome packaging, is the seventh substrate for AVP. In human AdV-2 (HAdV-2), it is 415-residues long with an AVP consensus cleavage site at the 351–352 position. It binds to the viral packaging ATPase IVa2 in vitro (Gustin et al., 1996; Ostapchuk et al., 2005; Perez-Romero et al., 2005) and contributes to the specificity of packaging, possibly via an interaction with capsid protein IIIa (Ma and Hearing, 2011; Wohl and Hearing, 2008). It has been reported to bind nonspecifically to DNA and to interact with pVII and its mature form, VII, in infected cells (Zhang and Arcos, 2005). However, as L1 52/55k does not bind to the DNA packaging sequence in vitro, its interaction with DNA may not be direct (Gustin et al., 1996; Ostapchuk et al., 2005; Perez-Romero et al., 2005). As virus particles do not contain AVP (Weber, 1976), the precursor versions of all AVP targets are contained by fully packaged, immature particles produced by the HAdV-2 thermosensitive mutant *ts1* under nonpermissive conditions.

The action of AVP has been found to be unique. On synthesis, the enzyme by itself is inactive (Mangel et al., 1993). It shows optimal activity in the presence of two cofactors namely an11-amino-acid peptide pVIc, derived from the C terminus of the polypeptide VI precursor pVI, and the viral DNA. Cleavage at the C-terminus releases pVIc, which then binds with AVP, and the AVP–pVIc complex slides along DNA, processing the virus precursor proteins also bound to the DNA (Blainey et al., 2013).

L1 52/55k is a potential substrate for AVP, as it contains an AVP consensus cleavage sequence. Regarding the cleavage of this substrate by AVP, Pérez-Berná et al. (2014) concluded the following:

1. AVP cleaves L1 52/55k at noncanonical sites.
2. Cleavage of L1 52/55k by AVP requires the presence of dsDNA.
3. Cleavage of L1 52/55k impairs its interactions with other proteins in the viral particle.
4. AVP needs cofactors for complete enzyme activity. One cofactor is the viral DNA, which stimulates AVP activity in vitro (Mangel et al., 1993) and the second cofactor is a plasmin-sensitive virion protein, which is an 11–amino

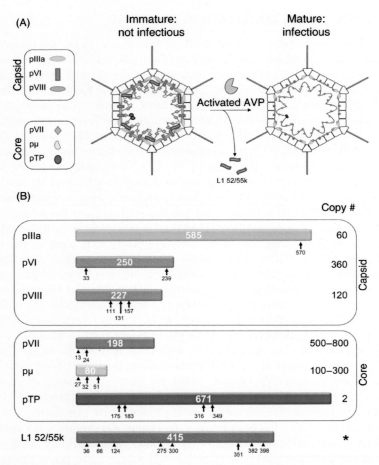

FIGURE 3.1 **Substrates of adenovirus protease (AVP).** (A) Representation of location of substrates in the viral particle. The internal location of L1 52/55k is inferred from its interactions with core elements. (B) Each human adenovirus (HAdV)-C2 precursor protein is represented as a *bar* with the polypeptide length in amino acids indicated in the center. *Arrows* indicate the consensus cleavage sites and the *arrowheads* indicate the nonconsensus sites by arrowheads. The prefix "p" denotes the unprocessed precursors. *Numerical digits* refer to copy numbers and an *asterisk* (*) for L1 52/55k indicates that its copy number varies depending on the assembly stage: 100 copies in empty particles; 50 in fully packaged, immature *ts1* particles; and 0 in mature virions. (*From Mangel, W.F., San Martin, C., May, E.R., 2014. Structure, function and dynamics in adenovirus maturation. Viruses 6 (11), 4536–4570.*)

acid peptide, pVIc (GVQSLKRRRCF), from the C-terminus of the precursor to virion protein VI, pVI (Mangel et al., 1993; Webster and Kemp, 1993). Actin was also considered a potential cytoplasmic cofactor for AVP because its C-terminal amino acid sequence (SGPSIVHRKCF) is highly homologous to the amino acid sequence of pVIc (GVQSLKRRRCF).

(A)

(B)

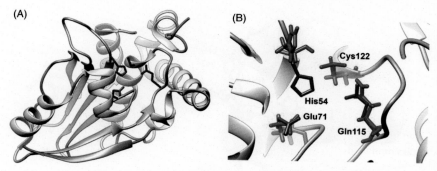

FIGURE 3.2   **Crystal structure of the AVP–pVIc complex and locations of the four amino acid residues involved in catalysis in AVP and in AVP–pVIc.** (A) Secondary structure of the AVP–pVIc complex with the four amino acid residues involved in catalysis in *blue* and the pVIc peptide in *green.* (B) The four amino acids involved in catalysis in AVP–pVIc *(blue)* and in AVP *(red)* are juxtaposed. Only His54 is in a different position in the two structures. *(From Mangel, W.F., San Martin, C., May, E.R., 2014. Structure, function and dynamics in adenovirus maturation. Viruses 6 (11), 4536–4570.)*

## 4   STRUCTURE OF AVP

Initially, when AVP was discovered it could not be classified in any category of proteases, but when its crystal structure complexed with pVIc was determined (Ding et al., 1996; McGrath et al., 1996, 2003), its structure could be defined. The complex structure was ovoid that appeared to consist of two domains (Fig. 3.2A), one containing a five-stranded β-sheet and the other mostly α-helices. When AVP–pVIc structure was compared with other unique protein molecules, AVP appeared to belong to a new family of proteases. A comparison of its structure with that of papain suggested that it was a cysteine proteinase. The four amino acids involved in catalysis by papain have identical counterparts at the same relative positions in the AVP–pVIc complex (Ding et al., 1996) (Fig. 3.2B). The positions of a helix and several β-strands within the central region of AVP appeared to be identical as those in papain (Mangel et al., 1997).

In the interaction of AVP with pVIc, it appeared that the N-terminus of pVIc is involved in the binding of pVIc to AVP and its C-terminus stimulates the AVP activity. While the N-terminus binds in a preexistent pocket, the C-terminus was observed to bind in an induced pocket (Mangel et al., 2014). The crucial amino acid residues that are involved in the binding of pVIc to AVP and in stimulating the activity of AVP are Gly1′, Cys10′, Val2′, and Phe11′. The first two can tolerate only homologous substitution and the latter two can tolerate only hydrophobic substitution.

AVP, pVIc, and AVP–pVIc complexes bind to DNA, where the nonsequence specific interaction between AVP–pVIc and DNA exhibits an electrostatic

(A) (B)

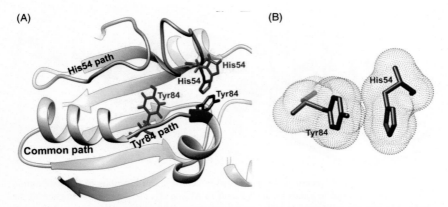

FIGURE 3.3  **AVP activation pathway.** (A) Binding of pVIc *(green)* to AVP, a series of contiguous conformational changes occur along a common path that bifurcates into upper and lower paths. At the end of the upper path, His54 drops down from its position in AVP *(red)* to a position in AVP–pVIc *(blue)* that is opposite to Cys122. At the end of the lower path, Tyr84 *(red)* moves 11 Å to a position in AVP–pVIc *(blue)* where it can form a cation–π interaction with His54. (B) Electron clouds of Tyr84 and His54 in their cation–π interaction. *(From Mangel, W.F., San Martin, C., May, E.R., 2014. Structure, function and dynamics in adenovirus maturation. Viruses 6 (11), 4536–4570.)*

interaction, involving positively charged groups on AVP–pVIc and negatively charged phosphate groups on DNA (Mangel et al., 1996; Record et al., 1976). There is also a substantial favorable nonelectrostatic component of the binding interaction of AVP–pVIc to DNA (McGrath et al., 2001). It was indicated that much of the binding free energy under physiological conditions resulted from nonspecific interactions between AVP–pVIc and base or sugar residues on the DNA, but the dominant factor driving the nonspecific interaction between AVP–pVIc and DNA was the entropic contribution from the release of counter ions.

Activation of AVP by pVIc results into a set of conformational changes in the neighboring regions. They occur along a common path that divides into upper and lower paths (Fig. 3.3). At the end of the upper path, His54 drops down from its position in AVP to a position in AVP–pVIc that is opposite to Cys122. At the end of the lower path, Tyr84 moves 11 Å to a position in AVP–pVIc where it can form a cation–π interaction with His54. The activation pathway is triggered when the three N-terminal amino acids of pVIc (Gly-Val-Gln) bind in a preformed, hydrophobic pocket, the N-terminal binding pocket (NT pocket), on AVP. When AVP is activated, it leads to the maturation of virions.

## 5  AVP INHIBITORS

As discussed earlier, the activation of AVP is an essential step in the production of the infectious virus. This activation is followed by the processing of virion precursor proteins. The protease, encoded in the late cassette L3 as a 23-kDa polypeptide, is also called adenain cysteine protease, or simply L3/p23 protease. AVP is activated by its two cofactors: (1) pVIc (Mangel et al., 1993; Webster et al., 1993), and (2) viral DNA (Mangel et al., 1993). The activation of AVP is an essential step in the production the infectious virus, and hence is an attractive target for antiviral therapy. As the structure of enzyme plays an important role in the design and development of enzyme inhibitors, McGrath et al. (2003) determined crystal structures of the AVP–pVIc complex at 1.6 Å resolution (PDB ID: 1NLN). Recently Baniecki et al. (2013) have determined the nascent form of the adenovirus proteinase structure with atomic resolution (PDB ID: 4EKF).

Based on the crystal structures of the AVP–pVIc complex and AVP, Mc-Grath et al. (2013) then attempted to design the first-generation AVP inhibitors. A number of differences in the structure of the two greatly facilitated the design and development of AVP inhibitors. It was described that on the surface of the AVP–pVIc complex, the N- and C-termini of pVIc were bound in pockets, while on the surface of AVP structure only the NT pocket for pVIc was fully formed. The C-terminal binding pocket (CT pocket) for pVIc was, therefore, assumed to be induced upon the binding of pVIc. McGrath et al. (2013) therefore predicted that a drug binding in the NT pocket would prevent the binding of pVIc to AVP and hence prevent the activation of AVP. However, the NT pocket was found to be a legitimate druggable site (McGrath et al., 2013) and structure-based drug design on the NT pocket in AVP yielded several hundred potential inhibitors. McGrath et al. (2013) identified a hit compound, NSC 36806 (1), which could bind with both the AVP and AVP–pVIc.

1: NSC 36806

As in the docking study, McGrath et al. (2013) found that a significant portion of NSC 36806 was not interacting with AVP and as its molecular weight was very high, they initiated a substructure search of the NCI repository and found eight compounds structurally related to NSC 36806, out of which two compounds, NSC 37248 (**2**) and NSC 37249 (**3**), were observed to be effective inhibitors. NSC 37248 (**2**) was structurally similar to NSC 36806 (**1**), hence like NSC 36806, it was also found to bind to both the NT pocket in AVP and to the active site in AVP–pVIc complexes. However, NSC 37249 (**3**) was predicted to bind only to the active site in AVP–pVIc complexes, and thus it was assumed and also proved to be a competitive inhibitor of AVP–pVIc complexes.

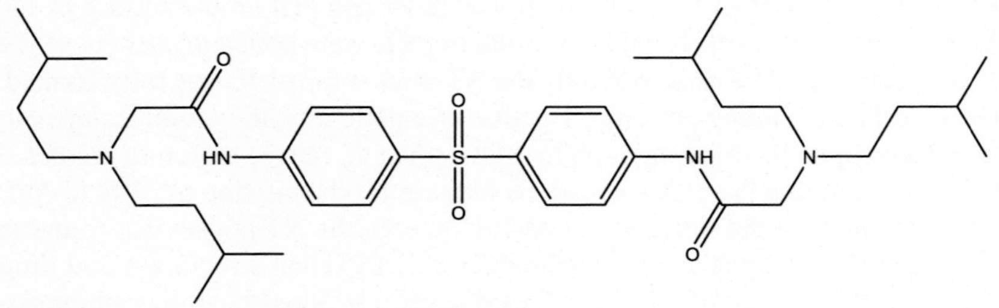

2: NSC 37248

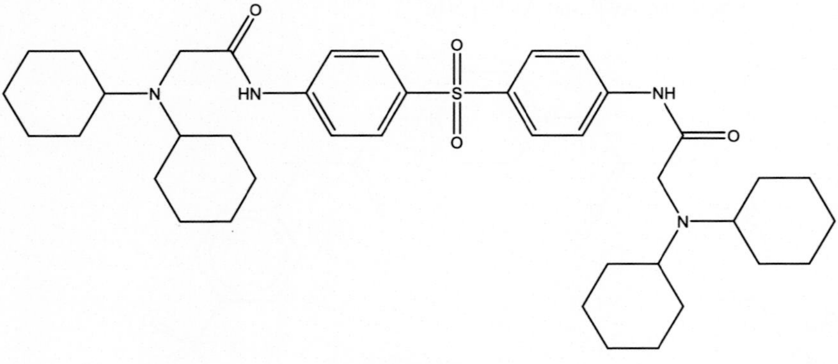

3: NSC 37249

Through a concise two-pronged hit discovery approach, Sweeney et al. (2014) recently identified two hits, a tetrapeptide nitrile (**4**) and a pyrimidine nitrile (**5**), as mutually exclusive starting points for investigating potent adenain inhibitors. These two compounds were highly potent against adenain with $IC_{50}$ values were 0.04 and 24 µM, respectively. X-ray cocrystal structures

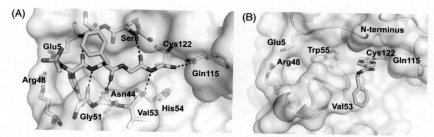

FIGURE 3.4  **X-ray cocrystal structures of adenain complexed with tetrapeptide nitrile (4) and pyrimidine nitrile (5).** (A) Tetrapeptide nitrile (**4**) bound in the active site of AVP2 (PDB code: 4PIE). (B) Pyrimidine nitrile (**5**) bound in the active site of AVP2 (PDB code: 4PID). *(From Sweeney, A.M., Grosche, P., Ellis, D., 2014. Discovery and structure-based optimization of adenain inhibitors. ACS Med. Chem. Lett. 5 (8), 937–941.)*

of both hits in complex with adenain were obtained (Sweeney et al., 2014). As shown in Fig. 3.4, both inhibitors were found to be covalently bound to the catalytic Cys122 residue of adenain through their nitrile group. For compound **4**, all amide groups were found to be involved in hydrogen bonding interactions with the protease. The P4 chlorophenyl group appeared to fill the hydrophobic S4 pocket of the substrate-binding site, which was involved in a cation−π interaction with the side chain of Arg48. However, the P3 phenyl group did not appear to have any interaction with the protease, rather the peptide chain expanded into the nonprime site (Fig. 3.4A). Fig. 3.4A also reveals the involvement of 3-chloro substituent of the P4 phenyl ring in halogen bonding with the carbonyl oxygen of Ala46 backbone residue.

Regarding inhibitor **5**, Fig. 3.4B shows that it has much less interaction than **4**. Unlike **4**, no portion of it extends into the substrate-binding pocket. The only interaction of it is through the thioimidate moiety, resulting from attack of the Cys122 SH-group on the nitrile group of the inhibitor, forming a hydrogen bond with Gln115.

**4**: Tetrapeptide nitrile                    **5**: Pyrimidine nitrile

However, despite the excellent interactions of compound **4** with the protease and consequent high potency in the biochemical assay, it was not found to be active in a viral replication assay. This anomaly could be attributed to its poor permeability due to its peptidic nature, and thus Sweeney et al. (2014) concentrated on the development of nonpeptidic inhibitors belonging to compound **5**. This led them to investigate, with various pros and cons, two highly active pyrimidine nitriles **6** and **7**, which had their $IC_{50}$ values against AVP8 as 0.004 and 0.003 µM, respectively.

**6**: $R_1$= H, $R_2$ = $CH_3OCH_2$,     $IC_{50}$ (AVP8) = 0.004 µM
**7**: $R_1$= H, $R_2$ = $CH_3CH(OH)$, $IC_{50}$ (AVP8) = 0.003 µM

As part of the ongoing efforts to optimize compound **4**, the Sweeney group aimed to further reduce the peptidic character of **4** by designing a novel replacement of the P4–P3 amide bond and by investigating irreversible inhibitors (Grosche et al., 2015). Due to the peptidic nature, compound **4** had poor permeability. Thus Grosche et al. (2015) first attempted to increase its permeability and hence its cellular potency. Based on crystal structure of adenain complexed with **4**, they designed a novel peptidomimetic scaffold, where the $NHCH(CH_2Ph)$ unit of the central Phe residue was replaced by a *meta*-substituted phenyl ring (**8**). They thought that introducing an ethanone linker between this phenyl moiety and the P4 phenyl group together with an *ortho*-substituent on the central phenyl ring would conserve the original P4–P3 binding mode. This led them to formulate a general structure as shown by **8**. The keto group of the ethanone linker of this structure was predicted to form a hydrogen bond with the amide NH of Glu5, while the *ortho*-substituent would force the carbonyl group out of the plane of the phenyl ring for optimal orientation. Further, it was assumed that gemdimethyl substitution at the benzylic position would be highly advantageous for proper positioning of the 3,5-dichlorophenyl moiety into the S4 subsite. Additionally, as already discussed by Sweeney et al. (2014), (3),5-dichloro substitution at the P4 phenyl ring may

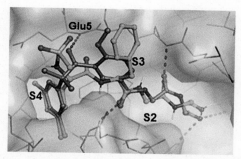

FIGURE 3.5 **Superposition of the putative docking pose of 8 (*blue*) with the X-ray cocrystal structure of 4 (*green*, PDB code 4PIE).** *(From Grosche, P., Sirockin, F., Sweeney, A.M., Ramage, P., Erbel, P., Melkko, S., Bernardi, A., Hughes, N., Ellis, D., Combrink, K.D., Jarousse, N., Altmann, E., 2015. Structure-based design and optimization of potent inhibitors of the adenoviral protease. Bioorg. Med. Chem. Lett. 25 (3), 438–443.)*

enhance hydrophobic interactions in S4. The binding mode of this new template (**8**) to adenain and its overlay with the X-ray cocrystal structure with **4** is shown in Fig. 3.5 (Grosche et al., 2015).

P4                 P3                          P2              P1

**8**

All structure–activity relationship (SAR) consideration led Grosche et al. (2015) to synthesize 4 analogs of **8** (**8a–d**), out of which the last two were found to be highly potent when tested against AVP8 and AVP5, but when tested in antiviral cytopathic effect (CPE) assay both compounds (**8c** and **8d**) exhibited only weak activity.

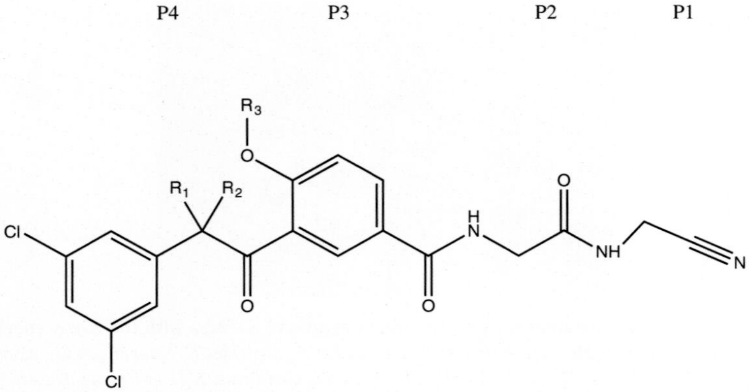

P4                    P3                    P2            P1

8a: $R_1 = R_2 = H$, $R_3 = CH_3$;                          $IC_{50}$ (AVP8) = 2.74 μM; $IC_{50}$ (AVP5) = 1.1 μM

8b: $R_1 = H$, $R_2 = R_3 = CH_3$;                         $IC_{50}$ (AVP8) = 0.95 μM; $IC_{50}$ (AVP5) = 0.31 μM

8c: $R_1 = R_2 = R_3 = CH_3$;                              $IC_{50}$ (AVP8) = 0.04 μM; $IC_{50}$ (AVP5) = 0.03 μM

8d: $R_1 = R_2 = CH_3$; $R_3 = (CH_3)_2$; $N (CH_3)_2$; $IC_{50}$ (AVP8) = 0.01 μM; $IC_{50}$ (AVP5) = 0.007 μM

Grosche et al. (2015) attempted to also optimize the pyrimidine nitrile scaffold (**5**), resulting into two achiral peptidomimetics, **9a** and **9b**, that were found to be exceptionally potent inhibitors of adenain. Both have only one amide bond. The binding of one of them (**9a**) when complexed with adenain (PDB code: 4WX4) was found to be as shown in Fig. 3.6 (Grosche et al., 2015). The figure shows that the compound (**9a**) could form four hydrogen bonds with the enzyme.

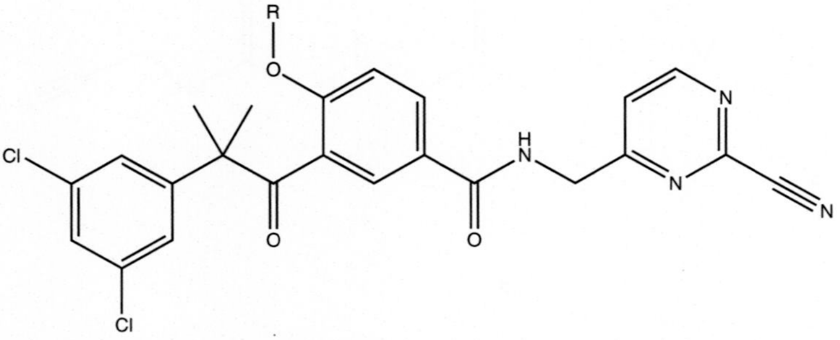

9a: R = $CH_3$,     $IC_{50}$ (AVP8) = 0.0006 μM; $IC_{50}$ (AVP5) = 0.0001 μM

9b: R = $(CH_3)_2N(CH_3)_2$, $IC_{50}$ (AVP8) = 0.0002 μM; $IC_{50}$ (AVP5) = 0.0001 μM

However, with all efforts so far, there are currently no approved antiviral therapies for HAdV infections. The only option is to treat patients with AdV infections

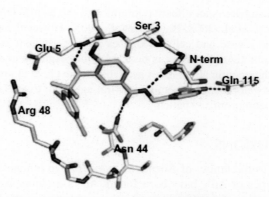

FIGURE 3.6   **X-ray co-crystal structure of 9a in adenain (PDB code: 4WX4).** *(From Grosche, P., Sirockin, F., Sweeney, A.M., Ramage, P., Erbel, P., Melkko, S., Bernardi, A., Hughes, N., Ellis, D., Combrink, K.D., Jarousse, N., Altmann, E., 2015. Structure-based design and optimization of potent inhibitors of the adenoviral protease. Bioorg. Med. Chem. Lett. 25 (3), 438–443.)*

with broad-acting antivirals, such as ribavirin (**10**), cidofovir (**11**), and ganciclovir (GCV) (**12**), which show variable results in all cases (Lenaerts et al., 2008). Ribavirin is a broad-spectrum antiviral agent showing in vitro activity against DNA and RNA viruses. It has shown variable activity against different HAdV types, displaying maximum activity against subgroup C (Morfin et al., 2005, 2009). Cidofovir displays efficacy against several DNA viruses and exhibits antiviral activity against all HAdV species, but has low oral bioavailability and significant toxicity (tubular necrosis), and does not confer long-term protection (Lujan-Zilbermann, 2008; Sanchez-Cespedes et al., 2014). However, currently even with its low bioavailability and nephrotoxicity, cidofovir is the anti-HAdV drug of choice for preemptive therapy (Lion et al., 2010; Matthes-Martin et al., 2012).

**10**: Ribavirin                **11**: Cidofovir                **12**: Ganciclovir

GCV is another antiviral drug that has been used for the treatment of HAdV infections with variable results (Bordigoni et al., 2001; Chen et al., 1997; Kurosaki et al., 2004).

GCV triphosphate is an effective inhibitor of the HAdV DNA polymerase in vitro (Lenaerts et al., 2008; Naesens et al., 2005). However, as HAdV has no viral thymidine kinase and cellular kinases are not efficient at phosphorylating GCV, its anti-HAdV activity is poor. Martinez-Aguado et al. (2015) have reviewed the therapeutic alternatives available in the clinic at present and prospects for the development of specific antiviral agents to treat HAdV infections in the immunocompromised patient.

## 6 CONCLUSIONS

AdVs belong to the family of Adenovirdae viruses. In humans, more than 50 distinct serotypes of AdV have been found. AdVs are nonenveloped viruses. The AVP plays an important role during the maturation of the virus, which is unique. A comparison of its structure with that of papain suggested that it can be classified as a cysteine proteinase.

AVP is also called adenain. It is an inactive enzyme at synthesis and it requires two cofactors for its maximal activity: (1) an 11–amino-acid peptide pVIc, derived from the C-terminus of the polypeptide VI precursor, pVI, and (2) viral DNA. Cleavage at the C-terminus releases pVIc, which then binds with AVP, and the AVP–pVIc complex slides along DNA, processing the virus precursor proteins, also bound to the DNA. The activation of AVP is an essential step in the production the infectious virus, and hence is an attractive target for antiviral therapy. As the structure of the enzyme plays an important role in the design and development of enzyme inhibitors, the crystal structures of the AVP–pVIc complex at 1.6 Å resolution was determined and the atomic resolution structure of the nascent form of AVP was also recently reported. On the basis of the crystal structures of AVP–pVIc complex and AVP, attempts were made to design first-generation AVP inhibitors and a hit compound, NSC 36806 (**1**), which could bind with both the AVP and AVP–pVIc. Through a concise two-pronged hit discovery approach, two hits, a tetrapeptide nitrile (**4**) and a pyrimidine nitrile (**5**), were identified. As part of the ongoing efforts to optimize compounds **4** and **5**, attempts were made to reduce their peptidic character, leading to more potent AVP inhibitors. However, with the efforts so far, there are currently no approved antiviral therapies for HAdV infections. The only option to treat patients with AdV infections include the use of broad-acting antivirals, such as ribavirin (**10**), cidofovir (**11**), and GCV (**12**), in all cases showing variable results.

## REFERENCES

Baniecki, M.L., McGrath, W.J., Mangel, W.F., 2013. Regulation of a viral proteinase by a peptide and DNA in one-dimensional space. III. Atomic resolution structure of the nascent form of the adenovirus proteinase. J. Biol. Chem. 288, 2081–2091.

Blainey, P.C., Graziano, V., Pérez-Berná, A.J., McGrath, W.J., Flint, S.J., San Martín, C., Xie, X.S., Mangel, W.F., 2013. Regulation of a viral proteinase by a peptide and DNA in one-dimensional space. IV. Viral proteinase slides along DNA to locate and process its substrates. J. Biol. Chem. 288, 2092–2102.

Blanche, F., Monegier, B., Faucher, D., Duchesne, M., Audhuy, F., Barbot, A., Bouvier, S., Daude, G., Dubois, H., Guillemin, T., Maton, L., 2001. Polypeptide composition of an adenovirus type 5 used in cancer gene therapy. J. Chromatogr. 921, 39–48.

Bordigoni, P., Anne-Sophie Carret, A.-S., Venard, V.R., Witz, F., Faou, A.L., 2001. Treatment of adenovirus infections in patients undergoing allogeneic hematopoietic stem cell transplantation. Clin. Infect. Dis. 32, 1290–1297.

Challberg, M.D., Kelly, Jr., T.J., 1981. Processing of the adenovirus terminal protein. J. Virol. 38, 272–277.

Chen, F.E., Liang, R.H.S., Yuen, K.Y., Chan, T.K., Peiris, M., 1997. Treatment of adenovirus-associated haemorrhagic cystitis with ganciclovir. Bone Marrow Transplant. 20, 997–999.

Ding, J., McGrath, W.J., Sweet, R.M., Mangel, W.F., 1996. Crystal structure of the human adenovirus proteinase with its 11 amino acid cofactor. EMBO J. 15, 1778–1783.

Grosche, P., Sirockin, F., Sweeney, A.M., Ramage, P., Erbel, P., Melkko, S., Bernardi, A., Hughes, N., Ellis, D., Combrink, K.D., Jarousse, N., Altmann, E., 2015. Structure-based design and optimization of potent inhibitors of the adenoviral protease. Bioorg. Med. Chem. Lett. 25, 438–443.

Guimet, D., Hearing, P., 2013. The adenovirus L4-22K protein has distinct functions in the post-transcriptional regulation of gene expression and encapsidation of the viral genome. J. Virol. 87, 7688–7699.

Gustin, K.E., Imperiale, M.J., 1998. Encapsidation of viral DNA requires the adenovirus L1 52/55-kilodalton protein. J. Virol. 72, 7860–7870.

Gustin, K.E., Lutz, P., Imperiale, M.J., 1996. Interaction of the adenovirus L1 52/55-kilodalton protein with the IVa2 gene product during infection. J. Virol. 70, 6463–6467.

Howitt, J., Anderson, C.W., Freimuth, P., 2003. Adenovirus interaction with its cellular receptor. Car. Curr. Top. Microbiol. Immunol. 272, 331–364.

Ishibashi, M., Maizel, Jr., J.V., 1974. The polypeptides of adenovirus. V. Young virions, structural intermediate between top components and aged virions. Virology 57, 409–424.

Kelkar, S.A., Pfister, K.K., Crystal, R.G., Leopold, P.L., 2004. Cytoplasmic dynein mediates adenovirus binding to microtubules. J. Virol. 78, 10122–10132.

Kurosaki, K., Miwa, N., Yoshida, Y., Kurokawa, M., Kurimoto, M., Endo, S., Shiraki, K., 2004. Therapeutic basis of vidarabine on adenovirus-induced haemorrhagic cystitis. Antiviral Chem. Chemother. 15, 281–285.

Lenaerts, L., De Clercq, E., Naesens, L., 2008. Clinical features and treatment of adenovirus infections. Rev. Med. Virol. 18, 357–374.

Lion, T., Kosulin, K., Landlinger, C., Rauch, M., Preuner, S., Jugovic, D., Pötschger, U., Lawitschka, A., Peters, C., Fritsch, G., Matthes-Martin, S., 2010. Monitoring of adenovirus load in stool by real-time PCR permits early detection of impending invasive infection in patients after allogeneic stem cell transplantation. Leukemia 24, 706–714.

Lujan-Zilbermann, J., 2008. Infections in hematopoietic stem cell transplant recipients. In: Long, S.S., Pickering, L.K., Prober, C.G. (Eds.), Principles and Practice of Pediatric Infectious Diseases. Churchill Livingstone, Philadelphia, PA, pp. 562–566.

Ma, H.C., Hearing, P., 2011. Adenovirus structural protein IIIa is involved in the serotype specificity of viral DNA packaging. J. Virol. 85, 7849–7855.

Mangel, W.F., Martin, S.C., May, E.R., 2014. Structure, function and dynamics in adenovirus maturation. Viruses 6, 4536–4570.

Mangel, W.F., McGrath, W.J., Toledo, D.L., Anderson, C.W., 1993. Viral DNA and a viral peptide can act as cofactors of adenovirus virion proteinase activity. Nature 361, 274–275.

Mangel, W.F., Toledo, D.L., Brown, M.T., Martin, J.H., McGrath, W.J., 1996. Characterization of three components of human adenovirus proteinase activity in vitro. J. Biol. Chem. 271, 536–543.

Mangel, W.F., Toledo, D.L., Ding, J., Sweet, R.M., McGrath, W.J., 1997. Temporal and spatial control of the adenovirus proteinase by both a peptide and the viral DNA. Trends Biochem. Sci. 22, 393–398.

Martinez-Aguado, P., Serna-Gallego, A., Marrugal-Lorenzo, J.A., Gomez-Marin, I., Sanchez-Cespedes, J., 2015. Antiadenovirus drug discovery: potential targets and evaluation methodologies. Drug Discov. Today 20, 1235–1242.

Martin, M.A., Knipe, D.M., Fields, B.N., Howley, P.M., Griffin, D., Lamb, R., 2007. Fields' virology. Wolters Kluwer Health/Lippincott Williams and Wilkins, Philadelphia, PA (p. 2395).

Matthes-Martin, S., Feuchtinger, T., Shaw, P.J., Engelhard, D., Hirsch, H.H., Cordonnier, C., Ljungman, P., 2012. European guidelines for diagnosis and treatment of adenovirus infection in leukemia and stem cell transplantation: summary of ECIL-4 (2011). Transpl. Infect. Dis. 14, 555–563.

McGrath, W.J., Baniecki, M.L., Li, C., Mc Whirter, S.M., Brown, M.T., Toledo, D.L., Mangel, W.F., 2001. Human adenovirus proteinase: DNA binding and stimulation of proteinase activity by DNA. Biochemistry 40, 13237–13245.

McGrath, W.J., Ding, J., Sweet, R.M., Mangel, W.F., 1996. Preparation and crystallization of a complex between human adenovirus serotype 2 proteinase and its 11-amino-acid cofactor pVIc. J. Struct. Biol. 117, 77–79.

McGrath, W.J., Ding, J., Sweet, R.M., Mangel, W.F., 2003. Crystallographic structure at 1.6 Å resolution of the human adenovirus proteinase in a covalent complex with its 11-amino-acid peptide cofactor: insights on a new fold. Biochim. Biophys. Acta. 1648, 1–11.

McGrath, W.J., Graziano, V., Zabrocka, K., Mangel, W.F., 2013. First generation inhibitors of the adenovirus proteinase. FEBS Lett. 587, 2332–2339.

Meier, O., Boucke, K., Hammer, S.V., Keller, S., Stidwill, R.P., Hemmi, S., Greber, U.F., 2002. Adenovirus triggers macropinocytosis and endosomal leakage together with its clathrin-mediated uptake. J. Cell Biol. 158, 1119–1131.

Morfin, F., Dupuis-Girod, S., Frobert, E., Mundweiler, S., Carrington, D., Sedlacek, P., Bierings, M., Cetkovsky, P., Kroes, A.C., van Tol, M.J., Thouvenot, D., 2009. Differential susceptibility of adenovirus clinical isolates to cidofovir and ribavirin is not related to species alone. Antivir. Ther. 14, 55–61.

Morfin, F., Dupuis-Girod, S., Mundweiler, S., Falcon, D., Carrington, D., Sedlacek, P., Bierings, M., Cetkovsky, P., Kroes, A.C., van Tol, M.J., Thouvenot, D., 2005. In vitro susceptibility of adenovirus to antiviral drugs is species-dependent. Antiviral Ther. 10, 225–229.

Naesens, L., Lenaerts, L., Andrei, G., Snoeck, R., Van Beers, D., Holy, A., Balzarini, J., De Clercq, E., 2005. Antiadenovirus activities of several classes of nucleoside and nucleotide analogues. Antimicrob. Agents Chemother. 49, 1010–1016.

Ostapchuk, P., Anderson, M.E., Chandrasekhar, S., Hearing, P., 2006. The L4 22-kilodalton protein plays a role in packaging of the adenovirus genome. J. Virol. 80, 6973–6981.

Ostapchuk, P., Almond, M., Hearing, P., 2011. Characterization of empty adenovirus particles assembled in the absence of a functional adenovirus IVa2 protein. J. Virol. 85, 5524–5531.

Ostapchuk, P., Yang, J., Auffarth, E., Hearing, P., 2005. Functional interaction of the adenovirus IVa2 protein with adenovirus type 5 packaging sequences. J. Virol. 79, 2831–2838.

Pérez-Berná, A.J., Mangel, F.B., McGrath, W.J., Graziano, V., Flint, J., Martin, C.S., 2014. Processing of the L1 52/55k protein by the adenovirus protease: a new substrate and new insights into virion maturation. J. Virol. 88, 1513–1524.

Perez-Romero, P., Tyler, R.E., Abend, J.R., Dus, M., Imperiale, M.J., 2005. Analysis of the interaction of the adenovirus L1 52/55-kilodalton and IVa2 proteins with the packaging sequence in vivo and in vitro. J. Virol. 79, 2366–2374.

Record, Jr., M.T., Lohman, M.L., de Haseth, P., 1976. Ion effects on ligand-nucleic acid interactions. J. Mol. Biol. 107, 145–158.

Rowe, W.P., Huebner, R.J., Gilmore, L.K., Parrott, R.H., Ward, T.G., 1953. Isolation of a cytopathogenic agent from human adenoids undergoing spontaneous degeneration in tissue culture. Proc. Soc. Exp. Biol. Med. 84, 570–573.

Sanchez-Cespedes, J., Moyer, C.L., Whitby, L.R., Boger, D.L., Nemerow, G.R., 2014. Inhibition of adenovirus replication by a trisubstituted piperazin-2-one derivative. Antiviral Res. 108, 65–73.

Stewart, P.L., Chiu, C.Y., Huang, S., Muir, T., Zhao, Y., Chait, B., Mathias, P., Nemerow, G.R., 1997. Cryo-EM visualization of an exposed RGD epitope on adenovirus that escapes antibody neutralization. EMBO J. 16, 1189–1198.

Sweeney, A.M., Grosche, P., Ellis, D., Combrink, Keith, Erbel, P., Hughes, N., Sirockin, F., Melkko, S., Bernardi, A., Ramage, P., Jarousse, N., Altmann, E., 2014. Discovery and structure-based optimization of adenain inhibitors. ACS Med. Chem. Lett. 5, 937–941.

Trotman, L.C., Mosberger, N., Fornerod, M., Stidwill, R.P., Greber, U.F., 2001. Import of adenovirus DNA involves the nuclear pore complex receptor CAN/Nup214 and histone H1. Nat. Cell Biol. 3, 1092–1100.

Weber, J., 1976. Genetic analysis of adenovirus type 2. III. Temperature sensitivity of processing viral proteins. J. Virol. 17, 462–471.

Weber, J.M., 1999. Role of endoprotease in adenovirus infection. In: Seth, P. (Ed.), Adenoviruses: Basic Biology to Gene Therapy. R.G. Landes, Austin, TX, pp. 79–83.

Webster, A., Kemp, G., 1993. The active adenovirus protease is the intact L3 23K protein. J. Gen. Virol. 74, 1415–1420.

Webster, A., Hay, R.T., Kemp, G., 1993. The adenovirus protease is activated by a virus-coded disulphide-linked peptide. Cell 72, 97–104.

Wohl, B.P., Hearing, P., 2008. Role for the L1-52/55K protein in the serotype specificity of adenovirus DNA packaging. J. Virol. 82, 5089–5092.

Wu, K., Guimet, D., Hearing, P., 2013. The adenovirus L4-33K protein regulates both late gene expression patterns and viral DNA packaging. J. Virol. 87, 6739–6747.

Zhang, W., Arcos, R., 2005. Interaction of the adenovirus major core protein precursor, pVII, with the viral DNA packaging machinery. Virology 334, 194–202.

Zhang, W., Imperiale, M.J., 2000. Interaction of the adenovirus IVa2 protein with viral packaging sequences. J. Virol. 74, 2687–2693.

# Chapter 4

# *Alphavirus* Nonstructural Proteases and Their Inhibitors

Akalabya Bissoyi\*, Subrat K. Pattanayak\*, Arindam Bit\*, Ashish Patel\*, Abhishek K. Singh\*\*, Sudhanshu S. Behera[†], Debabrata Satpathy\*

*\*National Institute of Technology, Raipur, Chhattisgarh, India; \*\*University of Allahabad, Allahabad, Uttar Pradesh, India; [†]DFARD, Bhubaneswar, Odisha, India*

## 1 INTRODUCTION

Alphaviruses belong to the Togaviridae (Group IV) family of viruses, and are categorized as C priority pathogens according to the US National Institute of Allergy and Infectious Diseases of United States of America. These viruses are known to be the cause of major fatality due to their high mortality and morbidity rates in underdeveloped regions of Africa, Asia, and some parts of Latin America (Strauss and Strauss, 1994). Alphaviruses are able to infect various vertebrates, such as human, horses, rodent, fish, and birds, as well as invertebrates and include numerous categories of viruses, such as Chikungunya virus (CHIKV), Barmah Forest virus, O'Nyong–Nyong virus, Mayaro virus, Rose River virus, Semiliki Forest virus (SFV), Sindbis virus (SINV), Una virus, Tonate virus, Eastern equine encephalitis, Western equine encephalitis, and Venezuelan equine viruses (Atkins, 2013). The infections caused by these viruses are of potential risk, and many of them have so far no treatment. The treatment, management, and vaccination procedure to combat diseases caused by these viruses have not been rewarding (Rayner et al., 2002). Alphaviruses are generally RNA viruses with a positive-sense, and single-stranded RNA genome (Koonin et al., 1993). These RNA viruses encode nonstructural (NS) proteins (NSPs) that can be used for replication to achieve a maturation state. These protein moieties can be targeted for the development and testing of potential antiviral drugs. The goal of computational study is to provide an alternative tool that is highly reliable and that may be complementary to the current experimental techniques. The two important theoretical approaches, molecular dynamics simulations (Pattanayak and Chowdhuri, 2011, 2012) and docking (Agarwal et al., 2015; Bissoyi et al., 2013), have been found to be quite reliable in this respect. Other computational procedures have been widely utilized

**Viral Proteases and Their Inhibitors. http://dx.doi.org/10.1016/B978-0-12-809712-0.00004-6**

to investigate the drug–receptor interactions in various types of biochemical problems (Garg et al., 1999; Gupta et al., 1983).

Recently, several protease inhibitors have been developed using computer-aided drug design methodologies (Gupta, 2013; Gupta et al., 1983; Wlodawer and Vondrasek, 1998), synthetic approaches, high-throughput screening method (Mayr and Bojanic, 2009), and drug reposition–based approaches (Sundberg, 2000), which could possibly target the NSPs responsible for virus replication. In addition, protease inhibitors targeting viral NSPs have also been isolated from various sources, such as plant extracts (Bhakat et al., 2014). These protease inhibitors have shown different degrees of inhibition of viral NSPs, which necessitates the development of selective and potent therapeutic agents against the devastating alphaviruses (De Francesco and Carfi, 2007). In this chapter, attempts have been made to discuss the development and design of antiviral drugs against various alphaviruses. In addition, the development of protease inhibitors targeting NSPs of alphaviruses has been also presented.

## 2 ALPHAVIRUSES

Alphaviruses have been established as a model system to examine the budding and assembly of enveloped viruses. The RNA genome of these viruses encodes multiple copies of capsid proteins (CPs) that assemble along with replicated RNA to form the nucleocapsid core (NC). The icosahedral budding and assembly pattern of these simple enveloped viruses are described in detail here.

## 2.1 Life Cycle of Virus

Alphaviruses invade the host cells through the clathrin-coated endocytic pathway. Acidic pH at the endosomal membrane enables fusion and results in the release of NC into the cytoplasm. Upon the shifting of CPs to ribosomes, NC becomes uncoated and discharges the 49S RNAs genome into the cytoplasm. These genomic RNAs are translated into the NSPs. The information of negative-sense replica of genomic RNA is transcribed by the NSPs. This RNA works as an archetype for the production of genomic and subgenomic RNA. The subgenomic RNA has greater capability to synthesize proteins, in comparison to the genomic RNA that also encodes the basic proteins of the virus. In the structural polyprotein, CP being at the N-terminus followed by PE2 (E3þE2), 6K, and E1. *Alphavirus* virions are formed by the assembly of 240 copies of CP, E2, and E1 (Fig. 4.1A). The transmembrane E1 glycoprotein mediates the coalescence between viral and endosomal membranes. On the other hand, the transmembrane segment E2 is responsible for the binding of the cell receptor. Endoplasmic reticulum (ER) and Golgi apparatus process the E1 and PE2, a precursor of E2 and E3 as a heterodimer, and transport them in the form of spikes to the cell surface. These spikes are constituted of three heterodimers of E1/E2 that promotes the correct folding of E2 and E3 which work as a chaperone and prevent the

(A)

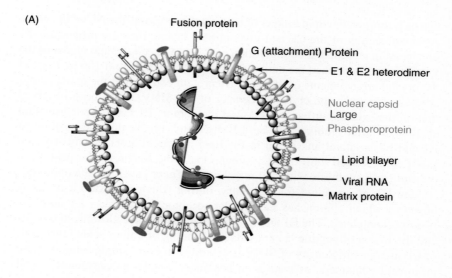

(B)

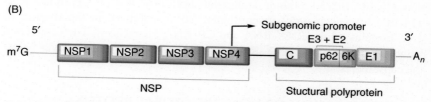

**FIGURE 4.1**   (A) Structure of *Alphavirus*. The hypothetical alpha capsid arrangement, genome, particle, and proteins, (B) Schematic representation of *Alphavirus* genome showing the RNA sequence. *C*, Capsid; *E*, envelope; *NSP*, nonstructural protein.

premature fusion of E1 with Golgi in an acidic pH environment. During virus entry, fusion of the glycoprotein spike complex produces E2 and E3 by cleavage of PE2 in the Golgi, with the help of a protease similar to furin. The function of 6K is not clear, but it promotes the pathogenic effect of the particle. In the infected cell, an icosahedral NC is formed in the cytoplasm by the encasing of the only copy of genome RNA with 240 copies of CP. Matured virus budded from the host cell membrane by the action of NC and E1/E2 trimeric spikes present in its plasma membrane.

## 2.2   *Alphavirus* and its Virion Structure

The conformation of alphaviruses (Fig. 4.1A) is clearly defined by the two popular techniques: (1) image reconstruction technique and (2) cryoelectron microscopy (cryo-EM). The outer protein layer of these viruses consists of glycoproteins E1 and E2 (Fig. 4.1). Membrane extension of these glycoproteins can pass through the lipid bilayer of the host that encloses the NC of the virus. After interaction of the CP and glycoprotein layers, they are proportionally organized in icosahedral configuration of T $^{1⁄4}$ 4. A pseudoatomic model of the virus is generated when the amino acid sequence of CP from 106 to 264 and the crystal arrangements of a part of E1 protein extend into the extracellular space and fit in to the cryo-EM density of the SINV. It has been observed that on the exterior surface of the viral membrane, E1 takes the icosahedral shell shape after binding with cryo-EM density. Practically, the position of E1 and E2 is lateral and radial to the lipid bilayer simultaneously. The top of E1 is populated with the E2, which cloaks the fusion of the peptide to prevent early synthesis with cell membranes. These envelope glycoproteins, E1 and E2, are required for cell entry, during which the exterior of E2 is unveiled, which approaches the cellular receptor and protects E1 until the fusion is completed. The E1 and E2 are separated in the endosome at low pH and the fusion of peptide is exposed. The binding of amino acids (106–264) of CP in cryo-EM density of SINV is exhibited in every single subunit of the prominent pentamers, as well as in hexamers (known as capsomeres), and detected in the NC layer made from the CP protease domain comprising of amino acids. Indeed a minor contact has been observed for CP amino acids 114–264 in capsomere(s). Accordingly, NC is stabilized in the absence of glycoproteins by the RNA–RNA and CP–RNA interactions. The connection between CP–RNA and RNA–RNA appears in the RNA–protein layer below the capsomeres.

## 3  GLOBAL VISION OF NONSTRUCTURAL PROTEINS

The NSPs are essential in the replication process of *Alphavirus*. In addition, these NSPs also play a vital role in the infection and interaction of the virus with host cells. Out of the four NSPs, NSP1 (537 aa), whose structure is shown in Fig. 4.2, binds with membrane proteins and activates guanylyl transferase and methyl transferase, which are involved in the covering of 26S and 42S RNAs (Ahola and Kääriäinen, 1995; Kääriäinen and Ahola, 2002; Koonin et al., 1993; Laakkonen et al., 1994). In the response of *Alphavirus* capping, the methyl group is transferred by NSP1 from S-adenosyl methionine (AdoMet) to GTP. Subsequently, before transferring 7-methyl-GMP (m7GMP) to virus RNA, a covalent complex is formed between m7GMP and NSP1. The response of capping of *Alphavirus* is somewhat different from the capping of the cellular mRNAs, where guanylyl transferase is first transmitted from GMP to RNA, followed by methylation of GMP-RNA (Ahola and Kääriäinen, 1995). The enzymatic actions of NSP1 are necessary for virus replication. A virus becomes noninfectious by single-point mutation because this transformation abolishes the enzymatic activities (Wang et al., 1996). NSP1 is strongly bound with the plasma membrane of infected cells, and when NSP1 is expressed alone then it functions as a single membrane anchor in virus replication complexes (Peränen et al., 1995; Salonen et al., 2003;

(A)                                            (B)

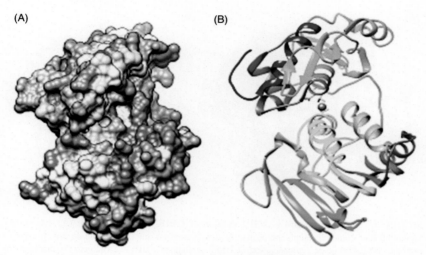

FIGURE 4.2    **3D Protein stucture of NSP1 domain of Chikungunya virus (CHIKV).** (A) 3D representations of surface charge distribution of respective proteins, *blue* color represents the positively charged regions and *orange*, the negatively charged regions, and (B) cartoon representation of the proteins. 3D protein structures of CHIKV NSP1 were designed using TASSER server (Zhang, 2008). The structure is drawn using Chimera visualization tool (Pettersen et al., 2004).

Žusinaite et al., 2007). Amphipathic alpha-helix is formed when amino acid 245–264 of NSP1 mediates the binding of plasma membrane (Ahola and Kääriäinen, 1995; Žusinaite et al., 2007). The binding activities of NSP1 with the membrane can be affected by a single-point mutation and it is lethal for the virus (Žusinaite et al., 2007). In SFV, the Cys 418–420 residues of NSP1 and in SINV, Cys420 of NSP1 tighten the membrane binding by posttranslational palmitoylation (Laakkonen et al., 1998; Wang et al., 1996; Žusinaite et al., 2007). Palmitoylation is required only for the formation of the replication complexes or sustainability of the virus, and not for enzymatic activities of NSP1 (Ahola and Kääriäinen, 1995; Žusinaite et al., 2007). However, removal or exchange of cysteine residues 418–420 in NSP1 of SFV significantly obstructs virus duplication and causes growth of expletive alterations (Žusinaite et al., 2007). It has been observed from the virus expression that palmitoylation-deficient NSP1 is not capable of causing the disease in mice; these viruses induce low levels of viruses in the blood stream, but no infection has been observed in the brain tissues (Ahola and Kääriäinen, 1995). NSP1 is responsible for the initiation of structure on the cell surface, similar to filopodia that is representative of *Alphavirus*-septic cells (Hardy and Strauss, 1989; Laakkonen et al., 1998; Rikkomen, 1996; Žusinaite et al., 2007). The functional activities and significance of filopodia-like structures and palmitoylation remain unidentified.

Of the four NSPs, NSP2 (799 aa in SFV) is known for different types of enzymatic actions and significant roles in virus infection. N-terminal domain of NSP2 (RNA helicase) comprises the activities of nucleoside triphosphatase

(A)                                           (B)

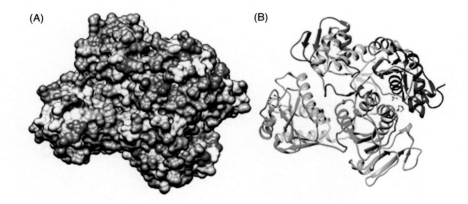

FIGURE 4.3  **Representation of NSP2 domain of CHIKV (PDB: 3TRK).** (A) The surface view with *blue color* represents the positively charged regions and *orange*, the negatively charged regions. (B) Ribbon-type representation of NSP2 domain of CHIKV. Structures are drawn using Chimera visualization tool (Pettersen et al., 2004).

and RNA triphosphatase as shown in Fig. 4.3 (Bouraï et al., 2012; Lastarza et al., 1994; Merits et al., 2001; Peränen et al., 1990; Rikkomen, 1996). The activities of RNA helicase and RNA triphosphatase are very important for the unwinding of RNA duplexes during replication and removal of 5′-phosphate, respectively, from viral RNA during capping reaction. The removal of 5′-phosphate before the guanylyl transferase activity of NSP1 is a very crucial step that facilitates the attachment of m7GMP.

The papain-like protease activity present at C-terminal domain of NSP2 is responsible for processing of NS polyprotein (Cruz et al., 2010; Fazakerley et al., 2002). Moreover, the protease participates in the formation of the replication complex and subsequently regulates the replication process. This proteolytic activity has also been shown to be hampered due to mutations that abolish its efficient activity and leads to the fatality of the virus (Cruz et al., 2010). The C-terminus of NSP2 contains latent methyl transference activity, which plays an important role in the regulation of minus-strand synthesis and the development of cytopathic effects (Cruz et al., 2010). It has been documented that approximately 50% of NSP2 is present in the nuclear fraction of infected cells, as evidenced by the cell fractionation studies (Peränen et al., 1990). The nuclear transportation of NSP2 starts at the earliest during the infection, and the nuclear localization is mediated through the pentapeptide PRRRV (aa 647–651) (Rikkomen, 1996). The nuclear translocation of NSP2 has been shown to be affected with mutations, which results in attenuated phenotypes with reduced cytotoxicity and transcriptional and translational shutdown of the host cells, yet renders the virus pathogenic in adult mice (Fazakerley et al., 2002; Rikkomen, 1996). In addition, apart from helicase- and methyltransferase-like domains, other sequences/domains within NSP2 are essential for virus-induced

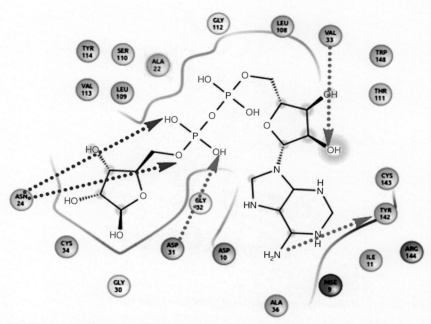

**FIGURE 4.4    A 2D representation of structure of NSP3 macro domain of CHIKV (PDB ID: 3GPO) in complex with ADP ribose.** *Arrows* represent the hydrogen bondings between NSP3 and ADP ribose. Structure drawn using Schrödinger (2011) software.

shutdown of cellular transcription (Akhrymuk et al., 2012; Firth et al., 2011; Gorchakov et al., 2005; Strauss and Strauss, 1994; Varadhachary et al., 2008) and suppression of antiviral responses (Akhrymuk et al., 2012; Gorchakov et al., 2008).

The primary structure of NSP3 (482 aa in SFV) is composed of three discrete domains that contain a small macro domain of 160 amino acids at the N-terminus, which is conserved among alphaviruses, rubella virus, hepatitis E virus, and coronaviruses (Fig. 4.4). This conserved macro domain also shares sequence similarity with the macro domains of proteins of eukaryotic organisms, archaea, and eubacteria (Koonin et al., 1993; Pehrson and Fuji, 1998). The middle domain of *Alphavirus* NSP3 is a conserved region and the third C-terminal domain (starting from Tyr324 in SFV) is hypervariable in length and sequence (Strauss and Strauss, 1994). However, the function of NSP3 is not clearly understood.

The study of replication complexes suggested that NSP3 is found in the cytoplasm in the form of aggregates with irregular boundaries of variable sizes. In addition, a small fraction of NSP3 is localized to the nuclear envelope (Gorchakov et al., 2008; Wang et al., 2006), suggesting that NSP3 might participate in RNA replication. It has also been demonstrated that the enzymatic

FIGURE 4.5    Ribbon-type representation of NSP4 domain of CHIKV.

activity only resides within the macro domain of NSP3 protein, which was identified as ADP-ribose-1-phosphate activity (Malet et al., 2009; Neuvonen and Ahola, 2009). *Alphavirus* macro domain has been found to have the capability of binding to poly-(ADP-ribose), RNA, and ADP-ribose as shown in Fig. 4.5 (Malet et al., 2009; Neuvonen and Ahola, 2009). Interestingly, poly-(ADP-ribose) polymerase-1 (PARP-1) interacts with SINV replication complexes through the C-terminal domain of NSP3 in neuronal cells (Park and Griffin, 2009).

Recent studies have suggested that only NSP3 of *Alphavirus* has the capability to get phosphorylated at serine/threonine residues at the C-terminus (Peränen et al., 1988; Vihinen and Saarinen, 2000; Vihinen et al., 2001). In SFV, 16 phosphorylation sites have been identified that are accumulated in the area of 50–amino acid residues. The complete removal of these sites led to reduced levels of RNA synthesis in cultured cells and subsequently reduced pathogenicity in mice (Vihinen et al., 2001). The phosphorylation of NSP3 is specifically important for minus-strand RNA synthesis, as evidenced by SINV-mutant studies (Vihinen et al., 2001). Thus the SFV and SINV studies have concluded that NSP3 plays a significant role in the modulation of neuropathogenesis in mice (Tuittila and Hinkkanen, 2003). Several studies reported that NSP3 also mediates the movement of *Alphavirus* replication complexes from the plasma membrane to intracellular vesicles. It has been confirmed by the analysis of subcellular localization of different forms of NSP intermediates (Hahn et al., 1989; Park and Griffin, 2009; Tanno et al., 2007).

Among NSPs of *Alphavirus*, NSP4 (614 aa in SFV) Fig. 4.5 represents the last catalytic subunit of *Alphavirus* that acts as RNA-dependent RNA polymerase (RdRP) (Hahn et al., 1989; Tanno et al., 2007). The C-terminal end of *Alphavirus* NSP4 contains the GDD motif (Gly315, Asp316, and Asp317) and shares sequence homology with RdRPs of other viruses (Argos et al., 1984). CHIKV NSP4 (EC 2.7.7.48) localizes to 2014–2462 sequence position in the polyprotein NSP 1234, which is constituted of 449 amino acids with the activity of polymerase. The potential catalytic site bearing a triad and two allosteric binding sites were identified to be similar to that in hepatitis C virus (HCV) NS5B polymerase. NSP4 of CHIKV included a highly conserved catalytic triad (GDD motif) responsible for polymerase activity.

The voluminous catalytic site is 2769.41 Å$^3$, whereas two allosteric binding sites with volume of 73.216 and 55.904 Å$^3$, were identified on both palm and thumb sites.

However, the N-terminal end of NSP4 is conserved in the alphaviruses. Although NSP4 is the major catalytic core of replication complex in *Alphavirus*, it still needs other NSPs for activity (Rubach et al., 2009). Mutation studies have suggested that the N-terminus of NSP4 interacts with host proteins and other NSPs, particularly NSP1, for its activity (Fata et al., 2002; Shirako et al., 2000). In addition, in vitro experiments have demonstrated that NSP4 also has terminal adenylyl transferase activity, suggesting its role in repair and maintenance of the poly(A)-tail of the viral genome (Shirako and Strauss, 1998; Tomar et al., 2006), which is an essential step for the survival of the virus. The level of NSP4 has been reported to be lower than the levels of other NSPs in the infected cells that are attained by several processes. First, the normal translation of different NS polyproteins is in the form of P123 due to the fact that most of the alpha viruses hold the opal termination codon in their genome. Moreover, the translation of P1234 and NSP4 is found with an efficiency of 15%–25% only. Second, the N-terminal amino acid in NSP4, which is a conserved tyrosine, directs the NSP4 to rapid proteasomal degradation. This destabilizing role of tyrosine has been confirmed by replacing it with a nonaromatic residue that leads to poor RNA replication (Shirako and Strauss, 1998).

## 3.1 NSP1 Protease Inhibitors

### 3.1.1 Bone Marrow Stromal Antigen-2–Based Inhibitors

Targeting N-terminal of NS1, which is responsible for methyl and guanyl transferase activities, would have an inhibitory effect on viral replication. Recently, NSP1 protein showed a crucial role in the downregulation of bone marrow stromal antigen-2 (BST-2) and thus paved the way for the development of BST-2–mediated chemicals acting against NSP1 (Tokarev et al., 2009). BST-2 is one of the important defense mechanisms of host cells, which occurs by the expression of interferon alpha (INFα). Its expression results in retaining the virus at the surface of the infected cells (Jones et al., 2013).

### 3.1.2 Small Hairpin RNA Molecules

Targeting gene silencers for specific viral proteins have been studied by several groups of researchers (Waterhouse et al., 2001). The small hairpin RNA (shRNA) of E1 and NSP1 show noncell-type specific antiviral effect and in broad-spectrum silencing against different types of strains of alphaviruses (Lam et al., 2012). The cell clones expressing shRNA against CHIKV E1 and NSP1 genes display inhibition activity of CHIKV production by degrading the viral single-stranded RNA.

## 3.2 NSP2 Inhibitors

With the advancement in the determination of crystal structures of viruses, NSP 2–3 macro domains were determined, which provided a platform for the development of new inhibitors for targeting alphaviruses. Recently, NSP2 is considered to be an important target for the development of antiviral drugs because of its role in viral replication and pathogenesis. Both C- and N-terminals have active sites.

### 3.2.1 Natural Inhibitors

In the past, the chemical constituents acting as drugs were extracted from natural products. Based on cellular high-throughput screening, 3040 natural compounds led to the discovery of natural NSP2 inhibitors (Lucas-Hourani et al., 2013). The compound ID1452-2 and its four analogs (Fig. 4.6) displayed potent and selective antiviral activity toward CHIKV. This result might initiate future efforts for the discovery of novel natural compounds as NSP2 selective inhibitors.

### 3.2.2 Druggable Targets Peptides

Recently, six peptide sequences were obtained and a global NSP2 sequence was constructed from various NSP2 strains of *Alphavirus* based on phylogenetic and global conservation (Singh et al., 2012). In silico analysis showed peptides sequences and the position of binding (Table 4.1).

### 3.2.3 NSP2-Based High-Throughput Screening

The molecular dynamics simulation and molecular docking are methods of high throughput screening. Singh et al. (2012) studied Chikungunya virus NSP2 protease protein to investigate the residues that are involved in the cleavage mechanism. Compounds with a finesse score more than 1.0 were considered for docking studies. The study identified four different compounds (ID27943,

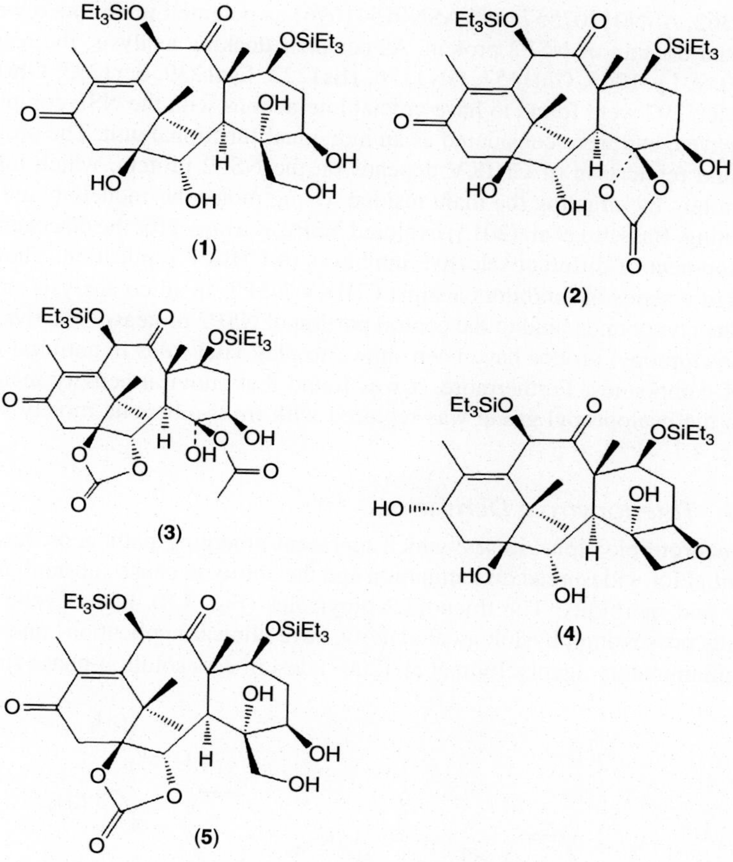

FIGURE 4.6 Chemical structures of compound ID1452-2 and its four analogs.

TABLE 4.1 Druggable Target Peptides With Tentative Binding Site at NSP2 Macro Domain

| Amino acid positions | Peptide sequences | Position of binding region |
|---|---|---|
| 22–28 | FPADRTT | 23, 28 |
| 142–159 | GTGGRLDSALPRIRIRAKN | 142–145, 154 |
| 235–249 | VRRRGVCVPLWNVTC | 236 |
| 445–459 | PKQSQRLLGELGPYP | 443, 458 |
| 488–505 | PENMHAHVWGGSRQRAIF | 504 |
| 639–655 | WVPSGHKHPHTFSHTPL | 642 |

ID21362, ASN01107557, and ASN01541696) as potential inhibitors toward C terminal domain of NSP2 protein. As per their docking analysis, the residues Gln1039, Lys1045, Glu1157, Gly1176, His1222, Lys1239, Ser1293, Glu1296, and Met1297 were found to have crucial interactions with the NSP complex to be cleaved, and were considered as an individual functional unit. The propagation and replication of CHIKV depends on the NSP2 protein, which inhibits the protein by targeting the main residue. Using molecular modeling and viral screening, Bassetto et al. (2013) explored NSP2 as a target for the discovery and development of different selective inhibitors of CHIKV replication. They discovered a series of inhibitors against CHIKV NSP2. In silico analysis showed that the compounds bind to the central portion of NSP2 protease. The hyrazone and cyclopropyl groups have been shown to play vital roles in antiviral activity of compounds. Furthermore, it was found that antiviral activity increased when the cyclopropyl group was replaced with *trans*-ethenylic moiety shown in Fig. 4.7.

### 3.2.4 Thienopyrrole Derivatives

The neurotropic alphaviruses, which represent emerging pathogens, have the potential for widespread dissemination and the ability to cause substantial mortality and morbidity. The thieno[3,2-b]pyrroles (Fig. 4.8), heterocyclic compounds possessing physiological activity, have clinical applications and act as antiinflammatory agents. Ilyin et al. (2007) described a solution-phase strategy

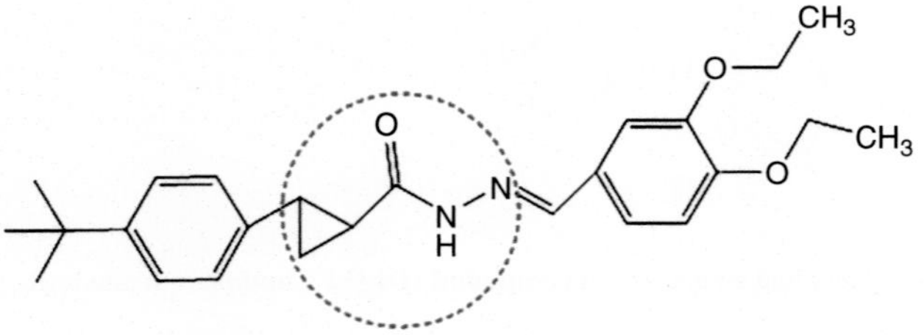

**(6)**

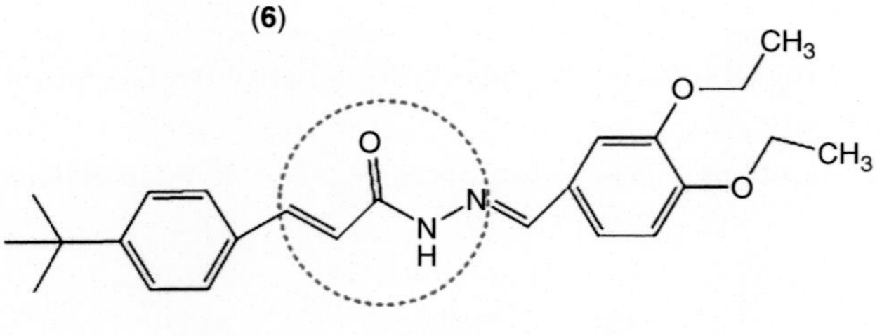

**(7)**

FIGURE 4.7  **Chemical structures of some CHIKV NSP2 protease inhibitors identified based on high-throughput screening.** *Red circle* indicates the active functional group involved in binding.

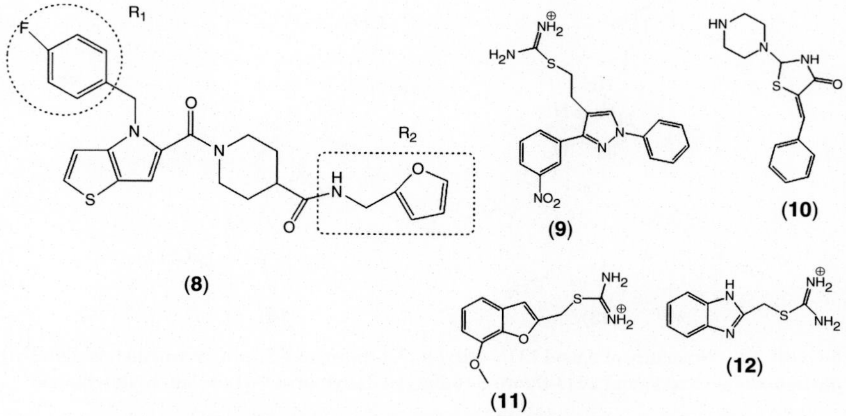

**FIGURE 4.8** The general structures of thienopyrroles and some of their analogs found active against Venezuelan equine encephalitis virus NSP2.

for the synthesis of novel combinatorial libraries containing a thieno[3,2-b] pyrrole core. Peng et al. (2009) identified thieno[3,2-6]pyrrole compounds with $IC_{50}$ values <10 μM/L and selectivity index >20. Based on structure–activity relationship analysis, 20 potential novel antiviral drugs out of 51,028 compounds were identified. Some of these compounds, shown in Fig. 4.8, were found to possess activity against neutrophilic *Alphavirus*, such as Venezuelan equine encephalitis virus. The thienopyrrole derivatives designed by homology modeling of protein sequences were shown to be effective inhibitors of NSP.

The NSP2 protein, which is responsible for CHIKV replication, causes a general host shutoff. By preventing IFN-induced gene expression, CHIKV suppresses the antiviral IFN response. Bora (2012) used homology modeling to predict the three-dimensional (3D) structure of NSP2 protease. With this model, a ligand [*N*-butyl-9-[3,4-dipropoxy-5-(propoxymethyl)oxolan-2-yl]purin-6-amine] was found to bind more efficiently than the ligand [3-fluoro-5-(6-hexylsulfanylpurin-9-yl)oxolan-2-yl]methanol. Both the ligands are shown in Fig. 4.9. One hydrogen bond was found to be formed between ligand **13** and NSP2. Carbonyl nitrogen atom in the pyrimidine ring of ligand **13** forms a hydrogen bond with the carbonyl oxygen (N–H...O) of Lys103 with a bond distance of 3.00 Å. The ligand **14** has higher gold score than ligand **13**. The former ligand is therefore supposed to be a promising inhibitor of NSP2 protease of CHIKV virus, and has been emerged as one of the promising antiviral drug candidates with potential symptomatic and disease-modifying effects.

Through molecular modeling by autodock and molecular dynamics simulation, Agarwal et al. (2015) elucidated the different structural aspects of NSP2 protein and different ligands, which are shown in the following Fig. 4.10.

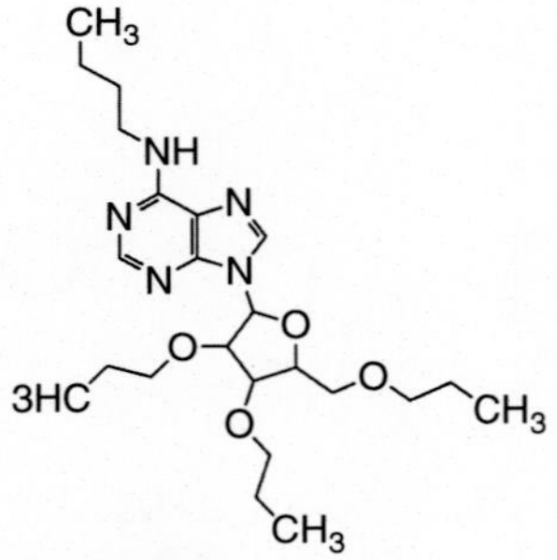

**(13)**

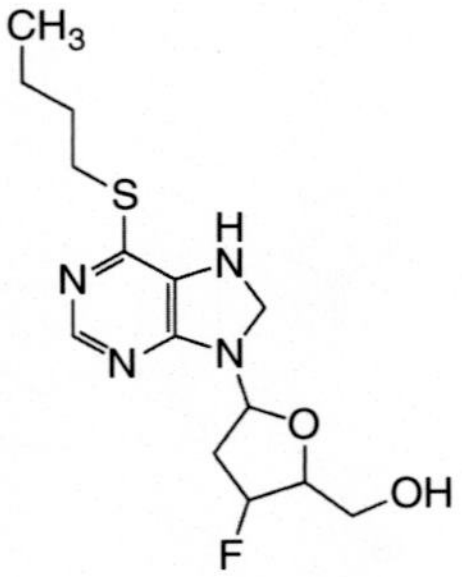

**(14)**

FIGURE 4.9 **Structure of ligand 13 [N-butyl-9-[3,4-dipropoxy-5-(propoxymethyl)oxolan-2-yl] purin-6-amine] and ligand 14 [3-fluoro-5-(6-hexylsulfanylpurin-9-yl)oxolan-2-yl]methanol.**

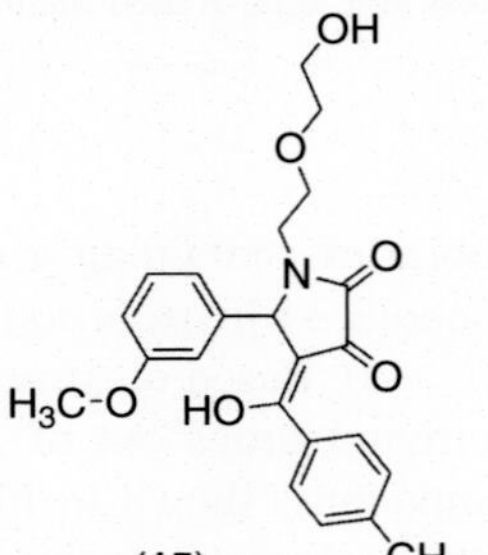

**(15)**

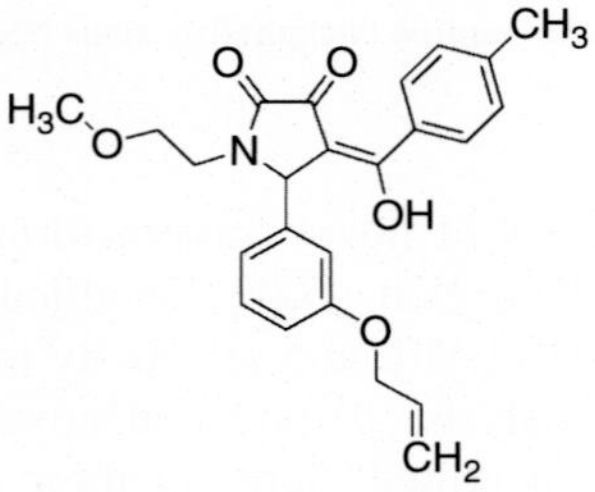

**(16)**

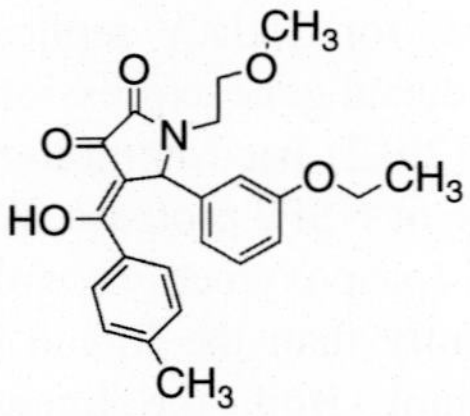

**(17)**

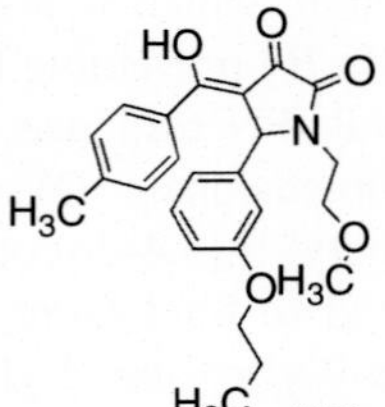

**(18)**

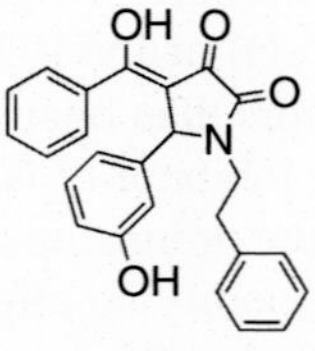

**(19)**

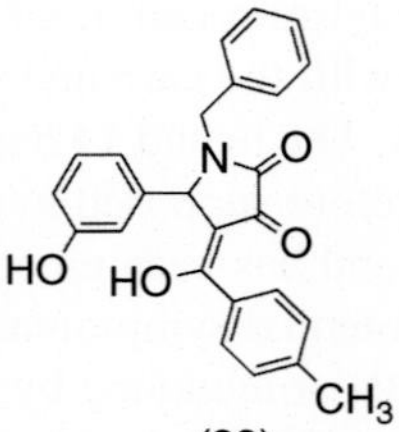

**(20)**

FIGURE 4.10 **Chemical structures of some in silico–predicted S-adenosyl methionine derivatives acting against NSP2 CHIKV protease.**

The ligand **20** (shown in Fig. 4.10) interacted with the protein with three hydrogen bonds having bond length of 2.86, 3.11, and 3.07 Å with Trp1084 and Tyr1047 amino acid residues. Ligand number **15** showed a better affinity toward CHIKV NSP2 protease having binding energy of 8.06 kcal/mol. It interacted with Trp1084 and Asp1246 amino acid residues of CNSP2 protease via two hydrogen bonds having bond length of 3.03 and 2.95 Å. The binding energies of other ligands, such as ligands **16–18**, were found in the range of −7.55 to −6.75 kcal/mol. Both ligand **15** and **20** have a common binding site at Trp1084. Ligand **15** interacted with catalytic amino acid residue Cys1013 hydrophobically. Catalytic amino acid residue His1083 was found to be in close proximity to the ligands with distance less than 5 Å, which is insufficient to form hydrophobic interactions. This indicates that the ligands might inhibit protease activity by blocking the binding of substrate to its active site. However, Trp1084 is a conserved residue of the NSP2 protease active site pocket and is present in close proximity to His1083. The catalytic activity of CNSP2 protease was reduced by interaction of the ligands with Trp1084. Interestingly, none of the ligands were found to be involved in π–π interaction with the CNSP2 protease. All the six ligands bonded with Ser1048, Tyr1079, and Gln1241 either through hydrogen bonding or hydrophobic interactions. These three amino acid residues create a microenvironment for the CNSP2 protease activity.

## 3.2.5   Thiazolidone Derivatives

The derivatives of pyrazole-like 1,3-thiazolidin-4-ones (Fig. 4.11) were synthesized by Jadav et al. (2015) and examined for the antiviral activity against CHIV (LR2006_OPY1) in Vero cell culture through CPE reduction assay. Three aralkylidene derivatives (compounds **6**, **7**, **8**) and *ortho*-methyl–substituted ligands

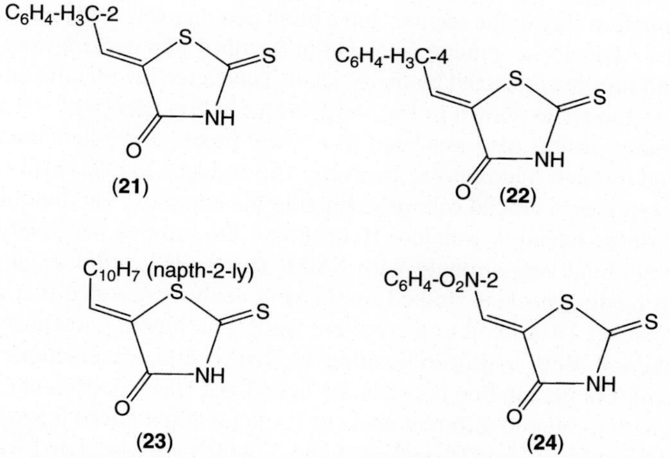

**FIGURE 4.11**   **Chemical structures of some thiazolidone derivatives.**

were found to be the most active among the rest. Furthermore, to understand the mechanism of protein and ligand interaction, molecular docking was carried out with NSP2 CHIKV protease (PDB: 3TRK). The results showed a strong interaction between aralkylidene with three amino acids (Tyr1047, Tyr1049, and Trp1084), while thiazolidinone showed a strong hydrophobic interaction with Cys1013, Tyr1047, Tyr1049, and Trp1084.

Compounds (**21–24**) were found to have antiviral activity at low concentration, particularly 0.42, 4.2, 3.6, 40.1, and 6.8 μM. Compound **21** was the most potent among the aforementioned compounds. Methyl substitution of *ortho*-position of compound **21** was favorable and was found to be 10 times more potent than its *para*-counterpart of compound **22**. The aralkylidene portion gets accommodated and exhibits two H-bonding interactions for compound **21**. Similarly, compound **22** showed interaction with residues in S2 and S1, but not in S3, which does not exhibit any H-bonding interaction with Tyr1047. Thiazolidone derivatives exhibited hydrogen-binding interactions with NSP2 protease. These interactions were between carbonyl oxygen of thiazolidone and the backbone amide hydrogen of Cys1013 and hydrogen of thiazolidone 2-amino nitrogen of the aforementioned compound.

## 3.3 NSP3 Protease Inhibitors

Like NSP2, the crystalline structure of N-terminal domain of NSP3 also became of great use, positioning NSP3 as a potential drug target for the development of antiviral drugs. To identify potential NSP3 inhibitors, Nguyen et al. (2014) screened 1541 compounds from NCI diverse sets against NSP3 of CHIKV and identified potential inhibitors that targeted CHIKV NSP3 complexed with ADP-ribose. They identified all possible binding pockets. The virtual screening was carried out using three steps. In the first step, docking was centered on the ADP-ribose–binding site, in the second step a blind docking was centered at the middle of the ADP-ribose–binding site, and in the third step docking was centered on the biding sites predicted by meatpacker. This led to the identification of the top 5 compounds as shown in Fig. 4.12, with binding energy of −8 kcal/mol. Furthermore, results also indicated that NSP3 protease residues interact with the ligand through hydrophobic contacts. The residues Val35, Val113, Tyr114, and Trp148 play a crucial role in stabilizing the complex. The binding energy of ADP-ribose having less than −10 kcal/mol, showed that negatively charged ligands can bind very strongly with NSP3. The binding affinities at the three different binding pockets studied by docking analysis revealed that all of the hits for pocket 1 can bind to the protein well, with binding energies less than −10 kcal/mol. Both hydrogen bonding and hydrophobic interactions play important roles in the binding between the ligand and NSP3. Different conformations of ligands bind to different pockets. Pocket 2 might not be a good location for binding, as compared to pockets 1 and 3, which are associated with higher binding affinities. Hydrogen-bonding interactions play an important role in the

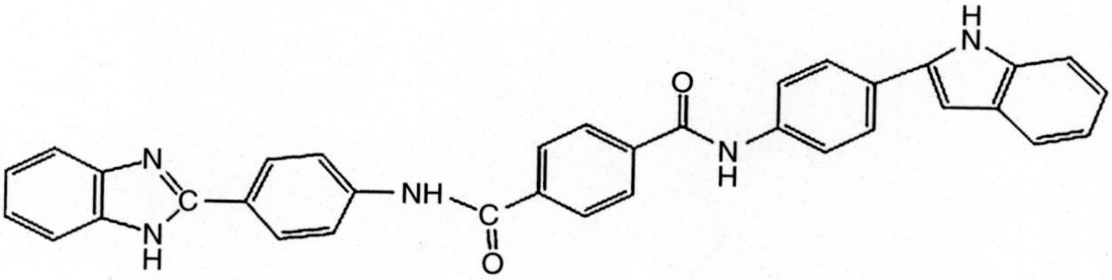

FIGURE 4.12   Chemical structure of in silico–screened ligands acting against NSP3 domain.

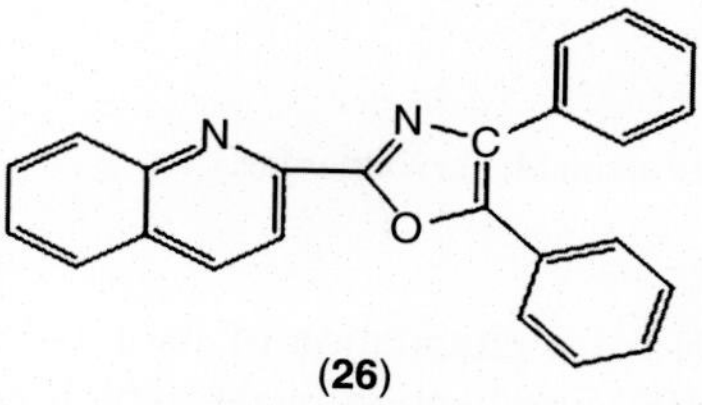

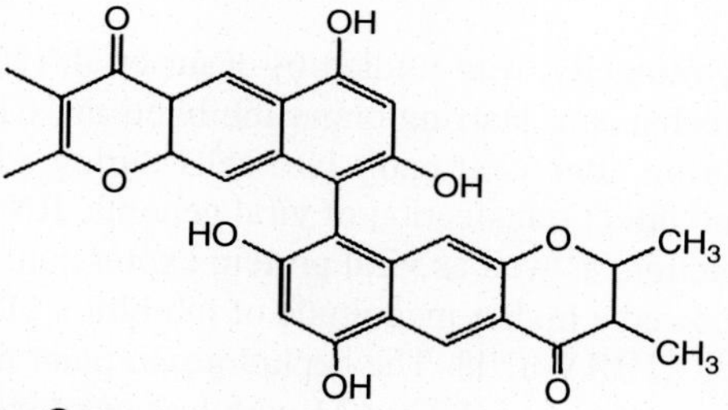

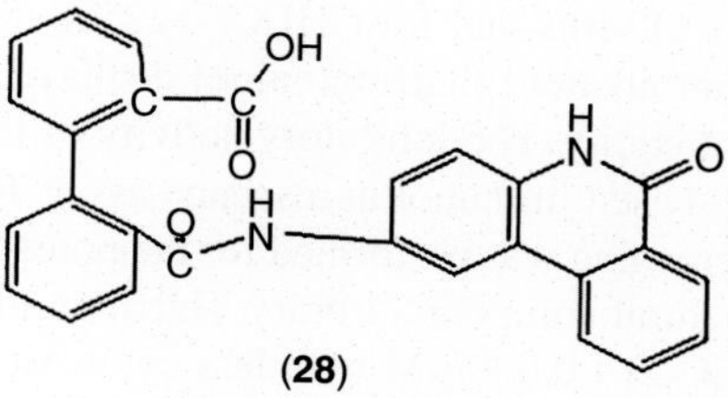

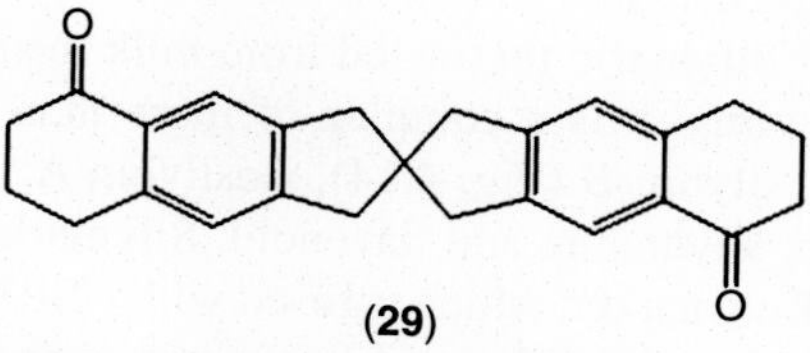

binding to pocket 1, whereas hydrophobic contacts are responsible for the interactions associated with binding to pocket 2.

### 3.3.1   Natural Inhibitors for NSP3 Protease

In an effort to identify novel inhibitors against NSP3 protease, Kaur et al. (2013) extensively studied the natural compound library. This study led to the identification of 44 compounds that exhibited more than 70% inhibition of CHIKV virus pathogenesis. Harringtonine, a cephalotaxine alkaloid (Fig. 4.13), which displayed potent inhibition activity against CHIKV infection with minimal

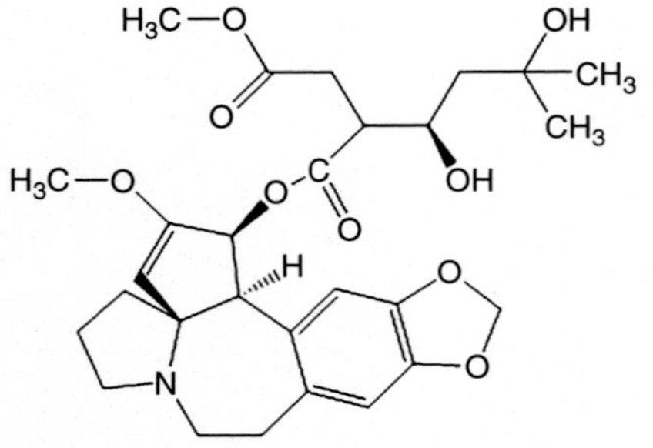

(30)

FIGURE 4.13  Chemical structure of a cephalotaxine alkaloid, harringtonine.

cytotoxicity, was studied by Kaur et al. (2013) for elucidation of its antiviral mechanism. Harringtonine inhibited the CHIKV replication cycle, which occurred after viral entry into cells during addition studies, cotreatment assays, and direct transfection of viral genomic RNA. It also affects CHIKV RNA production, as well as viral protein expression. At 5- and 10-μM concentrations, it showed a higher magnitude of inhibition of virus titer for CHIKV-122508 than for CHIKV-0708. The cepholotoxin plant alkaloid (harringtonine) displayed a strong $EC_{50}$ of 0.020 μM with less cytotoxicity. The inhibitory activity of theses compounds was tested on virus cell–based immunofluorescence assay. Recently, an immunofluorescence-based screening was performed for the potential inhibitors of NSP3 CHIKV using the natural compound library. Harringtonine (acephalotoxin alkaloid) displayed an $EC_{50}$ of 0.020 μM with less cytotoxicity (Fig. 4.13).

Silymarin (extracted from milk thistle) has been shown to inhibit hepatitis C virus. It is a complex of more than seven flavonolignans, such as silybin A, silybin B (Fig. 4.14), isosilybin A, isosilybin B, silychristin, isosilychristin, silydianin, and flavonoid. Silymarin is able to suppress the activity of the RLuc marker, which is fused with NSP3 protein of the virus (Lani et al., 2015). Plant-derived flavonoids are polyphenolic compounds endowed with a wide range of biological benefits to human health that not only includes antiinflammatory, antioxidant, antibacterial, and antifungal activities, but also antiviral activity. The increase in the number of drug-resistant microorganisms has brought natural compounds, such as flavonoids, to the forefront as an important natural resource to overcome this problem. A number of studies have successfully shown various types of flavonoids, such as rutin, naringin, baicalein, quercetin, and kaempferol, to be potential antiviral agents against a wide range of important viruses, including Dengue virus, HIV-1, H5N1 influenza A virus, coxsackievirus, and Japanese encephalitis virus (Seyedi et al., 2016). Baicalin is a metabolite of baicalein, which can be extracted from the root of Chinese

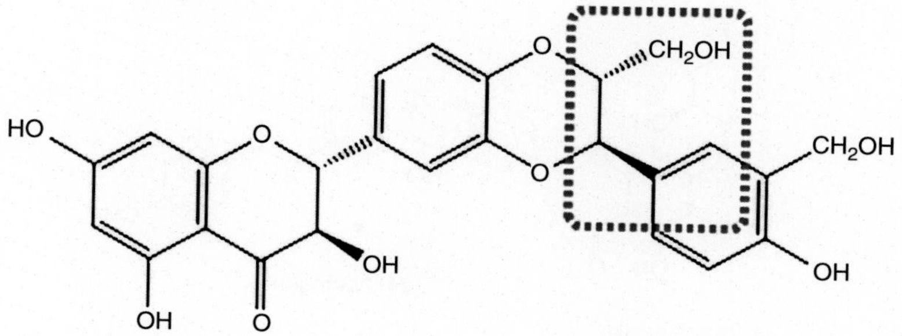

(31)

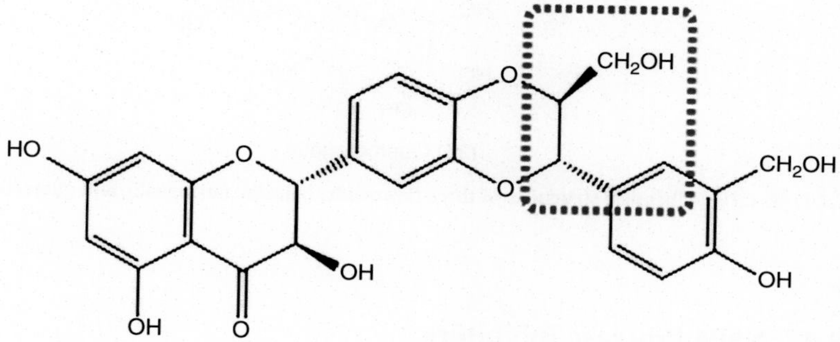

(32)

FIGURE 4.14   Chemical structures of silybin A and silybin B.

medicinal herbal plant *Scutellaria baicalensis*. Baicalin also serves as a potential antiviral agent against influenza viruses, where it acts as a neuraminidase inhibitor.

Seyedi et al. (2016) screened the potency of three flavonoids, baicalin, naringenin, and quercetagetin (Fig. 4.15), targeting NSP3 protein of CHIKV virus using computational approach. Docking results showed that of the three tested ligands, baicalin showed the highest binding affinity with −9.8 kcal/mol binding energy and low $K_i$ value of 0.064 μM for NSP3, followed by quercetagetin and naringenin, having binding energies −8.6 and −8.4 kcal/mol, respectively. The hydrogen-bond donors came from the protein residues, and the corresponding acceptors were derived from the ligands. The ligands interacted with the residues present in the active site of NSP3 (baicalin: Leu108, Tyr142, Ser110, and Thr111; naringenin: Ser110 and Thr111; quercetagetin: Cys34, Leu108, Arg144, and Asp145). One π–π interaction was found between baicalin and Tyr114 residue of NSP3.

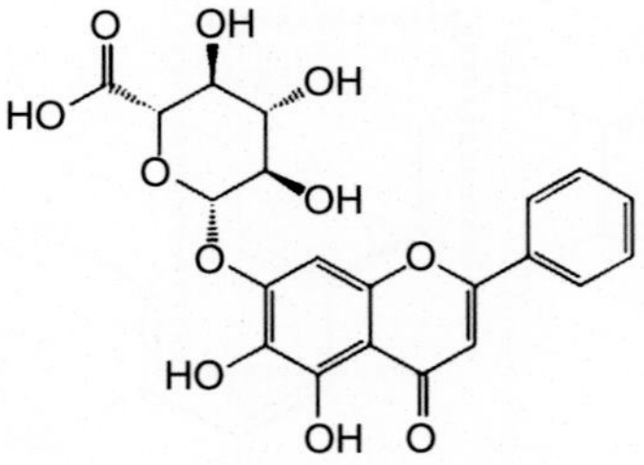

**(33)** Baicalin

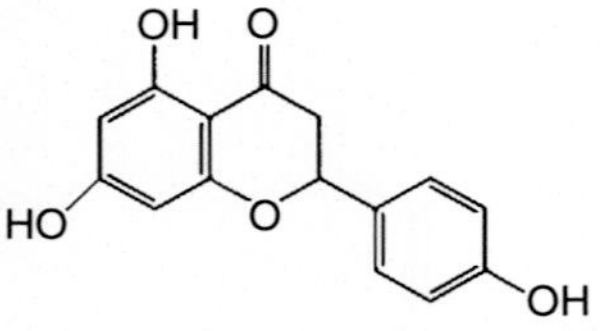

**(34)** Naringenin

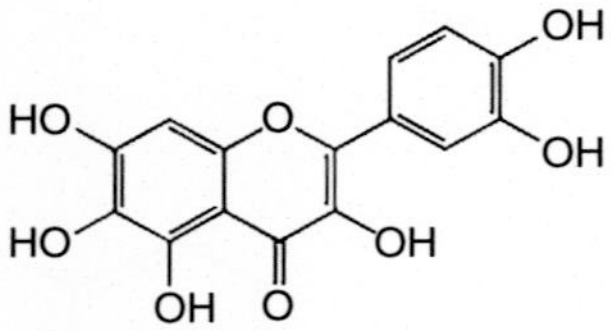

**(35)** Quercetagetin

FIGURE 4.15    **Chemical structures of three flavonoids: baicalin, naringenin, and quercetagetin.**

## 3.4    NSP4 Protease Inhibitors

### 3.4.1    Nucleoside and Nonnucleoside Inhibitors

In CHIKV NSP4 protease, amino acids Gly315, Asp316, and Asp317 form a catalytic triad and a nearby residue (proximity of 5 Å) acts as the active site. Kumar et al. (2012) identified one core binding site along with two allosteric sites situated in the palm and thumb domains. A docking result showed that nucleoside analogs had the best docking scores in this protease with binding energies in the range of −85.4173 to −78.049 kcal/mol (Kumar et al., 2012). A nucleoside NSP4 inhibitor, ribavirin (**36**, Fig. 4.16), as well as a nonnucleoside inhibitor, BILN 2061 (**37**, Fig. 4.16), could be docked in catalytic, as well as both the allosteric sites, localized in palm, as well as thumb domains, of NSP4, as shown in Fig. 4.17 (Patel et al., 2012). BILN 2061 showed the highest binding affinity with energy of −94.294 kcal/mol for the palm allosteric site.

Some indoles and a few benzimidazole derivatives were also found to better interact with the CHIKV NSP4 palm allosteric site, where the van der Waals interaction was dominant over the hydrogen-bonding interaction. The minimum core for catalytic specific (nucleoside) inhibitors was found to be four hydrogen-bond donors (HBDs), seven hydrogen-bond acceptors (HBAs), and one aromatic ring (R). The minimum core for nonnucleoside inhibitors related to palm allosteric site was found to be one HBD, four HBAs, two Rs, and three

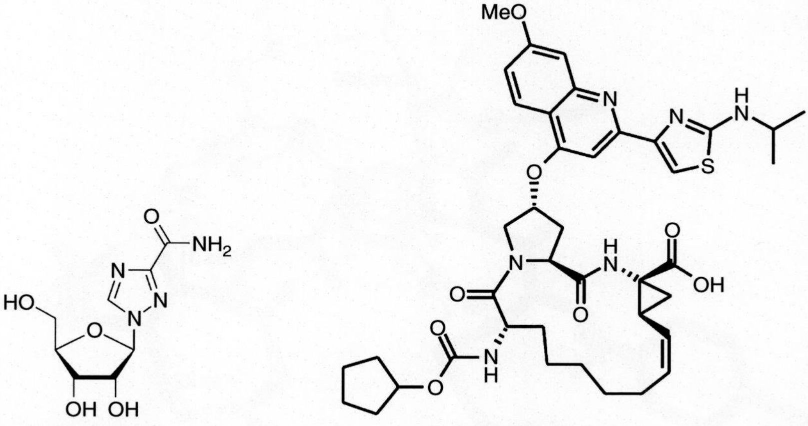

FIGURE 4.16 **Structures of a nucleoside inhibitor, ribavirin (36), and a nonnucleoside inhibitor, BILN 2061 (37), of NSP4 protease.**

hydrophobic regions. Ribavirin, a nucleoside analog (inhibitor), interacted with the catalytic site with significant binding energy. Nonnucleoside inhibitors, such as BILN 2061, and diketo acid derivatives, showed better interactions with the palm and thumb allosteric sites, respectively. The selective antiviral activity of favipiravir on the replication of CHIKV, alphaviruses, and favipiravir-resistant CHIKV variants, which all carry a K291R mutation in the RdRp NSP4, were studied by Delang et al. (2014). These authors also studied the effect of T-705 (favipiravir), a nucleotide prodrug that was developed for the treatment of influenza, on NSP4. The characterization of these virus variants in cell culture suggested that a highly conserved part of the viral polymerase of positive-strand RNA viruses is the target of favipiravir. The NSP2 (K49R and E622G) and NSP3 (Opal524W) mutations, which were also found in favipiravir-resistant CHIKV variants, did not result in phenotypic resistance to favipiravir, shown by reverse genetics. The lysine at position 291, which is located in a region of NSP4, possesses a high degree of conservation among alphaviruses. The structure of favipiravir (**38**, Fig. 4.18) bound in the active site of the Norwalk virus (NWV) polymerase structure was found to be superimposed over the cytosine base of CTP. Compared to original Lys291, the charged arginine side chain is much closer to the favipiravir-RTP inhibitor. No such specific interaction or any steric repulsion is observed between K291R and favipiravir-RTP or between Lys291 and favipiravir-RTP. T-705 and its active metabolite T-705 ribofuranosyl triphosphate (**39**, Fig. 4.18) have shown binding affinity for NSP4 protease domain. Metabolic experiment confirmed a direct effect of T-705 on CHIKV RNA synthesis.

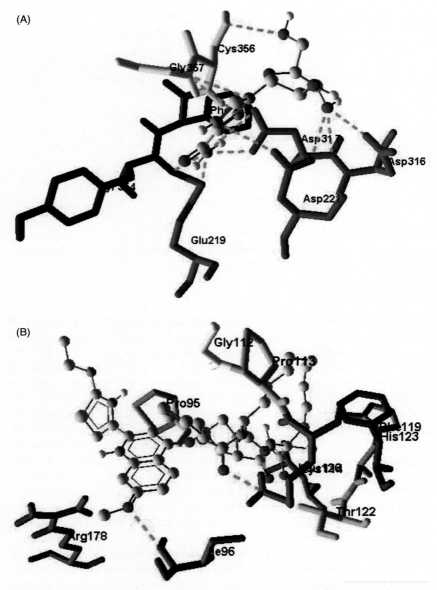

(A)

Cys356

Gly357

Phe

Asp317

Asp316

Asp22?

Glu219

(B)

Gly112

Pro113

Pro95

Phe119

His123

Cys120

Thr122

Arg178

Ile96

**FIGURE 4.17   Docked poses of ribavirin (A) and BILN2061 (B) interacting with the amino acids present in the catalytic and allosteric sites on the palm and thumb domains.** *(Reprinted from Patel, H., Jasarai, Y.T., Kapoora, R.G., 2012. Exploring the polymerase activity of Chikungunya viral non structural protein 4 (nsP4) using molecular modeling, epharmacophore and docking studies. Int. J. Pharm. Life Sci. 3, 1752–1765.)*

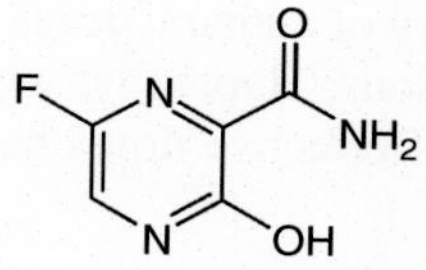

**FIGURE 4.18** Chemical structures of favipiravir or T-705 (38) and its metabolite, T-705 ribofuranosyl triphosphate (39) .

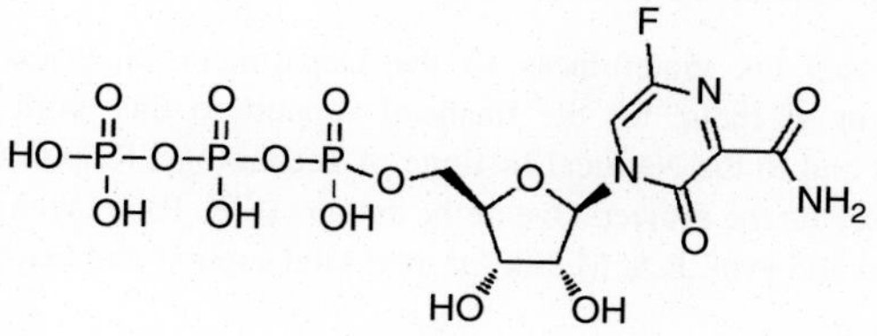

# 4 CONCLUDING REMARKS AND FUTURE PERSPECTIVES

Vaccines and therapeutic strategies are required to combat viruses and manage viral infections in all epidemic situations. The available vaccines and therapeutic strategies to combat viruses are very limited, which include viral polymerase inhibitors, recombinant interferons, and boosting host antiviral system. Recently, protease inhibitors were successfully developed to inhibit and manage hepatitis C and HIV viruses. In the present chapter, it has been shown that NSP2 from CHIKV can be effectively targeted by chemicals, that interact with numerous cellular proteins to control interferon signaling and gene transcription. Various forms of scaffold and chemical entities have already been considered for enhancing their capacity as drug molecules targeting NSPs. The crystallographic structures and functions of viral key proteins provide necessary information to develop potential targeted inhibitors against these viral proteins. These developed protease inhibitors ultimately combat diseases arising from different viruses, such as Mayaro virus, Ross River virus, CHIV, SINV, O'Nyong–Nyong virus, Eastern equine encephalitis virus, Semliki Forest virus, Una virus, Venezuelan equine encephalitis virus, Tonate virus, Western equine encephalitis virus, etc. Although significant improvement has been made in the direction of discovering novel protease inhibitors targeting viral NSP, the lack of clinically approved drug molecules is a limitation for protection against sudden disease occurrence. This chapter is equipped with useful information for developing chemical libraries to design *Alphavirus*-targeting inhibitors, and to construct pharmacophores for ligand-based drug design. This will increase the chances of discovery for small-molecule inhibitors. In conclusion, in near future

NSP inhibitor–based discovery of antiviral drugs need to be considered by both private and public organizations. Hopefully, with the grooming of business-friendly sectors, discovery of effective drugs based on NSP inhibitors would be a reality.

## ACKNOWLEDGMENTS

A. Bissoyi expresses his gratefulness to the Department of Science and Technology (DST), Government of India, for the financial support to this work through Grant No. YSS/2015/000618 and to the National Institute of Technology, Raipur, for providing necessary facilities to execute the project. One of the authors (S.K. Pattanayak) would like to thank Prof. S. Chowdhuri and Prof. B.S. Mallik for their kind support and helpful communications.

## REFERENCES

Agarwal, T., Asthana, S., Bissoyi, A., 2015. Molecular modeling and docking study to elucidate novel Chikungunya virus nsP2 protease inhibitors. Indian J. Pharmaceut. Sci. 77, 453–460.

Ahola, T., Kääriäinen, L., 1995. Reaction in *Alphavirus* mRNA capping: formation of a covalent complex of nonstructural protein nsP1 with 7-methyl-GMP. Proc. Nat. Acad. Sci. USA 92, 507–511.

Akhrymuk, I., Kulemzin, S.V., Frolova, E.I., 2012. Evasion of the innate immune response: the Old World *Alphavirus* nsP2 protein induces rapid degradation of Rpb1, a catalytic subunit of RNA polymerase II. J. Virol. 86, 7180–7191.

Argos, P., Kamer, G., Nicklin, M.J., Wimmer, E., 1984. Similarity in gene organization and homology between proteins of animal picomaviruses and a plant comovirus suggest common ancestry of these virus families. Nucleic Acids Res. 12, 7251–7267.

Atkins, G.J., 2013. The pathogenesis of alphaviruses. ISRN Virol. 2013, (Article ID 861912).

Bassetto, M., De Burghgraeve, T., Delang, L., Massarotti, A., Coluccia, A., Zonta, N., Gatti, V., Colombano, G., Sorba, G., Silvestri, R., 2013. Computer-aided identification, design and synthesis of a novel series of compounds with selective antiviral activity against Chikungunya virus. Antiviral Res. 98, 12–18.

Bhakat, S., Karubiu, W., Jayaprakash, V., Soliman, M.E., 2014. A perspective on targeting nonstructural proteins to combat neglected tropical diseases: dengue, West Nile and Chikungunya viruses. Eur. J. Med. Chem. 87, 677–702.

Bissoyi, A., Mahapatra, C., Chaudhuri, B., Mahajan, D., 2013. In silico prediction of novel drug molecule for migraine using blind docking. Global J. Biotechnol. Biochem. 8, 25–32.

Bora, L., 2012. Homology modeling and docking to potential novel inhibitor for Chikungunya (37997) protein nsP2 protease. J. Proteomics Bioinform. 5 (2).

Bouraï, M., Lucas-Hourani, M., Gad, H.H., Drosten, C., Jacob, Y., Tafforeau, L., Cassonnet, P., Jones, L.M., Judith, D., Couderc, T., 2012. Mapping of Chikungunya virus interactions with host proteins identified nsP2 as a highly connected viral component. J. Virol. 86, 3121–3134.

Cruz, C.C., Suthar, M.S., Montgomery, S.A., Shabman, R., Simmons, J., Johnston, R.E., Morrison, T.E., Heise, M.T., 2010. Modulation of type I IFN induction by a virulence determinant within the *Alphavirus* nsP1 protein. J. Virol. 399, 1–10.

De Francesco, R., Carfi, A., 2007. Advances in the development of new therapeutic agents targeting the NS3-4A serine protease or the NS5B RNA-dependent RNA polymerase of the hepatitis C virus. Adv. Drug Deliv. Rev. 59, 1242–1262.

Delang, L., Guerrero, N.S., Tas, A., Quérat, G., Pastorino, B., Froeyen, M., Dallmeier, K., Jochmans, D., Herdewijn, P., Bello, F., 2014. Mutations in the Chikungunya virus non-structural proteins cause resistance to favipiravir (T-705), a broad-spectrum antiviral. J. Antimicrob. Chemother. 69, 2770–2784.

Fata, C.L., Sawicki, S.G., Sawicki, D.L., 2002. Modification of Asn374 of nsP1 suppresses a Sindbis virus nsP4 minus-strand polymerase mutant. J. Virol. 76, 8641–8649.

Fazakerley, J.K., Boyd, A., Mikkola, M.L., Kääriäinen, L., 2002. A single amino acid change in the nuclear localization sequence of the nsP2 protein affects the neurovirulence of Semliki Forest virus. J. Virol. 76, 392–396.

Firth, A.E., Atasheva, S., Frolova, E.I., Frolov, I., 2011. Conservation of a packaging signal and the viral genome RNA packaging mechanism in *Alphavirus* evolution. J. Virol. 85, 8022–8036.

Garg, R., Gupta, S.P., Gao, H., Babu, M.S., Debnath, A.K., Hansch, C., 1999. Comparative quantitative structure-activity relationship studies on anti-HIV drugs. Chem Rev. 99, 3525–3602.

Gorchakov, R., Frolova, E., Frolov, I., 2005. Inhibition of transcription and translation in Sindbis virus-infected cells. J. Virol. 79, 9397–9409.

Gorchakov, R., Garmashova, N., Frolova, E., Frolov, I., 2008. Different types of nsP3-containing protein complexes in Sindbis virus-infected cells. J. Virol. 82, 10088–10101.

Gupta, S.P., 2013. QSAR and Molecular Modeling. Springer, the Netherlands.

Gupta, S.P., Singh, P., Bindal, M.C., 1983. QSAR studies on hallucinogens. Chem Rev. 83, 633–649.

Hahn, Y.S., Grakoui, A., Rice, C.M., Strauss, E.G., Strauss, J.H., 1989. Mapping of RNA-temperature-sensitive mutants of Sindbis virus: complementation group F mutants have lesions in nsP4. J. Virol. 63, 1194–1202.

Hardy, W.R., Strauss, J.H., 1989. Processing the nonstructural polyproteins of Sindbis virus: nonstructural proteinase is in the C-terminal half of nsP2 and functions both in *cis* and in *trans*. J. Virol. 63, 4653–4664.

Ilyin, A.P., Dmitrieva, I.G., Kustova, V.A., Manaev, A.V., Ivachtchenko, A.V., 2007. Synthesis of heterocyclic compounds possessing the 4 H-thieno[3, 2-b]pyrrole moiety. J. Comb. Chem. 9, 96–106.

Jadav, S. S., Nayan Sinha, B., Hilgenfeld, R., Pastorino, B., Lamballerie, X., Jayaprakash, V., 2015. Thiazolidone derivatives as inhibitors of Chikungunya virus. Eur. J. Med. Chem. 89, 172–178.

Jones, P.H., Maric, M., Madison, M.N., Maury, W., Roller, R.J., Okeoma, C.M., 2013. BST-2/tetherin-mediated restriction of Chikungunya (CHIKV) VLP budding is counteracted by CHIKV non-structural protein 1 (nsP1). J. Virol. 438, 37–49.

Kääriäinen, L., Ahola, T., 2002. Functions of *Alphavirus* nonstructural proteins in RNA replication. Prog. Nucleic Acid Res. Mol. Biol. 71, 187–222.

Kaur, P., Thiruchelvan, M., Lee, R.C.H., Chen, H., Chen, K.C., Ng, M.L., Chu, J.J.H., 2013. Inhibition of Chikungunya virus replication by harringtonine, a novel antiviral that suppresses viral protein expression. Antimicrob. Agents Chemother. 57, 155–167.

Koonin, E.V., Dolja, V.V., Morris, T.J., 1993. Evolution and taxonomy of positive-strand RNA viruses: implications of comparative analysis of amino acid sequences. Crit. Rev. Biochem. Mol. Biol. 28, 375–430.

Kumar, S., Kapopara, R.G., Patni, M.I., Pandya, H.A., Jasrai, Y.T., Patel, S.K., 2012. Int. J. Pharm. Life Sci. 3, 1631–1642.

Laakkonen, P., Auvinen, P., Kujala, P., Kääriäinen, L., 1998. *Alphavirus* replicase protein NSP1 induces filopodia and rearrangement of actin filaments. J. Virol. 72, 10265–10269.

Laakkonen, P., Hyvönen, M., Peränen, J., Kääriäinen, L., 1994. Expression of Semliki Forest virus nsP1-specific methyltransferase in insect cells and in *Escherichia coli*. J. Virol. 68, 7418–7425.

Lam, S., Chen, K.C., Ng, M.M.-L., Chu, J.J.H., 2012. Expression of plasmid-based shRNA against the E1 and nsP1 genes effectively silenced Chikungunya virus replication. PLoS One 7, e46396.

Lani, R., Hassandarvish, P., Chiam, C.W., Moghaddam, E., Chu, J.J.H., Rausalu, K., Merits, A., Higgs, S., Vanlandingham, D., Bakar, S.A., 2015. Antiviral activity of silymarin against Chikungunya virus. Sci. Rep. 5, 11421.

Lastarza, M.W., Grakoui, A., Rice, C.M., 1994. Deletion and duplication mutations in the C-terminal nonconserved region of Sindbis virus nsP3: effects on phosphorylation and on virus replication in vertebrate and invertebrate cells. J. Virol. 202, 224–232.

Lucas-Hourani, M., Lupan, A., Desprès, P., Thoret, S., Pamlard, O., Dubois, J., Guillou, C., Tangy, F., Vidalain, P.-O., Munier-Lehmann, H., 2013. A phenotypic assay to identify Chikungunya virus inhibitors targeting the nonstructural protein nsP2. J. Biomol. Screen. 18, 172–179.

Malet, H., Coutard, B., Jamal, S., Dutartre, H., Papageorgiou, N., Neuvonen, M., Ahola, T., Forrester, N., Gould, E.A., Lafitte, D., 2009. The crystal structures of Chikungunya and Venezuelan equine encephalitis virus nsP3 macro domains define a conserved adenosine binding pocket. J. Virol. 83, 6534–6545.

Mayr, L.M., Bojanic, D., 2009. Novel trends in high-throughput screening. Curr. Opin. Pharmacol. 9, 580–588.

Merits, A., Vasiljeva, L., Ahola, T., Kääriäinen, L., Auvinen, P., 2001. Proteolytic processing of Semliki Forest virus-specific non-structural polyprotein by nsP2 protease. J. Gen. Virol. 82, 765–773.

Neuvonen, M., Ahola, T., 2009. Differential activities of cellular and viral macro domain proteins in binding of ADP-ribose metabolites. J. Mol. Biol. 385, 212–225.

Nguyen, P.T., Yu, H., Keller, P.A., 2014. Discovery of in silico hits targeting the nsP3 macro domain of Chikungunya virus. J. Mol. Model. 20, 1–12.

Park, E., Griffin, D.E., 2009. The nsP3 macro domain is important for Sindbis virus replication in neurons and neurovirulence in mice. J. Virol. 388, 305–314.

Patel, H., Jasrai, Y.T., Kapopara, R.G., 2012. Exploring the polymerase activity of Chikungunya viral nonstructural protein 4 (nsP4) using molecular modeling, e-pharmacophore and docking studies. Int. J Pharm. Life Sci. 3, 1752–1765.

Pattanayak, S.K., Chowdhuri, S., 2011. Effect of water on solvation structure and dynamics of ions in the peptide bond environment: importance of hydrogen bonding and dynamics of the solvents. J. Phys. Chem. B 115, 13241–13252.

Pattanayak, S.K., Chowdhuri, S., 2012. A molecular dynamics simulations study on the behavior of liquid N-methylacetamide in presence of NaCl: structure, dynamics and H-bond properties. J. Mol. Liq. 172, 102–109.

Pehrson, J.R., Fuji, R.N., 1998. Evolutionary conservation of histone macroH2A subtypes and domains. Nucleic Acids Res. 26, 2837–2842.

Peng, W., Peltier, D.C., Larsen, M.J., Kirchhoff, P.D., Larsen, S.D., Neubig, R.R., Miller, D.J., 2009. Identification of thieno[3, 2-b]pyrrole derivatives as novel small molecule inhibitors of neurotropic alphaviruses. J. Infect. Dis. 199, 950–957.

Peränen, J., Laakkonen, P., Hyvönen, M., Kääriäinen, L., 1995. The *Alphavirus* replicase protein nsP1 is membrane-associated and has affinity to endocytic organelles. J. Virol. 208, 610–620.

Peränen, J., Rikkonen, M., Liljeström, P., Kääriäinen, L., 1990. Nuclear localization of Semliki Forest virus-specific nonstructural protein nsP2. J. Virol. 64, 1888–1896.

Peränen, J., Takkinen, K., Kalkkinen, N., Kääriäinen, L., 1988. Semliki Forest virus-specific nonstructural protein nsP3 is a phosphoprotein. J. Gen. Virol. 69, 2165–2178.

Pettersen, E.F., Goddard, T.D., Huang, C.C., Couch, G.S., Greenblatt, D.M., Meng, E.C., Ferrin, T.E., 2004. UCSF Chimera—a visualization system for exploratory research and analysis. J. Comput. Chem. 25, 1605–1612.

Rayner, J.O., Dryga, S.A., Kamrud, K.I., 2002. *Alphavirus* vectors and vaccination. Rev. Med. Virol. 12, 279–296.

Rikkomen, M., 1996. Functional significance of the nuclear-targeting and NTP-binding motifs of Semliki Forest virus nonstructural protein nsP2. J. Virol. 218, 352–361.

Rubach, J.K., Wasik, B.R., Rupp, J.C., Kuhn, R.J., Hardy, R.W., Smith, J.L., 2009. Characterization of purified Sindbis virus nsP4 RNA-dependent RNA polymerase activity in vitro. J. Virol. 384, 201–208.

Salonen, A., Vasiljeva, L., Merits, A., Magden, J., Jokitalo, E., Kääriäinen, L., 2003. Properly folded nonstructural polyprotein directs the Semliki Forest virus replication complex to the endosomal compartment. J. Virol. 77, 1691–1702.

Schrödinger, L., 2011. Schrödinger Software Suite. Schrödinger, LLC, New York, NY.

Seyedi, S.S., Shukri, M., Hassandarvish, P., Oo, A., Muthu, S.E., Abubakar, S., Zandi, K., 2016. Computational approach towards exploring potential anti-Chikungunya activity of selected flavonoids. Sci. Rep. 6.

Shirako, Y., Strauss, J.H., 1998. Requirement for an aromatic amino acid or histidine at the N terminus of Sindbis virus RNA polymerase. J. Virol. 72, 2310–2315.

Shirako, Y., Strauss, E.G., Strauss, J.H., 2000. Suppressor mutations that allow Sindbis virus RNA polymerase to function with nonaromatic amino acids at the N-terminus: evidence for interaction between nsP1 and nsP4 in minus-strand RNA synthesis. J. Virol. 276, 148–160.

Singh, K.D., Kirubakaran, P., Nagarajan, S., Sakkiah, S., Muthusamy, K., Velmurgan, D., Jeyakanthan, J., 2012. Homology modeling, molecular dynamics, e-pharmacophore mapping and docking study of Chikungunya virus nsP2 protease. J. Mol. Model. 18, 39–51.

Strauss, J.H., Strauss, E.G., 1994. The alphaviruses: gene expression, replication, and evolution. Microbiol. Rev. 58, 491–562.

Sundberg, S.A., 2000. High-throughput and ultra-high-throughput screening: solution-and cell-based approaches. Curr. Opin. Biotechnol. 11, 47–53.

Tanno, T., Bhanu, N.V., Oneal, P.A., Goh, S.-H., Staker, P., Lee, Y.T., Moroney, J.W., Reed, C.H., Luban, N.L., Wang, R.-H., 2007. High levels of GDF15 in thalassemia suppress expression of the iron regulatory protein hepcidin. Nat. Med. 13, 1096–1101.

Tokarev, A., Skasko, M., Fitzpatrick, K., Guatelli, J., 2009. Antiviral activity of the interferon-induced cellular protein BST-2/tetherin. AIDS Res. Hum. Retroviruses 25, 1197–1210.

Tomar, S., Hardy, R.W., Smith, J.L., Kuhn, R.J., 2006. Catalytic core of *Alphavirus* nonstructural protein nsP4 possesses terminal adenylyltransferase activity. J. Virol. 80, 9962–9969.

Tuittila, M., Hinkkanen, A.E., 2003. Amino acid mutations in the replicase protein nsP3 of Semliki Forest virus cumulatively affect neurovirulence. J. Gen. Virol. 84, 1525–1533.

Varadhachary, G.R., Wolff, R.A., Crane, C.H., Sun, C.C., Lee, J.E., Pisters, P.W., Vauthey, J.-N., Abdalla, E., Wang, H., Staerkel, G.A., 2008. Preoperative gemcitabine and cisplatin followed by gemcitabine-based chemoradiation for resectable adenocarcinoma of the pancreatic head. J. Clin. Oncol. 26, 3487–3495.

Vihinen, H., Saarinen, J., 2000. Phosphorylation site analysis of Semliki Forest virus nonstructural protein 3. J. Biol. Chem. 275, 27775–27783.

Vihinen, H., Ahola, T., Tuittila, M., Merits, A., Kääriäinen, L., 2001. Elimination of phosphorylation sites of Semliki Forest virus replicase protein nsP3. J. Biol. Chem. 276, 5745–5752.

Wang, Q.J., Ding, Y., Kohtz, S., Mizushima, N., Cristea, I.M., Rout, M.P., Chait, B.T., Zhong, Y., Heintz, N., Yue, Z., 2006. Induction of autophagy in axonal dystrophy and degeneration. J. Neurosci. 26, 8057–8068.

Wang, H.-L., O'Rear, J., Stollar, V., 1996. Mutagenesis of the Sindbis virus nsP1 protein: effects on methyltransferase activity and viral infectivity. J. Virol. 217, 527–531.

Waterhouse, P.M., Wang, M.-B., Lough, T., 2001. Gene silencing as an adaptive defence against viruses. Nature 411, 834–842.

Wlodawer, A., Vondrasek, J., 1998. Inhibitors of hiv-1 protease: a major success of structure-assisted drug design 1. Annu. Rev. Biophys. Biomol. Struct. 27, 249–284.

Zhang, Y., 2008. I-TASSER server for protein 3D structure prediction. BMC Bioinform. 9, 1.

Žusinaite, E., Tints, K., Kiiver, K., Spuul, P., Karo-Astover, L., Merits, A., Sarand, I., 2007. Mutations at the palmitoylation site of non-structural protein NSP1 of Semliki Forest virus attenuate virus replication and cause accumulation of compensatory mutations. J. Gen. Virol. 88, 1977–1985.

## FURTHER READING

Mills, N., 2006. ChemDraw Ultra 10.0. J. Am. Chem. Soc. 128, 13649–13650.

Chapter 5

# Structure and Function of *Alphavirus* Proteases

Shailly Tomar, Megha Aggarwal
*Indian Institute of Technology, Roorkee, Uttarakhand, India*

## 1   INTRODUCTION

Alphaviruses, belonging to the family Togaviridae, are enveloped, positive-sense, single-stranded RNA viruses (Griffin, 2007; Kuhn 2007; Strauss and Strauss, 1994). These are mosquitoborne viruses and cause a number of diseases, including rash, fever, arthritis, and encephalitis. The members of *Alphavirus* genus include Chikungunya virus (CHIKV); Eastern, Western, and Venezuelan equine encephalitis virus (VEEV); Aura virus (AURAV); Semliki Forest virus (SFV); Ross river virus (RRV); and Sindbis virus (SINV). Alphaviruses cause diseases in humans and livestock. VEEV has been classified as a select agent due to the existence of stockpiles of weapon-grade VEEV (Bronze et al., 2002). CHIKV infection causes fever, muscle pain, headache, rashes, neurological problems, and the severe condition might lead to persistent arthralgia (Larrieu et al., 2010). The reemergence of CHIKV covering large part of the world as an endemic, the high pathogenicity of the virus, which can result in persistent arthralgia, and the virus evolution that can be transmitted by more number of host vectors make it a noteworthy pathogen, which needs to be eradicated soon or against which preventive/therapeutic measures should be developed (Larrieu et al., 2010; Ligon, 2006; Powers and Logue, 2007; Schuffenecker et al., 2006).

The alphaviruses use a variety of receptors for cell attachment, including heparin sulfate, laminin receptors, $\alpha 1\beta 1$ integrin, and C-type lectins: DC-SIGN and L-SIGN to infect variety of hosts from humans to mammals to mosquitoes (Kielian, 1995; Klimstra et al., 2003, 1998). After receptor-mediated endocytosis, the acidification in the endosomes leads to the virus fusion through a mechanism in which E1–E2 heterodimer dissociates and E1 changes its conformation to expose the fusion peptide (Detulleo, 1998; Helenius, 1980; Justman et al., 1993; Kielian, 1985; Kondor-Koch, 1983; Mellman et al., 1986; Wahlberg, 1992). The fusion process follows the release

Viral Proteases and Their Inhibitors. http://dx.doi.org/10.1016/B978-0-12-809712-0.00005-8

**105**

of nucleocapsid into the cytoplasm. The nucleocapsid interacts with ribosomes and frees the genomic RNA (Singh and Helenius 1992). The genomic RNA itself acts as mRNA and translates to nonstructural (ns) polyproteins (P123 and P1234). P1234 is formed by the read through of the opal codon (Li and Rice 1989; Strauss et al., 1984). After polyprotein processing by nsP2 protease, four functional ns proteins (nsP1, nsP2, nsP3, and nsP4) are generated, which altogether, along with host factors, form the virus replication complexes. The nsP1 possesses guanine-7-methyltransferase and guanylyl transferase activities required for genome capping. Therefore, nsP1 has been considered as a capping enzyme (Ahola and Kääriäinen, 1995; Tomar et al., 2011). nsP1 is also associated with the membrane and plays a role in binding of the replication complex to the plasma membrane (Peränen et al., 1995). The nsP2 possesses helicase, NTase, RNA triphosphatase, and the proteolytic activities (Gomez de Cedrón et al., 1999; Rikkonen et al., 1994; Vasiljeva, 2001; Vasiljeva et al., 2000). The nsP3 contains ADP-ribose 1-phosphate phosphatase and RNA-binding activity (Malet et al., 2009; Park and Griffin, 2009). The nsP4 acts as an RNA-dependent RNA polymerase and hence is important for virus replication (Rubach et al., 2009; Tomar et al., 2006). Subsequently, the genomic RNA is transcribed into 26S subgenomic RNA through minus (−) strand RNA intermediate. The 26S subgenomic RNA gets translated into structural polyprotein, which is processed by capsid protease (CP; a serine protease) and the host proteases to form the functional structural proteins: capsid, E3, E2, 6K, and E1 glycoproteins. After the CP is released, it encapsidates the viral genomic RNA and assembles to form nucleocapsid assembly. Interestingly, the Aura virus capsid protein (AVCP) also encapsidates subgenomic RNA in virions (Rümenapf et al., 1995). The glycoproteins undergo posttranslational modifications in Golgi and endoplasmic reticulum, and are transported to the plasma membrane. The nucleocapsid core (NC) and cytoplasmic tails of E1 and E2 glycoproteins interact, leading to budding of the progeny virions at the host cell membrane. Thus the polyprotein processing by the proteases is the major step in the *Alphavirus* life cycle to initiate virus replication, budding, and assembly.

The cryoelectron microscopic structures are available for SINV, SFV, RRV, VEEV, and CHIKV at high resolutions (Cheng et al., 1995; Mancini et al., 2000; Mukhopadhyay et al., 2006; Sun et al., 2013; Zhang et al., 2002, 2011). The virions are of 70-nm diameter with icosahedral symmetry. The overall structure of the members of *Alphavirus* genus is almost similar. The 80 trimeric glycoprotein spikes of E1–E2 heterodimers penetrate through the lipid bilayer into the virion (Zhang et al., 2002). In total 240 copies of E1 and E2 glycoproteins are present individually. Similarly, 240 copies of the capsid protein assembles to form the NC, which is present just beneath the host-derived lipid layer in the virion. Both the spikes, as well as the NC, are arranged in $T = 4$ icosahedral symmetry (Paredes et al., 1992, 1993; Vogel et al., 1986; von Bonsdorff and Harrison, 1975).

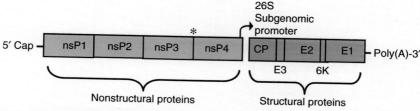

FIGURE 5.1 *Alphavirus* **genome.** The genome is (+) sense, single-stranded RNA, which includes two open reading frames (ORFs) and is capped at 5′ end. It also contains a 3′-poly(A) tail. The N-terminus, two-third region, encodes for the nonstructural (ns) proteins, which gets translated into nsP1, nsP2, nsP3, and nsP4. The *asterisk* (*) indicates the presence of a leaky stop codon. The carboxyl-terminus, one-third region, encodes for the structural proteins and gets translated by 26S subgenomic promoter into the capsid protease (CP) and 6K, and E1, E2, and E3 glycoproteins.

The *Alphavirus* genome is approximately 11.7-kb long plus-sense single-stranded RNA (Strauss and Strauss, 1994). The RNA consists of 5′-cap and 3′-poly(A) tail. It contains two open reading frames (ORFs): the ns ORF represents the 5′ two-third of the viral genome and the structural ORF represents the 3′ one-third of the genome. These two ORFs translate into ns and structural polyproteins, respectively. The structural, as well as ns proteins are translated as a single polyprotein, which further gets processed in a highly regulated fashion to produce functional viral proteins. The ns proteins perform various functions in the virus life cycle, such as replication, polyprotein processing, RNA transcription, and capping. The structural proteins form the virus core, encapsidate genomic RNA, and also are involved in virus assembly, budding, and disassembly. Fig. 5.1 shows the genome organization of alphaviruses.

## 2 *ALPHAVIRUS* PROTEASES

The positive-strand RNA viruses translate the viral RNAs into precursor polyproteins that get processed by either virus specific proteases, CP and nsP2, or by the host proteases (Fig. 5.2). The proteolytic processing of the polyprotein is an essential step in the virus life cycle for the production of functional viral proteins that regulates virus replication and assembly. The nsP2 protein is a multifunctional enzyme. The N-terminal region of nsP2 protein possesses ATPase and GTPase, RNA helicase, and RNA 5′ triphosphatase activities (Gomez de Cedrón et al., 1999; Rikkonen et al., 1994; Vasiljeva et al., 2000). The C-terminal region is involved in subgenomic RNA synthesis, nuclear transportation of nsP2, regulation of (−) strand RNA synthesis, and the proteolytic activity for ns polyprotein processing (Merits et al., 2001; Peränen et al., 1990; Sawicki and Sawicki, 1993; Sawicki et al., 2005; Suopanki et al., 1998; Vasiljeva, 2001; Vasiljeva et al., 2003). The nsp123 intermediate is transitory in the SINV nsP2 mutant (N614D) and cannot accomplish the complementary RNA synthesis (Lemm et al., 1994). Fig. 5.3 shows various enzymatic subdomains of nsP2 responsible for diverse functions of nsP2 in the virus life cycle.

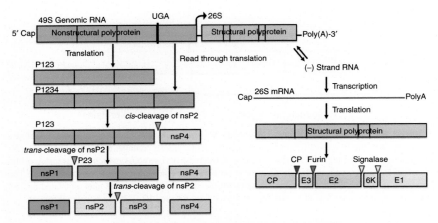

**FIGURE 5.2    Schematic representation of polyprotein processing in *Alphavirus*.** The ns poly-protein is translated by 49S genomic RNA and processed by *cis*- and *trans*-proteolysis of nsP2 protease. The structural proteins are translated by 26S subgenomic RNA via (−) strand RNA inter-mediate. These are processed by CP autoprotease activity, Furin, and Signalase.

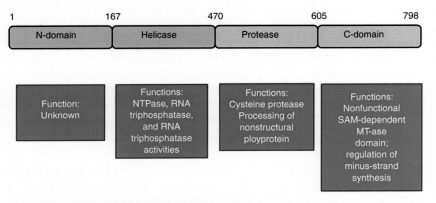

**FIGURE 5.3    Schematic of *Alphavirus* nsP2 domains.** The N-terminal domain is involved in NTPase, RNA, and RNA helicase activities, while the C-terminal domain consists of the protease domain. It also consists of a SAM-dependent methyltransferase (MTase)-like domain that is nonfunctional in nsP2. The numbering of different domains is done according to Chikungunya virus (CHIKV) nsP2.

The *Alphavirus* CP is a chymotrypsin-like serine protease and possesses *cis*-proteolytic activity for cleaving itself from the nascent structural polyprotein precursor (Aggarwal et al., 2012, 2014, 2015; Choi et al., 1991, 1996, 1997; Melancon and Garoff, 1987). The protease activity of capsid protein is essential for the maturation of structural polyprotein. After getting released from the polyprotein, the CP performs multiple functions, including RNA encapsidation, nucleocapsid assembly, and interaction with E1 and E2 glycoproteins for virus assembly and budding (Forsell et al., 1995; Geigenmüller-Gnirke et al., 1993;

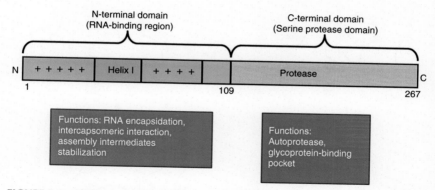

FIGURE 5.4 *Alphavirus* **capsid domains.** The CP comprises of the N-terminal domain (1–109) and C-terminal domain (110–267). The N-terminal region is positively charged and is involved in RNA encapsidation and capsid–capsid assembly. A helix I region is present in this domain, which is supposed to be involved in intercapsomeric interaction and assembly intermediates stabilization (Perera et al., 2001). The C-terminal domain consists of the protease domain, which also contains the glycoprotein-binding pocket. The numbering is done according to Aura virus capsid protein (AVCP).

Hahn et al., 1988; Hong et al., 2006; Metsikkö and Garoff, 1990; Owen and Kuhn, 1996; Tang et al., 2011; Vaux et al., 1988; Weiss et al., 1989; Zhao et al., 1994). The subdomains of the *Alphavirus* CP with different functions are shown in Fig. 5.4.

## 3 CAPSID PROTEASE

The CP autoproteolytic cleavage in alphaviruses takes place between the conserved Trp267 (the $P_1$ residue) and Ser268 (the $P_1'$ residue) (Aggarwal et al., 2012). After the autoproteolytic cleavage, the last C-terminus tryptophan remains bound in the active-site pocket of CP, which prevents further transproteolytic activity by blocking the entry of substrate in the CP active site. Therefore, the CP catalyzes only one enzymatic reaction in its life, as it becomes inactive afterward (Choi et al., 1991, 1996; Skoging and Liljeström, 1998). The active site of CP consists of His144, Ser218, and Asp166 residues similar to other serine proteases (Aggarwal et al., 2012; Hahn and Strauss, 1990). The CP of alphaviruses consists of the N- and the C-terminal regions. The N-terminal region, which is poorly conserved, predominantly contains basic amino acids and is intrinsically disordered similar to the N-terminal region of *Alphavirus* nsP4 RNA polymerase (Tomar et al., 2006). These disordered regions rich in basic residues have been postulated to be involved in protein–protein interactions and in interactions with nucleic acids (Forsell et al., 1995; Weiss et al., 1989). The C-terminal region consists of a chymotrypsin-like protease fold and possesses serine protease activity. Due to the disordered nature of the N-terminus, the full-length CP cannot be expressed, purified, or crystallized. Therefore, the

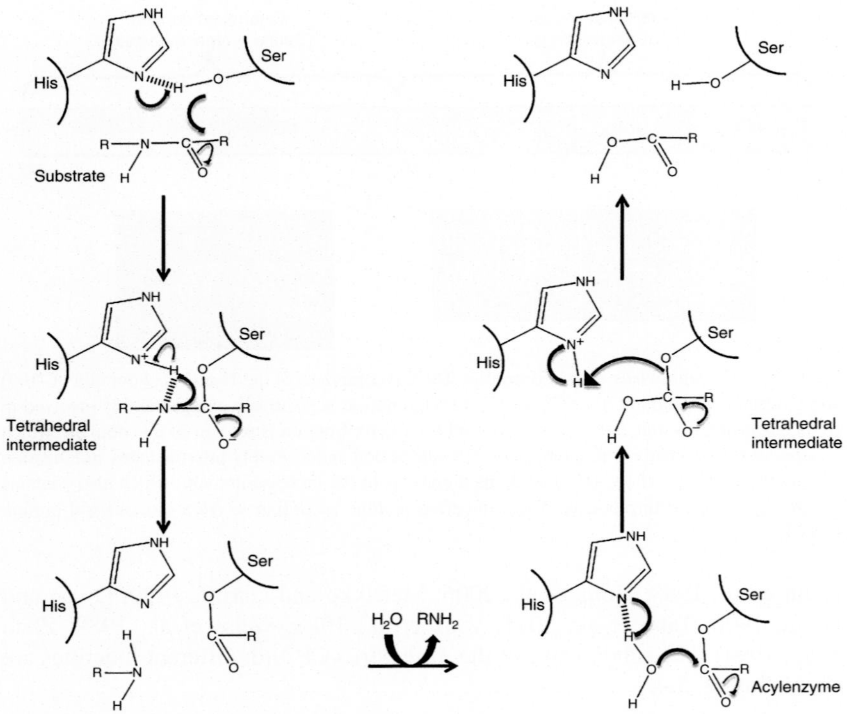

**FIGURE 5.5   Mechanism of serine protease action.** The two tetrahedral intermediates and the formation of acyl enzyme are shown. The product gets released from the reaction in two steps.

majority of structural work has been performed on the carboxyl terminal protease domain, and it has been well characterized. Even in the crystal structure of full-length CP, the electron density is visible for the protease domain only (Choi et al., 1991, 1997). This might be due to the disorderedness of N-terminus. Along with the active site, the C-terminal domain also contains a hydrophobic pocket that makes molecular contacts with the cytoplasmic tail of E2 glycoprotein (cdE2), which initiates the budding process of virus particles (Aggarwal et al., 2012; Metsikkö and Garoff, 1990; Tang et al., 2011; Zhao et al., 1994).

## 3.1   Proteolytic Mechanism

The complete proteolytic mechanism of CP is shown in Fig. 5.5. The proteolytic reaction starts with the formation of a hydrogen bond between the catalytic triad residues Asp166 and His144 (AVCP residue numbering). It makes the N-atom of histidine more electronegative. After the binding of the substrate, —OH group of active-site residue Ser acts as a nucleophile and attacks on the carbonyl carbon atom of the substrate. This leads to the acceptance of hydrogen by the electronegative N-atom of the His residue. Together this results in the shifting of the electron pair in the double bond of scissile bond carbonyl group toward the oxygen atom. This stage is termed as tetrahedral intermediate state.

The movement of the scissile covalent bond electrons toward the His H-atom results in the cleavage of the substrate peptide bond. Then the double bonds electrons move back from the carbonyl oxygen. This stage is called the acyl enzyme intermediate. Further, the water molecule enters and replaces the N-terminal part of the substrate. The similar reaction as before occurs again, which includes the electrons movement toward carbonyl oxygen and His accepts the proton from the water molecule, which finally generates one more tetrahedral intermediate. The oxyanion hole, consisting of conserved residues Gly216 and Ser218 (AVCP residue numbering), stabilizes the intermediates in the catalytic reaction by forming the H-bonds with the intermediates. Both the intermediates contain the negatively charged O-atom to fit into the oxyanion hole and have very low activation energies. The electron transfers from the substrate C and active-site serine O bond toward the H-atom of His, which makes the substrate carbonyl C more electron deficient. This results in the formation of double bond again, and the final step of cleaving the second half of the peptide substrate from Ser residue takes place.

## 3.2 Structure of Capsid Protease

The crystal structures of the protease domain of CP are available from different alphaviruses, including SINV, SFV, VEEV, and AURAV (Table 5.1). All these structures show the catalytically inactive postcleavage form of the capsid (Aggarwal et al., 2012; Choi et al., 1996, 1997; Tong et al., 1993). Also, the substrate-bound form of SCP structure is available in which the active-site residue Ser215 has been mutated to Ala to prevent the autocatalytic activity; after tryptophan two more residues are present at the C-terminus. Though the structure represents the inactive form of enzyme, it is informative with respect to the binding pattern of the substrate (Choi et al., 1996). Although the structural characterization was performed for the catalytic active SFV CP through nuclear magnetic resonance spectroscopy, fluorescence, and circular dichroism experiments (Morillas et al., 2008), the active SFV CP was found to be natively unfolded protein in the absence of last tryptophan residue, which is responsible for the stability of the capsid structure. Recently, the first crystal structure of active form of CP from AURAV was determined (Aggarwal et al., 2014) (Fig. 5.6). Till date, this is the only crystal structure available for the active form of *Alphavirus* CP. All these *Alphavirus* CP structures show chymotrypsin-like serine protease fold with Greek key motif (Fig. 5.6A). The active-site residues, Ser218, His144, and Asp166 (AVCP residue numbering), are conserved in *Alphavirus* CPs. The sequential position of the catalytic triad residues in various serine proteases might be different, but they have a very similar relative orientation in all the serine proteases. The C-terminal loop of CP brings the last C-terminal conserved $P_1$ tryptophan residue toward the active site for cleavage. In the active

**TABLE 5.1 CP Structural Data From Different Members of *Alphavirus* Genus**

| Viruses | PDB IDs | Features | Years | Resolutions (Å) | References |
|---|---|---|---|---|---|
| SINV | 2SNV | Refined structure | 1993 | 2.8 | Tong et al. (1993) |
| | 1KXE | (Y180S, E183G) | 1996 | 3.2 | Choi et al. (1996) |
| | 1KXD | (N222L) | 1996 | 3.0 | Choi et al. (1996) |
| | 1KXC | (N190K) | 1996 | 3.1 | Choi et al. (1996) |
| | 1KXB | (S215A) | 1996 | 2.9 | Choi et al. (1996) |
| | 1KXA | WT(106–264) | 1996 | 3.1 | Choi et al. (1996) |
| | 1KXF | (1–264) | 1996 | 2.38 | Choi et al. (1996) |
| | 2SNW | Type 3 crystal form | 1998 | 2.7 | Choi et al. (1996) |
| | 1WYK | (Dioxane bound) | 1998 | 2.0 | Lee et al. (1998) |
| | 1SVP | (Substrate bound) | 1996 | 2.0 | Lee et al. (1996) |
| SFV | 1VCP | Crystal form I | 1996 | 3.0 | Choi et al. (1997) |
| | 1VCQ | Crystal form II | 1996 | 3.1 | Choi et al. (1997) |
| VEEV | 1EP5 | WT | 2003 | 2.3 | Unpublished |
| | 1EP6 | WT | 2003 | 2.45 | Unpublished |
| AURAV | 4AGK | WT (110–267) | 2012 | 1.81 | Aggarwal et al. (2012) |
| | 4AGJ | (Dioxane bound) | 2012 | 1.98 | Aggarwal et al. (2012) |
| | 4UON | (Active form) | 2014 | 1.81 | Aggarwal et al. (2014) |
| | 5G4B | (Piperazine bound) | 2016 | 2.21 | Unpublished |

AURAV, Aura virus; SFV, Semliki Forest virus; SINV, Sindbis virus; VEEV, Venezuelan equine encephalitis virus.

AVCP, due to the absence of this Trp267, the loop is disordered and the density of the residues 262–265 is missing in the structure. The crystal structures of CP from different alphaviruses show the presence of crystallographic monomers, dimers, trimers, and also tetramers. However, the virus core assembly shows the monomeric nature of CP (Cheng et al., 1995). The solution form of CP is also monomeric in nature (Aggarwal et al., 2011, 2014; Morillas et al., 2008).

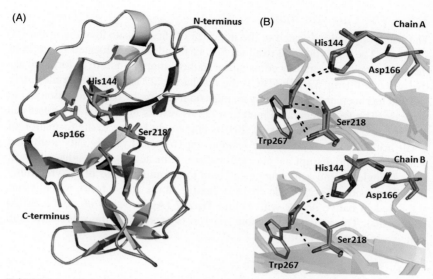

FIGURE 5.6  (A) Overall structure of a monomer of active AVCP (PDB ID: 4UON). The structure consists of a chymotrypsin-like fold with two subdomains. The active site residues *(green sticks)* are present at the interface of these two subdomains. (B) The conformational change in the active site of active and inactive (PDB ID: 4AGK) AVCP *(orange)* in both chain A *(green)* and chain B *(blue)* of active CP. The residues interactions are presented in *dashed black lines.*

Additionally, the mutation in the crystallographic dimer interface residues does not cause the reduction in virus replication or budding, indicating that these interactions are crystallographic and do not have any biological significance (Choi et al., 1996).

## 3.2.1  The Hydrophobic Pocket

The hydrophobic pocket in CP is present on the opposite side of the active site where cdE2 binds. Also, the N-arm residues of CP interact with the hydrophobic pocket of other symmetry-related molecules that mimic the interaction of CP and glycoprotein (Lee et al., 1996). Thus the hydrophobic pocket has been described as performing two different functions: capsid–capsid interaction for virus assembly and capsid–E2 interaction for virus budding. It is hypothesized that there is a conformational change in the hydrophobic pocket that leads to the switching of the function from virus assembly to virus budding (Choi et al., 1997). The conformational change has been found in the hydrophobic pocket of inactive (postcleavage form) and active (precleavage form) AVCP. In active AVCP, the N-arm binds to the hydrophobic pocket of other capsid molecules. However, this region is disordered in inactive AVCP and does not show this interaction. Therefore, the switch in the function of capsid could be a result of the conformational change of the pocket (Aggarwal et al., 2012, 2014).

## 3.2.2 The Catalytic Triad

The catalytic triad is located at the interface of the two subdomains in the CP structure. The catalytic triad residues Ser, His, and Asp are located far apart in the sequence; however, they come closer in the tertiary structure to form the active site of CP (Fig. 5.6). In CVCP, the residue Ile227 is located near the active site and has been proposed to play an important role in the catalytic action of the protease. The PCR-generated mutation I227K results in the loss of proteolytic activity of CVCP (Thomas et al., 2010). This might be due to the formation of a salt bridge between active-site residue Asp161 and Lys227. This additional salt bridge would disrupt the interactions within the active-site residues. Hence, even a single–amino acid change can lead to a complete loss of enzyme function. The residue Ser218 side chain is flexible in the active form, while it gets stabilized in native AVCP with bound Trp267 (Aggarwal et al., 2012, 2014) (Fig. 5.6B). The active-site residue Asp166 is exposed to the solvent in both inactive and active AVCP.

## 3.2.3 The Oxyanion Hole

The GDSG motif, found in many serine proteases, is highly conserved in *Alphavirus* CPs and it contains an active-site Ser residue. The glycine and serine residues in the motif form the oxyanion hole. The oxyanion hole is essential for proteolytic activity, as it stabilizes the tetrahedral intermediate through polar interactions (Cui et al., 2002; Rumthao et al., 2004). There is a large conformational variation in Gly216 residue present in the structures of active and inactive forms of AVCP. The residue Pro215 shows a flip in the peptide conformation between the two forms and creates steric hindrance for the substrate in the active form. Hence, it is possible that the conformation of Pro215 changes when the substrate binds to the pocket (Fig. 5.7). Also, the oxyanion hole of the active form contains a water molecule that is absent in the inactive forms of CP (Aggarwal et al., 2014). The substrate-bound form of SCP has two chains: chain A consists of a water molecule and chain B shows the presence of an H-bond between the oxyanion hole glycine and the scissile bond residues. Hence, chain A is described as the early enzymatic state with very loosely bound substrate, and chain B is involved in the late stage with properly bound substrate (Choi et al., 1996). Therefore, it can be concluded that the water molecule is present in the initial enzymatic state till the substrate can be accessed. As soon as the substrate enters in the specificity pocket, the water molecule is replaced by it and conformational changes occur to initiate new interactions with the bound substrate.

## 3.2.4 Specificity Pockets

The specificity pockets are referred to the pockets that are involved in binding the substrate. The scissile bond residues are called as $P_1$ and $P'_1$. At the C-terminus residues Ser268, Arg269, Ala270, and Ile271 are present, which are

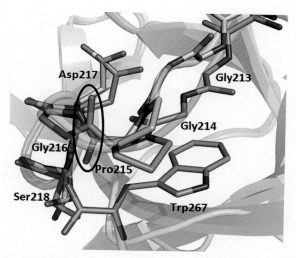

**FIGURE 5.7** Structural comparison of the active *(blue)* and inactive *(orange)* AVCP regarding the residues near oxyanion hole and $S_1$ pocket. The backbone peptide flip in Pro215 is highlighted in the *black circle*.

named as $P_1', P_2'$, $P_3'$, and $P_4'$, respectively (Table 5.2). The residues located at the N-terminus of Trp267: Glu266, Val265, and Thr264 are considered as $P_2$, $P_3$, and $P_4$ residues, respectively. The comparative studies of the *Alphavirus* CP with other chymotrypsin-like serine proteases show that different substrate residues bind to different specificity pockets and according to their binding, the specificity pockets are also assigned numbers (Tong et al., 1993) (Fig. 5.8A).

The $S_1$ specificity pocket is the primary substrate-binding site in serine proteases, as it is involved in substrate specificity along with other surface loops (Czapinska and Otlewski, 1999; Solivan et al., 2002). In *Alphavirus* CP, Trp residue binds to $S_1$ pocket, hence $S_1$ subsite should contain hydrophobic and small residues to give ample space and hydrophobicity for the binding of

**TABLE 5.2** The Substrate Residues for the Capsid Protease From Different Alphaviruses

|  | $P_4$ | $P_3$ | $P_2$ | $P_1$ | $P_1'$ | $P_2'$ | $P_3'$ | $P_4'$ |
|---|---|---|---|---|---|---|---|---|
| AURAV | T | V | E | W | S | R | A | I |
| CHIKV | A | E | E | W | S | L | A | I |
| VEEV | C | E | Q | W | S | L | V | T |
| SINV | T | E | E | W | S | A | A | P |
| SFV | S | E | E | W | S | A | P | L |

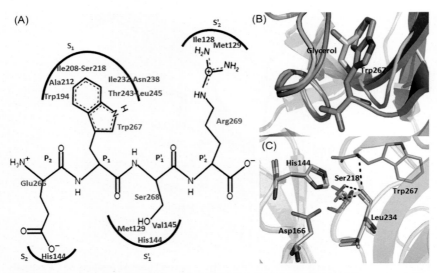

**FIGURE 5.8** (A) Schematic representation of the scissile bond residues and the corresponding specificity pocket residues. (B) The comparative analysis of active *(blue)* and inactive *(orange)* AVCP shows the presence of glycerol molecule at a position where Trp267 binds in inactive form. (C) Leu234 interacts with Ser218, as well as with Trp267 of inactive structure.

large tryptophan residue. The corresponding residue is serine in chymotrypsin, Val209 in SCP and Ala212 in AVCP. The $S_1$ pocket in AVCP consists of residues Trp194, Ile208 to Ser218, Ile232 to Asn238, and Thr243 to Leu245 (Fig. 5.8A). A glycerol molecule (used as a cryoprotectant) seems to be present in the $S_1$ pocket of the active AVCP exactly at the same position where Trp267 binds in the inactive form (Fig. 5.8B). The residue Ser214 is highly important in the catalysis in chymotrypsin, as it interacts with active-site residue Asp; however, the mutation in Ser214 does not influence the enzyme activity. Hence, it is possible that the backbone position of this residue is important (Krem et al., 2002). In AVCP, the corresponding residue is Leu234, which is conserved and forms hydrogen bonds with both the catalytic triad residue Ser218 and the scissile bond $P_1$ residue Trp267 (Fig. 5.8C). Therefore, the residue at this position is important for maintaining the conformational stability of the main catalytic residue, as well as in the binding of the substrate (Aggarwal et al., 2014). The $P_2$ and $P_3$ residues Glu266 and Val265 are present in an outward-oriented loop. Glu266 forms ionic interaction with active-site residue His144. The $S_4$ specificity pocket consists of Ala237, Ser246, and His261 residues to accommodate $P_4$ residue Thr264. The $S_1'$ pocket comprises residues Met129, His144, and Val145 to which $P_1'$ residue Ser268 binds (Fig. 5.8A). The $S_2'$ pocket and $S_4'$ pocket consists of residues Ile128 and Met129, and Asn123 and Lys127, respectively.

## 3.3 *trans*-Protease Activity and High Throughput Inhibitor Screening Assay

The *Alphavirus* CP is catalytically inactive after *cis*-autoproteolysis reaction due to the presence of the last tryptophan residue ($P_1$) into the specificity pocket. This prevents the binding of next or *trans*-substrate into the pocket (Choi et al., 1991). An attempt to detect the *trans*-protease activity of *Alphavirus* CP has been made, but the *trans*-activity could not be identified and the enzyme was pondered to be active for just one reaction in the complete virus life cycle (Hahn and Strauss, 1990). About 2 decades later, the studies were performed again to find out the *trans*-activity of SFV CP and the study concludes that SFV CP restores the activity after truncation of the last tryptophan that binds to the specificity pocket (Morillas et al., 2008). Different truncations from the C-terminus of the CP were made, including the deletion of the C-terminus tryptophan residue and all the constructs exhibit *trans*-esterase activity with the tyrosine- and tryptophan-based synthetic substrates. The kinetic parameters were very similar to those of other serine proteases ($k_{cat}/K_M = 5 \times 10^5$ M$^{-1}$s$^{-1}$) (Lottenberg et al., 1981; Odake et al., 1991; Powers and Kam, 1995). However, in SFV CP the *trans*-protease activity could not be determined.

In recent studies, AVCP shows *trans*-protease activity with the fluorescent peptide substrate containing the peptide sequence similar to the scissile bond (Aggarwal et al., 2014). Using the fluorescence resonance energy transfer (FRET)–based assay, the activity was assessed and the kinetic parameters were calculated. The same strategy has also been used to determine the *trans*-protease activity of CVCP (Aggarwal et al., 2015). The results show the high efficiency of the CP to cleave the synthetic fluorogenic peptide substrate in *trans* with $k_{cat}/K_m$ value of $1.11 \times 10^3$ M$^{-1}$s$^{-1}$. The peptide substrate DABCYL-GAEEWSLAIE-EDANS was used, which represents a sequence similar to that present in the scissile bond in CHIKV CP-E3. Furthermore, the protease activity of CVCP was performed in a 96-well plate in a high-throughput (HTP) format. The Z′ factor and coefficient of variance for CVCP activity are 0.64 and 8.68%, respectively. These results explain the adaptability of this assay as an HTP format for the screening of compound libraries. Thus the proteolytic assay has been developed for the HTP screening of protease inhibitors (PIs) against the enzymatic activity of *Alphavirus* CP (Aggarwal et al., 2014, 2015).

## 3.4 Capsid Protease Inhibition

The common serine PIs, including chloromethylketones, such as *N*-benzyloxy-carbonyl-leucine-tyrosine-chloromethylketone and *N*-benzyloxycarbonylphenylalanine-chloromethylketone, and phenylmethylsulfonyl fluoride (PMSF) (potent inhibitors of serine proteases) (Johnson and Moore, 2002, 2000;

Jung et al., 1995; Sidorowicz et al., 1980) have been tested for the inhibition of the esterase activity of the SFV CP. However, the serine PIs could not inhibit the *trans*-esterase activity of the CP (Morillas et al., 2008). Moreover, the other serine PI 4-(2-aminoethyl) benzenesulfonyl fluoride hydrochloride (AEBSF) also could not inhibit the *cis*-protease activity of CVCP (Thomas et al., 2010). Further, substrate-based compounds and small peptides were tested for the inhibition of CVCP activity (Aggarwal et al., 2015). The substrate-based candidate compounds and inhibitors, such as tryptophan, tryptamine, the common PI PMSF, benzamidine, *N*-*p*-tosyl-L-phenylalanine chloromethyl ketone (TPCK), and 3-acetylindole, and the peptide inhibitors, EW and EWS, were examined for the antiprotease activity of CVCP. However, none of these inhibitors were able to effectively inhibit CP activity. Moreover, in the crystal structure of the active form of AVCP, one glycerol molecule is present in the $S_1$ specificity pocket, which is exactly at the same position where $P_1$ residue Trp binds (Aggarwal et al., 2014) (Fig. 5.8B). Additionally, the CVCP activity data with increasing concentration of glycerol reveal the inhibitory effect of glycerol on CVCP activity. The enzyme was found to be inactivated in increasing concentrations of glycerol. On the basis of these results, there is a possibility to design glycerol-based inhibitors of *Alphavirus* CP. Hence, the availability of the crystal structure of the active form of *Alphavirus* CP and development of FRET-based HTP proteolytic assay for inhibitor screening will be beneficial for the identification and development of structure-based inhibitors targeting *Alphavirus*-specific CP activity (Aggarwal et al., 2015).

## 4 NSP2 PROTEASE

The dependence of *Alphavirus* replication on the activity of nsP2 protease to process the ns polyprotein into functional ns proteins makes the nsP2 protease an attractive target for the development of antiviral compounds. On the basis of the sequence analysis results, the nsP2 protease is considered to be a cysteine protease and it belongs to the C9 peptidase family of clan CA (Rawlings, 2006). Crystal structure and mutagenesis studies show that *Alphavirus* nsP2 is a papain-like protease having the catalytic diad consisting of residues Cys481 and His558 (SINV numbering). These active-site residues are completely conserved among the *Alphavirus* genus. The residue following the active site His558 in SINV is Trp 559 and this Trp is also conserved in all the members of the *Alphavirus* genus. This conserved Trp residue is essential for the proteolytic activity of SINV nsP2 (Strauss et al., 1992).

### 4.1 Mechanism

The nsP2 protease follows a mechanism similar to that of other cysteine proteases for proteolytic cleavage (Fig. 5.9). The very first step is the deprotonation of the thiol group by the basic side chain of His. The deprotonated cysteine attacks

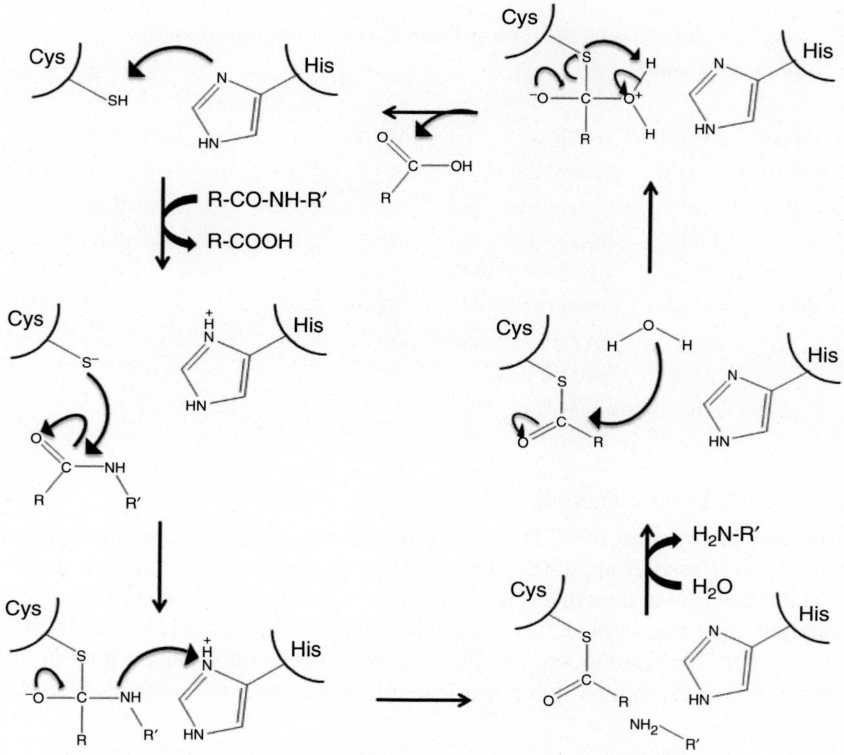

FIGURE 5.9    The cleavage mechanism of the cysteine protease.

via its anionic sulfur on the substrate carbonyl C-atom. Due to this nucleophilic attack, the N-terminal part of the substrate gets released and the His residue restores its deprotonated form. This generates a thioester intermediate, which is formed by the linkage of substrate C-terminus and cysteine thiol. Consequently, the hydrolysis of the thioester bond occurs and the remaining substrate fragment is separated from the cysteine thiol (Beveridge, 1996; Polgár, 1974; Russo et al., 2010).

## 4.2    Crystal Structures of nsP2 Protease

Crystal structures of nsP2 protease are available from VEEV (PDB ID: 2HWK, 5EZS, 5EZQ), SINV (PDB ID: 4GUA), and CHIKV (PDB ID: 3TRK, 4ZTB) [Hu et al., 2016; Russo et al., 2006; Shin et al., 2012; Narwal et al. (to be published)] (Table 5.3). Despite of the variation in the amino acid sequences, the 3D structures of nsP2 protease are highly conserved among all the three alphaviruses and are very similar to each other (Saisawang et al., 2015a).

**TABLE 5.3 nsP2 Crystal Structures From Different Members of the *Alphavirus* Genus**

| Viruses | PDB IDs | Features | Years | Resolutions (Å) | References |
|---|---|---|---|---|---|
| SINV | 4GUA | P23pro-ZBD | 2012 | 2.85 | Shin et al. (2012) |
| CHIKV | 3TRK, | Protease domain | 2011 | 2.4 | Unpublished |
| | 4ZTB | nsP2 protease–glycerol complex | 2016 | 2.59 | Unpublished |
| VEEV | 2HWK | Protease domain | 2006 | 2.45 | Russo et al. (2006) |
| | 5EZS | nsP2–E64d complex | 2006 | 2.16 | Hu et al. (2016) |
| | 5EZQ | nsP2 protease | 2016 | 1.66 | Hu et al. (2016) |

ZBD, Zinc-binding domain.

## 4.2.1 VEEV nsP2 Protease Structure

The crystal structure of VEEV nsP2 protease was determined at a resolution of 2.45 Å (Russo et al., 2006). The structure consists of two domains: the N- and the C-terminal domains (Fig. 5.10). The N-terminal domain consists of the catalytic diad that is made up of cysteine and histidine (Cys477 and His546) (Fig. 5.10). The residue Asp549 also comes closer to the catalytic diad in the crystal structure, though it is not oriented in a way to make contacts with the

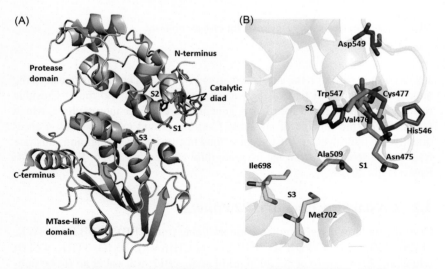

**FIGURE 5.10  Structure of the VEEV nsP2 protease (PDB ID: 2HWK).** (A) The protease domain *(blue)* and MTase-like domains *(pink)* are present at N- and C-terminus, respectively. The catalytic diad residues *(red)*, S₁ pocket residues *(green)*, S2 residues *(brown)*, and S3 residues *(yellow)* are shown as *sticks*. (B) The zoom-in view of the active site and specificity pockets.

active-site residues. Also, mutational analysis shows that this residue is not essential for nsP2 protease activity (Strauss et al., 1992). The C-terminal domain contains the SAM-dependent methyltransferase (MTase) domain that might be responsible for the recognition of the substrates and regulation of the substrate-binding cleft structure. The protease active site is located adjacent to the interface of these two domains. Though the catalytic diad conformation in nsP2 protease is similar to the papain-like cysteine protease, the overall fold of the nsP2 protease is unique. The conformation of the ns polyprotein allows the access of 3/4 site toward the protease and thus, this site first gets cleaved in *cis* and nsP4 is separated from the rest of the ns polyprotein (P123). The conformation of the nsP2 protease does not allow the *cis*-cleavage of 2/3 site. The inaccessibility of the cleavage site to the protease suggests the *trans*-cleavage of 2/3 site (Russo et al., 2006; Vasiljeva et al., 2003).

### 4.2.1.1  The Catalytic Diad

The catalytic diad is formed by residues Cys477 and His546, which form the thiolate–imidazolium ion pair in the activated form of nsP2 (Russo et al., 2010). However, this ion pair interaction is absent in the crystal structure of VEEV nsP2 (Russo et al., 2006). The structure analysis shows that there is a very ill-defined density for the His546 and the molecular dynamics studies suggests this residue is flexible, having different energetically favorable rotamers. The binding of the substrate helps in the stabilization of this interaction and precedes the reaction to proteolysis. The calculated $pK_a$ values for the catalytic diad in VEEV nsP2 show a value similar to that of His Cys thiolate–imidazolium ion pair. This too confirms the existence of this ion pair in nsP2 protease.

### 4.2.1.2  The Substrate-Binding Sites

The positions of different substrate-binding sites in the structure of VEEV nsP2 were revealed by molecular docking and fitting of the peptide corresponding to all three different cleavage sites into the active site of VEEV nsP2 (Russo et al., 2010). The two domains in nsP2 protease structure are separated by a long deep groove that is lined up by $S_1$, $S_2$, and $S_3$ sites (Fig. 5.10). These are present as a shallow depression at the surface of the protease. The shallow depression allows the binding of substrate having a small residue, such as glycine. From the structural and modeling studies, it was shown that five residues could bind to the substrate-binding site in VEEV nsP2 protease. If the peptide is longer than five amino acids, then it extends outside the binding pocket (Russo et al., 2010). The $S_1$ pocket residue Val476, along with Cys477, may be involved in the formation of oxyanion hole (Russo et al., 2006). The $S_2$ site consists of residue Trp547 (bulky aromatic residue) that is required for the binding of $P_2$ glycine to the enzyme. Hence, VEEV nsP2pro uses glycine specificity motif for the binding of specific substrates. The $S_3$ site is located in the C-terminal MTase domain in the protein structure.

In SFV nsP2, the $P_4$ residue is considered to play a major role in specificity determination (Lulla et al., 2006). The structure analysis of VEEV nsP2 shows the presence of a salt bridge between $S_4$ and $P_4$ residues, which supports and approves the importance of the $P_4$ residue as a molecular determinant of substrate specificity (Russo et al., 2010). Also, the cation–π interaction between the $P_1'$ and $S_1'$ residues endorses the preference of 3/4 cleavage over the other two nsP2 protease sites in the ns polyprotein. The covariance between the specificity pocket residues and the substrate residues exists for the maintenance of similar volume, charge, or hydrophobicity. Thus the variation in the residue of S sites among different *Alphavirus* nsP2 proteases should also contain the compatible variation in the corresponding substrate residues (Russo et al., 2010).

### 4.2.2 CHIKV nsP2

The structural, biochemical, and mutational analysis of the CHIKV nsP2 protease reveals very interesting features unique to CHIKV nsP2 protease. The investigations have shown that the CHIKV nsP2 protease is an altered papain-like protease and the catalytic Cys residue is interchangeable with a vicinal Ser residue that performs the proteolytic cleavage (Saisawang et al., 2015a) (Fig. 5.11). The mutation of either Cys478 or Ser482 does not result in activity loss, while

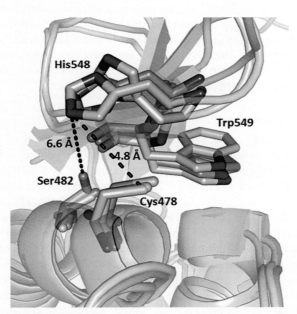

**FIGURE 5.11** **The structural alignment of nsP2 from VEEV** *(blue)* **(PDB ID: 2HWK), CHIKV** *(pink)* **(PDB ID: 3TRK), and SINV** *(sand)* **(PDB ID: 4GUA) exhibits the exchangeable Ser482 for Cys478 in CHIKV nsP2.** In other two nsP2 proteases, this residue is alanine. Conserved Trp549 of CHIKV nsP2 seems to be involved in the stabilization of the loop that brings His548 to the active site.

the mutation in both residues leads to a complete loss of protease activity in CHIKV. In the crystal structure of CHIKV nsP2, Ser482 seems far away from the catalytic diad (6.6 Å), from the catalytic residue His548 (Fig. 5.11). The inspection of the catalytic diad in the crystal structures of nsP2 from VEEV and SINV reveals a distance range of 5.6–8.1 Å between the catalytic diad. Hence, it is possible that a conformational movement is essential to bring the catalytic diad residues closer to cleave the substrate. These movements might also contribute to the mobility of Ser482 in CHIKV nsP2, to bring it closer to the cleavage site for carrying out proteolysis.

However, the recent study on CHIKV nsP2 demonstrates that Cys478 is essential to perform proteolysis as the mutant Cys478Ala completely abolishes the protease activity (Rausalu et al., 2016). The mutant also obliterates virus replication and release of infectious virus. In contrast, the Ser482Ala mutation shows no effect on proteolytic activity, RNA synthesis and infectious virus production.

Like the papain protease, along with the catalytic dyad, a conserved tryptophan residue is essential for the proteolytic activity of SINV and SFV nsP2 (Golubtsov et al., 2006; Strauss et al., 1992). However, structural analysis of CHIKV nsP2 shows that the orientation of this Trp549 does not permit it to take part in proteolysis. Instead, in CHIKV nsP2, Trp549 forms hydrophobic and van der Waals interactions with the pocket residues, and these interactions help in the stabilization of the loop containing the catalytic His548 residue (Saisawang et al., 2015a) (Fig. 5.11). Moreover, Trp549 is also involved in the glycine specificity motif selection of the substrate (Golubtsov et al., 2006). Furthermore, the mutation in the tryptophan residue does not influence the proteolytic activity of the enzyme. Unlike other *Alphavirus* nsP2 proteases, surprisingly, CHIKV nsP2 protease shows quite different characteristics. The mutation of Trp549 to either Ala or Phe shows nearly wildtype (WT) enzyme activity. Interestingly, these mutations have different effects with different substrates and the activity increases threefold for 2/3 substrate (Saisawang et al., 2015a). Thus it was concluded that Trp549 does not directly influence the activity; however, it maintains the appropriate conformation of the active site and substrate binding for efficient catalysis.

### 4.2.3 SINV nsP2

The SINV nsP2 structure is available as P23 complex bound to the nsP3 protein (Shin et al., 2012) (Fig. 5.12). It includes protease and MTase domains of nsP2 and macro and zinc-binding domains (ZBD) from nsP3 (Fig. 5.12A). nsP3 encircles the MTase domain of nsP2 and grasps it firmly; the dissociation mechanism of the two proteins after cleavage is not known. The superposition of the individual nsP2 and nsP3 structures with the complex nsP23 shows a major variation in the MTase domain of nsP2. Thus it is possible that the MTase domain movement (induced by either protein–protein interaction or RNA–protein interaction) might be involved in the dissociation of nsP2 and 3 after proteolytic cleavage. The nsP2/3 site is located in a narrow cleft at the interface of nsP2 and 3, and it is 40-Å apart from the active site of the protease, which confirms the *trans*-cleavage of 2/3 site.

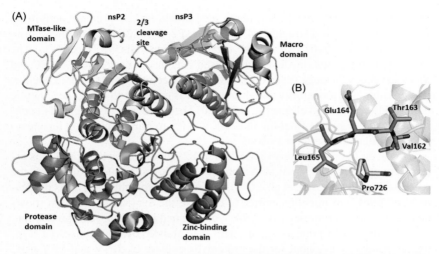

FIGURE 5.12   **The overall structure of SINV nsP23 (PDB ID: 4GUA).** (A) The cartoon diagram shows the presence of all the four domains: protease *(sand)* and MTase-like *(blue)* domains of nsP2 and macro domains *(pink)* and ZBDs *(teal)* domains of nsP3. The 2/3 cleavage site is present at the interface of nsP2 and 3. (B) Shows the Pro726 interaction with the nsP3 linker residues.

The *Alphavirus* infection causes the cytopathic effect in the infected cells by interfering with interferon function, which leads to cell death. The mutational studies show the role of nsP2 in disrupting host cell transcription and promoting cell cytopathogenicity (Frolova et al., 2002; Garmashova et al., 2006; Gorchakov et al., 2005). The mutations that lead to the noncytopathic effect of *Alphavirus* were recognized. The well-analyzed mutation was found in nsP2 region of the virus at position Pro726, which lies in its MTase domain. This mutation also influences virus replication and host cell transcription and translation, which in turn have a deleterious effect on viral replication and virus–host interaction (Dryga et al., 1997; Frolov and Schlesinger, 1994; Frolov et al., 1999; Weiss et al., 1980). The mutation of Pro726 in SINV nsP2 with the large side chain residue causes decrease in virus replication and production of noncytopathic virus; however, the small amino acid mutation shows WT virus characteristics. This is because of the steric clashes by the bulky amino acids with nsP3, as Pro726 interacts with the nsP3 linker region. Interestingly, the Pro726Gly mutation also shows the deleterious effect. It may be due to the distortion in the nsP2–3 interaction, as loop destabilization occurs if glycine is present. Interestingly, the Pro726 is found to be present at the interface of nsP2–3 in the structure (Shin et al., 2012). The structure shows Pro726 molecular contacts with the linker region of nsP3 (Fig. 5.12B). The residues surrounding Pro726 are either basic or hydrophobic. If the contacting nsP3 linker region residue Leu165 is changed to tryptophan, it does not influence the virus phenotype; however, the alteration of Leu165 to Asp or Glu leads to a reduction in virus infection and

P23 proteolysis. It is due to the stabilization of the structure in the presence of acidic amino acids, which reduces the interface flexibility required to change its conformation to access the P23 cleavage site (Shin et al., 2012).

In the early life cycle of virus, the replication complex synthesizes negative-strand RNA, but after the cleavage of the 2/3 site, the mature ns proteins are produced and the replication machinery changes the template for the production of positive-sense genomic, as well as subgenomic, RNA. The mechanism for the template switching is not well known. The structure of SINV nsP23 indicates the extensive RNA-binding surface that spans through both nsP2 and 3. Hence, it is a possibility that after the cleavage of nsP2/3, there is a change in the RNA-binding site that could be the cause of the switching of the RNA template by the replication complex (Shin et al., 2012).

## 4.3 The Proteolytic Action

The nsP2 possesses *cis*-, as well as *trans*-, proteolytic activity for the processing of nsP1234 polyprotein at three different cleavage sites (Fig. 5.2). The substrate residues for all the three different sites from different alphaviruses are tabulated in Table 5.4. The cleavage intermediate products having nsP2 protease are active and can perform the catalytic action efficiently (Vasiljeva et al., 2003). The processing of 2/3 occurs in *trans* (Russo et al., 2006). In VEEV nsP2 structure, the distance between the $P_1$ residue of the modeled substrate in the protease and the last visible residue in nsP2 shows a distance of ~42 Å, which clearly indicates that the cleavage of 2/3 site occurs in *trans*. This is confirmed by the nsP23 structure of SINV, where the structure analysis shows the presence of 2/3 cleavage site far apart from the active site, which is not accessible for the *cis* cleavage.

The proteolytic processing of ns polyprotein occurs in a highly regulated manner, which depends on the replication complex function and constituents (Shirako and Strauss, 1990). The nsP3 macrodomain also acts as a specificity determinant for the nsP2 protease substrates in SFV and SINV (Lulla et al., 2011). SFV nsP3 was shown to play certain role in the appropriate alignment of the 2/3 cleavage site. In the absence of the macro domain, the proteolytic cleavage of 2/3 site, in both these viruses, terminates. However, CHIKV nsP2 in the absence of macro domain is able to cleave the 2/3 site peptide (Saisawang et al., 2015b). The SINV nsP2 protease shows its higher efficiency for 1/2 and 2/3 sites, as compared to 3/4 site using native polyprotein (de Groot et al., 1990; Hardy and Strauss, 1989). SFV nsP2 protease data also show great variations in activity with different substrates and also between the full-length and protease domains of nsP2. Using the native polyprotein as the substrate, the 2/3 site cleavage is most efficient (Takkinen et al., 1991). However, when the synthetic substrate is used, 3/4 is the most promptly processed site (Golubtsov et al., 2006; Vasiljeva, 2001; Vasiljeva et al., 2003). The comparative analysis of nsP2 proteolytic activity of SINV, SFV, and VEEV shows decreased efficiency of the enzymes with the substrate MBP-NusG-His6 in which the cleavage site is located in the

**TABLE 5.4** $P_3'$–$P_4$ Residues for all Three Cleavage Sites of nsP2 From Different Alphaviruses

| | $P_4'$ | | | $P_3'$ | | | $P_2'$ | | | $P_1'$ | | | $P_1$ | | | $P_2$ | | | $P_3$ | | | $P_4$ | | |
|---|---|---|---|---|---|---|---|---|---|---|---|---|---|---|---|---|---|---|---|---|---|---|---|---|
| | ns 1/2 | ns 2/3 | ns 3/4 | ns 1/2 | ns 2/3 | ns 3/4 | ns 1/2 | ns 2/3 | ns 3/4 | ns 1/2 | ns 2/3 | ns 3/4 | ns 1/2 | ns 2/3 | ns 3/4 | ns 1/2 | ns 2/3 | ns 3/4 | ns 1/2 | ns 2/3 | ns 3/4 | ns 1/2 | ns 2/3 | ns 3/4 |
| CHIKV | E | Y | S | I | S | F | I | P | I | G | A | Y | A | C | G | G | G | G | A | A | A | R | R | R |
| VEEV | E | Y | S | V | S | F | S | P | I | G | A | Y | A | C | A | G | G | G | A | A | A | E | E | D |
| SINV | E | Y | S | V | S | F | L | P | I | A | A | Y | A | A | G | G | G | G | – | V | V | D | G | G |
| SFV | E | Y | S | V | S | F | V | P | I | G | A | Y | A | C | A | G | G | G | A | A | A | H | T | R |
| AURAV | E | Y | S | V | S | F | L | P | I | A | A | Y | A | A | G | G | G | G | A | S | V | D | G | G |

linker region of the fusion protein. However, all the proteases show comparatively higher efficiency in thioredoxin (Trx) fusion protein (Zhang et al., 2009). This might be due to the presence of specificity determinants in the residues beyond the scissile bond residues as well (Lulla et al., 2006). Also, the large size of nsP2 protease might halt it from cleaving the short linker region. The increasing length in the linker region improves protease efficiency. However, CHIKV nsP2 protease shows highly efficient protease activity with small peptide substrates, in contrast to other *Alphavirus* nsP2 proteases (Pastorino et al., 2008; Saisawang et al., 2015a). The CHIKV nsP2 also shows the maximum cleavage with the 3/4 site when fluorogenic substrates are used (Pastorino et al., 2008). The study shows the comparison of full-length and the protease domains of nsP2 for the proteolytic activity (Saisawang et al., 2015b). The protease domain has more activity than full-length nsP2; however, the proteolytic activity varies with different substrates. The effect of the metal ion and PI also varies between full-length and protease domain of nsP2 and also differs with different substrates (Saisawang et al., 2015b). The cross-reactivity experiment suggests that only VEEV nsP2 protease (not SINV and SFV nsP2) is able to cleave the heterologous substrate with high efficiency, even more than its native substrate (Zhang et al., 2009).

The two mutations Phe509Leu and Gly736Ser cause a defect in protease activity and demolish the downregulation of subgenomic RNA synthesis (Hahn et al., 1989; Suopanki et al., 1998). In the crystal structure of VEEV, the corresponding residues, Phe504 and Gly723, are buried into the hydrophobic pockets in N- and C-terminal domains of nsP2, respectively (Russo et al., 2006). The mutation in these residues might result in the disturbance of hydrophobicity, which destabilizes the hydrophobic core: the possible reason for the loss of activity. Also, the position of Gly723 in the structure shows that this position is unfavorable for the nonglycine residues in the Ramachandran Plot. The steric hindrance and the energetic barrier in Gly723Ser-mutated nsP2 could also be the other cause of the nonfunctional protease.

## 4.4 Inhibition Studies

A number of PIs were tested to inhibit the nsP2 protease activity of SFV (Vasiljeva, 2001). The nsP2 protease was completely resistant to PMSF (serine PI), pepstatin (aspartic PI), EDTA (metalloPI), and even the cysteine PIs: E-64 and leupeptin. However, another cysteine PI *N*-ethylmaleimide (NEM) could inhibit the protease action with 1/2 and 3/4 substrates. Interestingly, $Zn^{2+}$ and $Cu^{2+}$ could inhibit the enzyme activity. Likewise, different PIs were also tested for the CHIKV nsP2 protease inhibition activity using all three different substrates (Saisawang et al., 2015a). The inhibitors E-64 and leupeptin did not inhibit CHIKV nsP2 protease action. The effect of inhibitors and the metal ions differ with different alphaviruses, different mutated forms of CHIKV nsP2, and with different substrates. For example, cobalt usage in the protease activity assay of CHIKV nsP2 enhances enzymatic activity using 1/2 substrate, while it decreases enzymatic activity with 2/3 substrate. This suggests the complexity of the nsP2 catalytic mechanism, which needs to be studied extensively.

Few attempts have made to develop CHIKV nsP2 inhibitors on the structural basis using molecular docking, virtual screening, and molecular dynamics (Bora, 2012; Nguyen et al., 2015). The structure-based compounds were designed on the basis of the binding efficiency of the compounds to different pockets of nsP2 protease. Another study on the homology modeling on CHIKV nsP2 protease and virtual screening with around 5 million compounds identified 26 hits that could be effective against CHIKV (Bassetto et al., 2013). The selected 26 compounds were tested for the antiviral activities and two compounds **1** and **25** have an inhibitory effect on viral replication at a concentration that does not cause cell cytotoxicity. The $EC_{50}$ values from the CHIKV CPE assay and the dose–response curve are almost similar to each other. More computational studies were performed to identify and design the anti-CHIKV compounds and the nsP2 protease was used as a target. The molecular modeling and the e-pharmacophore mapping identified the features of nsP2 inhibitors (Singh et al., 2011, Dhindwal et al., 2016). The inhibitor should possess an aromatic ring with one hydrophobic and three H-bond–donating sites. By docking and e-pharmacophore mapping, four compounds were identified that could be used as the inhibitors of nsP2 protease because they could bind to the active site of the protease. Moreover, a potent inhibitor of VEEV, CID15997213, has been identified after HTP screening of 340,000 compound libraries, and this compound targets the N-terminal region of nsP2 protein (not the nsP2 protease); though the drug inhibition mechanism is not clear (Chung et al., 2014).

## 5 CONCLUSIONS

PIs are a class of antiviral inhibitor molecules that specifically inhibit virus-specific proteases and are currently widely used against HIV for the treatment of AIDS and hepatitis C virus (HCV) for the treatment of hepatitis. Similarly, the *Alphavirus* CP and nsP2 proteases are virus specific and essential viral enzymes required for the maturation of both structural, as well as ns, proteins, respectively. The activity of CP is indispensable for the initiation of structural polyprotein processing and the release of capsid, which performs numerous functions in the virus life cycle. The protease activity of nsP2 is necessary for regulated proteolytic processing of ns polyprotein, virus replication, host cell transcription, and host cell cytopathogenicity. Hence, these *Alphavirus*-specific proteases are considered to be the potential antiviral targets for drug development. In this chapter, we focused on the structural and functional aspects of the *Alphavirus* proteases: the CP (chymotrypsin-like serine protease) and nsP2 protease (papain-like cysteine protease). The structural features, including the active site, substrate-binding pockets, and oxyanion hole, have been discussed extensively. The structures from different members of the *Alphavirus* genus have been described and the comparison has been presented in detail. Different mutational studies show the role of different residues for the proteolytic activity and shed light on the mechanism of protease action, and its effects on direct catalysis or indirect substrate binding. The proteolytic activity of both the enzymes has been illustrated. The proteolytic activity of nsP2 is described with respect to its three different substrate sites in the ns polyprotein. Several attempts have been made to identify inhibitors against *Alphavirus* proteases, but little or no success has been achieved. Unfortunately, currently no *Alphavirus* inhibitors are commercially available.

This chapter focuses on the recent developments in the structure–activity relationship of *Alphavirus* nsP2 and CP proteases and also FRET-based CP assay for HTP screening of compound libraries. Henceforth, the study of structures and functions of *Alphavirus*-specific proteases will provide in-depth information, which can be exploited for the development of structure-based antivirals to combat *Alphavirus* infections.

# REFERENCES

Aggarwal, M., Dhindwal, S., Kumar, P., Kuhn, R.J., Tomar, S., 2014. *trans*-Protease activity and structural insights into the active form of the *Alphavirus* capsid protease. J. Virol. 88, 12242–12253.

Aggarwal, M., Dhindwal, S., Pratap, S., Kuhn, R.J., Kumar, P., Tomar, S., 2011. Crystallization, high-resolution data collection and preliminary crystallographic analysis of Aura virus capsid protease and its complex with dioxane. Acta Crystallogr. Sect. F 67, 1394–1398.

Aggarwal, M., Sharma, R., Kumar, P., Parida, M., Tomar, S., 2015. Kinetic characterization of trans-proteolytic activity of chikungunya virus capsid protease and development of a FRET-based HTS assay. Sci. Rep. 5, 14753.

Aggarwal, M., Tapas, S.P., Siwach, A., Kumar, P., Kuhn, R.J., Tomar, S., 2012. Crystal structure of Aura virus capsid protease and its complex with dioxane: new insights into capsid-glycoprotein molecular contacts. PLoS One 7, 0051288.

Ahola, T., Kääriäinen, L., 1995. Reaction in *Alphavirus* mRNA capping: formation of a covalent complex of nonstructural protein nsP1 with 7-methyl-GMP. Proc. Natl. Acad. Sci. USA 92, 507–511.

Bassetto, M., Burghgraeve, T.D., Delang, L., Massarotti, A., Coluccia, A., Zonta, N., Gatti, V., Colombano, G., Sorba, G., Silvestri, R., Tron, G.C., Neyts, J., Leyssen, P., Brancale, A., 2013. Computer-aided identification, design and synthesis of a novel series of compounds with selective antiviral activity against chikungunya virus. Antiviral Res. 98, 12–18.

Beveridge, A.J., 1996. A theoretical study of the active sites of papain and S195C rat trypsin: implications for the low reactivity of mutant serine proteinases. Protein Sci. 5, 1355–1365.

Bora, L., 2012. Homology modeling and docking to potential novel inhibitor for chikungunya (37997) protein nsP2 protease. J. Proteomics Bioinform. 05, 054–059.

Bronze, M.S., Huycke, M.M., Machado, L.J., Voskuhl, G.W., Greenfield, R.A., 2002. Viral agents as biological weapons and agents of bioterrorism. Am. J. Med. Sci. 323, 316–325.

Cheng, R.H., Kuhn, R.J., Olson, N.H., Choi, M.G.R.-K., Smith, T.J., Baker, T.S., 1995. Nucleocapsid and glycoprotein organization in an enveloped virus. Cell 80, 621–630.

Choi, H.K., Lu, G., Lee, S., Wengler, G., Rossmann, M.G., 1997. Structure of Semliki Forest virus core protein. Proteins Struct. Funct. Genet. 27, 345–359.

Choi, H.K., Lee, S., Zhang, Y.P., Mckinney, B.R., Wengler, G., Rossmann, M.G., Kuhn, R.J., 1996. Structural analysis of Sindbis Virus capsid mutants involving assembly and catalysis. J. Mol. Biol. 262, 151–167.

Choi, H.K., Tong, L., Minor, W., Dumas, P., Boege, U., Rossmann, M.G., Wengler, G., 1991. Structure of Sindbis virus core protein reveals a chymotrypsin-like serine proteinase and the organization of the virion. Nature 354, 37–43.

Chung, D.H., Jonsson, C.B., Tower, N.A., Chu, Y.-K., Sahin, E., Golden, J.E., Noah, J.W., Schroeder, C.E., Sotsky, J.B., Sosa, M.I., Cramer, D.E., Mckellip, S.N., Rasmussen, L., White, E.L., Schmaljohn, C.S., Julander, J.G., Smith, J.M., Filone, C.M., Connor, J.H., Sakurai, Y., Davey, R.A., 2014. Discovery of a novel compound with anti-Venezuelan equine encephalitis virus activity that targets the nonstructural protein 2. PLoS Pathog. 10, e1004213.

Cui, J., Marankan, F., Fu, W., Crich, D., Mesecar, A., Johnson, M.E., 2002. An oxyanion-hole selective serine protease inhibitor in complex with trypsin. Bioorg. Med. Chem. 10, 41–46.

Czapinska, H., Otlewski, J., 1999. Structural and energetic determinants of the S1-site specificity in serine proteases. Eur. J. Biochem. 260, 571–595.

de Groot, R.J., Hardy, W.R., Shirako, Y., Strauss, J.H., 1990. Cleavage-site preferences of Sindbis virus polyproteins containing the non-structural proteinase. Evidence for temporal regulation of polyprotein processing in vivo. EMBO J. 9, 2631–2638.

Detulleo, L., 1998. The clathrin endocytic pathway in viral infection. EMBO J. 17, 4585–4593.

Dhindwal, S., Kesari, P., Singh, H., Kumar, P., Tomar, S., 2016. Conformer and pharmacophore based identification of peptidomimetic inhibitors of chikungunya virus nsP2 protease. J. Biomol. Struct. Dyn., 1–18.

Dryga, S.A., Dryga, O.A., Schlesinger, S., 1997. Identification of mutations in a Sindbis virus variant able to establish persistent infection in BHK cells: the importance of a mutation in the nsP2 gene. Virology 228, 74–83.

Forsell, K., Suomalainen, M., Garoff, H., 1995. Structure-function relation of the NH2-terminal domain of the Semliki Forest virus capsid protein. J. Virol. 69, 1556–1563.

Frolov, I., Schlesinger, S., 1994. Comparison of the effects of Sindbis virus and Sindbis virus replicons on host cell protein synthesis and cytopathogenicity in BHK cells. J. Virol. 68, 1721–1727.

Frolov, I., Agapov, E., Hoffman, T.A., Prágai, B.M., Lippa, M., Schlesinger, S., Rice, C.M., 1999. Selection of RNA replicons capable of persistent noncytopathic replication in mammalian cells. J. Virol. 73, 3854–3865.

Frolova, E.I., Fayzulin, R.Z., Cook, S.H., Griffin, D.E., Rice, C.M., Frolov, I., 2002. Roles of nonstructural protein nsP2 and alpha/beta interferons in determining the outcome of Sindbis virus infection. J. Virol. 76, 11254–11264.

Garmashova, N., Gorchakov, R., Frolova, E., Frolov, I., 2006. Sindbis virus nonstructural protein nsP2 is cytotoxic and inhibits cellular transcription. J. Virol. 80, 5686–5696.

Geigenmüller-Gnirke, U., Nitschko, H., Schlesinger, S., 1993. Deletion analysis of the capsid protein of Sindbis virus: identification of the RNA binding region. J. Virol. 67, 1620–1626.

Golubtsov, A., Kääriäinen, L., Caldentey, J., 2006. Characterization of the cysteine protease domain of Semliki Forest virus replicase protein nsP2 by in vitro mutagenesis. FEBS Lett. 580, 1502–1508.

Gomez de Cedrón, M., Ehsani, N., Mikkola, M.L., García, J.A., Kääriäinen, L., 1999. RNA helicase activity of Semliki Forest virus replicase protein NSP2. FEBS Lett. 448, 19–22.

Gorchakov, R., Frolova, E., Frolov, I., 2005. Inhibition of transcription and translation in Sindbis virus-infected cells. J. Virol. 79, 9397–9409.

Griffin, D.E., 2007. In: Knipe, D.M. (Ed.), Alphaviruses. Fields Virology. Lippincott Williams and Wilkins, Philadelphia, PA, pp. 1023–1067.

Hahn, C.S., Strauss, J.H., 1990. Site-directed mutagenesis of the proposed catalytic amino acids of the Sindbis virus capsid protein autoprotease. J. Virol. 64, 3069–3073.

Hahn, C.S., Lustig, S., Strauss, E.G., Strauss, J.H., 1988. Western equine encephalitis virus is a recombinant virus. Proc. Natl. Acad. Sci. USA 85, 5997–6001.

Hahn, Y.S., Strauss, E.G., Strauss, J.H., 1989. Mapping of RNA-temperature-sensitive mutants of Sindbis virus: assignment of complementation groups A, B, and G to nonstructural proteins. J. Virol. 63, 3142–3150.

Hardy, W.R., Strauss, J.H., 1989. Processing the nonstructural polyproteins of Sindbis virus: nonstructural proteinase is in the C-terminal half of nsP2 and functions both in *cis* and in trans. J. Virol. 63, 4653–4664.

Helenius, A., 1980. On the entry of Semliki Forest virus into BHK-21 cells. J. Cell Biol. 84, 404–420.

Hong, E.M., Perera, R., Kuhn, R.J., 2006. *Alphavirus* capsid protein helix I controls a checkpoint in nucleocapsid core assembly. J. Virol. 80, 8848–8855.

Hu, X., Compton, J.R., Leary, D.H., Olson, M.A., Lee, M.S., Cheung, J., Ye, W., Ferrer, M., Southall, N., Jadhav, A., Glass, P.J., 2016. Kinetic, mutational, and structural studies of the venezuelan equine encephalitis virus nonstructural protein 2 cysteine protease. Biochemistry 55, 3007–3019.

Johnson, G., Moore, S.W., 2000. Cholinesterase-like catalytic antibodies: reaction with substrates and inhibitors. Mol. Immunol. 37, 707–719.

Johnson, G., Moore, S.W., 2002. Catalytic antibodies with acetylcholinesterase activity. J. Immunol. Methods 269, 13–28.

Jung, G., Ueno, H., Hayashi, R., Liao, T.H., 1995. Identification of the catalytic histidine residue participating in the charge-relay system of carboxypeptidase Y. Protein Sci. 4, 2433–2435.

Justman, J., Klimjack, M.R., Kielian, M.A., 1993. Role of spike protein conformational changes in fusion of Semliki Forest virus. J. Virol. 67, 7597–7607.

Kielian, M., 1985. pH-induced alterations in the fusogenic spike protein of Semliki Forest virus. J. Cell Biol. 101, 2284–2291.

Kielian, M., 1995. Membrane fusion And the *Alphavirus* life cycle. Adv. Virus Res., 113–151.

Klimstra, W.B., Nangle, E.M., Smith, M.S., Yurochko, A.D., Ryman, K.D., 2003. DC-SIGN and L-SIGN can act as attachment receptors for alphaviruses and distinguish between mosquito cell- and mammalian cell-derived viruses. J. Virol. 77, 12022–12032.

Klimstra, W.B., Ryman, K.D., Johnston, R.E., 1998. Adaptation of Sindbis virus to BHK cells selects for use of heparan sulfate as an attachment receptor. J. Virol. 72, 7357–7366.

Kondor-Koch, C., 1983. Expression of Semliki Forest virus proteins from cloned complementary DNA. I. The fusion activity of the spike glycoprotein. J. Cell Biol. 97, 644–651.

Krem, M.M., Prasad, S., Cera, E.D., 2002. Ser214 is crucial for substrate binding to serine proteases. J. Biol. Chem. 277, 40260–40264.

Kuhn, R.J., 2007. Togaviridae: the viruses and their replication. In: Knipe, D.M., Howley, P.M. (Eds.), Fields' Virology. fifth ed. Lippincott Williams and Wilkins, New York, NY, pp. 1001–1022.

Larrieu, S., Pouderoux, N., Pistone, T., Filleul, L., Receveur, M.C., Sissoko, D., Ezzedine, K., Malvy, D., 2010. Factors associated with persistence of arthralgia among Chikungunya virus-infected travellers: report of 42 French cases. J. Clin. Virol. 47, 85–88.

Lee, S., Kuhn, R.J., Rossmann, M.G., 1998. Probing the potential glycoprotein binding site of Sindbis virus capsid protein with dioxane and model building. Proteins Struct. Funct. Genet. 33, 311–317.

Lee, S., Owen, K.E., Choi, H.-K., Lee, H., Lu, G., Wengler, G., Brown, D.T., Rossmann, M.G., Kuhn, R.J., 1996. Identification of a protein binding site on the surface of the *Alphavirus* nucleocapsid and its implication in virus assembly. Structure 4, 531–541.

Lemm, J.A., Rümenapf, T., Strauss, E.G., Strauss, J.H., Rice, C.M., 1994. Polypeptide requirements for assembly of functional Sindbis virus replication complexes: a model for the temporal regulation of minus-and plus-strand RNA synthesis. EMBO J. 13, 2925–2934.

Li, G.P., Rice, C.M., 1989. Mutagenesis of the in-frame opal termination codon preceding nsP4 of Sindbis virus: studies of translational readthrough and its effect on virus replication. J. Virol. 63, 1326–1337.

Ligon, B.L., 2006. Reemergence of an unusual disease: the Chikungunya epidemic. Semin. Pediatr. Infect. Dis. 17, 99–104.

Lottenberg, R., Christensen, U., Jackson, C.M., Coleman, P.L., 1981. Assay of coagulation proteases using peptide chromogenic and fluorogenic substrates. Methods Enzymol. 80, 341–361.

Lulla, A., Lulla, V., Merits, A., 2011. Macromolecular assembly-driven processing of the 2/3 cleavage site in the *Alphavirus* replicase polyprotein. J. Virol. 86, 553–565.

Lulla, A., Lulla, V., Tints, K., Ahola, T., Merits, A., 2006. Molecular determinants of substrate specificity for Semliki Forest virus nonstructural protease. J. Virol. 80, 5413–5422.

Malet, H., Coutard, B., Jamal, S., Dutartre, H., Papageorgiou, N., Neuvonen, M., Ahola, T., Forrester, N., Gould, E.A., Lafitte, D., Ferron, F., Lescar, J., Gorbalenya, A.E., Lamballerie, X.D., Canard, B., 2009. The crystal structures of Chikungunya and Venezuelan equine encephalitis virus nsP3 macro domains define a conserved adenosine binding pocket. J. Virol. 83, 6534–6545.

Mancini, E.J., Clarke, M., Gowen, B.E., Rutten, T., Fuller, S.D., 2000. Cryo-electron microscopy reveals the functional organization of an enveloped virus, Semliki Forest virus. Mol. Cell. 5, 255–266.

Melancon, P., Garoff, H., 1987. Processing of the Semliki Forest virus structural polyprotein: role of the capsid protease. J. Virol. 61, 1301–1309.

Mellman, I., Fuchs, R., Helenius, A., 1986. Acidification of the endocytic and exocytic pathways. Annu. Rev. Biochem. 55, 663–700.

Merits, A., Kääriäinen, L., Vasiljeva, L., Auvinen, P., Ahola, T., 2001. Proteolytic processing of Semliki Forest virus-specific non-structural polyprotein by nsP2 protease. J. Gen. Virol. 82, 765–773.

Metsikkö, K., Garoff, H., 1990. Oligomers of the cytoplasmic domain of the p62/E2 membrane protein of Semliki Forest virus bind to the nucleocapsid in vitro. J. Virol. 64, 4678–4683.

Morillas, M., Eberl, H., Allain, F.-T., Glockshuber, R., Kuennemann, E., 2008. Novel enzymatic activity derived from the Semliki Forest virus capsid protein. J. Mol. Biol. 376, 721–735.

Mukhopadhyay, S., Zhang, W., Gabler, S., Chipman, P.R., Strauss, E.G., Strauss, J.H., Baker, T.S., Kuhn, R.J., Rossmann, M.G., 2006. Mapping the structure and function of the E1 and E2 glycoproteins in alphaviruses. Structure 14, 63–73.

Nguyen, P.T., Yu, H., Keller, P.A., 2015. Identification of chikungunya virus nsP2 protease inhibitors using structure-base approaches. J. Mol. Graph. Model. 57, 1–8.

Odake, S., Kam, C.M., Narasimhan, L., Poe, M., Blake, J.T., Krahenbuhl, O., Tschopp, J., Powers, J.C., 1991. Human and murine cytotoxic T lymphocyte serine proteases: subsite mapping with peptide thioester substrates and inhibition of enzyme activity and cytolysis by isocoumarins. Biochemistry 30, 2217–2227.

Owen, K.E., Kuhn, R.J., 1996. Identification of a region in the Sindbis virus nucleocapsid protein that is involved in specificity of RNA encapsidation. J. Virol. 70, 2757–2763.

Paredes, A.M., Brown, D.T., Rothnagel, R., Chiu, W., Schoepp, R.J., Johnston, R.E., Prasad, B.V., 1993. Three-dimensional structure of a membrane-containing virus. Proc. Natl. Acad. Sci. USA 90, 9095–9099.

Paredes, A.M., Simon, M.N., Brown, D.T., 1992. The mass of the Sindbis virus nucleocapsid suggests it has $T = 4$ icosahedral symmetry. Virology 187, 329–332.

Park, E., Griffin, D.E., 2009. Interaction of Sindbis virus non-structural protein 3 with poly(ADP-ribose) polymerase 1 in neuronal cells. J. Gen. Virol. 90, 2073–2080.

Pastorino, B.A., Peyrefitte, C.N., Almeras, L., Grandadam, M., Rolland, D., Tolou, H.J., Bessaud, M., 2008. Expression and biochemical characterization of nsP2 cysteine protease of chikungunya virus. Virus Res. 131, 293–298.

Peränen, J., Laakkonen, P., Hyvönen, M., Kääriäinen, L., 1995. The *Alphavirus* replicase protein nsP1 is membrane-associated and has affinity to endocytic organelles. Virology 208, 610–620.

Peränen, J., Rikkonen, M., Liljeström, P., Kääriäinen, L., 1990. Nuclear localization of Semliki Forest virus-specific nonstructural protein nsP2. J. Virol. 64, 1888–1896.

Perera, R., Owen, K.E., Tellinghuisen, T.L., Gorbalenya, A.E., Kuhn, R.J., 2001. *Alphavirus* nucleocapsid protein contains a putative coiled coil-helix important for core assembly. J. Virol. 75, 1–10.

Polgár, L., 1974. Mercaptide-imidazolium ion-pair: the reactive nucleophile in papain catalysis. FEBS Lett. 47, 15–18.

Powers, J.C., Kam, C.M., 1995. Peptide thioester substrates for serine peptidases and metalloendopeptidases. Methods Enzymol. 248, 3–18.

Powers, A.M., Logue, C.H., 2007. Changing patterns of chikungunya virus: re-emergence of a zoonotic arbovirus. J. Gen. Virol. 88, 2363–2377.

Rausalu, K., Utt, A., Quirin, T., Varghese, F.S., Žusinaite, E., Das, P.K., Ahola, T., Merits, A., 2016. Chikungunya virus infectivity, RNA replication and nonstructural polyprotein processing depend on the nsP2 protease's active site cysteine residue. Sci. Rep. 6, 37124.

Rawlings, N.D., 2006. MEROPS: the peptidase database. Nucleic Acids Res. 34, D270–D272.

Rikkonen, M., Peränen, J., Kääriäinen, L., 1994. ATPase and GTPase activities associated with Semliki Forest virus nonstructural protein nsP2. J. Virol. 68, 5804–5810.

Rubach, J.K., Wasik, B.R., Rupp, J.C., Kuhn, R.J., Hardy, R.W., Smith, J.L., 2009. Characterization of purified Sindbis virus nsP4 RNA-dependent RNA polymerase activity in vitro. Virology 384, 201–208.

Rümenapf, T., Brown, D.T., Strauss, E.G., König, M., Rameriz-Mitchel, R., Strauss, J.H., 1995. Aura *Alphavirus* subgenomic RNA is packaged into virions of two sizes. J. Virol. 69, 1741–1746.

Rumthao, S., Lee, O., Sheng, Q., Fu, W., Mulhearn, D.C., Crich, D., Mesecar, A.D., Johnson, M.E., 2004. Design, synthesis, and evaluation of oxyanion-hole selective inhibitor substituents for the S1 subsite of factor Xa. Bioorg. Med. Chem. Lett. 14, 5165–5170.

Russo, A.T., Malmstrom, R.D., White, M.A., Watowich, S.J., 2010. Structural basis for substrate specificity of *Alphavirus* nsP2 proteases. J. Mol. Graph. Model. 29, 46–53.

Russo, A.T., White, M.A., Watowich, S.J., 2006. The crystal structure of the Venezuelan equine encephalitis *Alphavirus* nsP2 protease. Structure 14, 1449–1458.

Saisawang, C., Saitornuang, S., Sillapee, P., Ubol, S., Smith, D.R., Ketterman, A.J., 2015a. Chikungunya nsP2 protease is not a papain-like cysteine protease and the catalytic dyad cysteine is interchangeable with a proximal serine. Sci. Rep. 5, 17125.

Saisawang, C., Sillapee, P., Sinsirimongkol, K., Ubol, S., Smith, D.R., Ketterman, A.J., 2015b. Full length and protease domain activity of chikungunya virus nsP2 differ from other *Alphavirus* nsP2 proteases in recognition of small peptide substrates. Biosci. Rep. 35, 1–9.

Sawicki, D.L., Sawicki, S.G., 1993. A second nonstructural protein functions in the regulation of *Alphavirus* negative-strand RNA synthesis. J. Virol. 67, 3605–3610.

Sawicki, D.L., Perri, S., Polo, J.M., Sawicki, S.G., 2005. Role for nsP2 proteins in the cessation of *Alphavirus* minus-strand synthesis by host cells. J. Virol. 80, 360–371.

Schuffenecker, I., Iteman, I., Michault, A., Murri, S., Frangeul, L., Vaney, M.C., Lavenir, R., Pardigon, N., Reynes, J.M., Pettinelli, F., Biscornet, L., Diancourt, L., Michel, S., Duquerroy, S., Guigon, G., Frenkiel, M.-P., Bréhin, A.C., Cubito, N., Desprès, P., Kunst, F., Rey, F.A., Zeller, H., Brisse, S., 2006. Genome microevolution of chikungunya viruses causing the indian ocean outbreak. PLoS Med. 3, e263.

Shin, G., Yost, S.A., Miller, M.T., Elrod, E.J., Grakoui, A., Marcotrigiano, J., 2012. Structural and functional insights into *Alphavirus* polyprotein processing and pathogenesis. Proc. Natl. Acad. Sci. USA 109, 16534–16539.

Shirako, Y., Strauss, J.H., 1990. Cleavage between nsP1 and nsP2 initiates the processing pathway of Sindbis virus nonstructural polyprotein P123. Virology 177, 54–64.

Sidorowicz, W., Jackson, G., Behal, F., 1980. Multiple molecular forms of human pancreas alanine aminopeptidase. Clin. Chim. Acta 104, 169–179.

Singh, I., Helenius, A., 1992. Role of ribosomes in Semliki Forest virus nucleocapsid uncoating. J. Virol. 66, 7049–7058.

Singh, K.D., Kirubakaran, P., Nagarajan, S., Sakkiah, S., Muthusamy, K., Velmurgan, D., Jeyakanthan, J., 2011. Homology modeling, molecular dynamics, e-pharmacophore mapping and docking study of chikungunya virus nsP2 protease. J. Mol. Model. 18, 39–51.

Skoging, U., Liljeström, P., 1998. Role of the C-terminal tryptophan residue for the structure-function of the *Alphavirus* capsid protein. J. Mol. Biol. 279, 865–872.

Solivan, S., Selwood, T., Wang, Z.M., Schechter, N.M., 2002. Evidence for diversity of substrate specificity among members of the chymase family of serine proteases. FEBS Lett. 512, 133–138.

Strauss, J.H., Strauss, E.G., 1994. The alphaviruses: gene expression, replication, and evolution. Microbiol. Rev. 58, 491–562.

Strauss, E.G., Groot, R.J.D., Levinson, R., Strauss, J.H., 1992. Identification of the active site residues in the nsP2 proteinase of Sindbis virus. Virology 191, 932–940.

Strauss, E.G., Rice, C.M., Strauss, J.H., 1984. Complete nucleotide sequence of the genomic RNA of Sindbis virus. Virology 133, 92–110.

Sun, S., Xiang, Y., Akahata, W., Holdaway, H., Pal, P., Zhang, X., Diamond, M.S., Nabel, G.J., Rossmann, M.G., 2013. Structural analyses at pseudo atomic resolution of Chikungunya virus and antibodies show mechanisms of neutralization. eLife 2, e00435.

Suopanki, J., Sawicki, D.L., Sawicki, S.G., Kääriäinen, L., 1998. Regulation of *Alphavirus* 26S mRNA transcription by replicase component nsP2. J. Gen. Virol. 79, 309–319.

Takkinen, K., Peränen, J., Kääriäinen, L., 1991. Proteolytic processing of Semliki Forest virus-specific non-structural polyprotein. J. Gen. Virol. 72, 1627–1633.

Tang, J., Jose, J., Chipman, P., Zhang, W., Kuhn, R.J., Baker, T.S., 2011. Molecular links between the E2 envelope glycoprotein and nucleocapsid core in Sindbis virus. J. Mol. Biol. 414, 442–459.

Thomas, S., Rai, J., John, L., Günther, S., Drosten, C., Pützer, B.M., Schaefer, S., 2010. Functional dissection of the *Alphavirus* capsid protease: sequence requirements for activity. Virol. J. 7, 327.

Tomar, S., Hardy, R.W., Smith, J.L., Kuhn, R.J., 2006. Catalytic core of *Alphavirus* nonstructural protein nsP4 possesses terminal adenylyltransferase activity. J. Virol. 80, 9962–9969.

Tomar, S., Narwal, M., Harms, E., Smith, J.L., Kuhn, R.J., 2011. Heterologous production, purification and characterization of enzymatically active Sindbis virus nonstructural protein nsP1. Protein Expr. Purif. 79, 277–284.

Tong, L., Wengler, G., Rossmann, M.G., 1993. Refined structure of Sindbis virus core protein and comparison with other chymotrypsin-like serine proteinase structures. J. Mol. Biol. 230, 228–247.

Vasiljeva, L., 2001. Site-specific protease activity of the carboxyl-terminal domain of Semliki Forest virus replicase protein nsP2. J. Biol. Chem. 276, 30786–30793.

Vasiljeva, L., Merits, A., Auvinen, P., Kaariainen, L., 2000. Identification of a novel function of the *Alphavirus* capping apparatus: RNA 5'-triphosphatase activity OF nsP2. J. Biol. Chem. 275, 17281–17287.

Vasiljeva, L., Merits, A., Golubtsov, A., Sizemskaja, V., Kaariainen, L., Ahola, T., 2003. Regulation of the sequential processing of Semliki Forest virus replicase polyprotein. J. Biol. Chem. 278, 41636–41645.

Vaux, D.J.T., Helenius, A., Mellman, I., 1988. Spike-nucleocapsid interaction in Semliki Forest virus reconstructed using network antibodies. Nature 336, 36–42.

Vogel, R.H., Provencher, S.W., Bonsdorff, C.H.V., Adrian, M., Dubochet, J., 1986. Envelope structure of Semliki Forest virus reconstructed from cryo-electron micrographs. Nature 320, 533–535.

Von Bonsdorff, C.H., Harrison, S.C., 1975. Sindbis virus glycoproteins form a regular icosahedral surface lattice. J. Virol. 16, 141–145.

Wahlberg, J.M., 1992. Membrane fusion process of Semliki Forest virus I: low pH-induced rearrangement in spike protein quaternary structure precedes virus penetration into cells. J. Cell Biol. 116, 339–348.

Weiss, B., Nitschko, H., Ghattas, I., Wright, R., Schlesinger, S., 1989. Evidence for specificity in the encapsidation of Sindbis virus RNAs. J. Virol. 63, 5310–5318.

Weiss, B., Rosenthal, R., Schlesinger, S., 1980. Establishment and maintenance of persistent infection by Sindbis virus in BHK cells. J. Virol. 33, 463–474.

Zhang, R., Hryc, C.F., Cong, Y., Liu, X., Jakana, J., Gorchakov, R., Baker, M.L., Weaver, S.C., Chiu, W., 2011. 4. 4 Å cryo-EM structure of an enveloped *Alphavirus* Venezuelan equine encephalitis virus. EMBO J. 30, 3854–3863.

Zhang, W., Mukhopadhyay, S., Pletnev, S.V., Baker, T.S., Kuhn, R.J., Rossmann, M.G., 2002. Placement of the structural proteins in Sindbis virus. J. Virol. 76, 11645–11658.

Zhang, D., Tözsér, J., Waugh, D.S., 2009. Molecular cloning, overproduction, purification and biochemical characterization of the p39 nsp2 protease domains encoded by three alphaviruses. Protein Expr. Purif. 64, 89–97.

Zhao, H., Lindqvist, B., Garoff, H., Von Bonsdorff, C.H., Liljeström, P., 1994. A tyrosine-based motif in the cytoplasmic domain of the *Alphavirus* envelope protein is essential for budding. EMBO J. 13, 4204–4211.

# Chapter 6

# *Flavivirus* Protease: An Antiviral Target

**Shailly Tomar, Rajat Mudgal, Benazir Fatma**
*Indian Institute of Technology, Roorkee, Uttarakhand, India*

## 1 INTRODUCTION

The genus *Flavivirus* consists of viruses which are known to cause morbidity and mortality worldwide. Dengue fever virus (DENV), West Nile virus (WNV), Yellow fever virus (YFV), Kyasanur Forest disease virus (KFDV), and recently notoriety-claimed Zika virus are few names that currently pose global health threats. They are confined mostly to the tropical and subtropical regions, particularly Southeast Asia, South America, and Australia, excluding the Polar Regions (Bhatt et al., 2013; Chancey et al., 2015; Rogers, 2006; Roth et al., 2014). Flaviviruses are mostly arthropod-borne viruses that are transmitted to host by mosquitoes or ticks. However, the vectors are not known for few members, such as Rio Bravo virus (RBV), which infects bats in America (Billoir et al., 2000), and Modoc virus (MODV) (Johnson, 1974). The pathogenic flaviviruses cause common health ailments, such as fever, headache, vomiting, nausea, rash, etc. However, in some cases *Flavivirus*, such as DENV, can cause fatal dengue hemorrhagic fever and dengue shock syndrome. Annually, DENV health calamity affects over 390 million people worldwide and rate of mortality can be 1.2%–3.5% (Bhatt et al., 2013).

YFV, the prototype of the genus *Flavivirus*, is known to cause mild illness with fever, headache, nausea, vomiting, fatigue, jaundice, and pain in muscles in humans, or in some cases to cause hemorrhagic fever that can progress and lead to organ failures (Monath, 2001). YFV is endemic to sub-Saharan Africa and tropical South America and is estimated to cause 200,000 cases of clinical disease and 30,000 deaths annually (Staples et al., 2010). Japanese encephalitis virus (JEV), another pathogenic *Flavivirus* that first emerged in the 1870s in Japan and has subsequently been reported across most of the Asia, has become the most important cause of epidemic encephalitis worldwide causing nearly 10,000 deaths annually (Solomon et al., 2003). WNV as the name suggests was first detected and isolated from the blood of the lady who was suffering

Viral Proteases and Their Inhibitors. http://dx.doi.org/10.1016/B978-0-12-809712-0.00006-X

from febrile illness, in the West Nile region of Uganda (Smithburn et al., 1940). WNV has become endemic in North America and is known for causing epidemic meningoencephalitis and arboviral encephalitis in the United States (Davis et al., 2006; DeBiasi et al., 2006). Tick-borne encephalitis virus (TBEV) causes tick-borne encephalitis (TBE), leading to fatal neurological manifestations in infected humans. More than 8000 cases of TBE have been reported between 1990–2007 in Europe and Russia (Mansfield et al., 2009).

DENV, the most prominent arbovirus, is further classified into four different serotypes (1–4) based on their interaction with the antibodies in the human serum. Recently, a fifth serotype of DENV has been reported (Normile, 2013). A vaccine designed for one serotype fails to generate neutralizing antibodies against other serotypes and progression leads to a more serious disease through a process known as antibody-mediated disease enhancement (Guzman and Kouri, 2008; Halstead, 1988). Hence, a recent approach for the development of vaccine against DENV focuses on the generation of a tetravalent vaccine, which can simultaneously immunize against all serotypes. Various traditional and molecular approaches are being undertaken in this direction (Chambers et al., 1997; Putnak, 1994).

Zika virus, one of the most recent *Flavivirus* that reemerged in Brazil in 2015, was first identified in Uganda in 1947 in rhesus monkeys suffering from yellow fever and was later identified in humans in 1952 in Uganda and the United Republic of Tanzania (Dick et al., 1952; Plourde and Bloch, 2016,). It has recently garnered infamy when WHO in February 2016 declared a "Public Health Emergency of International Concern" after its outbreak in Brazil, South America, Central America, Mexico, and the Caribbean in 2015 (Ventura et al., 2016). In 2015, Zika virus RNA was detected in the amniotic fluid of two pregnant women after detecting the symptom of microcephaly in the fetuses by prenatal ultrasound (Schuler and Faccini, 2016). Recently, a study in Northeastern Brazil has confirmed that vertical transmission of Zika virus is associated with severe fetal brain injury (Mlakar et al., 2016).

Fortunately and successfully, *Flavivirus* vaccines have been developed, and are available for YFV, JEV, and TBEV, for preventing and controlling these viruses. However, to date, no vaccine or virus-specific antivirals are available for WNV, DENV, and Zika virus (DeGroot et al., 2016; Heinz and Stiasny, 2012).

## 2 *FLAVIVIRUS* GENOME AND POLYPROTEIN PROCESSING

*Flavivirus* as the name suggests is derived from the term *flavus*, which means yellow in Latin (yellow fever caused by the prototype species YFV). They are positive-sense single-stranded (ss) RNA viruses. The genome of *Flavivirus* is ~10.5 kb in size. The 5'-terminal region of the viral genome is capped having a cap1-type structure and also contains the conserved dinucleotide sequence 5'-AG-3'. However, unlike cellular messenger RNA (mRNA), the 3'-end of the *Flavivirus* genome lacks a poly-A tail and terminates with the dinucleotide 5'-CU-3' (Brinton et al., 1986; Hahn et al., 1987; Wengler and Wengler, 1981).

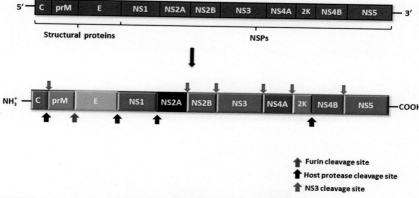

FIGURE 6.1    ***Flavivirus* genome organization and polyprotein processing.** *C*, Capsid; *E*, envelope; *NS*, nonstructural; *NSPs*, nonstructural proteins; *prM*, membrane precursor.

Though, in some strains of TBEV, including the prototype strain Neudoerfl, a poly-A tail has been identified on the 3′ end of the viral genome RNA (Mandl et al., 1991). The genome contains one large open reading frame of ∼10 kb, which encodes for a single polyprotein consisting of three structural proteins and seven nonstructural (NS) proteins (Fig. 6.1) involved in the replication of viral genome (Lindenbach and Rice, 2001). The viral RNA genome, being positive-sense, is translated by the host translation machinery as soon as the viral RNA is released in the host cytoplasm. These enveloped viruses infect the host cells by receptor-mediated endocytosis, and the mild acidic conditions in the lumen of endosomes trigger the fusion of the viral membrane with the endosomal membrane, which leads to the disassembly of nucleocapsid and the release of viral genomic RNA in the host cell cytoplasm. The released viral genomic RNA acts like a host mRNA and is translated into ∼370-kDa polyprotein precursor (Rice et al., 1985).

The order of proteins in the *Flavivirus* polyprotein as deduced from its nucleotide sequence analysis and polyprotein studies is $NH_3^+$-C-prM(M)-E-NS1-NS2A-NS2B-NS3-NS4A-NS4B-NS5-COOH (Bell et al., 1985; Rice et al., 1985, 1986). The translated polyprotein is directed into host endoplasmic reticulum (ER) membrane by signal sequences and cleaved into three structural proteins (capsid, precursor of membrane, and envelope) and seven nonstructural proteins (NSPs), NS1, NS2A, NS2B, NS3, NS4A, NS4B, and NS5, by the host proteases and the viral NS3 protease (NS3pro), which comprises of the N-terminal domain of NS3 (∼180 residues) (Figs. 6.1 and 6.2) (Chambers et al., 1991). NS3pro, a trypsin-like serine protease, in complex with its cofactor, NS2B, proteolytically processes the junctions of capsid-prM, NS2A–NS2B, NS2B–NS3, NS3–NS4A, and NS4B–NS5 (Fig. 6.3) (Chambers et al., 1989, 1991). The peptide bond between NS2B–NS3 is cleaved in *cis* by the heterodimeric complex of NS2B–NS3pro (Falgout et al., 1993).

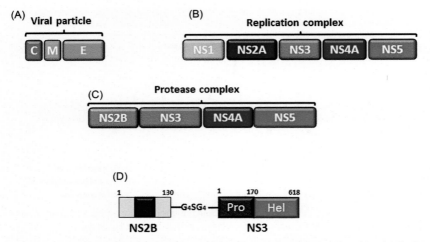

**FIGURE 6.2** **Different NSP complexes formed in host cell after polyprotein processing in the host cell.** (A–C) Various complexes reside in different cellular compartments and carry out assigned functions. (D) The construct used for expression of recombinant protease for biochemical and structural characterization (Assenberg et al., 2009).

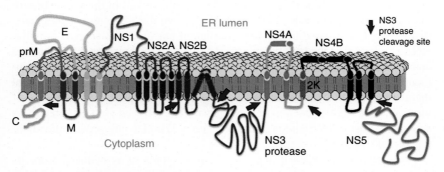

**FIGURE 6.3** **A model of endoplasmic reticulum (ER) membrane, the site for polyprotein processing in *Flavivirus*-infected cells.** *Brown arrows* indicate cleavage sites for NS3 protease within the embedded polyprotein.

The proteolytic processing of the polyprotein precursor also releases the viral polymerase NS5 that further participates in the replication process and transcribes negative-strand RNA using the viral RNA genome as a template, (Egloff et al., 2002). The negative strand in turn is transcribed back into the viral genomic RNA, which is packaged into the nucelocapsid cores and forms progeny virus particles (Filshie and Rehacek, 1968). Most of the *Flavivirus* NSPs play essential and multiple roles during the process of viral genome replication and assembly (Mackenzie et al., 1996; Murray et al., 2008; Smit et al., 2011) (Table 6.1).

**TABLE 6.1** *Flavivirus* Proteins and Their Functional Roles

| Proteins | Molecular weights (kDa) | Cleavage site at N-terminus (for YFV) | Proteases involved | Functions |
|---|---|---|---|---|
| C | ~11 | RR|S | NS2B–NS3 protease | Packaging of the viral RNA during RNA assembly, nucleocapsid formation (Chinmay et al., 2007) |
| E | ~54 | TGG|V | Signalase | Virus attachment to the host cell by cellular receptor, internalization of virion into the host cell via membrane-mediated fusion (Allison et al., 2001) |
| prM | ~26 | SRR|A | Furin | Glycosyated transmembrane protein acts like a chaperone protein during E protein synthesis, forms heterodimer with E proteins in immature virus particles (Zhang et al., 2003) |
| M | ~8 | AYS|A | Signalase | Transmembrane protein, cleaved part of prM (Zhang et al., 2003) |
| NS1 | ~46 | VGA|D | Signalase | RNA replication and immune invasion (Chung et al., 2006; Lindenbach and Rice, 2001; Mackenzie et al., 1996) |
| NS2A | ~22 | VTA|G | Membrane-bound host protease present in ER | Replication complex formation, virus assembly and evasion of host immune response (Kümmerer and Rice, 2002; Leung et al., 2008; Muñoz-Jordán et al., 2003) |
| NS2B | ~14 | RR|S | NS2B–NS3 protease | NS3 protease cofactor (Chambers et al., 1991) |
| NS3 | ~70 | RR|S | NS2B–NS3 protease | Trypsin-like serine protease (Chambers et al., 1989; Wengler et al., 1991), helicase NTPase and RNA triphosphatase activities (Wengler and Wengler, 1991) |
| NS4A | ~16 | RR|G | NS2B–NS3 protease | Replication complex formation and RNA replication (Mackenzie et al., 1998) |
| NS4B | ~28 | VAA|N | Signalase | Replication complex formation and RNA replication (Welsch et al., 2009) |
| NS5 | ~103 | RR|G | NS2B–NS3 protease | RNA-dependent RNA polymerase, methyltransferase (Egloff et al., 2002; Tan et al., 1996) |

M, Membrane; NTPase, nucleoside triphosphatase.

## 3   NS3 PROTEASE AND ITS ASSOCIATION WITH ITS COFACTOR NS2B

NS3 protein is a ~69 kDa protein and the N-terminal domain (DENV residue number 1–169) of the protein exhibits serine protease activity. The carboxy terminus of the protein has nucleoside triphosphatase (NTPase)–dependent RNA helicase activity, which aids in genome replication and in viral RNA synthesis. NS3 protease possesses a classic serine protease catalytic triad (His51, Asp75, and Ser135) (Aleshin et al., 2007; Polgar, 2005). The N-terminal domain of NS3 protease requires NS2B as its cofactor for the protease activity (Chambers et al., 1991). The NS3 protease domain is not very soluble in vitro and tends to form aggregates, suggesting that the NS3 protease is not catalytically active and may not fold properly without its cofactor NS2B. The engineered fusion gene of the NS2B cofactor connected to the NS3 protease domain via a flexible nine–amino acid long glycine-rich linker (Gly4-Ser-Gly4) was soluble, as well as catalytically active in vitro (Leung et al., 2001). The central region of NS2B has a hydrophilic part that acts as a chaperone for its binding with the NS3, by stabilizing the two terminal hydrophobic regions of NS3pro, which are predicted to be involved in membrane association of the NS2B/NS3pro complex. The NS2B–NS3pro complex carries out co- and posttranslational cleavage in the membrane of host ER (Clum et al., 1997; Westaway et al., 1997; Wichapong et al., 2010).

Mutagenesis studies elucidate that the central conserved hydrophilic region of NS2B is vital for the protease activity of NS3 protein and the hydrophobic region is involved in the membrane association of NS2B–NS3pro serine protease complex (Chambers et al., 1993; Droll et al., 2000). The protease activity of NS2B–NS3pro complex is dependent on the association of NS3pro with 47–amino acid (residues 49–95) long stretch of NS2B, which directly interacts with peptide substrate yielding a closed conformation. NS2B forms a β-hairpin that is tangled around NS3pro in its active form. NS2B wraps around the protease complex in a unique belt-like structure (Erbel et al., 2006). Nuclear magnetic resonance (NMR) and molecular dynamic studies have inferred the requirement of NS2B as a cofactor for NS3, which alone was shown to have disordered chymotrypsin-like fold (Gupta et al., 2015).

## 4   CATALYTIC MECHANISM

Virus-specific proteases can be found in nonenveloped ssRNA viruses (picornaviruses) (Matthews et al., 1994), enveloped dsRNA viruses (alphaviruses and flaviviruses) (Aggarwal et al., 2014, 2015), nonenveloped dsDNA viruses (adenoviruses) (Ding et al., 1996), and enveloped dsDNA viruses (herpesviruses) (Qiu et al., 1996). Proteases have been classified on the basis of principal active-site residue in the catalytic site. Most of the viral proteases belong to the class of cysteine or serine proteases. NS3 protease is a trypsin-like serine protease consisting of a signature serine protease catalytic triad comprising of residues

His51, Asp75, and Ser135 (amino acid residue number in DENV NS3pro) (Aleshin et al., 2007; Chambers et al., 1990b; Robin et al., 2009). *Flavivirus* NS2B–NS3pro complex follows a generally accepted mechanism for serine proteases. A serine protease cleaves the peptide bond in two phases: acylation and deacylation (Fig. 6.4). During acylation reaction, the protease activates the amide bond resulting in expulsion of an amine. Deacylation half of the reaction involves the replacement of leaving group with a hydroxyl ion from an activated water molecule, which is otherwise, a weak nucleophile.

The proteolysis reaction begins with the attack of Ser135 on the carbonyl carbon atom of the peptide substrate. This step is assisted by His51, which acts as a general base, to give rise to a tetrahedral intermediate. The resulting oxyanion is stabilized by hydrogen bond formation with the main chain NHs of oxyanion hole residues. The leaving group ($R_2NH_2$) departs after cleavage of scissile bond, giving rise to an acyl enzyme intermediate. The entry of water marks the beginning of the deacylation half of the reaction. Histidine accepts the proton from the incoming water molecule, leading to the formation of the second transition state or tetrahedral intermediate. The resulting protonated histidine is stabilized by the hydrogen bond with Asp75.

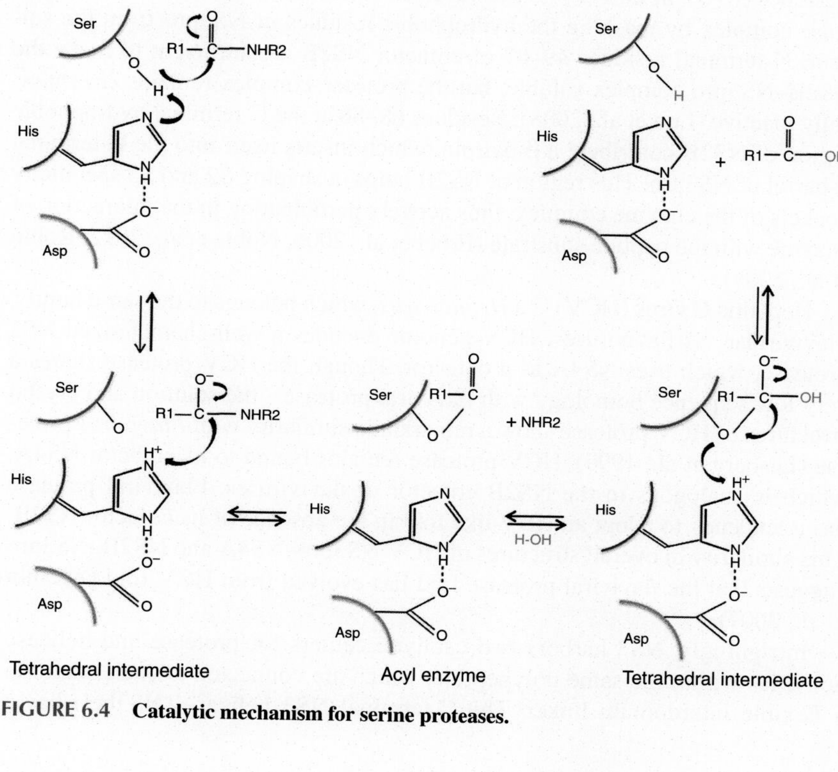

FIGURE 6.4  **Catalytic mechanism for serine proteases.**

The hydroxyl group replaces the leaving group in the peptide substrate resulting in the release of a carboxylic acid product ($R_1COOH$) (Valle and Falgout, 1998).

## 5  STRUCTURE OF NS2B–NS3PRO

For the development of NS2B–NS3pro complex inhibitors, determination of the three-dimensional structures of *Flavivirus* protease and its complexes with the potent inhibitors is required. Atleast 18 crystal structures of NS2B–NS3pro have been reported, providing insights into the enzymatic and inhibitor-binding activities of the protease (Table 6.2). The N-terminal domain of NS3 protease has been solved in the presence of the cofactor domain of NS2B for DENV, WNV, and Murray Valley encephalitis virus (MVEV) (Robin et al., 2009). A stretch of 35–48 amino acids present in the hydrophilic region of NS2B domain is involved in noncovalent interactions with the NS3 protease, which aids in the folding of the latter (Erbel et al., 2006). Crystal structures of NS2B–NS3pro of WNV and DENV (Fig. 6.5) exhibit chymotrypsin-like fold, which includes two β-barrels. Both the barrels are composed of six β-strands each, and the catalytic triad is located between the two β-barrels (Erbel et al., 2006; Luo et al., 2008). One β-strand in the N-terminal barrel is contributed by NS2B (residues 51–57 in DENV) (Erbel et al., 2006). This strand stabilizes the protease complex by masking the hydrophobic residues of NS3pro from the solvent. N-terminal residues 49–67 of cofactor NS2B are sufficient to make the NS2B–NS3pro complex soluble, but the protease complex remains enzymatically inactive (Luo et al., 2008). Residues 68–88 at the C-terminal hydrophobic region of NS2B contribute a β-hairpin, which inserts itself into the C-terminal β-barrel of NS3pro. This region of NS2B helps in shaping S2 and S3 specificity pockets of the enzyme complex, thus actively participating in the interaction of enzyme with the peptide substrate (Erbel et al., 2006; Noble et al., 2012; Robin et al., 2009).

Hepatitis C virus (HCV) is a *Hepacivirus*, which belongs to the same family, Flaviviridae, as flaviviruses. HCV genome encodes a well-characterized NS3 protease, which uses NS4A as a cofactor. Though the HCV protease shares a very low sequence homology with flaviviral proteases, the solution and crystal structures of HCV protease show a remarkable similarity with *Flavivirus* protease (Lesburg et al., 1999). HCV protease remains bound to its cofactor NS4A, which is analogous to the NS2B cofactor of flaviviruses. Flaviviral protease has a tendency to adopt an HCV-like fold in the absence of its cofactor NS2B. This similarity in overall structures of HCV NS3pro–NS4A and NS2B–NS3pro suggests that the flaviviral protease fold has evolved from HCV fold (Aleshin et al., 2007).

Intriguingly, NS3 harbors two catalytic centers for protease and helicase activities within the same polypeptide, which are connected to each other via a flexible interdomain linker. The C-terminal NS3 helicase (NS3hel) has a

**TABLE 6.2 PDB-Deposited Crystal Structures of *Flavivirus* Protease**

| Viral proteases | PDB entries | Ligands/ inhibitors | Resolution (Å) | References |
|---|---|---|---|---|
| JEV | 4R8T | Uncomplexed | 2.13 | Weinert et al. (2015) |
| Zika | 5LC0 | Boronate inhibitor | 2.7 | Lei et al. (2016) |
| DENV2 NS2B–NS3pro, A125C mutant | 4M9F (pH 8.5) | Uncomplexed | 2.7 | Yildiz et al. (2013) |
| | 4M9I (pH 5.5) | | 2.4 | |
| DENV2 NS2B–NS3pro | 4M9M (pH 8.5) | Uncomplexed | 1.5 | Yildiz et al. (2013) |
| | 4M9K (pH 5.5) | | 1.4 | |
| DENV2 NS2B–NS3pro | 4M9T | DTNB | 1.7 | Yildiz et al. (2013) |
| DENV3 NS2B–NS3pro | 3U1I | Covalently bound peptide inhibitor | 2.3 | Noble et al. (2012) |
| DENV3 NS2B–NS3pro | 3U1J | Aprotinin | 1.8 | Noble et al. (2012) |
| DENV4 NS2B–NS3 | 2WHX | Uncomplexed | 2.2 | Luo et al. (2010) |
| DENV4 NS2B–NS3, E173GP174 insertion | 2WZQ | Uncomplexed | 2.8 | Luo et al. (2010) |
| DENV1 NS2B–NS3 | 3L6P | Uncomplexed | 2.2 | Chandramouli et al. (2010) |
| DENV1 NS2B–NS3, S135A mutant | 3LKW | Uncomplexed | 2.0 | Chandramouli et al. (2010) |
| MVEV | 2WV9 | Uncomplexed | 2.75 | Assenberg et al. (2009) |
| DENV4 NS2B–NS3 | 2VBC | Uncomplexed | 3.1 | Luo et al. (2008) |
| DENV2 NS2B–NS3pro | 2FOM | Uncomplexed | 1.5 | Erbel et al. (2006) |
| WNV NS2B–NS3pro | 2FP7 | Covalently bound peptide inhibitor | 1.6 | Erbel et al. (2006) |
| WNV NS2B–NS3pro, H51A mutant | 2GGV | Uncomplexed | 1.8 | Aleshin et al. (2007) |
| WNV NS2B–NS3pro | 2IJO | Aprotinin | 2.3 | Aleshin et al. (2007) |
| WNV NS2B–NS3pro | 3E90 | Covalently bound peptide inhibitor | 2.4 | Robin et al. (2009) |

DENV, Dengue fever virus; DNTP, dithio-*bis*-nitrobenzoic acid; JEV, Japanese encephalitis virus; MVEV, Murray Valley encephalitis virus; WNV, West Nile virus.

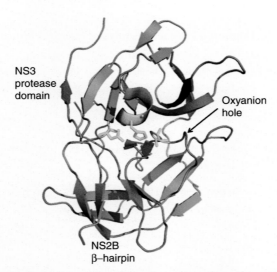

**FIGURE 6.5    Crystal structure of ligand-bound DENV-3 NS2B–NS3pro (Noble et al., 2012; PDB ID: 3U1I).** Peptide inhibitor has been omitted for clarity. β-Hairpin contributed by NS2B is shown in *orange*, while NS3 is shown in *cyan*. Catalytic triad (H51, D75, and S135) residues are shown as *yellow sticks*. Oxyanion hole loop in the protein has been shown in *magenta*.

Rec A–like NTPase catalytic center that utilizes the energy released from the hydrolysis of triphosphate moieties of nucleotides for its RNA unwinding activity (Chambers et al., 1990a; Luo et al., 2008). It is still not clear whether or not these two domains within the same polypeptide play a significant role in regulation of its biological activities. However, there is an evidence, which shows that NS3hel and NS5 (methyltransferase and RNA-dependent RNA polymerase) interact and that their individual activities are coordinated during viral genome replication (Kapoor et al., 1995; Tay et al., 2015; Yon et al., 2005).

Analysis of crystal structures of NS3pro in the different serotypes of DENV shows a hydrophobic loop (Gly29, Leu30, Phe31, and Gly32), which points toward the membrane during anchorage of protease to ER membrane (Aleshin et al., 2007). Residues that are evolutionary conserved among all serine proteases include Gly133, Gly136, Gly148, Gly149, Leu149, and Gly153. Apparently, these residues play an important role in maintaining the overall three-dimensional structure of the substrate-binding pocket of enzyme. Phylogenetic analysis suggests that Phe116, Ile123, Tyr161, and Val162 are also important for catalysis, and can be targeted for enzyme inhibition (Luo et al., 2008). These residues form a hydrophobic core that accommodates NS2B β-hairpin during the bound state of the enzyme. Crystal structure of

DENV-1 protease shows a metal-coordinated site consisting of His67 and His72 (of NS2B) and Glu-94 (of NS3), which is absent in any other flaviviral protease, as these two histidines are not conserved in any other *Flavivirus* NS2B. This unique site may play a role in stabilizing the binding of NS2B with NS3pro.

## 5.1    Catalytic Triad

In *Flavivirus* protease the active site is located between N- and C- terminal lobes of the NS2B–NS3pro. Upon binding to substrate peptide the catalytic site residues orient themselves to form a productive conformation for catalysis. The protease exhibits a trypsin-like fold, having two β-barrels where the Asp-His-Ser catalytic triad is confined in a cleft in between the two barrels (Assenberg et al., 2009). Residues making up the catalytic triad, His51, Asp75, and Ser135, appear distant in the sequence, but come close to each other in the three-dimensional structure of the enzyme (Erbel et al., 2006). The catalytic triad is involved in extensive hydrogen bond network in the substrate-bound form. In the active state, hydroxyl group of serine forms hydrogen bond with His–imidazole, which in turn interacts with the carboxyl group of aspartate (Fig. 6.6). In both, WNV and DENV protease structures aldehyde carbon of peptide inhibitors are bound to the hydroxyl group of catalytic Ser135 covalently, while aprotinin binds noncovalently to the same residue (Aleshin et al., 2007; Noble et al., 2012).

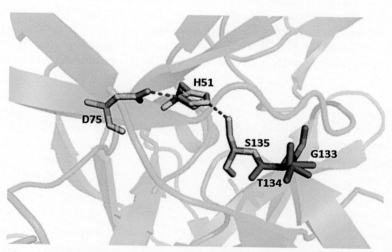

FIGURE 6.6    Detailed view of catalytic triad (shown as *yellow sticks*) and residues contributing to oxyanion hole (shown as *magenta sticks*). *Dashed lines* represent potential hydrogen bonds.

## 5.2 Oxyanion Hole

The oxyanion hole in flaviviral proteases is formed by NH groups of Gly133, Thr134, and Ser135. The oxyanion hole stabilizes the negative charge on the scissile peptide carbonyl oxygen formed during transition state (Fig. 6.6). In DENV-3, the oxyanion hole remains in an active conformation in the crystal structures of DENV NS3pro in complex with benzoyl-norleucine-Lys-Arg-Arg-H (Bz-nKRR-H) and aprotinin (Noble et al., 2012). However, in WNV protease structure the bond between Thr132 and Gly133 is flipped in substrate-free and Bz-nKRR-H–bound forms (Aleshin et al., 2007). Thus enzyme adopts a nonproductive conformation in the absence of a substrate. In case of WNV, it is possible that a productive form of enzyme is acquired only upon binding to an authentic substrate. This "induced fit" mechanism is energetically favorable for the enzyme in the absence of a substrate. In the DENV-3 protease–Bz-nKRR-H structure, a water molecule and sulfate group is also present in the oxyanion hole (Noble et al., 2012).

## 6 SUBSTRATE SPECIFICITY OF NS2B–NS3 PROTEASE

The catalytic triad of the protease is highly specific for substrates with dibasic P1 and P2 components and a small amino acid at P1′ (P, position). This dibasic recognition sequence in NS3 cleavage sites is highly conserved in the flavivirus genus (Table 6.3); although different flaviviruses prefer either Lys or Arg at P2. NS3 proteases of DENV and YFV recognize Arg at P2, and WNV protease recognizes Lys at P2 (Chappell et al., 2006). The crystal structure of *Flavivirus* protease NS2B–NS3pro–bound inhibitor complex predicts five well-formed substrate-binding pockets namely S1′, S1, S2, S3, and S4, incorporating P1′–P4 residues of substrate. S1′ pocket is lined by the catalytic residue His51 and a conserved Gly37 in WNV. This pocket is only large enough to accommodate small residues, such as glycine or serine. The S1 pocket is formed by nine

**TABLE 6.3 Specific Amino Acid Residues of Substrates Recognizable at Their Respective Positions by NS2B–NS3pro Enzyme**

| Viruses | P1′ | P1 | P2 | P3 | P4 | References |
|---------|-----|-----|-----|-----|-----|------------|
| DENV | G/S/A | R/K | R/Q | G/K/T/ | K/Q/ | Junaid et al. (2012) |
| WNV | G | R | K | T | Y | Chappell et al. (2006) |
| JEV | G | R | K | T | T | Jan et al. (1995) |
| YFV | S | R | R | A | G | Chambers et al. (1991) |
| Zika | S | R | K | G | T | — |
| MVEV | G | R | K | T | Y | — |

residues where P1 interacts via salt bridge with side chain of Asp129 and H-bond with the backbone carbonyl oxygen of the Tyr130 residue of NS3. The S1 pocket majorly comprises the bulky aromatic ring of residues: Tyr150 and Tyr161 of NS3pro. The S2 pocket of the WNV protease is lined by eight residues and is influenced by a negative electrostatic potential from backbone carbonyl oxygen atoms of Asp82 and Gly83 of NS2B, and Val71 and Lys72 of NS3pro. Acidic side chains of Asp82 of NS2B and Asp75 of the catalytic triad also contribute to this well-formed pocket. S3 and S4 pockets are not well defined in JEV protease and comprise of three and two uncharged amino acids, respectively, which leaves the substrate P4 and P3 residues being largely solvent exposed (Aleshin et al., 2007).

# 7 STRUCTURE OF NS2B–NS3PRO IN COMPLEX WITH INHIBITORS

Aprotinin/bovine pancreatic trypsin inhibitor–bound structures of WNV are similar to a Michaelis–Menten complex. Aprotinin occupies all major specificity pockets (S2–S2′) and mimics an enzyme–substrate precleavage complex. Aprotinin residues PCKARII (residues 13–19) are involved in antiparallel β-sheet interaction with E2B and E1 strands of WNV protease, while a second loop from residue 36 GGCR 39 interacts with another site of protease. A major difference between aprotinin-bound structures of WNV protease and DENV-3 protease lies in the interaction of NS2B cofactor with the aprotinin. Gln86 residue of WNV NS2B actively interacts with Arg39 of aprotinin (Aleshin et al., 2007), while in the case of DENV-3 protease–aprotinin–bound structure, NS2B adopts a different conformation (Noble et al., 2012) (Fig. 6.7). The two conformations of NS2B cofactor adopted in the crystal structures of WNV NS3 protease in its free and aprotinin-bound form might be a virus-specific mechanism to turn off the protease activity when it is not required (Fig. 6.8). Additionally, NS2B in DENV protease does not interact with aprotinin, still it acquires a closed belt structure conformation around DENV NS3pro in the peptide inhibitor–bound structure.

Both WNV and DENV proteases have been cocrystallized with Bz-nKRR-H, a peptide-based aldehyde inhibitor (Erbel et al., 2006; Noble et al., 2012). In the DENV NS3pro complex structure, Ser135 residue is covalently linked to the aldehyde carbon of Bz-nKRR-H and the resulting Bz-nKRR-H hydroxyl group interacts with His51 via a hydrogen bond. The P1 residue, Arg of the peptide inhibitor interacts via a salt bridge with Asp129 of NS3 and is involved in a π–cation interaction with Tyr161 residue, while P2 and P3 residues interact with Gly82 and Met84 of NS2B (Fig. 6.9). Crystal structure of WNV protease with another tripeptide inhibitor 2-naphthoyl-Lys-Lys-Arg-aldehyde (Naph-KKR-H) shows additional interactions of the peptide with another catalytic triad residue Asp75, and a water-mediated interaction with Asp129 and thus, it has been suggested that Naph-KKR-H could be a potent inhibitor (Robin et al., 2009).

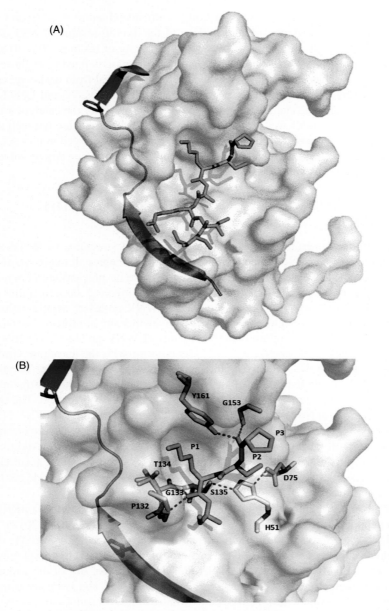

FIGURE 6.7   (A) Surface view of DENV-3 NS2B–NS3pro (Noble et al., 2012; PDB ID: 3UIJ) surface with selected aprotinin residues (shown as *green sticks*) 13 PCKARII 19 that interacts with the protease. The cartoon shows the invading NS2B β-hairpin. (B) Detailed interaction of P1, P2, and P3 residues of aprotinin (shown as *green sticks*) with selected residues of protease (shown as *gray sticks*). Potential hydrogen bonds are indicated by *dashed lines*. Only P1, P2, and P3 residues are shown for clarity. NS2B β-hairpin (shown in cartoon) does not participate in binding of aprotinin to DENV-3 protease.

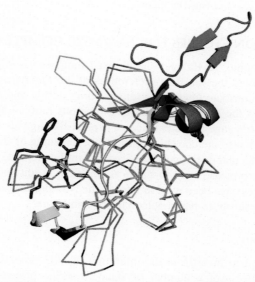

FIGURE 6.8 Comparison of complexed (Erbel et al., 2006; PDB entry 2FP7) and uncomplexed (H51A mutant, Aleshin et al., 2007; PDB ID: 2GGV) crystal structures of WNV NS2B–NS3pro. NS2B is shown in a cartoon representation, with the rest as *ribbon*. NS3pro region of complexed state is shown as a *green ribbon*, NS3pro in uncomplexed state is shown as a *cyan ribbon*, and ligand is shown as *blue sticks*. Remarkable plasticity can be seen in NS2B regions of bound (shown in *yellow*) and substrate-free *(magenta)* protease complex.

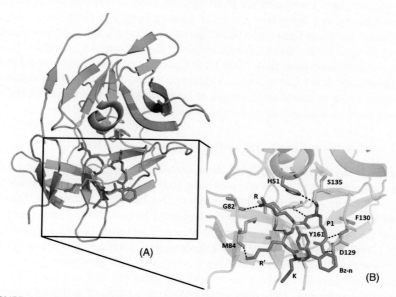

FIGURE 6.9 (A) Crystal structure of NS2B–NS3pro bound to peptide inhibitor Bz-nKRR-H (Noble et al., 2012; PDB ID: 3U1I). NS2B–NS3pro complex is shown in cartoon, while the peptide inhibitor is shown as *magenta sticks*. (B) Molecular interaction of peptide (shown as *magenta sticks*) with selected residues (shown as *sticks*) of DENV-3 NS2B–NS3pro. Potential hydrogen bonds are indicated by *dashed lines*.

## 8 INHIBITION STUDIES FOR NS3/NS2B COMPLEX

The crystallographic, NMR, and site-directed mutagenesis studies have analyzed and predicted the substrate preferences of the *Flavivirus* NS3 protease. These studies contribute toward the rational drug design and development of antivirals against the NS3pro active site. Numerous rational approaches have been employed to gauge the interaction of peptide-like ligands with the protein active site, which also includes high-throughput screening of chemical libraries and in silico molecular docking for successful identification and development of NS3pro-specific protease inhibitors (Yin et al., 2006). Additionally, fluorogenic peptide substrates have been used in vitro in the NS3 protease assays to tap the effect of certain substrate-based potential NS3 inhibitors.

### 8.1 Aprotinin

The first attempt to investigate the inhibition of NS3 protease was made by studying the effects of a number of standard serine protease inhibitors on NS2B–NS3pro complex of DENV, which failed except for "aprotinin" (Leung et al., 2001). In comparison to other peptidic inhibitors, aprotinin was shown to occupy more specificity pockets (S2–S2'). This basic pancreatic trypsin inhibitor was shown to inhibit the protease activity of NS3 at submicromolar concentrations by denying the substrate access to the protease active site by enveloping the enzyme, but in vivo efficacy of aprotinin in reduction of *Flavivirus* has not been reported yet. Aprotinin or bovine pancreatic trypsin inhibitor, an antifibrinolytic molecule used under the trade name Trasylol, was administered by injection as a medication to reduce bleeding during complex surgeries, but it was withdrawn worldwide in 2008 after reports suggested health hazard risks or sometimes death due to its use (Mangano et al., 2007). Thus the cause of its cytotoxicity and side effects needs to be addressed and investigated, before aprotinin or its analogs can be developed as potent DENV inhibitors.

### 8.2 Tripeptide Aldehydes and Cyclopentapeptides

In *Flavivirus*, the NS2B–NS3pro serine protease has substrate preferences for two basic residues to the N-terminal side, indicating a signal sequence for the cleavage of the *Flavivirus* polyprotein by the NS2B–NS3pro complex. This dibasic recognition sequence is the distinguished feature of the *Flavivirus* proteases and is unique in comparison to mammalian serine proteases (Shiryaev et al., 2007). This feature may help in designing of novel cationic potent inhibitors selective for *Flavivirus* proteases over host cellular proteases. As most of the cationic peptides possess antibacterial activity (as they are permeable to bacterial membranes) (Stoermer et al., 2008), this mode of action of cationic peptides can also be exploited for the lipid membrane of enveloped viruses, such as the flaviviruses. Studies based on molecular docking have identified that short cationic tripeptide that inhibits protease, as well as the virus, and might block the active site of the NS3pro (Schüller et al., 2011). The potent tripeptide (inhibition constant, $K_i$ = 9 nM) having N-and C-terminal modifications with a phenacetyl-cap and an aldehyde, respectively, displayed antiviral activity, having an $EC_{50}$ (concentration that gives 50% viral reduction) of 1.6 µM (Stoermer et al., 2008). Additionally some peptidic α-keto amide inhibitors have also been studied, and a cyclopeptide that binds to the active site of the NS2B–NS3pro complex of DENV-2, WNV, and YFV was shown to inhibit NS3 protease having $K_i$ in the nanomolar range (Brecher et al., 2013; Tambunan and Alamudi, 2010).

## 8.3 Plectasin

Plectasin, an antimicrobial cationic peptide, is the first fungus-derived defensin with therapeutic potential especially against Gram-positive bacteria (Rothan et al., 2013). Plectasin is a 40–amino acid long protein (MGFGCNGPWDED-DMQCHNHCKSIKGYKGGYCAK GGFVCKCY) which contains 6 cysteine residues that form 3 disulfides bridges in the folded protein molecule between Cys1 and Cys5, Cys2 and Cys4, and Cys3 and Cys6, thus stabilizing its α-helix–β-structural motif (Mandal et al., 2009). Plectasin was shown to have low cytotoxicity against mammalian cells, and can be exploited as one of the best inhibitory molecule against NS3 protease. The NS2B–NS3pro complex requires the interaction between negatively charged amino acids in NS2B cofactor and positively charged amino acids in NS3pro (Mygind et al., 2005). Plectasin being a cationic peptide has the potential to inhibit the activity of DENV protease by disrupting the binding between enzyme subunits. The recombinant plectasin inhibited DENV NS2B–NS3pro at $K_i$ value of approximately 5.0 µM. It is more potent than some of the other low–molecular weight inhibitors that showed a range of inhibitory activities against DENV NS2B–NS3pro complex (Rothan et al., 2013).

## 8.4 Guanidinylated 2,5-Dideoxystreptamine

Guanidinylated 2,5-dideoxystreptamine derivatives have been examined as competitive inhibitors for NS3 protease. They were shown to competitively inhibit the NS3 protease activity of DENV (1–4 serotypes), as well as of NS3 WNV protease, with $IC_{50}$ values in the range of 1–70 µM. The cytotoxicity level of the compound was detected to be low against vero cells, but the antiviral activity was also found to be low against DENV in infected vero cells. Inhibition kinetic studies of this class of compounds have shown a competitive behavior with a substrate, leading to the inference that the compounds only bind within the active site of the protease. Thus the multiple guanidine groups (G-groups) can be assumed to be an isosteric replacement for the nonprime-side arginine and lysine residues required for optimal substrate cleavage (Cregar et al., 2011).

The guanidinylated 2,5-dideoxystreptamine class of inhibitors has the potential of blocking NS3 protease activity of WNV and all four DENV strains.

Due to the lack of structural activity relationship studies for these compounds, it can only be speculated that the vital force for binding of these molecules in the protease active site is mainly electrostatic in nature. The specific binding interactions of the guanidinylated 2,5-dideoxystreptamine with the nonprime portion of the substrate-binding pocket of the protease that is highly negatively charged is perhaps because of the cationic sites of the G-group in the molecules. The most lipophilic (cLogP = −0.06) compound from this class was chosen for DENV-2 antiviral testing against vero cells at a maximum concentration of 100 μM. The compound showed no signs of cytotoxicity (Cregar et al., 2011).

## 8.5 Protegrins

Protegrins are a family of cationic peptides rich in arginine and cysteine (Kokryakov et al., 1993). Antiviral activity of protegrin-1 (PG-1) with a sequence of RGGRLCYCRRRFCVCVGR has been reported for DENV-infected Rhesus monkey kidney cells (MK2). However, the peptide PG-1 was found to be toxic for the MK2 cell line at concentrations greater than 12.5 μM. Additionally, it has also been reported that PG-1 at 25-μM concentration is also toxic to 293A human embryonic kidney cells. For human red blood cells, more than 50% cell lysis was observed at a concentration of 40 μM. However, the optimum concentration at which PG-1 is nontoxic to cells showed significant reduction in DENV viral replication at 24, 48, and 72 h after viral infection (Rothan et al., 2012). Fluorogenic peptide substrate *t*-butyloxycarbonylglycyl-L-arginyl-L-arginyl-L-4-methylcoumaryl-7-amide (Boc-Gly-Arg-Arg-MCA) has been used for assessing activities of potential protease inhibitor against DENV NS3pro. The kinetic assay that was performed to examine the inhibitory activity of PG-1 on the protease activity of NS2B–NS3 complex showed the inhibition potential of PG-1; by increasing the concentration of the inhibitor, around 95.7% inhibition was observed at 40-μM concentration. The study determined that NS2B–NS3pro activity is competitively inhibited by PG-1, with a inhibition constant value ($K_i$) of 5.85 μM (Rothan et al., 2012).

## 8.6 Chalcone

Chalcone is an aromatic ketone and an enone, and *Boesenbergia rotunda*, a common spice belonging to a member of the ginger family (Zingiberaceae), is the prime source of cyclohexenyl chalcone derivatives, 4-hydroxypanduratin A, and panduratin A. These compounds have shown promising inhibitory activities toward DENV NS3 protease with the $K_i$ values in the range of 21–25 μM (Kiat et al., 2006). The inhibition studies were performed by measuring the *trans*-cleavage activity of the protease using a fluorogenic peptide substrate Ac-Thr-Arg-Arg-MCA. The six compounds: pinostrobin, alpinetin, pinocembrin, cardamonin, panduratin A, and 4-hydroxypanduratin A that were extracted from *B. rotunda* showed inhibitory activity against DENV protease. These six extracted compounds having anti-Dengue activity were identified by NMR and GCMS.

# 9   CONCLUSIONS

In this chapter we have discussed various structural and functional aspects of flaviviral proteases, and strategies being undertaken for the development of compounds targeting the virus-specific protease. Till now the tarmac laid for the identification of potent inhibitors for the NS3 protease mostly aimed at the active site. Development of active-site inhibitors against HCV failed because it has a similar shallow active site and is hydrophilic in nature (solvent exposed). The fact that the active site is flat and charged makes the search for small-molecule inhibitors against NS3 protease cumbersome. Thus the inhibitors have lesser interaction with the enzyme, compromising the accessibility and binding efficiency. Solved X-ray and NMR structures of NS2B–NS3pro complex, showing NS2B as the essential cofactor for NS3 protease, provide alternative sites for the development and identifications of compounds selectively targeting the allosteric site of NS2B.

The European Union Viral Enzymes Involved in Replication (VIZIER) project, launched in 2004, targets the most conserved viral enzymes (of replication machinery), including flaviviral NS3 and NS5. The VIZIER project aims at bridging the gap between structural biology and virology laboratory expertise. It involves crystallizing enzymes, defining functional domains, and identifying small-molecule inhibitors through virtual docking. Within VIZIER consortium, validity of potent *Flavivirus* inhibitors is being assessed by studying their effects on the binding site of the cofactor NS2B (Bollati et al., 2010). In a recent study, a novel approach of investigating potential allosteric sites has been developed (Yildiz et al., 2013). Identified allosteric sites have been analyzed for inhibition by potential inhibitors from a small-molecule library.

Efficacy of potential inhibitory compounds is constrained by a number of factors, including cytotoxicity, cell permeability, and serum stability. Despite the availability of detailed information regarding the structures and binding mechanisms of proteases from several members of *Flavivirus* genus, identification of druggable sites and utilizing them for inhibition of this two-component protease is still challenging. As structures of proteases from different members of *Flavivirus* family share remarkable similarity, studies can be undertaken for designing antiflaviviral compounds, which have broad-spectrum effects against all DENV serotypes, as well as other *Flavivirus* members, such as YFV, WNV, and JEV.

# REFERENCES

Aggarwal, M., Dhindwal, S., Kumar, P., Kuhn, R.J., Tomar, S., 2014. *trans*-Protease activity and structural insights into the active form of the alphavirus capsid protease. J. Virol. 88, 12242–12253.

Aggarwal, M., Sharma, R., Kumar, P., Parida, M., Tomar, S., 2015. Kinetic characterization of trans-proteolytic activity of Chikungunya virus capsid protease and development of a FRET-based HTS assay. Sci. Rep. 5, 14753.

Aleshin, A.E., Shiryaev, S.A., Strongin, A.Y., Liddington, R.C., 2007. Structural evidence for regulation and specificity of flaviviral proteases and evolution of the Flaviviridae fold. Protein Sci. 16, 795–806.

Allison, S.L., Schalich, J., Stiasny, K., Mandl, C.W., Heinz, F.X., 2001. Mutational evidence for an internal fusion peptide in flavivirus envelope protein E. J. Virol. 75, 4268–4275.

Assenberg, R., Mastrangelo, E., Walter, T.S., Verma, A., Milani, M., Owens, R.J., Stuart, D.I., Grimes, J.M., Mancini, E.J., 2009. Crystal structure of a novel conformational state of the *Flavivirus* NS3 protein: implications for polyprotein processing and viral replication. J. Virol. 83, 12895–12906.

Bell, J.R., Kinney, R.M., Trent, D.W., Lenches, E.M., Dalgarno, L., Strauss, J.H., 1985. Amino-terminal amino acid sequences of structural proteins of three flaviviruses. Virology 143, 224–229.

Bhatt, S., Gething, P.W., Brady, O.J., Messina, J.P., Farlow, A.W., Moyes, C.L., Drake, J.M., Brownstein, J.S., Hoen, A.G., Sankoh, O., Myers, M.F., 2013. The global distribution and burden of dengue. Nature 496, 504–507.

Billoir, F., de Chesse, R., Tolou, H., de Micco, P., Gould, E.A., de Lamballerie, X., 2000. Phylogeny of the genus *Flavivirus* using complete coding sequences of arthropod-borne viruses and viruses with no known vector. J. Gen. Virol. 81, 781–790.

Bollati, M., Alvarez, K., Assenberg, R., Baronti, C., Canard, B., Cook, S., Coutard, B., Decroly, E., de Lamballerie, X., Gould, E.A., Grard, G., 2010. Structure and functionality in *Flavivirus* NS-proteins: perspectives for drug design. Antiviral Res. 87, 125–148.

Brecher, M., Zhang, J., Li, H., 2013. The *Flavivirus* protease as a target for drug discovery. Virol. Sin. 28, 326–336.

Brinton, M.A., Fernandez, A.V., Dispoto, J.H., 1986. The 3′-nucleotides of *Flavivirus* genomic RNA form a conserved secondary structure. Virology 153 (1), 113–121.

Chambers, T.J., Grakoui, A.R.A.S.H., Rice, C.M., 1991. Processing of the yellow fever virus nonstructural polyprotein: a catalytically active NS3 proteinase domain and NS2B are required for cleavages at dibasic sites. J. Virol. 65, 6042–6050.

Chambers, T.J., Hahn, C.S., Galler, R., Rice, C.M., 1990a. *Flavivirus* genome organization, expression, and replication. Ann. Rev. Microbiol. 44, 649–888.

Chambers, T.J., McCourt, D.W., Rice, C.M., 1989. Yellow fever virus proteins NS2A, NS213, and NS4B: identification and partial N-terminal amino acid sequence analysis. Virology 169, 100–109.

Chambers, T.J., Nestorowicz, A., Amberg, S.M., Rice, C.M., 1993. Mutagenesis of the yellow fever virus NS2B protein: effects on proteolytic processing, NS2B-NS3 complex formation, and viral replication. J. Virol. 67, 6797–6807.

Chambers, T.J., Tsai, T.F., Pervikov, Y., Monath, T.P., 1997. Vaccine development against dengue and Japanese encephalitis: report of a World Health Organization meeting. Vaccine 15 (14), 1494–1502.

Chambers, T.J., Weir, R.C., Grakoui, A., McCourt, D.W., Bazan, J.F., Fletterick, R.J., Rice, C.M., 1990b. Evidence that the N-terminal domain of nonstructural protein NS3 from yellow fever virus is a serine protease responsible for site-specific cleavages in the viral polyprotein. Proc. Natl. Acad. Sci. USA 87, 8898–8902.

Chancey, C., Grinev, A., Volkova, E., Rios, M., 2015. The global ecology and epidemiology of West Nile virus. BioMed Res. Int., 376230.

Chandramouli, S., Joseph, J.S., Daudenarde, S., Gatchalian, J., Cornillez-Ty, C., Kuhn, P., 2010. Serotype-specific structural differences in the protease-cofactor complexes of the dengue virus family. J. Virol. 84 (6), 3059–3067.

Chappell, K.J., Stoermer, M.J., Fairlie, D.P., Young, P.R., 2006. Insights to substrate binding and processing by West Nile Virus NS3 protease through combined modeling, protease mutagenesis, and kinetic studies. J. Biol. Chem. 281 (50), 38448–38458.

Chinmay, G.P., Christopher, T.J., Yu-hsuan, C., et al., 2007. Functional requirements of the yellow fever virus capsid protein. J. Virol. 81 (12), 6471–6481.

Chung, K.M., Liszewski, M.K., Nybakken, G., Davis, A.E., 2006. West Nile virus nonstructural protein NS1 inhibits complement activation by binding the regulatory protein factor H. Proc. Natl. Acad. Sci. USA 103, 19111–19116.

Clum, S., Ebner, K.E., Padmanabhan, R., 1997. Cotranslational membrane insertion of the serine proteinase precursor NS2B-NS3 (Pro) of dengue virus type 2 is required for efficient in vitro processing and is mediated through the hydrophobic regions of NS2B. J. Biol. Chem. 272, 30715–30723.

Cregar, Hernandez, L., Jiao, G.S., Johnson, A.T., Lehrer, A.T., Wong, T.A., Margosiak, S.A., 2011. Small molecule pan-Dengue and West Nile virus NS3 protease inhibitors. Antiviral Chem. Chemother. 21, 209–217.

Davis, L.E., DeBiasi, R., Goade, D.E., Haaland, K.Y., Harrington, J.A., Harnar, J.B., Pergam, S.A., King, M.K., DeMasters, B.K., Tyler, K.L., 2006. West Nile virus neuroinvasive disease. Ann. Neurol. 60, 286–300.

DeBiasi, Tyler, DeBiasi, R.L., Tyler, K.L., 2006. West Nile virus meningo encephalitis. Nat. Clin. Pract. Neurol. 2, 264–275.

DeGroot, A.S., Moise, L., Olive, D., Einck, L., Martin, W., 2016. Agility in adversity vaccines on demand. Exp. Rev. Vaccines 15, 1087–1091.

Dick, G.W., Kitchen, S.F., Haddow, A.J., 1952. Zika virus (I). Isolation and serological specificity. Trans. R. Soc. Trop. Med. Hyg. 46, 509–520.

Ding, J., McGrath, W.J., Sweet, R.M., Mangel, W.F., 1996. Crystal structure of the human adenovirus proteinase with its 11 amino acid cofactor. EMBO J. 15, 1778.

Droll, D.A., Murthy, H.K., Chambers, T.J., 2000. Yellow fever virus NS2B–NS3 protease: charged-to-alanine mutagenesis and deletion analysis define regions important for protease complex formation and function. Virology 275, 335–347.

Egloff, M.P., Benarroch, D., Selisko, B., Romette, J.L., Canard, B., 2002. An RNA cap (nucleoside-2′-O-)-methyltransferase in the *Flavivirus* RNA polymerase NS5: crystal structure and functional characterization. EMBO J. 21, 2757–2768.

Erbel, P., Schiering, N., D'Arcy, A., Renatus, M., Kroemer, M., Lim, S.P., Yin, Z., Keller, T.H., Vasudevan, S.G., Hommel, U., 2006. Structural basis for the activation of flaviviral NS3 proteases from Dengue and West Nile virus. Nat. Struct. Mol. Biol. 13, 372–373.

Falgout, B., Miller, R.H., Lai, C.J., 1993. Deletion analysis of Dengue virus type 4 nonstructural protein NS2B: identification of a domain required for NS2B-NS3 protease activity. J. Virol. 67, 2034–2042.

Filshie, B., Rehacek, J., 1968. Studies of the morphology of Murray Valley encephalitis and Japanese encephalitis Viruses growing in cultured mosquito cells. Virology 34, 435–443.

Gupta, G., Lim, L., Song, J., 2015. NMR and MD studies reveal that the isolated Dengue NS3 Protease is an intrinsically disordered chymotrypsin fold which absolutely requests NS2B for correct folding and functional dynamics. PloS One 10, e0134823.

Guzman, M.G., Kouri, G., 2008. Dengue haemorrhagic fever integral hypothesis: confirming observations 1987-2007. Trans. R. Soc. Trop. Med. Hyg. 102, 522–523.

Hahn, C.S., Hahn, Y.S., Rice, C.M., Lee, E., Dalgarno, L., Strauss, E.G., Strauss, J.H., 1987. Conserved elements in the 3′ untranslated region of *Flavivirus* RNAs and potential cyclization sequences. J. Mol. Biol. 198, 33–41.

Halstead, S.B., 1988. Pathogenesis of dengue: challenges to molecular biology. Science 239, 476.

Heinz, F.X., Stiasny, K., 2012. Flaviviruses and *Flavivirus* vaccines. Vaccine 30, 4301–4306.

Jan, L.R., Yang, C.S., Trent, D.W., et al., 1995. Processing of Japanese encephalitis virus non-structural proteins: NS2B-NS3 complex and heterologous proteases. J Gen Virol. 76 (Pt3), 573–580.

Johnson, M.K., 1974. The primary biochemical lesion leading to the delayed neurotoxic effects of some organophosphorus esters. J. Neurochem. 23, 785–789.

Junaid, M., Chalayut, C., Torrejon, A.S., Angsuthanasombat, C., Shutava, I., Lapins, M., Wikberg, J.E., Katzenmeier, G., 2012. Enzymatic analysis of recombinant Japanese encephalitis virus NS2B (H)-NS3pro protease with fluorogenic model peptide substrates. PloS One 7, e36872.

Kapoor, M., Zhang, L., Ramachandra, M., Kusukawa, J., Ebner, K.E., Padmanabhan, R., 1995. Association between NS3 and NS5 proteins of Dengue virus type 2 in the putative RNA replicase is linked to differential phosphorylation of NS5. J. Biol. Chem. 270, 19100–19106.

Kiat, T.S., Pippen, R., Yusof, R., Ibrahim, H., Khalid, N., Rahman, N.A., 2006. Inhibitory activity of cyclohexenylchalcone derivatives and flavonoids of Fingerroot, *Boesenbergia rotunda* (L.), towards Dengue-2 virus NS3 protease. Bioorg. Med. Chem. Lett. 16, 3337–3340.

Kokryakov, V.N., Harwig, S.S., Panyutich, E.A., Shevchenko, A.A., Aleshina, G.M., Shamova, O.V., Korneva, H.A., Lehrer, R.I., 1993. Protegrins: leukocyte antimicrobial peptides that combine features of corticostaticdefensins and tachyplesins. FEBS Lett. 327, 231–236.

Kümmerer, B.M., Rice, C.M., 2002. Mutations in the yellow fever virus nonstructural protein NS2A selectively block production of infectious particles. J. Virol. 76 (10), 4773–4784.

Lei, J., Hansen, G., Nitsche, C., Klein, C.D., Zhang, L., Hilgenfeld, R., 2016. Crystal structure of Zika virus NS2B-NS3 protease in complex with a boronate inhibitor. Science 353 (6298), 503–505.

Lesburg, C.A., Cable, M.B., Ferrari, E., Hong, Z., Mannarino, A.F., Weber, P.C., 1999. Crystal structure of the RNA-dependent RNA polymerase from hepatitis C virus reveals a fully encircled active site. Nat. Struct. Mol. Biol. 6, 937–943.

Leung, J.Y., Pijlman, G.P., Kondratieva, N., Hyde, J., Mackenzie, J.M., Khromykh, A.A., 2008. Role of nonstructural protein NS2A in *Flavivirus* assembly. J. Virol. 82, 4731–4741.

Leung, D., Schroder, K., White, H., Fang, N.X., Stoermer, M.J., Abbenante, G., Martin, J.L., Young, P.R., Fairlie, D.P., 2001. Activity of recombinant Dengue 2 virus NS3 protease in the presence of a truncated NS2B co-factor, small peptide substrates, and inhibitors. J. Biol. Chem. 276, 45762–45771.

Lindenbach, B.D., Rice, C.M., 2001. Flaviviridae: the viruses and their replication. Fields Virol. 1, 991–1041.

Luo, D., Wei, N., Doan, D.N., Paradkar, P.N., Chong, Y., Davidson, A.D., Kotaka, M., Lescar, J., Vasudevan, S.G., 2010. Flexibility between the protease and helicase domains of the dengue virus NS3 protein conferred by the linker region and its functional implications. J. Biol. Chem. 285 (24), 18817–18827.

Luo, D., Xu, T., Hunke, C., Grüber, G., Vasudevan, S.G., Lescar, J., 2008. Crystal structure of the NS3 protease-helicase from Dengue virus. J. Virol. 82, 73–83.

Mackenzie, J.M., Jones, M.K., Young, P.R., 1996. Immunolocalization of the Dengue virus nonstructural glycoprotein NS1 suggests a role in viral RNA replication. Virology 220, 232–240.

Mackenzie, J.M., Khromykh, A.A., Jones, M.K., Westaway, E.G., 1998. Subcellular localization and some biochemical properties of the flavivirusKunjin nonstructural proteins NS2A and NS4A. Virology 245 (2), 203–215.

Mandal, K., Pentelute, B.L., Tereshko, V., Thammavongsa, V., Schneewind, O., Kossiakoff, A.A., Kent, S.B., 2009. Racemic crystallography of synthetic protein enantiomers used to determine the X-ray structure of Plectasin by direct methods. Protein Sci. 18, 1146–1154.

Mandl, C.W., Kunz, C.H., Heinz, F.X., 1991. Presence of poly (A) in a *Flavivirus*: significant differences between the 3′ noncoding regions of the genomic RNAs of tick-borne encephalitis virus strains. J. Virol. 65, 4070–4077.

Mangano, D.T., Miao, Y., Vuylsteke, A., Tudor, I.C., Juneja, R., Filipescu, D., Hoeft, A., Fontes, M.L., Hillel, Z., Ott, E., Titov, T., 2007. Mortality associated with aprotinin during 5 years following coronary artery bypass graft surgery. JAMA 297 (5), 471–479.

Mansfield, K.L., Johnson, N., Phipps, L.P., Stephenson, J.R., Fooks, A.R., Solomon, T., 2009. Tick-borne encephalitis virus—a review of an emerging zoonosis. J. Gen. Virol. 90, 1781–1794.

Matthews, D.A., Smith, W.W., Ferre, R.A., Condon, B., Budahazi, G., Slsson, W., Villafranca, J.E., Janson, C.A., McElroy, H.E., Gribskov, C.L., Worland, S., 1994. Structure of human rhinovirus 3C protease reveals a trypsin-like polypeptide fold, RNA-binding site, and means for cleaving precursor polyprotein. Cell 77, 761–771.

Mlakar, J., Korva, M., Tul, N., Popović, M., Poljšak-Prijatelj, M., Mraz, J., Kolenc, M., Resman Rus, K., Vesnaver Vipotnik, T., Fabjan Vodušek, V., Vizjak, A., 2016. Zika virus associated with microcephaly. N. Engl. J. Med. 374, 951–958.

Monath, T.P., 2001. Yellow fever: an update. Lancet Infect. Dis. 1 (1), 11–20.

Muñoz-Jordán, J.L., Sánchez-Burgos, G.G., Laurent-Rolle, M., García-Sastre, A., 2003. Inhibition of interferon signaling by dengue virus. Proc. Natl. Acad. Sci. 100 (24), 14333–14338.

Murray, C.L., Jones, C.T., Rice, C.M., 2008. Architects of assembly: roles of flaviviridae non-structural proteins in virion morphogenesis. Nat. Rev. Microbiol. 6, 699–708.

Mygind, P.H., Fischer, R.L., Schnorr, K.M., Hansen, M.T., Sönksen, C.P., Ludvigsen, S., Raventós, D., Buskov, S., Christensen, B., Maria, L., Taboureau, O., 2005. Plectasin is a peptide antibiotic with therapeutic potential from a saprophytic fungus. Nature 437, 975–980.

Noble, C.G., Seh, C.C., Chao, A.T., Shi, P.Y., 2012. Ligand-bound structures of the Dengue virus protease reveal the active conformation. J. Virol. 86, 438–446.

Normile, D., 2013. Surprising new dengue virus throws a spanner in disease control efforts. Science 342, 415.

Plourde, A.R., Bloch, E.M., 2016. A Literature review of Zika virus. Emerg. Infect. Dis. 22, 1185–1192.

Polgar, L., 2005. The catalytic triad of serine peptidases. Cell. Mol. Life Sci. 62, 2161–2172.

Putnak, R., 1994. Progress in the development of recombinant vaccines against dengue and other arthropod-borne flaviviruses. Springer, New York, NY, (pp. 231–252).

Qiu, X., Culp, J.S., DiLella, A.G., Hellmig, B., Hoog, S.S., Janson, C.A., Smith, W.W., Abdel-Meguid, S.S., 1996. Unique fold and active site in cytomegalovirus protease. Nature 383, 275–279.

Rice, C.M., Aebersold, R., Teplow, D.B., Pata, J., Bell, J.R., Vorndam, A.V., Trent, D.W., Brandriss, M.W., Schlesinger, J.J., Strauss, J.H., 1986. Partial N-terminal amino acid sequences of three nonstructural proteins of two flaviviruses. Virology 151, 1–9.

Rice, C.M., Lenches, E.M., Eddy, S.R., Shin, S.J., Sheets, R.L., Strauss, J.H., 1985. Nucleotide sequence of Yellow fever virus: implications for *Flavivirus* gene expression and evolution. Science 229, 726–733.

Robin, G., Chappell, K., Stoermer, M.J., Hu, S.H., Young, P.R., Fairlie, D.P., Martin, J.L., 2009. Structure of West Nile virus NS3 protease: ligand stabilization of the catalytic conformation. J. Mol. Biol. 385, 1568–1577.

Rogers, D.J., 2006. The global distribution of Yellow fever and Dengue. Adv. Parasitol. 62, 81–220.

Roth, A., Mercier, A., Lepers, C., Hoy, D., Duituturaga, S., Benyon, E., Guillaumot, L., Souares, Y., 2014. Concurrent outbreaks of Dengue, Chikungunya and Zika virus infections-an unprecedented epidemic wave of mosquito-borne viruses in the Pacific 2012-2014. Eurosurveillance 19, 1–29.

Rothan, H.A., Abdulrahman, A.Y., Sasikumer, P.G., Othman, S., Abd, Rahman, N., Yusof, R., 2012. Protegrin-1 inhibits Dengue NS2B-NS3 serine protease and viral replication in MK2 cells. J. Biomed. Biotech., 1–6.

Rothan, H.A., Mohamed, Z., Suhaeb, A.M., Rahman, N.A., Yusof, R., 2013. Antiviral cationic peptides as a strategy for innovation in global health therapeutics for Dengue virus: high yield production of the biologically active recombinant Plectasin peptide. J. Integ. Biol. 17, 560–567.

Schuler, Faccini, L., 2016. Possible Association between Zika virus infection and microcephaly-Brazil. Morbid. Mortal. Wkly Rep. 65, 59–62.

Schüller, A., Yin, Z., Brian Chia, C.S., Doan, D.N., Kim, H.K., Shang, L., Loh, T.P., Hill, J., Vasudevan, S.G., 2011. Tripeptide inhibitors of dengue and West Nile virus NS2B-NS3 protease. Antiviral Res. 92, 96–101.

Shiryaev, S.A., Kozlov, I.A., Ratnikov, B.I., Smith, J.W., Lebl, M., Strongin, A.Y., 2007. Cleavage preference distinguishes the two-component NS2B–NS3 serine proteinases of Dengue and West Nile Viruses. Biochem. J. 401, 743–752.

Smit, J.M., Moesker, B., Rodenhuis, Zybert, I., Wilschut, J., 2011. *Flavivirus* cell entry and membrane fusion. Viruses 3, 160–171.

Smithburn, K.C., Hughes, T.P., Burke, A.W., Paul, J.H., 1940. A neurotropic virus isolated from the blood of a native of Uganda. Am. J. Trop. Med. 20, 471–472.

Solomon, T., Ni, H., Beasley, D.W., Ekkelenkamp, M., Cardosa, M.J., Barrett, A.D., 2003. Origin and evolution of Japanese encephalitis virus in southeast Asia. J. Virol. 77, 3091–3098.

Staples, J.E., Gershman, M., Fischer, M., 2010. Yellow fever vaccine: recommendations of the Advisory Committee on Immunization Practices (ACIP). Morbid. Mortal. Wkly Rep. 59, 1–27.

Stoermer, M.J., Chappell, K.J., Liebscher, S., Jensen, C.M., Gan, C.H., Gupta, P.K., Xu, W.J., Young, P.R., Fairlie, D.P., 2008. Potent cationic inhibitors of West Nile virus NS2B/NS3 protease with serum stability, cell permeability and antiviral activity. J. Med. Chem. 51, 5714–5721.

Tambunan, U.S., Alamudi, S., 2010. Designing cyclic peptide inhibitor of Dengue virus NS3-NS2B protease by using molecular docking approach. Bioinformation 5, 250–254.

Tan, B.H., Fu, J., Sugrue, R.J., Yap, E.H., Chan, Y.C., Tan, Y.H., 1996. Recombinant dengue type 1 virus NS5 protein expressed in *Escherichia coli* exhibits RNA-dependent RNA polymerase activity. Virology 216, 317–325.

Tay, M.Y., Saw, W.G., Zhao, Y., Chan, K.W., Singh, D., Chong, Y., Forwood, J.K., Ooi, E.E., Grüber, G., Lescar, J., Luo, D., 2015. The C-terminal 50 amino acid residues of Dengue NS3 protein are important for NS3-NS5 interaction and viral replication. J. Biol. Chem. 290, 2379–2394.

Valle, R.P., Falgout, B., 1998. Mutagenesis of the NS3 protease of Dengue virus type 2. J. Virol. 72 (1), 624–632.

Ventura, C.V., Maia, M., Bravo, Filho, V., Góis, A.L., Belfort, R., 2016. Zika virus in Brazil and macular atrophy in a child with microcephaly. Lancet 387, 228.

Weinert, T., Olieric, V., Waltersperger, S., Panepucci, E., Chen, L., Zhang, H., Zhou, D., Rose, J., Ebihara, A., Kuramitsu, S., Li, D., 2015. Fast native-SAD phasing for routine macromolecular structure determination. Nature Met. 12 (2), 131–133.

Welsch, S., Miller, S., Romero-Brey, I., Merz, A., Bleck, C.K., Walther, P., Fuller, S.D., Antony, C., Krijnse-Locker, J., Bartenschlager, R., 2009. Composition and three-dimensional architecture of the dengue virus replication and assembly sites. Cell Host Microbe. 5 (4), 365–375.

Wengler, G., Czaya, G., Färber, P.M., Hegemann, J.H., 1991. In vitro synthesis of West Nile virus proteins indicates that the amino-terminal segment of the NS3 protein contains the active centre of the protease which cleaves the viral polyprotein after multiple basic amino acids. J. Gen. Vir. 72 (4), 851–858.

Wengler, G., Wengler, G., 1981. Terminal sequences of the genome and replication-form RNA of the *Flavivirus* West Nile virus: absence of poly (A) and possible role in RNA replication. Virology 113, 544–555.

Wengler, G., Wengler, G., 1991. The carboxy-terminal part of the NS 3 protein of the West Nile flavivirus can be isolated as a soluble protein after proteolytic cleavage and represents an RNA-stimulated NTPase. Virology 184 (2), 707–715.

Westaway, E.G., Mackenzie, J.M., Kenney, M.T., Jones, M.K., Khromykh, A.A., 1997. Ultrastructure of Kunjin virus-infected cells: colocalization of NS1 and NS3 with double-stranded RNA, and of NS2B with NS3, in virus-induced membrane structures. J. Virol. 71, 6650–6661.

Wichapong, K., Pianwanit, S., Sippl, W., Kokpol, S., 2010. Homology modeling and molecular dynamics simulations of Dengue virus NS2B/NS3 protease: insight into molecular interaction. J. Mol. Recognit. 23, 283–300.

Yildiz, M., Ghosh, S., Bell, J.A., Sherman, W., Hardy, J.A., 2013. Allosteric inhibition of the NS2B-NS3 protease from Dengue virus. ACS Chem. Biol. 8, 2744–2752.

Yin, Z., Patel, S.J., Wang, W.L., Chan, W.L., Rao, K.R., Wang, G., Ngew, X., Patel, V., Beer, D., Knox, J.E., Ma, N.L., 2006. Peptide inhibitors of Dengue virus NS3 protease. Part 2: SAR study of tetrapeptide aldehyde inhibitors. Bioorg. Med. Chem. Lett. 16, 40–43.

Yon, C., Teramoto, T., Mueller, N., Phelan, J., Ganesh, V.K., Murthy, K.H., Padmanabhan, R., 2005. Modulation of the nucleoside triphosphatase/RNA helicase and 5′-RNA triphosphatase activities of Dengue virus type 2 nonstructural protein 3 (NS3) by interaction with NS5, the RNA-dependent RNA polymerase. J. Biol. Chem. 280, 27412–27419.

Zhang, Y., Corver, J., Chipman, P.R., Zhang, W., Pletnev, S.V., Sedlak, D., Baker, T.S., Strauss, J.H., Kuhn, R.J., Rossmann, M.G., 2003. Structures of immature flavivirus particles. EMBO J. 22 (11), 2604–2613.

## FURTHER READING

Chambers, T.J., Nestorowicz, A., Rice, C.M., 1995. Mutagenesis of the yellow fever virus NS2B/3 cleavage site: determinants of cleavage site specificity and effects on polyprotein processing and viral replication. J. Virol. 69, 1600–1605.

Ishikawa, T., Yamanaka, A., Konishi, E., 2014. A review of successful *Flavivirus* vaccines and the problems with those flaviviruses for which vaccines are not yet available. Vaccine 32, 1326–1337.

Li, J., Lim, S.P., Beer, D., Patel, V., Wen, D., Tumanut, C., Tully, D.C., Williams, J.A., Jiricek, J., Priestle, J.P., Harris, J.L., 2005. Functional profiling of recombinant NS3 proteases from all four serotypes of Dengue virus using tetrapeptide and octapeptide substrate libraries. J. Biol. Chem. 280, 28766–28774.

Lindenbach, B.D., Rice, C.M., 1997. *trans*-Complementation of yellow fever virus NS1 reveals a role in early RNA replication. J. Virol. 71, 9608–9617.

Niyomrattanakit, P., Yahorava, S., Mutule, I., Mutulis, F., Petrovska, R., Prusis, P., Katzenmeier, G., Wikberg, J.E., 2006. Probing the substrate specificity of the dengue virus type 2 NS3 serine protease by using internally quenched fluorescent peptides. Biochem. J. 397, 203–211.

Patkar, C.G., Jones, C.T., Chang, Y.H., Warrier, R., Kuhn, R.J., 2007. Functional requirements of the Yellow fever virus capsid protein. J. Virol. 81, 6471–6481.

WHO, 2016. Emergency Committee on Zika virus and observed increase in neurological disorders and neonatal malformations. WHO Statement.

Chapter 7

# Flavivirus NS2B/NS3 Protease: Structure, Function, and Inhibition

**Zhong Li\*, Jing Zhang\*, Hongmin Li\*\***
*\*Wadsworth Center, New York State Department of Health, Albany, NY, United States;*
*\*\*State University of New York, Albany, NY, United States*

## 1 INTRODUCTION

The genus *Flavivirus* is composed of more than 70 viruses. Many flaviviruses cause serious and deadly human diseases. The four serotypes of Dengue virus (DENV), yellow fever virus (YFV), West Nile virus (WNV), Zika virus (ZIKV), St. Louis encephalitis virus (SLEV), Japanese encephalitis virus (JEV), Powassan virus (POWV), and tickborne encephalitis virus (TBEV) are categorized as global emerging or reemerging pathogens. The World Health Organization (WHO) has estimated annual human cases of more than 390,000,000, 200,000, and 68,000 for DENV, YFV, and JEV, respectively. Approximately 3.9 billion people are at risk of DENV infection. ZIKV is a new reemerging mosquitoborne *Flavivirus*. Significant outbreaks initially occurred at Yap Island in 2007, French Polynesia in 2013, Easter Island in 2014, and most recently Brazil in 2015 (Calvet et al., 2016; Chen and Hamer, 2016). Due to global travels, the virus was quickly imported to many new territories, such as the United Kingdom, Canada, USA, etc. (Attar, 2016; Bogoch et al., 2016; Chen, 2016; Korhonen et al., 2016). Recently, it was also reported that ZIKV can be transmitted through sexual activities and blood transfusions (Foy et al., 2011; Musso et al., 2014, 2015; Patino-Barbosa et al., 2015). Importantly, increasing evidence suggests that ZIKV infections are linked to Guillain–Barré syndrome, as well as an increase in babies born with microcephaly (Calvet et al., 2016; Li et al., 2016; Martines et al., 2016; Rodriguez-Morales, 2016; Thomas et al., 2016). These associations strongly suggest that ZIKV infection during pregnancy might cause severe neurological damage in neonates. WHO has declared ZIKV as a global public health emergency (Gulland, 2016). Although effective vaccines exist for YFV, JEV, and TBEV, there are currently no safe and

Viral Proteases and Their Inhibitors. http://dx.doi.org/10.1016/B978-0-12-809712-0.00007-1

**163**

effective vaccines for WNV, DENV, and ZIKV. Furthermore, due to the dangers and difficulties inherent in mass vaccination of large at-risk populations, it is desirable to be able to treat severe *Flavivirus* infections with antiviral therapeutics that could be administered. This chapter will review recent advances in *Flavivirus* drug development targeting the essential viral protease.

## 1.1 The Flaviviral Genome Structure

The *Flavivirus* genome RNA is single stranded and of positive (i.e., mRNA-sense) polarity with a size of approximately 11 kb. The viral genome is composed of a 5′ untranslated region, a single long open reading frame, and a 3′ untranslated region (Fig. 7.1) (Rice et al., 1985; Shi et al., 2001). A cap is present at the 5′ end, followed by the conserved dinucleotide sequence 5′-AG-3′ (Cleaves and Dubin, 1979). The 3′ end of the genome does not terminate with a poly(A) tract, but with 5′-$CU_{OH}$-3′ (Wengler, 1981). The single open reading frame of *Flavivirus* encodes a polyprotein precursor of about 3430 amino acids (aa) (Fig. 7.1). The polyprotein is co- and posttranslationally processed by viral and cellular proteases into three structural proteins [capsid (C), pre-membrane (prM) or membrane (M), and envelope (E)] and seven nonstructural (NS) proteins (NS1, NS2A, NS2B, NS3, NS4A, NS4B, and NS5) (Chambers et al., 1990). The structural proteins participate in the formation of the viral particle and are involved in viral fusion with host cells (Li et al., 2008; Lindenbach et al., 2007; Marianneau et al., 1999; Tassaneetrithep et al., 2003). Low pH in the endosomal compartment triggers fusion of the viral and host cell membrane, leading to the release of the nucleocapsid and viral RNA into the cytoplasm. This process is mediated by the viral E protein that is able to switch among different oligomeric states: as a trimer of prM–E heterodimers in immature particles, as a dimer in mature virus, and as a trimer when fusing with a host cell (Bressanelli et al., 2004; Modis et al., 2004). The virus prM glycoprotein can be cleaved by host furin protease to release the N-terminal (Nter) "pr" residues during maturation, leaving only the ectodomain and C-terminal (Cter) transmembrane region of "M" in the virion. The pr peptide protects immature virions against premature fusion with the host membrane (Guirakhoo et al., 1992; Li et al., 2008; Zhang et al., 2003).

The NS proteins are involved in RNA replication, virion assembly, and evasion of innate immune responses (Lindenbach et al., 2007). The majority of the

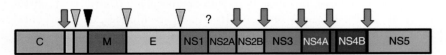

FIGURE 7.1 *Flavivirus* **polyprotein cleavage sites by viral protease** *(green arrows)*, **cellular signal peptidase** *(gray triangles)*, **and Golgi protease** *(black arrow)*. *C*, Capsid, *E*, envelope, *M*, membrane, *NS*, nonstructural.

*Flavivirus* NS proteins are multifunctional. NS1 is a large glycoprotein that is required for negative-strand RNA synthesis (Lindenbach and Rice, 1997, 1999; Muylaert et al., 1997). NS2A was reported to function in the generation of virus-induced membranes during virus assembly and/or release of infectious *Flavivirus* particles (Kummerer and Rice, 2002; Leung et al., 2008). NS2B is a required cofactor for the protease activity of NS3 (Arias et al., 1993; Chambers et al., 1991, 1993; Falgout et al., 1993). NS3 is a large multifunctional protein with the activities of a serine protease (with NS2B as a cofactor), a 5′ RNA triphosphatase, a nucleoside triphosphatase, and a helicase (Li et al., 1999; Warrener et al., 1993; Wengler, 1991). NS4A is an integral membrane protein involved in membrane rearrangements required to form the viral replication complex (Miller et al., 2007; Roosendaal et al., 2006). NS4B has been reported to inhibit the type I interferon response of host cells, and might modulate viral RNA synthesis (Grant et al., 2011; Munoz-Jordan et al., 2005; Umareddy et al., 2006). NS5 is the largest flaviviral protein with multiple enzymatic activities, namely the RNA-dependent RNA polymerase (Ackermann and Padmanabhan, 2001; Guyatt et al., 2001; Tan et al., 1996), the *N*-7 guanine and 2′-O ribose methyltransferase (Dong et al., 2012; Egloff et al., 2002; Koonin, 1993; Ray et al., 2006; Zhou et al., 2007), and the RNA guanylyltransferase (Issur et al., 2009). Several NS proteins, such as NS2A, NS4A, NS4B, and NS5, are thought to interfere with host immune responses (Ashour et al., 2009; Best et al., 2005; Daffis et al., 2010; Guo et al., 2005; Munoz-Jordan et al., 2003, 2005).

## 2 THE NS3/NS2B PROTEASE

Among the viral proteins, the NS3 protein (~618 aa) is the second largest protein encoded by *Flavivirus*. NS3 is a multifunctional protein with activities of a serine protease, an RNA triphosphatase, a nucleoside triphosphatase, and a helicase (Brecher et al., 2013). The viral protease is a complex with two components: the Nter 184 aa of viral NS3 protein and a hydrophobic core of about 40 aa in length within viral NS2B protein as an essential cofactor (Chambers et al., 1990, 1991; Falgout et al., 1991). The *Flavivirus* NS2B–NS3 protease is a highly conserved and replication-critical enzyme (Fig. 7.2) (Falgout et al., 1991). Fig. 7.2A shows a sequence alignment of the NS3 protease domains of selected flaviviruses that are significant human pathogens. Sequence conservations between DENVs, WNV, JEV, SLEV, ZIKV, and YFV range from 50% to 75%. POWV and TBEV have sequence identity of 71%. They share about 40% sequence identities with the DENV group. As shown in Fig. 7.2, several regions of NS3 are highly conserved: (1) the NS3 residues within the protease active site (Fig. 7.2B, left panel) and (2) the NS3 residues that form pockets to accommodate essential NS2B residues (Fig. 7.2B, middle and right panels). The sequence conservation of the core region of cofactor NS2B is also high (Brecher et al., 2013).

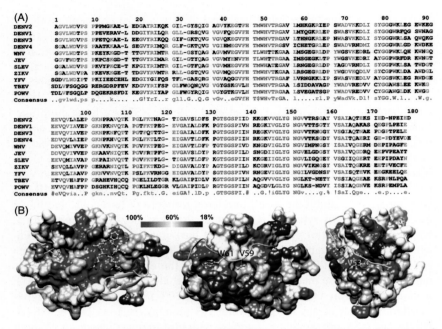

FIGURE 7.2 **Sequence conservation of the NS2B–NS3 protease complex.** (A) Alignment of sequences of the NS3 protease domains of representative flaviviruses that cause significant human diseases. Residues were colored according to the extent of their sequence conservation: >90% conserved *(red)*, 50%–90% conservation *(green)*, less or not conserved (<50%) *(black)*. (B) Mapping of sequence conservation to surface of the Dengue virus 3 (DENV3) NS3 protease domain (PDB code: 3U1I) (Noble et al., 2012). The structure is viewed from three angles, showing features in different regions. Left: Active-site surface with bound substrate analog inhibitor, represented in *ball-and-stick representation*. Images in middle and right show the NS3 pockets holding essential NS2B residues in *ribbon representation*. The NS2B residues V53, V59, and W61 are in *ball-and-stick representation*. The surface is colored according to sequence conservation (most conserved: *red*; least conserved: *blue*), resulting from the multiple sequence alignment.

This viral protease complex is required to work together with host proteases to cleave the polyprotein precursor produced by the viral genome. As shown in Fig. 7.1, the peptide bonds within capsid and between NS2A–NS2B, NS2B–NS3, NS3–NS4A, and NS4B–NS5 are cleaved by the NS2B/NS3 protease complex, leading to the release of mature individual NS proteins.

The NS2B/NS3 protease complex is essential for the *Flavivirus* replication and virion assembly, as evidenced by the lack of production of infectious virions in mutants carrying inactivating viral proteases (Chambers et al., 1993).

## 3 ENZYMATIC CHARACTERIZATION

The *Flavivirus* protease belongs to the trypsin serine protease superfamily with a catalytic triad (e.g., His51-As75-Ser135 for the DENV2 NS3) (Bazan and Fletterick, 1989; Chambers et al., 1991). Up to date, numerous studies have been published on enzymatic characterizations of *Flavivirus* proteases (for reviews, see Brecher et al., 2013; Nitsche et al., 2014; Sampath and Padmanabhan, 2009). Majority of these assays used covalently linked NS2B–NS3 constructs, in which the core region of NS2B was covalently linked to the NS3 protease domain through a flexible Gly-rich linker (Nitsche et al., 2014). Noncovalently linked NS2B–NS3 heterocomplexes through coexpression of individual NS2B and NS3 plasmids or with autocleavable linkers have also been used (de la Cruz et al., 2014; Kim et al., 2013). Alternatively, separately expressed NS2B and NS3 components were explored (Phong et al., 2011). Moreover, a construct of full-length NS2B–NS3 was investigated (Huang et al., 2013). It is worth noting that some active inhibitors against truncated NS2B–NS3 complex became inactive for the full-length NS2B–NS3 complex.

Besides construct differences, the assay conditions varied significantly. Although the protease assays were performed with pH ranging from 7.5 to 9.5, it was found that in general the NS2B–NS3 proteases were more active at high pH conditions, such as pH 9.0 or 9.5. Addition of 20%–30% glycerol could also stabilize the protease and improve the assay performance. Furthermore, various detergents at low concentrations were found to be beneficial for inhibitor screens.

In addition to constructs and assay condition variations, a wide range of substrates was used in these assays. Except the *cis*-cleavage site between NS2B and NS3 that shows a QR sequence at positions P1 and P2 for the DENV proteases, the natural cleavage sites of the *Flavivirus* proteases have a strong preference of dibasic residues at these two positions. In general, the NS2B/NS3 protease complex prefers a substrate with basic residues (Arg or Lys) at the P1 and P2 sites and a short side-chain aa (Gly, Ser, or Ala) at the P1′ site (Chambers et al., 1990; Gouvea et al., 2007). Investigations using combinatory peptide libraries not only confirmed the preference of dibasic sequence requirement at the P1 and P2 positions, but also revealed a strong preference for a third basic residue at the P3 position, as well as preference for small and polar residues at the P1′ and P3′ positions (Li et al., 2005). In contrast, P2′ and P4′ do not appear to have influence on substrate specificity.

Based on these studies, numerous synthetic substrates have been developed with chromo- or fluorogenic properties to facilitate inhibitor screenings (Nitsche et al., 2014). In general, these synthetic substrates fall into two categories. (1) Peptides with nonnature chromogenic or fluorogenic moieties at P1′. Protease cleavage releases the chromogenic or fluorogenic group from the peptide, leading to significant absorption in the UV/vis range or fluorescence. (2) Peptides with natural cleavage sites in the middle and nonnatural fluorophore moieties at both N- and C-termini that constitute internally quenched systems due to fluorescence resonance energy transfer. Protease cleavage releases the fluorophore from its quencher, resulting in significant fluorescence increase.

Although both types of substrates have been broadly used in various assays, substrates within group (1) have bulky chromophore or fluorophore in P1′, which is not as ideal as the protease complex prefers a small/noncharged aa in

P1′. Therefore, results from studies with these substrates may not fully represent the natural cleavage by the protease complex. In contrast, fluorescence resonance energy transfer substrates within group (2) can cover the entire peptide cleavage site from P4 to P4′, leading to not only high-affinity recognition, but also to a more precise mimicking of the natural peptide cleavage.

Due to significant variations of constructs, assay conditions, and substrates, the kinetic parameters determined for the *Flavivirus* NS2B–NS3 protease complexes show drastic differences (Nitsche et al., 2014). The $K_m$ values were observed from as low as 3.6 μM to as high as 4239 μM (Leung et al., 2001; Nitsche et al., 2014; Prusis et al., 2008). The $K_{cat}$ values range from 0.0022 to 300 s$^{-1}$. The resulting catalytic efficiencies as calculated by $K_{cat}/K_m$ vary from 1.6 to 50 × 10E6 M$^{-1}$s$^{-1}$.

Nevertheless, using various experimental systems, extensive mutagenesis studies have been performed on both NS2B and NS3 (Chappell et al., 2005, 2006, 2008a; Niyomrattanakit et al., 2004; Radichev et al., 2008; Salaemae et al., 2010; Shiryaev et al., 2007; Valle and Falgout, 1998; Zuo et al., 2009). Mutations of the NS3 catalytic residues (H51, D75, and S135) and those adjacent to these residues, such as V52, Y150, and G151, resulted in complete or almost complete loss of protease activity (Fig. 7.3A). Mutations of other adjacent residues, such as F130, G133, and N152 of NS3, led to a significant reduction in the protease activity, whereas mutations of V154, V155, I139, I140, and G144 caused moderate decrease in the protease activity. V154 and V155 are not as close to the active center as other residues. They may be involved in recognition of the P3 residue of substrate. Residues I139, I140, and G144 are located behind the active site. They may be important to contact NS2B cofactor (Fig. 7.3B).

On the NS2B side, mutagenesis studies indicated that two regions are critical for protease function. Region 1 is located between residues D58 and W61 (Fig. 7.3B). Region 2 is located within the hairpin structure of NS2B, involving NS2B residues from L74 to I86 (Fig. 7.3A). Mutations of region 1 residues of the WNV NS2B at positions equivalent to those of the DENV3 NS2B residues, D58, V59, T60, and W61, as well as residues V53 near region 1, significantly decreased the protease activity (Chappell et al., 2008a). Mutations of region 2 residues of the WNV NS2B at positions equivalent to those of the DENV3 NS2B residues, L74, I76, V78, D79, D81, G82, T83, M84, and I86, also drastically reduced protease activity. The NS2B residues within region 2 participate in the active-site formation. As shown in Fig. 7.3A, the NS2B residues at region 2, which are important for protease function, all face down to NS3 and make significant contacts with NS3 residues. In contrast, mutations of other residues that are exposed to solvent and do not contact NS3 showed little or no effects on protease activity. As shown in Figs. 7.2B and 7.3B, three out of five critical NS2B residues (V53, V59, and W61) within region 1 are held in deep pockets of NS3, suggesting that these residues are important for NS2B binding to NS3, as well as stabilization of NS3. Of these NS2B residues, W61 and G82, are strictly

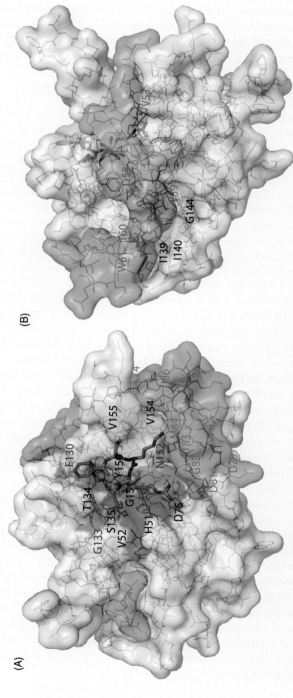

FIGURE 7.3    **Mapping of mutagenesis results to surface of the DENV3 NS2B–NS3 protease (PDB code: 3U1I) (Noble et al., 2012).** (A) *View from active site.* (B) *View from behind the active site. The surfaces are shown in* gray *for NS3 and* cyan *for NS2B. The NS3 residues with mutations that led to completely or almost completely abolished the protease activity are shown in* red, *the NS3 residues with mutations that significantly reduced protease activity are shown in* green, *the NS3 residues that moderately affect the protease activity are shown in* yellow. *The NS2B residues that are essential for the protease activity are shown in* orange.

conserved among flaviviruses (Brecher et al., 2013). In general, residues within region 1 are more conserved than those in region 2 of NS2B, suggesting that region 1 is critical for NS2B–NS3 recognition, whereas region 2 is important for fine substrate specificity.

## 4  STRUCTURE OF THE NS3/NS2B PROTEASE COMPLEX

The *Flavivirus* proteases belong to the trypsin-like serine protease family and display a general trypsin-like fold (Erbel et al., 2006; Nitsche et al., 2014). Crystal structures have been determined in both apo and inhibitor-bound forms for the NS2B–NS3 proteases of many flaviviruses (Aleshin et al., 2007; Assenberg et al., 2009; Chandramouli et al., 2010; Erbel et al., 2006; Luo et al., 2008a,b, 2010; Noble and Shi, 2012; Robin et al., 2009; Sampath and Padmanabhan, 2009). Majority of these crystallized protease constructs used covalently linked forms (e.g., NS2B-G$_4$SG$_4$linker-NS3). In all these structures, the NS2B fragment is composed of about 44–47 aa, providing an essential co-factor function (Chambers et al., 1990, 1991; Falgout et al., 1990). In the absence of substrates or active-site inhibitors, the NS2B is in a so-called "open" conformation (Fig. 7.4A). The Nter portion (residues 51–66) of NS2B, including essential NS2B residues within region 1, is tightly associated with NS3. This portion of NS2B forms a β-strand and is an integral part of the β-barrel of NS3 (Erbel et al., 2006). Structural comparison indicates that the NS2B residues within the Nter portion display similar conformations in all structures, regardless of substrate or inhibitor binding (Fig. 7.4B). The conformational similarity of the NS2B Nter portions is consistent with the important structural role of this part of NS2B. For example, it was known that the Nter portion of NS2B (aa 49–66 only) is sufficient to bind and stabilize the NS3 conformation (Luo et al., 2008a, 2010), although such a complex lacks protease activity (Luo et al., 2008b, 2010; Phong et al., 2011). The NS2B residues within this region are highly conserved, especially for several hydrophobic residues at positions 51, 53, 59, and 61 (in DENV2 order), with Trp61 strictly conserved (Brecher et al., 2013). Three of these residues (53, 59, and 61) were found essential for the protease function (Chappell et al., 2008a). These conserved hydrophobic residues bind deeply into several pockets of NS3 (Fig. 7.2B).

Residues within the Cter portion (aa 67–95) of the NS2B cofactor, including essential residues within region 2 (often referred as NS2Bc), display greater sequence variation than those within region 1. The NS2B region 2 may contribute to the fine substrate specificities, as region 2 is part of the protease active site (Fig. 7.3). In addition, in contrast to the Nter region, which shows similar conformations, the Cter portion (beyond aa 66) of NS2B shows significantly large conformational differences between inhibitor-bound and inhibitor-free structures, and even between inhibitor-free structures (Fig. 7.4B). These results suggest that the Nter portion, but not the Cter portion, of NS2B is essential for NS2B to bind and stabilize NS3.

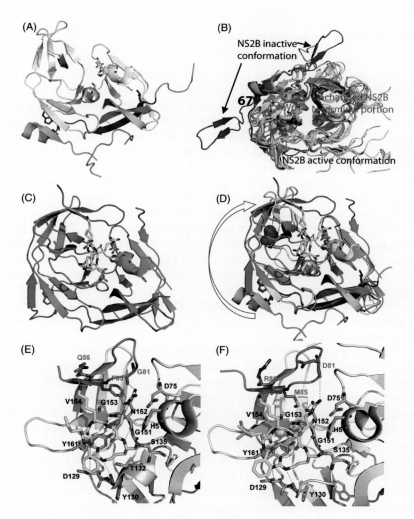

FIGURE 7.4 **Crystal structures of** *Flavivirus* **NS2B–NS3 proteases.** (A) Crystal structure of the DENV2 NS2B–NS3 protease in the absence of substrate analogs (PDB: 2FOM) (Erbel et al., 2006); NS3 in *gray* color, N-terminal (Nter) of NS2B in *blue*, and NS2Bc in *cyan*. (B) Superposition of all available crystal structures of the NS2B–NS3 protease complex, in the absence or presence of inhibitors. All NS3 chains were colored *gray*, with NS2B in different colors. PDB codes: 2FP7 (West Nile Virus; WNV, with peptide inhibitor, *green*) (Erbel et al., 2006), 2FOM (DENV2, apo form, *cyan*) (Erbel et al., 2006), 2GGV (WNV, apo form, *red*) (Aleshin et al., 2007), 2IJO (WNV, aprotinin bound, *yellow*) (Aleshin et al., 2007), 3E90 (WNV, with peptide inhibitor, *blue*) (Robin et al., 2009), 2WV9 (Murray Valley encephalitis virus; MVEV, NS3 full-length, apo form, *orange*) (Assenberg et al., 2009), 3LKW (DENV1, apo form, *brown*) (Chandramouli et al., 2010), 2WHX (DENV4, NS3 full-length, apo form, *gray*) (Luo et al., 2010), and 3U1I (DENV3, with peptide inhibitor, *magenta*) (Noble et al., 2012). L51 and W61 were labeled and shown in *stick representation*. (C) Crystal structure of the DENV3 NS2B–NS3 protease in complex with substrate analog inhibitor (Noble et al., 2012). The structure is shown in *green* for NS3 and *magenta* for NS2B. The bound inhibitor is in stick representation. (D) Structure superimposition between inhibitor-bound (3U1I) and inhibitor-free (2FOM) proteases. The colors are the same as in panels (A) and (C). Residue A125 is in *red sphere representation*. (E) Close-up view of the active site of the WNV NS2B–NS3 protease (2FP7) (Erbel et al., 2006). (F) Close-up view of the active site of the DENV3 NS2B–NS3 protease (3U1I) (Noble et al., 2012). Hydrogen bonds are shown in *dashed lines*, NS3 residues are in *cyan* color, and NS2B residues in *magenta*.

The Cter portion of NS2B is part of the protease active site in WNV and DENV. Although the conformations of NS2Bc vary considerably in apo crystal structures (Fig. 7.4B), the NS2Bc shows remarkable conformational similarity upon binding of substrate analogs or the protease inhibitor aprotinin (Fig. 7.4C). After the binding of inhibitor or substrate to the NS3 active site, the NS2Bc "wraps around" the NS3, closing the NS3 active site with the so-called active "closed" conformation (Noble et al., 2012) (Fig. 7.4D). It was noticed that the conformation of the Nter portion of NS2B remains the same as that of apo form (Noble et al., 2012). NS2B binding and conformational change are required for NS3 function; mutations that abrogate NS2B binding greatly reduce the proteolytic activity of the complex (Chappell et al., 2008a; Niyomrattanakit et al., 2004). In the structure of inhibitor-bound form, the NS2Bc forms a β-hairpin and extends toward the NS3 active site. NS2B residues within this region make direct interactions, including hydrogen bonds and salt bridges, with substrate analogs or aprotinin inhibitors. Mutagenesis studies confirmed the functional significance of this region of NS2B (Chappell et al., 2008a; Niyomrattanakit et al., 2004), likely due to its structural role in the formation of the protease active site.

The active site of the *Flavivirus* NS2B/NS3 protease complex is quite flat and hydrophilic (Fig. 7.2B) and requires several basic residues as substrates. Fig. 7.4E–F shows the close-up views of the active sites of the WNV and DENV3 proteases, respectively. The irreversible peptide-based aldehyde inhibitors covalently attaches to the active-site residue S135. Majority of the contacts are not specific. For examples, only three out of eight hydrogen bonds between the substrate analog and the WNV protease complex involve the side-chain atoms of the protease between OD1 of NS3 D129 and NE1 of the P1 Arg of the substrate analog, between OD1 of NS3 N152 and NE of the P2 Arg of substrate, and between OE1 of NS3 Q86 and NZ of the P3 Lys of substrate, whereas all the rest are made between the main-chain atoms of NS3 and substrates (Fig. 7.4E). Similar interactions were observed between the DENV3 protease and its substrate analog (Fig. 7.4F). For both structures, the P1 Arg of the substrate analog is specifically recognized by the NS3 D129 side chain. NS3 Asn152 is likely responsible for the recognition of the P2 Arg of the substrate. Although specific contact exists between the P3 Lys of substrate and NS3 Q86 side chain of the WNV protease, it is not present in the structure of the DENV protease–inhibitor complex.

The X-ray crystal structures discussed differ especially with regard to the localization of the Cter portion of NS2B (NS2Bc), depending on the presence or absence of a cocrystallized inhibitor. This essential NS2Bc is wrapped around the active site of NS3 as a short β-hairpin when it is in its catalytically active form. As discussed earlier, this part of the protease–cofactor complex is vital for the recognition of the substrate and adequate catalytic efficiency. The crystal structures show that without any cocrystallized ligand, a flexible NS2Bc domain was observed (open state) (Fig. 7.4A). Additionally, for the cocrystal structures of the *Flavivirus* NS2B–NS3 proteases with substrate analogs, the "closed" NS2Bc wrapped around the protease domain was observed (Fig. 7.4C). However, the structure of the 3U1J (DENV3) cocrystallized with aprotinin (BPTI) lacked the expected electron density in the Cter domain of NS2B. This would thus indicate an "open" state with a flexible cofactor. Therefore, these elaborate crystallization studies, and thus the ambiguous results, were unable to reveal the relevant conformational state of the protease–cofactor complex. This ambiguity may be due to the crystallization artifacts used or the artificial protease constructs with a permanent, nonnatural covalent linkage between the two domains. To complement the X-ray structural studies, NMR was used to identify the predominant conformation in solution for the various protease constructs.

Research and reports about the commonly employed covalently linked truncated construct NS2B(H)-glyNS3pro provided some of the first suitable NMR spectra and signal assignments for Dengue protease (Bi et al., 2013; Bodenreider et al., 2009; Chen et al., 2014; de la Cruz et al., 2011, 2014; Kim et al., 2013; Li et al., 2014; Su et al., 2009a,b). NMR studies using various techniques indicated that the "closed" conformation of NS2Bc was found to be the predominant state in solution, regardless of the presence or absence of a ligand for both unlinked and linked NS2B–NS3 constructs. Conversely, the "open" conformation is sparsely populated when in solution.

However, it was also observed that as pH and salt concentrations increased, a more pronounced conformational exchange occurred, which indicated that electrostatic interactions are crucial for the association of NS2Bc to the active site (de la Cruz et al., 2014). Of note, in contrast to the linked construct, which is stable at room temperature over longer incubation times, the unlinked protease precipitation occurred during the experiments. This would indicate a total dissociation of the complete cofactor (with Nter and Cter domains) from the protease. This effect was even more distinct at high ionic strengths. As isolated NS3pro was previously reported to be practically insoluble, and because dissociation leads to precipitation of at least the NS3 domain, there is some value of this construct in performing enzymatic assays and related purposes.

Additional evidence of the conformational flexibility came from mutagenesis studies by Yildiz et al. (2013). They tried to use cysteine mutagenesis approach to identify allosteric inhibition sites for the NS2B–NS3 protease complex. Cysteine mutants were introduced to the NS3 surface residues. Thiol-reactive probes, such as DTNB, were introduced to react with the cysteine mutants, leading to the identification of a region around the Ala125 residue as an allosteric site, which appeared sensitive toward the binding of covalent inhibitors. The A125 of NS3 is located beneath the hairpin structure of NS2Bc in the active protease conformation (Fig. 7.4D). DTNB covalently attached to A125C could block NS2Bc from forming the "closed" conformation. The results indicated that NS2Bc can be easily displaced from the active site.

## 5 INHIBITORS FOR THE NS3/NS2B PROTEASE

Viral proteases are proven antiviral targets. Numerous inhibitors have been successfully developed against the HIV virus and used in treatment of AIDS (Menendez-Arias, 2010). Additionally, a few Hepatitis C virus (HCV) protease inhibitors, telaprevir, simeprevir, paritaprevir, and grazoprevir, have recently received FDA approval to treat chronic HCV infections (Kowdley et al., 2014; Lawitz et al., 2015; Lin et al., 2006a,b, 2009) (Fig. 7.5). This success involving protease inhibitors with other viruses has made the *Flavivirus* protease the focal point of development for anti-*Flavivirus* therapy. Research into this has used both high-throughput screening (HTS) and structure-based drug design to explore and identify inhibitors against *Flavivirus* protease.

Leung et al. (2001) reported the first inhibition studies using a recombinant covalently linked NS2B/NS3 protease complex of DENV2. Of the 16 standard serine protease inhibitors tested, only aprotinin, a basic pancreatic trypsin inhibitor, was shown to inhibit the enzyme with nanomolar $IC_{50}$ (Leung et al., 2001; Mueller et al., 2007). Aprotinin binds the NS2B/NS3 proteases of all four serial types of DENV with high affinity (picomolar) (Li et al., 2005), but the in vivo efficacy of aprotinin in reducing *Flavivirus* has not been reported. Despite the fact that aprotinin is a potent inhibitor for the *Flavivirus* NS3 protease, there are several safety issues in using it as a drug. Aprotinin is a small protein that inhibits trypsin and other related proteolytic enzymes. Under the trade name Trasylol (Bayer), it was administered by injection as a medication to reduce bleeding during complex surgery, such as heart and liver surgery, before 2007. However, in 2008 the drug was permanently withdrawn after studies suggested that its use increased the risk of complications or death (Mangano et al., 2006, 2007).

Besides standard serine protease inhibitors, several peptidic α-keto amide inhibitors were also investigated (Leung et al., 2001). Two peptidic inhibitor candidates showed promising protease inhibition activity with low micromolar $IC_{50}$. Several similar peptidic inhibitor candidates, including cyclopeptides (Gao et al., 2010; Xu et al., 2012), were also found to be active for the NS2B/NS3 protease complex of DENV2, WNV, and YFV with the absolute inhibition constant or $K_i$ as low as 43 nM (Chanprapaph et al., 2005; Knox et al., 2006;

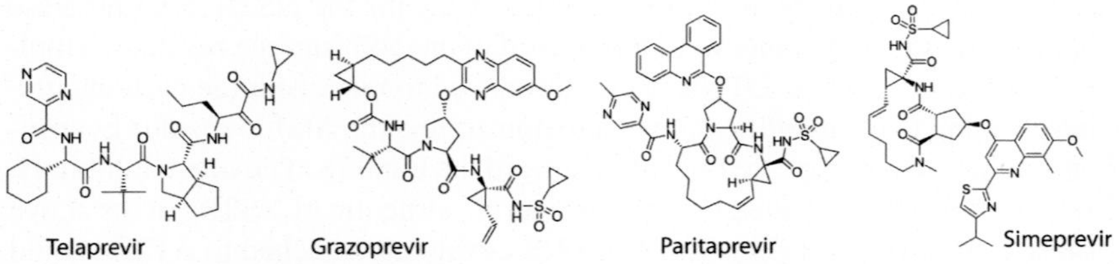

Telaprevir    Grazoprevir    Paritaprevir    Simeprevir

FIGURE 7.5    **Some FDA-approved Hepatitis C virus (HCV) protease inhibitors.**

Nall et al., 2004; Nitsche et al., 2012; Schuller et al., 2011; Yin et al., 2006a,b). Though the in vivo efficacy of these peptidic inhibitor candidates has not yet been verified, they are highly charged and thus may have poor bioavailability. Many studies verified this notion, for example, Shiryaev et al. (2006) reported that D-arginine–based peptides are potent inhibitors for the WNV NS3 protease, with $K_i$ as low as 1 nM in an in vitro biochemical protease assay. However, when a cell-based virus reduction assay was used, the inhibitor only showed micromolar inhibitory activity against the WNV (Shiryaev et al., 2006). In addition, Stoermer et al. (2008) reported a peptidic inhibitor candidate that had high potency ($K_i$ = 9 nM) for the WNV protease. The inhibitor was composed of cationic tripeptide (KKR) with a phenacetyl-cap at the Nter and an aldehyde at the Cter and was cell permeable and stable in serum. However, it displayed a comparatively much reduced antiviral activity ($EC_{50}$ = 1.6 µM) (Stoermer et al., 2008). The less-than-ideal activities of these peptide-based inhibitors in cell-based assays could be explained by the poor penetration of charged peptides across the cell membrane. Despite this, the low bioavailability of these substrate inhibitors could limit their potential as effective chemotherapeutics (Chappell et al., 2008b; Noble et al., 2010).

As far as the standard inhibitors based on substrates are concerned, attempts have been made to exploit proteins as inhibitors (Rothan et al., 2012). In one study, Rothan et al. (2012) reported that retrocyclin-1 (RC-1) inhibits the NS2B/NS3 protease activity in vitro with $IC_{50}$ in micromolar range. However, there were only moderate reductions in the growth of the virus even at a 150-µM concentration.

Nonsubstrate-based inhibitors were also investigated, though they revealed only moderate inhibition activity ($IC_{50}$ in low micromolar range) (Cregar-Hernandez et al., 2011; Ganesh et al., 2005; Jia et al., 2010; Kiat et al., 2006). To explore more small-molecular inhibitors for the protease, both in silico–based and protein-based HTS have been developed (Aravapalli et al., 2012; Deng et al., 2012; Ekonomiuk et al., 2009a,b; Ezgimen et al., 2012; Gao et al., 2013; Johnston et al., 2007; Knehans et al., 2011; Lai et al., 2013a,b; Mueller et al., 2008; Nitsche et al., 2011; Samanta et al., 2012; Steuer et al., 2011; Tiew et al., 2012; Tomlinson and Watowich, 2012; Tomlinson et al., 2009). Using these techniques, several such small-molecule inhibitors were identified that had low micromolar or high nanomolar inhibition activities for the WNV and DENV proteases (Bodenreider et al., 2009; Cregar-Hernandez et al., 2011; Ekonomiuk et al., 2009a; Johnston et al., 2007; Knehans et al., 2011; Lai et al., 2013b; Mueller et al., 2008; Sidique et al., 2009; Tomlinson and Watowich, 2011; Tomlinson et al., 2009; Yang et al., 2011). Though these compounds were potent inhibitors ($IC_{50}$ up to 0.105 µM) for the *Flavivirus* NS3 protease, some of them showed poor stability and some had a half-life of only 1–2 h in solution (Johnston et al., 2007). However, the assay of these compounds, except a few, was not cell based (Mueller et al., 2008; Tomlinson et al., 2009; Yang et al., 2011). In two studies (Mueller et al., 2008; Tomlinson et al., 2009), several compounds were discovered to inhibit the growth of WNV and DENV with $EC_{50}$ in the low micromolar range. Moreover, Yang et al. (2011) showed that a compound could inhibit the DENV NS3 protease with $IC_{50}$ of 15 μM. This compound appeared to be much more potent in a replicon-based antiviral assay ($EC_{50}$ = 0.17 μM), as compared to the enzyme-based protease assay, possibly due to additional cellular targets.

All current efforts of inhibitor identification for the NS3 protease are focused on the protease active site. However, only limited success has been achieved with this method. This may be due to the fact that the active site of the *Flavivirus* NS3 protease is not only flat, but also highly charged (Aleshin et al., 2007; Assenberg et al., 2009; Chandramouli et al., 2010; Erbel et al., 2006; Luo et al., 2008a,b, 2010; Robin et al., 2009), making it difficult to find small-molecule inhibitors for the NS2B/NS3 protease. Therefore, alternative approaches should be considered. Most notably, the *Flavivirus* NS3 protease requires NS2B as a cofactor for function. Therefore, the NS2B–NS3 association site may present a possible target for identification and development of compounds that inhibit *Flavivirus* NS3 protease function via blocking the NS2B–NS3 association. The crystal structures of the NS2B/NS3 complex (Aleshin et al., 2007; Assenberg et al., 2009; Chandramouli et al., 2010; Erbel et al., 2006; Luo et al., 2008a, 2010) and sample data from functional studies (Chambers et al., 2005; Chappell et al., 2006, 2008a; Niyomrattanakit et al., 2004; Radichev et al., 2008) provide solid bases for HT screening of compound libraries to identify various allosteric inhibitors. This approach has not been extensively explored to date. Only two reports indicated that a noncompetitive inhibitor was identified with high potency against the NS3 protease, through a protein-based HTS assay. Though one of the identified compounds was very unstable in solution (Johnston et al., 2007; Pambudi et al., 2013), docking experiments suggested that the compound binds in such a way to the NS3 surface that it would interfere with the binding of NS3 and its cofactor NS2B (Johnston et al., 2007; Pambudi et al., 2013). Although a crystal structure of the inhibitor–NS3 complex is required to confirm the mechanism of action of this type of inhibitor, in vitro inhibition studies have indicated that the compound identified by Pambudi et al. (2013) targeting the NS2B–NS3 interactions could efficiently inhibit all four serotypes of DENV with $EC_{50}$ of 0.74–4.92 μM. This compound has also shown moderate inhibition activity toward YFV in addition to the aforementioned viruses, indicating a potentially broad antiviral spectrum. Mutagenesis studies further showed that mutations of DENV4 and YFV residues that were predicted to interact with the inhibitor candidate would affect the sensitivity of viruses to this compound (Pambudi et al., 2013). These results strongly supported the hypothesis that the NS2B–NS3 interaction was a valid therapeutic target for anti-DENV drugs and thus suggested that efforts should be made to develop allosteric inhibitors targeting the NS2B–NS3 interaction.

# 6 FUTURE DIRECTIONS

The NS2B–NS3 protease represents a promising drug target, as viral proteases have proven to be great therapeutic targets for the development of drugs against many diseases, such as AIDS and Hepatitis. However, most attempts to develop *Flavivirus* protease inhibitors focused on the NS3 active site have had very limited success (Brecher et al., 2013; Noble and Shi, 2012; Sampath and Padmanabhan, 2009), possibly due to two of the site features (Lim et al., 2013; Luo et al., 2015; Sampath and Padmanabhan, 2009). First, the active site is flat and featureless, which makes specific inhibitors very unlikely. Second, the active site preferentially binds substrates with basic (positively charged) residues in its P2 and P1 positions, meaning that effective inhibitors in biochemical assays must be similarly charged; this charge requirement results in poor bioavailability in vivo (Fig. 7.2B).

Alternative strategies to identify inhibitors that don't bind at the NS3 active site, but to other regions of the protein to inhibit its function, may overcome these restrictions. One such strategy is to target the NS2B–NS3 interactions, which are known to be essential for viral replication. NS2B as a cofactor is essential for the NS3 protease function (Chappell et al., 2008a; Niyomrattanakit et al., 2004). In the *Flavivirus* protease, it has been shown that two regions of NS2B, Nter constituted of aa 53–61 and Cter constituted of aa 74–86, are essential for the protease function (Chappell et al., 2008a; Niyomrattanakit et al., 2004; Phong et al., 2011; Radichev et al., 2008). NS2B Nter residues display similar conformations in all structures (Brecher et al., 2013). Further, in NS2B Cter only aa 49–66 peptide would be sufficient to stabilize the NS3 conformation (Luo et al., 2008a, 2010). Moreover, unlike the flat and featureless active site (Fig. 7.2B), the NS3 pockets holding the NS2B Nter residues (such as key contact residues L51, V53, V59, and W61) are deep and hydrophobic (Fig. 7.2B).

The development of the protease inhibitor began with the determination of the three-dimensional structures of the NS3 protease, the NS2B/NS3 protease complex, and the protease–inhibitor complexes (Aleshin et al., 2007; Assenberg et al., 2009; Chandramouli et al., 2010; Erbel et al., 2006; Hammamy et al., 2013; Luo et al., 2008a,b, 2010; Noble et al., 2012; Robin et al., 2009). At this point, 14 different crystal structures of the NS2B/NS3 complex are available that include the apo structures of proteases of WNV, DENV1, DENV2, DENV4, and Murray Valley encephalitis virus (MVEV); the structures of proteases of WNV and DENV3 in complex peptide substrate–based inhibitors; and the broad-spectrum serine protease inhibitor aprotinin-bound structures of proteases of WNV and DENV3.

Historically, the most straightforward approach to develop inhibitors of an enzyme target has been to screen the compounds that bind competitively to the active site and thus displace the native substrate. The advantage of this traditional approach is that characterization of the properties of a particular enzyme's substrate is often a sufficient starting point for selecting compounds that match or exceed the substrate in its affinity for the enzyme. Unfortunately, for several reasons this approach has not been very successful to yield effective inhibitors of *Flavivirus* NS2B/NS3 protease. One of the reasons is the flat and hydrophilic nature of active site of NS2B/NS3, which does not permit any effective interaction of the inhibitor. The second reason is that the NS2B/NS3 active site is similar enough to those of host serine proteases, so that many compounds turn to be very toxic to the host, as has been observed in the case of aprotinin. The third reason is that the active site preferentially binds with positively charged moieties and thus can have deleterious effects on compound bioavailability.

However, there are many things that can be learnt from the development of active-site inhibitors of HCV protease. Although two HCV protease substrate–based inhibitors were developed, resistant mutations arose quickly (Wyles, 2013). This could be explained by the active site of the HCV protease being very shallow and solvent exposed. Thus the featureless property of the active site of the HCV protease implies that inhibitors would rely on relatively few interactions with the enzyme for tight binding, which results in a low barrier to resistance and extensive cross-resistance (Romano et al., 2010; Wegzyn and Wyles, 2012). It has been reported that a single key mutation resulted in a significant loss of inhibition and cross-resistance of the inhibitor (Romano et al., 2010; Wyles, 2012, 2013). Similarly, the active site of *Flavivirus* NS2B/NS3 protease complex is also flat and featureless, as well as hydrophilic in nature. Therefore, potential drug resistance should be taken into account, especially when the development of active-site inhibitors for *Flavivirus* protease complex is considered.

Fortunately, the solved crystal structures of the *Flavivirus* protease in both substrate-bound and unbound states have provided great mechanistic insight into the protease function. Details of the interaction of the NS2B cofactor with NS3 have suggested that an allosteric approach to inhibition via disruption of NS2B/NS3 binding would be very promising. Thus lead compounds developed by this approach are less likely to have any of the drawbacks previously discussed and observed with active-site inhibitors. Such compounds would also be more amenable to both computational and HTS screening methods. This "structure-guided" approach may suggest additional allosteric sites in *Flavivirus* protease for future studies, and has the potential to open new and broad avenues to drug discovery in other disease target proteins.

# 7 ACKNOWLEDGMENTS

This research was partially supported by a grant (AI094335) from the National Institutes of Health.

# REFERENCES

Ackermann, M., Padmanabhan, R., 2001. De novo synthesis of RNA by the dengue virus RNA-dependent RNA polymerase exhibits temperature dependence at the initiation but not elongation phase. J. Biol. Chem. 276, 39926–39937.

Aleshin, A., Shiryaev, S., Strongin, A., Liddington, R., 2007. Structural evidence for regulation and specificity of flaviviral proteases and evolution of the Flaviviridae fold. Pro. Sci. 16, 795–806.

Aravapalli, S., Lai, H., Teramoto, T., Alliston, K.R., Lushington, G.H., Ferguson, E.L., Padmanabhan, R., Groutas, W.C., 2012. Inhibitors of Dengue virus and West Nile virus proteases based on the aminobenzamide scaffold. Bioorg. Med. Chem. 20, 4140–4148.

Arias, C.F., Preugschat, F., Strauss, J.H., 1993. Dengue 2 virus NS2B and NS3 form a stable complex that can cleave NS3 within the helicase domain. Virology 193, 888–899.

Ashour, J., Laurent-Rolle, M., Shi, P.Y., Garcia-Sastre, A., 2009. NS5 of dengue virus mediates STAT2 binding and degradation. J. Virol. 83, 5408–5418.

Assenberg, R., Mastrangelo, E., Walter, T.S., Verma, A., Milani, M., Owens, R.J., Stuart, D.I., Grimes, J.M., Mancini, E.J., 2009. Crystal structure of a novel conformational state of the *Flavivirus* NS3 protein: implications for polyprotein processing and viral replication. J. Virol. 83, 12895–12906.

Attar, N., 2016. ZIKA virus circulates in new regions. Nat. Rev. Microbiol. 14, 62.

Bazan, J.F., Fletterick, R.J., 1989. Detection of a trypsin-like serine protease domain in flaviviruses and pestiviruses. Virology 171, 637–639.

Best, S.M., Morris, K.L., Shannon, J.G., Robertson, S.J., Mitzel, D.N., Park, G.S., Boer, E., Wolfinbarger, J.B., Bloom, M.E., 2005. Inhibition of interferon-stimulated JAK-STAT signaling by a tick-borne *Flavivirus* and identification of NS5 as an interferon antagonist. J. Virol. 79, 12828–12839.

Bi, Y., Zhu, L., Li, H., Wu, B., Liu, J., Wang, J., 2013. Backbone (1)H, (1)(3)C and (1)(5)N resonance assignments of dengue virus NS2B-NS3p in complex with aprotinin. Biomol. NMR Assign. 7, 137–139.

Bodenreider, C., Beer, D., Keller, T.H., Sonntag, S., Wen, D., Yap, L., Yau, Y.H., Shochat, S.G., Huang, D., Zhou, T., Caflisch, A., Su, X.C., Ozawa, K., Otting, G., Vasudevan, S.G., Lescar, J., Lim, S.P., 2009. A fluorescence quenching assay to discriminate between specific and nonspecific inhibitors of dengue virus protease. Analyt. Biochem. 395, 195–204.

Bogoch, I.I., Brady, O.J., Kraemer, M.U., German, M., Creatore, M.I., Kulkarni, M.A., Brownstein, J.S., Mekaru, S.R., Hay, S.I., Groot, E., Watts, A., Khan, K., 2016. Anticipating the international spread of Zika virus from Brazil. Lancet 387, 335–336.

Brecher, M., Zhang, J., Li, H., 2013. The *Flavivirus* protease as a target for drug discovery. Virol. Sin. 28, 326–336.

Bressanelli, S., Stiasny, K., Allison, S.L., Stura, E.A., Duquerroy, S., Lescar, J., Heinz, F.X., Rey, F.A., 2004. Structure of a *Flavivirus* envelope glycoprotein in its low-pH-induced membrane fusion conformation. EMBO J. 23, 728–738.

Calvet, G., Aguiar, R.S., Melo, A.S., Sampaio, S.A., de Filippis, I., Fabri, A., Araujo, E.S., de Sequeira, P.C., de Mendonca, M.C., de Oliveira, L., Tschoeke, D.A., Schrago, C.G., Thompson, F.L., Brasil, P., Dos Santos, F.B., Nogueira, R.M., Tanuri, A., de Filippis, A.M., 2016. Detection and sequencing of Zika virus from amniotic fluid of fetuses with microcephaly in Brazil: a case study. Lancet Infect. Dis. 16, 653–660.

Chambers, T.J., Droll, D.A., Tang, Y., Liang, Y., Ganesh, V.K., Murthy, K.H., Nickells, M., 2005. Yellow fever virus NS2B-NS3 protease: characterization of charged-to-alanine mutant and revertant viruses and analysis of polyprotein-cleavage activities. J. Gen. Virol. 86, 1403–1413.

Chambers, T.J., Grakoui, A., Rice, C.M., 1991. Processing of the yellow fever virus nonstructural polyprotein: a catalytically active NS3 proteinase domain and NS2B are required for cleavages at dibasic sites. J. Virol. 65, 6042–6050.

Chambers, T.J., Hahn, C.S., Galler, R., Rice, C.M., 1990. *Flavivirus* genome organization, expression, and replication. Annu. Rev. Microbiol. 44, 649–688.

Chambers, T.J., Nestorowicz, A., Amberg, S.M., Rice, C.M., 1993. Mutagenesis of the yellow fever virus NS2B protein: effects on proteolytic processing, NS2B-NS3 complex formation, and viral replication. J. Virol. 67, 6797–6807.

Chandramouli, S., Joseph, J.S., Daudenarde, S., Gatchalian, J., Cornillez-Ty, C., Kuhn, P., 2010. Serotype-specific structural differences in the protease-cofactor complexes of the dengue virus family. J. Virol. 84, 3059–3067.

Chanprapaph, S., Saparpakorn, P., Sangma, C., Niyomrattanakit, P., Hannongbua, S., Angsuthanasombat, C., Katzenmeier, G., 2005. Competitive inhibition of the dengue virus NS3 serine protease by synthetic peptides representing polyprotein cleavage sites. Biochem. Biophys. Res. Commun. 330, 1237–1246.

Chappell, K.J., Nall, T.A., Stoermer, M.J., Fang, N.X., Tyndall, J.D., Fairlie, D.P., Young, P.R., 2005. Site-directed mutagenesis and kinetic studies of the West Nile Virus NS3 protease identify key enzyme-substrate interactions. J. Biol. Chem. 280, 2896–2903.

Chappell, K.J., Stoermer, M.J., Fairlie, D.P., Young, P.R., 2006. Insights to substrate binding and processing by West Nile virus NS3 protease through combined modeling, protease mutagenesis, and kinetic studies. J. Biol. Chem. 281, 38448–38458.

Chappell, K.J., Stoermer, M.J., Fairlie, D.P., Young, P.R., 2008a. Mutagenesis of the West Nile virus NS2B cofactor domain reveals two regions essential for protease activity. J. Gen. Virol. 89, 1010–1014.

Chappell, K.J., Stoermer, M.J., Fairlie, D.P., Young, P.R., 2008b. West Nile Virus NS2B/NS3 protease as an antiviral target. Curr. Med. Chem. 15, 2771–2784.

Chen, L.H., 2016. Zika virus infection in a Massachusetts resident after travel to Costa Rica: a case report. Ann. Intern. Med. 164, 574–576.

Chen, L.H., Hamer, D.H., 2016. Zika virus: rapid spread in the western hemisphere. Ann Intern Med 164, 613–615.

Chen, W.N., Loscha, K.V., Nitsche, C., Graham, B., Otting, G., 2014. The dengue virus NS2B-NS3 protease retains the closed conformation in the complex with BPTI. FEBS Lett. 588, 2206–2211.

Cleaves, G.R., Dubin, D.T., 1979. Methylation status of intracellular dengue type 2 40 S RNA. Virology 96, 159–165.

Cregar-Hernandez, L., Jiao, G.S., Johnson, A.T., Lehrer, A.T., Wong, T.A., Margosiak, S.A., 2011. Small molecule pan-dengue and West Nile virus NS3 protease inhibitors. Antivir. Chem. Chemother. 21, 209–217.

Daffis, S., Szretter, K.J., Schriewer, J., Li, J., Youn, S., Errett, J., Lin, T.Y., Schneller, S., Zust, R., Dong, H., Thiel, V., Sen, G.C., Fensterl, V., Klimstra, W.B., Pierson, T.C., Buller, R.M., Gale, Jr., M., Shi, P.Y., Diamond, M.S., 2010. 2′-O methylation of the viral mRNA cap evades host restriction by IFIT family members. Nature 468, 452–456.

de la Cruz, L., Chen, W.N., Graham, B., Otting, G., 2014. Binding mode of the activity-modulating C-terminal segment of NS2B to NS3 in the dengue virus NS2B-NS3 protease. FEBS J. 281, 1517–1533.

de la Cruz, L., Nguyen, T.H., Ozawa, K., Shin, J., Graham, B., Huber, T., Otting, G., 2011. Binding of low molecular weight inhibitors promotes large conformational changes in the dengue virus NS2B-NS3 protease: fold analysis by pseudocontact shifts. J. Am. Chem. Soc. 133, 19205–19215.

Deng, J., Li, N., Liu, H., Zuo, Z., Liew, O.W., Xu, W., Chen, G., Tong, X., Tang, W., Zhu, J., Zuo, J., Jiang, H., Yang, C.G., Li, J., Zhu, W., 2012. Discovery of novel small molecule inhibitors of dengue viral NS2B-NS3 protease using virtual screening and scaffold hopping. J. Med. Chem. 55, 6278–6293.

Dong, H., Chang, D.C., Hua, M.H., Lim, S.P., Chionh, Y.H., Hia, F., Lee, Y.H., Kukkaro, P., Lok, S.M., Dedon, P.C., Shi, P.Y., 2012. 2′-O methylation of internal adenosine by *Flavivirus* NS5 methyltransferase. PLoS Pathog. 8, e1002642.

Egloff, M.P., Benarroch, D., Selisko, B., Romette, J.L., Canard, B., 2002. An RNA cap (nucleoside-2′-O-)-methyltransferase in the *Flavivirus* RNA polymerase NS5: crystal structure and functional characterization. EMBO J. 21, 2757–2768.

Ekonomiuk, D., Su, X.C., Ozawa, K., Bodenreider, C., Lim, S.P., Otting, G., Huang, D., Caflisch, A., 2009a. Flaviviral protease inhibitors identified by fragment-based library docking into a structure generated by molecular dynamics. J. Med. Chem. 52, 4860–4868.

Ekonomiuk, D., Su, X.C., Ozawa, K., Bodenreider, C., Lim, S.P., Yin, Z., Keller, T.H., Beer, D., Patel, V., Otting, G., Caflisch, A., Huang, D., 2009b. Discovery of a non-peptidic inhibitor of West Nile virus NS3 protease by high-throughput docking. PLoS Negl. Trop. Dis. 3, e356.

Erbel, P., Schiering, N., D'Arcy, A., Renatus, M., Kroemer, M., Lim, S., Yin, Z., Keller, T., Vasudevan, S., Hommel, U., 2006. Structural basis for the activation of flaviviral NS3 proteases from dengue and West Nile virus. Nat. Struct. Mol. Biol. 13, 372–373.

Ezgimen, M., Lai, H., Mueller, N.H., Lee, K., Cuny, G., Ostrov, D.A., Padmanabhan, R., 2012. Characterization of the 8-hydroxyquinoline scaffold for inhibitors of West Nile virus serine protease. Antiviral Res. 94, 18–24.

Falgout, B., Bray, M., Schlesinger, J.J., Lai, C.J., 1990. Immunization of mice with recombinant vaccinia virus expressing authentic dengue virus nonstructural protein NS1 protects against lethal dengue virus encephalitis. J. Virol. 64, 4356–4363.

Falgout, B., Miller, R.H., Lai, C.J., 1993. Deletion analysis of dengue virus type 4 nonstructural protein NS2B: identification of a domain required for NS2B-NS3 protease activity. J. Virol. 67, 2034–2042.

Falgout, B., Pethel, M., Zhang, Y.M., Lai, C.J., 1991. Both nonstructural proteins NS2B and NS3 are required for the proteolytic processing of dengue virus nonstructural proteins. J. Virol. 65, 2467–2475.

Foy, B.D., Kobylinski, K.C., Chilson Foy, J.L., Blitvich, B.J., Travassos da Rosa, A., Haddow, A.D., Lanciotti, R.S., Tesh, R.B., 2011. Probable non-vector-borne transmission of Zika virus, Colorado, USA. Emerg. Infect. Dis. 17, 880–882.

Ganesh, V.K., Muller, N., Judge, K., Luan, C.H., Padmanabhan, R., Murthy, K.H., 2005. Identification and characterization of nonsubstrate based inhibitors of the essential dengue and West Nile virus proteases. Bioorg. Med. Chem. 13, 257–264.

Gao, Y., Cui, T., Lam, Y., 2010. Synthesis and disulfide bond connectivity-activity studies of a kalata B1-inspired cyclopeptide against dengue NS2B-NS3 protease. Bioorg. Med. Chem. 18, 1331–1336.

Gao, Y., Samanta, S., Cui, T., Lam, Y., 2013. Synthesis and in vitro evaluation of West Nile Virus protease inhibitors based on the 1,3,4,5-tetrasubstituted 1H-pyrrol-2(5H)-one scaffold. ChemMedChem 8, 1554–1560.

Gouvea, I.E., Izidoro, M.A., Judice, W.A., Cezari, M.H., Caliendo, G., Santagada, V., dos Santos, C.N., Queiroz, M.H., Juliano, M.A., Young, P.R., Fairlie, D.P., Juliano, L., 2007. Substrate specificity of recombinant dengue 2 virus NS2B-NS3 protease: influence of natural and unnatural basic amino acids on hydrolysis of synthetic fluorescent substrates. Arch. Biochem. Biophys. 457, 187–196.

Grant, D., Tan, G.K., Qing, M., Ng, J.K., Yip, A., Zou, G., Xie, X., Yuan, Z., Schreiber, M.J., Schul, W., Shi, P.Y., Alonso, S., 2011. A single amino acid in nonstructural protein NS4B confers virulence to dengue virus in AG129 mice through enhancement of viral RNA synthesis. J. Virol. 85, 7775–7787.

Guirakhoo, F., Bolin, R.A., Roehrig, J.T., 1992. The Murray Valley encephalitis virus prM protein confers acid resistance to virus particles and alters the expression of epitopes within the R2 domain of E glycoprotein. Virology 191, 921–931.

Gulland, A., 2016. Zika virus is a global public health emergency, declares WHO. BMJ 352, i657.

Guo, J., Hayashi, J., Seeger, C., 2005. West Nile virus inhibits the signal transduction pathway of alpha interferon. J. Virol. 79, 1343–1350.

Guyatt, K.J., Westaway, E.G., Khromykh, A.A., 2001. Expression and purification of enzymatically active recombinant RNA-dependent RNA polymerase (NS5) of the *Flavivirus* Kunjin. J. Virol. Methods 92, 37–44.

Hammamy, M.Z., Haase, C., Hammami, M., Hilgenfeld, R., Steinmetzer, T., 2013. Development and characterization of new peptidomimetic inhibitors of the West Nile virus NS2B-NS3 protease. ChemMedChem 8, 231–241.

Huang, Q., Li, Q., Joy, J., Chen, A.S., Ruiz-Carrillo, D., Hill, J., Lescar, J., Kang, C., 2013. Lyso-myristoyl phosphatidylcholine micelles sustain the activity of Dengue non-structural (NS) protein 3 protease domain fused with the full-length NS2B. Protein Expr. Purif. 92, 156–162.

Issur, M., Geiss, B.J., Bougie, I., Picard-Jean, F., Despins, S., Mayette, J., Hobdey, S.E., Bisaillon, M., 2009. The *Flavivirus* NS5 protein is a true RNA guanylyltransferase that catalyzes a two-step reaction to form the RNA cap structure. RNA 15, 2340–2350.

Jia, F., Zou, G., Fan, J., Yuan, Z., 2010. Identification of palmatine as an inhibitor of West Nile virus. Arch. Virol. 155, 1325–1329.

Johnston, P.A., Phillips, J., Shun, T.Y., Shinde, S., Lazo, J.S., Huryn, D.M., Myers, M.C., Ratnikov, B.I., Smith, J.W., Su, Y., Dahl, R., Cosford, N.D., Shiryaev, S.A., Strongin, A.Y., 2007. HTS identifies novel and specific uncompetitive inhibitors of the two-component NS2B-NS3 proteinase of West Nile virus. Assay Drug Dev. Technol. 5, 737–750.

Kiat, T.S., Pippen, R., Yusof, R., Ibrahim, H., Khalid, N., Rahman, N.A., 2006. Inhibitory activity of cyclohexenyl chalcone derivatives and flavonoids of fingerroot, *Boesenbergia rotunda* (L.), towards dengue-2 virus NS3 protease. Bioorg. Med. Chem. Lett. 16, 3337–3340.

Kim, Y.M., Gayen, S., Kang, C., Joy, J., Huang, Q., Chen, A.S., Wee, J.L., Ang, M.J., Lim, H.A., Hung, A.W., Li, R., Noble, C.G., Lee le, T., Yip, A., Wang, Q.Y., Chia, C.S., Hill, J., Shi, P.Y., Keller, T.H., 2013. NMR analysis of a novel enzymatically active unlinked dengue NS2B-NS3 protease complex. J. Biol. Chem. 288, 12891–12900.

Knehans, T., Schuller, A., Doan, D.N., Nacro, K., Hill, J., Guntert, P., Madhusudhan, M.S., Weil, T., Vasudevan, S.G., 2011. Structure-guided fragment-based in silico drug design of dengue protease inhibitors. J. Comput. Aided Mol. Des. 25, 263–274.

Knox, J.E., Ma, N.L., Yin, Z., Patel, S.J., Wang, W.L., Chan, W.L., Ranga Rao, K.R., Wang, G., Ngew, X., Patel, V., Beer, D., Lim, S.P., Vasudevan, S.G., Keller, T.H., 2006. Peptide inhibitors of West Nile NS3 protease: SAR study of tetrapeptide aldehyde inhibitors. J. Med. Chem. 49, 6585–6590.

Koonin, E.V., 1993. Computer-assisted identification of a putative methyltransferase domain in NS5 protein of flaviviruses and lambda 2 protein of reovirus. J. Gen. Virol. 74, 733–740.

Korhonen, E.M., Huhtamo, E., Smura, T., Kallio-Kokko, H., Raassina, M., Vapalahti, O., 2016. Zika virus infection in a traveller returning from the Maldives, June 2015. Euro Surveill. 21, 2.

Kowdley, K.V., Lawitz, E., Poordad, F., Cohen, D.E., Nelson, D.R., Zeuzem, S., Everson, G.T., Kwo, P., Foster, G.R., Sulkowski, M.S., Xie, W., Pilot-Matias, T., Liossis, G., Larsen, L.,

Khatri, A., Podsadecki, T., Bernstein, B., 2014. Phase 2b trial of interferon-free therapy for hepatitis C virus genotype 1. N. Engl. J. Med. 370, 222–232.

Kummerer, B.M., Rice, C.M., 2002. Mutations in the yellow fever virus nonstructural protein NS2A selectively block production of infectious particles. J. Virol. 76, 4773–4784.

Lai, H., Dou, D., Aravapalli, S., Teramoto, T., Lushington, G.H., Mwania, T.M., Alliston, K.R., Eichhorn, D.M., Padmanabhan, R., Groutas, W.C., 2013a. Design, synthesis and characterization of novel 1,2-benzisothiazol-3(2H)-one and 1,3,4-oxadiazole hybrid derivatives: potent inhibitors of Dengue and West Nile virus NS2B/NS3 proteases. Bioorg. Med. Chem. 21, 102–113.

Lai, H., Sridhar Prasad, G., Padmanabhan, R., 2013b. Characterization of 8-hydroxyquinoline derivatives containing aminobenzothiazole as inhibitors of Dengue virus type 2 protease in vitro. Antiviral Res. 97, 74–80.

Lawitz, E., Gane, E., Pearlman, B., Tam, E., Ghesquiere, W., Guyader, D., Alric, L., Bronowicki, J.P., Lester, L., Sievert, W., Ghalib, R., Balart, L., Sund, F., Lagging, M., Dutko, F., Shaughnessy, M., Hwang, P., Howe, A.Y., Wahl, J., Robertson, M., Barr, E., Haber, B., 2015. Efficacy and safety of 12 weeks versus 18 weeks of treatment with grazoprevir (MK-5172) and elbasvir (MK-8742) with or without ribavirin for hepatitis C virus genotype 1 infection in previously untreated patients with cirrhosis and patients with previous null response with or without cirrhosis (C-WORTHY): a randomised, open-label phase 2 trial. Lancet 385, 1075–1086.

Leung, J.Y., Pijlman, G.P., Kondratieva, N., Hyde, J., Mackenzie, J.M., Khromykh, A.A., 2008. Role of nonstructural protein NS2A in *Flavivirus* assembly. J. Virol. 82, 4731–4741.

Leung, D., Schroder, K., White, H., Fang, N.X., Stoermer, M.J., Abbenante, G., Martin, J.L., Young, P.R., Fairlie, D.P., 2001. Activity of recombinant dengue 2 virus NS3 protease in the presence of a truncated NS2B co-factor, small peptide substrates, and inhibitors. J. Biol. Chem. 276, 45762–45771.

Li, H., Clum, S., You, S., Ebner, K.E., Padmanabhan, R., 1999. The serine protease and RNA-stimulated nucleoside triphosphatase and RNA helicase functional domains of dengue virus type 2 NS3 converge within a region of 20 amino acids. J. Virol. 73, 3108–3116.

Li, J., Lim, S.P., Beer, D., Patel, V., Wen, D., Tumanut, C., Tully, D.C., Williams, J.A., Jiricek, J., Priestle, J.P., Harris, J.L., Vasudevan, S.G., 2005. Functional profiling of recombinant NS3 proteases from all four serotypes of dengue virus using tetrapeptide and octapeptide substrate libraries. J. Biol. Chem. 280, 28766–28774.

Li, L., Lok, S.M., Yu, I.M., Zhang, Y., Kuhn, R.J., Chen, J., Rossmann, M.G., 2008. The *Flavivirus* precursor membrane-envelope protein complex: structure and maturation. Science 319, 1830–1834.

Li, C., Xu, D., Ye, Q., Hong, S., Jiang, Y., Liu, X., Zhang, N., Shi, L., Qin, C.F., Xu, Z., 2016. Zika virus disrupts neural progenitor development and leads to microcephaly in mice. Cell Stem Cell 19, 672.

Lim, S.P., Wang, Q.Y., Noble, C.G., Chen, Y.L., Dong, H., Zou, B., Yokokawa, F., Nilar, S., Smith, P., Beer, D., Lescar, J., Shi, P.Y., 2013. Ten years of Dengue drug discovery: progress and prospects. Antiviral Res. 100, 500–519.

Li, H., Zhu, L., Hou, S., Yang, J., Wang, J., Liu, J., 2014. An inhibition model of BPTI to unlinked Dengue virus NS2B-NS3 protease. FEBS Lett. 588, 2794–2799.

Lin, C., Kwong, A.D., Perni, R.B., 2006a. Discovery and development of VX-950, a novel, covalent, and reversible inhibitor of hepatitis C virus NS3.4A serine protease. Infect. Disord. Drug Targets 6, 3–16.

Lin, T.I., Lenz, O., Fanning, G., Verbinnen, T., Delouvroy, F., Scholliers, A., Vermeiren, K., Rosenquist, A., Edlund, M., Samuelsson, B., Vrang, L., de Kock, H., Wigerinck, P., Raboisson, P.,

Simmen, K., 2009. In vitro activity and preclinical profile of TMC435350, a potent hepatitis C virus protease inhibitor. Antimicrob. Agents Chemother. 53, 1377–1385.

Lin, K., Perni, R.B., Kwong, A.D., Lin, C., 2006b. VX-950, a novel hepatitis C virus (HCV) NS3-4A protease inhibitor, exhibits potent antiviral activities in HCV replicon cells. Antimicrob. Agents Chemother. 50, 1813–1822.

Lindenbach, B.D., Rice, C.M., 1997. trans-Complementation of yellow fever virus NS1 reveals a role in early RNA replication. J. Virol. 71, 9608–9617.

Lindenbach, B.D., Rice, C.M., 1999. Genetic interaction of *Flavivirus* nonstructural proteins NS1 and NS4A as a determinant of replicase function. J. Virol. 73, 4611–4621.

Lindenbach, B.D., Thiel, H.-J., Rice, C.M., 2007. Flaviviridae: The Virus and Their Replication, fourth ed. Lippincott William and Wilkins, Philadelphia, PA.

Luo, D., Vasudevan, S.G., Lescar, J., 2015. The *Flavivirus* NS2B-NS3 protease-helicase as a target for antiviral drug development. Antiviral Res. 118, 148–158.

Luo, D., Wei, N., Doan, D.N., Paradkar, P.N., Chong, Y., Davidson, A.D., Kotaka, M., Lescar, J., Vasudevan, S.G., 2010. Flexibility between the protease and helicase domains of the dengue virus NS3 protein conferred by the linker region and its functional implications. J. Biol. Chem. 285, 18817–18827.

Luo, D., Xu, T., Hunke, C., Gruber, G., Vasudevan, S.G., Lescar, J., 2008a. Crystal structure of the NS3 protease-helicase from Dengue virus. J. Virol. 82, 173–183.

Luo, D., Xu, T., Watson, R.P., Scherer-Becker, D., Sampath, A., Jahnke, W., Yeong, S.S., Wang, C.H., Lim, S.P., Strongin, A., Vasudevan, S.G., Lescar, J., 2008b. Insights into RNA unwinding and ATP hydrolysis by the *Flavivirus* NS3 protein. EMBO J. 27, 3209–3219.

Mangano, D.T., Miao, Y., Vuylsteke, A., Tudor, I.C., Juneja, R., Filipescu, D., Hoeft, A., Fontes, M.L., Hillel, Z., Ott, E., Titov, T., Dietzel, C., Levin, J., 2007. Mortality associated with aprotinin during 5 years following coronary artery bypass graft surgery. JAMA 297, 471–479.

Mangano, D.T., Tudor, I.C., Dietzel, C., 2006. The risk associated with aprotinin in cardiac surgery. N. Engl. J. Med. 354, 353–365.

Marianneau, P., Steffan, A.M., Royer, C., Drouet, M.T., Jaeck, D., Kirn, A., Deubel, V., 1999. Infection of primary cultures of human Kupffer cells by Dengue virus: no viral progeny synthesis, but cytokine production is evident. J. Virol. 73, 5201–5206.

Martines, R.B., Bhatnagar, J., Keating, M.K., Silva-Flannery, L., Muehlenbachs, A., Gary, J., Goldsmith, C., Hale, G., Ritter, J., Rollin, D., Shieh, W.J., Luz, K.G., Ramos, A.M., Davi, H.P., Kleber de Oliveria, W., Lanciotti, R., Lambert, A., Zaki, S., 2016. Notes from the field: evidence of Zika virus infection in brain and placental tissues from two congenitally infected newborns and two fetal losses—Brazil, 2015. Morb. Mort. Wkly. Rep. 65, 159–160.

Menendez-Arias, L., 2010. Molecular basis of human immunodeficiency virus drug resistance: an update. Antiviral Res. 85, 210–231.

Miller, S., Kastner, S., Krijnse-Locker, J., Buhler, S., Bartenschlager, R., 2007. The non-structural protein 4A of dengue virus is an integral membrane protein inducing membrane alterations in a 2K-regulated manner. J. Biol. Chem. 282, 8873–8882.

Modis, Y., Ogata, S., Clements, D., Harrison, S.C., 2004. Structure of the dengue virus envelope protein after membrane fusion. Nature 427, 313–319.

Mueller, N.H., Pattabiraman, N., Ansarah-Sobrinho, C., Viswanathan, P., Pierson, T.C., Padmanabhan, R., 2008. Identification and biochemical characterization of small-molecule inhibitors of west nile virus serine protease by a high-throughput screen. Antimicrob. Agents Chemother. 52, 3385–3393.

Mueller, N.H., Yon, C., Ganesh, V.K., Padmanabhan, R., 2007. Characterization of the West Nile virus protease substrate specificity and inhibitors. Int. J. Biochem. Cell. Biol. 39, 606–614.

Munoz-Jordan, J.L., Laurent-Rolle, M., Ashour, J., Martinez-Sobrido, L., Ashok, M., Lipkin, W.I., Garcia-Sastre, A., 2005. Inhibition of alpha/beta interferon signaling by the NS4B protein of flaviviruses. J. Virol. 79, 8004–8013.

Munoz-Jordan, J.L., Sanchez-Burgos, G.G., Laurent-Rolle, M., Garcia-Sastre, A., 2003. Inhibition of interferon signaling by Dengue virus. Proc. Natl. Acad. Sci. USA 100, 14333–14338.

Musso, D., Nhan, T., Robin, E., Roche, C., Bierlaire, D., Zisou, K., Shan Yan, A., Cao-Lormeau, V.M., Broult, J., 2014. Potential for Zika virus transmission through blood transfusion demonstrated during an outbreak in French Polynesia, November 2013 to February 2014. Euro Surveill. 19, 20761.

Musso, D., Roche, C., Robin, E., Nhan, T., Teissier, A., Cao-Lormeau, V.M., 2015. Potential sexual transmission of Zika virus. Emerg. Infect. Dis. 21, 359–361.

Muylaert, I.R., Galler, R., Rice, C.M., 1997. Genetic analysis of the yellow fever virus NS1 protein: identification of a temperature-sensitive mutation which blocks RNA accumulation. J. Virol. 71, 291–298.

Nall, T.A., Chappell, K.J., Stoermer, M.J., Fang, N.X., Tyndall, J.D., Young, P.R., Fairlie, D.P., 2004. Enzymatic characterization and homology model of a catalytically active recombinant West Nile virus NS3 protease. J. Biol. Chem. 279, 48535–48542.

Nitsche, C., Behnam, M.A., Steuer, C., Klein, C.D., 2012. Retro peptide-hybrids as selective inhibitors of the Dengue virus NS2B-NS3 protease. Antiviral Res. 94, 72–79.

Nitsche, C., Holloway, S., Schirmeister, T., Klein, C.D., 2014. Biochemistry and medicinal chemistry of the dengue virus protease. Chem. Rev. 114, 11348–11381.

Nitsche, C., Steuer, C., Klein, C.D., 2011. Arylcyanoacrylamides as inhibitors of the Dengue and West Nile virus proteases. Bioorg. Med. Chem. 19, 7318–7337.

Niyomrattanakit, P., Winoyanuwattikun, P., Chanprapaph, S., Angsuthanasombat, C., Panyim, S., Katzenmeier, G., 2004. Identification of residues in the dengue virus type 2 NS2B cofactor that are critical for NS3 protease activation. J. Virol. 78, 13708–13716.

Noble, C.G., Shi, P.Y., 2012. Structural biology of dengue virus enzymes: towards rational design of therapeutics. Antiviral Res. 96, 115–126.

Noble, C.G., Chen, Y.L., Dong, H., Gu, F., Lim, S.P., Schul, W., Wang, Q.Y., Shi, P.Y., 2010. Strategies for development of Dengue virus inhibitors. Antiviral Res. 85, 450–462.

Noble, C.G., Seh, C.C., Chao, A.T., Shi, P.Y., 2012. Ligand-bound structures of the dengue virus protease reveal the active conformation. J. Virol. 86, 438–446.

Pambudi, S., Kawashita, N., Phanthanawiboon, S., Omokoko, M.D., Masrinoul, P., Yamashita, A., Limkittikul, K., Yasunaga, T., Takagi, T., Ikuta, K., Kurosu, T., 2013. A small compound targeting the interaction between nonstructural proteins 2B and 3 inhibits dengue virus replication. Biochem. Biophys. Res. Commun. 25, 393–398.

Patino-Barbosa, A.M., Medina, I., Gil-Restrepo, A.F., Rodriguez-Morales, A.J., 2015. Zika: another sexually transmitted infection? Sex. Transm. Infect. 91, 359.

Phong, W.Y., Moreland, N.J., Lim, S.P., Wen, D., Paradkar, P.N., Vasudevan, S.G., 2011. Dengue protease activity: the structural integrity and interaction of NS2B with NS3 protease and its potential as a drug target. Biosci. Rep. 31, 399–409.

Prusis, P., Lapins, M., Yahorava, S., Petrovska, R., Niyomrattanakit, P., Katzenmeier, G., Wikberg, J.E., 2008. Proteochemometrics analysis of substrate interactions with dengue virus NS3 proteases. Bioorg. Med. Chem. 16, 9369–9377.

Radichev, I., Shiryaev, S.A., Aleshin, A.E., Ratnikov, B.I., Smith, J.W., Liddington, R.C., Strongin, A.Y., 2008. Structure-based mutagenesis identifies important novel determinants of the NS2B cofactor of the West Nile virus two-component NS2B-NS3 proteinase. J. Gen. Virol. 89, 636–641.

Ray, D., Shah, A., Tilgner, M., Guo, Y., Zhao, Y., Dong, H., Deas, T., Zhou, Y., Li, H., Shi, P., 2006. West nile virus 5′-cap structure is formed by sequential guanine N-7 and ribose 2′-O methylations by nonstructural protein 5. J. Virol. 80, 8362–8370.

Rice, C.M., Lenches, E.M., Eddy, S.R., Shin, S.J., Sheets, R.L., Strauss, J.H., 1985. Nucleotide sequence of yellow fever virus: implications for *Flavivirus* gene expression and evolution. Science 229, 726–733.

Robin, G., Chappell, K., Stoermer, M.J., Hu, S.H., Young, P.R., Fairlie, D.P., Martin, J.L., 2009. Structure of West Nile virus NS3 protease: ligand stabilization of the catalytic conformation. J. Mol. Biol. 385, 1568–1577.

Rodriguez-Morales, A.J., 2016. Zika and microcephaly in Latin America: an emerging threat for pregnant travelers? Travel Med. Infect. Dis. 14, 5.

Romano, K.P., Ali, A., Royer, W.E., Schiffer, C.A., 2010. Drug resistance against HCV NS3/4A inhibitors is defined by the balance of substrate recognition versus inhibitor binding. Proc. Natl. Acad. Sci. USA 107, 20986–20991.

Roosendaal, J., Westaway, E.G., Khromykh, A., Mackenzie, J.M., 2006. Regulated cleavages at the West Nile virus NS4A-2K-NS4B junctions play a major role in rearranging cytoplasmic membranes and Golgi trafficking of the NS4A protein. J. Virol. 80, 4623–4632.

Rothan, H.A., Han, H.C., Ramasamy, T.S., Othman, S., Rahman, N.A., Yusof, R., 2012. Inhibition of dengue NS2B-NS3 protease and viral replication in Vero cells by recombinant retrocyclin-1. BMC Infect. Dis. 12, 314.

Salaemae, W., Junaid, M., Angsuthanasombat, C., Katzenmeier, G., 2010. Structure-guided mutagenesis of active site residues in the dengue virus two-component protease NS2B-NS3. J. Biomed. Sci. 17, 68.

Samanta, S., Cui, T., Lam, Y., 2012. Discovery, synthesis, and in vitro evaluation of West Nile virus protease inhibitors based on the 9,10-dihydro-3H,4aH-1,3,9,10a-tetraazaphenanthren-4-one scaffold. ChemMedChem 7, 1210–1216.

Sampath, A., Padmanabhan, R., 2009. Molecular targets for *Flavivirus* drug discovery. Antiviral Res. 81, 6–15.

Schuller, A., Yin, Z., Brian Chia, C.S., Doan, D.N., Kim, H.K., Shang, L., Loh, T.P., Hill, J., Vasudevan, S.G., 2011. Tripeptide inhibitors of dengue and West Nile virus NS2B-NS3 protease. Antiviral Res. 92, 96–101.

Shi, P.Y., Kauffman, E.B., Ren, P., Felton, A., Tai, J.H., Dupuis, A.P., Jones, S.A., Ngo, K.A., Nicholas, D.C., Maffei, J., Ebel, G.D., Bernard, K.A., Kramer, L.D., 2001. High-throughput detection of West Nile virus RNA. J. Clin. Microbiol. 39, 1264–1271.

Shiryaev, S.A., Ratnikov, B.I., Aleshin, A.E., Kozlov, I.A., Nelson, N.A., Lebl, M., Smith, J.W., Liddington, R.C., Strongin, A.Y., 2007. Switching the substrate specificity of the two-component NS2B-NS3 *Flavivirus* proteinase by structure-based mutagenesis. J. Virol. 81, 4501–4509.

Shiryaev, S., Ratnikov, B., Chekanov, A., Sikora, S., Rozanov, D., Godzik, A., Wang, J., Smith, J., Huang, Z., Lindberg, I., Samuel, M., Diamond, M., Strongin, A., 2006. Cleavage targets and the D-arginine-based inhibitors of the West Nile virus NS3 processing proteinase. Biochem. J. 393, 503–511.

Sidique, S., Shiryaev, S.A., Ratnikov, B.I., Herath, A., Su, Y., Strongin, A.Y., Cosford, N.D., 2009. Structure-activity relationship and improved hydrolytic stability of pyrazole derivatives that are allosteric inhibitors of West Nile Virus NS2B-NS3 proteinase. Bioorg. Med. Chem. Lett. 19, 5773–5777.

Steuer, C., Gege, C., Fischl, W., Heinonen, K.H., Bartenschlager, R., Klein, C.D., 2011. Synthesis and biological evaluation of alpha-ketoamides as inhibitors of the Dengue virus protease with antiviral activity in cell-culture. Bioorg. Med. Chem. 19, 4067–4074.

Stoermer, M.J., Chappell, K.J., Liebscher, S., Jensen, C.M., Gan, C.H., Gupta, P.K., Xu, W.J., Young, P.R., Fairlie, D.P., 2008. Potent cationic inhibitors of West Nile virus NS2B/NS3 protease with serum stability, cell permeability and antiviral activity. J. Med. Chem. 51, 5714–5721.

Su, X.C., Ozawa, K., Qi, R., Vasudevan, S.G., Lim, S.P., Otting, G., 2009a. NMR analysis of the dynamic exchange of the NS2B cofactor between open and closed conformations of the West Nile virus NS2B-NS3 protease. PLoS Negl. Trop. Dis. 3, e561.

Su, X.C., Ozawa, K., Yagi, H., Lim, S.P., Wen, D., Ekonomiuk, D., Huang, D., Keller, T.H., Sonntag, S., Caflisch, A., Vasudevan, S.G., Otting, G., 2009b. NMR study of complexes between low molecular mass inhibitors and the West Nile virus NS2B-NS3 protease. FEBS J. 276, 4244–4255.

Tan, B.H., Fu, J., Sugrue, R.J., Yap, E.H., Chan, Y.C., Tan, Y.H., 1996. Recombinant dengue type 1 virus NS5 protein expressed in *Escherichia coli* exhibits RNA-dependent RNA polymerase activity. Virology 216, 317–325.

Tassaneetrithep, B., Burgess, T.H., Granelli-Piperno, A., Trumpfheller, C., Finke, J., Sun, W., Eller, M.A., Pattanapanyasat, K., Sarasombath, S., Birx, D.L., Steinman, R.M., Schlesinger, S., Marovich, M.A., 2003. DC-SIGN (CD209) mediates Dengue virus infection of human dendritic cells. J. Exp. Med. 197, 823–829.

Thomas, D.L., Sharp, T.M., Torres, J., Armstrong, P.A., Munoz-Jordan, J., Ryff, K.R., Martinez-Quinones, A., Arias-Berrios, J., Mayshack, M., Garayalde, G.J., Saavedra, S., Luciano, C.A., Valencia-Prado, M., Waterman, S., Rivera-Garcia, B., 2016. Local transmission of Zika virus—Puerto Rico, November 23, 2015-January 28, 2016. Morb. Mort. Wkly. Rep. 65, 154–158.

Tiew, K.C., Dou, D., Teramoto, T., Lai, H., Alliston, K.R., Lushington, G.H., Padmanabhan, R., Groutas, W.C., 2012. Inhibition of Dengue virus and West Nile virus proteases by click chemistry-derived benz[d]isothiazol-3(2H)-one derivatives. Bioorg. Med. Chem. 20, 1213–1221.

Tomlinson, S.M., Watowich, S.J., 2011. Anthracene-based inhibitors of dengue virus NS2B-NS3 protease. Antiviral Res. 89, 127–135.

Tomlinson, S.M., Watowich, S.J., 2012. Use of parallel validation high-throughput screens to reduce false positives and identify novel dengue NS2B-NS3 protease inhibitors. Antiviral Res. 93, 245–252.

Tomlinson, S.M., Malmstrom, R.D., Russo, A., Mueller, N., Pang, Y.P., Watowich, S.J., 2009. Structure-based discovery of Dengue virus protease inhibitors. Antiviral Res. 82, 110–114.

Umareddy, I., Chao, A., Sampath, A., Gu, F., Vasudevan, S.G., 2006. Dengue virus NS4B interacts with NS3 and dissociates it from single-stranded RNA. J. Gen. Virol. 87, 2605–2614.

Valle, R.P., Falgout, B., 1998. Mutagenesis of the NS3 protease of dengue virus type 2. J. Virol. 72, 624–632.

Warrener, P., Tamura, J.K., Collett, M.S., 1993. RNA-stimulated NTPase activity associated with yellow fever virus NS3 protein expressed in bacteria. J. Virol. 67, 989–996.

Wegzyn, C.M., Wyles, D.L., 2012. Antiviral drug advances in the treatment of human immunodeficiency virus (HIV) and chronic hepatitis C virus (HCV). Curr. Opin. Pharmacol. 12, 556–561.

Wengler, G., 1981. Terminal sequences of the genome and replicative-from RNA of the *Flavivirus* West Nile virus: absence of poly(A) and possible role in RNA replication. Virology 113, 544–555.

Wengler, G., 1991. The carboxy-terminal part of the NS 3 protein of the West Nile *Flavivirus* can be isolated as a soluble protein after proteolytic cleavage and represents an RNA-stimulated NTPase. Virology 184, 707–715.

Wyles, D.L., 2012. Beyond telaprevir and boceprevir: resistance and new agents for hepatitis C virus infection. Top. Antivir. Med. 20, 139–145.

Wyles, D.L., 2013. Antiviral resistance and the future landscape of hepatitis C virus infection therapy. J. Infect. Dis. 207 (Suppl. 1), S33–S39.

Xu, S., Li, H., Shao, X., Fan, C., Ericksen, B., Liu, J., Chi, C., Wang, C., 2012. Critical effect of peptide cyclization on the potency of peptide inhibitors against Dengue virus NS2B-NS3 protease. J. Med. Chem. 55, 6881–6887.

Yang, C.C., Hsieh, Y.C., Lee, S.J., Wu, S.H., Liao, C.L., Tsao, C.H., Chao, Y.S., Chern, J.H., Wu, C.P., Yueh, A., 2011. Novel dengue virus-specific NS2B/NS3 protease inhibitor, BP2109, discovered by a high-throughput screening assay. Antimicrob. Agents Chemother. 55, 229–238.

Yildiz, M., Ghosh, S., Bell, J.A., Sherman, W., Hardy, J.A., 2013. Allosteric inhibition of the NS2B-NS3 protease from dengue virus. ACS Chem. Biol. 8, 2744–2752.

Yin, Z., Patel, S.J., Wang, W.L., Wang, G., Chan, W.L., Rao, K.R., Alam, J., Jeyaraj, D.A., Ngew, X., Patel, V., Beer, D., Lim, S.P., Vasudevan, S.G., Keller, T.H., 2006a. Peptide inhibitors of Dengue virus NS3 protease. Part 1: warhead. Bioorg. Med. Chem. Lett. 16, 36–39.

Yin, Z., Patel, S.J., Wang, W.L., Chan, W.L., Ranga Rao, K.R., Wang, G., Ngew, X., Patel, V., Beer, D., Knox, J.E., Ma, N.L., Ehrhardt, C., Lim, S.P., Vasudevan, S.G., Keller, T.H., 2006b. Peptide inhibitors of dengue virus NS3 protease. Part 2: SAR study of tetrapeptide aldehyde inhibitors. Bioorg. Med. Chem. Lett. 16, 40–43.

Zhang, Y., Corver, J., Chipman, P.R., Zhang, W., Pletnev, S.V., Sedlak, D., Baker, T.S., Strauss, J.H., Kuhn, R.J., Rossmann, M.G., 2003. Structures of immature *Flavivirus* particles. EMBO J. 22, 2604–2613.

Zhou, Y., Ray, D., Zhao, Y., Dong, H., Ren, S., Li, Z., Guo, Y., Bernard, K.A., Shi, P.Y., Li, H., 2007. Structure and function of *Flavivirus* NS5 methyltransferase. J. Virol. 81, 3891–3903.

Zuo, Z., Liew, O.W., Chen, G., Chong, P.C., Lee, S.H., Chen, K., Jiang, H., Puah, C.M., Zhu, W., 2009. Mechanism of NS2B-mediated activation of NS3pro in dengue virus: molecular dynamics simulations and bioassays. J. Virol. 83, 1060–1070.

Chapter 8

# Design and Development of HCV NS5B Polymerase Inhibitors

**Debasis Das**
*Sree Chaitanya College, Habra, West Bengal, India*

## 1 INTRODUCTION

Hepatitis C virus (HCV) was first characterized in 1989 as the major cause of non-A and non-B hepatitis (Choo et al., 1989). HCV is recognized as the principal agent of hepatitis C infection and the cause of hepatocellular carcinoma and chronic liver cirrhosis (Alter et al., 1992, 1999). It is a burden on public health worldwide and a principal cause of liver transplantation. An estimated population of 170–200 million people worldwide, 2%–3% of total world population, is believed to suffer from chronic HCV infection. Approximately 3–4 million people are newly infected in this disease each year (Alter, 2007; Rickheim et al., 2002). HCV is commonly spread by direct contact with infected blood and blood products, such as blood transfusions, hemodialysis, and intravenous drug use. A small certain percentage of infected people can resolve HCV infection by their natural immunity (Meyer et al., 2007) and approximately 8,000–10,000 deaths result annually from the disease in the United States alone. The symptoms can be mild or nonexistent for years after initial infection and patients can be asymptomatic for decades before developing liver cirrhosis and/or liver cancer (Hoofnagle, 2002).

### 1.1 Current Standard of Care

The goal of therapy for Hepatitis C is to eradicate the virus completely or prevent the development of complications. To date, there is no vaccine that can be used to prevent HCV infection in human. It has been suggested that developing effective HCV vaccines will be difficult for any genotype of HCV (Farci, 2001). The current standard of care (SOC) for HCV-infected patients consists of a combined therapy of pegylated interferon-$\alpha$ (pegIFN-$\alpha$) in combination with ribavirin (RBV) for 24 or 48 weeks, depending on the HCV genotype. This combination therapy has been the SOC for patients with chronic hepatitis C, regardless the strain of virus genotype1-6 (Ghany et al., 2009). Both the components of this therapy are associated with side effects including headache, fever, depression,

*Viral Proteases and Their Inhibitors.* http://dx.doi.org/10.1016/B978-0-12-809712-0.00008-3

**189**

myalgia, and hemolytic anemia (Elloumi et al., 2007; Manns et al., 2006). The success rate for achieving a sustained virologic response (SVR) for genotype 1 patients in the United States, Europe, and Japan remains low, with a SVR of 40%–50% .To address these deficiencies, extensive efforts have been made toward the identification potent direct-acting antiviral (DAAs) agents against HCV (Au and Pockros, 2014; Casey and Lee, 2013). Several clinical trials have shown various combinations of agents, including interferon-free regimens, to be highly effective in the clearance or SVR of chronic hepatitis C infection. Several structural motifs of HCV inhibitors have been identified and many pharmaceutical companies are competing to identify new drugs (Haudecoeur et al., 2013).

## 2 HCV VIRUS

HCV is a small, spherically enveloped, single-stranded, positive (+)-sense RNA virus of 40–60 nm in diameter. It belongs to Flaviviridae family (Rosenberg, 2001; Tan et al., 2002). HCV infects liver cells at the level of $10^8$–$10^{11}$ copies of HCV RNA per gram of tissue. Some recent reports indicate that HCV can infect T-cells, B lymphocytes, dendritic cells, and cells of the central nervous system (Laskus et al., 2007). At least seven major genotypes (1–7) of HCV have been discovered with various subtypes (a, b, c, d, etc.).

- *Genotype 1*: is the most common genotype worldwide—46.2% of the population of the world is infected with hepatitis C. The reason that genotype 1 is the most common worldwide is that it is contaminated through blood transfusions, blood products, and organ transplantation and unsafe injection drug use.
- *Genotype 2*: accounts for about 9.1% of the cases of hepatitis C worldwide. Genotype 2 has always been one of the easiest genotypes to treat.
- *Genotype 3*: is the second most prevalent genotype worldwide, effecting about 30% of the population worldwide. At this time, genotype 3 is the most difficult to treat.
- *Genotype 4*: is mostly confined to Africa. It accounts for 90% of the genotypes in Egypt, which has the highest prevalence of HCV in the World.
- *Genotype 5*: has almost exclusively been found in South Africa and it accounts for 0.8% of the worldwide population of hepatitis C.
- *Genotype 6*: accounts for 5.4% of the worldwide population of hepatitis C and is mostly found in Southeast Asia.
- *Genotype 7*: A small number of people are identified with this genotype. They have all been from the Democratic Republic of the Congo.

## 2.1 HCV Genome

The HCV genome is an uncapped, small, positive-sense, single-strand RNA virus with a 9.6 kb genome (Choo et al., 1991; Takamizawa et al., 1991). The RNA genome harbors a single open reading frame that is flanked by 5′ and 3′ nontranslated RNA segments (NTR's). The open reading frame encodes a single polyprotein of approximately 3010 amino acids that is further processed by host peptidases and viral proteases to provide at least 10 viral proteins. These include three HCV structural proteins. These include three HCV structural proteins (C, E1, and E2), and seven nonstructural (NS) proteins (NS2, p7, NS3, NS4A, NS4B, NS5A, and NS5B). Due to their necessary roles in viral replication, viral enzymes, such as the HCV protease NS3/NS4A and NS5B have been the most studied viral proteins and recognized as the most viable protein targets for small molecule HCV drug discovery.

## 2.2 NS5B Polymerase

The HCV nonstructural 5B (NS5B) is a RNA-dependent RNA polymerase (RdRp) that resides at the C-terminus of the polyprotein. It is one of the central enzymes in the viral replication cycle. The NS5B is responsible for viral RNA synthesis and genome replication. It is a validated target for antiviral therapy (Lindenbach and Rice, 2005). As HCV NS5B has no counterpart in mammalian cells, it is expected that inhibition of the enzyme will not cause target-related side effects. The 65-kDa proteins have the common fold of other viral polymerases with characteristic thumb, finger, and palm domains (Fig. 8.1) and functions inside the cell as a component of the HCV replicase complex. The palm domain contains the active site. Four allosteric sites have been identified: palm I, palm II, thumb I, and thumb II (Ago et al., 1999; Bressanelli et al., 1999; Lesburg

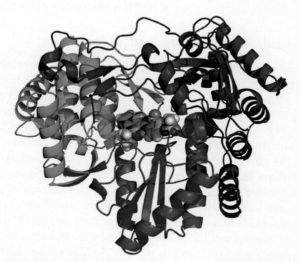

FIGURE 8.1   **Ribbon structure of HCV RdRp and its three subdomains.** The palm, fingers, and thumb subdomains of HCV polymerase are colored *red, blue,* and *green,* respectively. We consider the palm site to include residues 188–227 and 287–370, the fingers site to include residues 1–187 and 228–286, and the thumb site to include residues 371–563. The figure was generated using PyMOL Molecular Graphics System, Schrödinger, LLC. *(Reprinted with permission from Das, D., Hong, J., Chen, S.-H., Wang, G., Beigelman, L., Seiwert, S.D., Buckman, B.O., 2011. Recent advances in drug discovery of benzothiadiazine and related analogs as HCV NS5B polymerase inhibitors. Bioorg. Med. Chem. 19, 4690–4703. Copyright 2011 Elsevier)*

et al., 1999). HCV polymerase is characterized by a β-hairpin loop protruding in the central RNA-binding channel and two additional loops that extend from the fingers to the surface of the thumb domain, all of which are thought to be involved in the regulation of RNA synthesis (Bressanelli et al., 1999). Thus in the past decades, attentions have been paid to discover many NS5B polymerase inhibitors.

## 3   HCV NS5B INHIBITORS

HCV NS5B became an attractive target for drug discovery for HCV therapy. NS5B inhibitors interfere the functions of the enzyme at each individual step that is associated with distinct structure of the replicating complex. Inhibitors of HCV NS5B can be broadly classified into three classes, as described below, based on their chemical structure and/or mode of action:

1. *Nucleoside or nucleotide analogs inhibitors (NIs):* The NIs bind at the enzyme active site. They act as competitors of nucleotide triphosphates (NTPs) and interfere with any step during RNA synthesis.
2. *Nonnucleoside inhibitors:* These inhibitors allosterically target the NS5B, and the majority of them inhibit the initiation stage of RNA synthesis.
3. *Small molecule inhibitors:* They have a distinct mechanism of inhibition in that they covalently modify the residues near the active site of NS5B and inhibit its activity. Some of them act by chelating the divalent metal ions needed by NS5B. These classes of NS5B inhibitors are discussed here.

### 3.1   Nucleoside or Nucleotide Analogs Inhibitors

The nucleoside analogue inhibitors (NIs) are considered as the prodrugs. These inhibitors are normally synthesized in the nonphosphorylated form and converted in the active triphosphate form in the cytoplasm of infected cells by cellular kinases (Carroll and Olsen, 2006). This class includes the purine analog ribavirin (**1**) (Fig. 8.2), a small molecule drug that was approved by the Food and Drug Administration (FDA) for the treatment of HCV. Ribavirin has a minor and transient effect on viral load on monotherapy to treat patients with HCV. However, ribavirin in combination with pegylated interferon eradicates HCV in more than 50% of treated patients. The action mechanisms of ribavirin, to enhance response with interferon, are not clearly known (Ward et al., 2008).

The first nucleoside analogs to demonstrate activity against HCV NS5B were the class of 2/C-methyl modified ribonucleosies. 2'-C-Methylcytidine **NM107**, (**2**, $EC_{50}$ (replicon) = 2.5 µM) and its prodrug **NM283** (**3**, Lawitz et al., 2007) were identified as the active analogs of this series (Fig. 8.2). The efficacy of **NM107** was hampered by its low oral bioavailability. So, 3'-O-L-valinyl ester analog (valopicitabine, **3**) was developed with a notable improvement in the pharmacokinetic profile.

The prodrug **NM283** (**3**) was the first HCV polymerase inhibitor to advance into clinical trial by Idenix Phramaceuticals and Novartis (Toniutto et al., 2007).

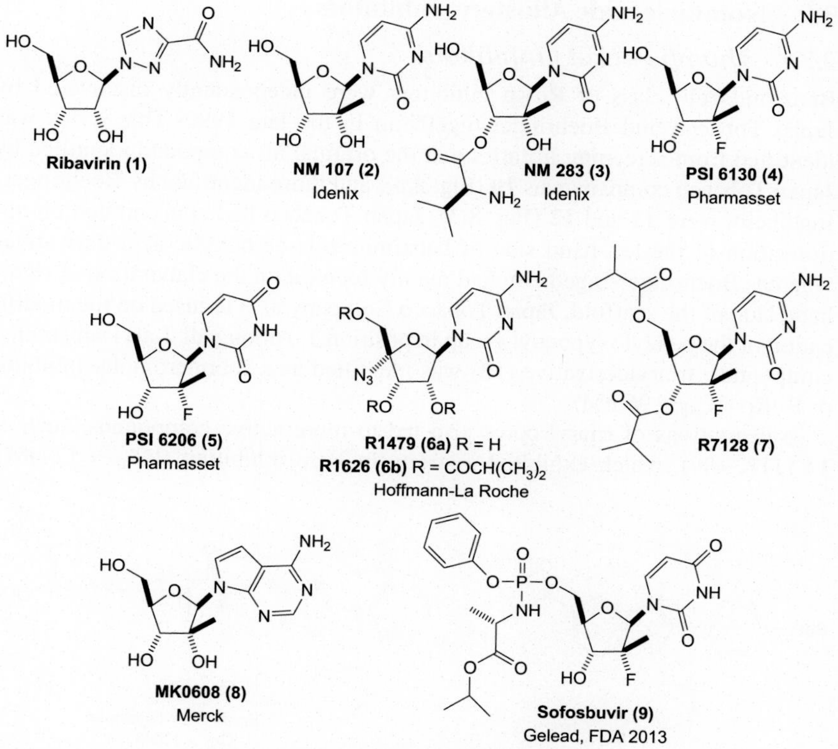

FIGURE 8.2   Top nucleoside inhibitors (NIs) of HCV NS5B.

Pharmasset pushed two compounds into clinical study **PSI-6130** (**4**) [EC$_{90}$ (replicon) = 5–6 mM] and **PSI 6206** (**5**) (Wang et al., 2009a). Two derivatives **R1479** (**6a**) (EC$_{50}$, replicon = 1.3 μM) and **R1626** (**6b**), the prodrugs of **R1479** were discovered as very effective and pushed to clinical trial by Hoffnan-La Roche. But the clinical trial of **R1626** was stopped at Phase IIb study due to unexpected safety problems. **R-7128** (**7**), the *bis*-isobutyl ester prodrug of PSI-6130 (2′-deoxy-2′-fluoro-2′-C-methyl cytidine) was the most advanced active site NI in clinical development. The results of a combination of **R-7128** (500 or 1500 mg BID for 28 days) with PEG-IFN/RBV showed 85% rapid virologic response and good tolerability. The compound, 7-deaza-adenosine derivative **MK-0608** (**8**), was discovered as the most effective derivative by Merck co. Recently, **Sofosbuvir** (**9**) has been discovered by Gilead scientist as a very effective drug. **Sofosbuvir** plus pegylated interferon plus ribavirin can be used to treat HCV genotypes 1 and 4. The cure rates are up to 90% in people with HCV genotype 1 and up to 96% in people with HCV genotype 4. The treatment duration is 12 weeks.

## 3.2 Nonnucleoside Allosteric Inhibitors

### 3.2.1 Thumb Pocket I Inhibitors

Benzimidazole class of RdRp inhibitors were independently discovered by Japan Tobacco and Boehringer-Ingelheim in the late 1990s.This series was identified from screening libraries and the original hit compound identified by Japan Tobacco company was **10** (Fig. 8.3), and those identified by Boehringer-Ingelheim were **11** and **12** (Fig. 8.3). Japan Tobacco had concentrated on optimization of the left-hand side of benzimidazole carboxylic acid derivatives whereas Boehringer-Ingelheim had mainly focused on the elaboration of right-hand side of the scaffold. Japan Tobacco Company also focused on the modification of the benzyloxyphenyl group in position 2. Among all 2-aryl substituted compounds, triarylderivative (**13**) was identified as a submicromolar inhibitor of RdRp ($IC_{50}$ = 96 nM).

Substitutions of triaryl cores also led to more active compounds, such as **14** (JTK-109), which exhibited a strong in vitro inhibition ($IC_{50}$ = 17 nM).

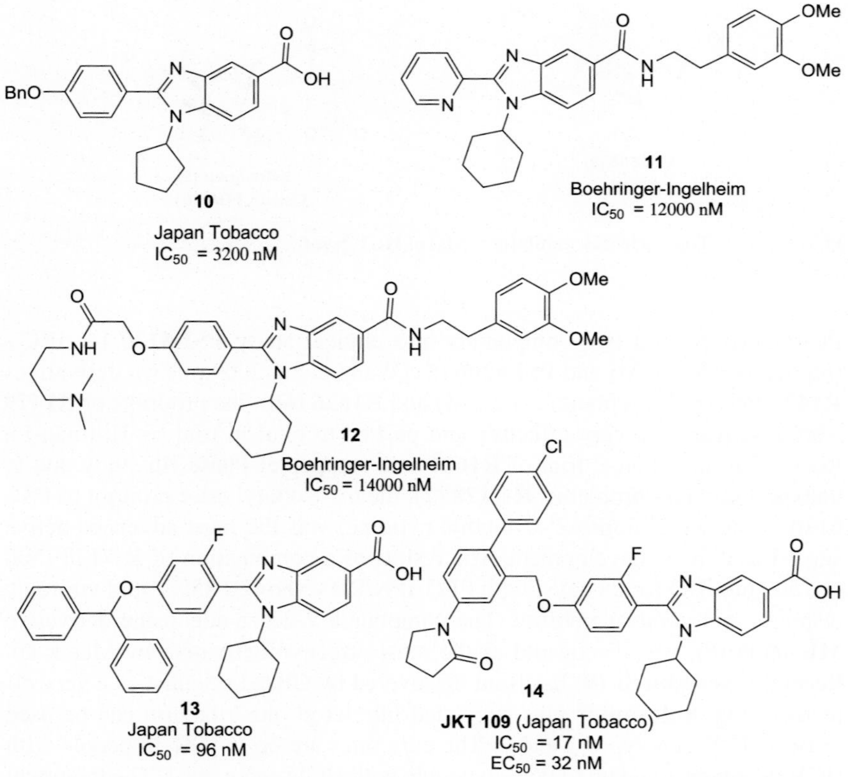

FIGURE 8.3   Screened hits from benzimidazole series as HCV NS5B and JKT inhibitors.

Additionally, compound **14** showed good inhibition activity on genotypes 1a and 3a ($IC_{50}$ = 62 and 61 nM). JTK-109 (**14**) was an advanced candidate into clinical studies but was discontinued by Japan Tobacco (Hirashima et al., 2006).

Boehringer-Ingelheim focused on the elaboration of the right-hand side of the scaffold through formation of amide derivative. Boehringer-Ingelheim reported the discovery and preliminary optimization of benzimidazole scaffold. It was also shown that cyclohexyl or cyclopentyl ring at N1-position, a small aromatic or heteroaromatic ring at C-2 and a carboxylic acid function at C-5 position were essential to show the activity **15** (Fig. 8.4) (Beaulieu et al., 2004). Significant improvement of potencies (58-fold) was attained by elaborating the right hand side of inhibitors through arrangement of functionalized amino acid residues (**16**).

The physicochemical properties of the molecules were adjusted by replacing the ionizable carboxylic acid moiety with thiazole moiety (**17**), which led to the discovery of inhibitor active in cell culture at low micromolar concentration (Beaulieu et al., 2004).

The replacement study of the 5-carboxylic acid substituent by various amido moieties was under taken by Boehringer-Ingelheim. It was found that an amido

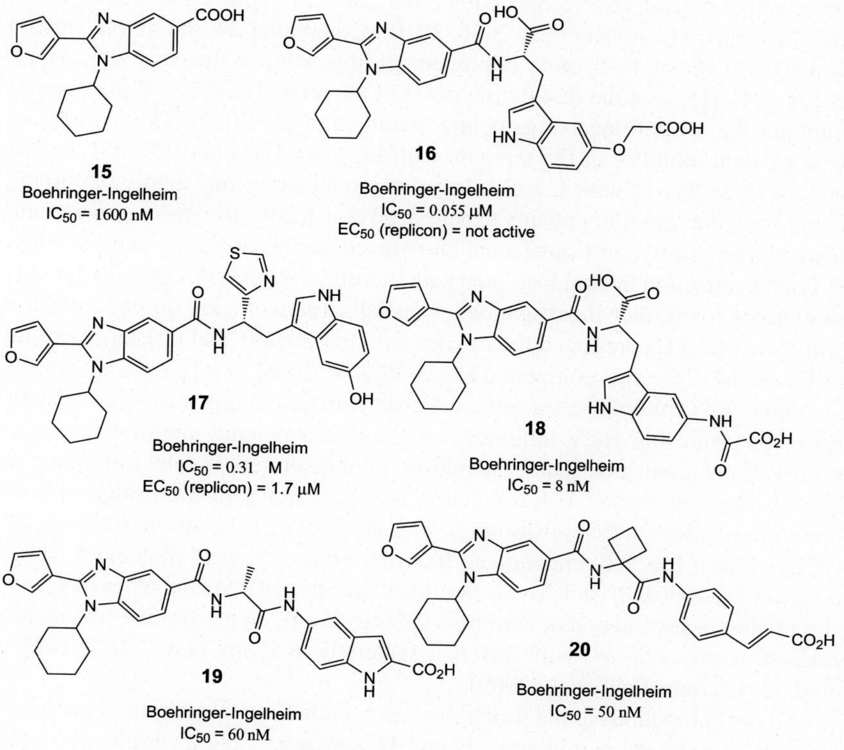

**15**
Boehringer-Ingelheim
$IC_{50}$ = 1600 nM

**16**
Boehringer-Ingelheim
$IC_{50}$ = 0.055 μM
$EC_{50}$ (replicon) = not active

**17**
Boehringer-Ingelheim
$IC_{50}$ = 0.31 M
$EC_{50}$ (replicon) = 1.7 μM

**18**
Boehringer-Ingelheim
$IC_{50}$ = 8 nM

**19**
Boehringer-Ingelheim
$IC_{50}$ = 60 nM

**20**
Boehringer-Ingelheim
$IC_{50}$ = 50 nM

FIGURE 8.4   A few active imidazole-NS5B inhibitors.

coupled with tryptophan nucleus had strong influence on the activity of the compound (**18**) ($IC_{50}$ = 8 nM, Fig. 8.4). Additionally, other classes of aromatic substituents, such as N-substituted tyrosine derivatives, were also evaluated, but were found to have much low activity on the isolated enzyme (Beaulieu et al., 2004). Introduction of alanine-based spacers between the carboxyl group and the aromatic part of the chain provided compounds with good in vitro activity (**19**, $IC_{50}$ = 60 nM) (Goulet et al., 2010). Modifications at positions 2 and 5 confirmed the strong effect of 2-furyl and 5-amido chain substitutions on the biological activity of these molecules. Cyclobutyl derivative (**20**) was found to inhibit RdRp with an $IC_{50}$ of 50 nM.

### 3.2.2 Indole 2-Carboxylic Acids

In order to improve the lipophilicity of imidazole molecules, replacement of a nitrogen atom by a carbon was tried, leading to several indole derivatives (Fig. 8.5). The core-refining from imidazole to indole scaffold was done systematically by several groups (Zhao et al., 2015). Among them, some compounds showed very good in vitro activities, such as **21** and **22** ($IC_{50}$ = 16 and 44 nM, respectively). Further optimizations of the 5-amido substitution pattern afforded more active derivatives, such as compound **23**, which strongly inhibited RdRp ($IC_{50}$ = 4 nM) (Beaulieu et al., 2006, 2010). Compound **24**, the indole version amide of **17** showed a twofold improvement in potency in the enzymatic assay. BILB 1941 (**25**) was the first thumb pocket I allosteric HCV NS5B polymerase inhibitor from Boehringer-Ingelheim) (Beaulieu et al., 2012). This compound was a potent inhibitor of the replicon with $EC_{50}$ = 84 nM and 153 nM against genotypes 1b and 1a with 153 nM. In 5-day monotherapy and a dose of 450 mg administered every 8 h, patients experienced ≥ 1 log10 decrease in mean viral load. Unfortunately, gastrointestinal intolerance precluded testing at higher doses but the reduction in viral load seen with this class of inhibitors provided proof-of-concept for further investigations. Genelabs Technologies (in collaboration with Novartis AG) presented the in vitro antiviral activity and resistance profile of **GL60667** (**26**). This compound has an $EC_{50}$ = 75 nM in a 1b replicon assay.

Since 2006, indole-based tetracyclic derivatives emerged in several studies to be promising HCV inhibitors. A significant potency improvement was observed by conformational restriction of substituted indole inhibitors at Thumb Domain Pocket I. Introduction of heteroatoms in the bridge offered opportunities for further substitution. The discovery of the strong inhibitor **27** ($IC_{50}$ = 9 nM, Fig. 8.6) encouraged focusing on the cyclized molecules (Ikegashira et al., 2006). Bristol-Myers Squibb, Genelabs, IRBM, and other laboratories identified several potent inhibitors having seven-, eight-, or nine-membered rings structures (**28–34**) with different substitutions (Ding et al., 2012; Narjes et al., 2011; Stansfield et al., 2009).

All the compounds in the series showed good biological activity in nanomolar scale. Among them, inhibitors **31** and **32** were more potent with $IC_{50}$ = 4 nM (Narjes et al., 2011).

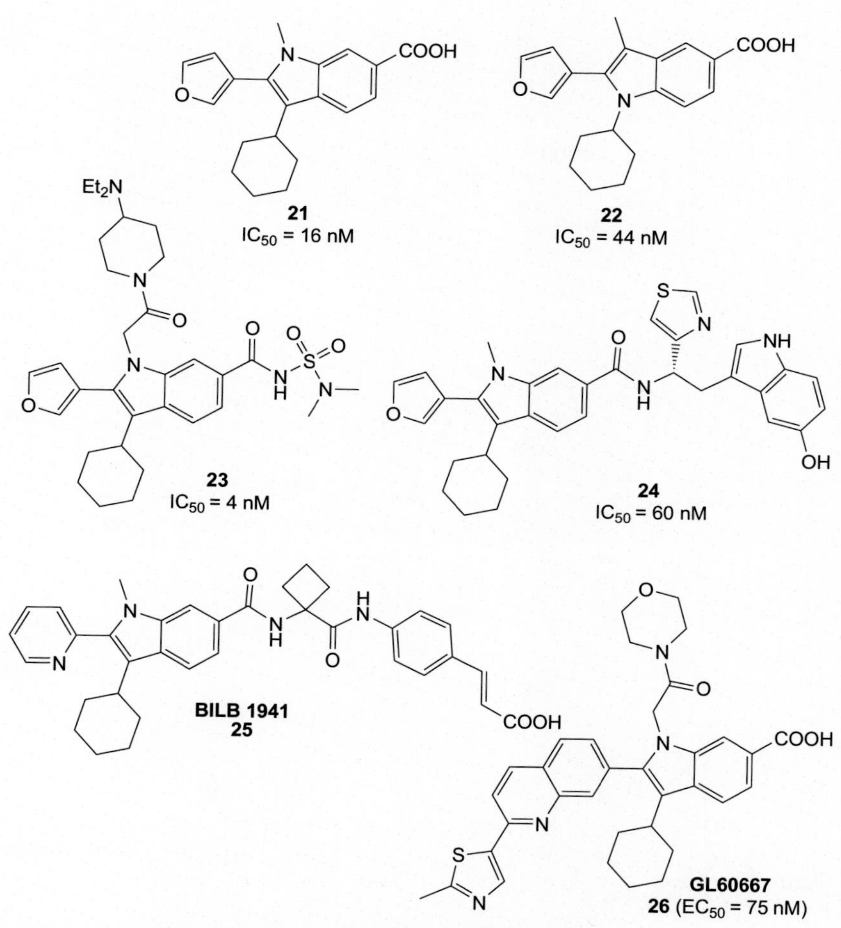

FIGURE 8.5 Some indole derivatives as HCV NS5B inhibitors.

Macrocyclic derivatives were investigated for the thumb pocket 1 inhibitors (Fig. 8.7). The compound **35** ($IC_{50}$ = 31 nM) was a potent compound (Vendeville et al., 2010). Optimization of the series led to the potent drug TMC 647055 (**36**). This was a nonzwitterionic 17-membered macrocycle possessing good potency ($EC_{50}$ = 82 nM) and good pharmacokinetic profiles in rats and dogs with minimal cell toxicity ($CC_{50}$ = >20 µM) (Vendeville et al., 2012). Phase I trial of the drug was done and it is being studied in Phase II.

## 3.2.3 Thumb Pocket II Inhibitors

Thumb Pocket II is a hydrophobic cavity situated at the base of the thumb domain of NS5B, near to the enzyme active site (~ 35 Å away). The thiophene

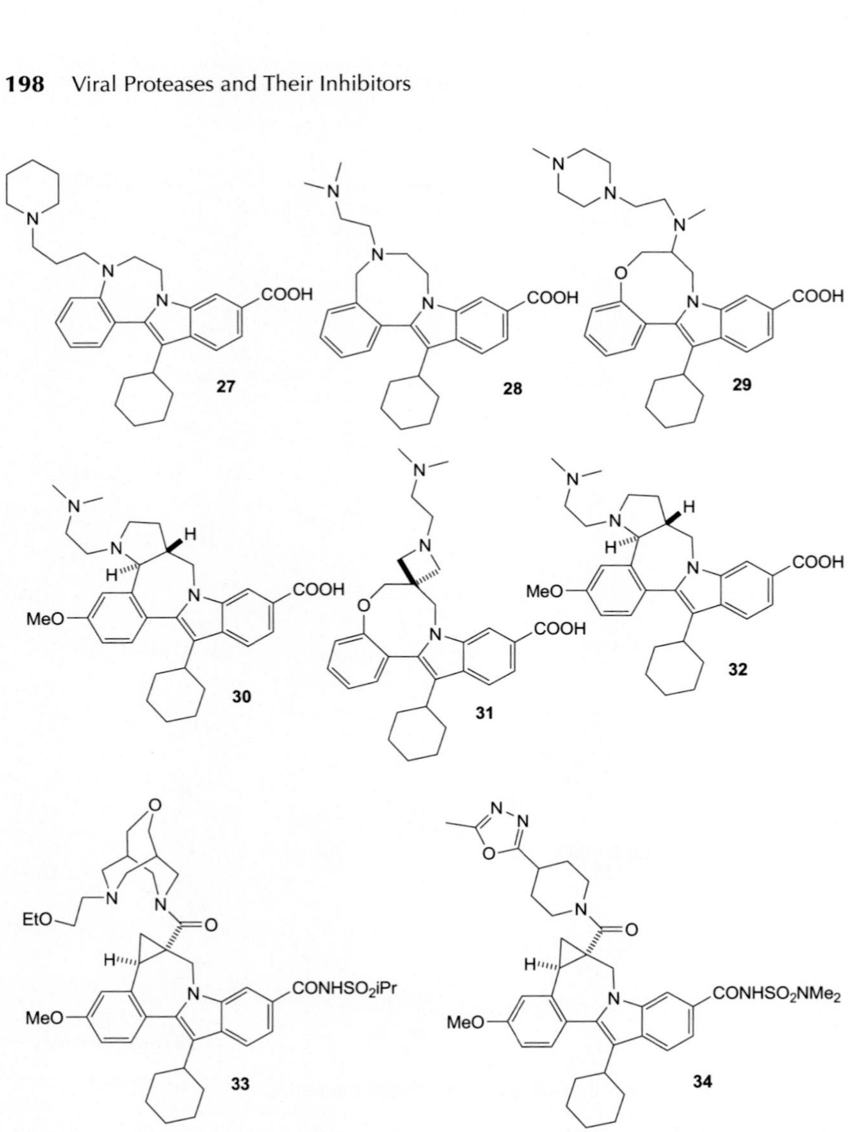

**FIGURE 8.6** **Tetracyclic, pentacyclic, and macrocycle indoles thumb pocket I inhibitors.**

carboxylic acid series was identified by Shire Biochem as the inhibitors of the site. Since the original Shire disclosures, several other companies discovered potent thiophenecarboxylic acids. Among them compound **37** (Fig. 8.8) was identified to have good potency ($IC_{50}$ = 1.5 µM) (Chan et al., 2004).

In late 2006, ViroChem Pharma identified more potent thiophenecarboxylic acids as thumb pocket II inhibitors and initiated a Phase I proof-of-concept clinical study of VCH-759 (**38**). This compound had submicromolar antiviral activity in both 1a and 1b replicons ($EC_{50}$ = 0.34 and 0.27 µM, respectively) and was well tolerated in HCV RNA with 800 mg BID and TID doses (Cooper et al., 2009). VCH-916 (**39**) was another highly protein bound (> 98%),

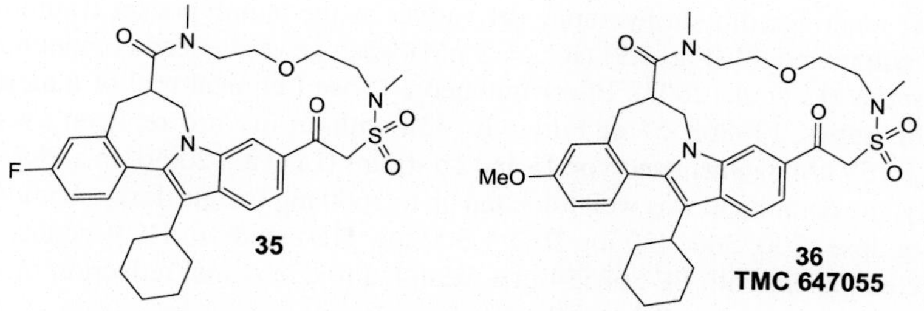

FIGURE 8.7    Macrocycle derivatives for Thumb Pocket I.

submicromolar inhibitor for both 1a and 1b replicons (1a/1b $EC_{50}$ = 79/110 nM) with an attractive preclinical ADME-PK profile.

Several other categories of chemicals were known to bind in Thumb Pocket II weakly. One of them was a group of pyranoindole derivatives from Viropharma/ Wyeth, in which a member HCV-371 (**40**, Fig. 8.9) was found to be a potent inhibitor of thumb pocket II ($IC_{50}$ = 0.33 μM, $EC_{50}$ = 4.8 μM) but failed to demonstrate any significant antiviral effects in HCV infected patients at the doses studied and both were discontinued from further development (LaPorte et al., 2008). It was highlighted that the R-enantiomer of the tested compounds were always better inhibitors than the S-enantiomer (Gopalsamy et al., 2004). Compound **41** was highly potent in the series ($IC_{50}$ = 2 nM) (LaPorte et al., 2008).

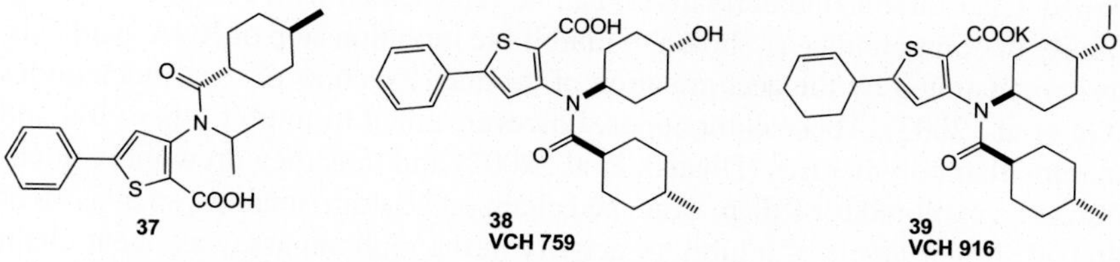

FIGURE 8.8    Thumb Pocket II inhibitors of HCV NS5B.

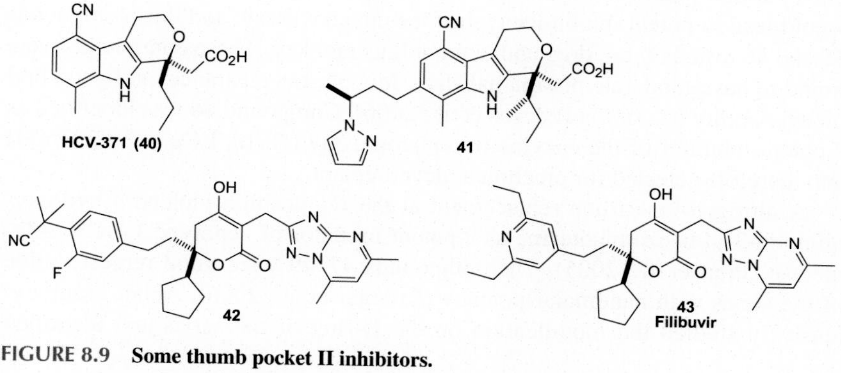

FIGURE 8.9    Some thumb pocket II inhibitors.

Pfizer identified dihydropyranone derivatives as the thumb pocket II inhibitor. The compound **42** (Fig. 8.9) ($IC_{50} = 3$ nM) was one of the lead compounds in the series (Li et al., 2007). Pfizer initiated a Phase I clinical trial of a member of this series, PF-868554 or Filibuvir (**43**), with an in vitro replicon average $EC_{50} = 59$ nM against genotype 1a and 1b strains (Li et al., 2009). In a Phase I a study, the compound was well tolerated up to 1600 mg (single dose). A multiple rising dose (100, 300, 450 mg BID + 300 mg TID) study in HCV genotype 1 infected patient with PF-868554 in a mean led to a maximal reduction in viral load ranging from 1 to 2.1 log10.

## 3.3 Palm Site I Inhibitors

The junction of the thumb and palm domains of NS5B in the close proximity to the active site of the enzyme was earlier referred to as Pocket III. Study on shapes of the compounds acting with Palm Site I and II had shown that they were distinctly different.

### 3.3.1 Benzothiadiazine-Based Inhibitors

Benzo-1,2,4-thiadiazines class of RdRp inhibitors were discovered from HTS of the GlaxoSmithKline (GSK) proprietary compounds, where this class of compounds were reported as nonnucleoside RdRp inhibitors (Dhanak et al., 2002). In the past few years, benzo-1,2,4-thiadiazine scaffold was extensively studied and were reported as HCV-RNA polymerase inhibitors (Das et al., 2011). The benzo-1,2,4-thiadiazine derivatives, can be represented by **44** (Fig. 8.10), This family of compounds was shown to inhibit the initiation step of RNA synthesis, as demonstrated by the accumulation of products, such as di- or trinucleotides (Gu et al., 2003). These compounds, however, failed to inhibit other viral and mammalian polymerases (Dhanak et al., 2002) and thus they drew more attention to be explored for RdRp. Branched chain (at N-atom) analogs, such as **45** or **46** had similar levels of inhibitory activity in the replicon assay as linear chain analogs, but had sixfold enhancement in enzyme inhibition. Structure activity studies suggested that substitution on A-ring of **44** with alkyl chains, halogens, nitro-, amino-, and carboxylic acid functionalities at the 5, 6, 7, or 8 positions would lead to potent inhibitors against the enzyme assay, and thus compounds **45** and **46** tested on rat, dog, and cynomolgus monkey. These compounds were found to have moderate plasma half-live, low plasma clearance, and good oral bioavailability ($\geq 35\%$) (Tedesco et al., 2006). Compound **46** was identified as a potent inhibitor of the enzyme in replicon assay (1b/1a, $EC_{50} = 38/207$ nM) and therefore selected for preclinical development.

A nitrogen-for-carbon replacement at the N1 position yielded 4-hydroxy-quinolin-3-yl benzothiadiazine as a potent inhibitor of genotype 1 HCV polymerase (Pratt et al., 2005). The compounds **47–49** were good representative of the series with nanomolar potency (Krueger et al., 2006). Abbot group extensively studied the modification on the D-ring of the series and identified

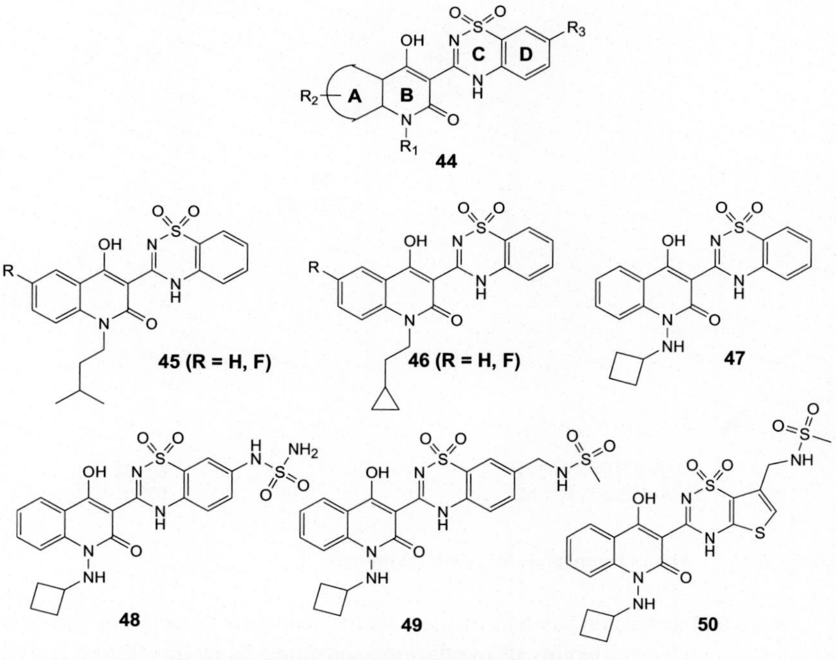

FIGURE 8.10    **4-Hydroxyquinolin-2(1H)-ones as Palm I inhibitors.**

compound **50** for animal studies. This group postulated that the π-stacking of the aromatic rings in the benzothiadiazine analogs might be the cause of their lower solubility (Rosen et al., 1988), as they had observed that the solubility of fluoroquinolone antibacterial agents could be improved by replacing nitrogen atom of dihydroquinolinedione ring system with a carbon atom. Following a similar approach, dialkyl-chains were introduced in the B-ring of benzothiadiazines.

Abbott laboratory reported the antiviral activity and resistance profile of **A-782759 (51)** (Fig. 8.11) (Mo et al., 2005), an analog of the benzothiadiazine series. Further optimization of **A-782759** series led to identify the compound **A-837093 (52)** with significantly improved metabolism and pharmacokinetic properties. This compound displayed excellent activity in HCV replicon with $EC_{50}$ of 6 nM and 11 nM against HCV genotype 1b-N and 1a-77, respectively. The rat, dog, and monkey PK profiling of **A-837093 (52)** and another compound **A-848837 (53)** were demonstrated (Wagner et al., 2009). Both the drugs were found at high concentration in rat liver tissue. The liver to plasma ratios at the 6 h time point were 17-fold for both compounds and increased slightly to 20-fold at the 12 h time point. **A-837093** was tested in chimpanzee HCV model (Chen et al., 2007). The pK profile of **A-848837** was reported (Molla et al., 2007).

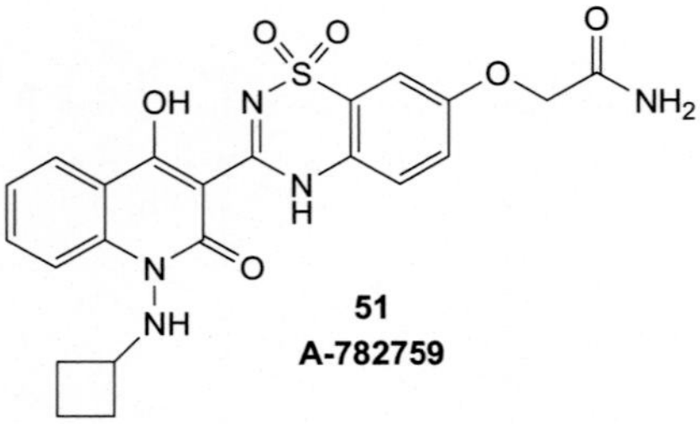


**51**
**A-782759**

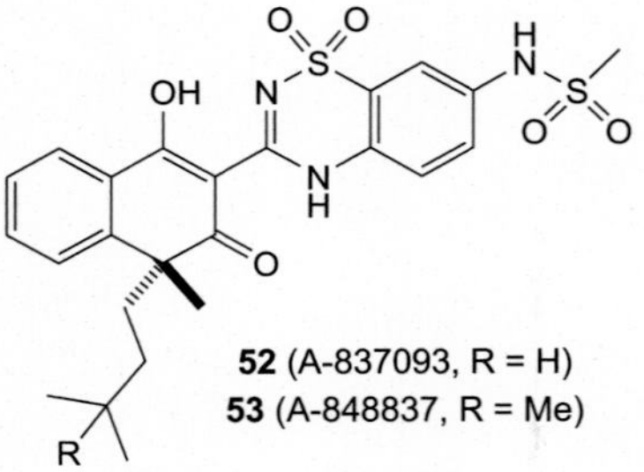


**52** (A-837093, R = H)
**53** (A-848837, R = Me)

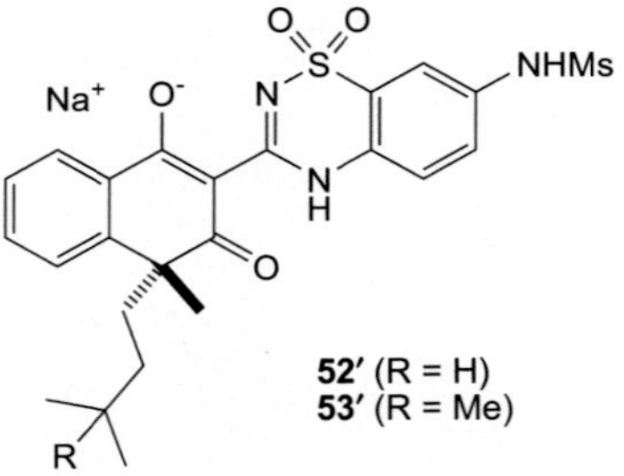


**52′** (R = H)
**53′** (R = Me)

FIGURE 8.11 **Abbott compounds for NS5B inhibitors.**

Anadys Pharmaceuticals had focused on the reduction of polar surface area to improve the bioavailability of pyridazione-substitued benzothiadiazine derivatives. The pyrrolo[1,2-*b*]pyridazinone moiety **54**, a fused five membered ring with pyridazinone ring, was assessed by Anadys Pharmaceuticals. Some of the interesting compounds in the group were **54** and **55** (Fig. 8.12), which had shown good inhibition activity ($IC_{50}$ < 10 nM) (Ruebsam et al., 2008). Reduction of the pyrrole core into pyrrolidine afforded **56**, a structural analog of compound **57** sharing the same potency ($IC_{50}$ < 10 nM). Removal of the pyrrolidine nitrogen led to a new chiral subclass of molecules bearing a fused cyclopentanylanalog, such as **58**. Partially saturated quinolinone core with a methylene or an ethylene bridge on ring A gave **ANA598** (**59**), which was a potent inhibitor of the replicon (1b/1a $EC_{50}$ = 5 and 50 nM). ANA 598, named as **Setrobuvir** was an experimental drug for the treatment of hepatitis C from Anadys and developed by Roche. It was in Phase IIb clinical trials, used in combination with interferon and ribavirin, targeting hepatitis C patients with genotype 1. **Setrobuvir** works by inhibiting the hepatitis C enzyme NS5B (Ruebsam et al., 2009).

InterMune team shifted the N-atom from the B-ring to the junction of A–B rings and made a series of azolo[1,5-*a*]pyridine benzo-thiadiazines and azolo[1,5-*a*]-pyrimidine-benzothiadiazines (Das et al., 2011). Most of the compounds with these scaffolds were demonstrated very potent NS5B inhibition activity ($IC_{50}$ = <5 nM) and exhibited promising replicon potency values (Wang et al., 2009b, c, d). Among them, compounds **60–64** (Fig. 8.13) were identified as the most potent compounds in the series and the X-ray structure of **62** bound to the polymerase was described. The most active compound in the replicon system was compound **64c** with an $EC_{50}$ value of 2.3 nM and $CC_{50}$ value

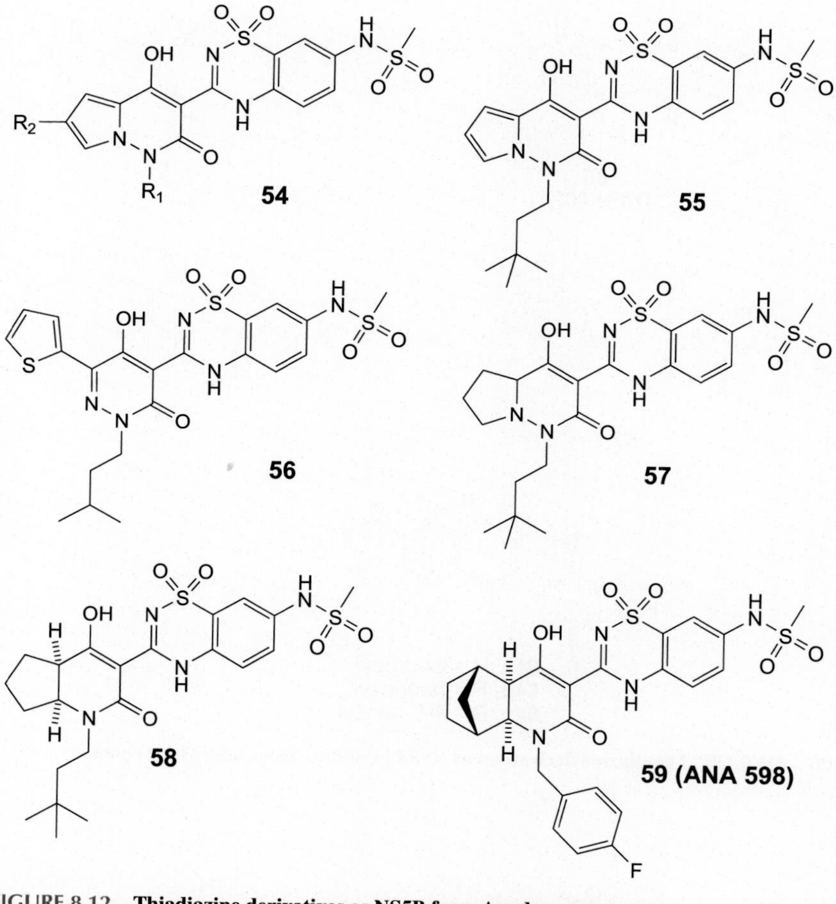

FIGURE 8.12    Thiadiazine derivatives as NS5B from Anadys company.

greater than 100 μM. In vitro combination studies of HCV protease inhibitor danoprevir with benzothaidiazine ITMN-8020 (**60**) had shown enhanced antiviral activity and suppression of drug-resistant variants (Tan et al., 2009).

### 3.3.2  Aryl Dihydrouracils and Acyl Pyrrolidines

A high-throughput screening campaign by Abbott identified an aryl dihydrouracil fragment that was subsequently modified to produce the potent, nonnucleoside inhibitor known as dasabuvir (**65**, Fig. 8.14). It is also known as ABT-333 ($IC_{50}$ levels of 2.2 nM against HCV genotypes 1 and 2) (Maring et al., 2009). Dasabuvir was similar to those selected by the benzothiadiazine class of inhibitors to bind to the palm I site of NS5B (Liu et al., 2012). The results from single ascending dose/multiple ascending dose (SAD/MAD) with ABT-333 at

**60**
**ITMN 8020**

**61**

**62**

**63**

**64a**: R = isopropyl
**64b**: R = neohexyl
**64c**: R = 4-F-benzyl

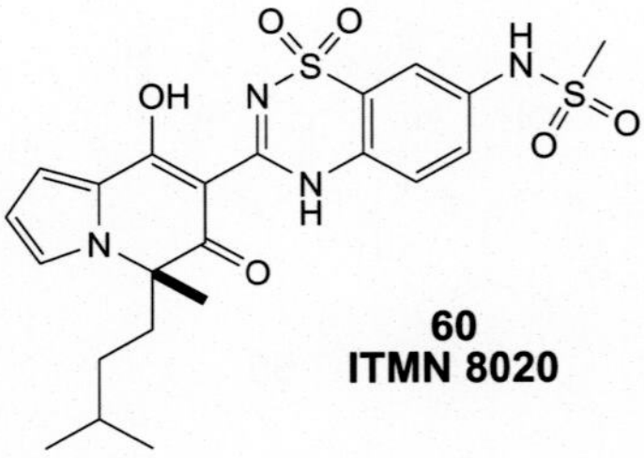

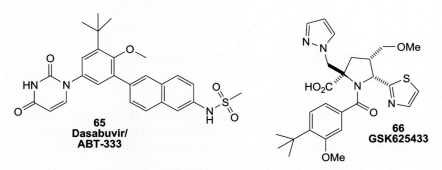

FIGURE 8.13 **Thiadiazine derivatives as NS5B inhibitosr from InterMune company.**

**65**
**Dasabuvir/**
**ABT-333**

**66**
**GSK625433**

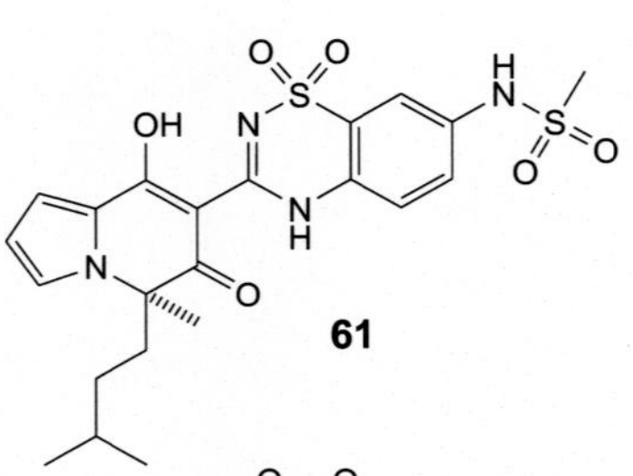

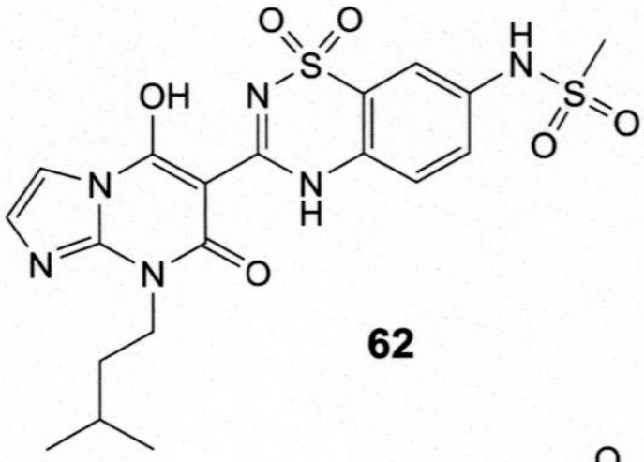

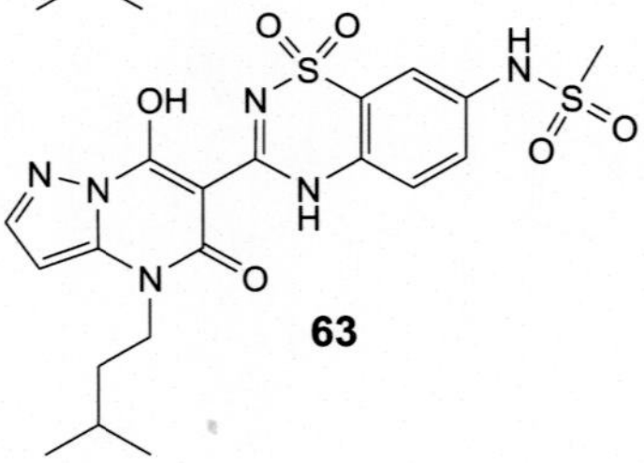

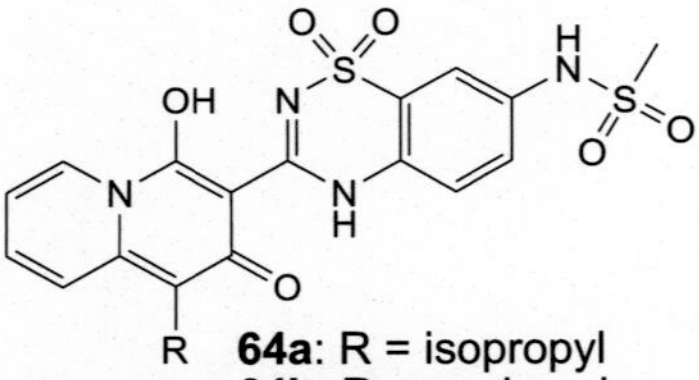

FIGURE 8.14 **Aryl dihydrouracils and acyl pyrrolidines as palm site I inhibitors.**

doses 200–400 mg BID to healthy volunteers were promising. Dasabuvir was approved in the United States by FDA (Gentile et al., 2014).

Through the HTS process, GSK identified chiral acylpyrrolidine class of inhibitors as the Palm Site I binding compound (Slater et al., 2007).

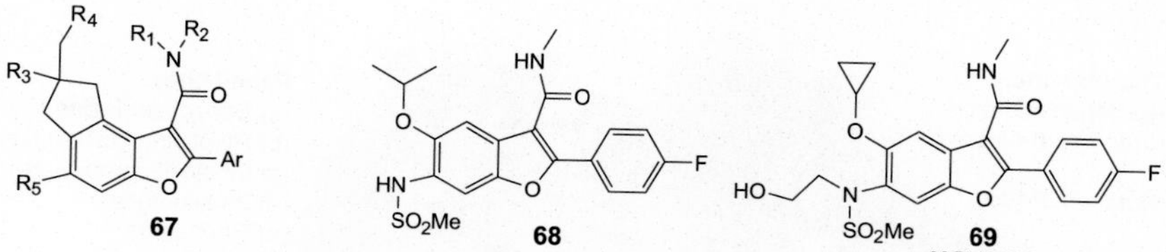

FIGURE 8.15   **Benzofuran derivatives as Palm site II inhibitors.**

Thiophene-based chiral compound with a *tert*-butyl substitution on the phenyl ring and a modified carboxylic acid moiety improved the potency. The cumulated and successful efforts led to the discovery of compound GSK625433 (**66**) ($IC_{50} = 28$ nM) (Gray et al., 2007). This compound was successfully studied in Phase I clinical trials both in healthy volunteers and infected adult patients. Unfortunately, the clinical trial was halted due the observation of long-term mouse carcinogenicity studies.

## 3.4   Palm Site II Inhibitors

### 3.4.1   Benzofuran Class of Inhibitors

Palm Site II is located at the junction of palm and thumb domains. Palm site II partially overlaps with Palm site I, in proximity to the active site of enzyme. This binding pocket was identified by the ViroPharma and Wyeth group during the study of benzofuran derivatives as the NS5B inhibitors (Fig. 8.15). Benzofuran derivatives of general structure **67** were studied and many promising candidates were found. ViroPharma/Wyeth identified two potent compounds **68** and **69** in this series. Almost a complete study on inhibition, pharmacokinetics, resistance, or mechanisms studies was performed on a compound, HCV-796 (**69**) ($IC_{50} = 40$ nM), in the series with very promising result (Howe et al., 2008; Kneteman et al. 2009). Based on favorable preclinical profile, HCV-796 was pushed into Phase I clinical trial. In 2006, it was administered as 1000 mg BID for 14 days to genotype 1 HCV-infected patients. Although Phase I clinical trials led to encouraging results, the program was stopped for safety concerns (elevation of liver enzymes) at an early stage of Phase II (ViroPharma Press Release, 2007).

## 4   MECHANISM OF THE NS5B POLYMERASE AND INHIBITOR INTERACTIONS

It has been suggested that different nonnucleoside inhibitors bind at different site of NS5B protease as shown in Fig. 8.16 (Mayhoub, 2012). Mayhoub, 2012 has pointed out that current research focuses on three points: (1) enhancing the efficacy of the IFNα formulas to improve the patient tolerance, (2) finding alternative ribavirin-like molecules to reduce its toxicity, and (3) investigating DAAs that target specific key steps of the viral life cycle (Chart 8.1).

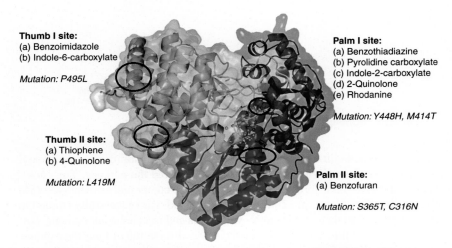

**Thumb I site:**
(a) Benzoimidazole
(b) Indole-6-carboxylate

*Mutation: P495L*

**Palm I site:**
(a) Benzothiadiazine
(b) Pyrolidine carboxylate
(c) Indole-2-carboxylate
(d) 2-Quinolone
(e) Rhodanine

*Mutation: Y448H, M414T*

**Thumb II site:**
(a) Thiophene
(b) 4-Quinolone

*Mutation: L419M*

**Palm II site:**
(a) Benzofuran

*Mutation: S365T, C316N*

**FIGURE 8.16  Summary of NNIs and their binding sites within HCV NS5B.** *(Reprinted with permission from Mayhoub, A.S., 2012. Hepatitis C RNA-dependent RNA polymerase inhibitors: a review of structure-activity and resistance relationships, different scaffolds and mutations. Bioorg. Med. Chem. 20, 3150–3161. Copyright 2012 Elsevier.)*

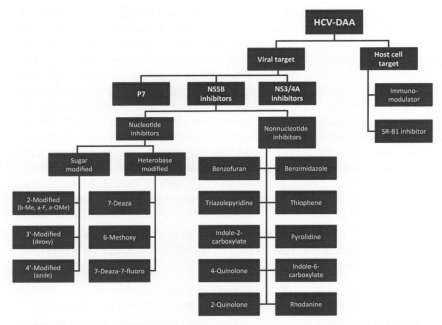

**CHART 8.1  Diagrammatic chart for HCV–DAA categories.** *(Reprinted with permission from Mayhoub, A.S., 2012. Hepatitis C RNA-dependent RNA polymerase inhibitors: a review of struc-ture-activity and resistance relationships, different scaffolds and mutations. Bioorg. Med. Chem. 20, 3150–3161. Copyright 2012 Elsevier.)*

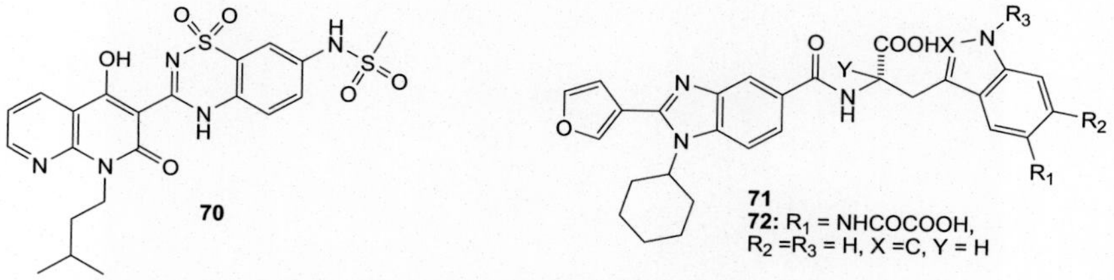

FIGURE 8.17   Structure of thiadiazine and imidazole.

Several quantitative structure–activity relationship studies (QSAR) and docking studies performed on various classes of NS5B inhibitors by Gupta and coworkers (Gupta et al., 2013; Patil et al., 2012a,b, 2013, 2014; Samanta et al., 2012) pointed out the essential physicochemical properties of a given class of compounds that may enhance their inhibition potency. In a docking study on a series benzothiadiazine derivatives as genotype 1 HCV polymerase inhibitors, Patil et al. (2012a) pointed out as to how the most active compound of the series (**70**) could interact with the NS5B polymerase. It was observed that while docked in NS5B polymerase (PDB id:2DXS), the OH moiety of this compound acted as hydrogen bond donor and formed a hydrogen bond with CO moiety of Leu492 residue of the enzyme. Further, the isopentyl group of the compound at N-1 was engulfed in a hydrophobic pocket of the enzyme formed by Ala396 and Trp397, where it could have hydrophobic interaction. Other hydrophobic interactions observed were of benzothiadiazine nucleus with Pro494, Trp500, and His428 residues of the enzyme (Fig. 8.17).

Some benzimidazole derivatives acting as NS5B polymerase inhibitors were also subjected to QSAR and molecular modeling studies by some authors (Patel et al., 2008; Patil et al., 2012b). When Patel et al. (2008) made docking studies on a series of compounds, such as shown by **71**, they found that there were significant hydrophobic and hydrogen-bond interactions between the molecules and the polymerase (PDB ID: 2dxs). With reference to a member of the series (**72**), the interaction between the inhibitor and the enzyme was as shown in Fig. 8.18. The figure shows that the benzimidazole ring forms hydrophobic contacts with residues identified as His428, Trp500, Pro495, and Val494. The cyclohexyl group attached to the N1 position of the benzimidazole ring is located in a deep hydrophobic pocket formed by Leu392, Ala395, Ala396, Ile424, Leu425, and Phe429. A total of five hydrogen bonds are shown to be formed between compound **72** and the HCV NS5B polymerase as shown by yellow dotted line in the figure. Similar binding interactions between the NS5B polymerase and another series of benzimidazoles, were observed by Patil et al. (2012b). Patil et al. had also observed that their representative compound when docked in active site of the enzyme when bound with the ligand (2dxs) overlapped with the bound ligand. This suggested that the benzimidazole nucleus successfully interacts at the active of the enzyme.

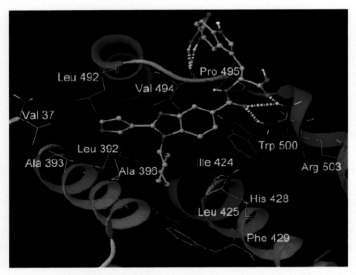

**FIGURE 8.18  A representation of binding of a benzimidazole inhibitor (72) in the allosteric site of NS5B polymerase.** *(Reprinted with the permission from Patel, P.D., Patel, M.R., Basu, N.K., Talele T.T., 2008. 3D QSAR and molecular docking studies of benzimidazole derivatives as hepatitis C virus NS5B polymerase inhibitors. J. Chem. Inform. Model. 48, 42–55. Copyright 2008 American Chemical Society.)*

## 5   DAA AND FUTURE HEPATITIS C DRUGS

In 2012, the DAA medications have been approved for the HCV treatment (Asselah and Marcellin, 2012; Bartenschlager et al., 2013). Since then, there have been many clinical trials with combination of DAAs by Abbott, Boehringer-Ingelheim, Bristol-Myers Squibb, and Gilead with cure rates of 90%–100%. DAAs can eradicate HCV infection in almost all patients with chronic liver disease, including advanced fibrosis, and inhibit continuous inflammation in the liver (Tamori et al., 2016). DAAs are classified into three groups–NS3/4A serine protease inhibitors, NS5A inhibitors, and NS5B polymerase inhibitors. DAA of NS5B are classified into two subclasses: nucleos(/t)ide competitive polymerase inhibitors and allosteric inhibitors of RNA polymerase [nonnucleos(/t)ide polymerase inhibitors, NNPIs]. NNPIs induce conformational changes in the NS5B polymerase enzyme by binding to its various allosteric sites. These agents have a lower barrier of resistance and appear to be genotype specific. A combination of therapies with multiple drugs against different HCV targets may be used to inhibit HCV replication effectively (http://www.hepatitiscnewdrugresearch.com). The problem associated with DAA treatment is that DAA regimens are too expensive to be accessible to all patients with HCV. Future research may get a direction to eradicate the HCV infection with new drugs and less expensive treatment for the HCV related diseases.

Some compounds from natural sources were also found to have NS5B polymerase inhibition activity. Ma et al. (2009) had found that triterpenes isolated

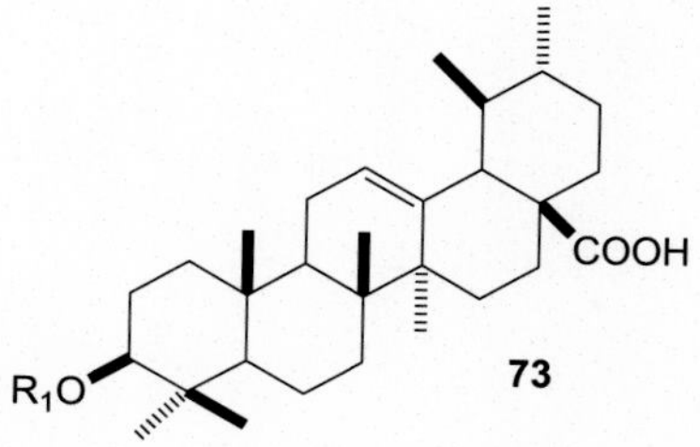

| R$_1$ | IC$_{50}$ |
|---|---|
| H | 35 |
| COCH$_3$ | 22 |
| COCOOH | 15 |
| COCH$_2$COOH | 06 |
| CO(CH$_2$)$_2$COOH | 09 |
| CO(CH$_2$)$_3$COOH | 18 |

FIGURE 8.19  Compounds from *Cynomorium songaricum* as HCV protease inhibitors.

from *Cynomorium songaricum* had HCV PR inhibition activity as shown in Fig. 8.19. The compound **73**, malonylursolic acid hemiester, was found to be the most active. These triterpenes can be developed to be therapeutically useful to treat HCV inhibition. Besides triterpenes, Patil et al. (2011) also discussed in a recent review that many other natural compounds, such as naringenin, proanthocyanidin, curcumin, epigallocatechin-3-gallate, quercetin, and abrogates also have diverse anti-HCV properties.

# 6  CONCLUSIONS

HCV infection, causing chronic hepatitis C, becomes the main cause of liver transplantation in the Western world. High annual mortality rate related to HCV infection and related diseases have enforced to discover novel therapeutic compounds for HCV treatment. In the past several years, GSK, Boehringer-Ingelheim, Anadys, Abbott, InterMune, Roche, and other pharmaceutical companies have discovered several compounds, mainly acting as HCV NS3/4A protease and HCV NS5B polymerase inhibitors, that have been pushed to clinical trial. However, due to the lack of "proof reading" function of NS5B, the HCV RNA genome can rapidly mutate to escape antiviral drug therapy. The mutations and consequent viral resistance will likely be significant concern for the development of small molecule HCV inhibitors. The viral resistance will be less expected to appear if protease inhibitors are administered with viral enzymes NS5B RNA-dependent RNA-polymerase inhibitors. In the past decade, a tremendous amount of work has been done to unravel the structure and function of the HCV RdRp, which has led to the more clear understanding of inhibitors and allosteric site interactions (Table 8.1). At the same time, a number of drugs

**TABLE 8.1 A List of Advanced NS5B Inhibitors in Clinical Trials**

| Compounds | Trade name | Originator | Developer | Class of Compound | Status | References |
|---|---|---|---|---|---|---|
| Mericitabine | R7128 | Pharmasset | Roche | Pyrimidine nucleosides | Phase II | Wedemeyer et al. (2016) |
| Balapiravir | R1626 | Pharmasset | Roche | Pyrimidine nucleosides | Phase II in 2008 | Nelson et al. (2012) |
| PSI-6130 | PSI-6130 | Pharmasset | Pharmasset | Pyrimidine nucleosides | | Ma et al. (2007) |
| PSI 7851 | PSI 7851 | Pharmasset | Pharmasset | Pyrimidine nucleosides | Phase I in 2010 | Lam et al. (2010) |
| Valopicitabine | NM283 | Idenix Pharmaceuticals | Idenix/Novartis | Nucleosides | Phase II in 2007 | Toniutto et al. (2007) |
| IDX 184 | IDX 184 | Idenix Pharmaceuticals | Idenix Pharmaceuticals | Nucleotides | Phase I/II in 2013 | |
| Nesbuvir | HCV-796 | ViroPharma | ViroPharma, Wyeth | Benzofurans | Phase I in 2008 | Kneteman et al. (2009); Howe et al. (2008) |
| ABT-072 | ABT-072 | Abbott | AbbVie | Small molecule | Phase II | |
| Dasabuvir | ABT-333 | Abbott | AbbVie | Pyrimidines | *Marketed*, FDA approved 2014 | Maring et al. (2009) |
| VCH-759 | VCH-759 | Shire Pharmaceuticals | Vertex | Small molecules | Phase I in 2014 | Cooper et al. (2009) |
| VCH 916 | VCH-916 | Shire Pharmaceuticals | Vertex | Small molecules | Phase I in 2010 | Tan and He (2011); Proulx et al. (2008) |

| | | | | | | |
|---|---|---|---|---|---|---|
| Filibuvir | PF-868554 | Pfizer | Pfizer | Pyrimidines | Phase II in 2013 | Jiao et al. (2014) |
| BILB-1941 | BILB-1941 | Boehringer-Ingelheim | Boehringer-Ingelheim | Small molecule | Phase II in 2014 | |
| IDX 375 | IDX 375 | Idenix | Idenix | Small molecule | Phase II in 2011 | |
| Tegobuvir | GS9190 | Gilead Sciences | Gilead Sciences | Small molecule | Phase II in 2011 | |
| MK-3281 | MK-3281 | Merck | Merck | | Phase I in 2009 | |
| MK 3682 | IDX 21437 | Idenix Pharmaceuticals | Merck | Uracil nucleotides | Phase II | |
| MK 0608 | MK 0608 | Merck | Merck | Small molecule | Phase I in 2014 | |
| Deleobuvir | BI 207127 | Boehringer-Ingelheim | Boehringer-Ingelheim | Indoles | Phase I | |
| Lomibuvir | VCH-222/VX-222 | Virochem Pharma | Vertex | Small molecule | Phase II in 2014 | |
| TMC647055 | TMC647055 | Janssen | Janssen | Macrocyclic compound | Phase II | |
| Setrobuvir | ANA 598 | Anadys Pharmaceuticlas | Anadys Pharmaceuticlas, Roche | Benzothiadiazines | Phase II | Ruebsam et al. (2009) |
| JTK-109 | JTK-109 | Japan Tobacco Co. | Japan Tobacco Co. | Imidazoles | Discontinued at Phase I in 2003 | |
| GSK 625433 | GSK 625433 | GlaxoSmithKline | GlaxoSmithKline | Pyrrolidines | Discontinued at Phase I in 2008 | |
| Sofosbuvir | GS 7977 | Gilead Sciences | Gilead Sciences | Small molecules | Marketed, completes a phase III trial in hepatitis C (treatment-naive) in Estonia and Russia | |

have been stopped at phase I or Phase II stage of clinical trial due to safety issue. The structural diversity of identified inhibitors provides a unique opportunity to perform QSAR. The large number of identified potent inhibitors of RdRp will be the source for clinical development in the near future.

## REFERENCES

Ago, H., Adachi, T., Yoshida, A., Yamamoto, M., Habuka, N., Yatsunami, K., Miyano, M., 1999. Crystal structure of the RNA-dependent-RNA polymerase of hepatitis C virus. Structure 7, 1417–1426.

Alter, M.J., 2007. Epidemiology of hepatitis C virus infection. World J. Gastroenterol. 13, 2436–2441.

Alter, M.J., Kruszon-Moran, D., Nainan, O.V., McQuillan, G.M., Gao, F., Moyer, L.A., Kaslow, R.A., Margolis, H.S., 1999. The prevalence of hepatitis C virus infection in the United States, 1988 through 1994. New Eng. J. Med. 341, 556–562.

Alter, M.J., Margolis, H.S., Krawczynski, K., Judson, F.N., Mares, A., Alexander, W.J., Hu, P.Y., Miller, J.K., Gerber, M.A., Sampliner, R.E., 1992. The natural history of community-acquired hepatitis C in the United states. New Engl. J. Med. 327, 1899–1950.

Asselah, T., Marcellin, P., 2012. Direct-acting antivirals for the treatment of chronic hepatitis C: one pill a day for tomorrow. Liver Int. 32, 88–102.

Au, J.S., Pockros, P.J., 2014. Novel therapeutic approaches for hepatitis C. Clin. Pharmacol. Ther. 95, 78–88.

Bartenschlager, R., Lohmann, V., Penin, F., 2013. The molecular and structural basis of advanced antiviral therapy for hepatitis C virus infection. Nat. Rev. Microbiol. 11, 482–496.

Beaulieu, P.L., Bös, M., Bousquet, Y., Fazal, G., Gauthier, J., Gillard, J., Goulet, S., LaPlante, S., Poupart, M.-A., Lefebvre, S., McKercher, G., Pellerin, C., Austel, V., Kukolj, G., 2004. Non-nucleoside inhibitors of the hepatitis C virus NS5B polymerase: discovery and preliminary SAR of benzimidazole derivatives. Bioorg. Med. Chem. Lett. 14, 119–124.

Beaulieu, P.L., Bös, M., Cordingley, M.G., Chabot, C., Fazal, G., Garneau, M., Gillard, J.R., Jolicoeur, E., LaPlante, S., McKercher, G., Poirier, M., Poupart, M.A., Tsantrizos, Y.S., Duan, J., Kukolj, G., 2012. Discovery of the first thumb pocket 1 NS5B polymerase inhibitor (BILB 1941) with demonstrated antiviral activity in patients chronically infected with genotype 1 hepatitis C virus (HCV). J. Med. Chem. 55, 7650–7666.

Beaulieu, P.L., Gillard, J., Bykowski, D., Brochu, C., Dansereau, N., Duceppe, J.-S., Haché, B., Jakalian, A., Lagacé, L., LaPlante, S., McKercher, G., Moreau, E., Perrault, S., Stammers, T., Thauvette, L., Warrington, J., Kukolj, G., 2006. Improved replicon cellular activity of non-nucleoside allosteric inhibitors of HCVNS5B polymerase: From benzimidazole to indole scaffolds. Bioorg. Med. Chem. Lett. 16, 4987–4993.

Beaulieu, P.L., Jolicoeur, E., Gillard, J., Brochu, C., Coulombe, R., Dansereau, N., Duan, J., Garneau, M., Jakalian, A., Kühn, P., Lagacé, L., LaPlante, S., McKercher, G., Perrault, S., Poirier, M., Poupart, M.-A., Stammers, T., Thauvette, L., Thavonekham, B., Kukolj, G., 2010. N-Acetamideindolecarboxylic acid allosteric 'finger-loop' inhibitors of the hepatitis C virus NS5B polymerase: Discovery and initial optimization studies. Bioorg. Med. Chem. Lett. 20, 857–861.

Bressanelli, S., Tomei, L., Roussel, A., et al., 1999. Crystal structure of the RNA-dependent RNA polymerase of hepatitis C virus. Proc. Natl. Acad. Sci. USA 96, 13034–13039.

Carroll, S.S., Olsen, D.B., 2006. Nucleoside analog inhibitors of hepatitis C virus replication. Infect. Disord. Drug. Targets 6, 17–29.

Casey, L.C., Lee, W.M., 2013. Hepatitis C virus therapy update. Curr. Opin. Gastroenterol. 29, 243–249.

Chan, L., Pereira, O., Reddy, T.J., et al., 2004. Discovery of thiophene-2-carboxylic acids as potent inhibitors of HCV NS5B polymerase and HCV subgenomic RNA replication. Part 2: tertiary amides. Bioorg. Med. Chem. Lett. 14, 797–800.

Chen, C.-M., He, Y., Lu, L., Lim, H.B., Tripathi, R.L., Middleton, T., Hernandez, L.E., Beno, D.W.A., Long, M.A., Kati, W.M., Bosse, T.D., Larson, D.P., Wagner, R., Lanford, R.E., Kohlbrenner, W.E., Kempf, D.J., Pilot-Matias, T.J., Molla, A., 2007. Activity of a potent hepatitis C virus polymerase inhibitor in the chimpanzee model. Antimicrob. Agents Chemother. 51, 4290–4296.

Choo, Q.L., Kuo, G., Weiner, A.J., Overby, L.R., Bradley, D.W., Houghton, M., 1989. Isolation of a cDNA clone derived from a blood-borne non-A, non-B viral hepatitis genome. Science 244, 359–362.

Choo, Q.L., Richman, K.H., Han, J.H., Berger, K., Lee, C., Dong, C., Gallegos, C., Coit, D., Medina-Selby, A., Barr, P.J., Weiner, A.J., Bradley, D.W., Kuo, G., Houghton, M., 1991. Genetic organization and diversity of the hepatitis C virus. Proc. Natl. Acad. Sci. USA 88, 2451–2455.

Cooper, C., Lawitz, E.J., Ghali, P., Roodriguez-Torres, M., Anderson, F.H., Lee, S.S., Bedard, J., Chauret, N., Thibert, R., Boivin, I., Nicolas, O., Proulx, L., 2009. Evaluation of VCH-759 monotherapy in hepatitis C infection. J. Hepatol. 51, 39–46.

Das, D., Hong, J., Chen, S.-H., Wang, G., Beigelman, L., Seiwert, S.D., Buckman, B.O., 2011. Recent advances in drug discovery of benzothiadiazine and related analogs as HCV NS5B polymerase inhibitors. Bioorg. Med. Chem. 19, 4690–4703.

Dhanak, D., Duffy, K.J., Johnston, V.K., Lin-Goerke, J., Darcy, M., Shaw, A.N., Gu, B., Silverman, C., Gates, A.T., Nonnemacher, M.R., Earnshaw, D.L., Casper, D.J., Kaura, A., Baker, A., Greenwood, C., Gutshall, L.L., Maley, D., DelVecchio, A., Macarron, R., Hofmann, G.A., Alnoah, Z., Cheng, H.-Y., Chan, G., Khandekar, S., Keenan, R.M., Sarisky, R.T., 2002. Identification and biological characterization of heterocyclic inhibitors of the hepatitis C virus RNA-dependant RNA polymerase. J. Biol. Chem. 277, 38322–38327.

Ding, M., He, F., Hudyma, T.W., Zheng, X., Poss, M.A., Kadow, J.F., Beno, B.R., Rigat, K.L., Wang, Y.-K., Fridell, R., Lemm, J.A., Qiu, D., Liu, M., Voss, S., Pelosi, L., Roberts, S.B., Gao, M., Knipe, J., Gentles, R.G., 2012. Synthesis and SAR studies of novel heteroaryl fused tetracyclic indole-diamide compounds:Potent allosteric inhibitors of the hepatitis C virus NS5B polymerase. Bioorg. Med. Chem. Lett. 22, 2866–2871.

Elloumi, H., Houissa, F., Hadj, N.B., Gargouri, D., Romani, M., Kharrat, J., Ghorbel, A., 2007. Sudden hearing loss associated with peginterferon and rivavirin combination therapy during hepatitis C treatment. World J. Gastroenterol. 13, 5411–5412.

Farci, P., 2001. Hepatic tumors in children. Clin. Liver Dis. 5, 895–916.

Gentile, I., Buonomo, A.R., Borgia, G., 2014. Dasabuvir: a non-nucleoside inhibitor of NS5B for the treatment of hepatitis C virus infection. Rev. Recent Clin. Trials 9 (2), 115–123.

Ghany, M.G., Strader, D.B., Thomas, D.L., 2009. Diagnosis, management, and treatment of hepatitis C: an update. Hepatology 49, 1335–1374.

Gopalsamy, A., Lim, K., Ciszewski, G., Park, K., Ellingboe, J.W., Bloom, J., Insaf, S., Upeslacis, J., Mansour, T.S., Krishnamurthy, G., Damarla, M., Pyatski, Y., Ho, D., Howe, A.Y.M., Orlowski, M., Feld, B., O'Connell, J., 2004. Discovery of pyrano[3,4-b]indoles as potent and selective HCV NS5B polymerase inhibitors. J. Med. Chem. 47, 6603–6608.

Goulet, S., Poupart, M.-A., Gillard, J., PoirierM, Kukolj, G., Beaulieu, P.L., 2010. Discovery of benzimidazolediamide finger loop (Thumb I Pocket) allosteric inhibitors ofHCVNS5B polymerase: implementing parallel synthesis for rapid linker optimization. Bioorg. Med. Chem. 20, 196–200.

Gray, F., Amphlett, E., Bright, H., Chambers, L., Cheasty, A., Fenwick, R., Haigh, D., Hartley, D., Howes, P., Jarvest, R., Mirzai, F., Nerozzi, F., Parry, N., Slater, M., Smith, S., Thommes, P., Wilkinson, C., Williams, E., 2007. GSK625433; A novel and highly potent inhibitor of the HCVNS5B polymerase. J. Hepatol. 46, S225–S1225.

Gu, B., Johnston, V.K., Lester, L., Gutshall, L.L., Nguyen, T.T., Gontarek, R.R., Darcy, M.G., Tedesco, R., Dhanak, D., Duffy, K.J., Kao, C., Sarisky, R.T., 2003. Arresting initiation of hepatitis C virus RNA synthesis using heteocyclic derivatives. J. Biol. Chem. 278, 16602–16607.

Gupta, S.P., Samanta, S., Masand, N., Patil, V.M., 2013. k-Nearest neighbor—molecular field analysis on human HCV NS5B polymerase inhibitors: 2, 5-disubstituted imidazo[4,5-c]pyridines. Med. Chem. Res. 22, 330–339.

Haudecoeur, R., Peuchmaur, M., Ahmed-Belkeacem, A., Pawlotsky, J., Boumendjel, A., 2013. Structure–activity relationships in the development of allosteric hepatitis C virus RNA-dependent RNA polymerase inhibitors: ten years of research. Med. Res. Rev. 33, 934–984.

Hirashima, S., Suzuki, T., Ishida, T., et al., 2006. Benzimidazole derivatives bearing substituted biphenyls as hepatitis C virus NS5B RNA-dependent RNA polymerase inhibitors: structure-activity relationship studies and identification of a potent and highly selective inhibitor JTK-109. J. Med. Chem. 49, 4721–4736.

Hoofnagle, J.H., 2002. Course and outcome of hepatitis C. Hepatology 36, S21–S29.

Howe, A.Y., Cheng, H., Johann, S., Mullen, S., Chunduru, S.K., Young, D.C., Bard, J., Chopra, R., Krishnamurthy, G., Mansour, T., O'Connell, J., 2008. Molecular mechanism of hepatitis C virus replicon variants with reduced susceptibility to a benzofuran inhibitor, HCV-796. Antimicrob. Agents Chemother. 52, 3327–3338.

Ikegashira, K., Oka, T., Hirashima, S., Noji, S., Yamanaka, H., Hara, Y., Adachi, T., Tsuruha, J.-I., Doi, S., Hase, Y., Noguchi, T., Ando, I., Ogura, N., Ikeda, S., Hashimoto, H., 2006. Discovery of conformationally constrained tetracyclic compounds as potent hepatitis C virus NS5B RNA polymerase inhibitors. J. Med. Chem. 49, 6950–6953.

Jiao, P., Xue, W., Shen, Y., Jin, N., Liu, H., 2014. Understanding the drug resistance mechanism of hepatitis C Virus NS5B to PF-00868554 due to mutations of the 423 site: a computational study. Mol. Biosyst. 10, 767–777.

Kneteman, N.M., Howe, A.Y.M., Gao, T., Lewis, J., Pevear, D., Lund, G., Douglas, D., Mercer, D.F., Tyrrell, D.L.J., Immermann, F., Chaudhary, I., Speth, J., Villano, S.A., O'Connell, J., Collett, M., 2009. HCV796: A selective nonstructural protein 5B polymerase inhibitor with potent anti-hepatitis C virus activity in vitro, in mice with chimeric human livers, and in humans infected with hepatitis C virus. Hepatology 49, 745–752.

Krueger, A.C., Madigan, D.L., Jiang, W.W., Kati, W.M., Liu, D., Liu, Y., Maring, C.J., Masse, S., McDaniel, K.F., Middleton, T., Mo, H., Molla, A., Montgomery, D., Pratt, J.K., Rockway, T.W., Zhang, R., Kempf, D.J., 2006. Inhibitors of HCV NS5B polymerase: synthesis and structure-activity relationships of N-alkyl-4-hydroxyquinolon-3-yl-benzothiadiazine sulfamides. Bioorg. Med. Chem. Lett. 16, 3367–3370.

Lam, A.M., Murakami, E., Espiritu, C., Steuer, H.M., Niu, C., Keilman, M., Bao, H., Zennou, V., Bourne, N., Julander, J.G., Morrey, J.D., Smee, D.F., Frick, D.N., Heck, J.A., Wang, P., Nagarathnam, D., Ross, B.S., Sofia, M.J., Otto, M.J., Furman, P.A., 2010. PSI-7851 a pronucleotide of b-D-2'-Deoxy-2'-fluoro-2'C-methyluridine monophosphate is a potent and pan-genotype inhibitor of hepatitis C virus replication. Antimicrob. Agents Chemother. 54, 3187–3196.

LaPorte, M.G., Jackson, R.W., Draper, T.L., Gaboury, J.A., Galie, K., Herbertz, T., Hussey, A.R., Rippin, S.R., Benetatos, C.A., Chunduru, S.K., Christensen, J.S., Coburn, G.A., Rizzo, C.J., Rhodes, G., O'Connell, J., Howe, A.Y.M., Mansour, T.S., Collett, M.S., Pevear, D.C., Young, D.C., Gao, T., Tyrrell, D.L.J., Kneteman, N.M., Burns, C.J., Condon, S.M., 2008. The discovery of pyrano[3,4-b]indole-based allosteric inhibitors of HCV NS5B polymerase with in vivo activity. ChemMedChem 3, 1508–1515.

Laskus, T., Operskalski, E.A., Radkowski, M., Wilkinson, J., Mack, W.J., de Giacomo, M., Alharthi, L., Chen, Z., Xu, J., Kovacs, A., 2007. Negative-strand Hepatitis C Virus (HCV) RNA in peripheral blood mononuclear cells from anti-HCV-positive/ HIV – infected women. J. Infect. Diseases 195, 124–133.

Lawitz, E., Nguyen, T., Younes, Z., Santoro, J., Gitlin, N., McEniry, D., et al., 2007. Clearance of HCV RNA with valopicitabine (NM283) plus peginterferon in treatment-naïve patients with HCV-1 infection: results at 24 and 48 weeks. J. Hepatol. 46 (Suppl. 1), S9.

Lesburg, C.A., Cable, M.B., Ferrari, E., Hong, Z., Mannarino, A.F., Weber, P.C., 1999. Crystal structure of the RNA-dependent RNA polymerase from hepatitis C virus reveals a fully encircled active site. Nat Struct. Biol. 6, 937–943.

Li, H., Linton, A., Tatlock, J., Gonzalez, J., Borchardt, A., Abreo, M., Patel, L., Drowns, M., Ludlum, S., Goble, M., Yang, M., Blazel, J., Rahavendran, S.V., Skor, H., Shi, S.T., Lewis, C., Fuhrman, S., 2007. Allosteric inhibitors of hepatitis C polymerase: discovery of potent and orally bioavailable carbon-linked dihydropyranones. J. Med. Chem. 50, 3969–3972.

Li, H., Tatlock, J., Linton, A., Gonzalez, J., Jewell, T., Patel, L., Ludlum, S., Drowns, M., Rahavendran, S.V., Skor, H., Hunter, R., Shi, S.T., Herlihy, K.J., Parge, H., Hickey, M., Yu, X., Chau, F., Nonomiya, J., Lewis, C., 2009. Discovery of (R)-6-cyclopentyl-6-2-(2,6-diethylpyridin-4-yl) ethyl)-3-((5,7-dimethyl-[1,2,4]triazolo[1,5-a]pyrimidin-2-yl)methyl)-4-hydroxy-5,6-dihydropyrano- 2-one (PF-00868554) as a potent and orally available hepatitis C virus polymerase inhibitor. J. Med. Chem. 52, 1255–1258.

Lindenbach, B.D., Rice, C.M., 2005. Unraveling hepatitis C virus replication from genome to function. Nature 436, 933–938.

Liu, Y., Lim, B.H., Jiang, W.W., Flentge, C.A., Hutchinson, D.K., Madigan, D.L., Randolph, J.T., Wagner, R., Maring, C.J., Kati, W.M., Molla, A., 2012. Identification of aryl dihydrouracil derivatives as palm initiation site inhibitors of HCV NS5B polymerase. Bioorg. Med. Chem. Lett. 22, 3747–3750.

Ma, H., Jiang, W.R., Robledo, N., Leveque, V., Ali, S., Lara-Jaime, T., Masjedizadeh, M., Smith, Cammack, N., Klumpp, K., Symons, J., 2007. Characterization of the metabolic activation of hepatitis C virus nucleoside inhibitor β-D-2′-deoxy-2′-fluoro-2′-C-methylcytidine (PSI-6130) and identification of a novel active 5'-triphosphate species. J. Biol. Chem. 282, 29812–29820.

Ma, C.M., Wei, Y., Wang, Z.G., Hattori, M., 2009. Triterpenes from *Cynomorium songaricum*— analysis of HCV protease inhibitory activity, quantification, and content change under the influence of heating. J. Nat. Med. 63, 9–14.

Manns, M.P., Wedemeyer, H., Cornberg, M., 2006. Treating viral hepatitis C: efficacy, side effects and compilcations. Gut 55, 1350–1359.

Maring, C, Wagner, R., Hutchinson, D., Flentge, C., Kati, W., Koev, G. Liu, Y., Beno, D., Shen, J., Lau, Y., Gao, Y., Fischer, J., Vaidyanathan, S., Lim, H., Beyer, J., Mondal, R., Molla, A., 2009. Preclinical Potency, Pharmacokinetic and ADME Characterization of ABT-333, A Novel Non-Nucleoside HCV Polymerase Inhibitor EASL, 44th Annual Meeting Copenhagen, Denmark, April 22–26, 2009.

Mayhoub, A.S., 2012. Hepatitis C RNA-dependent RNA polymerase inhibitors: a review of structure-activity and resistance relationships, different scaffolds and mutations. Bioorg. Med. Chem. 20, 3150–3161.

Meyer, M.F., Lehmann, M., Cornberg, M., Wiegand, J., Manns, M.P., Klade, C., Wedemeyer, H., 2007. Clearance of low levles of HCV viremiain the absence of a strong adaptive immune responce. Virol. J. 4, 1–11.

Mo, H., Lu, L., Piloy-Matias, T., Pittawalla, R., Mondal, R., Masse, S., Dekhtyar, T., Ng, T., Koev, G., Stoll, V., Stewart, K.D., Pratt, J., Donner, P., Rockway, T.W., Maring, C., Molla, A., 2005. Mutations conferring resistanceto a hepatitis C virus (HCV) RNA-dependant RNA polymerase inhibitor alone or in combination with an HCV serine protease inhibitor in vitro. Antimicrob. Agents Chemother. 49, 4305–4314.

Molla, A., Wagner, R., Lu, L., He, D., Chen, C.M., Koev, G., Masse, S., Cai, Y., Klein, C., Beno, D., Hernandez, L., Krishnan, P., Pithawalla, R., Pilot-Matias, T., Middleton, T., Lanford, R., Kati, W., Kempf, D., 2007. Characterization of pharmacokinetic/pharmacodynamic parameters for the novel HCV polymerase inhibitor A-848837. J. Hepatol. 46 (Suppl. 1), S234–S235.

Narjes, F., Crescenzi, B., Ferrara, M., Habermann, J., Colarusso, S., del Rosario Rico Ferreira, M., Stansfield, I., Mackay, A.C., Conte, I., Ercolani, C., Zaramella, S., Palumbi, M.-C., Meuleman, P., Leroux-Roels, G., Giuliano, C., Fiore, F., DiMarco, S., Baiocco, P., Koch, U., Migliaccio, G., Altamura, S., Laufer, R., De Francesco, R., Rowley, M., 2011. Discovery of (7R)-14-cyclohexyl-7-{[2-(dimethylamino)ethyl](methyl)-amino}-7,8-dihydro-6H-indolo[1,2-e][1,5]benzoxazocine-11-carboxylic acid (MK-3281), a potentand orally bioavailable finger-loop inhibitor of the hepatitis C virus NS5B polymerase. J. Med. Chem. 54, 289–301.

Nelson, D.R., Zeuzem, S., Andreone, P., Ferenci, P., Herring, R., Jensen, D.M., Marcellin, P., Pockros, P.J., Rodríguez-Torres, M., Rossaro, L., Rustgi, V.K., Sepe, T., Sulkowski, M., Thomason, I.R., Yoshida, E.M., Chan, A., Hill, G., 2012. Balapiravir plus peginterferon alfa-2a (40KD)/ribavirin in a randomized trial of hepatitis C genotype 1 patients. Ann. Hepatol. 11 (1), 15–31.

Patel, P.D., Patel, M.R., Basu, N.K., Talele, T.T., 2008. 3D QSAR and molecular docking studies of benzimidazole derivatives as hepatitis C virus NS5B polymerase inhibitors. J. Chem. Inform. Model. 48, 42–55.

Patil, V.M., Gupta, S.P., Samanta, S., Masand, N., 2011. Current perspective of HCV NS5B inhibitors: a review. Curr. Med. Chem. 18, 5564–5597.

Patil, V.M., Gupta, S.P., Samanta, S., Masand, N., 2012a. 3D-QSAR and docking studies on a series of benzothiadiazine derivativesas genotype 1 HCV polymerase inhibitors. Med. Chem. 8, 1099–1107.

Patil, V.M., Gupta, S.P., Samanta, S., Masand, N., 2013. Virtual Screening of Imidazole Analogs as potential hepatitis C virus NS5B polymerase inhibitors. Chem. Papers 67, 236–244.

Patil, V.M., Gupta, S.P., Samanta, S., Masand, N., 2014. QSAR analysis of some heterocyclic compounds as HCV NS5B polymerase inhibitors. J. Eng. Sci. Manag. Educ. 7, 215–219.

Patil, V.M., Gurukumar, K.R., Chudayeu, M., Gupta, S.P., Samanta, S., Masand, N., Basu, K., 2012b. Synthesis, In vitro and in silico NS5B polymerase inhibitory activity of benzimidazole derivatives. Med. Chem. 8, 629–635.

Pratt, J.K., Donner, P., McDaniel, K.F., Maring, C.J., Kati, W.K., Mo, H., Middleton, T., Liu, Y., Ng, T., Xie, Q., Zhang, R., Montgomery, D., Molla, A., Kempf, D.J., Kohlbrenner, W., 2005. Inhibitors of HCV NS5B polymerase: synthesis and structure-activity relationships of N-1 heteroalkyl- 4-hydroxyquinolon-3-yl-benzothiadiazines. Bioorg. Med. Chem. Lett. 15, 1577–1582.

Proulx L, Bourgault B, Chauret N, et al., 2008. Results of a safety, tolerability and pharmacokinetic phase I study of VCH-916, a novel polymerase inhibitor for HCV, following single ascending doses in healthy volunteers. 43rd Annual Meeting of the European Association for the Study of the Liver, Milan, Italy, April 23–27, 2008.

Rickheim, D.G., Nelsen, C.J., Fassett, J.T., Timchenko, N.A., Hansen, L.K., Albrecht, J.H., 2002. Differntialregulation of cyclins D1 and D3 in hepatocyte proliferation. Hepatology 36, 30–38.

Rosen, T., Chu, D.T.W., Lico, I.M., Fernandes, P.B., Marsh, K., Shen, L., Cepa, V.G., Pernet, A.G., 1988. Deign, synthesis and properties of (4S)-7-(4-amino-2-substituted-pyrrolidine-1-yl)quinolone-2-carboxylic acids. J. Med. Chem. 31, 1598–1611.

Rosenberg, S., 2001. Recent advances in the molecular biology of hepatitis C virus. J. Mol. Biol. 313, 451–464.

Ruebsam, F., Murphy, D.E., Tran, C.V., Li, L.S., Zhao, J., Dragovich, P.S., McGuire, H.M., Xiang, A.X., et al., 2009. Discovery of tricyclic 5,6-dihydro-1H-pyridin-2-ones as novel, potent, and orally bioavailable inhibitors of HCV NS5B polymerase. Bioorg. Med. Chem. Lett. 19 (22), 6404–6412.

Ruebsam, F., Webber, S.E., Tran, M.T., Tran, C.V., Murphy, D.E., Zhao, J., Dragovich, P.S., Kim, S.H., Li, L.-S., Zhou, Y., Han, Q., Kissinger, C.R., Showalter, R.E., Lardy, M., Shah, A.M., Tsan, M., Patel, R., LeBrun, L.A., Kamran, R., Sergeeva, M.V., Bartkowski, D.M., Nolan, T.G., Norris, D.A., Kirkovsky, L., 2008. Pyrrolo[1,2-b]pyridazin-2-ones as potent inhibitors of HCV NS5B polymerase. Bioorg. Med. Chem. Lett. 18, 3616–3621.

Samanta, S., Gupta, S.P., Masand, N., Basu, N.K., Patil, V.M., 2012. Docking directed kNN MFA study of pyrimidine nucleosides as HCV NS5B inhibitors. Int. J. Drug Des. Discov. 3, 809–813.

Slater, M.J., Amphlett, E.M., Andrews, D.M., Bravi, G., Burton, G., Cheasty, A.G., Corfield, J.A., Ellis, M.R., Fenwick, R.H., Fernandes, S., Guidetti, R., Haigh, D., Hartley, C.D., Howes, P.D., Jackson, D.L., Jarvest, R.L., Lovegrove, V.L.H., Medhurst, K.J., Parry, N.R., Price, H., Shah, P., Singh, O.M.P., Stocker, R., Thommes, P., Wilkinson, C., Wonacott, A., 2007. Optimization of novel acyl pyrrolidine inhibitors of hepatitis C virus RNA-dependent RNA polymerase leading to a development candidate. J. Med. Chem. 50, 897–900.

Stansfield, I., Ercolan, i, C., Mackay, A., Conte, I., Pompei, M., Koch, U., Gennari, N., Giuliano, C., Rowley, M., Narjes, F., 2009. Tetracyclic indole inhibitors of hepatitis C virus NS5B-polymerase. Bioorg. Med. Chem. Lett. 19, 627–632.

Takamizawa, A., Mori, C., Fuke, I., Manabe, S., Murakami, S., Fujita, J., Onihi, E., Anodh, T., Yoshida, I., Okayama, H., 1991. Structure and organization of the hepatitis C virus genome isolated from human carriers. J. Virol. 65, 1105–1113.

Tamori, A., Enomoto, M., Kawada, N., 2016. Recent advances in antiviral therapy for chronic hepatitis C. Mediators Inflamm. 2016, 1–11.

Tan, S.-L., He, Y., 2011. Hepatitis C Antiviral Drug Discovery and Development. Caister Academic Press, Norfolk, UK, 2011.

Tan, S.L., Pause, A., Shi, Y., Sonenberg, N., 2002. Hepatitis C therapeutics: current status and emerging strategies. Nat. Rev. Drug Discov. 1, 867–881.

Tan, H., Wang, G., Moy, C., Kossen, K., Rajagopalan, R., Misialek, S., Ruhrmund, D., Hooi, L., Snarskaya, N., Stoycheva, A., Buckman, B., Seiwert, S., Beigelman, L., 2009. Forty fourth EASL Copenhagen April 22–26, 2009 Poster 936. Available from: http://www.kenes.com/easl2009/posters/Abstract 817.htm

Tedesco, R., Shaw, A.N., Bambal, R., Chai, D., Concha, N.O., Darcy, M.G., Danak, D., Fitch, D.M., Gates, A., Gerhardt, W.G., Halegoua, D.L., Han, C., Hofmann, G.A., Johnston, V.K., Kaura, A.C., Liu, N., Keenan, R.M., Lin-Goerke, J., Sarisky, R.T., Wiggall, K.J., Zimmerman, M.N., Duffy, K.J., 2006. 3-(1,1-dioxo-2H-(1,2,4)-benzothiadiazin-3-yl)-4-hydroxy-2(1H)-quinolinones, potent inhibitors of hepatitis C virus RNA-dependant RNA polymerase. J. Med. Chem. 49, 971–983.

Toniutto, P., Fabris, C., Bitetto, D., Fornasiere, E., Rapetti, R., Pirisi, M., 2007. Valocipitabine dihydrochloride: a specific polymerase inhibitor of hepatitis C virus. Curr. Opin. Investig. Drugs 8, 150–158.

Vendeville, S., Lin, T.-I., Hu, L., et al., 2012. Finger loop inhibitors of the HCV NS5B polymerase. Part II optimization of teteracyclic indole-ased macrocycle leading to the discovery of TMC647055. Bioorg. Med. Chem. Lett. 22, 4437–4443.

Vendeville, S., Raboisson, P., Lin, T.-I., 2010. Macrocyclic indole derivatives useful as hepatitis C virus inhibitors. PCT International Application 2010, WO2010003658.

ViroPharma Press Release, August 10th, 2007.

Wagner, R., Larson, D.P., Beno, D.W.A., Bosse, T.D., Darbyshire, J.F., Gao, Y., Gates, B.D., He, W., Henry, R.F., Hernandez, L.E., Douglas, K., Hutchinson, Jiang, W.W., Kati, W.M., Klein, L.L., Koev, G., Kohlbrenner, W., Krueger, C.A., Liu, J., Liu, Y., Long, M.A., Maring, C.L., Masse, S.V., Middleton, T., Montgomery, D.A., Pratt, J.K., Stuart, P., Molla, A., Kempf, D.J., 2009. Inhibitors of hepatitis C virus polymerase: synthesis and biological characterization of unsymmetrical dialkyl-hydroxynaphthalenoyl-benzothiadiazines. J. Med. Chem. 52, 1659–1669.

Wang, P.Y., Chun, B.K., Rachakonda, S., Du, J.F., Khan, N., Shi, J.X., Stee, W., Cleary, D., Ross, B.S., Sofia, M.J., 2009a. An efficient and diastereoselective synthesis of PSI-6130: a clinically efficacious inhibitor of HCV NS5B polymerase. J. Org. Chem. 74, 6819–6824.

Ward, C.L., Dev, A., Rigby, S., Symonds, W.T., Patel, K., Zekry, A., Pawlotsky, J.G., McHutchison, J.G., 2008. Interferon and ribavirin therapy does not select for resistance mutations in hepatitis C virus polymerase. J. Viral Hepatitis 15, 571–577.

Wang, G., He, Y., Sun, J., Das, D., Hu, M., Huang, J., Ruhrmund, D., Hooi, L., Misialek, S., Ravi Rajagopalan, T.V., Stoycheva, A., Buckman, B., Kossen, K., Seiwert, S.D., Beigelman, L., 2009b. HCV NS5B polymerase inhibitors 1: synthesis and in vitro activity of 2-(1,1-dioxo-2H-[1,2,4]benzothiadiazin-3-yl)-1-hydroxynaphthalene derivatives. Bioorg. Med. Chem. Lett. 19, 4476–4479.

Wang, G., Lei, H., Wang, X., Das, D., Hong, J., Mackinnon, C.H., Coulter, T.S., Montalbetti, C.A.G.N., Meras, R., Gai, X., Bailey, S.E., Ruhrmund, D., Hooi, L., Misialek, S., Ravi Rajagopalan, T.V., Cheng, R.K.Y., Barker, J., Felicetti, B., Schonfeld, D.L., Stoycheva, A., Buckman, B., Kossen, K., Seiwert, S.D., Beigelman, L., 2009c. HCV NS5B polymerase inhibitors 2: synthesis and in vitro activity of (1,1-dioxo-2H-[1,2,4]benzothiadiazin-3-yl) azolo[1,5-a]pyridine and azolo[1,5-a]pyrimidine derivatives. Bioorg. Med. Chem. Lett. 19, 4480–4483.

Wang, G., Zhang, L., Wu, X., Das, D., Ruhrmund, D., Hooi, L., Misialek, S., Ravi Rajagopalan, T.V., Buckman, B., Kossen, K., Seiwert, S.D., Beigelman, L., 2009d. HCV NS5B polymerase inhibitors 3: synthesis and in vitro activity of 3-(1,1-dioxo-2H-[1,2,4]benzothiadiazin-3-yl)-4-hydroxy-2H-quinolizin-2-one derivatives. Bioorg. Med. Chem. Lett. 19, 4484–4487.

Wedemeyer, H., Forns, X., Hézode, C., Lee, S.S., Scalori, A., Voulgari, A., Le Pogam, S., Nájera, I., Thommes, J.A., 2016. Mericitabine and Either Boceprevir or Telaprevir in combination with Peginterferon Alfa-2a plus Ribavirin for patients with chronic Hepatitis C Genotype 1 infection and prior null response: the randomized DYNAMO 1 and DYNAMO 2 studies. PLoS One 11 (1), e0145409.

Zhao, F., Liu, N., Zhan, P., Jiang, X., Liu, X., 2015. Discovery of HCV NS5B thumb site I inhibitors: core-refining from benzimidazole to indole scaffold. Eur. J. Med. Chem. 94, 218–228.

## FURTHER READING

Beaulieu, P.-L., 2009. Recent advances in the developmenton NS5B polymerase. Expert. Opin. Ther. Patents 19, 145–169.

Beaulieu, P.L., 2010. Filibuvir, a non-nucleoside NS5B polymerase inhibitor for the potential oral treatment of chronic HCV infection. Investig. Drugs J. 13, 938–948.

Burton, G., Ku, T.W., Carr, T.J., KiesowT, SariskyRT, Lin-Goerke, J., Baker, A., EarnshawDL, Hofmann, G.A., Keenan, R.M., Dhanak, D., 2005. Identification of small molecule inhibitors of the hepatitis C virus RNA-dependant RNA polymerase from a pyrrolidine combinatorial mixture. Bioorg. Med. Chem. Lett. 15, 1553–1556.

Ding, M., He, F., Poss, M.A., Rigat, K.L., Wang, Y.-K., Roberts, S.B., Qiu, D., Fridell, R.A., Gao, M., Gentles, R.G., 2011. The synthesis of novel heteroaryl-fused 7,8,9,10-tetrahydro-6H-azepino[1,2-a]indoles, 4-oxo-2,3-dihydro-1H-[1,4]diazepino[1,7-a]indoles and 1,2,4,5-tetra-hydro-[1,4]oxazepino[4,5-a]indoles. Effectiveinhibitors of HCV NS5B polymerase. Org. Bio-mol. Chem. 9, 6654–6662.

Rockway, T.W., Zhang, R., Liu, D., Betebenner, D.A., McDaniel, K.F., Pratt, J.K., Beno, D., Mont-gomery, D., Jiang, W.W., Masse, S., Kati, W.M., Middleton, T., Molla, A., Maring, C.J., Kempf, D.J., 2006. Inhibitors of HCV NS5B polymerase: synthesis and structure-activity relationships of N-1-benzyl and N-1-[3- methylbutyl]-4-hydroxy-1,8-naphthyridon-3-yl benzothiadiazine analogs containing substituents on the aromatic ring. Bioorg. Med. Chem. Lett. 16, 3833–3838.

Tedesco, R., Chai, D., Concha, N.O., Darcy, M.G., Dhanak, D., Fitch, D.M., Gates, A., Johnston, V.K., Keenan, R.M., Lin-Goerke, J., Sarisky, R.T., Shaw, A.N., Valko, K.L., Wiggal, K.J., Zim-merman, M.N., Duffy, K.J., 2009. Synthesis and biological activity of heteroaryl 3-(1,1- di-oxo-2H-(1,2,4)-benzothiadiazin-3-yl)-4-hydroxy-2(1H)-quinolinone derivatives as hepatitis C virus NS5B polymerase inhibitors. Bioorg. Med. Chem. Lett. 19, 4354–4358.

Chapter 9

# Studies on HIV-1 Protease and its Inhibitors

Sonal Dubey
*Krupanidhi College of Pharmacy, Bengaluru, Karnataka, India*

## 1 INTRODUCTION

Protease (PR), proteinase, or peptidase refers to a group of proteolytic enzymes whose major function is to catalyze the hydrolysis of peptide bonds that link amino acids together in a polypeptide chain with high sequence selectivity and catalytic proficiency. PRs can be found in animals, plants, bacteria, archaeans, and viruses. Depending on their catalytic mechanism, PRs can be classified into seven different categories: (1) serine PRs contain a serine alcohol, (2) cysteine PRs contain a cysteine thiol, (3) threonine PRs contain a threonine secondary alcohol, (4) aspartic PRs contain an aspartate carboxylic acid, (5) glutamic PRs contain a glutamate carboxylic acid, (6) metalloPRs contain a metal (usually zinc), and (7) asparagine peptide lyases that contain an asparagine to perform an elimination reaction (not requiring water) (Oda, 2012; Rawlings and Barrett, 1993; Rawlings et al., 2011).

Human immunodeficiency viruses (HIVs) are retroviruses and among them HIV of type 1 (HIV-1) has been found to be the causative agent of acquired immunodeficiency syndrome (AIDS). It contains an aspartyl PR that plays an essential role in the life cycle of HIV. It cleaves newly synthesized polyproteins at the appropriate places to create mature protein components of an infectious HIV virion. Without an effective HIV PR, HIV virions remain noninfectious (Kohl et al., 1988; Kräusslich et al., 1989). Thus mutation of HIV PR's active site or inhibition of its activity disrupts HIV's ability to replicate and infect additional cells, making HIV PR inhibition the subject of considerable pharmaceutical research.

Before going into the HIV-1 PR in detail, let us have a look at the structure of the HIV-1 (Fig. 9.1), its life cycle (Fig. 9.2), and the role of its PR. The HIV is a RNA retrovirus belonging to the subfamily of Lentivirus. It contains two major envelope proteins: gp120 and gp41; three enzymes: reverse transcriptase (RT), integrase (IN), and PR; and two copies of RNA.

As shown in Fig. 9.2, HIV-1 tends to infect CD4$^+$ T cells, as CD4 receptors have high affinity for gp120a glycoprotein present on the envelope of virus. This

Viral Proteases and Their Inhibitors. http://dx.doi.org/10.1016/B978-0-12-809712-0.00009-5

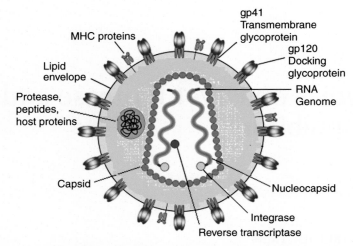

FIGURE 9.1   **Structure of human immunodeficiency virus type 1 (HIV-1).** *MHC*, Major histocompatibility complex.

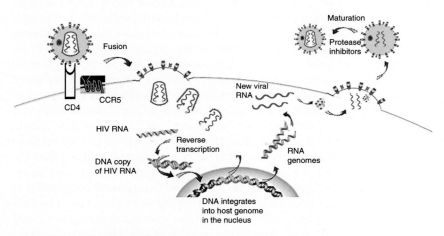

FIGURE 9.2   **Life cycle of HIV-1.** *CCR5*, Chemokine receptor 5.

interaction between the viral envelope proteins and CD4 receptor is followed by gp120 binding to chemokine receptor 5 (CCR5) coreceptor. Both these glycoproteins mediate the fusion of viral cell with the host cell and diffuse the viral contents into the host cell. As the viral contents enter a cell, its RNA is reverse transcribed to DNA by a virally encoded enzyme, RT. The viral DNA then enters the nucleus of the cell, where it is integrated into the cellular DNA by a second virally encoded enzyme, IN. Activation of the host cell turns on the transcription of the viral DNA into messenger RNA. This mRNA comes out of the cell nucleus through nuclear pores and gets attached to ribosome, which then translates the mRNA into viral proteins. HIV-1 PR, the third virally encoded enzyme, is required in this step to cleave a viral polyprotein precursor into individual mature proteins. The viral RNA and viral proteins assemble at the cell surface into new virions, which then bud from the cell. The cell capsid is formed, acquires the envelope glycoproteins, and is released to infect another cell. The budding and release of multiple virions causes extensive damage to the cell, leading to cell death (Barré-Sinoussi et al., 2013).

## 2   3D STRUCTURE OF HIV-1 PROTEASE

The first crystal structure of HIV-1 PR was obtained by Navia et al. (1989), followed by a little more accurate structure by Wlodawer et al. (1989) in the same year. Till date over 771 structures of HIV-1 PR, its mutants, and enzyme complexed with various inhibitors have been reported in the protein data bank (PDB). The HIV-1 PR or the retroviral PRs are unique proteins, as they are homodimeric enzymes containing a single, symmetric active site, which involves two adjacent aspartic acid side chains. Protonation of one aspartic acid forms a hydrogen bond with the other one and stabilizes the complex (Pearl, 1987). HIV-1, as well as HIV-2, PRs are aspartyl PRs, each consisting of two identical 99–amino acid subunit (Figs. 9.3 and 9.4). They function as a symmetric homodimer with one active site, which is C2 symmetric in the absence of ligands. The two monomer subunits are referred to as chain A and chain B, with residues numbered from 1 to 99 in chain A and from 101 to 199 in chain B. Each monomer unit has a flap, which is an extended β-sheet region (a loop rich in Gly and having residue 45–55 in each monomer), and plays an important role in the substrate binding by acting as a part of the substrate-binding site. The aspartyl residue, Asp25, is present at

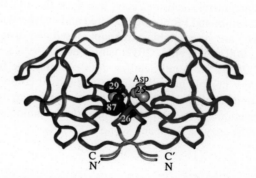

**FIGURE 9.3   The structure of homodimeric HIV-1 protease (PR).** The two identical subunits are shown by *blue* and *red ribbons* with their flaps (residues 45–55) and the N- and C-termini (residues 1–4 and 96–99). The catalytic Asp25 is represented by *gray van der Waals spheres*. Residues T26, D29, and R87 are represented by *blue van der Waals spheres*. Mutating these residues to Ala, Asn, and Lys, respectively, yields a stable folded monomer.

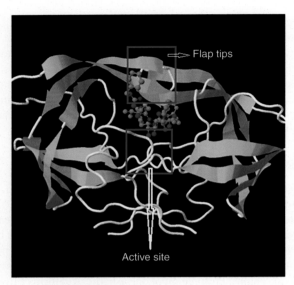

**FIGURE 9.4    Structure of HIV-1 PR.** The structure 3TLH (PDB) presented using the "cartoon" option in Pymol (http://www.pymol.org), with chain A shown in *gray* and chain B in *yellow*. This highlights the secondary structure of the protein, which is largely β-sheet. The heavy atoms of the inhibitor TL-3 are shown as *ball and stick model* between the active site and the β-hairpin flap tips (Heal et al., 2012).

the bottom of the cavity. The active site is situated within the core and consists of three residues Asp-Thr-Gly from each of the monomers. This triad is a form of active site common to aspartic PRs. In PR, it is located at residues 25–27 and, due to symmetry, at residues 125–127 (Hong et al., 1997). The active-site region is capped by two identical β-hairpin loops, the flaps (residues 45–55 in each monomer), which regulates a substrate entry into the active site. While the flap β-hairpins in the ligand-bound PR are well ordered and interact with the substrate, in the free PR they are very flexible and adopt closed and open conformations (Freedberg et al., 2002; Heal et al., 2012; Ishima et al., 1999; Levy et al., 2004; Perryman et al., 2004; Scott and Schiffer, 2000). In the free form of the PR, these flaps are very flexible and adopt open and closed conformations (Harte et al., 1992; Swaminathan et al., 1991). The substrate binds to the enzyme in an extended conformation and its interaction with different amino acids determines the specificity of the enzyme (Brik and Wong, 2003).

## 2.1    Subsites

There are two subsites each of S1, S2, and S3 in the enzyme. The S1 subsites and the S3 subsites, located next to S1 subsites, are quite hydrophobic in nature; although S2 subsites are also hydrophobic, their Asp29 and Asp30 residues are not. The two subunits of PR interact through four-stranded antiparallel β-sheets consisting of residues 1–4 and 96–99 of N-and C-terminal monomers, respectively (Fig. 9.5).

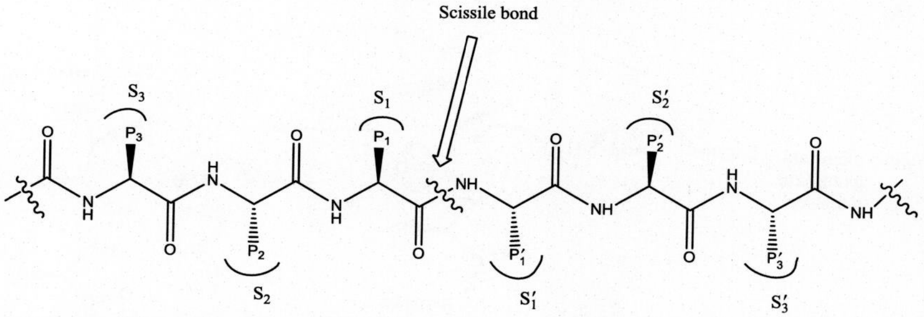

FIGURE 9.5    Standard nomenclature: $P_1...P_n$, $P'_1...P'_n$ is used to designate amino acid residues of peptide structures. The corresponding binding sites on the PR are referred to as $S_1...S_n$, $S'_1...S'_n$ subsites.

## 3   MECHANISM OF BINDING WITH HIV-1 PROTEASE

PR is an essential enzyme for many biological processes. PRs catalyze the hydrolysis of peptide bonds with high sequence selectivity and proficiency. The catalytic hydrolysis takes place by various mechanisms. Based on the catalytic residue used by the enzyme, PRs have been classified into seven different classes. PRs generally use an activated water molecule to attack the amide bond carbonyl of the substrate's scissile bond. This activated water molecule then performs a nucleophilic attack on the amide bond carbonyl of the scissile bond to generate an oxyanion tetrahedral intermediate. Protonation followed by rearrangement of the scissile amide N-atom results in the breakdown of the tetrahedral intermediate to hydrolyze the peptide. When the activation of water molecule is achieved by the two aspartyl β-carboxy groups at the active site, they are classified as aspartate PRs. The HIV-1 PR belongs to the aspartate family (Fig. 9.6).

In all aspartic PRs, including HIV-1 PR, there is a conserved sequence Asp-Thr-Gly. This catalytic triad is located at residues 25–27 and their symmetric counterparts 125–127. As mentioned previously these active sites are beneath the flaps. The mutation of Asp25 with any other similar amino acid, such as Asn (Darke et al., 1989), Thr (Seelmeier et al., 1988), or Ala (Mous et al., 1988), renders the molecule without any proteolytic activity.

The mechanism of HIV-1 aspartyl PR (Fig. 9.7) is a one-step hydrolysis reaction, in which the nucleophilic water molecule and the acidic proton attacks the scissile peptide bond and breaks the protein molecule into the desired units. This figure explains the mechanism at the molecular level (Jaskólski et al., 1991). The two catalytic aspartates can exist in four different protonation states: the dianionic form ($-1, -1$), two monoanionic forms ($-1, 0$) and ($0, -1$), and in diprotonated or neutral form ($0, 0$). This protonation state of the two aspartyl groups is dependent on the local environment and hence is different for different inhibitors. Aspartyl PR has a unique ability

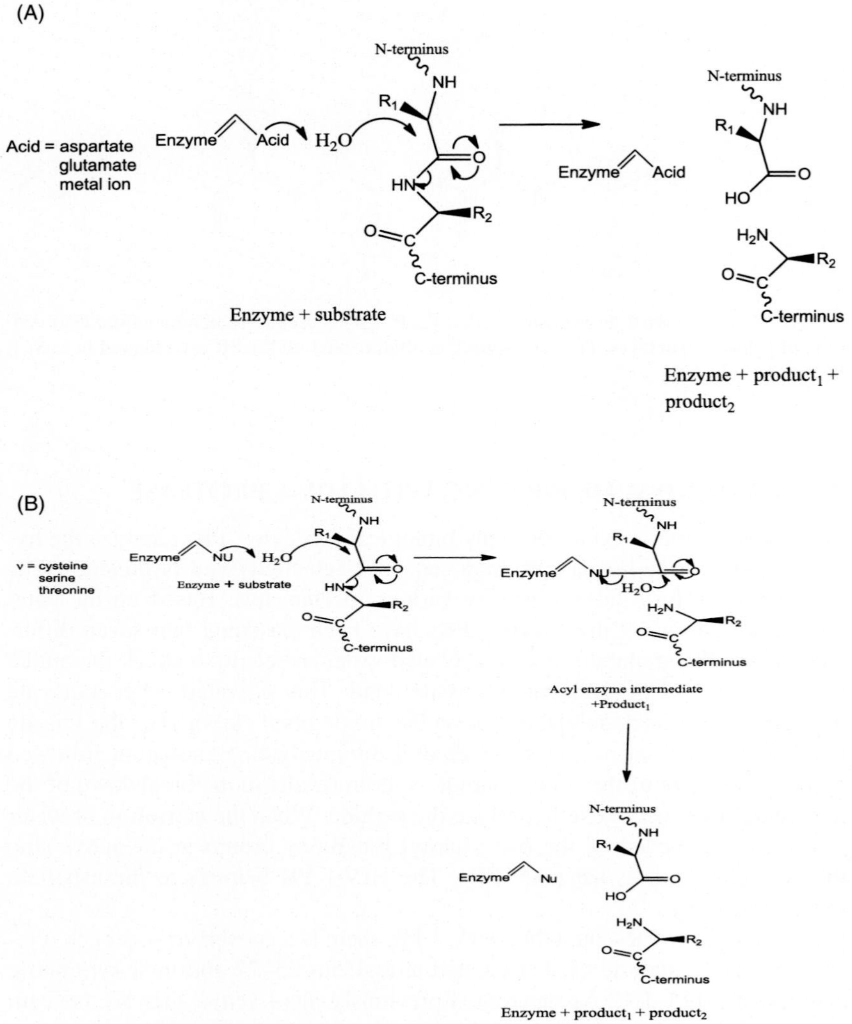

(A)

(B)

FIGURE 9.6 **Two proposed catalytic mechanisms of the PR enzyme based on the catalytic residue used by the enzyme.** (A) Mechanism 1 is for aspartate, glutamate, and metalloion residues. (B) Mechanism 2 is for cysteine, serine, and threonine residues.

to function over a wide range of pH (2.0–7.4) (Harte and Beveridge, 1993; Jaskólski et al., 1991).

There is a direct H-bonding between the flap and the substrate. The carbonyl group between the $P_1$ and $P_2$, and the one between $P_1'$ and $P_2'$, both make hydrogen bonds with the same water molecule. This water molecule creates a tetrahedral geometry by making two more hydrogen bonds with the flaps (Fig. 9.8) (Gustchina and Weber, 1990).

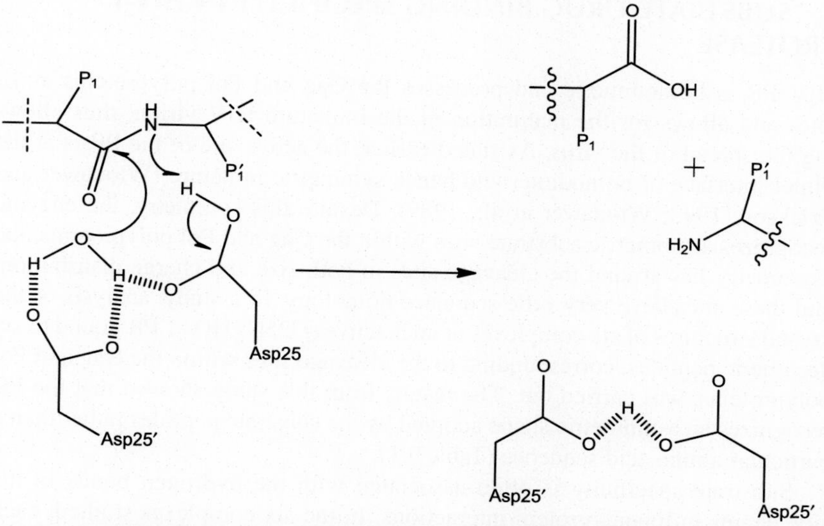

**FIGURE 9.7** **Proposed concerted catalytic mechanism for HIV-1 PR showing the involvement of various amino acids.**

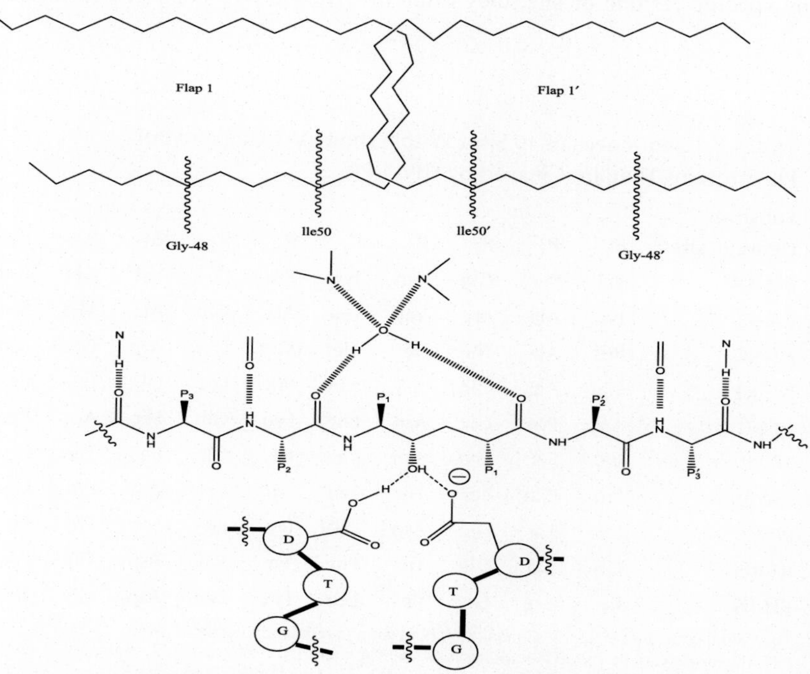

**FIGURE 9.8** **Schematic representation of the two flaps of HIV PR and their hydrogen bonds with the water molecule.** The same water molecule makes another two hydrogen bonds with the peptidyl inhibitor. The figure also shows the interaction of the central hydroxyl group of inhibitor with the catalytic aspartates of HIV PR.

## 4  SUBSTRATE/DRUG BINDING SPECIFICITY IN HIV-1 PROTEASE

HIV PR is homodimeric and processes the Gag and Pol polyproteins at 10 sites and allows for the maturation of the immature HIV virion, thus allowing the spread of the virus. As stated earlier, the active site of the PR is at the dimer interface of homodimer and hence symmetric in nature (Wlodawer and Erickson, 1993; Wlodawer et al., 1989). Despite this symmetry, the enzyme recognizes asymmetric substrate sites within the Gag and Pol polyproteins, the asymmetry lies around the cleavage sites in both size and charge distribution, and these sites have very little sequence homology. In a study, analysis of the crystal structures of six complexes of an inactive (D25N) HIV-1 PR bound to six decameric peptides, corresponding to the cleavage sites within the Gag and Pol polyproteins, was carried out. The results from this study showed that the PR recognizes an asymmetric shape adopted by the substrate peptides rather than a particular amino acid sequence (Table 9.1).

Substrate specificity is often associated with the hydrogen bonds in the side chains in ligand–protein interactions. In the six complexes studied, each substrate peptide formed between one and three side chains hydrogen bonds with the PR (Table 9.2). Although these hydrogen bonds did not form between any specific residue or site, they often involve the $P_2/P_2$ or $P_3/P_3$ residues of

**TABLE 9.1 Sequences of 10 Sites Within the HIV-1 Gag and Pol Polyproteins That are Cleaved by HIV-1 PR**

| Substrate cleavage sites[a] | $P_5$ | $P_4$ | $P_3$ | $P_2$ | $P_1$ | $P_1$ | $P_2$ | $P_3$ | $P_4$ | $P_5$ |
|---|---|---|---|---|---|---|---|---|---|---|
| MA-CA | Val | Ser | Gln | Asn | Tyr | Pro | Ile | Val | Gln | Asn |
| CA-p2 | Lys | Ala | Arg | Val | Leu | Ala | Glu | Ala | Met | Ser |
| p2-NC | Pro | Ala | Thr | Ile | Met | Met | Gln | Arg | Gly | Asn |
| NC-p1 | Glu | Arg | Gln | Ala | Asn | Phe | Leu | Gly | Lys | Ile |
| p1-p6 | Arg | Pro | Gly | Asn | Phe | Leu | Gln | Ser | Arg | Pro |
| TF-PR | Val | Ser | Phe | Asn | Phe | Pro | Gln | Ile | Thr | Leu |
| AutoP | Pro | Gln | Ile | Thr | Leu | Trp | Lys | Arg | Pro | Leu |
| PR-RT | Cys | Thr | Leu | Asn | Phe | Pro | Ile | Ser | Pro | Ile |
| RT-RH | Gly | Ala | Glu | Thr | Phe | Tyr | Val | Asp | Gly | Ala |
| RH-IN | Ile | Arg | Lys | Ile | Leu | Phe | Leu | Asp | Gly | Ile |

The crystal structures of the six sequences in bold were determined in complex with an inactive HIV-1 protease variant. AutoP, Autoproteolysis site; CA, capsid; IN, integrase; MA, matrix; NC, nucleocapsid; RH, RNAse H; RT, reverse transcriptase; TF, *trans*-frame peptide.
[a]The cleavage sites are identified by the proteins released once the site is cleaved.

**TABLE 9.2 Observed Side Chain to Side Chain and Side Chain to Backbone Hydrogen Bonds**

| Cleavage sites | Substrate atoms | Protein atoms | Distances | Substrate atoms | Protein atoms | Distances (Å) |
|---|---|---|---|---|---|---|
| MA-CA (dimer 1) | SerP4 OG | Asp30 OD2 | 2.4 | AsnP2 OD1 | Asp29 N | 2.9 |
| | GlnP3 OE1 | Arg8 NH2 | 3.0 | AsnP2 OD1 | Asp30 N | 2.8 |
| MA-CA (dimer 2) | SerP4 OG | Asp30 OD2 | 2.5 | AsnP2 OD1 | Asp29 N | 2.7 |
| | GlnP3 OE1 | Arg8 NH2 | 3.1 | AsnP2 OD1 | Asp30 N | 2.9 |
| Ca-p2 | GluP2 OE1 | Asp30 OD2 | 2.5 | GluP2 OE2 | Asp29 N | 2.9 |
| | | | | GluP2 OE2 | Asp30 N | 3.3 |
| p2-NC | GlnP2 NE2 | Asp30 OD2 | 2.9 | GlnP2 OE1 | Asp29 N | 3.1 |
| | ThrP3 OG1 | Asp29 OD2 | 3.4 | GlnP2 OE1 | Asp30 N | 2.7 |
| | | Arg8 NH2 | 2.8 | — | — | — |
| | ArgP3[b] NE | Arg8 NH1 | 3.0 | — | — | — |
| | | Arg8 NH2 | 3.2 | — | — | — |
| p1-p6 | GlnP2 NE2 | Asp30 OD2 | 3.0 | GlnP2 OE1 | Asp29 N | 2.9 |
| | ArgP4 NE | Asp30 OD1 | 3.2 | GlnP2 OE1 | Asp30 N | 2.7 |
| | ArgP4 NE | Asp30 OD2 | 3.0 | — | — | — |
| | ArgP4 NH2 | Gln58 OE1 | 3.4 | — | — | — |
| RT-RH | TyrP1 OH | Arg8 NH1 | 3.2 | GlnP3 NE2 | Gly48 O | 3.0 |
| | AspP3 OD1 | Arg8 NH1 | 3.1 | — | — | — |
| RH-IN | AspP3 OD1 | Arg8 NH1 | 3.4 | — | — | — |

Superscript b denotes one of the two observed conformations of ArgP3.

the peptide. Only four PR side chains (Arg8/8, Asp29, Asp30/30, and Gln58) formed direct side chain hydrogen bonds with the substrate peptides. When the $P_2/P_2$ residue is either Glu or Gln, as in CA-p2, p2-NC, and p1-p6, the side chain makes a hydrogen bond with Asp30, OD2, as well as with the amide nitrogen atoms of aspartic acid residues 29 and 30. The latter hydrogen bonds were also formed when $P_2/P_2$ was either Asp or Asn. The absence of a polar side chain at the $P_2/P_2$ position in the structures MA-CA, RT-RH, and RH-IN prevents them from forming this kind hydrogen bond with the "catalytic" Asn25ND2.

Certain conserved hydrogen bonds are formed in the PR. These conserved hydrogen bonds between the peptides and water molecules primarily involve backbone atoms. Five water molecules were completely conserved between the

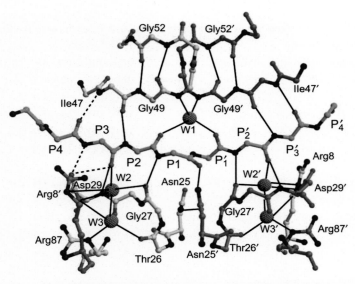

**FIGURE 9.9**   **A diagrammatic representation of conserved hydrogen bonds between the five water molecules and the peptide (shown in *green*), as well as the two monomers of the PR that are shown in *cyan* and *magenta* colors.** While binding with the substrate, there are certain conformational changes in the PR, which break the symmetry, but the overall pattern of hydrogen bonds and water structure is conserved.

five structures that were determined at a 2.0-Å resolution (Fig. 9.9). These conserved water molecules (W's) include the three molecules that directly contact the peptide (Baldwin et al., 1995; Mahalingam et al., 2001) and two that stabilize them. W1 tethers the flaps of the PR at Ile50 (N) and Ile50′ (N) with $P_2$ (O) and $P_2'$ (O). W2/W2′ hydrogen bond to the peptide at $P_2$ (O)/$P_1'$ (O) and to the PR at Arg8′ (NE), Gly27 (O), Asp29 (OD1)/Arg8 (NE), Gly27′ (O), Asp29′ (OD1), as well as to two other water molecules, W3/W3′. Although W3/W3′ do not directly contact the substrate peptides, they do stabilize their conformation by binding to the PR at Thr26 (O) and Asp29 (OD1)/Arg87 (NE). Thus these five water molecules stabilize the extended conformation of the peptides by hydrogen bonding with the carbonyl oxygen atoms of the peptide. This extended conformational stabilization is not possible in PR in the absence of water, as no direct hydrogen bonding would be involved. It can thus be concluded that for the numerous backbone–backbone hydrogen bonds necessary for substrate binding, the peptide must be able to form a relatively extended conformation to partake in. The specificity of the enzyme to the substrate is achieved when $P_1$ and $P_3$ (or the $P_3/P_4$ region) contact each other on the unprimed side of the enzyme cleavage site to form a toroid. PR recognizes a shape rather than a particular amino acid sequence. On the other side of the enzyme cleavage site, that is, $P_1'$ and $P_3'$, there must be enough space for water to access the peptide. The role of this water molecule for substrate recognition or for product release is still not clear. The extension in the enzyme structure should be enough to allow the formation of hydrogen bonds, but not so much also that it prevents the formation of either a toroid or a cluster of water.

The hydrophobic toroid at $P_1–P_3$ indicates a newer avenue for drug design. Most of the currently marketed drugs developed against HIV-1 PR bind predominantly to the PR between the $P_2 – P_2'$ positions, as these sites directly bridge the active site. Even though the compounds themselves being nonsymmetric, they cover the same region in both monomers of the PR. When a drug-resistant mutation occurs, it affects the inhibitor's binding at two independent sites. These kinds of resistance of PI can be avoided or decreased by designing an inhibitor that, rather than covering the region from $P_2 – P_2'$, covers the region from $P_3 – P_1'$.

The HIV-1 PR undergoes frequent mutation due to the infidelity of its RT. The most studied mutations in HIV-1 PR are the ones that arise in the infected patients with some drug-resistant characteristics undergoing treatment. In spite of mutations, however, the enzyme continues to function and recognize the various substrate sites within the Gag and Pol polyproteins. The secondary structure of HIV-1 PR is primarily β-sheet, the majority of the tertiary hydrogen bonds are found between the backbone atoms. A total of only 4 side chains are involved in protein–substrate hydrogen bonds, 17 side chains are involved in protein–protein hydrogen bonds, and 12 are involved in water-mediated protein–protein hydrogen bonds (Prabu-Jeyabalan et al., 2002).

## 5  DESIGN AND DEVELOPMENT OF HIV-1 PR INHIBITORS

### 5.1  First-Generation Protease Inhibitors

Initial therapies for HIV/AIDS patients consisted of nucleoside RT inhibitor monotherapy, such as zidovudine or AZT. The first protease inhibitor (PI) to be introduced in the market in 1995 was saquinavir (SQV) (**1**, Fig. 9.10). With this drug began the era of combination therapy for HIV/AIDS patients known as highly active antiretroviral therapy (HAART). It consists of a combination of at least three drugs. HAART therapy with a RT inhibitor greatly improved the disease condition by reducing viral loads, improving CD4 cell counts, and halting the progression to AIDS (Ghosh and Chapsal, 2013; Gulick et al., 1997; Hammer et al., 1997).

The first-generation PIs were based on hydroxyethylene and hydroxyethylamine isosteres (Ghosh et al., 2005). The central hydroxyl group mimics the transition state of the hydrolysis step by binding with the catalytic aspartic acid residues. SQV is a very potent inhibitor with a $K_i$ of 0.12 nM (Roberts et al., 1990). A crystal structure of SQV-bound HIV PR showed that the inhibitor binds in an

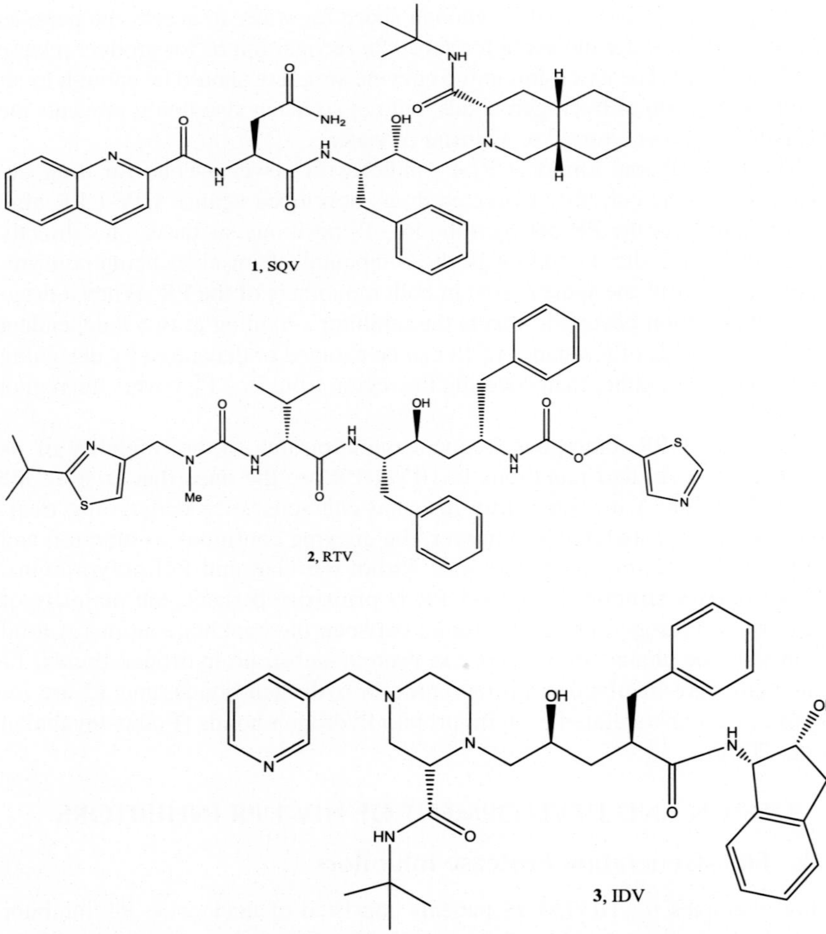

**FIGURE 9.10** **First-generation HIV protease inhibitors (PIs).** *APV*, Amprenavir; *IDV*, indinavir; *NFV*, nelfinavir; *RTV*, ritonavir; *SQV*, saquinavir.

extended conformation (Krohn et al., 1991). Further research in this field led to the development of other PIs, such as ritonavir (RTV) (**2**, Fig. 9.10), developed by Abbott Laboratories, which was approved by the FDA in 1996 and had a very potent $K_i$ of 0.015 nM (Kempf et al., 1995). However, it was later found out that RTV is a potent inhibitor of cytochrome P450 3A, a major metabolic enzyme for PR inhibition, which made this drug to be used as a pharmacokinetic booster rather than a PI (Kumar et al., 1996). Indinavir (IDV) (**3**, Fig. 9.10) was developed by Merck and was the next PI to be approved by the FDA in 1996. IDV is

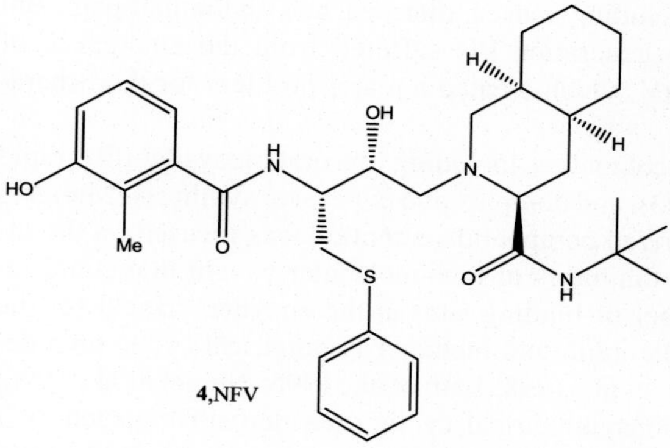

**FIGURE 9.10** *(Cont.)*

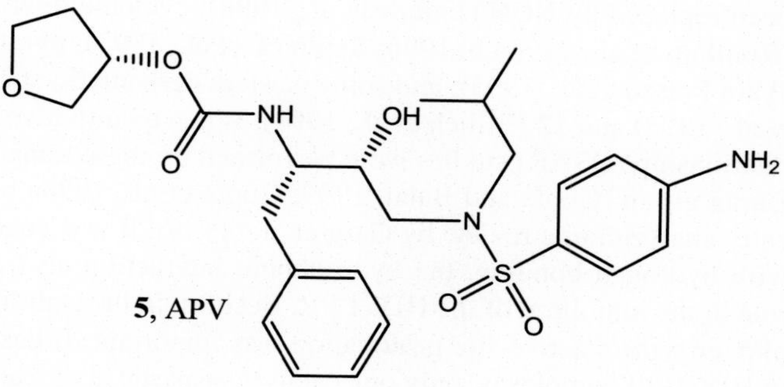

a very potent molecule, with a $K_i$ of 0.36 nM (Vacca et al., 1994), but it has a very short half-life (1.8 h), which poses a compliance problem with the patient due to frequent dosing. Nelfinavir (NFV) (**4**, Fig. 9.10), developed by Agouron Pharmaceuticals in collaboration with Eli Lilly, was approved in 1997 (Kaldor et al., 1997). It has a $K_i$ of 2 nM. Amprenavir (APV) (**5**, Fig. 9.10), approved in 1999, also contains a hydroxyethylamine isostere and exhibits a $K_i$ of 0.6 nM (Kim et al., 1995).

## 5.2 Second-Generation PR Inhibitors

The first-generation PIs suffered from limited efficacy. As many of these compounds were peptidic in structure, it resulted in their largely poor oral bioavailability, high metabolic clearance, low half-life, and requirement of more frequent dosing. Furthermore, they suffered from side effects, such as gastrointestinal distress, including nausea, diarrhea, and abdominal pain. But most importantly, the first-generation PIs suffered from the emergence of drug-resistant strains of HIV, which created a major problem for their therapeutic utility.

Due to the poor pharmacokinetics, including low oral bioavailability, rapid excretion (Olson et al., 1993), and complex and expensive synthesis (Maligres et al., 1995) of peptide-derived compounds, attention was focused on the investigation of nonpeptidic inhibitors of low–molecular weight that could interact with a limited number of binding sites at the enzyme, critical for the inhibition. Such nonpeptidic inhibitors included symmetrical cyclic urea derivatives, such as **6** (Jadhav et al., 1998; Lam et al., 1996; Nugiel et al., 1996; Wilkerson et al., 1996); nonsymmetrical cyclic urea derivatives, such as **7** (Wilkerson et al., 1997); cyclic cyanoguanidines, such as **8** where 1-O has been replaced by NCN (Jadhav et al., 1998); cycloalkylpyranones, such as **9** (Romines et al., 1995a,b, 1996; Skulnick et al., 1997); pyranones, such as **10** (Vara Prasad et al., 1995); and some isostere derivatives, such as **11** (Vazquez et al., 1995) and **12** (Billich et al., 1994). When quantitative structure activity relationship (QSAR) studies were performed on these nonpeptidic PIs by the Gupta group (Gupta and Babu, 1999; Gupta et al., 1998a,b, 1999) and critically analyzed in a review by Garg et al. (1999), it was concluded that along with hydrogen bonding, the hydrophobic interaction also plays a dominant role in the inhibition of the HIV-1 PR. Besides, in the stabilization of PI complex enzyme's active-site protonation was important. It has been shown that in HIV-1–PI complexes, only one catalytic aspartic acid residue is protonated (Chen and Tropsha, 1995; Ferguson et al., 1991). The PI has a binding pocket that has a considerable number of hydrophobic residues. In most of the QSAR equations obtained, the hydrophobic parameter (ClogP) has shown significance on the activity bearing a positive coefficient, and some had parabolic relation with ClogP having an optimum value of ClogP ($ClogP_O$), which ranged from 4.49 to 6.96 corresponding well to ClogP value of FDA-approved PIs. An X-ray structure of a cyclic urea in an HIV-1 PR pocket indicated that $S_2/S_2'$ pockets in HIV-1 PR are essentially hydrophobic (Lam et al., 1996). The residues that make up these pockets are Val32, Ile47, Ile50, and Ile84 in each monomer, and the phenyl group of the cyclic urea makes hydrophobic interactions with them. When any residue is mutated, it loses its contact with the inhibitor. The importance of hydrophobic residues in the binding pocket reconfirms the contribution of the hydrophobicity of inhibitors on anti-HIV activity. While a substrate-based PI can bind with the enzyme as shown in Fig. 9.11 (Babine and Bender, 1997), a nonsymmetrical nonpeptidic PI, an arylthiomethane, was found to bind with the enzyme as shown in Fig. 9.12 (Vara Prasad et al., 1995).

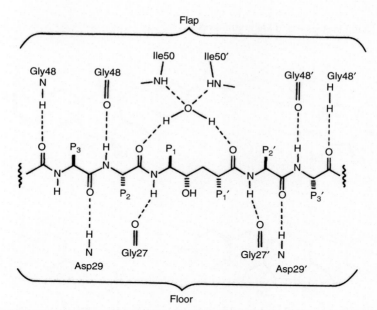

**FIGURE 9.11** **A representation of binding of a substrate-based HIV-1–PI with the enzyme.** *(Reprinted with permission from Babine, R.E., Bender, S.L., 1997. Molecular recognition of protein–ligand complexes: applications to drug design. Chem. Rev. 97 (5), 1359–1472. Copyright 1997 American Chemical Society.)*

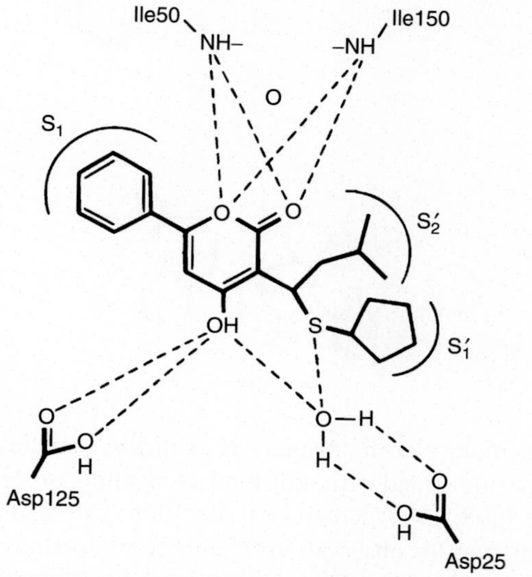

**FIGURE 9.12** **Model of binding of arylthiomethanes with HIV-1 PR based on X-ray crystallographic studies.** $S_1$, $S'_1$, $S'_2$, etc., can be van der Waals or hydrophobic sites. *(Reprinted with permission from Prasad, J.V., Para, K.S., Tummino, P.J., Ferguson, D., McQuade, T.J., Lunney, E.A., Rapundalo, S.T., Batley, B.L., Hingorani, G., Domagala, J.M., 1995. Nonpeptidic potent HIV-1 protease inhibitors: (4-hydroxy-6-phenyl-2-oxo-2H-pyran-3-yl) thiomethanes that span P1-P2' subsites in a unique mode of active site binding. J. Med. Chem. 38 (6), 898–905. Copyright 1995 American Chemical Society.)*

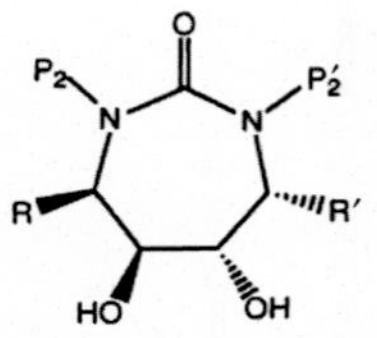

6

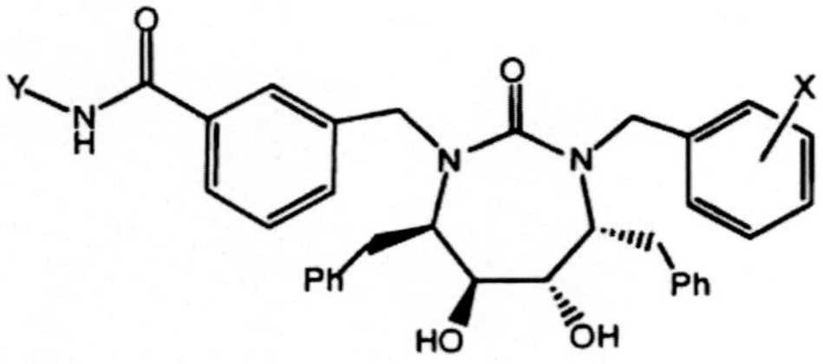

7

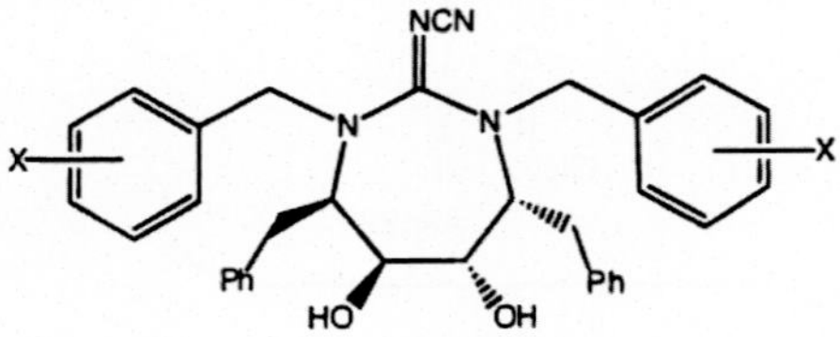

8

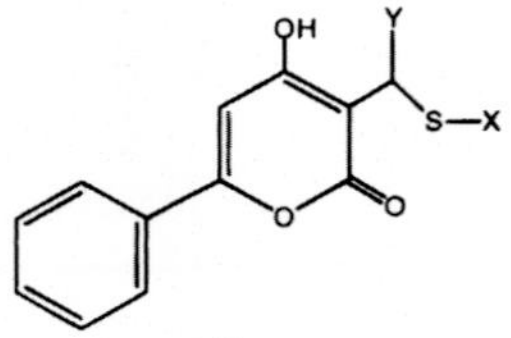

9

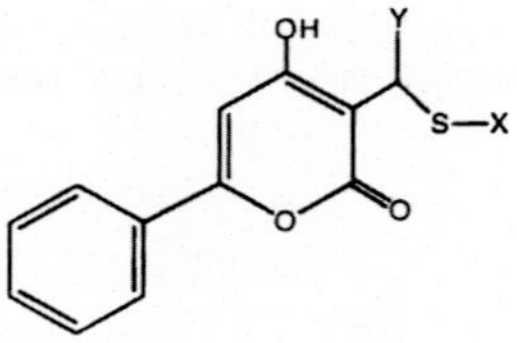

10

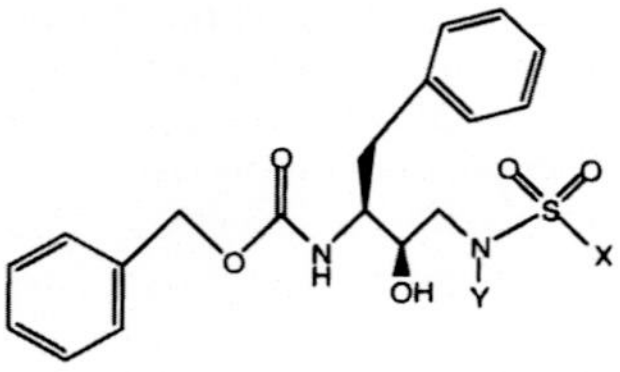

11

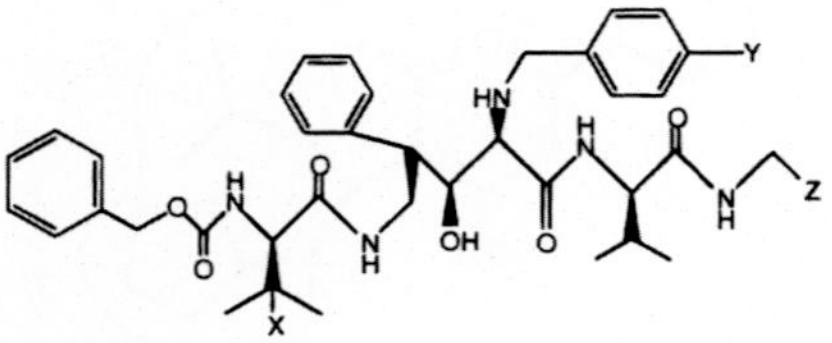

12

Comparative molecular field analysis (CoMFA) studies were also performed on cyclic ureas and different kind of peptide isosteres by some authors (Debnath, 1998, 1999; Kroemer et al., 1995; Oprea et al., 1994; Waller et al., 1993) to support the observations that these compounds adopt a common mode of binding and involve steric and electrostatic interactions. The sincere efforts to develop second-generation PIs led to development of lopinavir (LPV) (**13**, Fig. 9.13) by Abbott Laboratories that had better properties than RTV. Unable to be dosed alone, LPV became the first PI available as a combination pill with RTV as a pharmacokinetic booster (Sham et al., 1998). Atazanavir (ATV) (**14**, Fig. 9.13) approved in 2003, had a longer half-life and became the first

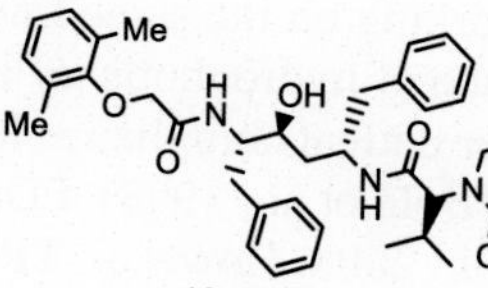

**13, LPV**

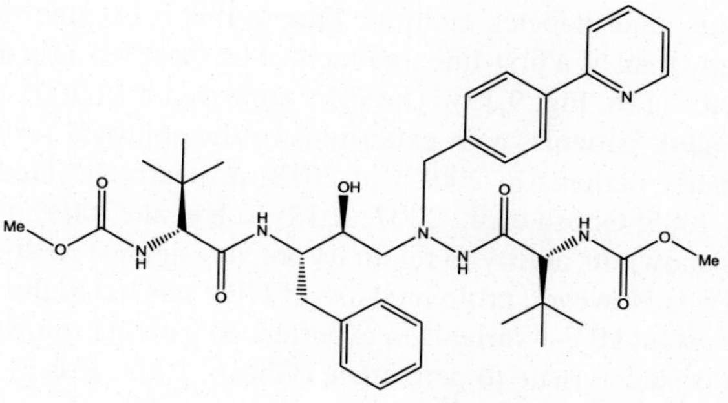

**14, ATV**

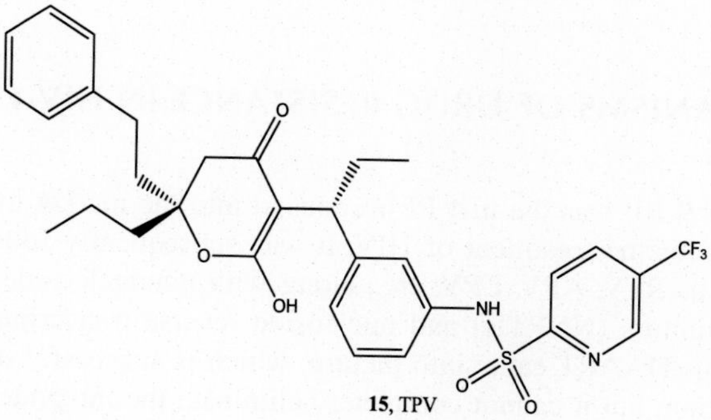

**15, TPV**

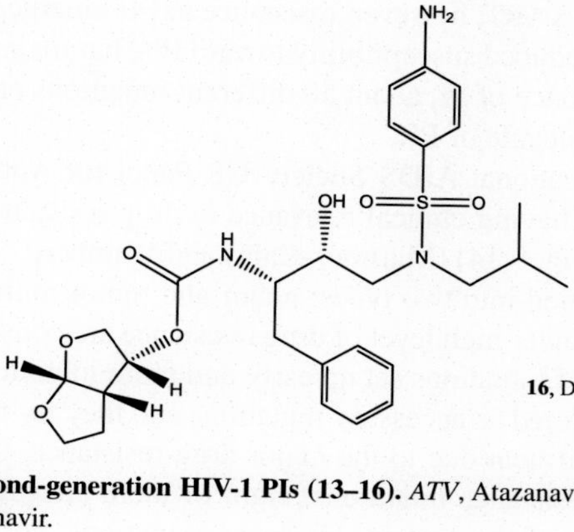

**16, DRV**

FIGURE 9.13  **Second-generation HIV-1 PIs (13–16).** *ATV*, Atazanavir; *DRV*, darunavir; *LPV*, lopinavir; *TPV*, tipranavir.

PI to be electively dosed once daily. Working on the same line to enhance the half-life of the PIs by achieving a balance of hydrophobicity and hydrophilicity, having a lower pill burden, and better patient compliance led to the discovery of tipranavir (TPV) (**15**, Fig. 9.13) (Bold et al., 1998). FDA approved it in 2005 with an extension to pediatric use in 2008. However, TPV suffered from more severe side effects compared to other PIs, such as intracranial hemorrhage, hepatitis, and diabetes mellitus. Due to this it became more useful in salvage therapy than as a first-line treatment. The most recently approved PI is darunavir (DRV) (**16**, Fig. 9.13). The FDA approved it in 2006 for treatment-experienced adult patients, with extensions to the approval for treatment-naive and pediatric patients in 2008 and 2013, respectively (Deeks, 2014; De Meyer et al., 2005; Ghosh et al., 2007, 2012; Koh et al., 2003). DRV is a good drug and has shown its ability to retain its potency against multidrug-resistant strains of HIV-1. However, prolonged use of DRV has led to the emergence of multidrug-resistant HIV-1 variants in experienced patients, showing the ability of the drug resistance issue to perpetuate (Ghosh, 2009; Koh et al., 2010; Tie et al., 2004).

## 6 MECHANISMS OF DRUG RESISTANCE IN HIV-1 PROTEASE

SQV (1, Fig. 9.10) was the first PI introduced into the market in 1995, which raised hopes for the treatment of HIV. It was subsequently followed by other PIs, such as RTV, APV, LPV, etc., along with nonnucleoside reverse transcriptase inhibitors (NNRTIs) and nucleoside reverse transcriptase inhibitors (NRTIs). Thus HAART came into picture, which is successful in suppressing viral replication, but it cannot completely eliminate the integrated viral DNA, which makes the long-term use of these drugs essential. Resistance can be seen for all the drugs in HAART; however, resistance to PIs emerges quickly. The resistant strains have reduced susceptibility toward PIs, but maintain the PR function. Even in the absence of PI, about 50 different mutations can be observed in a relatively small molecule of PR.

As per the International AIDS Society US Panel for Antiretroviral Resistance, the mutations having clinical relevance to drug resistance lie in 37 of 99 residues of HIV-1 (Fig. 9.14) (Martinez-Cajas and Wainberg, 2007). These mutations can be classified into two types: major and minor mutations. Seventeen mutation sites that render high level of drug resistance are considered major mutations, these kinds of mutations set up early and more inhibitor specific. Minor mutations are considered as accessory mutations and they try to compensate for the replication impairment due to the major drug-resistance mutations. Among the major sites of resistance mutations, eight of them alter residues that form the active-site cavity of the protein. However, mutations, not in the active-site cavity or flap, but those in the distal region, are also at times selected as major

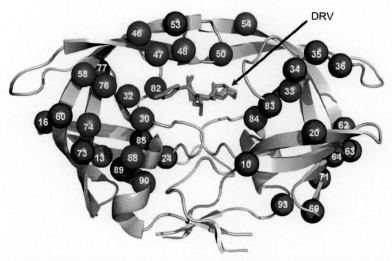

**FIGURE 9.14** **Sites of the resistance mutations in PR dimer.** The PR dimer is in *pink ribbons* with DRV in *green sticks*. Major and minor resistance mutations are colored as *red* and *blue spheres*, respectively. Mutations are distributed on both the monomers to increase visibility.

mutations by PIs. Resistance of particular drugs and their specific mutations are reported in Table 9.3.

There are several molecular mechanisms associated with resistance arising from specific mutations. These mechanisms are dependent on the type of mutation and/or drugs used in HAART therapy. This complicates the analysis of drug-specific resistance mechanism. However, there are three distinct mechanisms, as described here, by which the virus can evade the PIs.

1. *Mutation at the active site that alters direct interactions of the PR with inhibitor or substrate, leading to reduced inhibition.*

   Residues Leu23, Asp30, Val32, Met46, Ile47, Gly48, Ile50, Val82, and Ile84 interact with substrates or inhibitors in the active-site cavity and undergo mutations in drug resistance (Table 9.3). The V82A mutation shows a shift in the main chain atoms of residues 81 and 82 that partially compensate for the loss of interactions due to substitution of a smaller side chain (Mahalingam et al., 2004). Structural studies of V82A in the context of the active wildtype (wt) PR show similar small changes with IDV, DRV, and SQV (Kozisek et al., 2007; Tie et al., 2004, 2005, 2007). In contrast, in studies with inactive PR containing mutation D25N and V82A in complex with SQV, the PR was found to loose almost all the van der Waals interactions along with the hydrophobic interactions and some H-bonds with the inhibitor (Prabu-Jeyabalan et al., 2003). However, the crystal structure of mutant V82A in the catalytically active PR more or less retains all the van der Waals

**TABLE 9.3 Resistance Mutations of HIV-1 PR**

Minor mutations columns are headed by position number and wild-type residue (shown as "position residue").

| Drugs | Major mutations | 10 L | 11 V | 13 I | 16 G | 20 K | 24 L | 30 D | 32 V | 33 L | 34 E | 35 E | 36 M | 43 K | 46 M | 47 I | 48 G | 50 I | 53 F | 54 I | 58 Q | 60 D | 62 I | 63 L | 64 I | 69 H | 71 A | 73 G | 74 T | 76 L | 77 V | 82 V | 83 N | 84 I | 85 I | 88 N | 89 L | 90 L | 93 I |
|---|---|---|---|---|---|---|---|---|---|---|---|---|---|---|---|---|---|---|---|---|---|---|---|---|---|---|---|---|---|---|---|---|---|---|---|---|---|---|---|
| Saquinavir/r | G48V, L90M | I/R/V | | | | | I | | | | | | | | | | | | | V/L | | | V | | | | V/T | S | | | I | A | | V | | | | L | |
| Indinavir/r | M46I/L, V82A/F/T, I84V | I/R/V | | | | M/R | I | | I | | | | I | | | | | | | V | | | | | | | V/T | S/A | | | I | | | | | | | M | |
| Nelfinavir/r | D30N, L90M | F/I | | | | | | | | | | | I | | I/L | | | | | | | | | | | | V/T | | | | V | A/F/T/S | | V | | D/S | | | |
| Fosamprenavir/r | I50V, I84V | F/I/R/V | | | | | | | I | | | | | | I/L | V | | | | L/V/M | | | | | | | | S | | V | | A/F/T/S | | | | | | M | |
| Lopinavir/r | V32I, I47V/A, V82A/F/T/S | F/I/R/V | | | | M/R | I | | | F | | | | | I/L | | | V | L | V/L/A/M/T/S | | | | P | | | V/T | S | | V | | | | V | | | | M | |

Residues in the substrate-binding site are in blue. Resistance mutations for which structural information is available are in red.

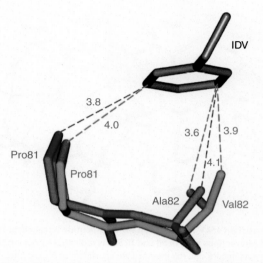

**FIGURE 9.15    The V82A mutation shows a shift in the main chain atoms of residues 81 and 82 that partially compensates for the loss of interactions due to substitution of a smaller side chain.** The wildtype (wt) Val82 (PDB ID: 1SDT) and Ala mutant (PDB ID: 1SDV) are shown as *green* and *magenta sticks*, respectively.

contacts with IDV (Mahalingam et al., 2004) (Fig. 9.15) that partially compensate for the loss of interactions due to substitution of a smaller side chain.

2. *Mutations at the dimer interface that alter PR stability.*

   Residues Leu24, Ile50, and Phe53 lie at the interface between the two subunits. Leu24 does not interact directly with the substrate or inhibitor, although this residue is adjacent to the catalytic Asp25. Ile50 and Phe53 are located in the flap. L24I is observed as a minor mutation in resistance to SQV, IDV, NFV, LPV, and ATV. I50V is a major resistance mutation for Fosamprenavir and DRV, and a minor mutation for LPV. F53L appears as a minor resistance mutation selected by LPV and ATV.

3. *Distal mutations showing a variety of effects.*

   This category of mutations includes mutations of flap residues and other residues without direct contact with inhibitors. Distal mutations are often observed together with other resistance mutations. Flap mutant I54M is selected as a major drug-resistant mutation in treatment with DRV, although residue 54 has no direct interactions with inhibitors. The structure of the I54M variant has been analyzed with DRV and SQV (Liu et al., 2008). Mutation of residue 54 induces changes in residues 80–82 (80's loop) that interact with inhibitors. In case of variant I54M, the 80's loop is shifted away from residue 54 due to increased side chain length resulting in weaker interactions with DRV. In contrast, the I54V variant has no significant change in interactions with DRV or SQV. Another flap residue Ile54 does not interact directly with inhibitor, but its mutations, including I54V, are

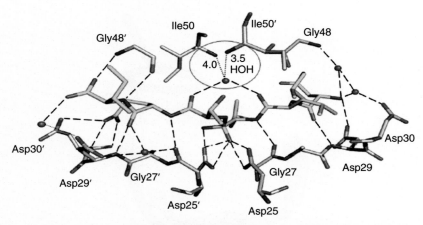

**FIGURE 9.16**   **The variant I54V (PDB ID: 2B80) with peptide tetrahedral intermediate has lost water-mediated interactions with Ile50 and Ile50′ as indicated by *red circles*.** The hydrogen bond interactions between the PR and the intermediate are shown as *broken lines*.

attributed to resistance against all PIs, except for NFV. The crystal structure of mutant without inhibitor has shown that the loss of water mediated hydrogen bond interactions with the tetrahedral reaction intermediate of a peptide and the flap residues Ile50 and Ile50′ (Fig. 9.16). The mutant structure of I54V–DRV complex reveals a shift in the 80's loop, although no significant overall change occurs in the interactions between PR and the inhibitor (Liu et al., 2008). These studies indicate that weaker binding of the inhibitor due to the I54V substitution may contribute to drug resistance (Weber and Agniswamy, 2009).

## 7   INHIBITION

The knowledge gained over the last 3 decades regarding the structure and mechanism of HIV PR has paved the path toward the development of effective drugs for this enzyme. Presently there are seven FDA-approved HIV PIs. The development of these inhibitors can be categorized into two broad classes. First, the inhibitors developed using the basic core of nonhydrolyzable hydroxyethylene or hydroxyethylamine moiety. Other noncleavable transition state isosteres that have been explored successfully, for example, statine, norstatine, phosphinate, reduced amide, dihydroxyethylene (Wlodawer and Erickson, 1993), α-keto amide (Munoz et al., 1994), and silicon-based inhibitors (Chen et al., 2001). The idea of using the basic core is to isosterically replace the scissile bond, which is believed to mimic the tetrahedral transition state of the proteolytic reaction. SQV, developed by Hoffman–La Roche in 1995 was the first FDA-approved drug for the treatment of AIDS, NFV (Lilly and Agouron, 1997), and APV (Vertex Pharmaceuticals, 1999) are all the examples of drugs having hydroxyethylamine isostere replacement. Similar examples having nonhydro-lyzable different isostere core are IDV, RTV, and LPV.

The second category of HIV PIs can be classified based on their effects on the rigidity of the β-hairpin "flaps": those that restrict the flexibility and those that do not. It has been usually observed that the molecules that retain flexibility of flaps permit the "flexibilty-assisted catalysis," a mechanism introduced by Piana et al. (2002). However, tripnavir, a nonpeptidic inhibitor, is distinct due to its strength and its effect of rigidifying the flaps. This observation suggests that their mode of action is quite different from other PIs, such as DRV, which interacts with the active site rather than the flaps. Rigidity analysis indicates two different modes of action for HIV-1 PIs: direct interaction with the active site or interaction with the flaps to rigidify them in a closed conformation and thus indirectly prevent access to the active site. For example, RTV is often administered in combination with other PIs during antiretroviral therapy. Such type of complementary approach of using one inhibitor that targets the flaps of the PR and one that may not account for the efficacy of TPV–RTV combination reported by Hicks et al. (2006). These observations suggest that to develop a therapeutically useful PI, the molecule should be able to interact with the active site and the flaps simultaneously, which will make it difficult for the virus to evade by mutation.

## 8 STRUCTURE-GUIDED DESIGN OF PIS FOR RESISTANT HIV PROTEASE

Molecular mechanisms of drug resistance and structure and activity of PIs have laid a solid foundation for the design of antiviral PIs targeting resistant strains of HIV. The earlier PIs, designed to target wt HIV-1 PR, retain a peptide-like chemical backbone. PIs developed later, such as APV, TPV, and DRV, include sulfonamide, which replace the peptide-like carbonyl group (Koh et al., 2003; Mehandru and Markowitz, 2003). The designing of inhibitors that fit within the envelope formed by bound substrates, as PR cleavage of its substrates is essential for viral replication, resulted in tight-binding (subnanomolar) inhibitors of HIV-1 PR variants (Altman et al., 2008; Chellappan et al., 2007; Prabu-Jeyaba-lan et al., 2006). Targeting the flaps in the open conformation of the unliganded PR dimer is another approach in which molecules, such as pyrrolidine-based inhibitors, bind between the open flaps in the PR (Bottcher et al., 2008). Inor-ganic metallocarboranes also bind to the open flaps of the PR and inhibit drug-resistant variants (Kozisek et al., 2008).

The most successful strategy to date for targeting resistant HIV-1 infections is exemplified by DRV in 2006 (Weber and Agniswamy, 2009). Interactions with the main chain amide and carbonyl oxygen are important, as they are not easily altered by mutation; at the same time the introduction of hydrogen bond donor or acceptor groups in the inhibitor to form hydrogen bond interactions with conserved regions of HIV PR makes the molecule less susceptible to the development of resistance (Fig. 9.17).

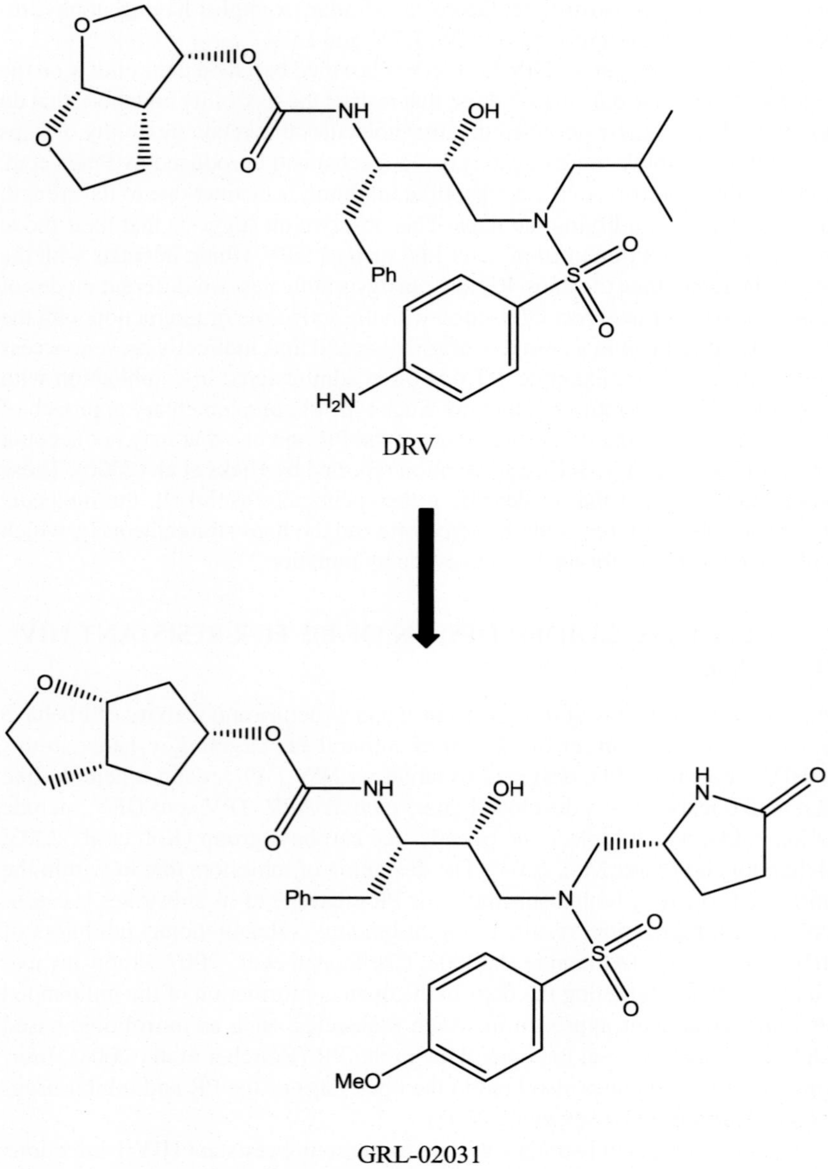

DRV

GRL-02031

FIGURE 9.17   Chemical structures of DRV and the new antiviral inhibitor GRL-02031.

# 9 DESIGN OF DEFECTIVE VARIANTS FOR HIV-1 PR INHIBITION

Defective variants of HIV-1 PR were attempted to inhibit wt HIV-1 PR activity. These variants lead to the formation of heterodimers and destabilize the formation of inactive variant homodimers of HIV-1 PR through substitutions at Asp25, Ile49, and Gly50 (Babé et al., 1995; McPhee et al., 1996). The defective monomers were refolded in vitro in the presence of wt HIV-1 PR and showed dose-dependent inhibition of proteolytic activity. The formation of inactive heterodimers between defective and wt HIV PR monomers made the PR an active target for inhibition. With the design of several such defective HIV PR monomers, with one or more substitutions at Asp25, Gly49, and/or Ile50, several defective PR monomers were reported, which demonstrated inhibition in cell culture (Babé et al., 1995; McPhee et al., 1996). Coexpression of the inactive PR variant with wt HIV PR inhibited viral maturation in both transiently transfected and stably integrated cell lines in a dose-dependent manner, consistent with dominant-negative inhibition of the viral PR (Babé et al., 1995). A very potent dominant-negative inhibitor could be constituted of a variant monomer containing three substituents, D25K/G49W/I50W (KWW). A heterodimer composed of wt HIV PR and HIV PR KWW is shown in Fig. 9.18 (Rozzelle et al., 2000). The substitution of lysine for the catalytic Asp25 may

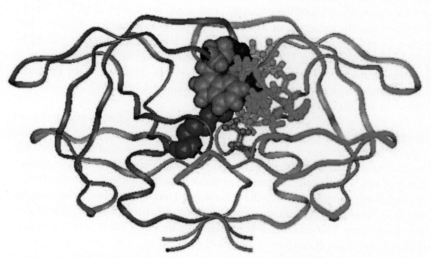

FIGURE 9.18    **Model of the HIV PR heterodimer with the D25K, G49W, and I50W (KWW) substitutions.** The KWW monomer is shown in *blue* and the wt monomer in *green*. In KWW monomer, space-filling representation with *red* refers to D25K, *light blue* to G49W, and *dark blue* to I50W. (*Reprinted with permission from Rozzelle, J.E., Dauber, D.S., Todd, S., Kelley, R., Craik, C.S., 2000. Macromolecular inhibitors of HIV-1 protease characterization of designed heterodimers. J. Biol. Chem. 275 (10), 7080–7086. Copyright 2000 American Society for Biochemistry and Molecular Biology, Inc.*)

stabilize the heterodimer by hydrophobic interactions between the methylene groups of the lysine side chain and the hydrophobic substrate-binding pocket, as well as by favorable charge–charge interactions with the aspartate of the wt monomer (McPhee et al., 1996). Further, favorable hydrophobic interactions with the S1 and S29 subsites of an associated monomer via a pseudosubstrate-like mechanism could be introduced by replacing the residues Gly49 and Ile50 with tryptophans.

Rozzelle et al. (2000) examined the three defective monomers, HIV PR D25K (K), HIV PR G49W/I50W (WW), and HIV PR KWW, with single, double, and triple mutations, respectively, and demonstrated that dominant-negative inhibition was attributed to heterodimer formation and that the dimer interface of HIV PR could be engineered to stabilize these inactive heterodimer complexes relative to homodimeric wt HIV PR. These authors also concluded that the potency of inhibition could be increased through the rational design of a more stable heterodimeric interface.

## 10   RECENT PROGRESS TOWARD HIV-1 PROTEASE INHIBITORS

The second-generation PIs, particularly DRV (Fig. 9.13), are potent nonpeptidyl inhibitors of the mature PR dimer (Hayashi et al., 2014; Koh et al., 2011). As for the acquisition of the catalytic proteolytic properties of HIV-1 PR, it is essential that dimerization of PR monomers should take place. Thus the inhibition of PR dimerization by DRV represents a novel approach to halt HIV-1 progression. Except for DRV and TPV, no other PIs exhibit dimerization inhibition properties. These types of PIs possessing dual inhibitory properties may exhibit a high genetic barrier toward resistance (Ghosh and Chapsal, 2013; Ghosh et al., 2010a).

For further developments in this field, production of next-generation PIs needs to be focused on the improvement of drug-resistance profiles having high genetic barrier toward resistance. Efforts need to be directed for the development of new PIs using molecular design, with the focus on the design and synthesis of novel $P_2$ ligands promoting enhanced backbone binding interactions to combat drug resistance. Research is also needed on the ways to develop nonpeptide PIs containing different structural scaffolds other than hydroxyethylsulfonamide isosteres. The current on-going research on PIs is based on the following types of chemicals (Ghosh et al., 2016).

1. Inhibitors containing bis-tetrahydrofuran as a $P_2$ ligand (e.g., **17**) (Yedidi et al., 2013).
2. Inhibitors containing substituted bis-tetrahydrofuran as a $P_2$ ligand (e.g., **18**) (Ghosh et al., 2015b).

3. Inhibitors containing cyclopentyltetrahydrofuran as a $P_2$ ligand (e.g., **19**) (Ghosh et al., 2006b).
4. Inhibitors containing tetrahydropyranyltetrahydrofuran as a $P_2$ ligand (e.g., **20**) (Ghosh et al., 2011).
5. Inhibitors containing tris-tetrahydrofuran as a $P_2$ ligand (e.g., **21**) (Ghosh et al., 2010b).
6. Inhibitors containing fused tricyclic derivatives as $P_2$ ligands (e.g., **22**) (Ghosh et al., 2015a).
7. Inhibitors with extended $P_1$ ligands (e.g., **23**) (Ghosh et al., 2010a).
8. Benzamide-based inhibitors (e.g., **24**) (Bonini et al., 2005).
9. Inhibitors containing a proline derivative as a $P_2$ ligand (e.g., 25) (Gao et al., 2011).
10. Oxazolidinone-based $P_2$ ligands (e. g., **26**) (Ali et al., 2010).
11. Urea- and triazole-derived inhibitors (e.g., **27**) (Brik et al., 2005).
12. ATV-based compounds (e.g., **28**) (Öhrngren et al., 2011).
13. Macrocyclic inhibitors (e.g., **29**) (Ghosh et al., 2014).
14. Nonsulfonamide inhibitors (e.g., **30**) (Ghosh et al., 2006a).
15. Allophenylnorstatine-containing inhibitors (e.g., **31**) (Ami et al., 2007).
16. Diol and pseudosymmetric inhibitors (e.g., **32**) (Clemente et al., 2008).
17. Nontransition state inhibitors (e.g., **33**) (Garino et al., 2005).
18. Cyclic sulfamide inhibitors (e.g., **34**) (Ganguly et al., 2011).
19. Lysinol-containing inhibitor (e.g., **35**) (Jones et al., 2010).
20. Cyclic urea–derived inhibitors (e.g., **36**) (Zhao et al., 2005).
21. Natural product–based inhibitors (e.g., **37**) (Wei et al., 2009).
22. Pentacycloundecane-containing inhibitors (e.g., **38**) (Dufau et al., 2012).
23. Dimerization inhibitors (e.g., **39**) (Bannwarth et al., 2006).

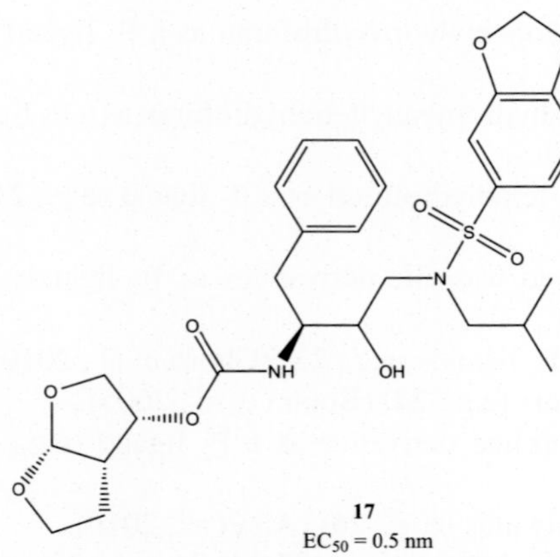

**17**
$EC_{50} = 0.5$ nm

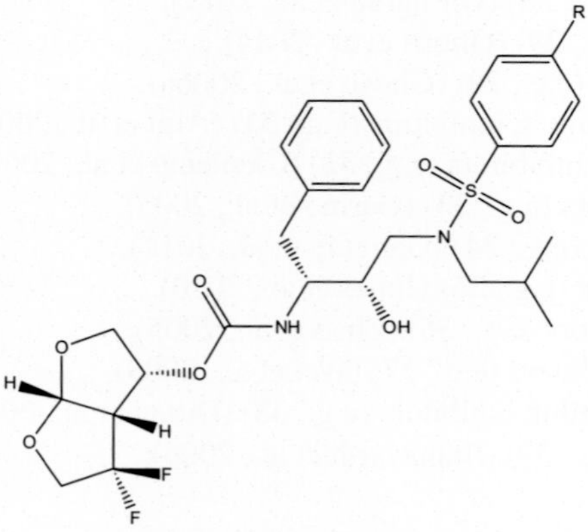

**18**, R = NH$_2$; $K_i = 0.0058$ nM; $EC_{50} = 0.003$ μM; $CC_{50} = 37$ μM

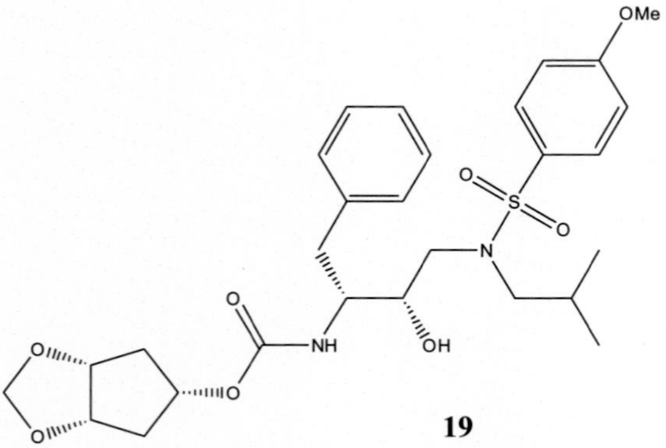

**19**

$K_i = 0.11$ nm; $IC_{50} = 0.0038$ μm

**20,** $K_i = 0.0027$ nM; $IC_{50} = 0.0005$ μM; $EC_{50} = 0.2$ nM

**21,** $K_i = 5.9$ pM; $IC_{50} = 1.8$ nM

**22**

$K_i = 0.01$ nM; $IC_{50} = 1.9$ nM; $EC_{50} = 0.0019$ μM

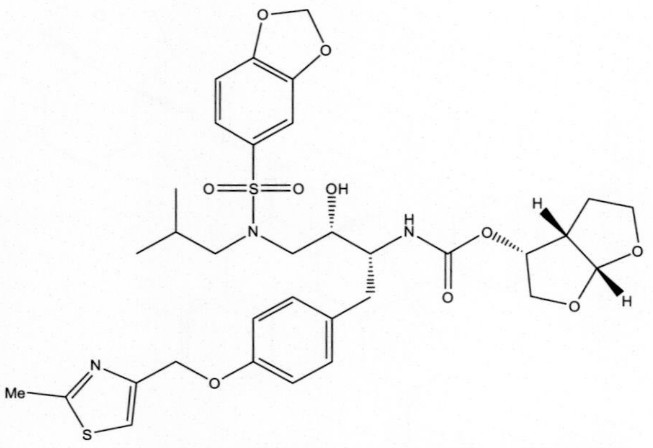

**23** (Brecanavir), EC$_{50}$ = 0.2 nM, IC$_{50}$ = 0.7 nM

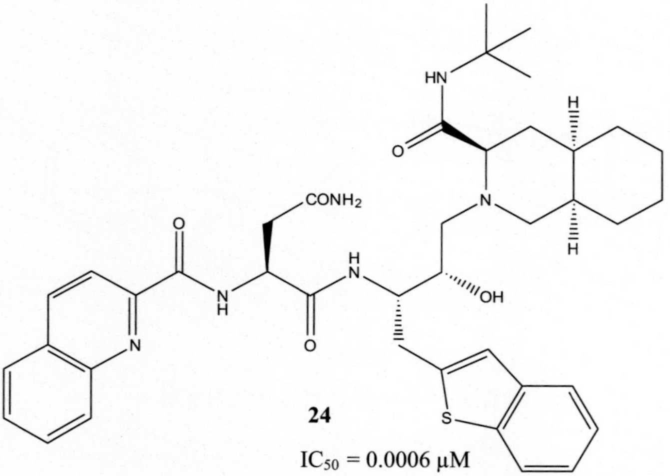

**24**

IC$_{50}$ = 0.0006 μM

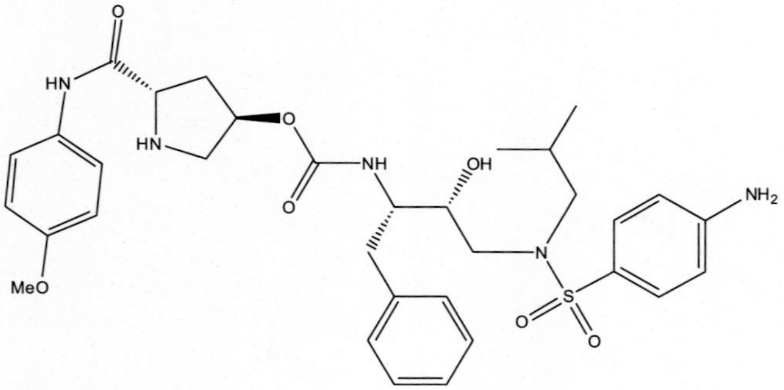

**25,** IC$_{50}$ = 15.4 nM

**26**, IC$_{50}$ = 5 nM; EC$_{50}$ = 0.8 μM

**27**, $K_i$ = 1.7 nM, IC$_{50}$ = 6 nM

**28**, $K_i$ = 3.6 nM, EC$_{50}$ = 1.0 μM

## 11  CONCLUSIONS

Notwithstanding the sustained efforts and advances in the antiretroviral therapy in last 3 decades, HIV is perceived as a chronic disease. HIV-1 PR is an important enzyme that helps in the maturation of the HIV-1 after its replication in the nucleus of the host cell. Inhibition of HIV-1 PR causes a major block in the life cycle of the virus. PIs were among the first drugs approved by US FDA for the treatment of AIDS. Structure-based drug design has resulted in several PIs that have increased the treatment options and improved viral suppression. However, the emergence of drug resistance has affected the long-term potency of clinical drugs.

Thus the question is: can PIs help get rid of HIV-1? These inhibitors cannot cure AIDS, as they cannot destroy all the viruses; however, they can reduce the number of viruses by almost 99%. The problem remains with the dormant virus present in latently infected cells, which are not active at the time of the treatment, but can activate themselves after a period of time. Thus there is a huge doubt on the complete elimination of the virus by any kind of treatment.

Although HAART, which uses the combination of several (three or more) antiretroviral drugs, is more effective than using just one medicine (monotherapy) to fight HIV, development of new therapies to cure AIDS is going on in several directions, for example, development of new PIs that are not cross-resistant to current drugs, studies on heterodimers of PR, inhibition of HIV-1 entry, and targeting conserved HIV RNA sequences with small molecules, etc. However, there is still a long way to go, as far as a total cure of HIV is concerned. Among all these, HIV-1 PR remains the most studied and explored target for AIDS treatment.

## REFERENCES

Ali, A., Reddy, G.S., Nalam, M.N., Anjum, S.G., Cao, H., Schiffer, C.A., Rana, T.M., 2010. Structure-based design, synthesis, and structure-activity relationship studies of HIV-1 protease inhibitors incorporating phenyloxazolidinones. J. Med. Chem. 53, 7699–7708.

Altman, M.D., Ali, A., Reddy, G.S., Nalam, M.N., Anjum, S.G., Cao, H., Chellappan, S., Kairys, V., Fernandes, M.X., Gilson, M.K., Schiffer, C.A., Rana, T.M., Tidor, B., 2008. HIV-1 protease inhibitors from inverse design in the substrate envelope exhibit sub nanomolar binding to drug resistant variants. J. Am. Chem. Soc. 130, 6099–6113.

Ami, E., Nakahara, K., Sato, A., Nguyen, J.T., Hidaka, K., Hamada, Y., Nakatani, S., Kimura, T., Hayashi, Y., Kiso, Y., 2007. Synthesis and antiviral property of allophenylnorstatine-based HIV protease inhibitors incorporating D-cysteine derivatives as P2/P3 moieties. Bioorg. Med. Chem. Lett. 17, 4213–4217.

Babé, L.M., Roset, J., Craik, C.S., 1995. Trans-dominant inhibitory human immunodeficiency virus type 1 protease monomers prevent protease activation and virion maturation. Proc. Natl. Acad. Sci. USA 92, 10069–10073.

Babine, R.E., Bender, S.L., 1997. Molecular recognition of protein-ligand complexes: applications to drug design. Chem. Rev. 97, 1359.

Baldwin, E.T., Bhat, T.N., Gulnik, S., Liu, B., Topol, I.A., Kiso, Y., Mimoto, T., Mitsuya, H., Erickson, J.W., 1995. 45 Structure of HIV-1 protease with KNI-272, a tight-binding transition-state analog containing allophyenylnorstatine. Structure 3, 581–590.

Bannwarth, L., Kessler, A., Pèthe, S., Collinet, B., Merabet, N., Boggetto, N., Sicsic, S., Reboud-Ravaux, M., Ongeri, S., 2006. Molecular tongs containing amino acid mimetic fragments: new inhibitors of wild-type and mutated HIV-1 protease dimerization. J. Med. Chem. 49, 4657–4664.

Barré-Sinoussi, F., Ross, A.L., Delfraissy, J.F., 2013. Past, present and future: 30 years of HIV research. Nat. Rev. Microbiol. 11, 877–883.

Billich, A., Charpiot, B., Fricker, G., Gstach, H., Lehr, P., Peichl, P., Scholz, D., Rosenwirth, B., 1994. HIV proteinase inhibitors containing 2-aminobenzylstatine as a novel scissile bond replacement: biochemical and pharmacological characterization. Antiviral Res. 25, 215.

Bold, G., Fässler, A., Capraro, H.G., Cozens, R., Klimkait, T., Lazdins, J., Mestan, J., Poncioni, B., Rösel, J., Stover, D., Tintelnot-Blomley, M., Acemoglu, F., Beck, W., Boss, E., Eschbach, M., Hürlimann, T., Masso, E., Roussel, S., Ucci-Stoll, K., Wyss, D., Lang, M., 1998. New aza-dipeptide analogues as potent and orally absorbed HIV-1protease inhibitors: candidates for clinical development. J. Med. Chem. 41, 3387–3401.

Bonini, C., Chiummiento, L., Bonis, M.D., Funicello, M., Lupattelli, P., Suanno, G., Berti, F., Campaner, P., 2005. Synthesis, biological activity and modelling studies of two novel anti HIV PR inhibitors with a thiophene containing hydroxyethylamino core. Tetrahedron 61, 6580–6589.

Bottcher, J., Blum, A., Dorr, S., Heine, A., Diederich, W.E., Klebe, G., 2008. Targeting the open-flap conformation of HIV-1 protease with pyrrolidine-based inhibitors. ChemMedChem 3, 1337–1344.

Brik, A., Wong, C.-H., 2003. HIV-1 protease: mechanism and drug discovery. Org. Biomol. Chem. 1, 5–14.

Brik, A., Alexandratos, J., Lin, Y.C., Elder, J.H., Olson, A.J., Wlodawer, A., Goodsell, D.S., Wong, C.H., 2005. 1,2,3-Triazole as a peptide surrogate in the rapid synthesis of HIV-1 protease inhibitors. ChemBioChem 6, 1167–1169.

Chellappan, S., Kairys, V., Fernandes, M.X., Schiffer, C., Gilson, M.K., 2007. Evaluation of the substrate envelope hypothesis for inhibitors of HIV-1 protease. Proteins 68, 561–567.

Chen, X., Tropsha, A., 1995. J. Med. Chem. 38, 42.

Chen, A., Sieburth, S.M., Glekas, A., Hewitt, G.W., Trainor, G.L., Viitanen, S.E., Garber, S.S., Cordova, B., Jeffry, S., Klabe, R., 2001. Drug design with a new transition state analog of the hydrated carbonyl: silicon-based inhibitors of the HIV protease. Chem. Biol. 8, 1161.

Clemente, J.C., Robbins, A., Graña, P., Paleo, M.R., Correa, J.F., Villaverde, M.C., Sardina, F.J., Govindasamy, L., Agbandje-McKenna, M., McKenna, R., Dunn, B.M., Sussman, F., 2008. Design, synthesis, evaluation, and crystallographic-based structural studies of HIV-1 protease inhibitors with reduced response to the V82A mutation. J. Med. Chem. 51, 852–860.

Darke, P.L., Leu, C.T., Davis, L.J., Heimbach, J.C., Diehl, R.E., Hill, W.S., Dixon, R.A., Sigal, I.S., 1989. Human immunodeficiency virus protease. Bacterial expression and characterization of the purified aspartic protease. J. Biol. Chem. 264, 2307.

Debnath, A.K., 1998. Comparative molecular field analysis (CoMFA) of a Series of symmetrical bis-benzamide cyclic urea derivatives as HIV-1 protease inhibitors. J. Chem. Inf. Comput. Sci. 38, 761–767.

Debnath, A.K., 1999. Three-dimensional quantitative structure-activity relationship study on cyclic urea derivatives as HIV-1 protease inhibitors: application of comparative molecular field analysis. J. Med. Chem. 42, 249–259.

Deeks, E.D., 2014. Darunavir: a review of its use in the management of HIV-1 infection. Drugs 74, 99–125.

De Meyer, S., Azijn, H., Surleraux, D., Jochmans, D., Tahri, A., Pauwels, R., Wigerinck, P., de Béthune, M.P., 2005. TMC114, a novel human immunodeficiency virus type 1 protease inhibitor active against protease inhibitor-resistant viruses, including a broad range of clinical isolates. Antimicrob. Agents Chemother. 49, 2314–2321.

Dufau, L., Marques Ressurreicao, A.S., Fanelli, R., Kihal, N., Vidu, A., Milcent, T., Soulier, J.L., Rodrigo, J., Desvergne, A., Leblanc, K., Bernadat, G., Crousse, B., Reboud-Ravaux, M., Ongeri, S., 2012. Carbonylhydrazide-based molecular tongs inhibit wild-type and mutated HIV-1 protease dimerization. J. Med. Chem. 55, 6762–6775.

Ferguson, D.M., Radmer, R.J., Kollman, P.A., 1991. Determination of the relative binding free energies of peptide inhibitors to the HIV-1 protease. J. Med. Chem. 34, 2654–2659.

Freedberg, D.I., Ishima, R., Jacob, J., Wang, Y.X., Kustanovich, I., Louis, J.M., Torchia, D.A., 2002. Rapid structural fluctuations of the free HIV protease flaps in solution: Relationship to crystal structures and comparison with predictions of dynamics calculations. Protein Sci. 11, 221–232.

Ganguly, A.K., Alluri, S.S., Caroccia, D., Biswas, D., Wang, C.H., Kang, E., Zhang, Y., McPhail, A.T., Carroll, S.S., Burlein, C., Munshi, V., Orth, P., Strickland, C., 2011. Design, synthesis, and X-ray crystallographic analysis of a novel class of HIV-1 protease inhibitors. J. Med. Chem. 54, 7176–7183.

Gao, B.L., Zhang, C.M., Yin, Y.Z., Tang, L.Q., Liu, Z.P., 2011. Design and synthesis of potent HIV-1 protease inhibitors incorporating hydroxyprolinamides as novel P2 ligands. Bioorg. Med. Chem. Lett. 21, 3730–3733.

Garg, R., Gupta, S.P., Gao, H., Babu, M.S., Debnath, A.K., Corwin Hansch, C., 1999. Comparative quantitative structure-activity relationship studies on anti-HIV drugs. Chem. Rev. 99, 3525–3601.

Garino, C., Bihel, F., Pietrancosta, N., Laras, Y., Quelever, G., Woo, I., Klein, P., Bain, J., Boucher, J.L., Kraus, J.L., 2005. New 2-bromomethyl-8-substituted-benzo[c]chromen-6-ones. Synthesis and biological properties. Bioorg. Med. Chem. Lett. 15, 135–138.

Ghosh, A.K., 2009. Harnessing nature's insight: design of aspartyl protease inhibitors from treatment of drug-resistant HIV to Alzheimer's disease. J. Med. Chem. 52, 2163–2176.

Ghosh, A.K., Chapsal, B.D., 2013. Design of the anti-HIV protease inhibitor darunavir. In: Ganellin, C.R., Jefferis, R., Roberts, S. (Eds.), Introduction to Biological and Small Molecule Drug Research and Development. Elsevier, Amsterdam, pp. 355–385.

Ghosh, A.K., Anderson, D.D., Mitsuya, H., 2010a. The FDA approved HIV-1 protease inhibitors for treatment of HIV/AIDS. Abraham, D.J., Rotella, D.P. (Eds.), Burger's Medicinal Chemistry and Drug Discovery, vol. 7, seventh ed. John Wiley & Sons, Hoboken, NJ, pp. 1–74.

Ghosh, A.K., Anderson, D.D., Weber, I.T., Mitsuya, H., 2012. Enhancing protein backbone binding–a fruitful concept for combating drug-resistant HIV. Angew. Chem. Int. Ed. 51, 1778–1802.

Ghosh, A.K., Chapsal, B.D., Baldridge, A., Steffey, M.P., Walters, D.E., Koh, Y., Amano, M., Mitsuya, H., 2011. Design and synthesis of potent HIV-1 protease inhibitors incorporating hexahydrofuropyranol derived high affinity P(2) ligands: structure-activity studies and biological evaluation. J. Med. Chem. 54, 622–634.

Ghosh, A.K., Dawson, Z.L., Mitsuya, H., 2007. Darunavir, a conceptually new HIV-1 protease inhibitor for the treatment of drug resistant HIV. Bioorg. Med. Chem. 15, 7576–7580.

Ghosh, A.K., Ostwald, H.L., Prato, G., 2016. Recent progress in the development of HIV1 protease inhibitors for the treatment of HIV/AIDS. J. Med. Chem. 59, 5172–5208.

Ghosh, A.K., Swanson, L.M., Cho, H., Leshchenko, S., Hussain, K.A., Kay, S., Walters, D.E., Koh, Y., Mitsuya, H., 2005. Structure-based design: synthesis and biological evaluation of a series of novel cycloamide-derived HIV-1 protease inhibitors. J. Med. Chem. 48, 3576–3585.

Ghosh, A.K., Sridhar, P.R., Leshchenko, S., Hussain, A.K., Li, J., Kovalevsky, A.Y., Walters, D.E., Wedekind, J.E., Grum-Tokars, V., Das, D., Koh, Y., Maeda, K., Gatanaga, H., Weber, I.T., Mitsuya, H., 2006a. Structure-based design of novel HIV-1 protease inhibitors to combat drug resistance. J. Med. Chem. 49, 5252–5261.

Ghosh, A.K., Schiltz, G., Perali, R.S., Leshchenko, S., Kay, S., Walters, D.E., Koh, Y., Maeda, K., Mitsuya, H., 2006b. Design and synthesis of novel HIV-1 protease inhibitors incorporating oxyindoles as the P2′-ligands. Bioorg. Med. Chem. Lett. 16, 1869–1873.

Ghosh, A.K., Schiltz, G.E., Rusere, L.N., Osswald, H.L., Walters, D.E., Amano, M., Mitsuya, H., 2014. Design and synthesis of potent macrocyclic HIV-1 protease inhibitors involving P1-P2 ligands. Org. Biomol. Chem. 12, 6842–6854.

Ghosh, A.K., Xu, C.X., Osswald, H.L., 2015a. Enantioselective synthesis of dioxatriquinane structural motifs for HIV-1 protease inhibitors using a cascade radical cyclization. Tetrahedron Lett. 56, 3314–3317.

Ghosh, A.K., Xu, C.X., Rao, K.V., Baldridge, A., Agniswamy, J., Wang, Y.F., Weber, I.T., Aoki, M., Miguel, S.G., Amano, M., Mitsuya, H., 2010b. Probing multidrug-resistance and protein-ligand interactions with oxatricyclic designed ligands in HIV-1 protease inhibitors. ChemMedChem 5, 1850–1854.

Ghosh, A.K., Yashchuk, S., Mizuno, A., Chakraborty, N., Agniswamy, J., Wang, Y.F., Aoki, M., Gomez, P.M., Amano, M., Weber, I.T., Mitsuya, H., 2015b. Design of gem-difluoro-bis-tetrahydrofuran as P2 ligand for HIV-1 protease inhibitors to improve brain penetration: synthesis, X-ray studies, and biological evaluation. ChemMedChem 10, 107–115.

Gulick, R.M., Mellors, J.W., Havlir, D., Eron, J.J., Gonzalez, C., McMahon, D., Richman, D.D., Valentine, F.T., Jonas, L., Meibohm, A., Emini, E.A., Chodakewitz, J.A., Deutsch, P., Holder, D., Schleif, W.A., Condra, J.H., 1997. Treatment with indinavir, zidovudine, and lamivudine in adults with human immunodeficiency virus infection and prior antiretroviral therapy. N. Engl. J. Med. 337, 734–739.

Gupta, S.P., Babu, M.S., 1999. Quantitative structure-activity relationship studies on cyclic cyano-guanidines acting as HIV-1 protease inhibitors. Bioorg. Med. Chem. 7, 2549–2553.

Gupta, S.P., Babu, M.S., Garg, R., Sowmya, S., 1998a. Quantitative structure activity relationship studies on cyclic urea-based HIV protease inhibitors. J. Enzyme Inhib. 13, 399.

Gupta, S.P., Babu, M.S., Kaw, N., 1999. Quantitative structure-activity relationships of some HIV-protease inhibitors. J. Enzyme Inhib. 14, 109.

Gupta, S.P., Babu, M.S., Sowmya, S., 1998b. A quantitative structure-activity relationship study on some sulfolanes and arylthiomethanes acting as HIV-1 protease inhibitors. Bioorg. Med. Chem. 6, 2185.

Gustchina, A., Weber, I.T., 1990. Comparison of inhibitor binding in HIV-1 protease and in non-viral aspartic proteases: the role of the flap. FEBS Lett. 269, 269–272.

Hammer, S.M., Squires, K.E., Hughes, M.D., Grimes, J.M., Demeter, L.M., Currier, J.S., Eron, Jr., J.J., Feinberg, J.E., Balfour, Jr., H.H., Deyton, L.R., Chodakewitz, J.A., Fischl, M.A., Phair, J.P., Pedneault, L., Nguyen, B.Y., Cook, J.C., 1997. A controlled trial of two nucleoside analogues plus indinavir in persons with human immunodeficiency virus infection and CD4 cell counts of 200 per cubic millimeter or less. AIDS Clinical Trials Group 320 Study Team. N. Engl. J. Med. 337, 725–733.

Harte, W.E., Beveridge, D.L., 1993. Mechanism for the destabilization of the dimer interface in a mutant HIV-1 protease: a molecular dynamics study. J. Am. Chem. Soc. 115, 1231–1234.

Harte, W.E., Swaminathan, S., Beveridge, D.L., 1992. Molecular dynamics of HIV-1 protease. Proteins 13, 175–194.

Hayashi, H., Takamune, N., Nirasawa, T., Aoki, M., Morishita, Y., Das, D., Koh, Y., Ghosh, A.K., Misumi, S., Mitsuya, H., 2014. Dimerization of HIV-1 protease occurs through two steps relating to the mechanism of protease dimerization inhibition by darunavir. Proc. Natl. Acad. Sci. USA 111, 12234–12239.

Heal, J.W., Jimenez-Roldan, J.E., Wells, S.A., Freedman, R.B., Römer, R.A., 2012. Inhibition of HIV-1 protease: the rigidity perspective. Bioinformatics 28 (3), 350–357.

Hicks, C.B., Cahn, P., Cooper, D.A., Walmsley, S.L., Katlama, C., Clotet, B., Lazzarin, A., Johnson, M.A., Neubacher, D., Mayers, D., Valdez, H., 2006. Durable efficacy of tipranavir-ritonavir in combination with an optimised background regimen of antiretroviral drugs for treatment-experienced HIV-1-infected patients at 48 weeks in the randomized evaluation of strategic intervention in multi-drug resistant patients with tipranavir (RESIST) studies: an analysis of combined data from two randomised open-label trials. Lancet 368, 466–475.

Hong, L., Zhang, X.J., Foundling, S., Hartsuck, J.A., Tang, J., 1997. Structure of a g48h mutant of HIV-1 protease explains how glycine-48 replacements produce mutants resistant to inhibitor drugs. FEBS J. 420, 11–16.

Ishima, R., Freedberg, D.I., Wang, Y.X., Louis, J.M., Torchia, D.A., 1999. Flap opening and dimer interface flexibility in the free and inhibitor-bound HIV protease, and their implications for function. Structure 7, 1047–1055.

Jadhav, P.K., Woerner, F.J., Lam, P.Y.S., Hodge, C.N., Eyermann, C.J., Man, H.-W., Daneker, W.F., Bacheler, L.T., Rayner, M.M., Meek, J.L., Erickson-Viitanen, S., Jackson, D.A., Calabrese, J.C., Schadt, M., Chang, C.-H., 1998. Nonpeptide cyclic cyanoguanidines as HIV-1 protease inhibitors: synthesis, structure-activity relationships, and X-ray crystal structure studies. J. Med. Chem. 41, 1446–1455.

Jaskólski, M., Tomasselli, A.G., Sawyer, T.K., Staples, D.G., Heinrikson, R.L., Schneider, J., Kent, S.B.H., Wlodawer, A., 1991. Structure at 2.5-A resolution of chemically synthesized human immunodeficiency virus type 1 protease complexed with a hydroxyethylene-based inhibitor. Biochemistry 30, 1600–1609.

Jones, K.L., Holloway, M.K., Su, H.P., Carroll, S.S., Burlein, C., Touch, S., DiStefano, D.J., Sanchez, R.I., Williams, T.M., Vacca, J.P., Coburn, C.A., 2010. Epsilon substituted lysinol derivatives as HIV-1 protease inhibitors. Bioorg. Med. Chem. Lett. 20, 4065–4068.

Kaldor, S.W., Kalish, V.J., Davies, J.F., Shetty, B.V., Fritz, J.E., Appelt, K., Burgess, J.A., Campanale, K.M., Chirgadze, N.Y., Clawson, D.K., Dressman, B.A., Hatch, S.D., Khalil, D.A., Kosa, M.B., Lubbehusen, P.P., Muesing, M.A., Patick, A.K., Reich, S.H., Su, K.S., Tatlock, J.H., 1997. Viracept (nelfinavir mesylate, AG1343): a potent, orally bioavailable inhibitor of HIV-1 protease. J. Med. Chem. 40, 3979–3985.

Kempf, D.J., Marsh, K.C., Denissen, J.F., McDonald, E., Vasavanonda, S., Flentge, C.A., Green, B.E., Fino, L., Park, C.H., Kong, X.P., Wideburg, N.E., Saldivar, A., Ruiz, L., Kati, W.M., Sham, H.L., Robins, T., Stewart, K.D., Hsu, A., Plattner, J.J., Leonard, J.M., Norbeck, D.W., 1995. ABT-538 is a potent inhibitor of human immunodeficiency virus protease and has high oral bioavailability in humans. Proc. Natl. Acad. Sci. USA 92, 2484–2488.

Kim, E.E., Baker, C.T., Dwyer, M.D., Murcko, M.A., Rao, B.G., Tung, R.D., Navia, M.A., 1995. Crystal structure of HIV-1 protease in complex with VX-478, a potent and orally bioavailable inhibitor of the enzyme. J. Am. Chem. Soc. 117, 1181–1182.

Koh, Y., Aoki, M., Danish, M.L., Aoki-Ogata, H., Amano, M., Das, D., Shafer, R.W., Ghosh, A.K., Mitsuya, H., 2011. Loss of protease dimerization inhibition activity of darunavir is associated with the acquisition of resistance to darunavir by HIV-1. J. Virol. 85, 10079–10089.

Koh, Y., Amano, M., Towata, T., Danish, M., Leshchenko-Yashchuk, S., Das, D., Nakayama, M., Tojo, Y., Ghosh, A.K., Mitsuya, H., 2010. In vitro selection of highly darunavir-resistant and replication competent HIV-1 variants by using a mixture of clinical HIV-1 isolates resistant to multiple conventional protease inhibitors. J. Virol. 84, 11961–11969.

Koh, Y., Nakata, H., Maeda, K., Ogata, H., Bilcer, G., Devasamudram, T., Kincaid, J.F., Boross, P., Wang, Y.F., Tie, Y., Volarath, P., Gaddis, L., Harrison, R.W., Weber, I.T., Ghosh, A.K., Mitsuya, H., 2003. Novel bis-tetrahydrofuranylurethane-containing nonpeptidic protease inhibitor (PI) UIC-94017 (TMC114) with potent activity against multi-PI-resistant human immunodeficiency virus in vitro. Antimicrob. Agents Chemother. 47, 3123–3129.

Kohl, N.E., Emini, E.A., Schleif, W.A., 1988. Active human immunodeficiency virus protease is required for viral infectivity. Proc. Natl. Acad. Sci. USA 85 (13), 4686–4690.

Kozisek, M., Bray, J., Rezacova, P., Saskova, K., Brynda, J., Pokorna, J., Mammano, F., Rulisek, L., Konvalinka, J., 2007. Molecular analysis of the HIV-1 resistance development: enzymatic activities, crystal structures, and thermodynamics of nelfinavir-resistant HIV protease mutants. J. Mol. Biol. 374, 1005–1016.

Kozisek, M., Cigler, P., Lepsik, M., Fanfrlik, J., Rezacova, P., Brynda, J., Pokorna, J., Plesek, J., Gruner, B., Grantz, S.K., Vaclavikova, J., Kral, V., Konvalinka, J., 2008. Inorganic polyhedral metallacarborane inhibitors of HIV protease: a new approach to overcoming antiviral resistance. J. Med. Chem. 51, 4839–4843.

Kräusslich, H.G., Ingraham, R.H., Skoog, M.T., Wimmer, E., Pallai, P.V., Carter, C.A., 1989. Activity of purified biosynthetic proteinase of human immunodeficiency virus on natural substrates and synthetic peptides. Proc. Natl. Acad. Sci. USA 86 (3), 807–811.

Kroemer, R.T., Ettmayer, P., Hecht, P., 1995. 3D-quantitative structure-activity relationships of human immunodeficiency virus type-1 proteinase inhibitors: comparative molecular field analysis of 2-heterosubstituted statine derivatives-implications for the design of novel inhibitors. J. Med. Chem. 38, 4917–4928.

Krohn, A., Redshaw, S., Ritchie, J.C., Graves, B.J., Hatada, M.H., 1991. Novel binding mode of highly potent HIV-proteinase inhibitors incorporating the (R)-hydroxyethylamine isostere. J. Med. Chem. 34, 3340–3342.

Kumar, G.N., Rodrigues, A.D., Buko, A.M., Denissen, J.F., 1996. Cytochrome P450-mediated metabolism of the HIV-1 protease inhibitor ritonavir (ABT-538) in human liver microsomes. J. Pharmacol. Exp. Ther. 277, 423–431.

Lam, P.Y.S., Ru, Y., Jadhav, P.K., Aldrich, P.E., DeLucca, G.V., Eyerman, C.J., Chang, C.-H., Emmett, G., Holler, E.R., Danekar, W.F., Li, L., Confalone, P.N., McHugh, R.J., Han, Q., Li, R., Markwalder, J.A., Seitz, S.P., Sharpe, T.R., Bacheler, L.T., Rayner, M.M., Klabe, R.M., Shum, L., Winslow, D.L., Kornhauser, D.M., Jackson, D.A., Erickson-Viitanen, S., Hodge, C.N., 1996. Cyclic HIV protease inhibitors: synthesis, conformational analysis, P2/P2′ structure-activity relationship, and molecular recognition of cyclic ureas. J. Med. Chem. 39, 3514–3525.

Levy, Y., Caflisch, A., Onuchic, J.N., Wolynes, P.G., 2004. The folding and dimerization of HIV-1 protease: evidence for a stable monomer from simulations. J. Mol. Biol. 340, 67–79.

Liu, F., Kovalevsky, A.Y., Tie, Y., Ghosh, A.K., Harrison, R.W., Weber, I.T., 2008. Effect of flap mutations on structure of HIV-1 protease and inhibition by saquinavir and darunavir. J. Mol. Biol. 381, 102–115.

Mahalingam, B., Louis, J., Hung, J., Harrison, R., Weber, I., 2001. Structural implications of drug-resistant mutants of HIV-1 protease: high-resolution crystal structures of the mutant protease/substrate analogue complexes. Proteins 43, 455–464.

Mahalingam, B., Wang, Y.F., Boross, P.I., Tozser, J., Louis, J.M., Harrison, R.W., Weber, I.T., 2004. Crystal structures of HIV protease V82A and L90M mutants reveal changes in the indinavir binding site. Eur. J. Biochem. 271, 1516–1524.

Maligres, P.E., Upadhyay, V., Rossen, K., Cianciosi, S.J., Purick, R.M., Eng, K.K., Reamer, R.A., Askin, D., Volante, R.P., Reider, P.J., 1995. Diastereoselective syn-epoxidation of 2-alkyl-4-enamides to epoxyamides: Synthesis of the Merck HIV-1 protease inhibitor epoxide intermediate. Tetrahedron Lett. 36, 2195.

Martinez-Cajas, J.L., Wainberg, M.A., 2007. Protease inhibitor resistance in HIV-infected patients: molecular and clinical perspectives. Antiviral Res. 76, 203–221.

Mcphee, F., Good, A.C., Kuntz, I.D., Craik, C.S., 1996. Engineering human immunodeficiency virus 1 protease heterodimers as macromolecular inhibitors of viral maturation. Proc. Natl. Acad. Sci. USA 93, 11477–11481.

Mehandru, S., Markowitz, M., 2003. Tipranavir: a novel non-peptidic protease inhibitor for the treatment of HIV infection. Expert Opin. Investig. Drugs 12, 1821–1828.

Mous, J., Heimer, E.P., Le Grice, S.J.J., 1988. Processing protease and reverse transcriptase from human immunodeficiency virus type I polyprotein in *Escherichia coli*. J. Virol. 62, 1433.

Munoz, B., Giam, C.Z., Wong, C.H., 1994. Alpha-ketoamide Phe-Pro isostere as a new core structure for the inhibition of HIV protease. Bioorg. Med. Chem. 2, 1085–1090.

Navia, M.A., Fitzgerald, M.D.P., Mckeever, B.M., Leu, C.T., Heimbach, J.C., Herber, W.K., Sigal, I.S., Darke, P.L., Spronger, J.P., 1989. Three-dimensional structure of aspartyl protease from human immunodeficiency virus HIV-1. Nature 337, 615–620.

Nugiel, D.A., Jacobs, K., Worley, T., Patel, M., Kaltenbach, III., R.F., Meyer, D.T., Jadhav, P.K., DeLucca, G.V., Smyser, T.E., Klabe, R.M., Bacheler, L.T., Rayner, M.M., Seitz, S.P., 1996. Preparation and structure-activity relationship of novel P1/P1'-substituted cyclic urea-based human immunodeficiency virus type-1 protease inhibitors. J. Med. Chem. 39, 2156–2169.

Oda, K., 2012. New families of carboxyl peptidases: serine-carboxyl peptidases and glutamic peptidases. J. Biochem. 151 (1), 13–25.

Öhrngren, P., Wu, X., Persson, M., Ekegren, J.K., Wallberg, H., Vrang, L., Rosenquist, Å., Samuelsson, B., Unge, T., Larhed, M., 2011. HIV-1 protease inhibitors with a tertiary alcohol containing transition-state mimic and various P2 and P1' substituents. MedChemComm 2, 701–709.

Olson, L.G., Bolin, D.R., Bonner, M.P., Bos, M., Cook, C.M., Fry, D.C., Graves, B.J., Hatada, M., Hill, D.E., 1993. Concepts and progress in the development of peptide mimetics. J. Med. Chem. 36, 3039–3049.

Oprea, T.I., Waller, C.L., Marshall, G.R., 1994. Three-dimensional quantitative structure-activity relationship of human immunodeficiency virus (I) protease inhibitors. 2. Predictive power using limited exploration of alternate binding modes. J. Med. Chem. 37, 2206.

Pearl, L., 1987. The catalytic mechanism of aspartic protinases. FEBS Lett. 214, 8.

Perryman, A.L., Lin, J.H., McCammon, J.A., 2004. HIV-1 protease molecular dynamics of a wild-type and of the V82F/I84V mutant: possible contributions to drug resistance and a potential new target site for drugs. Protein Sci. 13, 1108–1123.

Piana, S., Carloni, P., Rothlisberger, U., 2002. Drug resistance in HIV-1 protease: flexibility-assisted mechanism of compensatory mutations. Protein Sci. 11, 2393–2402.

Prabu-Jeyabalan, M., King, N.M., Nalivaika, E.A., Heilek-Snyder, G., Cammack, N., Schiffer, C.A., 2006. Substrate envelope and drug resistance: crystal structure of RO1 in complex with wild-type human immunodeficiency virus type 1 protease. Antimicrob. Agents Chemother. 50, 1518–1521.

Prabu-Jeyabalan, M., Nalivaika, E.A., King, N.M., Schiffer, C.A., 2003. Viability of a drug-resistanthuman immunodeficiency virus type 1 protease variant: structural insights for better antiviral therapy. J. Virol. 77, 1306–1315.

Prabu-Jeyabalan, M., Nalivaika, E., Schiffer, C.A., 2002. Substrate shape determines specificity of recognition for HIV-1 protease: analysis of crystal structures of six substrate complexes. Structure 10, 369–381.

Rawlings, N.D., Barrett, A.J., 1993. Evolutionary families of peptidases. Biochem. J. 290 (Pt. 1), 205–218.

Rawlings, N.D., Barrett, A.J., Bateman, A., 2011. Asparagine peptide lyases: a seventh catalytic type of proteolytic enzymes. J. Biol. Chem. 286, 38321–38328.

Roberts, N.A., Martin, J.A., Kinchington, D., Broadhurst, A.V., Craig, J.C., Duncan, I.B., Galpin, S.A., Handa, B.K., Kay, J., Kröhn, A., Lambert, R.W., Merrett, J.H., Mills, J.S., Parkes, K.E., Redshaw, S., Ritchie, A.J., Taylor, D.L., Thomas, G.J., Machin, P.J., 1990. Rational design of peptide-based HIV proteinase inhibitors. Science 248, 358–361.

Romines, K.R., Morris, J.K., Howe, W.J., Tomich, P.K., Horng, M.-M., Chong, K.-T., Hinshaw, R.R., Anderson, D.J., Strohbach, J.W., Turner, S.R., Mizsak, S.A., 1996. Cycloalkylpyranones and cycloalkyldihydropyrones as HIV protease inhibitors: exploring the impact of ring size on structure-activity relationships. J. Med. Chem. 39, 4125–4130.

Romines, K.R., Watenpaugh, K.D., Tomick, P.K., Howe, W.J., Morris, J.K., Lovasz, K.D., Mulichak, A.M., Finzel, B.C., Lynn, J.C., Horng, M.-M., Schwende, F.J., Ruwart, M.J., Zipp, G.L., Chong, K.-T., Dolak, L.A., Toth, L.N., Howard, G.M., Rush, B.D., Wilkinson, K.F., Possert, P.L., Dalga, R.J., Hinshaw, R.R., 1995a. Use of medium-sized cycloalkyl rings to enhance secondary binding: discovery of a new class of human immunodeficiency virus (HIV) protease inhibitors. J. Med. Chem. 38, 1884–1891.

Romines, K.R., Watenpaugh, K.D., Howe, W.J., Tomich, P.K., Lovasz, K.D., Morris, J.K., Janakiraman, M.N., Lynn, J.C., Horng, M.-M., Chog, K.-T., Hinshaw, R.R., Dolak, L.A., 1995b. Structure-based design of nonpeptidic HIV protease inhibitors from a cyclooctylpyranone lead structure. J. Med. Chem. 38, 4463–4473.

Rozzelle, J.E., Dauber, D.S., Todd, S., Kelly, R., Craik, C.S., 2000. Macromolecular inhibitors of HIV-1 protease: characterization of designed heterodimers. J. Biol. Chem. 275, 7080–7086.

Scott, W.R.P., Schiffer, C.A., 2000. Curling of flap tips in HIV-1 protease as a mechanism for substrate entry and tolerance of drug resistance. Structure 8, 1259–1265.

Seelmeier, S., Schmidt, H., Turk, V., Von der Helm, K., 1988. Human immunodeficiency virus has an aspartic-type protease that can be inhibited by pepstatin A. Proc. Natl. Acad. Sci. USA 85, 6612–6616.

Sham, H.L., Kempf, D.J., Molla, A., Marsh, K.C., Kumar, G.N., Chen, C.M., Kati, W., Stewart, K., Lal, R., Hsu, A., Betebenner, D., Korneyeva, M., Vasavanonda, S., McDonald, E., Saldivar, A., Wideburg, N., Chen, X., Niu, P., Park, C., Jayanti, V., Grabowski, B., Granneman, G.R., Sun, E., Japour, A.J., Leonard, J.M., Plattner, J.J., Norbeck, D.W., 1998. ABT-378, a highly potent inhibitor of the human immunodeficiency virus protease. Antimicrob. Agents Chemother. 42, 3218–3224.

Skulnick, H.I., Johnson, P.D., Aristoff, P.A., Morris, J.K., Lovasz, K.D., Howe, W.J., Watenpaugh, K.D., Janakiraman, M.N., Anderson, D.J., Reischer, R.J., Schwartz, T.M., Banitt, L.S., Tomich, P.K., Lynn, J.C., Horng, M.-M., Chong, K.-T., Hinshaw, R.R., Dolak, L.A., Seest, E.P., Schwende, F.J., Rush, B.D., Howard, G.M., Toth, L.N., Wilkinson, K.R., Kakuk, T.J., Johnson, C.W., Cole, S.L., Zaya, R.M., Zipp, G.L., Possert, P.L., Dalga, R.J., Zhong, W.-Z., Williams, M.G., Romines, K.R., 1997. Structure-based design of nonpeptidic HIV protease inhibitors: the sulfonamide-substituted cyclooctylpyramones. J. Med. Chem. 40, 1149–1164.

Swaminathan, S., Harte, W.E., Beveridge, D.L., 1991. Investigation of domain structure in proteins via molecular dynamics simulation: application to HIV-1 protease dimer. J. Am. Chem. Soc. 113, 2717–2721.

Tie, Y., Boross, P.I., Wang, Y.F., Gaddis, L., Hussain, A.K., Leshchenko, S., Ghosh, A.K., Louis, J.M., Harrison, R.W., Weber, I.T., 2004. High resolution crystal structures of HIV-1 protease with a potent nonpeptide inhibitor (UIC-94017) active against multi-drug-resistant clinical strains. J. Mol. Biol. 338, 341–352.

Tie, Y., Boross, P.I., Wang, Y.F., Gaddis, L., Liu, F., Chen, X., Tozser, J., Harrison, R.W., Weber, I.T., 2005. Molecular basis for substrate recognition and drug resistance from 1.1 to 1.6 angstroms resolution crystal structures of HIV-1 protease mutants with substrate analogs. FEBS J. 272, 5265–5277.

Tie, Y., Kovalevsky, A.Y., Boross, P., Wang, Y.F., Ghosh, A.K., Tozser, J., Harrison, R.W., Weber, I.T., 2007. Atomic resolution crystal structures of HIV-1 protease and mutants V82A and I84V with saquinavir. Proteins 67, 232–242.

Vacca, J.P., Dorsey, B.D., Schleif, W.A., Levin, R.B., Mcdaniel, S.L., Darke, P.L., Zugay, J., Quintero, J.C., Blahy, O.M., Roth, E., Sardana, V.V., Schlabach, A.J., Graham, P.I., Condra, J.H., Gotlib, L., Holloway, M.K., Lin, J., Chen, I.W., Vastag, K., Ostovic, D., Anderson, P.S., Emini, E.A., Huff, J.R., 1994. L-735,524—An orally bioavailable human-immunodeficiency-virus type-1 protease inhibitor. Proc. Natl. Acad. Sci. USA 91, 4096–4100.

Vara Prasad, J.V.N., Para, K.S., Tummino, P.J., Ferguson, D., Mcquade, E.J., Lunney, E.A., Rapundalo, S.T., Batley, B.L., Hingorani, G., Domagala, J.M., Gracheck, S.J., Bhat, T.N., Liu, B., Baldwin, E.T., Erickson, J.W., Sawyer, T.K., 1995. Nonpeptidic potent HIV-1 protease inhibitors: (4-hydroxy-6-phenyl-2-oxo-2H- pyran-3-yl)thiomethanes that span P1-P2' subsites in a unique mode of active site binding. J. Med. Chem. 38, 898–905.

Vazquez, M.L., Bryant, M.L., Clare, M., Decrescenzo, G.A., Doherty, E.M., Freskos, J.N., Getman, D.P., Houseman, K.A., Julien, J.A., Kocan, G.P., Mueller, R.A., Shieh, H.-S., Stallings, W.C., Stegeman, R.A., Talley, J.J., 1995. Inhibitors of HIV-1 protease containing the novel and potent (R)-(hydroxyethyl)sulfonamide isostere. J. Med. Chem. 38, 581–584.

Waller, C.L., Oprea, T.I., Giolitti, A., Marshall, G.R., 1993. Three-dimensional QSAR of human immunodeficiency virus (I) protease inhibitors. 1. A CoMFA study employing experimentally-determined alignment rules. J. Med. Chem. 36, 4152–41560.

Weber, I.T., Agniswamy, J., 2009. HIV-1 protease: structural perspectives on drug resistance. Viruses 1, 1110–1136.

Wei, Y., Ma, C.M., Hattori, M., 2009. Synthesis of dammarane-type triterpene derivatives and their ability to inhibit HIV and HCV proteases. Bioorg. Med. Chem. 17, 3003–3010.

Wilkerson, W.W., Akamike, E., Cheatham, W.W., Hollis, A.Y., Collins, R.D., DeLucca, I., Lam, P.Y.S., Ru, Y., 1996. HIV protease inhibitory bis-benzamide cyclic ureas: a quantitative structure-activity relationship analysis. J. Med. Chem. 39, 4299–4312.

Wilkerson, W.W., Dax, S., Cheatham, W.W., 1997. Nonsymmetrically substituted cyclic urea HIV protease inhibitors. J. Med. Chem. 40, 4079–4088.

Wlodawer, A., Erickson, J.W., 1993. Structure-based inhibitors of HIV-1 protease. Annu. Rev. Biochem. 62, 543–585.

Wlodawer, A., Miller, M., Jaskolski, M., Sathyanarayana, B.K., Baldwin, E., Weber, I.T., Selk, L.M., Clawson, L., Schneider, J., Kent, S.B.H., 1989. Science 245, 616.

Yedidi, R.S., Maeda, K., Fyvie, W.S., Steffey, M., Davis, D.A., Palmer, I., Aoki, M., Kaufman, J.D., Stahl, S.J., Garimella, H., Das, D., Wingfield, P.T., Ghosh, A.K., Mitsuya, H., 2013. P2' Benzene carboxylic acid moiety is associated with decrease in cellular uptake: evaluation of novel nonpeptidic HIV-1 protease inhibitors containing P2 bis-tetrahydrofuran moiety. Antimicrob. Agents Chemother. 57, 4920–4927.

Zhao, C., Sham, H.L., Sun, M., Stoll, V.S., Stewart, K.D., Lin, S., Mo, H., Vasavanonda, S., Saldivar, A., Park, C., McDonald, E.J., Marsh, K.C., Klein, L.L., Kempf, D.J., Norbeck, D.W., 2005. Synthesis and activity of N-acyl azacyclic urea HIV-1 protease inhibitors with high potency against multiple drug resistant viral strains. Bioorg. Med. Chem. Lett. 15, 5499–5503.

## FURTHER READING

Ghosh, A.K., Chapsal, B.D., 2010. Second-generation approved HIV protease inhibitors for the treatment of HIV/AIDS. Ghosh, A.K. (Ed.), Aspartic Acid Proteases as Therapeutic Targets—Methods and Principles in Medicinal Chemistry, vol. 45, Wiley-VCH, Weinheim, pp. 169–204.

King, N.M., Prabu-Jeyabalan, M., Nalivaika, E.A., Wigerinck, P., de Bethune, M.P., Schiffer, C.A., 2004. Structural and thermodynamic basis for the binding of TMC114, a next-generation human immunodeficiency virus type 1 protease inhibitor. J. Virol. 78, 12012–12021.

Makatini, M.M., Petzold, K., Sriharsha, S.N., Soliman, M.E., Honarparvar, B., Arvidsson, P.I., Sayed, Y., Govender, P., Maguire, G.E., Kruger, H.G., Govender, T., 2011. Pentacycloundec-ane-based inhibitors of wild-type C-South African HIV-protease. Bioorg. Med. Chem. Lett. 21, 2274–2277.

Mimoto, T., Imai, J., Kisanuki, S., Enomoto, H., Hattori, N., Akaji, K., Kiso, Y., 1992. Kynostatin (KNI)-227 and -272, highly potent anti-HIV agents: conformationally constrained tripeptide in-hibitors of HIV protease containing allophenylnorstatine. Chem. Pharm. Bull. 40, 2251–2253.

Toh, H., Ono, M., Saigo, K., Miyata, T., 1985. Synthetic non-peptide inhibitors of HIV protease. Nature 315, 691–692.

Weber, I.T., Kovalevsky, A.Y., Harrison, R.W., 2007. Structures of HIV protease guide inhibi-tor design to overcome drug resistance. In: Cladwell, G.W., Atta-ur-Rahman, Player, M.R., Choudhry, M.I. (Eds.), Frontiers in Drug Design, Discovery: Structure-Based Drug Design in the 21st Century. third ed. Bentham Science Publishers Ltd., Emirate of Sharja, UAE, pp. 45–62.

# Chapter 10

# Studies on Picornaviral Proteases and Their Inhibitors

**Vaishali M. Patil\*, Satya P. Gupta\*\***

*\*Kharvel Subharti College of Pharmacy, Swami Vivekanand Subharti University, Meerut, Uttar Pradesh, India; \*\*National Institute of Technical Teachers' Training and Research, Bhopal, Madhya Pradesh, India*

## 1 INTRODUCTION

The word Picornavirus (PV) was derived from words "pico" and "RNA" describing small and the single stranded ribonucleic acid present in it, respectively. PVs, an important class of human and animal pathogens, belong to the family of Picornaviridae (Ehrenfeld et al., 2010) consisting of 285 different PV types. The 29 species of PVs are classified in 13 genera with 5 proposed genera. The PV classification is detailed in Table 10.1 (Knowles et al., 2010; Stanley et al., 2005). PVs are pervasive in nature and are small icosahedral positive single stranded RNA viruses. PV infection causes variety of benign (i.e., common cold) to fatal diseases (i.e., polio myelitis in human and foot-and-mouth disease in animals) (Melnick, 1983). Among the various known genera, the best known PVs are enteroviruses (EVs) which include Picornavirus (PV), human rhinovirus (HRV), foot-and-mouth disease virus (FMDV), and hepatitis A virus (HAV) (Dussart et al., 2005; Le Guyader et al., 2008; Sapkal et al., 2009; Zhang et al., 2009).

PVs are also divided on the basis of physicochemical properties, such as acid stability, and density, with the aid of advances at molecular level (Brown et al., 2009; Greninger et al., 2009; Holtz et al., 2008; Kapoor et al., 2008; McErlean et al., 2008; Zoll et al., 2009). The number of identified PVs is increasing and vaccine is available only against PO (poliovirus), HAV, and FMDV, while no prophylaxis cast is available for other PV infections (Xu et al., 2010). Thus to control these infections, development of effective vaccine, as well as antiviral compound should complement each other.

Viral Proteases and Their Inhibitors. http://dx.doi.org/10.1016/B978-0-12-809712-0.00010-1

263

**TABLE 10.1 Detailed Classification of the Virus Family Picornaviridae, Genus, and Virus Types**

| Genera | Species | Number of virus types |
|---|---|---|
| *Enterovirus* | *Human enterovirus A* (HEV-A)<br>*Human enterovirus B* (HEV-B)<br>Human enterovirus C (HEV-C)<br>Human enterovirus D (HEV-D) | 21<br>59<br>19<br>3 |
| | Simian enterovirus A (SEV-A) | 3 |
| | Bovine enterovirus | 2 |
| | Porcine enterovirus A (PEV-A)<br>Porcine enterovirus B (PEV-B) | 1<br>2 |
| | Human rhinovirus A (HRV-A)<br>Human rhinovirus B (HRV-B)<br>Human rhinovirus C (HRV-C) | 74<br>25<br>10 |
| *Hepatovirus* | Hepatitis A virus | 1 |
| *Parechovirus* | Human parechovirus | 14 |
| | Ljunganvirus | 1 |
| *Kobuvirus* | Aichi virus | 1 |
| | Bovine kobuvirus | 1 |
| | Porcine kobuvirus | 1 |
| *Cardiovirus* | Encephalomyocarditis virus | 1 |
| | Theilovirus | 12 |
| *Cosavirus* | Human Cosavirus (HcoSV A-E) | 5 |
| *Aphthovirus* | *Foot-and-mouth disease virus* | 7 |
| | *Equine rhinitis A virus* | 1 |
| | *Bovine rhinovirus* | 3 |
| *Erbovirus* | *Equine rhinitis B virus* | 3 |
| *Teschovirus* | *Porcine teschovirus* | 11 |
| *Sapelovirus* | | |
| *Senecavirus* | *Seneca valley virus* | 1 |
| *Tremovirus* | *Avian encephalomyelitis-like viruses* | 1 |
| *Avihepatovirus* | *Duck hepatitis virus* | 3 |
| *Seal picornavirus* | | |

## 1.1 Picornaviruses

The *picornaviridae* family consists of six genera of pathogens infecting humans and they are *Enterovirus, Parechovirus, Hepatovirus, Cardiovirus, Kobuvirus*, and *Cosavirus*.

## 1.1.1   Enterovirus

*Enterovirus* genus is circulated worldwide and is the largest among the known virus types and species. Seven species are known as EVs which infect humans or other animals. In human, EV infection is mild or asymptometric with other respiratory tract symptoms and may be associated with severe illnesses. EV infection is also observed in patients with a gammaglobulinemia chronic fatigue syndrome (Chapman and Kim 2008; Huber 2008). The human EV and PV may cause poliomyelitis or polio or infantile paralysis and is characterized by acute flaccid paralysis (AFP). In such patients, an ongoing inflammatory process has been reported with increased mRNA expression for proinflammatory cytokines in the central nervous system (Gonzalez et al., 2002). In addition to this, it has always been a threat to children and some of the annual virus out breaks have been observed in different parts of the world (Table 10.2) (Lee et al., 2010; Narkeviciute and Vaiciuniene 2004; Wong et al., 2010; Yang et al., 2009; Zhao et al., 2005).

**TABLE 10.2 Picornavirus Diseases and Receptors**

| Genera | Representative species | Diseases | Receptors |
|---|---|---|---|
| *Aphthovirus* | Foot-and-mouth disease virus | Foot-and-mouth disease | Integrin |
| *Cardiovirus* | Encephalomyocarditis virus | Myocarditis, encephalitis | VCAM-1 |
| | Theiler's murine encephalomyelitis virus | Neurological disease | Sialic acid |
| *Enterovirus* | Coxsackievirus B1-6 | Hand, foot and mouth disease | CAR |
| | Enterovirus 71 | Hand, foot, and mouth disease | PSGL-1, SCARB2 |
| | Poliovirus 1–3 | Poliomyelitis | CD155 (PVR) |
| | Major group rhinovirus | Respiratory disease | ICAM-1 |
| | Minor group rhinovirus | Respiratory disease | LDL |
| *Hepatovirus* | Hepatitis A virus | Hepatitis | HAVcr-1 |
| *Kobuvirus* | Aichi virus | Gastroenteritis | — |

CAR, Coxsackie and adenovirus receptor; HAVcr-1, hepatitis A virus cell receptor; ICAM-1, intercellular adhesion molecule; LDL, low-density lipoprotein; PVR, poliovirus receptor; PSGL, P-selectin glycoprotein ligand; SCARB, scavenger receptor class B member; VCAM, vascular cell adhesion molecule.

## 1.1.2 Human Rhinovirus

The HRV has three species and is the major cause of upper respiratory infection (Palmenberg et al., 2009), hospital born respiratory infection, and asthma in children, as well as adults (Arden and Mackay, 2010; De Almeida et al., 2010; Dougherty and Fahy, 2009; Gern, 2010).

## 1.1.3 Human Parechoviruses (HPeV)

Human parechovirus (HPeV) also has three species, which mainly infects young children. HPeV leads to gastrointestinal and respiratory tract infections and also causes alpha-fetoprotein, encephalitis, aseptic meningitis, myocarditis, neonatal sepsis, and Reye syndrome (Wolthers et al., 2008). Recently many of the HPeV serotypes have been identified and new HPeVs are emerging with the advances of molecular methods.

## 1.1.4 Hepatitis A Virus

In the genus *Hepatovirus*, HAV is the only species that affects humans and primates. Among the HAV infected individuals, 70% have clinical hepatitis while only 1% of infected children are symptomatic. With improved hygiene, fewer cases of HAV infection have been reported (Brundage and Fitzpatrick, 2006). The major reasons behind HAV outbreak are contaminated food and water, intravenous drug users, and homosexual men (Gervelmeyer et al., 2006; Hauri et al., 2006; Lee et al., 2008; O'Donovan et al., 2001; Pontrelli et al., 2007; Sattar et al., 2000; Stene-Johansen et al., 2007).

## 1.1.5 Kobuvirus

In 1989, the first species *Kobuvirus*, the cause of oyster associated gastroenteritis has been isolated (Yamashita et al., 1991). This genus consists of *Aichivirus*, bovine *Kobuvirus*, and porcine *Kobuvirus*. Aichi virus has been observed as a common cause of gastroenteritis in Japan, Asia, Europe, and South America (Goyer et al., 2008; Le Guyader et al., 2008; Oh et al., 2006).

## 1.1.6 Human Cosavirus (HCoSV)

The human cosavirus (HCoSV) has been isolated from healthy children, as well as those with AFP and acute diarrhea. The HCoSVs are categorized into five genomic groups, HCoSV-A to HCoSV-E, but the virus has not been found to be associated with any particular disease (Holtz et al., 2008; Kapoor et al., 2008).

## 1.1.7 Cardiovirus

Cardioviruses are of two types namely, theilovirus (ThV) and encephalomyocarditis virus (EMCV). Theiloviruses are further subclassified into five types. Vilyuisk human encephalomyelitis and scaffold virus infect humans causing respiratory diseases, gastroenteritis, myocarditis, and acute and chronic encephalitis. The isolated scaffold viruses, that is, SAFV1-8, have been found in upper respiratory or gastroenteritis patients (Blinkova et al., 2009; Zoll et al., 2009). EMCV infection in other mammalian species is associated with myocarditis, encephalitis, and is fatal.

## 2    PICORNAVIRUS: LIFE CYCLE AND STRUCTURE

In viral life cycle, viral proteins have important roles in viral replication, translation, as well as in altering various host functions, such as cellular gene expression, protein localization, signal transduction, and membrane rearrangement. Various important steps in replication cycle of PV are: (1) PV binds to the host cell via a cellular receptor; (2) after entry and uncoating, viral RNA enters the cell cytoplasm as positive-sense single-stranded RNA and are translated via cap-independent internal ribosomal entry site (IRES) translation; (3) formation of virus-induced membranous vesicles in the nuclear periphery; (4) the original positive-strand acts as a template for the synthesis of a negative-strand intermediate and viral RNA replication takes place; (5) synthesis of multiple positive-strand RNAs from the negative-strand RNAs; and (6) translation or encapsidation and release from the cell (Fig. 10.1).

In PVs, the encapsulated genome forms a virion ($\sim$30 nm size) without an envelope and consists of a positive-stranded RNA molecule (6700–8800 nucleotides). As shown in Fig. 10.2A, the viral genome has a single open reading frame and structured untranslated regions (UTR) at the highly structured 5′- and 3′-end, and a 3′-poly(A) tail. The uncapped viral genome has viral protein genome linked (VPg) to the viral protein 3B, instead of the 5′-end covalently coupled to the protein, and this VPg is used as a primer during RNA replication. The cap-independent translation is mediated by the 5′-UTR with an IRES. In all PVs,

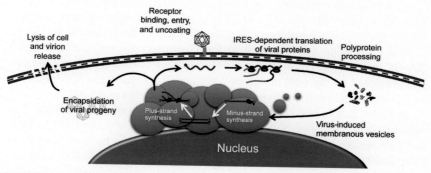

**FIGURE 10.1    The picornavirus replication cycle.** *IRES,* Internal ribosomal entry site. *(Reprinted with permission from Cathcart, A.L., Baggs, E.L., Semler, B.L., 2015. Picornaviruses: pathogenesis and molecular biology. Reference Module in Biomedical Research, third ed. Elsevier. Copyright 2015 Elsevier.)*

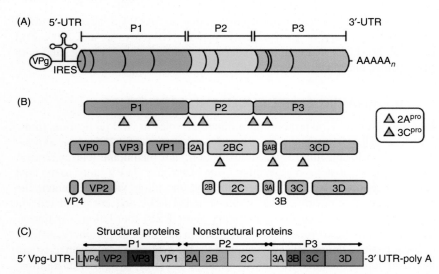

**FIGURE 10.2** Schematic representation of *Enterovirus* (A–B). Open access publication; and picornavirus genome (C). *UTR*, Untranslated regions. *(Parts A–B, Reprinted from van der Linden, L., Vives-Adrián, L., Seliško, B., Ferrer-Orta, C., Liu, X., Lanke, K., Ulferts, R., De Palma, A. M., Tanchis, F., Goris, N., et al., 2015. The RNA template channel of the RNA-dependent RNA polymerase as a target for development of antiviral therapy of multiple genera within a virus family. PLoS Pathog. 11, e1004733; Part C, Reprinted with permission from Norder, H., De Palma, A. M., Seliško, B., Costenaro, L., Papageorgiou, N., Arnan, C., Coutard, B., Lantez, V., De Lamballerie, X., Baroni, C., Sola, M., Tan, J., Neyts, J., Canard, B., Coll, M., Gorbalenya, A. E., Hilgenfeld, R., 2011. Picornavirus non-structural proteins as targets for new anti-virals with broad activity. Antivir. Res. 89, 204–218. Copyright 2011 Elsevier.)*

the open reading frame is organized in a similar pattern with some differences between genera. The open reading frame of EV encodes three areas (P1, P2, and P3) of a single polyprotein. The P1 region contains structural proteins (VP1, VP2, VP3, and VP4) and P2 and P3 regions contain nonstructural proteins, 2A–2C and 3A–3D, respectively (Fig. 10.2B–C) (van der Linden et al., 2015). P2 and P3 regions get processed to form nonstructural replication proteins (2A–2C and 3A–3D, respectively) and cleavage intermediates. The polyprotein gets processed by the viral proteases $2A^{pro}$ and 3C proteinases ($3C^{pro}$) into the viral proteins and some stable precursors (van der Linden et al., 2015).

The host factors and capsid proteins are involved in receptor binding. The capsid proteins constitute 60 copies of 4 P1-encoded polypeptides (VP1 to VP4) and they initiate infection by binding to a receptor on the host membrane. In PVs, similar type of receptor molecules have been observed and that the extracellular regions are made up of 2–5 amino-terminal immunoglobulin-like domains (Coyne and Bergelson, 2006; He et al., 2000; Kolatkar et al., 1999; Shafren et al., 1997a,b,c). The amino-terminal domain D1 contributes in the binding with the conserved amino acid residues of the PV canyon and triggers viral instability and uncoating. In case of poliovirus and rhinovirus a single receptor suffices for viral entry, but in some viruses nonimmunoglobulin superfamily (non-IgSF) cell surface receptors [low-density lipoprotein receptor (LDLR) and decay-accelerating factor (DAF or CD55)] are utilized (Bergelson et al., 1994; Coyne and Bergelson, 2006; Hewat et al., 2000; Rossmann et al., 2002). Apart from the virus-receptor interaction factors, cellular factors and viral genome elements interact with the 5'-UTR to influence the efficiency of translation initiation and virus replication.

The PV-encoded proteases, such as 2A and 3C proteases ($L^{pro}$, $2A^{pro}$, and $3C^{pro}$) play an important role in viral polyprotein processing (Strebel and Beck, 1986; Svitkin et al., 1979; Toyoda et al., 1986) by cleavage of viral polypeptides and inhibiting various host machineries. PV infection affects viral replication by altering host membrane permeability along with production of membranous structures (Bienz et al., 1983; Guinea and Carrasco, 1990; Lacal and Carrasco, 1982; Schlegel et al., 1996). The viral replication complex has been found to be associated with virus-induced membranous vesicle (Bienz et al., 1992) and replication-associated viral proteins (2B, 2BC, 3A, and 3D) (Bienz et al., 1994; Schlegel et al., 1996; Suhy et al., 2000). Some of the host cellular proteins and affecting PV proteins with their function are as summarized in Table 10.3.

## 2.1 Picornaviral Particles

In virion, the icosahedral shell is formed by VP1-3 and the inner surface of the particle by the VP4 (Racaniello, 2007), while for its assembly and maturation, cleavage of P1 is required and the products remain intact. VP1-3 possesses common nuclei of eight wedge-shaped and linked β–strands. The details of connecting loops in β–strands determine the surface topology and control the receptor specificity, as well as antigenic properties. In EV, "canyon"—a deep depression around the fivefold axes below VP1—separates the surface-oriented protrusions of VP1 from VP-2 and -3. Canyon also acts as engagement site having cellular receptors (Colonno et al., 1988; Olson et al., 1993). In *Cardiovirus*, the canyon is partially filled-in with series of depressions, that is, "pits," involved in receptor binding (Grant et al., 1994; Hertzler et al., 2000; Luo et al., 1987; Toth et al., 1993). The FMDV surface is comparatively smooth and cellular receptors are specific toward flexible loop of the particle surface (Acharya et al., 1989; Logan et al., 1993). The viral polyproteins on proteolysis of P1 yields VP0, VP1, and VP3, which are not physically separated. The formed pentameric subunits along with viral RNA genome copy are required for "provirion" generation (Hoey and Martin, 1974; Guttman and Baltimore 1977; Lee et al., 1993). Finally, VP0 cleaves into VP2 and VP4 which coincides to addition of RNA into the immature particle (Arnold et al., 1987; Curry et al., 1997; Hindiyeh et al., 1999). For poliovirus, the X-ray structure shows an outward, trefoil-shaped depression similar to the binding site of the genomic RNA (Basavappa

**TABLE 10.3 Alteration of Host Cell Proteins and Functions**

| Cellular proteins | Picornavirus proteins | Function |
| --- | --- | --- |
| GBF1, BIG1/2 | PV 3A, PV 3CD | Recruits Arf-1 to membranes |
| ACBD3, PI4KIIIβ | Aichi 3A, CVB 3A, HRV 3A, PV 3A | May be involved in formation of membranous vesicles |
| eIF4G | CVB 2A$^{pro}$, FMDV L$^{pro}$, HRV 2A$^{pro}$, PV 2A$^{pro}$ | Cleavage shuts off cap-dependent translation |
| PABP | CVB 2A$^{pro}$, PV 2A$^{pro}$, PV 3C | Cleavage contributes to inhibition of host cell translation |
| 4E-BP1 | EMCV, PV () | Dephosphorylation inhibits host cell translation |
| Histone H3 | FMDV 3C | Inhibits host cell transcription |
| TBP, CREB, Oct-1, TFIIIC | PV 3C | Inhibits host cell transcription |
| RIG-1 | EMCV 3C, HRV 3C, PV 3C | Cleavage inhibits cellular immune response |
| NF-κB | FMDV L$^{pro}$, PV 3C | Cleavage disrupts IFN signaling |
| IPS-1 | HAV 3ABC, HRV 3C, HRV 2A$^{pro}$ | Cleavage disrupts IFN signaling pathway |
| IRF-3 | TMEV L | Inhibition of IRF-3 dimerization blocks IFN-β transcription |
| Gemin3 | PV 2A$^{pro}$ | Cleavage inhibits cellular splicing complex assembly |
| Nucleoporins | EMCV L, HRV 2A$^{pro}$, PV 2A$^{pro}$ | Cleavage (2A) or phosphorylation (L) disrupts nucleocytoplasmic transport |
| Dystrophin | CVB 2A$^{pro}$ | Cleavage disrupts cytoskeleton |
| G3BP | PV 3C | Disrupts cellular stress-granule formation |

et al., 1994; Chen et al., 1989). This structural RNA perfectly docks in the empty capsid structures. These results propose cleavage mechanism where RNA polarizes the carbonyl oxygen and is engaged by the histidine side chain with its water (Basavappa et al., 1994). His195 mutations inhibit VP0 cleavage to accumulate noninfectious virions (Hindiyeh et al., 1999).

## 2.2 Picornavirus Receptors

The various important functions of receptor include entry of virus, cell attachment, signaling, endocytosis, and initiation of capsid alterations. Most of the PV receptors belong to different families, such as immunoglobulin-like LDLR, the complement control, the integrin and the T cell immunoglobulin domain mucin-like receptors. The different types of receptors, types of interactions, plasticity of receptor use, and their roles are summarized further.

### 2.2.1 Immunoglobulin Superfamily Receptors (VCAM-1, ICAM-1, PVR, CAR)

Many well characterized receptors belonging to the immunoglobulin superfamily have been investigated. For HRVs, intercellular adhesion molecule-1 (ICAM-1) (Greve et al., 1989; Staunton et al., 1989; Tomassini et al., 1989) and PV receptor (PVR i.e., CD155) (Gromeier et al., 2000; Takai et al., 2008), have been identified. For EMCV, a murine vascular cell adhesion molecule-1 (VCAM-1) receptor (Huber, 1994) and in the epithelial cells coxsackie and adenovirus receptor (CAR) have been identified (Coyne and Bergelson 2005; Freimuth et al., 2008). In this class of receptors, the virus binding to cell surface and initiation of genomic delivery to the cytoplasm is contributed by the Ig-like domain.

### 2.2.2 Decay Accelerating Factor Receptor

For EVs, the DAF or CD55 receptor is a primary receptor and is comprised of the extracellular modules attached to the membrane with different biding sites on capsid, as well as receptor (Lea, 2002; Powell et al., 1999). The infectious entry is further triggered by interaction with a second or coreceptor.

### 2.2.3 Low-Density Lipoprotein Receptors

In HRVs, depending on the virus cell–binding inhibition ability, two receptors, namely LDLR-related protein (LRP) and very low density lipoprotein receptor (VLDLR), have been identified (Hofer et al., 1994; Marlovits et al., 1998a,b,c). In this class, an extracellular domain, a transmembrane domain, and a cytoplasmic domain have been observed.

### 2.2.4 Integrins/Cell Adhesion Receptors

Integrins subunit is constituted of a large extracellular globular domain, a transmembrane domain, and a small cytoplasmic domain. Integrins $\alpha_v\beta_1$, $\alpha_v\beta_3$, $\alpha_v\beta_6$ and $\alpha_v\beta_8$ act as entry sites for *Aphthovirus* FMDV and initiate internalization of endocytosis without any structural changes (Duque et al., 2004; Jackson et al., 2000, 2002, 2004). Most of the viruses, containing integrin receptors, bind through the Arg-Gly-Asp (RGD) motif (Hendry et al., 1999) with echovirus 1 as an exception (Xing et al., 2004).

## 2.2.5  HAV Cellular Receptor (TIM-1)

Class I integral membrane glycoproteins, that is, primate cellular receptor (HAVcr-1) and the human analog (huHAVcr-1) with an extracellular N-terminal membrane distal Ig-like and mucin-like domains (TIM), have been reported (Thompson et al., 1998). A study investigated that, HAV infection in childhood stimulates TIM-1 and in later phase of life protects from allergy. This concludes that the increased allergy and asthma cases in western countries are due to the reduced HAV incidences (McIntire et al., 2004).

## 2.2.6  EV71 Receptors

In 2009, two EV71 receptors, that is, human P-selectin glycoprotein ligand-1 and scavenger receptor class B member 2, acting as leukocyte infecting EV71 and endocytosis receptor, respectively, have been reported (Nishimura et al., 2009; Yamayoshi et al., 2009).

## 2.2.7  Sialic Acid

The crystallographic structure complexed with sialic acid and Theiler's murine encephalomyelitis virus (TMEV) has proposed sialic acid as a receptor (Lipton et al., 2006; Zhou et al., 2000). Use of sialic acid-mediated entry in EMCV (Guy et al., 2009), *Aphthovirus*, and equine rhinitis A virus (ERAV) (Stevenson et al., 2004; Warner et al., 2001) has been observed.

## 2.2.8  Coreceptors

The DAF-binding EVs, such as coxsackieviruses, CAV21, and CBV3, use core-ceptors instead of DAF for binding to the "canyon" and altering capsid (Shafren et al., 1997a,b,c).

# 2.3  Picornaviral Proteinases

According to the active site nucleophiles, cellular proteinases have been cat-egorized into four classes, that is, serine, cysteine (thiol), aspartyl (acid), and metallo-proteinases, as well as exo- and endovariety. The substrate binding pocket directs the specificity of exo-proteinases for removing the N- or C-termi-nal residues and the endo-proteinases to cleave the internal sites. The proteinase subsite binding residues ($S_1$, $S_2$, etc.) of the substrate binds to the N-terminal of scissile bond and the residues of substrates are termed as $P_1$, $P_2$, etc. While the proteinase site binding substrate residues ($S_1'$, $S_2'$, etc.) progressively forms scissile bond with the C-terminal of the substrate and the residues of the sub-strates are termed as $P_1'$, $P_2'$, etc. (Schechter and Berger, 1967). The studies have investigated some important features of RNA proteinases, that is, (1) role as discrete proteins or proteolytic domains of larger forms; (2) activity depends on the type of larger form of the proteinase domain; (3) binding of other virus pro-teins or RNA controls activity and substrate specificity; (4) cleavage of host-cell proteins alter its function; (5) activation requires special environment; and (6) catalytic cleavage may be intra- or intermolecular through same mechanism.

The virus-encoded proteinases contribute toward specificity and regulate proteolysis of the other virus protein precursors and control biogenesis. Cleavage of specific cellular protein target modifies cellular macromolecular process which promotes replication cycle. Their important aspects include regulation of interaction with other viral proteins and/or nucleic acid, structural and mechanistic properties (Ryan and Flint, 1997). The picornaviral protein encodes a single, long open reading frame. The full-length translation products are absent due to rapid primary cleavage (Fig. 10.3) and the formed products are in *cis* or *trans* form. The polyprotein constitutes three types of proteinases, that is, L, 2A, and 3C (L$^{pro}$, 2A$^{pro}$, and 3C$^{pro}$, respectively).

### 2.3.1 Aphthovirus *L Proteinase*

The *Aphthovirus* or FMDV polyproteins have been processed by proteinases. The *Aphthovirus*, equine rhinovirus (ERV) 1 and 2, shows presence of L$^{pro}$ at the N-terminus. The L$^{pro}$ undergoes cotranslational cleavage at C-terminus forming Lab$^{pro}$ and Lb$^{pro}$ (Clarke et al., 1985; Sangar et al., 1987; Strebel and Beck, 1986). A relationship proposed between L$^{pro}$ and thiol proteinases has been confirmed by site-directed mutagenesis (Gorbalenya et al., 1991; Piccone et al., 1995; Roberts and Belsham, 1995).

### 2.3.2 Enterovirus *and Rhinovirus 2A Proteinase*

In EV and rhinovirus, virus-encoded proteinase (2A$^{pro}$) mediates the primary cleavage between the P1 capside protein precursor and the replicative domains of the polyprotein (P2, P3) (Fig. 10.3) at the Tyr-Gly pair (Sommergruber et al., 1989; Toyoda et al., 1986). It also processes a P3 precursor within 3D

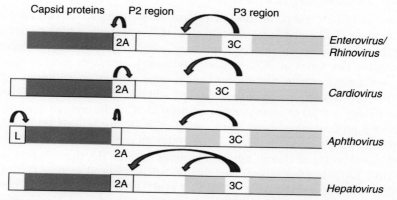

FIGURE 10.3 **Primary cleavage of picornaviral proteinases.** *Rectangular box,* Single, open reading frame; *blue arrows,* cleavage activity source.

to yield 3C′ and 3D′ (Hanecak et al., 1982; McLean et al., 1976; Rueckert et al., 1979). Mutation in the Tyr-Gly pair produces a parental phenotype and does not affect poliovirus growth (Lee and Wimmer, 1988). Thiol proteinase inhibitors, such as iodoaceamide and *N*-ethylmaleimide show 2A$^{pro}$ inhibition (Konig and Rosenwirth, 1988). 2A$^{pro}$ also blocks host-cell protein synthesis which is mediated by 220 kDa polypeptide cleavages (Etchison et al., 1982).

### 2.3.3 Aphthovirus *and* Cardiovirus *2A Protein*

In *Aphthovirus* and *Cardiovirus*, the primary polyprotein cleaves between P1 and 2A, as well as the capsid protein precursor of the C-terminal of 2A and 2B (Fig. 10.2). The 2A/2B cleavage has no sequence similarity with 2A$^{pro}$ (EV and RV) but in Theiler's murine encephalomyelitis (TME) and encephalomyocarditis (EMC) viruses the 2A proteins are conserved. In *Aphthovirus* and *Cardiovirus* 2A proteins the three C-terminal residues and in 2B proteins the N-terminal residues are conserved. The proteinases are not required for the 2A/2B site cleavage of FMDV or TMEV polyproteins (Clarke and Sangar, 1988; Roos et al., 1989; Ryan et al., 1989).The mechanism derived for picornaviral 2A mediated cleavage involves: (1) cleavage of Gly-Pro peptide bond by 2A; (2) substrate specificity represented by the 2A sequence; or (3) inhibition of formation of Gly-Pro peptide bond by interrupting the protein synthesis (Ryan and Flint, 1997).

### 2.3.4 Hepatovirus *2A Protein*

In *Hepatovirus* (hepatitis A virus), the 3C$^{pro}$ mediate the capsid-replicative protein primary cleavage and no function has been ascribed for 2A (Fig. 10.2) (Jia et al., 1993; Schultheiss et al., 1994).

### 2.3.5 *Picornavirus 3C Proteinase*

In case of EMCV and poliovirus, the proteolytic activities are associated with 3C proteins (P22 and P3-7c, respectively). 3C$^{pro}$ possesses a high degree of sequence similarity and mediates primary (single, 2A/2B) and secondary cleavages. The series of 3C$^{pro}$-mediated secondary cleavages process capsid and replicative protein precursors. Unlike other genera, in poliovirus, the 3CD$^{pro}$ mediates processing of the capsid protein precursor. 3C retains proteolytic activity and 3D alters substrate specificity of 3C$^{pro}$. Incorporation of purified 3AB protein enhances auto-processing of 3CD$^{pro}$ to 3C and 3D, as well as protein 2BC processing (Molla et al., 1994). In addition to this, 3C$^{pro}$ also mediates cleavage of various host-cell proteins (histone H3, transcription factor IIIC, TATA-binding protein, and microtubule-associated protein 4) (Clark et al., 1991; Das and Dasgupta, 1993; Falk et al., 1990; Joachims et al., 1995; Tesar and Marquardt, 1990). 3C$^{pro}$ has been characterized using proteinase inhibitors. The serine and thiol proteinase inhibitors possess proteolytic activity

(Baum et al., 1991; Gorbalenya and Svitkin, 1983; Korant, 1972, 1973; Korant et al., 1985; Pelham, 1978; Summers et al., 1972). Sequence alignment has detected the similarities between the large subclass cellular proteinases and 3C$^{pro}$, such as chymo-trypsin-like fold and the identity of the residues forming a catalytic triad analogous to that of serine proteinases. The predicted triad structure shows His$^{40}$, an acidic residue (Asp$^{85}$/Asp$^{71}$/Glu$^{71}$), the nucleophilic residue (Cys$^{147}$). The mutagenesis results established the role in catalysis, as well as predicted Gorbalenya (Cheah et al., 1990; Grubman et al., 1995; Hämmerle et al., 1991; Kean et al., 1991, 1993; Lawson and Semler, 1990). The mutation suppressing property of *cis* form is mapped within 3C$^{pro}$ (Andino et al., 1990b). Such mutations affect 3CD$^{pro}$ binding and are mapped to multiple sites of 3C$^{pro}$ (Andino et al., 1990a,b, 1993; Leong et al., 1993; Walker et al., 1995). The host-cell proteins (EF-1α or p36) regulate the RNA binding properties of 3CD$^{pro}$.

## 2.4 Picornaviral IRES-Mediated Translation

The picornaviral IRESs have been categorized based on homology, secondary structure, and other properties as:

- *Type I IRES*: poliovirus, rhinovirus, coxackievirus, EV
- *Type II IRES*: FMDV, *Cardiovirus* (e.g, EMCV, paraechoviruses, *Kobuvirus*)
- *Type III IRES*: HAV

The IRES-mediated initiation requires canonical initiation factors and IRES transactivating factors (ITAFs) as per viral requirements. ITAFs may act as IRES chaperones which stabilize the entire IRES in appropriate configuration for binding with canonical translation factors and finally ribosomes (Pilipenko et al., 2000). The various proteins involved in picornaviral IRES-mediated translation are summarized in Table 10.4. During translation, involvement of some canonical translation factors (eIF4G, eIF4B, eIF3, and eIF2) and noncanonical translation factors (polypyrimidine tract-binding protein, PTB/p57/hnRNPI) has been reported (Back et al., 2002; Hellen and Sarnow, 2001; Michael et al., 1995).

## 2.5 Role of Picornaviral Proteinases

The picornaviral proteins are encoded in a single, long open reading frame and the three types of polyproteins are L (L$^{pro}$ or FMDV L), 2A (2A$^{pro}$), and 3C$^{pro}$. The proteinases have been characterized based on the catalytic mechanism, substrate specificity, and 3D-structure.

In PVs, proteinases play significant role in the viral life cycle and virus–host interaction, such as viral polyprotein processing, initiation of viral RNA synthesis. 3C$^{pro}$, 2A$^{pro}$, and other proteinases mediate release of mature and functional proteins from the polyprotein. 3C$^{pro}$ is involved in multiple roles

**TABLE 10.4 Cellular Proteins Involved in Picornaviral IRES-Mediated Translation**

| Proteins | Binding sites/associated viral proteins | Viruses |
|---|---|---|
| PTB | IRES, 3C | PV, HRV, EMCV, FMDV, TMEV, HAV |
| nPTB | IRES | PV, TMEV |
| PCBP1 | Clover leaf, IRES | PV, HRV |
| PCBP2 | Clover leaf, IRES, 2A, 3CD, 3C | PV, HRV, HAV, CVB3 |
| Unr | IRES | PV, HRV |
| La | IRES, 3C | PV, HRV, HAV, CVB3, EMCV |
| ITAF45 | IRES | FMDV |
| HnRNP A1 | IRES | HRV2, EV71 |
| Nucleolin/C23 | IRES | PV, HRV |
| DRBP 76: NF45 heterodimer | IRES | HRV2 |
| GAPDH | IRES | HAV |
| FBP2 | IRES | EV71 |
| PABP | IRES | PV, EMCV |

Source: Lin, J.Y., Chen, T.C., Weng, K.F., Chang, S.C., Chen, L.L., Shih, S.R., 2009. Viral and host proteins involved in picornavirus life cycle. J. Biomed. Sci. 16, 103. Copyright permission from Elsevier.

in host cells to viral pathogenesis. In host cells, $3C^{pro}$ plays an important role in shutting off transcription, inhibiting protein synthesis and nucleo-cytoplasmic transport, and cell death. During cellular transcription, $3C^{pro}$ targets host cell proteins namely, $TAF_{110}$, TBP, CREB-1, Oct-1, $P_{53}$, TFIIIC (PV), CstF64 (EV71), and Histone 3 (FMDV) (Banerjee et al., 2005; Falk et al., 1990; Kundu et al., 2005; Shen et al., 1996; Weidman et al., 2001; Yalamanchili et al., 1997b). $3C^{pro}$ is also involved in cleavage of translation-related proteins in host cells namely, eIF4A-I (FMDV) (Li et al., 2001), eIF4G-I (FMDV) (Belsham et al., 2000), eIF5B (PV, CV, HRV) (De Breyne et al., 2008), G3BP1 (CV), PCBP2 (HAV, PV) (Perera et al., 2007; Zhang et al., 2007a), PABP (PV, HAV, EMCV) (Kobayashi et al., 2012; Kuyumcu-Martinez et al., 2004; Zhang et al., 2007b), and Sam68 (FMDV) (Lawrence et al., 2012).

## 2.5.1   Role of 3C$^{pro}$ in the Viral Life Cycle

In the viral life cycle, 3C$^{pro}$ has main role in: (1) processing of the viral polyprotein; (2) initiation of viral RNA synthesis; and (3) switching translation to RNA replication. Among the 29 genera of PV family, 3C$^{pro}$ regulates primary, as well as secondary polyprotein cleavages and generates multiple viral proteins. In *Cardiovirus, Kobuvirus, Teschovirus, Sapelovirus*, and *Senecavirus*, the primary function is controlled by 3C$^{pro}$ (Guarné et al., 2000; Salas and Ryan, 2010). In EVs, 2A$^{pro}$ mediates proteolytic cleavage between P1 and 2A. In aphthoviruses, cardioviruses, teschoviruses, cosaviruses, sencaviruses, and erboviruses, the 2A–2B cleavage is induced through separation of capsid proteins by ribosome, and in other cases the cleavage is mediated by 3C$^{pro}$ (Castelló et al., 2011; Palmenberg et al., 2010). When L and 2A$^{pro}$ are not proteolytic (e.g., *aichi virus*), 3C$^{pro}$ and 3CD precursors complete the cleavages. It has been reported that 3CD$^{pro}$ is more efficient than 3C$^{pro}$ for cleavage at VP2–VP3 and VP3–VP1 in EVs and apthoviruses (Amineva et al., 2004; Ryan et al., 1989; Ypma-Wong et al., 1988). In PVs, synthesis of positive- and negative-stranded RNA depends on RNA-dependent RNA polymerase (RdRp). The 3C$^{pro}$ and its precursors (3ABC, 3BC, 3BCD, and P3) stimulate RNA synthesis in vitro (Kusov et al., 1997; Pathak et al., 2007; Yin et al., 2003). For RNA replication, 3C$^{pro}$ and its precursors interact with the three RNA structures, that is, cloverleaf-like structure in the 5′-NCR (5′CL), *cis*-acting replication RNA element (*cre*), and 3′NTR-poly (A) tail (Andino et al., 1990a,b, 1993; Yin et al., 2003). Some of the viral and cellular proteins, such as 3AB, 3CD (in PV), and 3AB, 3ABC (in HAV) bind with 3′NCR of genomic RNA (Harris et al., 1994; Kusov et al., 1997). The identified 3C$^{pro}$ regions are N-terminal region (K12 and R13), central region (KFRDI86) and C-terminal region (T153, G154, and K155) (Blair et al., 1998; Franco et al., 2005; Moustafa et al., 2015). The RNA synthesis mechanism of 3C$^{pro}$ is similar to that of the RNA-binding. 3C$^{pro}$ and 3CD$^{pro}$ are different in terms of protease specificities and RNA-binding, but translation and replication by binding to RNA sequences is primarily regulated by 3C$^{pro}$ (Chan and Boehr, 2015).

## 2.5.2   Host Cells Intervention by 3C$^{pro}$

In PV, 3C$^{pro}$ mediated cleavage of host factors alters transcription, initiation of protein synthesis, nucleocytoplasmic transport, and favoring replication cycle, which are essential for cell viability.

In PV infected host cells, RNA synthesis is blocked due to DNA-dependent RNA polymerase cleavage by 3C$^{pro}$. 3C$^{pro}$ shows rapid shut-off of transcription by inducing cleavage of some factors (Banerjee et al., 2005; Kundu et al., 2005; Shen et al., 1996; Weidman et al., 2001; Yalamanchili et al., 1997a,b).The 3C$^{pro}$ interrupts cap-dependent translation by cleavage of nuclear histone H3 and CstF64 in FMDV and EV71, respectively (Falk et al., 1990; Weng et al., 2009), as well as inhibits transcription and alters gene expression in host cells. In host cells, protein synthesis depends on the 5' terminal cap structure and cap-binding proteins, while for viral translation mRNA requires IRES. In addition to eIFs, 3C$^{pro}$ also cleaves IRES-transacting factors to initiate translation. In PV proteinases, L, 2A, and 3C$^{pro}$ have essential role in the cleavage of host translation initiation factors. The function of 3C$^{pro}$ in cleavage of translation-related proteins in host cells infected by different PVs is as follows:

FMDV: binds capped *mRNA* to 40S ribosomal subunit and unwinds double-stranded RNA (eIF4A I); initiates translation by bringing *m*RNA to the 40S ribosome (eIF4G I); and cellular differentiation and proliferation (Sam68).

PV, CV, HRV: positioning of methionine *t*RNA on the condon of *m*RNA (eIF5B)

CV: offers RNA-binding protein to interact with Ras-GAP (G3BP1)

HAV, PV: activates translation and controls gene expression (PCBP2)

PV, HAV, EMCV: shortening of poly (A) and initiates translation (PABP)

In cytoplasm, the initiation of translation and replication depends on nuclear-resident proteins. PVs have been reported to promote viral replication by interrupting the nucleocytoplasmic trafficking of cells by targeting five types of host proteins, that is, nucleoporins Nup62, Nup98, Nup153, Nup214, and Nup358 (Caly et al., 2015; Flather and Semler, 2015; Ghildyal et al., 2009; Park et al., 2008, 2010; Walker et al., 2013; Watters and Palmenberg, 2011).

The cleavage of transcription and translation initiation factors required for cell survival and death by 3C$^{pro}$ induces apoptosis in PVs. 3C$^{pro}$ expression causes degradation of cellular DNA, as well as generation of apoptotic bodies (Ricour et al., 2009).The p53, transcriptional activator, gets degraded by 3C$^{pro}$ and promotes apoptosis through transcription-independent mechanism (Weidman et al., 2001) in addition to multiple pathways, thus releasing the virus.

3C$^{pro}$ also mediates cleavage of innate immune-related proteins, such as NEMO in FMDV; IRF7, TRIF, TAK1, and TAB2 in EV71; MAVS and TRIF in CVB3; and NEMO, TRIF, and MAVS in HAV (Lei et al., 2011, 2014; Mukherjee et al., 2011; Qu, 2010; Qu et al., 2011; Wang et al., 2014, 2012). PVs suppress host immune and antiviral response using interfering signaling pathways.

The crystal structure for 3C$^{pro}$ has been reported since 1990s (Bergmann et al., 1999; Birtley et al., 2005; Matthews et al., 1994; Mosimann et al., 1997) with two equivalent β-barrel domains, composed of six antiparallel strands and a shallow substrate binding groove (Matthews et al., 1999). In various picornaviral 3C$^{pro}$, a flexible surface loop (β-ribbon) has been identified (Matthews et al., 1999; Yin et al., 2005) that accommodates a part of N-terminal peptide substrate by folding over the peptide-binding groove of the enzyme and plays an important role in substrate recognition (Cui et al., 2011). The β-ribbon improves substrate accessibility by adopting an open conformation and exposing substrate binding clefts. The enzyme-substrate complex is stabilized by interactions between the β-ribbon and N-terminal end. Extensive sequence analysis has been performed to investigate the cleavage specificity of 3C$^{pro}$. 3C$^{pro}$ cleaves polypeptides in PV (at P1-Gln/P1'-Gly junctions), FMDV (at P1', P1-Gln, or

P1-Glu), and in HAV it shows preference for small hydrophilic residue at P2 (Curry et al., 2007; Seipelt et al., 1999).

3C$^{pro}$ acts on host cells by cleavage of essential factors required for cell viability, transcription, initiation of protein synthesis, nucleocytoplasmic transport, and growth which favor viral replication cycle. It induces cleavage of cellular factor required for transcription, translation, and nucleocytoplasmic trafficking which help in viral replication by modulating cell physiology. These interactions make 3C$^{pro}$ an important target in viral pathogenesis.

## 2.6 Nonstructural Protein Target

The picornaviral nonstructural protein has been characterized in different types and they offer suitable targets for antiviral drugs (Table 10.5).

## 2.7 3D$^{pol}$—Structure and Function

In picornviruses, the highly conserved amino acid sequences of 3D$^{pol}$ are not homologous but share basic structure and binding motif similarity to some extent (O'Reilly and Kao, 1998). The structure of 3D$^{pol}$ in poliovirus, rhinovirus, and FMDV is typical with finger, palm, and thumb domains (Ferrer-Orta et al., 2004; Hansen et al., 1997; Love et al., 2004; Thompson and Peersen, 2004).With amino acids present at N-terminal, 3D$^{pol}$ possesses important functionalities, encircles active site of palm domain, and on mutation disrupts 3D$^{pol}$ activity (Ferrer-Orta et al., 2004; Spagnolo et al., 2010; Thompson and Peersen, 2004). In *Coxsackievirus* 3D$^{pol}$, hydrophobicity at residue 5 contributes toward activity (Campagnola et al., 2008).

**TABLE 10.5 Investigated Nonstructural Region/Proteins**

| Genus | Number of types sequenced | Nonstructural proteins with solved structure | | | |
|---|---|---|---|---|---|
| | | 2A$^{pro}$ | 3C$^{pro}$ | 3D$^{pol}$ | 3CD$^{pro}$ |
| *Enterovirus* | 14 | CV-B4, HRV2 | PV, HRV-2, HRV-14 | PV, HRV-1, HRV-14, HRV-16 | PV |
| *Parechovirus* | 2 | — | — | — | — |
| *Hepatovirus* | 1 | — | HAV | — | — |
| *Cardiovirus* | 1 | — | — | — | — |
| *Aphthovirus* | 1 | — | FMDV | FMDV | — |
| *Erbovirus* | 1 | — | — | — | — |
| *Kobuvirus* | 1 | — | — | — | — |

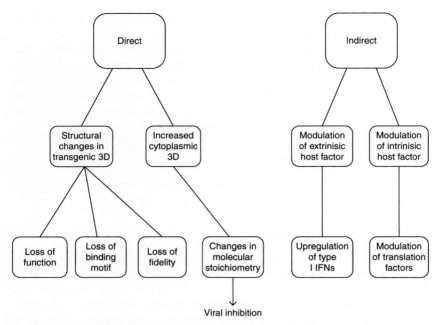

**FIGURE 10.4  Potential mechanism of 3D$^{pol}$ antiviral action on viruses.** *(Reprinted from Kerkvliet, J., Edukulla, R., Rodriguez, M., 2010. Novel roles of the picornaviral 3D polymerase in viral pathogenesis. Adv. Virol. 2010, 9. Open access publication.)*

The important role of 3D$^{pol}$ in picornaviral replication has provided many strategies to target function of 3D$^{pol}$ to slow down picornaviral replication. The nucleoside analog, ribavirin, has transferred 3D$^{pol}$ into a lethal mutagenesis state. Most of the nucleoside analogs get incorporated into host cell RNA and that is a major factor behind frequent side effects, as well as emergence of resistance to the treatment. The possible mechanism of 3D as antiviral protein is represented in Fig. 10.4. It is proposed that transgenic 3D itself oligomerizes to form a lattice. In host cell, transgenic 3D disrupts interaction with other viral molecules by disturbing one/more binding motifs. Studies have been carried out in transgenic and nontransgenic mice infected with TMEV, EMCV, and PRV to hypothesize the competitive or interfering binding site interactions by transgenic 3D. The broad spectrum antiviral properties of transgenic 3D do not support the dominant negative properties of 3D reported previously. The function and binding mode of transgenic 3D and viral 3D$^{pol}$ are quiet similar. 3D reduces the replication due to molecular stoichiometric differences in the cytoplasm. In in vitro models, increased concentration of 3D$^{pol}$ has been found to inhibit the stimulating effect of 3AB in 3D polymerization (Paul et al., 1994; Richards and Ehrenfeld, 1998; Rodriguez-Wells et al., 2001). Indirect mechanism has been proposed based on the inhibition of dsDNA virus by transgenic mice. For efficient RNA synthesis by transgenic 3D, it requires transcription and processed viral proteins forming replication complex. Transgenic 3D can regulate host proteins by replicating host mRNA transcript encoding the required proteins. In the poliovirus 3D$^{pol}$, translocation of 3CD$^{pro}$ by nuclear localization sequence near the N-terminus prevents transcription in host (Amineva et al., 2004; Sharma et al., 2004). Thus if the secondary structure is available, transgenic 3D may get translocated in the nucleus and replicate gene transcripts. In transgenic mice, the antiviral effect is observed in DNA and RNA viruses and further studies will provide interesting facts on this hypothesis.

## 2.8 Picornaviral Inhibitors

Over the past decades significant progress has been made toward the development of suitable drugs for the treatment of picornaviral infections. Some virus-specific compounds have been evaluated clinically for their efficacy against rhinovirus (Hayden et al., 1995; Hayden and Turner, 2012; Turner and Hayden, 1992) and EV infections (Schiff et al., 1992, 1996). Approaches including interferon and bradykinin antagonists were found to be unsuccessful (Higgins et al., 1990; Monto et al., 1989; Turner et al., 1989). Based on the mechanism adopted for inhibiting picornaviral infection, various developments are discussed here.

## 3 INHIBITORS OF THE NONSTRUCTURAL *ENTEROVIRUS* PROTEIN 3C OR PROTEASE INHIBITORS

Rupintrivir (AG7088, **1**) and AG7404 (**2**) are the only EV protease inhibitors (Binford et al., 2005; Matthews et al., 1999) that have entered into clinical studies for inhibition of 3C$^{pro}$. Rupintrivir (**1**), the irreversible peptidomimetic with $\alpha$, $\beta$–unsaturated ester, inhibits CV-B3 enzyme (EC$_{50}$ = 1.3 $\mu$M). SAR findings concluded improved enzyme affinity by replacement of the ethyl ester in position P2$'$ of Rupintrivir (**1**) by large aromatic moieties (Lee et al., 2007) due to hydrophobic interaction with residue Tyr22 of the 3C$^{pro}$. Repetitive cultivation of CV-B3 in the presence of Rupintrivir (**1**) raised three resistance mutations in the 3C$^{pro}$, T68A and N126Y. Clinical trials for Rupintrivir (**1**) had been discontinued due to limited activity against natural RV infections. Its analog, AG7404 (**2**), has improved oral bioavailability, as well as in vitro antiviral activity, but the clinical studies were discontinued (Dragovich et al., 2003; Patick et al., 2005).

1, Rupintrivir (AG-7088)                                       2, AG 7404

The three-dimensional structural analysis of the observed mutations has shown the presence of mutations in loops at remote location, major structural rearrangements in N126Y, and structural similarity of T68A mutant with the wild-type enzyme (Tan et al., in preparation). Recombinant viruses, harboring these mutations, have been demonstrated to be resistant against Rupintrivir (**1**).

The 3C proteases recognize the special structural features near the 5′ terminus of the genome for viral RNA binding. However, the proteolytic substrate pocket of 3C$^{pro}$ is quite well conserved in enteroviruses and rhinoviruses. The 3C$^{pro}$ binds viral RNA mainly through a conserved loop of the sequence Lys-Phe-Arg-Asp-Ile that connects the two β-barrel domains and is located on the side of the molecule opposite to the protease active site (Gorbalenya et al., 1990; Hämmerle et al., 1991; Matthews et al., 1994). Residues within the FMDV 3C protein have shown to be involved in viral RNA interaction, with VPg uridylylation activity and virus replication (Nayak et al., 2006). Once the interaction between the 3C$^{pro}$ and the 5′-terminal RNA of the virus is characterized, it could be utilized as a target for structure-based drug discovery.

A novel class of macrocyclic transition state inhibitors (**3**) effective against 3C$^{pro}$, SARS-3CL$^{pro}$23, and NV 3CL$^{pro}$24 has been designed based on following assumptions: (1) ligand is recognized by proteases in the β-strand conformation (Madala et al., 2010; Tyndall et al., 2005), (2) macrocyclization preorganizes a peptidyl transition state mimic in a suitable β-strand conformation which can bind to the protease active site (Gilon et al., 1991; Glenn et al., 2002; Tyndall and Fairlie, 2001), (3) macrocyclization can improve affinity, cellular permeability, proteolytic stability, as well as drug-like properties (Adessi and Soto, 2002; Edwards and Price, 2010; Gleeson, 2008; Lipinski, 2000; Marsault and Peterson, 2011; McCreary and Fairlie, 1998; Meanwell, 2011; Ritchie et al., 2011; Veber et al., 2002), (4) the plasticity of S$_3$ subsite can be utilized in the 3C and 3CL proteases by tethering the P$_1$ Gln side chain to the P$_3$ residue side chain, and (5) computational data suggested the ring size with $n = 3$ may have lesser intraligand strain and better receptor binding. The synthesized inhibitor **3** has shown IC$_{50}$ of 5.1, 1.8, and 15.5 μM against NV 3CL$^{pro}$, CVB3 Nancy strain 3C$^{pro}$, and SRAS-CoV 3CL$^{pro}$, respectively (Mandadapu et al., 2013).

**3** (R = isobutyl, n = 3, X = CHO)

## 3.1 Assembly Inhibitors

The assembly inhibitors can interrupt the multiple steps involved in assembly of new virus particles. Hsp90 on interaction with P1 capsid precursor enables cleavage and/or recognition of 3CD$^{pro}$ by ensuring correct folding (Geller et al., 2007; Tsou et al., 2013). Geldanamycin (**4**) and its analog 17-AAG (17-allyamino-17-demethoxygeldanamycin, **5**) impair morphogenesis by allowing Hsp90 and P1 interactions and altering P1 processing. In the assembly of EV subgroup, the role of glutathione has been identified by the glutathione synthesis inhibitor (BSO, buthionine sulfoxime, **6**) and glutathione scavenger (TP219) (Ma et al., 2014; Mikami et al., 2004; Smith and Dawson, 2006; Thibaut et al., 2014). BSO (**6**) and glutathione-depleting compound were found to be well tolerated but the in vivo antiviral activity has not been studied (Bailey et al., 1994, 1997; O'Dwyer et al., 1996).

**4, Geldanamycin**                **5, 17-AAG**          **6,BSO (Buthionine sulfoxime)**

## 3.1.1 Receptor Binding Inhibitors

### 3.1.1.1 Soluble ICAM-1 (sICAM)

ICAM-1 is a ligand for lymphocyte function-associated antigen (LFA-1) (Dustin and Springer, 1988). In various immunological interactions, adhesion of cells, expressing LFA-1, to those expressing ICAM-1 is an important step. The availability of 3D-strucutures of the capsid of several rhinovirus serotypes and nonpolio EV has identified the presence of deep trough or "canyons" on the viral shell surface acting as binding site for ICAM-1 (Giranda et al., 1990; Kim et al., 1989; Oliviera et al., 1993; Olson et al., 1993; Muckelbauer et al., 1995; Rossmann et al., 1985; Rossmann and Palmenberg, 1988; Zhao et al., 1996). The canyon acts as a virion receptor-binding site as supported by its properties: (1) as compared to solvent-exposed residues, the amino acid residues in the canyon are highly conserved; and (2) the canyon does not come under host immune system pressure and due to dimensions it is not accessible for antibody molecule binding (Rossmann and Palmenberg, 1988).

The soluble form of ICAM-1 (sICAM-1) has shown broad-spectrum activity against major rhinovirus serotypes (IC$_{50}$ = 0.1 to 41.1 µg/mL) (Arruda et al., 1992; Crump et al., 1993; Greve et al., 1991; Martin et al., 1990; Ohlin et al., 1994).

HRV-1A and -2 belonging to the minor receptor-binding group were not inhibited by ICAM-1 (Crump et al., 1993). Fusion of the five domains of ICAM-1 with IgA1, IgM, and IgG1 has been found to significantly improve inhibition.

### 3.1.2 Capsid Binding Inhibitors

The ligand binding site location and atomic interactions within the binding site have been revealed by X-ray crystallography studies of various small molecule inhibitors of PVs (Bibler-Muckelbauer et al., 1994; Kim et al., 1993; Smith et al., 1986). The capsid-binding compounds show selectivity and specificity for the serotypes due to variation in size and amino acid composition of the hydrophobic binding pockets (Kim et al., 1993; Mallamo et al., 1992). Compounds with similar mechanism of inhibition of viral uncoating and/or adsorption show selectivity, irrespective of their structural categories (Zhang et al., 1992). Compounds, such as pirodavir (**7**), Pleconaril (**8**), SCH 38057 (**9a**), and SDZ 35-682 (**9b**) bind to the capsid proteins in the same hydrophobic pocket of HRV-14. Among these, pirodavir (R 77975) (**7**) is a pyridazine derivative and has reached clinical phase of development due to improved potency and spectrum of activity as compared to its previous analog R 61837 (Andries et al., 1991, 1992). The inactive acid derivative obtained from ester hydrolysis of Pirodavir has shown inhibitory effect against rhinovirus infections but its clinical studies are halted.

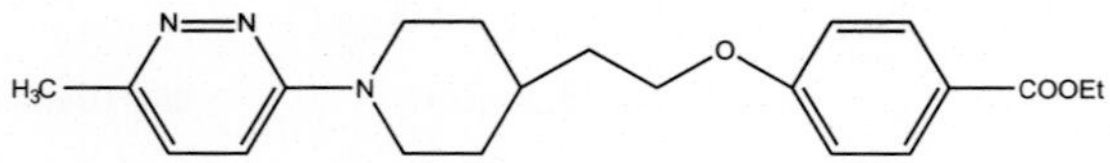

**7, Pirodavir(R 77975)**

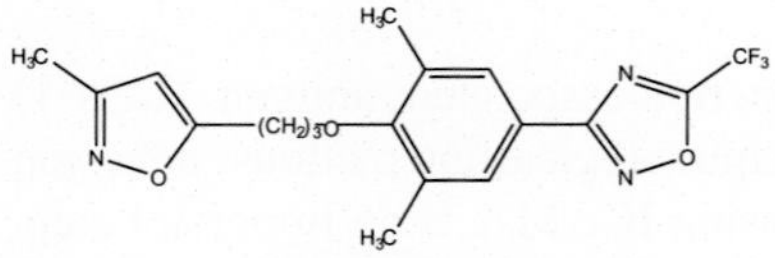

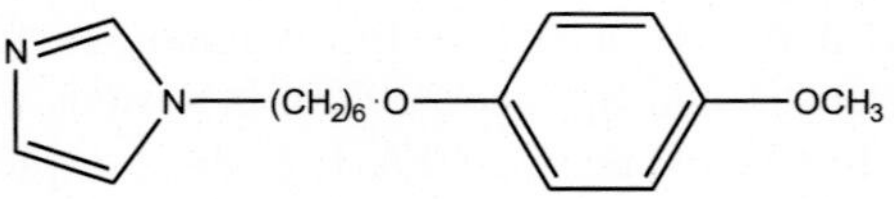

**8, Pleconaril (WIN 63843)**

**9a, SCH 38057**

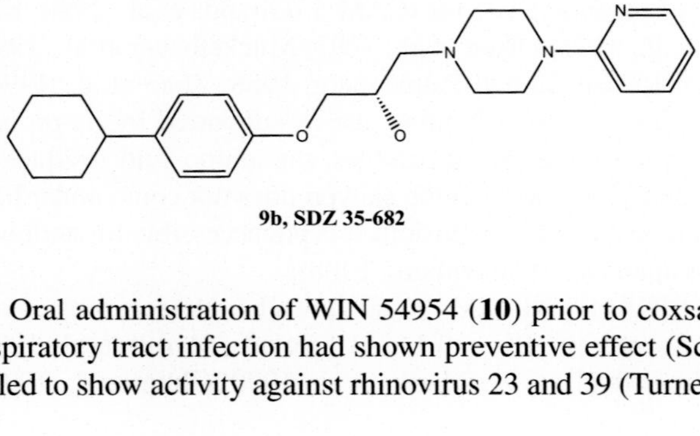

**9b, SDZ 35-682**

Oral administration of WIN 54954 (**10**) prior to coxsackievirus A21 upper respiratory tract infection had shown preventive effect (Schiff et al., 1992), but failed to show activity against rhinovirus 23 and 39 (Turner and Hayden, 1992).

WIN 54954 (**10**) has been withdrawn from clinical trials due to short half life in vivo, rapid metabolism, as well as association of reversible hepatitis with drug metabolites. Further, SAR studies have been extended to address the key problems, such as potency, chemical and metabolic stability, pharmacokinetics, and safety by targeting the oxazoline ring which appears to be chemically and metabolically unstable. The oxazoline ring was replaced by various heterocycles, leading to identification of a promising, more potent and stable oxadiazole, WIN 61893 (**11**) (Diana et al., 1994).

10, WIN 54954                                    11, WIN 61893

Evaluation of metabolic stability and metabolic products for both WIN 54954 (**10**) and WIN 61893 (**11**) concluded improved stability of WIN 61893 with approximately 8 metabolic products. For the metabolic products, terminal methyl group was hydrolyzed (3:1 ratio) and prevention of hydroxylation at this site reduced the extent of metabolism. Corresponding trifluoromethyl analogs did not show reduction to that extent of metabolism as WIN 61893 (**11**) and similarly replacement of the methyl group on the oxadiazole ring (Pleconaril, **8**) had only two metabolic products. Thus the trifluoromethyl group protects the molecule by preventing metabolism.

Clinical studies have been reported for Pleconaril (**8**), BTA798 (**13**, Biota Pharmaceuticals, Alpharetta, USA), and Pocapavir (V-073, **12**) that binds to the capsid in the canyon, a trough involved in receptor binding. Thus these compounds impair first step, that is, receptor attachment and/or viral uncoating) of viral replication (Linden et al., 2015). Pleconaril (VP 63843, **8**) is a predecessor of WIN 54954 (**10**) from the series of WIN compounds with broad spectrum of activity against various rhinovirus serotypes. Pleconaril has shown inhibition of 95% isolates of nonpolio EVs (100 ng/mL). Echovirus 11 has been observed as the most sensitive serotype with inhibition of 90% clinical isolates (6 ng/mL). It has shown 18-fold improvement in half-life, > 4-fold improvement in terminal half-life and 2- to 3-fold decreased plasma clearance (oral dose 10 mg/kg) (Diana et al., 1995). In the first Phase trials, Pleconaril has been well tolerated up to 400 mg/kg oral dose and in the second Phase trials oral 200 mg/kg 3 times dosage was found to be effective against an upper respiratory tract infection with *Coxsackievirus* A21 (Schiff et al., 1996). Despite the broad spectrum of activity, Pleconaril (**8**) has been rejected by FDA due to safety issues (Schiff and Sherwood, 2000; Senior, 2002). It had no significant effect for asthma prevention and common cold symptoms in asthma patients (Merck and Dohme, 2015).

BTA798 (**13**) was found to lead to significantly reduced symptoms in asthma patients with naturally acquired RV infections during clinical phase IIb trials. Similarly Pocapavir (**12**) had been evaluated clinically against PV and EV (Buontempo et al., 1997; Oberste et al., 2009). It had shown good tolerance but no apparent antiviral effect. The identified reasons behind failure of capsid binders in clinical trials were the lack of activity against all EVs and quick emergence of resistance (De Palma et al., 2008a,b; Kaiser et al., 2000; Ledford et al., 2004; Pevear et al., 1999).

**12, Pocapavir (V-073)/SCH 48973**          **13, BTA 798**

WIN compounds and SCH 38057 (**14**, Pocapavir, V-073) bind at the same hydrophobic pocket (Rozhon et al., 1993; Zhang et al., 1992). The successors of SCH 38057 (**14**), SCH 47802 (**15**) (in vitro $IC_{50}$ = 0.03-10 μg/mL, in vivo oral dose 60 mg/kg) (Buontempo et al., 1995; Cox et al., 1995, 1996) and SCH 48973 (**12**) (dose 3–20 mg/kg) (Wright-Minogue et al., 1996), had shown oral absorption of 78% and an oral bioavailability of 52% (Veals et al., 1996).

**14, SCH38057** (1-[6-(2-chloro-4-methoxyphenoxy)-hexyl]imidazole hydrochloride))

**15, SCH 47802**

### 3.1.3 SDZ 35-682

Structurally related to Pirodavir (**7**), SDZ 35-682 (**16**) had shown inhibition of several strains of rhinoviruses (0.1 µg/mL). SDZ 35-682 has shown in vitro and in vivo effect, as well as protection against echovirus 9 (Rosenwirth et al., 1995).

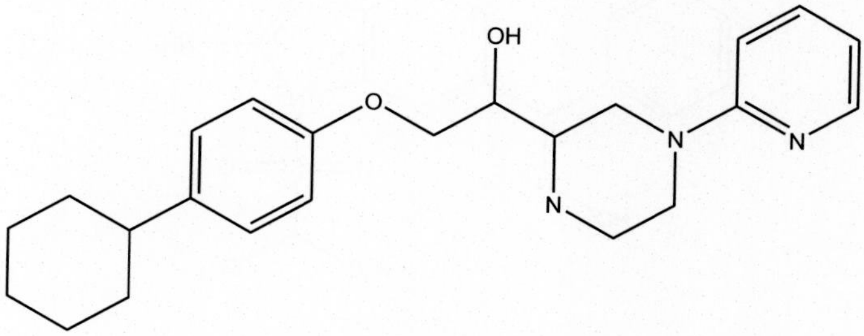

**16, (SDZ 35-682)**

### 3.1.4 Chalcone Amides

The antirhinoviral activity of chalcones has been well reported (Ishitsuka et al., 1982a,b; Ninomiya et al., 1984, 1985). Ro 09-0410 (**17**) has been effective against a variety of rhinoviruses but has limited activity against the EVs. Absence of its activity against HRV infections is attributed to its lack of aqueous solubility and bioavailability (Al-Nakib et al., 1987; Phillpotts et al., 1984). Further, amide analogs have been added to this category which posses 4.5–12 times higher inhibition of 12 rhinovirus serotypes (Ninomiya et al., 1990). The binding site is same for all the analogs and is in close proximity to that of Pleconaril. Improved aqueous solubility has been observed for compounds with ethoxy and methylamino substitutions, but there has been no further studies on such series.

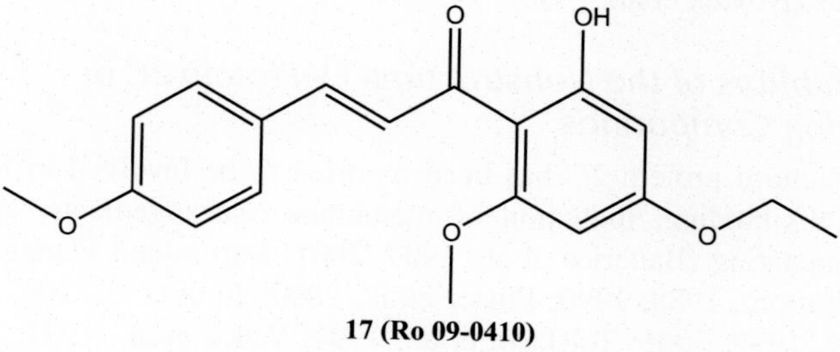

**17 (Ro 09-0410)**

### 3.1.5 3A Coding Region

A potent inhibitor of rhinoviruses and EVs, Enviroxime (**18**), was reported (DeLong et al., 1980) but that failed in clinical studies due to serious side effects on oral administration (Hayden and Gwaltney, 1982; Miller et al., 1985; Phillpotts et al., 1983). Enviroxime (**18**) targets 3A-coding region of the virus to block synthesis of plus-strand viral RNA during viral RNA replication (Heinz and Vance, 1995). The data supports 3A coding region as an additional target to inhibit replication of PVs.

**18, Enviroxime**

### 3.1.6 Zinc Salts

In 1984, the efficacy of zinc gluconate as lozenges to reduce the established cold symptoms had been reported for the first time (Eby et al., 1984). The study was, however, discontinued due to mouth irritation and palatability problems (Farr and Gwaltney, 1987). Any Further study failed to produce beneficial effects of zinc acetate or zinc gluconate (Douglas et al., 1987; Farr et al., 1987; Geist et al., 1987). Ingestion of zinc (300 mg/day for 6 weeks) had shown declined lymphocyte proliferative responses, polymorphonuclear phagocytic function and high-density lipoprotein concentration (Chandra, 1984). The proposed mechanism for inhibition of PVs is: (1) inhibition of viral peptide cleavage (Korant and Butterworth, 1976); (2) complexation of $Zn^{2+}$ with ICAM-1-binding sites on the viral capsid protein surface and thus inhibiting the binding of cells to rhinovirus (Novick et al., 1996).

### 3.1.7 Inhibitors of the Nonstructural EV Protein 2C or 2C-Targeting Compounds

The nonstructural protein 2C has been reported to be involved in RNA replication, RNA-binding, induction of membrane rearrangements, encapsidation, and uncoating (Banerjee et al., 1997, 2001; Barton and Flanegan, 1997; Li and Baltimore, 1988, 1990; Pfister et al., 2000; Tang et al., 2007; Teterina et al., 1992, 1997, 2006; Tolskaya et al., 1994; Vance et al., 1997; Verlinden et al., 2000). The nucleoside triphosphate-binding motif of 2C has been found to be involved in ATPase activity (Mirzayan and Wimmer, 1994; Rodríguez and Carrasco, 1993). Thiazolobenzimidazoles is the first class of replication inhibitors of various EVs in a dose-dependent manner. Chemical modifications with specific substituents have been identified for improving antiviral activity. For TBZE-029 (**19**), the mechanism of action has been investigated. During time of drug-addition study, this category of inhibitors were found to affect the viral replication cycle during attachment, entry, uncoating, and release (De Palma et al., 2008a,b,c) but not to affect polyprotein synthesis and/or processing or ATPase activity of 2C. Similarly, none of the previously reported anti-EV compounds, targeting 2C, were found to inhibit enzymatic activity. This class of 2C inhibitors could inhibit enzymatic ATPase activity or interfere with 2C function which differs from its ATPase activity. Inhibition of enzymatic ATPase cannot be mimicked using the purified 2C protein as the conditions differ for the complete replication complex (Gorbalenya et al., 1990; Pfister and Wimmer, 1999). Some other compounds targeting 2C were guanidine hydrochloride, fluoxetine (**20**), HBB (**21**), and MRL-1237 (**22**) (De Palma et al., 2008c; Hadaschik et al., 1999; Pincus et al., 1986; Sadeghipour et al., 2012; Shimizu et al., 2000; Ulferts et al., 2013). Lack of crystal structure availability for 2C is one of the major reasons behind limited knowledge toward its mechanism and thus these inhibitors failed to enter in vivo studies in human.

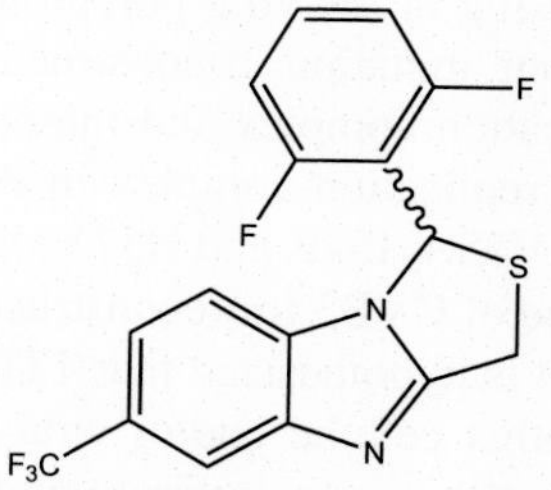

**19, TBZE-029,** $IC_{50} = 7 \mu M$ (SI > 83)

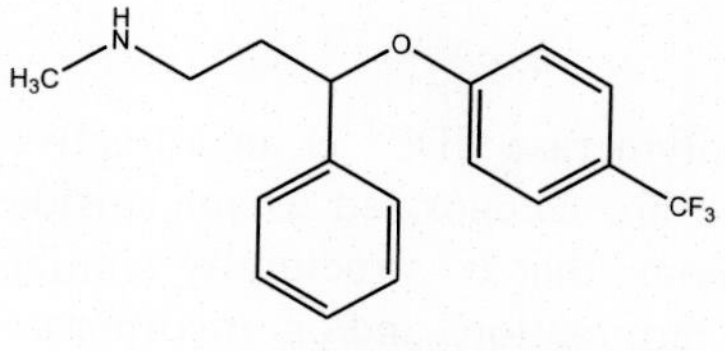

**20, Fluoxetine**

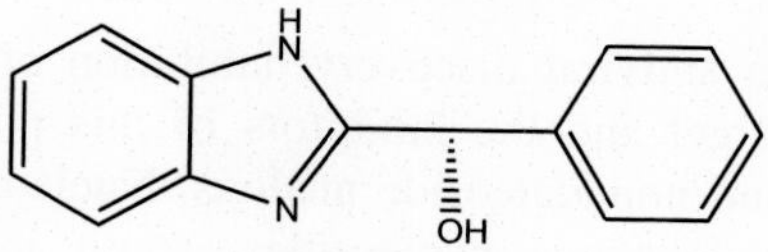

**21, HBB**

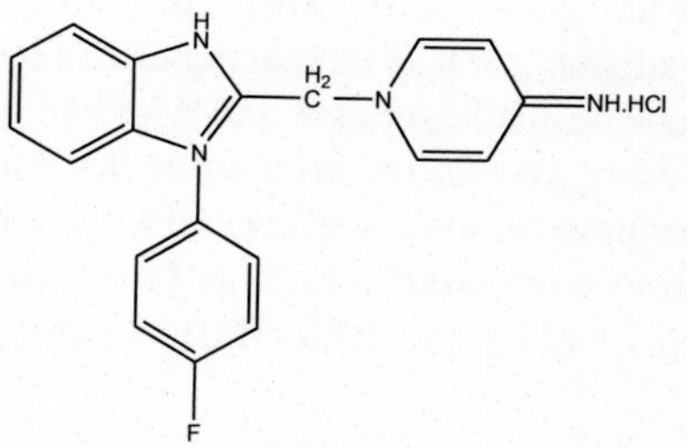

**22, MRL-1237**

## 3.1.8 Inhibitors of the Nonstructural Enterovirus Protein 3A

**23, TTP-8307** (IC$_{50}$ = 1.2 µM, CVB3; SI > 83)

Enviroxime (**18**) reported 20 years ago as inhibitor of protein 3A EV replication is associated with development of cross-resistance (Heinz and Vance, 1995, 1996). TTP-8307 (**23**) has shown selective inhibition of replication in various entero- and rhinoviruses in a dose-dependent manner but do not alter synthesis and processing of the viral polyprotein. All TTP-8307-resistant variants possess at least one mutation in nonstructural protein 3A which contributes in the viral replication complex and interacts with several other viral proteins to assemble the replication complex in the right conformation. The identified mutations (I8T, V45A, I54F, and H57Y) have been reintroduced in an infectious full-length clone of CVB3 for reconstructing the resistant phenotype (De Palma et al., 2009). It is hypothesized that TTP-8307 (**23**) disrupts certain interactions of 3A with other cellular and/or viral proteins. Thus viral inhibition targeting 3A needs further investigation to understand the role of proteins involved in replication complex formation.

## 3.1.9 Polymerase Inhibitors

In antiviral discovery, inhibition of viral polymerase 3D$^{pol}$ is an attractive target and the inhibitors of this polymerase are categorized as nucleoside and nonnucleoside analogs. Nucleoside analogs, that is, structurally similar endogenous nucleosides, result in chain termination and/or incorporate incorrect nucleosides leading to mutation (e.g., Ribavirin causes lethal mutagenesis) (Crotty et al., 2000, 2001), while the other category of inhibitors adopt different mechanisms, such as increasing error rate of 3D$^{pol}$, competing with incoming nucleoside triphosphates and Mg$^{2+}$ (e.g., Amiloride) (Gazina et al., 2011), and altering productive binding of the template-primer to 3D$^{pol}$ by binding at the template-binding site (ex. GPC-N114, **24**) (van der Linden et al., 2015). For some compounds, such as Gliotoxin (**25**), aurintricarboxylic acid (**26**), DTrip-22 (**27**), and BPR-3P0128 (**28**), the mechanism is not reported (Chen et al., 2009; Hung et al., 2010; Miller et al., 1968; Rodriguez and Carrasco, 1992; Velu et al., 2014).

**24, GPC N114**          **25, Gliotoxin**          **26, Aurintricarboxilic acid**

**27, DTrip 22**                    **28, BPR3P0128**

## 3.2 PI4KIIIβ Inhibitors

Enviroxime, a potent inhibitor of EVs (Wikel et al., 1980) has been recently identified as PI4KIIIβ inhibitor (Hsu et al., 2010; van der Schaar et al., 2012). During clinical phase, Enviroxime was found to successfully reduce the symptoms but unable to produce any in vivo effect (Hayden and Gwaltney, 1982; Higgins et al., 1988; Levandowski et al., 1982; Miller et al., 1985; Phillpotts et al., 1981, 1983). It was withdrawn from further trial due to gastrointestinal side effects and poor pharmacokinetic profile (Phillpotts et al., 1983). Other reported PI4KIIIβ inhibitors are GW5074 (**29**), T-00127-HEV1 (**30**), and AG 7404 (**2**). Among these inhibitors, AG 7404 (**2**) has shown strong inhibition of in vivo CVB4 replication and CVB4-induced pathology (van der Schaar et al., 2013). T-00127-HEV1 (**30**) is a small molecule inhibiting PI4KIIIβ and lethal to mice (Spickler et al., 2013). Some of the PI4KIIIβ inhibitors have shown interference with lymphocyte proliferation and chemokine secretion in a mixed leukocyte reaction (Lamarche et al., 2012). This has potentiated the studies toward investigating role of PI4KIIIβ inhibitors for the treatment of EV infections.

29, GW 5074                    30, T-00127-HEV1

## 3.3 OSBP Inhibitors

Oxysterol-binding protein-1 (OSBP) is essential for viral replication and has been identified as a target for some compounds, such as Itraconazole (**31**), OSW-1 (**32**), AN-12-H5 (**33**), 25-hydroxycholesterol, AN-11-E6 (**34**), AN-22-A6 (**35**), AN-23-F6 (**36**, anti-EV71 compound) and T-00127-HEV2 (**37**) (Albulescu et al., 2015; Arita et al., 2010, 2013; Strating et al., 2015). In 1992, itraconazole (**31**) was approved by FDA as an antifungal drug and later as anticancer reagent (Shim and Liu, 2014). Such reports have encouraged the antiviral studies of itraconazole (**31**) in mouse models.

**31, Itraconazole**

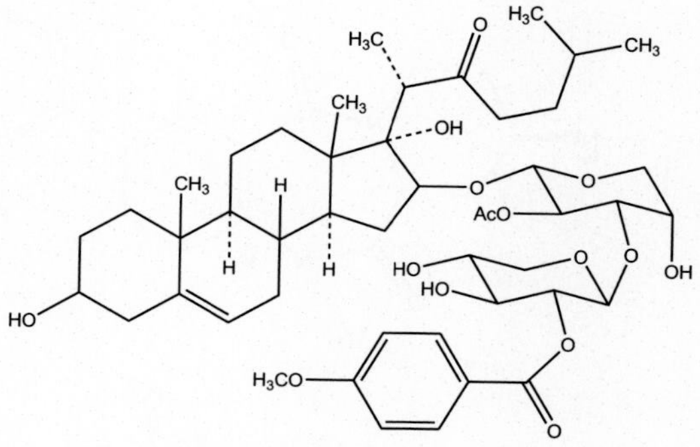

**32, OSW-1**

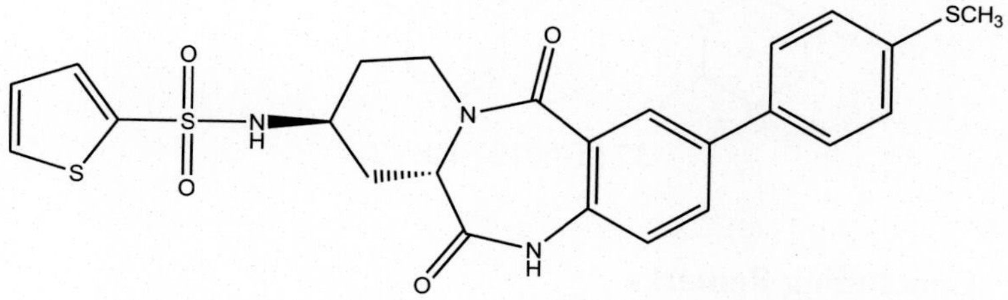

**33**

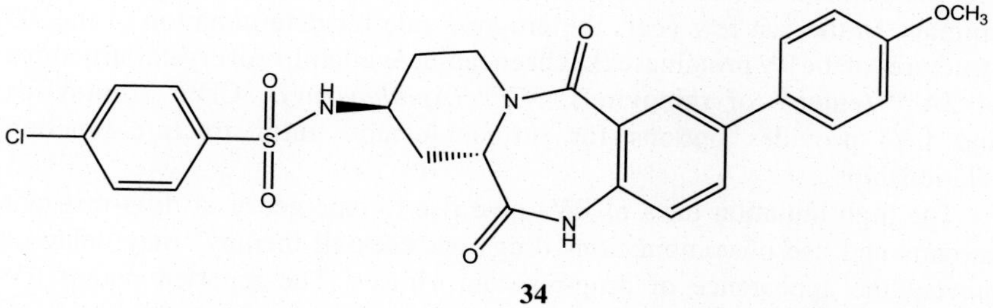

**34**

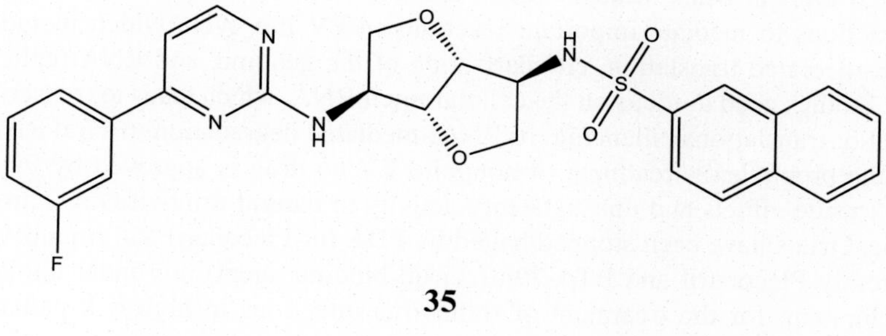

**35**

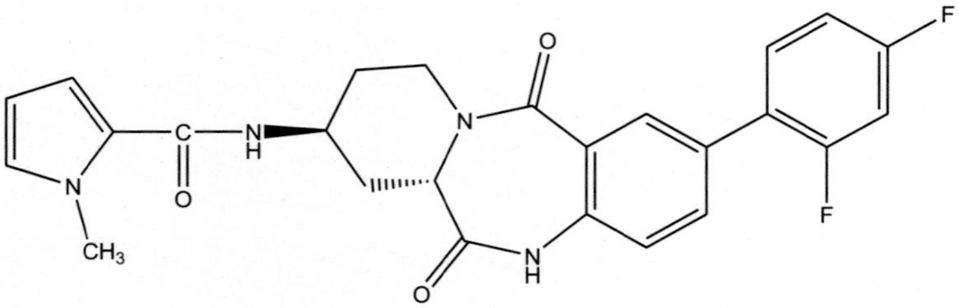

**36**

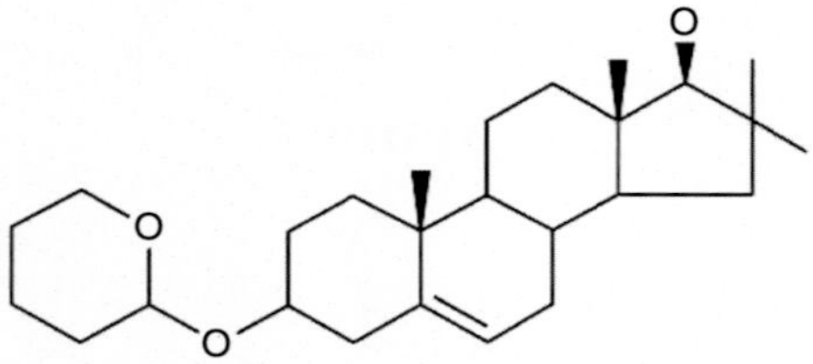

**37** (T-00127-HEV2)

## 3.4 Concluding Remarks

Picornaviridae is one of the largest viral families that affect both humans and animals. In the last few years the progress rate for determination of the 3D structures of the PV proteinases has been rapid. Availability of crystal structures of 2A$^{pro}$ (entero- or rhinoviral), 3C$^{pro}$ (*Aphthovirus*), 3CD$^{pro}$ (rhinovirus and EV) provides options for further insights into structure–function relationships.

The high mutation rates of EVs give rise to emergence of drug-resistant variants and use of combinations drugs or "cocktail therapy" may delay or prevent the appearance of drug-resistant viruses. The reported potent EV inhibitors target various stages of viral replication cycle and some of them have entered clinical trials. Studies have identified use of multiple RNA–protein interactions to mediate important reactions in PV life cycle which includes IRES-mediated translation, circularization of the genome, and RNA replication. Studies need to focus on the cellular microRNA, which leads to transcriptional or translational silencing, or RNAi-mediated degradation of viral RNA. For the prophylaxis/treatment of nonpolio EV no drug is approved by FDA. Due to side effects and unsatisfactory activity in natural rhinovirus infection, clinical trials have been stopped/halted by FDA for Pleconaril and rupintrivir. Currently Pleconaril and BTA-798 (capsid binding agent) are under clinical development for the treatment of rhinovirus infections in high-risk patients with chronic lung diseases.

The progress toward the discovery of antipicornaviral compounds is remarkable and has attempted to address many critical issues related to potency, chemical and metabolic stability, bioavailability, and safety. The clinical efficacy of some compounds has been established against RV (intranasal route) and EV (oral route) infections. The efforts will hopefully provide a cure for common cold and as therapeutic or prophylactic options for preventing asthma, COPD and other PV infections. Thus demand for the development of drugs for the treatment of picornaviral infection is essential along with exploring many other potential interesting targets.

## REFERENCES

Acharya, R., Fry, E., Stuart, D., Fox, G., Rowlands, D., Brown, F., 1989. The three-dimensional structure of foot-and-mouth disease virus at 2.9 Å resolution. Nature 337, 709–716.

Adessi, C., Soto, C., 2002. Converting a peptide into a drug: strategies to improve stability and bioavailability. Curr. Med. Chem. 9, 963–978.

Albulescu, L., Strating, J.R., Thibaut, H.J., van der Linden, L., Shair, M.D., Neyts, J., van Kuppeveld, F.J., 2015. Broad-range inhibition of enterovirus replication by OSW-1, a natural compound targeting OSBP. Antivir. Res. 117, 110–114.

Al-Nakib, W., Higgins, P., Barrow, I., Tyrell, D.A.J., Lenox-Smith, I., Ishitsuka, H., 1987. Intranasal chalcone Ro 0410, as prophylaxis against rhinovirus infections in human volunteers. J. Antimicrob. Chemother. 20, 887–892.

Amineva, S.P., Aminev, A.G., Palmenberg, A.C., Gern, J.E., 2004. Rhinovirus 3C protease precursors 3CD and 3CD'localize to the nuclei of infected cells. J. Gen. Virol. 85, 2969–2979.

Andino, R., Rieckhof, G.E., Achacoso, P.L., Baltimore, D., 1993. Poliovirus RNA synthesis utilizes an RNP complex formed around the 5'-end of viral RNA. EMBO J 12, 3587–3598.

Andino, R., Rieckhof, G.E., Baltimore, D., 1990a. A functional ribonucleoprotein complex forms around the 5' end of poliovirus RNA. Cell 63, 369–380.

Andino, R., Rieckhof, G.E., Trono, D., Baltimore, D., 1990b. Substitutions in the protease (3C$^{pro}$) gene of poliovirus can suppress a mutation in the 5'-noncoding region. J. Virol. 64, 607–612.

Andries, K., Dewindt, B., Snoeks, J., Willebrords, K., Van Eemeren, K., Stokbroekx, K., 1991. Antirhinovirus spectrum and mechanism of action of R77975. Antivir. Res. 15 (Suppl. 1), Abstract 101.

Andries, K., Stokbroekx, R., Dewindt, B., Snoeks, J., 1992. In vitro activity of pirodavir (R77975), a substituted phenoxy-pyridazinamine with broad spectrum antipicornaviral activity. Antimicrob. Agents Chemother. 36, 100–107.

Arden, K.E., Mackay, I.M., 2010. Newly identified human rhinoviruses: molecular methods heat up the cold viruses. Rev. Med. Virol. 20, 156–176.

Arita, M., Kojima, H., Nagano, T., Okabe, T., Wakita, T., Shimizu, H., 2013. Oxysterol-binding protein family I is the target of minor enviroxime-like compounds. J. Virol. 87, 4252–4260.

Arita, M., Takede, Y., Wakita, T., Shimizu, H., 2010. A bifunctional anti-enterovirus compound that inhibitis replication and early stage of *Enterovirus* 71 infection. J. Gen. Virol. 91, 2734–2744.

Arnold, E., Luo, M., Vriend, G., Rossmann, M.G., Palmenberg, A.C., Parks, G.D., Nicklin, M.J., Wimmer, E., 1987. Implications of the picornavirus capsid structure for polyprotein processing. Proc. Natl. Acad. Sci. USA 84, 21–25.

Arruda, E., Crump, C.E., Martin, S.D., Merluzzi, V., Hayden, F.G., 1992. In vitro studies of the antirhinovirus activity of soluble intercellular adhesion molecule-L. Antimicrob. Agents Chemother. 36, 1186–1191.

Back, S.H., Kim, Y.K., Kim, W.J., Cho, S., Oh, H.R., Kim, J.E., Jang, S.K., 2002. Translation of polioviral mRNA is inhibited by cleavage of polypyrimidine tract-binding proteins executed by polioviral 3C(pro). J. Virol. 76, 2529–2542.

Bailey, H.H., Mulcahy, R.T., Tutsch, K.D., Arzoomanian, R.Z., Alberti, D., Tombes, M.B., Wilding, G., Pomplun, M., Spriggs, D.R., 1994. Phase I clinical trial of intravenous L-buthionine sulfoximine and melphalan: an attempt at modulation of glutathione. J. Clin. Oncol. 12, 194–205.

Bailey, H.H., Ripple, G., Tutsch, K.D., Arzoomanian, R.Z., Alberti, D., Feierabend, C., Mahvi, D., Schink, J., Pomplun, M., Mulcahy, R.T., Wilding, G., 1997. Phase I study of continuous-infusion L-S,R-buthionine sulfoximine with intravenous melphalan. J. Natl. Cancer Inst. 89, 1789–1796.

Banerjee, R., Echeverri, A., Dasgupta, A., 1997. Poliovirus-encoded 2C polypeptide specifically binds to the 3'-terminal sequences of viral negative-strand RNA. J. Virol. 71, 9570–9578.

Banerjee, R., Tsai, W., Kim, W., Dasgupta, A., 2001. Interaction of poliovirus-encoded 2C/2BC polypeptides with the 3' terminus negative-strand cloverleaf requires an intact stem-loop B. Virology 280, 41–51.

Banerjee, R., Weidman, M.K., Navarro, S., Comai, L., Dasgupta, A., 2005. Modifications of both selectivity factor and upstream binding factor contribute to poliovirus-mediated inhibition of RNA polymerase I transcription. J. Gen. Virol. 86, 2315–2322.

Barton, D.J., Flanegan, J.B., 1997. Synchronous replication of poliovirus RNA: initiation of negative-strand RNA synthesis requires the guanidine-inhibited activity of protein 2C. J. Virol. 71, 8482–8489.

Basavappa, R., Syed, R., Flore, O., Icenogle, J.P., Filman, D.J., Hogle, J.M., 1994. Role and mechanism of the maturation cleavage of VP0 in poliovirus assembly: structure of the empty capsid assembly intermediate at 2.9 Å resolution. Protein Sci. 3, 1651–1669.

Baum, E.Z., Bebernitz, G.A., Palant, O., Mueller, T., Plotch, S.J., 1991. Purification, properties and mutagenesis of poliovirus 3C protease. Virology 185, 140–150.

Belsham, G.J., McInerney, G.M., Ross-Smith, N., 2000. Foot-and-mouth disease virus 3C protease induces cleavage of translation initiation factors eIF4A and eIF4G within infected cells. J. Virol. 74, 272–280.

Bergelson, J.M., Chan, M., Solomon, K.R., St. John, N.F., Lin, H., Finberg, R.W., 1994. Decay-accelerating factor (CD55), a glycosylphosphatidylinositol-anchored complement regulatory protein, is a receptor for several echoviruses. Proc. Natl. Acad. Sci. USA 91, 6245–6248.

Bergmann, E.M., Cherney, M.M., McKendrick, J., Frormann, S., Luo, C., Malcolm, B.A., Vederas, J.C., James, M.N., 1999. Crystal structure of an inhibitor complex of the 3C proteinase from hepatitis A virus (HAV) and implications for the polyprotein processing in HAV. Virology 265, 153–163.

Bibler-Muckelbauer, J.K., Kremer, M.J., Rossmann, M.G., Diana, G.D., Dutko, F.J., Pevear, D.C., McKinlay, M.A., 1994. Human rhinovirus 14 complexed with fragments of antiviral compounds. Virology 202, 360–369.

Bienz, K., Egger, D., Pfister, T., 1994. Characteristics of the poliovirus replication complex. Arch. Virol. (Suppl. l9), 147–157.

Bienz, K., Egger, D., Pfister, T., Troxler, M., 1992. Structural and functional characterization of the poliovirus replication complex. J. Virol. 66, 2740–2747.

Bienz, K., Egger, D., Rasser, Y., Bossart, W., 1983. Intracellular distribution of poliovirus proteins and the induction of virus-specific cytoplasmic structures. Virology 131, 39–48.

Birtley, J.R., Knox, S.R., Jaulent, A.M., Brick, P., Leatherbarrow, R.J., Curry, S., 2005. Crystal structure of foot-and-mouth disease virus 3C protease. New insights into catalytic mechanism and cleavage specificity. J. Biol. Chem. 280, 11520–11527.

Binford, S.L., Maldonado, F., Brothers, M.A., Weady, P.T., Zalman, L.S., Meador, III, J.W., Matthews, D.A., Patick, A.K., 2005. Conservation of amino acids in human rhinovirus 3C protease correlates with broad-spectrum antiviral activity of Rupintrivir, a novel human rhinovirus 3C protease inhibitor. Antimicrob. Agents Chemother. 49, 619–626.

Blair, W.S., Parsley, T.B., Bogerd, H.P., Towner, J.S., Semler, B.L., Cullen, B.R., 1998. Utilization of a mammalian cell-based RNA binding assay to characterize the RNA binding properties of picornavirus 3C proteinases. RNA 4, 215–225.

Blinkova, O., Kapoor, A., Victoria, J., Jones, M., Wolfe, N., Naeem, A., Shaukat, S., Sharif, S., Alam, M.M., Angez, M., Zaidi, S., Delwart, E.L., 2009. Cardioviruses are genetically diverse and cause common enteric infections in South Asian children. J. Virol. 83, 4631–4641.

Brown, B.A., Maher, K., Flemister, M.R., Naraghi-Arani, P., Uddin, M., Oberste, M.S., Pallansch, M.A., 2009. Resolving ambiguities in genetic typing of human enterovirus species C clinical isolates and identification of *Enterovirus* 96, 99 and 102. J. Gen. Virol. 90, 1713–1723.

Brundage, S.C., Fitzpatrick, A.N., 2006. Hepatitis A. Am. Fam. Physician 73, 2162–2168.

Buontempo, P.J., Cox, S., Wright-Minogue, J., DeMartino, J.L., Skelton, A.M., Ferrari, E., Albin, R., Rozhon, E.J., Girijavallabhan, V., Modlin, J.F., O'Connell, J.F., 1997. SCH 48973: a potent, broad-spectrum, antienterovirus compound. Antimicrob. Agents Chemother 41, 1220–1225.

Buontempo, P., Wright-Minogue, J., Cox, S., DeMartino, J., Girijavallablan, V., Rozhon, E., Schwartz, J., O'Connell, J., 1995. Mechanism of action of SCH 47802: an antipicornavirus molecule. Antivir. Res. 26 (Suppl. 3), A346.

Caly, L., Ghildyal, R., Jans, D.A., 2015. Respiratory virus modulation of host nucleocytoplasmic transport, target for therapeutic intervention? Front. Microbiol. 6, 848.

Campagnola, G., Weygandt, M., Scoggin, K., Peersen, O., 2008. Crystal structures of coxsackievirus B3 3Dpol highlights the functional importance of residue 5 in picornavirus polymerases. J. Virol. 82 (19), 9458–9464.

Castelló, A., Alvarez, E., Carrasco, L., 2011. The multifaceted poliovirus 2A protease: Regulation of gene expression by picornavirus proteases. J. Biomed. Biotech. 2011, 369648.

Cathcart, A.L., Baggs, E.L., Semler, B.L., 2015. Picornaviruses: pathogenesis and molecular biology. In: Reference Module in Biomedical Research, third ed. Elsevier.

Chan, Y.M., Boehr, D.D., 2015. Allosteric functional switch in poliovirus 3C protease. Biophysical J. 108 (Suppl. 1), 528a.

Chandra, R.K., 1984. Excessive intake of zinc impairs immune responses. J. Am. Medi. Assoc. 252, 1143–1145.

Chapman, N.M., Kim, K.S., 2008. Persistent coxsackievirus infection, *Enterovirus* persistence in chronic myocarditis and dilated cardiomyopathy. Curr. Top. Microbiol. Immunol. 323, 275–292.

Cheah, K.-C., Leong, L.E.-C., Porter, A.G., 1990. Site-directed mutagenesis suggests close functional relationship between human rhinovirus 3C cysteine protease and cellular trypsin-like serine proteases. J. Biol. Chem. 265, 7180–7187.

Chen, T.-C., Chang, H.-Y., Lin, P.-F., Chern, J.-H., Hsu, J.T.-A., Chang, C.-Y., Shih, S.R., 2009. Novel antiviral agent DTriP-22 targets RNA-dependent RNA polymerase of *Enterovirus* 71. Antimicrob. Agents Chemother. 53, 2740–2747.

Chen, Z.G., Stauffacher, C., Li, Y., Schmidt, T., Bomu, W., Kamer, G., Shanks, M., Lomonossoff, G., Johnson, J.E., 1989. Protein-RNA interactions in an icosahedral virus at 3.0 Å resolution. Science 245, 154–159.

Clark, M.E., Hammerle, T., Wimmer, E., Dasgupta, A., 1991. Poliovirus proteinase 3C converts an active form of transcription factor IIIC to an inactive form: a mechanism for inhibition of host cell polymerase III transcription by poliovirus. EMBO J. 10, 2941–2947.

Clarke, B.E., Sangar, D.V., 1988. Processing and assembly of foot-and-mouth disease virus proteins using subgenomic RNA. J. Gen. Virol. 69, 2313–2325.

Clarke, B.E., Sangar, D.V., Burroughs, J.N., Newton, S.E., Carroll, A.R., Rowlands, D.J., 1985. Two initiation sites for foot-and-mouth disease virus polyprotein in vivo. J. Gen. Virol. 66, 2615–2626.

Colonno, R.J., Condra, J.H., Mizutani, S., Callahan, P.L., Davies, M.E., Murcko, M.A., 1988. Evidence for the direct involvement of the rhinovirus canyon in receptor binding. Proc. Natl. Acad. Sci. USA 85, 5449–5453.

Cox, S., Buontempo, P.J., Wright-Minogue, J., DeMartino, J.L., Skelton, A.M., Ferrari, E., Schwartz, J., Rozhon, E.J., Linn, C.C., Girijavallabhan, V., O'Connell, J.F., 1996. Antipicornavirus activity of SCH 47802 and analogs: In vitro and in vivo studies. Antiviral Res. 32, 71–79.

Cox, S., Buontempo, J., Wright-Minogue, J., Skelton, A., DeMartino, J., Roshon, E., Linn, C., Girijavallablan, V., Schwartz, J., O'Connell, J., 1995. Antipicornavirus activity of SCH 47802: in vitro and in vivo studies. Antivir. Res. 26 (Suppl. 3), A345.

Coyne, C.B., Bergelson, J.M., 2005. CAR: a virus receptor within the tight junction. Adv. Drug Deliv. Rev. 57, 869–882.

Coyne, C.B., Bergelson, J.M., 2006. Virus-induced Abl and Fyn kinase signals permit coxsackievirus entry through epithelial tight junctions. Cell 124, 119–131.

Crotty, S., Cameron, C.E., Andino, R., 2001. RNA virus error catastrophe: direct molecular test by using ribavirin. Proc. Natl. Acad. Sci. USA 98, 6895–6900.

Crotty, S., Maag, D., Arnold, J.J., Zhong, W., Lau, J.Y., Hong, Z., Andino, R., Cameron, C.E., 2000. The broad-spectrum antiviral ribonucleoside ribavirin is an RNA virus mutagen. Nat. Med. 6, 1375–1379.

Crump, C.E., Arruda, E., Hayden, F.G., 1993. In vitro inhibitory activity of soluble ICAM-1 for the numbered sero- types of human rhinovirus. Antivir. Chem. Chemother. 4, 323–327.

Cui, S., Wang, J., Fan, T., Qin, B., Guo, L., Lei, X., Wang, J., Wang, M., Jin, Q., 2011. Crystal structure of human enterovirus 71 3C protease. J. Mol. Biol. 408, 449–461.

Curry, S., Fry, E., Blakemore, W., Abu-Ghazaleh, R., Jackson, T., King, A., Lea, S., Newman, J., Stuart, D., 1997. Dissecting the roles of VP0 cleavage and RNA packaging in picornavirus capsid stabilization: the structure of empty capsids of foot-and-mouth disease virus. J. Virol. 71, 9743–9752.

Curry, S., Roqué-Rosell, N., Zunszain, P.A., Leatherbarrow, R.J., 2007. Foot-and-mouth disease virus 3C protease: Recent structural and functional insights into an antiviral target. Int. J. Biochem. Cell Biol. 39, 1–6.

Das, S., Dasgupta, A., 1993. Identification of the cleavage site and determinants required for poliovirus 3C$^{pro}$-catalysed cleavage of human TATA-binding protein transcription factor TBP. J. Virol. 67, 3326–3331.

De Almeida, M.B., Zerbinati, R.M., Tateno, A.F., Oliveira, C.M., Romão, R.M., Rodrigues, J.C., Pannuti, C.S., da Silva Filho, L.V., 2010. Rhinovirus C and respiratory exacerbations in children with cystic fibrosis. Emerg. Infect. Dis. 16, 996–999.

De Breyne, S., Bonderoff, J.M., Chumakov, K.M., Lloyd, R.E., Hellen, C.U., 2008. Cleavage of eukaryotic initiation factor eIF5B by *Enterovirus* 3C proteases. Virology 378, 118–122.

De Palma, A.M., Heggermont, W., Lanke, K., Coutard, B., Bergmann, M., Monforte, A.M., Canard, B., De Clercq, E., Chimirri, A., Pürstinger, G., Rohayem, J., van Kuppeveld, F., Neyts, J., 2008a. The thiazolobenzimidazole TBZE-029 inhibits *Enterovirus* replication by targeting a short region immediately downstream from motif C in the nonstructural protein 2C. J. Virol. 82, 4720–4730.

De Palma, A.M., Pürstinger, G., Wimmer, E., Patick, A.K., Andries, K., Rombaut, B., De Clercq, E., Neyts, J., 2008b. Potential use of antiviral agents in polio eradication. Emerg. Infect. Dis. 14, 545–551.

De Palma, A.M., Thibaut, H., Lanke, K., Heggermont, W., Ireland, S., Andrews, R., Arimilli, M., Altel, T., De Clercq, E., van Kuppeveld, F., Neyts, J., 2009. Mutations in the non-structural protein 3A confer resistance to the novel *Enterovirus* inhibitor TTP-8307. Antimicrob. Agents Chemother. 53, 1850–1857.

De Palma, A.M., Vliegen, I., De Clercq, E., Neyts, J., 2008c. Selective inhibitors of picornavirus replication. Med. Res. Rev. 28, 823–884.

DeLong, D.C., Nelson, J.D., Wu, C.Y.E., Warren, B., Wikel, J., Chamberlin, J., Montgomery, D., Paget, C.J., 1980. Virus inhibition studies with AR-336. 1. Tissue culture activity. In abstracts of the 78th Annual Meeting of the American Society for Microbiology, 1978, Washington, DC, (Abstract S-128, p. 234).

Diana, G.D., Volkots, D.L., Nitz, T., Bailey, T.R., Long, M.A., Vescio, N., Aldous, S., Pevear, D.C., Dutko, F., 1994. Oxadiazoles as ester bioisosteric replacements in compounds related to disoxaril. Antirhinovirus activity. J. Med. Chem. 37, 2421–2436.

Diana, G.D., Rudewicz, P., Pevear, D.C., Nitz, T.J., Aldous, S.C., Aldous, D.J., Robinson, D.T., Draper, T., Dutko, F.T., Aldi, C., Gendron, G., Oglesby, R.C., Volkots, D.L., Reuman, M., Bailey, T.R., Czerniak, R., Block, T., Roland, R., Oppermann, J., 1995. Picornavirus inhibitors: trifluoromethyl substitution provides a global protective effect against hepatic metabolism. J. Med. Chem. 38, 1355–1371.

Dougherty, R.H., Fahy, J.V., 2009. Acute exacerbations of asthma, epidemiology, biology and the exacerbation-prone phenotype. Clin. Exp. Allergy 39, 193–202.

Douglas, R.M., Miles, H.B., Moore, B.W., Ryan, P., Pinnock, C.B., 1987. Failure of effervescent zinc acetate lozenges to alter the course of upper respiratory tract infections in Australian adults. Antimicrob. Agents Chemother. 31, 1263–1265.

Dragovich, P.S., Prins, T.J., Zhou, R., Johnson, T.O., Hua, Y., Luu, H.T., Sakata, S.K., Brown, E.L., Maldonado, F.C., Tuntland, T., Lee, C.A., Fuhrman, S.A., Zalman, L.S., Patick, A.K., Matthews, D.A., Wu, E.Y., Guo, M., Borer, B.C., Nayyar, N.K., Moran, T., Chen, L., Reijto, P.A., Rose, P.W., Guzman, M.C., Dovalsantos, E.Z., Lee, S., McGee, K., Mohajeri, M., Liese, A., Tao, J., Kosa, M.B., Liu, B., Batugo, M.R., Gleeson, J.P., Wu, Z.P., Liu, J., Meador, 3rd, J.W., Ferre, R.A., 2003. Structure-based design, synthesis, and biological evaluation of irreversible human rhinovirus 3C protease inhibitors. 8. Pharmacological optimization of orally bioavailable 2-pyridone-containing peptidomimetics. J. Med. Chem. 46, 4572–4585.

Duque, H., LaRocco, M., Golde, W.T., Baxt, B., 2004. Interactions of foot-and-mouth disease virus with soluble bovine alphaVbeta3 and alphaVbeta6 integrins. J. Virol. 78, 9773–9781.

Dussart, P., Cartet, G., Huguet, P., Lévêque, N., Hajjar, C., Morvan, J., Vanderkerckhove, J., Ferret, K., Lina, B., Chomel, J.J., Norder, H., 2005. Outbreak of acute hemorrhagic conjunctivitis in French Guiana and West Indies caused by coxsackievirus A24variant: phylogenetic analysis reveals Asian import. J. Med. Virol. 75, 559–565.

Dustin, M.L., Springer, T.A., 1988. Lymphocyte function associated antigen-1 (LFA- 1) interaction with intercellular adhesion molecule-1 (rCAM-1) is one of at least three mechanisms for lymphocyte adhesion to cultured endothelial cells. J. Cell Biol. 107, 321–331.

Eby, G.A., Davis, D.R., Halcomb, W.W., 1984. Reduction in duration in common colds by zinc gluconate lozenges in a double blind study. Antimicrob. Agents Chemother. 25, 20–24.

Edwards, M.P., Price, D.A., 2010. Role of physicochemical properties and ligand lipophilicity efficiency in addressing drug safety risk. Annu. Rep. Med. Chem. 45, 381–391.

Ehrenfeld, E., Domingo, E., Roos, R.P. (Eds.), 2010. The Picornaviruses. ASM Press, Washington, p. 493.

Etchison, D., Milburn, S.C., Edery, I., Sonenberg, N., Hershey, J.W.B., 1982. Inhibition of HeLa cell protein synthesis following poliovirus infection correlates with the proteolysis of a 200,000-dalton polypeptide associated with eukaryotic initiation factor 3 and cap binding protein complex. J. Biol. Chem. 257, 14806–14810.

Falk, M.M., Grigera, P.R., Bergmann, I.E., Zibert, A., Multhaup, G., Beck, E., 1990. Foot-and-mouth disease virus protease 3C induces specific proteolytic cleavage of host cell histone H3. J. Virol. 64, 748–756.

Farr, B.M., Conner, E.M., Betts, R.F., Oleske, J., Minnifer, A., Gwaltney, Jr., J.M., 1987. Two randomized controlled trials of zinc gluconate lozenge therapy of experimentally induced rhinovirus colds. Antimicrob. Agents Chemother. 8, 1183–1187.

Farr, B.M., Gwaltney, Jr., J.M., 1987. The problems of taste in placebo matching an evaluation of zinc gluconate for the common cold. J. Chronic Dis. 40, 875–879.

Ferrer-Orta, C., Arias, A., Perez-Luque, R., Escarmís, C., Domingo, E., Verdaguer, N., 2004. Structure of foot-and-mouth disease virus RNA-dependent RNA polymerase and its complex with a template-primer RNA. J. Biol. Chem. 279, 47212–47221.

Flather, D., Semler, B.L., 2015. Picornaviruses and nuclear functions: targeting a cellular compartment distinct from the replication site of a positive-strand RNA virus. Front. Microbiol. 6, 594.

Franco, D., Pathak, H.B., Cameron, C.E., Rombaut, B., Wimmer, E., Paul, A.V., 2005. Stimulation of poliovirus synthesis in a HeLa cell-free in vitro translation-RNA replication system by viral protein 3CD$^{pro}$. J. Virol. 79, 6358–6367.

Freimuth, P., Philipson, L., Carson, S.D., 2008. The coxsackievirus and adenovirus receptor. Curr. Top. Microbiol. Immunol. 323, 67–87.

Gazina, E.V., Smidansky, E.D., Holien, J.K., Harrison, D.N., Cromer, B.A., Arnold, J.J., Parker, W.W., Cameron, C.E., Petrou, S., 2011. Amiloride is a competitive inhibitor of coxsackievirus B3 RNA polymerase. J. Virol. 85, 10364–10374.

Geist, F.C., Bateman, J.A., Hayden, F.G., 1987. In vitro activity of zinc salts against human rhinovirus. Antimicrob. Agents Chemother 31, 622–624.

Geller, R., Vignuzzi, M., Andino, R., Frydman, J., 2007. Evolutionary constraints on chaperone-mediated folding provide an antiviral approach refractory to development of drug resistance. Genes Dev. 21, 195–205.

Gern, J.E., 2010. The ABCs of rhinoviruses, wheezing, and asthma. J. Virol. 84, 7418–7426.

Gervelmeyer, A., Nielsen, M.S., Frey, L.C., Sckerl, H., Damberg, E., Mølbak, K., 2006. An outbreak of hepatitis A among children and adults in Denmark, August 2002 to February 2003. Epidemiol. Infect. 134, 485–491.

Ghildyal, R., Jordan, B., Li, D., Dagher, H., Bardin, P.G., Gern, J.E., Jans, D.A., 2009. Rhinovirus 3C protease can localize in the nucleus and alter active and passive nucleocytoplasmic transport. J. Virol. 83, 7349–7352.

Gilon, C., Halle, D., Chorev, M., Selinger, Z., Byk, G., 1991. Backbone cyclization: a new method for conferring conformational constraint on peptides. Biopolymers 31, 745–750.

Giranda, V.L., Chapman, M.S., Rossmann, M.G., 1990. Modeling of the human intercellular adhesion molecule-1, the human rhinovirus major group receptor. Proteins 7, 227–233.

Gleeson, M.P., 2008. Generation of a set of simple, interpretable ADMET rules of thumb. J. Med. Chem. 51, 817–834.

Glenn, M.P., Pattenden, L.K., Reid, R.C., Tyssen, D.P., Tyndall, J.D., Birch, C.J., Fairlie, D.P., 2002. Beta-strand mimicking macrocyclic amino acids: template for protease inhibitors with antiviral activity. J. Med. Chem. 45, 371–381.

Gonzalez, R.H., Khademi, M., Andersson, M., Wallstrom, E., Borg, K., Olsson, T., 2002. Prior poliomyelitis-evidence of cytokine production in the central nervous system. J. Neurol. Sci. 205, 9–13.

Gorbalenya, A.E., Koonin, E.V., Lai, M.M.C., 1991. Putative papain-related thiol protease of posi-tive strand RNA viruses: identification of rubi- and *Aphthovirus* proteases and delineation of a novel conserved domain associated with proteases of rubi-, alpha- and coronaviruses. FEBS Lett. 288, 201–205.

Gorbalenya, A.E., Koonin, E.V., Wolf, Y.I., 1990. A new superfamily of putative NTP binding do-mains encoded by genomes of small DNA and RNA viruses. FEBS Lett. 262, 145–148.

Gorbalenya, A.E., Svitkin, Y.V., 1983. Encephalomyocarditis virus protease: purification and role of the SH groups in processing of the precursor of structural proteins. Biochemistry 48, 385–395.

Goyer, M., Aho, L.S., Bour, J.B., Ambert-Balay, K., Pothier, P., 2008. Seroprevalence distribution of *Aichi virus* among a French population in 2006-2007. Arch. Virol. 153, 1171–1174.

Grant, R.A., Hiremath, C.N., Filman, D.J., Syed, R., Andries, K., Hogle, J.M., 1994. Structures of poliovirus complexes with anti-viral drugs: implications for viral stability and drug design. Curr. Biol. 4, 784–797.

Greninger, A.L., Runckel, C., Chiu, C.Y., Haggerty, T., Parsonnet, J., Ganem, D., DeRisi, J.L., 2009. The complete genome of klassevirus—a novel picornavirus in pediatric stool. Virol. J. 6, 82.

Greve, J.M., Davis, G., Meyer, A.M., Forte, C.P., Yost, S.C., Marlor, C.W., Kamarck, M.E., McClel-land, A., 1989. The major human rhinovirus receptor is ICAM-1. Cell 56, 839–847.

Greve, J.M., Fortes, C.P., Marlor, C.W., Meyer, A.M., Hoover-Litty, H., Wunderlich, D., McClel-land, A., 1991. Mechanism of receptor mediated rhinovirus neutralization defined by two solu-ble forms of ICAM-1. J. Virol. 65, 6015–6023.

Gromeier, M., Solecki, D., Patel, D.D., Wimmer, E., 2000. Expression of the human poliovirus receptor/CD155 gene during development of the central nervous system: implications for the pathogenesis of poliomyelitis. Virology 273, 248–257.

Grubman, M.J., Zellner, M., Bablanian, G., Mason, P.W., Piccone, M.E., 1995. Identification of the active-site residues of the 3C proteinase of foot-and-mouth disease virus. Virology 213, 581–589.

Guarné, A., Hampoelz, B., Glaser, W., Carpena, X., Tormo, J., Fita, I., Skern, T., 2000. Structural and biochemical features distinguish the foot-and-mouth disease virus leader proteinase from other papain-like enzymes. J. Mol. Biol. 302, 1227–1240.

Guinea, R., Carrasco, L., 1990. Phospholipid biosynthesis and poliovirus genome replication, two coupled phenomena. EMBO J. 9, 2011–2016.

Guttman, N., Baltimore, D., 1977. Morphogenesis of poliovirus. IV. existence of particles sediment-ing at 150S and having the properties of provirion. J. Virol. 23, 363–367.

Guy, M., Chilmonczyk, S., Cruciere, C., Eloit, M., Bakkali-Kassimi, L., 2009. Efficient infection of buffalo rat liver-resistant cells by encephalomyocarditis virus requires binding to cell surface sialic acids. J. Gen. Virol. 90, 187–196.

Hadaschik, D., Klein, M., Zimmermann, H., Eggers, H.J., Nelsen-Salz, B., 1999. Dependence of echovirus 9 on the *Enterovirus* RNA replication inhibitor 2-(α-Hydroxybenzyl)-benzimidazole maps to nonstructural protein 2C. J. Virol. 73, 10536–10539.

Hammerle, T., Hellen, C.U.T., Wimmer, E., 1991. Site-directed mutagenesis of the putative catalytic triad of poliovirus 3C proteinase. J. Biol. Chem. 266, 5412–5416.

Hanecak, R., Semler, B.L., Anderson, C.W., Wimmer, E., 1982. Proteolytic processing of poliovirus polypeptides: antibodies to poly polypeptide P3-7c inhibit cleavage at glutamine glycine pairs. Proc. Natl. Acad. Sci. USA 79, 3973–3977.

Hansen, J.L., Long, A.M., Schultz, S.C., 1997. Structure of the RNA-dependent RNA polymerase of poliovirus. Structure 5, 1109–1122.

Harris, K.S., Xiang, W., Alexander, L., Lane, W.S., Paul, A.V., Wimmer, E., 1994. Interaction of poliovirus polypeptide 3CD$^{pro}$ with the 5′ and 3′ termini of the poliovirus genome. Identification of viral and cellular cofactors needed for efficient binding. J. Biol. Chem. 269, 27004–27014.

Hauri, A.M., Fischer, E., Fitzenberger, J., Uphoff, H., Koenig, C., 2006. Active immunization during an outbreak of hepatitis A in a German day-care centre. Vaccine 24, 5684–5689.

Hayden, F.G., Gwaltney, Jr., J.M., 1982. Prophylactic activity of intranasal enviroxime against experimentally induced rhinovirus type 39 infection. Antimicrob. Agents Chemother. 21, 892–897.

Hayden, F.G., Hipskind, G.J., Woerner, D.H., Eisen, G.F., Janssens, M., Janssen, P.A., Andries, K., 1995. Intranasal pirodavir (R77,975) treatment of rhinovirus colds. Antimicrob. Agents Chemother. 39, 290–294.

Hayden, F.G., Turner, R.B., 2012. Rhinovirus genetics and virulence: looking for needles in a haystack. Am. J. Resp. Crit. Care Med. 186, 818–820.

He, Y., Bowman, V.D., Mueller, S., Bator, C.M., Bella, J., Peng, X., Baker, T.S., Wimmer, E., Kuhn, R.J., Rossmann, M.G., 2000. Interaction of the poliovirus receptor with poliovirus. Proc. Natl. Acad. Sci. USA 97, 79–84.

Heinz, B.A., Vance, L.M., 1995. The antiviral compound enviroxime targets the 3A coding region of rhinovirus and poliovirus. J. Virol. 69, 4189–4197.

Heinz, B.A., Vance, L.M., 1996. Sequence determinants of 3A-mediated resistance to enviroxime in rhinoviruses and enteroviruses. J. Virol. 70, 4854–4857.

Hellen, C.U., Sarnow, P., 2001. Internal ribosome entry sites in eukaryotic mRNA molecules. Genes Dev. 15, 1593–1612.

Hendry, E., Hatanaka, H., Fry, E., Smyth, M., Tate, J., Stanway, G., Santti, J., Maaronen, M., Hyypia, T., Stuart, D., 1999. The crystal structure of coxsackievirus A9: new insights into the uncoating mechanisms of enteroviruses. Structure 7, 1527–1538.

Hertzler, S., Luo, M., Lipton, H.L., 2000. Mutation of predicted virion pit residues alters binding of Theiler's murine encephalomyelitis virus to BHK-21 cells. J. Virol. 74, 1994–2004.

Hewat, E.A., Neumann, E., Conway, J.F., Moser, R., Ronacher, B., Marlovits, T.C., Blaas, D., 2000. The cellular receptor to human rhinovirus 2 binds around the 5-fold axis and not in the canyon: a structural view. EMBO J. 19, 6317–6325.

Higgins, P.G., Barrow, G.I., Al-Nakib, W., Tyrrell, D.A., DeLong, D.C., Lenox-Smith, I., 1988. Failure to demonstrate synergy between interferon-alpha and a synthetic antiviral, enviroxime, in rhinovirus infections in volunteers. Antivir. Res. 10, 141–149.

Higgins, P.G., Barrow, G.I., Tyrell, D.A.J., 1990. A study of the efficacy of the bradykinin antagonist, NPC 567, in rhinovirus infections in human volunteers. Antivir. Res. 14, 339–344.

Hindiyeh, M., Li, Q.H., Basavappa, R., Hogle, J.M., Chow, M., 1999. Poliovirus mutants at histidine 195 of VP2 do not cleave VP0 into VP2 and VP4. J. Virol. 73, 9072–9079.

Hoey, E.M., Martin, S.J., 1974. A possible precursor containing RNA of a bovine *Enterovirus*: the provirion 11. J. Gen. Virol. 24, 515–524.

Hofer, F., Gruenberger, M., Kowalski, H., Machat, H., Huettinger, M., Kuechler, E., Blaas, D., 1994. Members of the low density lipoprotein receptor family mediate cell entry of a minor-group common cold virus. Proc. Natl. Acad. Sci. USA 91, 1839–1842.

Holtz, L.R., Finkbeiner, S.R., Kirkwood, C.D., Wang, D., 2008. Identification of a novel picornavirus related to cosaviruses in a child with acute diarrhea. Virol. J. 5, 159.

Hsu, N.-Y., Ilnytska, O., Belov, G., Santiana, M., Chen, Y.-H., Takvorian, P.M., Pau, C., van der Schaar, H., Kaushik-Basu, N., Balla, T., Cameron, C.E., Ehrenfeld, E., van Kuppeveld, F.J., Altan-Bonnet, N., 2010. Viral reorganization of the secretory pathway generates distinct organelles for RNA replication. Cell 141, 799–811.

Huber, S.A., 1994. VCAM-1 is a receptor for encephalomyocarditis virus on murine vascular endothelial cells. J. Virol. 68, 3453–3458.

Huber, S., 2008. Host immune responses to coxsackievirus B3. Curr. Top. Microbiol. Immunol. 323, 199–221.

Hung, H.C., Chen, T.C., Fang, M.Y., Yen, K.J., Shih, S.R., Hsu, J.T.A., Tseng, C.P., 2010. Inhibition of *Enterovirus* 71 replication and the viral 3D polymerase by aurintricarboxylic acid. J. Antimicrob. Chemother. 65, 676–683.

Ishitsuka, H., Ninomiya, Y.T., Ohsawa, C., Fujiu, M., Suhara, Y., 1982a. Direct and specific inactivation of rhinovirus by chalcone Ro 09-0410. Antimicrob. Agents Chemother. 22, 617–621.

Ishitsuka, H., Ohsawa, C., Ohiva, T., Umeda, I., Suhara, Y., 1982b. Antipicornavirus flavone Ro 09-0179. Antimicrob. Agents Chemother. 22, 611–616.

Jackson, T., Clark, S., Berryman, S., Burman, A., Cambier, S., Mu, D., Nishimura, S., King, A.M., 2004. Integrin alphavbeta8 functions as a receptor for foot-and-mouth disease virus: role of the beta-chain cytodomain in integrin-mediated infection. J. Virol. 78, 4533–4540.

Jackson, T., Mould, A.P., Sheppard, D., King, A.M., 2002. Integrin alphavbeta1 is a receptor for foot- and-mouth disease virus. J. Virol. 76, 935–941.

Jackson, T., Sheppard, D., Denyer, M., Blakemore, W., King, A.M., 2000. The epithelial integrin alphavbeta6 is a receptor for foot-and-mouth disease virus. J. Virol. 74, 4949–4956.

Jia, X.Y., Summers, D.F., Ehrenfeld, E., 1993. Primary cleavage of the HAV capsid protein precursor in the middle of the proposed 2A coding region. Virology 193, 515–519.

Joachims, M., Harris, K.S., Etchison, D., 1995. Poliovirus protease 3C mediates cleavage of microtubule-associated protein 4. Virology 211, 451–461.

Kaiser, L., Crump, C.E., Hayden, F.G., 2000. In vitro activity of pleconaril and AG7088 against selected serotypes and clinical isolates of human rhinoviruses. Antivir. Res. 47, 215–220.

Kapoor, A., Victoria, J., Simmonds, P., Slikas, E., Chieochansin, T., Naeem, A., Shaukat, S., Sharif, S., Alam, M.M., Angez, M., Wang, C., Shafer, R.W., Zaidi, S., Delwart, E., 2008. A highly prevalent and genetically diversified Picornaviridae genus in South Asian children. Proc. Natl. Acad. Sci. USA 105, 20482–20487.

Kean, K.M., Howell, M.T., Grunert, S., Girard, M., Jackson, R.J., 1993. Substitution mutations at the putative catalytic triad of the poliovirus 3C protease have differential effects on cleavage at different sites. Virology 194, 360–364.

Kean, K.M., Tetrina, N.L., Marc, D., Girad, M., 1991. Analysis of putative active site residues of the poliovirus 3C protease. Virology 163, 330–340.

Kim, S., Smith, T.J., Chapman, M.S., Rossmann, M.G., Pevear, D.C., Dutko, F.J., Feck, P.T., Diana, G.D., McKinlay, M.A., 1989. Crystal structure of human rhinovirus serotype 1A (HRV 1A). J. Mol. Biol. 210, 91–111.

Kim, S., Willingmann, P., Gong, Z.X., Kremer, V.I.J., Chapman, M.S., Minor, I., Oliviera, M.A., Rossmann, M.G., Andries, K., Diana, G.D., Dutko, F.J., McKinlay, M.A., Pevear, D.C., 1993. A comparison of the antirhinoviral drug binding pocket in HRV 14 and HRV 1A. J. Mol. Biol. 230, 206–227.

Knowles, N.J., Hovi, T., King, A.M.Q., Stanway, G., 2010. Overview of Taxonomy. In: Ehrenfeld, E., Domingo, E., Roos, R.P. (Eds.), The Picornaviruses. ASM Press, Washington, pp. 19–32.

Kobayashi, M., Arias, C., Garabedian, A., Palmenberg, A.C., Mohr, I., 2012. Site-specific cleavage of the host poly(A) binding protein by the encephalomyocarditis virus 3C proteinase stimulates viral replication. J. Virol. 86, 10686–10694.

Kolatkar, P.R., Bella, J., Olson, N.H., Bator, C.M., Baker, T.S., Rossmann, M.G., 1999. Structural studies of two rhinovirus serotypes complexed with fragments of their cellular receptor. EMBO J. 18, 6249–6259.

Konig, H., Rosenwirth, B., 1988. Purification and partial characterizationof poliovirus protease 2A by means of a functional assay. J. Virol. 62, 1243–1250.

Korant, B.D., 1972. Cleavage of viral precursor proteins in vivo and in vitro. J. Virol. 10, 751–759.

Korant, B.D., 1973. Cleavage of poliovirus-specific polypeptide aggregates. J. Virol. 12, 556–563.

Korant, B.D., Butterworth, B.E., 1976. Inhibition by zinc of rhinovirus protein cleavage: interaction of zinc with capsid polypeptides. J. Virol. 18, 298–306.

Korant, B.D., Brzin, J., Turk, V., 1985. Cystatin, a protein inhibitor of cysteine proteases alters viral protein cleavages in infected human cells. Biochem. Biophys. Res. Commun. 127, 1072–1076.

Kundu, P., Raychaudhuri, S., Tsai, W., Dasgupta, A., 2005. Shutoff of RNA polymerase II transcription by poliovirus involves 3C protease-mediated cleavage of the TATA-binding protein at an alternative site: Incomplete shutoff of transcription interferes with efficient viral replication. J. Virol. 79, 9702–9713.

Kusov, Y.Y., Morace, G., Probst, C., Gauss-Müller, V., 1997. Interaction of hepatitis A virus (HAV) precursor proteins 3AB and 3ABC with the 51 and 31 termini of the HAV RNA. Virus Res. 51, 151–157.

Kuyumcu-Martinez, N.M., Van Eden, M.E., Younan, P., Lloyd, R.E., 2004. Cleavage of poly (A)-binding protein by poliovirus 3C protease inhibits host cell translation: A novel mechanism for host translation shutoff. Mol. Cell Biol. 24, 1779–1790.

Lacal, J.C., Carrasco, L., 1982. Relationship between membrane integrity and the inhibition of host translation in virus-infected mammalian cells. Comparative studies between encephalomyocarditis virus and poliovirus. Eur. J. Biochem. 127, 359–366.

Lamarche, M.J., Borawski, J., Bose, A., Capacci-Daniel, C., Colvin, R., Dennehy, M., Ding, J., Dobler, M., Drumm, J., Gaither, L.A., Gao, J., Jiang, X., Lin, K., McKeever, U., Puyang, X., Raman, P., Thohan, S., Tommasi, R., Wagner, K., Xiong, X., Zabawa, T., Zhu, S., Wiedmann, B., 2012. Anti-hepatitis C virus activity and toxicity of type III phosphatidylinositol-4-kinase beta inhibitors. Antimicrob. Agents Chemother. 56, 5149–5156.

Lawrence, P., Schafer, E.A., Rieder, E., 2012. The nuclear protein Sam68 is cleaved by the FMDV 3C protease redistributing Sam68 to the cytoplasm during FMDV infection of host cells. Virology 425, 40–52.

Lawson, M.A., Semler, B.L., 1990. Picornavirus protein processing - enzymes, substrates, and genetic regulation. Curr. Top. Microbiol. Immunol. 161, 49–87.

Le Guyader, F.S., Le Saux, J.C., Ambert-Balay, K., Krol, J., Serais, O., Parnaudeau, S., Giraudon, H., Delmas, G., Pommepuy, M., Pothier, P., Atmar, R.L., 2008. Aichi virus, norovirus, astrovirus, *Enterovirus*, and rotavirus involved in clinical cases from a French oyster-related gastroenteritis outbreak. J. Clin. Microbiol. 46, 4011–4017.

Lea, S., 2002. Interactions of CD55 with non-complement ligands. Biochem. Soc. Trans. 30, 1014–1019.

Ledford, R.M., Patel, N.R., Demenczuk, T.M., Watanyar, A., Herbertz, T., Collett, M.S., Pevear, D.C., 2004. VP1 sequencing of all human rhinovirus serotypes: Insights into genus phylogeny and susceptibility to antiviral capsid-binding compounds. J. Virol. 78, 3663–3674.

Lee, C.S., Lee, J.H., Kwon, K.S., 2008. Outbreak of hepatitis A in Korean military personnel. Jap. J. Infect. Dis. 61, 239–241.

Lee, M.S., Lin, T.Y., Chiang, P.S., Li, W.C., Luo, S.T., Tsao, K.C., Liou, G.Y., Huang, M.L., Hsia, S.H., Huang, Y.C., Chang, S.C., 2010. An investigation of epidemic *Enterovirus* 71 infection in Taiwan, 2008: clinical, virologic, and serologic features. Pediatr. Infect. Dis. J. 29, 1030–1034.

Lee, W.M., Monroe, S.S., Rueckert, R.R., 1993. Role of maturation cleavage in infectivity of picornaviruses: activation of an infectosome. J. Virol. 67, 2110–2122.

Lee, V.S., Nimmanpipug, P., Aruksakunwong, O., Promsri, S., Sompornpisut, P., Hannongbua, S., 2007. Structural analysis of lead fullerene-based inhibitor bound to human immunodeficiency virus type 1 protease in solution from molecular dynamics simulations. J. Mol. Graph. Model. 26, 558–570.

Lee, C.K., Wimmer, E., 1988. Proteolytic processing of poliovirus polyprotein: elimination of 2A$^{pro}$-mediated, alternative cleavage of polypeptide 3CD by in vitro mutagenesis. Virology 166, 405–414.

Lei, X., Han, N., Xiao, X., Jin, Q., He, B., Wang, J., 2014. *Enterovirus* 71 3C inhibits cytokine expression through cleavage of the TAK1/TAB1/TAB2/TAB3 complex. J. Virol. 88, 9830–9841.

Lei, X., Sun, Z., Liu, X., Jin, Q., He, B., Wang, J., 2011. Cleavage of the adaptor protein TRIF by *Enterovirus* 71 3C inhibits antiviral responses mediated by toll-like receptor 3. J. Virol. 85, 8811–8818.

Leong, L.E.C., Walker, P.A., Porter, A.G., 1993. Human rhinovirus-14 protease 3C (3C$^{pro}$) binds specifically to the 5′-noncoding region of the viral RNA. J. Biol. Chem. 268, 25735–25739.

Levandowski, R.A., Pachucki, C.T., Rubenis, M., Jackson, G.G., 1982. Topical enviroxime against rhinovirus infection. Antimicrob. Agents Chemother. 22, 1004–1007.

Li, J.P., Baltimore, D., 1988. Isolation of poliovirus 2C mutants defective in viral RNA synthesis. J. Virol. 62, 4016–4021.

Li, J.P., Baltimore, D., 1990. An intragenic revertant of a poliovirus 2C mutant has an uncoating defect. J. Virol. 64, 1102–1107.

Li, X., Lu, H.H., Mueller, S., Wimmer, E., 2001. The C-terminal residues of poliovirus proteinase 2A(pro) are critical for viral RNA replication but not for cis- or trans-proteolytic cleavage. J. Gen. Virol. 82 (Pt 2), 397–408.

Linden, L., Vives-Adrián, L., Selisko, B., Ferrer-Orta, C., Liu, X., Lanke, K., Ulferts, R., De Palma, A.M., Tanchis, F., Goris, N., Lefebvre, D., De Clercq, K., Leyssen, P., Lacroix, C., Pürstinger, G., Coutard, B., Canard, B., Boehr, D.D., Arnold, J.J., Cameron, C.E., Verdaguer, N., Neyts, J., van Kuppeveld, F.J.M., 2015. The RNA Template Channel of the RNA-Dependent RNA Polymerase as a Target for Development of Antiviral Therapy of Multiple Genera within a Virus Family. PLoS Pathog 11, e1004733.

Lipinski, C.A., 2000. Drug-like properties and the causes of poor solubility and poor permeability. J. Pharmacol. Toxicol. Methods 44, 235–249.

Lipton, H.L., Kumar, A.S., Hertzler, S., Reddi, H.V., 2006. Differential usage of carbohydrate co-receptors influences cellular tropism of Theiler's murine encephalomyelitis virus infection of the central nervous system. Glycoconj. J. 23, 39–49.

Logan, D., Abu-Ghazaleh, R., Blakemore, W., Curry, S., Jackson, T., King, A., Lea, S., Lewis, R., Newman, J., Parry, N., Rowlands, D., Stuart, D., Fry, E., 1993. Structure of a major immunogenic site on foot-and-mouth disease virus. Nature 362, 566–568.

Love, R.A., Maegley, K.A., Yu, X., Ferre, R.A., Lingardo, L.K., Diehl, W., Parge, H.E., Dragovich, P.S., Fuhrman, S.A., 2004. The crystal structure of the RNA-dependent RNA polymerase from human rhinovirus: a dual function target for common cold antiviral therapy. Structure 12, 1533–1544.

Luo, M., Vriend, G., Kamer, G., Minor, I., Arnold, E., Rossmann, M.G., Boege, U., Scraba, D.G., Duke, G.M., Palmenberg, A.C., 1987. The atomic structure of Mengo virus at 3.0 Å resolution. Science 235, 182–191.

Ma, H.-C., Liu, Y., Wang, C., Strauss, M., Rehage, N., Chen, Y.-H., Altan-Bonnet, N., Hogle, J., Wimmer, E., Mueller, S., Paul, A.V., Jiang, P., 2014. An interaction between glutathione and the capsid is required for the morphogenesis of C-cluster enteroviruses. PLoS Pathog. 10, e1004052.

Madala, P.K., Tyndall, J.D., Nall, T., Fairlie, D.P., 2010. Update 1 of: Proteases universally recognize beta strands in their active sites. Chem. Rev. 110, PR1–PR31.

Mallamo, J.P., Diana, G.D., Pevear, D.C., Dutko, F.J., Chapman, M.S., Kim, K.H., Minor, I., Oliviera, M., Rossmann, M.G., 1992. Conformationally restricted analogues of disoxaril: a comparison of the activities against human rhinovirus types 14 and 1A. J. Med. Chem. 35, 4690–4695.

Mandadapu, S.R., Weerawama, P.M., Prior, A.M., Uy, R.A.Z., Aravapalli, S., Alliston, K.R., Lushington, G.H., Kim, Y., Hua, D.H., Chang, K.O., Groutas, W.C., 2013. Macrocyclic inhibitors of 3C and 3C-like proteases of picornavirus, norovirus, and coronairus. Bioorg. Med. Chem. Lett. 23, 3709–3712.

Marlovits, T.C., Abrahamsberg, C., Blaas, D., 1998a. Soluble LDL minireceptors. Minimal structure requirements for recognition of minor group human rhinovirus. J. Biol. Chem 273, 33835–33840.

Marlovits, T.C., Abrahamsberg, C., Blaas, D., 1998b. Very-low-density lipoprotein receptor fragment shed from HeLa cells inhibits human rhinovirus infection. J. Virol. 72, 10246–10250.

Marlovits, T.C., Zechmeister, T., Schwihla, H., Ronacher, B., Blaas, D., 1998c. Recombinant soluble low-density lipoprotein receptor fragment inhibits common cold infection. J. Mol. Recognit. 11, 49–51.

Marsault, E., Peterson, M.L., 2011. Macrocycles are great cycles: Applications, opportunities, and challenges of synthetic macrocycles in drug discovery. J. Med. Chem. 54, 1961–2004.

Martin, S.D., Staunton, D.E., Springer, T.A., Stratowa, C., Sommergruber, W., Merluzzi, V.I., 1990. A soluble form of intercellular adhesion molecule-1 inhibits rhinovirus infection. Nature 344, 70–72.

Matthews, D.A., Dragovich, P.S., Webber, S.E., Fuhrman, S.A., Patick, A.K., Zalman, L.S., Hendrickson, T.F., Love, R.A., Prins, T.J., Marakovits, J.T., Zhou, R., Tikhe, J., Ford, C.E., Meador, J.W., Ferre, R.A., Brown, E.L., Binford, S.L., Brothers, M.A., DeLisle, D.M., Worland, S.T., 1999. Structure-assisted design of mechanism-based irreversible inhibitors of human rhinovirus 3C protease with potent antiviral activity against multiple rhinovirus serotypes. Proc. Natl. Acad. Sci. USA 96, 11000–11007.

Matthews, D.A., Smith, W.W., Ferre, R.A., Condon, B., Budahazi, G., Sisson, W., Villafranca, J.E., Janson, C.A., McElroy, H.E., Gribskov, C.L., Worland, S., 1994. Structure of human rhinovirus 3C protease reveals a trypsin-like polypeptide fold, RNA-binding site, and means for cleaving precursor polyprotein. Cell 77, 761–771.

McCreary, R.P., Fairlie, D.P., 1998. Macrocyclic peptidomimetics: potential for drug development. Curr. Opin. Drug Disc. Deliv. 1, 208–217.

McErlean, P., Shackelton, L.A., Andrews, E., Webster, D.R., Lambert, S.B., Nissen, M.D., Sloots, T.P., Mackay, I.M., 2008. Distinguishing molecular features and clinical characteristics of a putative new rhinovirus species, human rhinovirus C (HRV C). PLoS One 3, e1847.

McIntire, J.J., Umetsu, D.T., DeKruyff, R.H., 2004. TIM-1, a novel allergy and asthma susceptibility gene. Springer Semin. Immunopathol. 25, 335–348.

McLean, C., Matthews, T.J., Rueckert, R.R., 1976. Evidence of ambiguous processing and selective degradation in the noncapsid proteins of rhinovirus 1a. J. Virol. 19, 903–914.

Meanwell, N.A., 2011. Improving drug candidates by design: a focus on physicochemical properties as a means of improving compound disposition and safety. Chem. Res. Toxicol. 24, 1420–1456.

Melnick, J.L., 1983. Portraits of viruses: the picornaviruses. Intervirology 20, 61–100.

Merck Sharp & Dohme Corp. Effects of Pleconaril Nasal Spray on Common Cold Symptoms and Asthma Exacerbations Following Rhinovirus Exposure. Available online: https://clinicaltrials.gov/ct2/show/NCT00394914

Michael, W.M., Siomi, H., Choi, M., Pinol-Roma, S., Nakielny, S., Liu, Q., Dreyfuss, G., 1995. Signal sequences that target nuclear import and nuclear export of pre-mRNA-binding proteins. Cold Spring Harb. Symp. Quant. Biol. 60, 663–668.

Mikami, T., Satoh, N., Hatayama, I., Nakane, A., 2004. Buthionine sulfoximine inhibits cytopathic effect and apoptosis induced by infection with human echovirus 9. Arch. Virol. 149, 1117–1128.

Miller, P.A., Milstrey, K.P., Trown, P.W., 1968. Specific inhibition of viral ribonucleic acid replication by Gliotoxin. Science 159, 431–432.

Miller, F.D., Monto, A.S., DeLong, D.C., Exelby, A., Bryan, E.R., Srivastava, S., 1985. Controlled trial of enviroxime against natural rhinovirus infections in a community. Antimicrob. Agents Chemother. 27, 102–106.

Mirzayan, C., Wimmer, E., 1994. Biochemical studies on poliovirus polypeptide 2C: Evidence for ATPase activity. Virology 199, 176–187.

Molla, A., Harris, K.S., Paul, A.V., Shin, S.H., Mugavero, J., Wimmer, E., 1994. Stimulation of poliovirus proteinase 3C$^{pro}$-related proteolysis by the genome-linked protein VPg and its precursor 3AB. J. Biol. Chem. 269, 27015–27020.

Monto, A.S., Schwartz, S.A., Albrecht, J.K., 1989. Ineffectiveness of post exposure prophylaxis of rhinovirus infection with low-dose intranasal alpha interferon in families. Antimicrob. Agents Chemother. 33, 387–390.

Mosimann, S.C., Cherney, M.M., Sia, S., Plotch, S., James, M.N., 1997. Refined X-ray crystallographic structure of the poliovirus 3C gene product. J. Mol. Biol. 273, 1032–1047.

Moustafa, I.M., Gohara, D.W., Uchida, A., Yennawar, N., Cameron, C.E., 2015. Conformational ensemble of the poliovirus 3CD precursor observed by md simulations and confirmed by saxs: a strategy to expand the viral proteome? Viruses 7, 5962–5986.

Muckelbauer, J.K., Kremer, M., Minor, I., Diana, G., Dutko, F.J., Groarke, J., Pevear, D.C., Rossmann, M.G., 1995. The structure of coxsackievirus B3 at 3.5 A resolution. Structure 15, 653–667.

Mukherjee, A., Morosky, S.A., Delorme-Axford, E., Dybdahl-Sissoko, N., Oberste, M.S., Wang, T., Coyne, C.B., 2011. The Coxsackievirus B 3C protease cleaves MAVS and TRIF to attenuate host type I interferon and apoptotic signaling. PLoS Pathog. 7, e1001311.

Narkeviciute, I., Vaiciuniene, D., 2004. Outbreak of echovirus 13 infection among Lithuanian children. Clin. Microbiol. Infect. 10, 1023–1025.

Nayak, A., Goodfellow, I.G., Woolaway, K.E., Birtley, J., Curry, S., Belsham, G.J., 2006. Role of RNA structure and RNA binding activity of foot-and-mouth disease virus 3C protein in VPg uridylylation and virus replication. J. Virol. 80, 9865–9875.

Ninomiya, Y., Aoyama, M., Umeda, I., Suhara, Y., Ishitsuka, H., 1985. Comparative studies on the modes of action of the antirhinovirus agent Ro 09-0410, Ro 09-0179, RJ'VII-15,731, 4'6-dichloroflavin and enviroxirne. Antimicrob. Agents Chemother. 27, 595–599.

Ninomiya, Y., Ohasawa, C., Aoyama, M., Umeda, I., Suhara, Y., Ishitsuka, H., 1984. Antivirus agent Ro 09-0410 binds to rhinovirus specifically and stabilizes the virus conformation. Virology 134, 269–276.

Ninorniya, Y., Shimma, N., Ishitsuka, H., 1990. Comparative studies on the antirhinovirus activityand mode of action of the rhinovirus capsid binding agents, chalcone amides. Antivir. Res. 13, 61–64.

Nishimura, Y., Shimojima, M., Tano, Y., Miyamura, T., Wakita, T., Shimizu, H., 2009. Human P-selectin glycoprotein ligand-1 is a functional receptor for *Enterovirus* 71. Nat. Med. 15, 794–797.

Norder, H., De Palma, A.M., Selisko, B., Costenaro, L., Papageorgiou, N., Arnan, C., Coutard, B., Lantez, V., De Lamballerie, X., Baroni, C., Sola, M., Tan, J., Neyts, J., Canard, B., Coll, M., Gorbalenya, A.E., Hilgenfeld, R., 2011. Picornavirus non-structural proteins as targets for new anti-virals with broad activity. Antivir. Res. 89, 204–218.

Novick, S.G., Godfroy, J.C., Godfrey, N.J., Wilder, H.R., 1996. How does zinc modify the common cold? Clinical observation and implications regarding mechanisms of action. Medical Hypothesis 46, 295–302.

O'Donovan, D., Cooke, R.P., Joce, R., Eastbury, A., Waite, J., Stene-Johansen, K., 2001. An outbreak of hepatitis A amongst injecting drug users. Epidemiol. Infect. 127, 469–473.

O'Dwyer, P.J., Hamilton, T.C., LaCreta, F.P., Gallo, J.M., Kilpatrick, D., Halbherr, T., Brennan, J., Bookman, M.A., Hoffman, J., Young, R.C., Comis, R.L., Ozols, R.F., 1996. Phase I trial of buthionine sulfoximine in combination with melphalan in patients with cancer. J. Clin. Oncol. 14, 249–256.

O'Reilly, E.K., Kao, C.C., 1998. Analysis of RNA-dependent RNA polymerase structure and function as guided by known polymerase structures and computer predictions of secondary structure. Virology 252, 287–303.

Oberste, M.S., Moore, D., Anderson, B., Pallansch, M.A., Pevear, D.C., Collett, M.S., 2009. In vitro antiviral activity of V-073 against polioviruses. Antimicrob. Agents Chemother. 53, 4501–4503.

Oh, D.Y., Silva, P.A., Hauroeder, B., Diedrich, S., Cardoso, D.D., Schreier, E., 2006. Molecular characterization of the first Aichi viruses isolated in Europe and in South America. Arch. Virol. 151, 1199–1206.

Ohlin, A., Hoover-Litty, H., Sanderson, G., Paessens, A., Johnston, S.L., Holgate, S.T., Huguenel, E., Greve, J.M., 1994. Spectrum of activity of soluble intercellular adhesion molecule-1 against rhinovirus reference strains and field isolates. Antimicrob. Agents Chemother. 38, 1413–1415.

Oliviera, M.A., Zhao, R., Lee, W.M., Kremer, M.J., Minor, I., Rueckert, R.R., Diana, G.D., Pevear, D.C., Dutko, F.J., McKinlay, M.A., Rossmann, M.G., 1993. The structure of human rhinovirus 16. Structure 1, 51–68.

Olson, N.H., Kolatkar, P.R., Oliviera, M.A., Cheng, R.H., Greve, J.M., McClelland, A., Baker, T.S., Rossmann, M.G., 1993. Structure of a human rhinovirus complexed with its receptor molecule. Proc. Natl. Acad. Sci. USA 90, 507–511.

Palmenberg, A., Neubauer, D., Skern, T., 2010. Genome organization and encoded proteins. In: Ehrenfeld, E., Domingo, E., Roos, R. (Eds.), The Picornaviruses. ASM Press, Washington, DC, USA, pp. 3–17.

Palmenberg, A.C., Spiro, D., Kuzmickas, R., Wang, S., Djikeng, A., Rathe, J.A., Fraser-Liggett, C.M., Liggett, S.B., 2009. Sequencing and analyses of all known human rhinovirus genomes reveal structure and evolution. Science 324, 55–59.

Park, N., Katikaneni, P., Skern, T., Gustin, K.E., 2008. Differential targeting of nuclear pore complex proteins in poliovirus-infected cells. J. Virol. 82, 1647–1655.

Park, N., Skern, T., Gustin, K.E., 2010. Specific cleavage of the nuclear pore complex protein Nup62 by a viral protease. J. Biol. Chem. 285, 28796–28805.

Pathak, H.B., Arnold, J.J., Wiegand, P.N., Hargittai, M.R., Cameron, C.E., 2007. Picornavirus genome replication: assembly and organization of the VPg uridylylation ribonucleoprotein (initiation) complex. J. Biol. Chem. 282, 16202–16213.

Patick, A.K., Brothers, M.A., Maldonado, F., Binford, S., Maldonado, O., Fuhrman, S., Petersen, A., Smith, 3rd, G.J., Zalman, L.S., Burns-Naas, L.A., Tran, J.Q., 2005. In vitro antiviral activity and single-dose pharmacokinetics in humans of a novel, orally bioavailable inhibitor of human rhinovirus 3C protease. Antimicrob. Agents Chemother. 49, 2267–2275.

Paul, A.V., Cao, X., Harris, K.S., Lama, J., Wimmer, E., 1994. Studies with poliovirus polymerase $3D^{pol}$. Stimulation of poly(U) synthesis in vitro by purified poliovirus protein 3AB. J. Biol. Chem. 269, 29173–29181.

Pelham, H.R.B., 1978. Translation of encephalomyocarditis virus RNA in vitro yields an active proteolytic processing enzyme. Eur. J. Chem. 85, 457–462.

Perera, R., Daijogo, S., Walter, B.L., Nguyen, J.H., Semler, B.L., 2007. Cellular protein modification by poliovirus: the two faces of poly (rC)-binding protein. J. Virol. 81, 8919–8932.

Pevear, D.C., Tull, T.M., Seipel, M.E., Groarke, J.M., 1999. Activity of pleconaril against enteroviruses. Antimicrob. Agents Chemother. 43, 2109–2115.

Pfister, T., Jones, K.W., Wimmer, E., 2000. A cysteine-rich motif in poliovirus protein 2C(ATPase) is involved in RNA replication and binds zinc in vitro. J. Virol. 74, 334–343.

Pfister, T., Wimmer, E., 1999. Characterization of the nucleoside triphosphatase activity of poliovirus protein 2C reveals a mechanism by which guanidine inhibits poliovirus replication. J. Biol. Chem. 274, 6992–7001.

Phillpotts, R.J., Higgins, P.G., Willman, J.S., Tyrell, D.J.A., LenoxSmith, I., 1984. Evaluation of antirhinovirus chalcone Ro 09-0415 given orally to volunteers. Antimicrob. Agents Chemother. 14, 403–409.

Phillpotts, R.J., Jones, R.W., Delong, D.C., Reed, S.E., Wallace, J., Tyrrell, D.A., 1981. The activity of enviroxime against rhinovirus infection in man. Lancet 1, 1342–1344.

Phillpotts, R.J., Wallace, J., Tyrell, D.A.J., Tagart, V.B., 1983. Therapeutic activity of enviroxime against rhinovirus infection in volunteers. Antimicrob. Agents Chemother. 23, 671–675.

Piccone, M.E., Zellner, M., Kumosinski, T.F., Mason, P.W., Grubman, M.J., 1995. Identification of the active-site residues of the L proteinase of foot-and-mouth disease virus. J. Virol. 69, 4950–4956.

Pilipenko, E.V., Pestova, T.V., Kolupaeva, V.G., Khitrina, E.V., Poperechnaya, A.N., Agol, V.I., Hellen, C.U., 2000. A cell cycle-dependent protein serves as a template-specific translation initiation factor. Genes Dev. 14, 2028–2045.

Pincus, S.E., Diamond, D.C., Emini, E.A., Wimmer, E., 1986. Guanidine-selected mutants of poliovirus: Mapping of point mutations to polypeptide 2C. J. Virol. 57, 638–646.

Pontrelli, G., Boccia, D., DI Renzi, M., Massari, M., Giugliano, F., Celentano, L.P., Taffon, S., Genovese, D., DI Pasquale, S., Scalise, F., Rapicetta, M., Croci, L., Salmaso, S., 2007. Epidemiological and virological characterization of a large communitywide outbreak of hepatitis A in southern Italy. Epidemiol. Infect. 136, 1027–1034.

Powell, R.M., Ward, T., Goodfellow, I., Almond, J.W., Evans, D.J., 1999. Mapping the binding domains on decay accelerating factor (DAF) for haemagglutinating enteroviruses: implications for the evolution of a DAF-binding phenotype. J. Gen. Virol. 80 (Pt 12), 3145–3152.

Qu, L., 2010. Evasion of RIG-I/MDA5 and TLR3-Mediated Innate Immunity by Hepatitis A Virus. Ph.D. Thesis, University of Texas Medical Branch, Galveston, TX, USA.

Qu, L., Feng, Z., Yamane, D., Liang, Y., Lanford, R.E., Li, K., Lemon, S.M., 2011. Disruption of TLR3 signaling due to cleavage of TRIF by the hepatitis A virus protease-polymerase processing intermediate, 3CD. PLoS Pathog. 7, e1002169.

Racaniello, V.R., 2007. Picornaviridae: the viruses and their replication. In: Knipe, D.M., Howley, P.M., Griffin, D.E., Lamb, R.A., Martin, M.A., Roizman, B., Strauss, S.E. (Eds.), Fields Virology. fifth ed. Lippincott Williams & Wilkins, USA, pp. 795–838.

Richards, O.C., Ehrenfeld, E., 1998. Effects of poliovirus 3AB protein on 3D polymerase-catalyzed reaction. J. Biol. Chem. 273 (21), 12832–12840.

Ricour, C., Delhaye, S., Hato, S.V., Olenyik, T.D., Michel, B., van Kuppeveld, F.J., Gustin, K.E., Michiels, T., 2009. Inhibition of MRNA export and dimerization of interferon regulatory factor 3 by theiler's virus leader protein. J. Gen. Virol. 90, 177–186.

Ritchie, T.J., Ertl, P., Lewis, R., 2011. The graphical representation of ADME-related molecule properties for medicinal chemists. Drug Disc. Today 16, 65–72.

Roberts, P.J., Belsham, G.J., 1995. Identification of critical amino acids within the foot-and-mouth disease virus leader protein, a cysteine protease. Virology 213, 140–146.

Rodriguez, P.L., Carrasco, L., 1992. Gliotoxin: inhibitor of poliovirus RNA synthesis that blocks the viral RNA polymerase 3D$^{pol}$. J. Virol. 66, 1971–1976.

Rodríguez, P.L., Carrasco, L., 1993. Poliovirus protein 2C has ATPase and GTPase activities. J. Biol. Chem. 268, 8105–8110.

Rodriguez-Wells, V., Plotch, S.J., DeStefano, J.J., 2001. Primerdependent synthesis by poliovirus RNA-dependent RNA polymerase (3D$^{pol}$). Nucleic Acids Res. 29, 2715–2724.

Roos, R.P., Kong, W., Semler, B.L., 1989. Polyprotein processing of Theilers murine encephalomyelitis virus. J. Virol. 63, 5344–5353.

Rosenwirth, B., Kis, Z.L., Eggers, H.J., 1995. In vivo efficacy of SDZ 35-682, a new picornavirus capsid-binding agent. Antivir. Res. 26, 55–64.

Rossmann, M.G., Arnold, E., Erickson, J.W., Frankinberger, E.A., Griffith, J.P., Hecht, H.J., Johnson, J.E., Kamer, G., Luo, M., Mosser, A.G., Rueckert, R.R., Sherry, B., Vriend, G., 1985. Structure of a human common cold virus and functional relationship to other picornaviruses. Nature 317, 145–153.

Rossmann, M.G., He, Y., Kuhn, R.J., 2002. Picornavirus-receptor interactions. Trends Microbiol. 10, 324–331.

Rossmann, M.G., Palmenberg, A.C., 1988. Conservation of the putative receptor attachment site in picornaviruses. Virology 164, 373–382.

Rozhon, E., Cox, S., Buontempo, P., O'Connell, J., Slater, W., DeMartino, J., Schwartz, J., Miller, G., Arnold, E., Zhang, A., Viorrow, C., Jablonski, S., Pinto, P., Versace, R., Duelfer, T., Girijavallabhan, V., 1993. SCH 38057: a picornavirus capsid-binding molecule with antiviral activity after the initial stage of viral uncoating. Antivir. Res. 21, 15–35.

Rueckert, R.R., Matthews, T.J., Kew, O.M., Pallansch, M.A., McLean, C., Omilianowski, D.R., 1979. In: Perez-Bercoff, R. (Ed.), The Molecular Biology of Picornaviruses. Plenum Press, New York, pp. 113–125, Chapter 6.

Ryan, M.D., Belsham, G.J., King, A.M.Q., 1989. Specificity of substrate-enzyme interactions in foot-and-mouth disease virus polyprotein processing. Virology 173, 35–45.

Ryan, M.D., Flint, M., 1997. Virus-encoded proteinases of the picornavirus super-group. J. Gen. Virol. 78, 699–723.

Sadeghipour, S., Bek, E.J., McMinn, P.C., 2012. Selection and characterisation of guanidine-resistant mutants of human enterovirus 71. Virus Res. 169, 72–79.

Salas, E.M., Ryan, M.D., 2010. Translation and protein process. In: Ehrenfeld, E., Domingo, E., Roos, R. (Eds.), The Picornaviruses. ASM Press, Washington, DC, USA, pp. 141–150.

Sangar, D.V., Newton, S.E., Rowlands, D.J., Clarke, B.E., 1987. All FMDV serotypes initiate protein synthesis at two separate AUGs. Nucleic Acids Res. 15, 3305–3315.

Sapkal, G.N., Bondre, V.P., Fulmali, P.V., Patil, P., Gopalkrishna, V., Dadhania, V., Ayachit, V.M., Gangale, D., Kushwaha, K.P., Rathi, A.K., Chitambar, S.D., Mishra, A.C., Gore, M.M., 2009. Enteroviruses in patients with acute encephalitis, Uttar Pradesh, India. Emerg. Infect. Dis. 15, 295–298.

Sattar, S.A., Jason, T., Bidawid, S., Farber, J., 2000. Foodborne spread of hepatitis A: recent studies on virus survival, transfer and inactivation. Can. J. Infect. Dis. 11, 159–163.

Schechter, I., Berger, A., 1967. On the size of the active site in proteases I papain. Biochem. Biophys. Res. Commun. 27, 157–162.

Schiff, G.M., McKinlay, M.A., Sherwood, J.R., 1996. Oral efficacy of VP 63843 in coxsackievirus A21 infected volunteers. Interscience Conference on Antimicrobial Agents and Chemotherapy, New Orleans, 16–19 September, 1996, (Abstract H43).

Schiff, G.M., Sherwood, J.R., 2000. Clinical activity of pleconaril in an experimentally induced coxsackievirus A21 respiratory infection. J. Infect. Dis. 181, 20–26.

Schiff, G.M., Sherwood, J.R., Young, E.C., Mason, L.J., Gamble, I.N., 1992. Prophylactic efficacy of WIN 54954 in prevention of experimental human coxsackie A21 infection and illness. Antivir. Res. 17 (Suppl. 1), 92.

Schlegel, A., Giddings, Jr., T.H., Ladinsky, M.S., Kirkegaard, K., 1996. Cellular origin and ultrastructure of membranes induced during poliovirus infection. J. Virol. 70, 6576–6588.

Schultheiss, T., Kusov, Y.Y., Gauss-Muller, V., 1994. Proteinase 3C of hepatitis A virus (HAV) cleaves the HAV polyprotein P2-P3 at all sites including VP1/2A and 2A/2B. Virology 198, 275–281.

Seipelt, J., Guarné, A., Bergmann, E., James, M., Sommergruber, W., Fita, I., Skern, T., 1999. The structures of picornaviral proteinases. Virus Res. 62, 159–168.

Senior, K., 2002. FDA panel rejects common cold treatment. Lancet Infect. Dis. 2, 264.

Shafren, D.R., Dorahy, D.J., Greive, S.J., Burns, G.F., Barry, R.D., 1997a. Mouse cells expressing human intercellular adhesion molecule-1 are susceptible to infection by coxsackievirus A21. J. Virol. 71, 785–789.

Shafren, D.R., Dorahy, D.J., Ingham, R.A., Burns, G.F., Barry, R.D., 1997b. Coxsackievirus A21 binds to decay-accelerating factor but requires intercellular adhesion molecule 1 for cell entry. J. Virol. 71, 4736–4743.

Shafren, D.R., Williams, D.T., Barry, R.D., 1997c. A decay-accelerating factor-binding strain of coxsackievirus B3 requires the coxsackievirus-adenovirus receptor protein to mediate lytic infection of rhabdomyosarcoma cells. J. Virol. 71, 9844–9848.

Sharma, R., Raychaudhuri, S., Dasgupta, A., 2004. Nuclear entry of poliovirus protease-polymerase precursor 3CD: implications for host cell transcription shut-off. Virology 320, 195–205.

Shen, Y., Igo, M., Yalamanchili, P., Berk, A.J., Dasgupta, A., 1996. DNA binding domain and subunit interactions of transcription factor IIIC revealed by dissection with poliovirus 3C protease. Mol. Cell. Biol. 16, 4163–4171.

Shim, J.S., Liu, J.O., 2014. Recent advances in drug repositioning for the discovery of new anticancer drugs. Int. J. Biol. Sci. 10, 654–663.

Shimizu, H., Agoh, M., Agoh, Y., Yoshida, H., Yoshii, K., Yoneyama, T., Hagiwara, A., Miyamura, T., 2000. Mutations in the 2C region of poliovirus responsible for altered sensitivity to benzimidazole derivatives. J. Virol. 74, 4146–4154.

Smith, A.D., Dawson, H., 2006. Glutathione is required for efficient production of infectious picornavirus virions. Virology 353, 258–267.

Smith, T.J., Kremer, M.J., Luo, M., Vriend, G., Arnold, E., Kamer, G., Rossmann, M.G., McIGnlay, M.A., Diana, G.D., Otto, M.J., 1986. The site of attachment in human rhinovirus 14 of antiviral agents that inhibit uncoating. Science 233, 1286–1293.

Sommergruber, W., Zorn, M., Blaas, D., Fessl, F., Volkmann, P., Maurer-Fogy, I., Pallai, P., Merluzzi, V., Matteo, M., Skern, T., Keuchler, E., 1989. Polypeptide 2A of human rhinovirus type 2: identification as a protease and characterization by mutational analysis. Virology 169, 68–77.

Spagnolo, J.F., Rossignol, E., Bullitt, E., Kirkegaard, K., 2010. Enzymatic and nonenzymatic functions of viral RNAdependent RNA polymerases within oligomeric arrays. RNA 16, 382–393.

Spickler, C., Lippens, J., Laberge, M.-K., Desmeules, S., Bellavance, É., Garneau, M., Guo, T., Hucke, O., Leyssen, P., Neyts, J., et al., 2013. Phosphatidylinositol 4-kinase III beta is essential for replication of human rhinovirus and its inhibition causes a lethal phenotype in vivo. Antimicrob. Agents Chemother. 57, 3358–3368.

Stanley, J., Bisaro, D.M., Briddon, R.W., Brown, J.K., Fauquet, C.M., Harrison, B.D., Rybicki, E.P., Stenger, D.C., 2005. Geminiviridae. In: Fauquet, C.M., Mayo, M.A., Maniloff, J., Desselberger, U., Ball, L.A. (Eds.), Virus Taxonomy, Eighth Report of the International Committee on the Taxonomy of Viruses. Elsevier, London, pp. 301–306.

Staunton, D.E., Merluzzi, V.J., Rothlein, R., Barton, R., Marlin, S.D., Springer, T.A., 1989. A cell adhesion molecule, ICAM-1, is the major surface receptor for rhinoviruses. Cell 56, 849–853.

Stene-Johansen, K., Tjon, G., Schreier, E., Bremer, V., Bruisten, S., Ngui, S.L., King, M., Pinto, R.M., Aragonès, L., Mazick, A., Corbet, S., Sundqvist, L., Blystad, H., Norder, H., Skaug, K., 2007. Molecular epidemiological studiesshowthat hepatitisAvirus is endemic among active homosexual men in Europe. J. Med. Virol. 79, 356–365.

Stevenson, R.A., Huang, J.A., Studdert, M.J., Hartley, C.A., 2004. Sialic acid acts as a receptor for equine rhinitis A virus binding and infection. J. Gen. Virol. 85, 2535–2543.

Strating, J.R., van der Linden, L., Albulescu, L., Bigay, J., Arita, M., Delang, L., Leyssen, P., van der Schaar, H.M., Lanke, K.H., Thibaut, H.J., Ulferts, R., Drin, G., Schlinck, N., Wubbolts, R.W., Sever, N., Head, S.A., Liu, J.O., Beachy, P.A., De Matteis, M.A., Shair, M.D., Olkkonen, V.M., Neyts, J., van Kuppeveld, F.J., 2015. Itraconazole inhibits *Enterovirus* replication by targeting the Oxysterol-binding protein. Cell Rep. 10, 600–615.

Strebel, K., Beck, E., 1986. A second protease of foot-and-mouth disease virus. J. Virol. 58, 893–899.

Suhy, D.A., Giddings, Jr., T.H., Kirkegaard, K., 2000. Remodeling the endoplasmic reticulum by poliovirus infection and by individualviral proteins: an autophagy-like origin for virus-induced vesicles. J. Virol. 74, 8953–8965.

Summers, D.F., Shaw, E.N., Stewart, M.L., Maizel, J.V., 1972. Inhibition of cleavage of large poliovirus-specific precursor proteins in infected HeLa cells by inhibitors of proteolytic enzymes. J. Virol. 10, 880–884.

Svitkin, Y.V., Gorbalenya, A.E., Kazachkov, Y.A., Agol, V.I., 1979. Encephalomyocarditis virus-specific polypeptide p22 possessing a proteolytic activity: preliminary mapping on the viral genome. FEBS Lett. 108, 6–9.

Takai, Y., Miyoshi, J., Ikeda, W., Ogita, H., 2008. Nectins and nectin-like molecules: roles in contact inhibition of cell movement and proliferation. Nat. Rev. Mol. Cell Bio. 9, 603–615.

Tang, W.F., Yang, S.Y., Wu, B.W., Jheng, J.R., Chen, Y.L., Shih, C.H., Lin, K.H., Lai, H.C., Tang, P., Horng, J.T., 2007. Reticulon 3 binds the 2C protein of *Enterovirus* 71 and is required for viral replication. J. Biol. Chem. 282, 5888–5898.

Tesar, M., Marquardt, O., 1990. Foot-and-mouth disease virus protease 3C inhibits cellular transcription and mediates cleavage of histone H3. Virology 174, 364–374.

Teterina, N.L., Gorbalenya, A.E., Egger, D., Bienz, K., Ehrenfeld, E., 1997. Poliovirus 2C protein determinants of membrane binding and rearrangements in mammalian cells. J. Virol. 71, 8962–8972.

Teterina, N.L., Kean, K.M., Gorbalenya, A.E., Agol, V.I., Girard, M., 1992. Analysis of the functional significance of amino acid residues in the putative NTP-binding pattern of the poliovirus 2C protein. J. Gen. Virol. 73 (Pt 8), 1977–1986.

Teterina, N.L., Levenson, E., Rinaudo, M.S., Egger, D., Bienz, K., Gorbalenya, A.E., Ehrenfeld, E., 2006. Evidence for functional protein interactions required for poliovirus RNA replication. J. Virol. 80, 5327–5337.

Thibaut, H.J., van der Linden, L., Jiang, P., Thys, B., Canela, M.D., Aguado, L., Rombaut, B., Wimmer, E., Paul, A., Pérez-Pérez, M.J., van Kuppeveld, F.J., Neyts, J., 2014. Binding of glutathione to *Enterovirus* capsids is essential for virion morphogenesis. PLoS Pathog. 10, e1004039.

Thompson, P., Lu, J., Kaplan, G.G., 1998. The Cys-rich region of hepatitis A virus cellular receptor 1 is required for binding of hepatitis A virus and protective monoclonal antibody 190/4. J. Virol. 72, 3751–3761.

Thompson, A.A., Peersen, O.B., 2004. Structural basis for proteolysis-dependent activation of the poliovirus RNA dependent RNA polymerase. EMBO J. 23, 3462–3471.

Tolskaya, E.A., Romanova, L.I., Kolesnikova, M.S., Gmyl, A.P., Gorbalenya, A.E., Agol, V.I., 1994. Genetic studies on the poliovirus 2C protein, an NTPase. A plausible mechanism of guanidine effect on the 2C function and evidence for the importance of 2C oligomerization. J. Mol. Biol. 236, 1310–1323.

Tomassini, J.E., Graham, D., DeWitt, C.M., Lineberger, D.W., Rodkey, J.A., Colonno, R.J., 1989. cDNA cloning reveals that the major group rhinovirus receptor on HeLa cells is intercellular adhesion molecule. Proc. Natl. Acad. Sci. USA 86, 4907–4911.

Toth, K.S., Zhang, C.X., Lipton, H.L., Luo, M., 1993. Crystallization and preliminary X-ray diffraction studies of Theiler's virus (GDVII strain). J. Mol. Biol. 231, 1126–1129.

Toyoda, H., Nicklin, M.J.H., Murray, M.G., Anderson, C.W., Dunn, J.J., Studier, F.W., Wimmer, E., 1986. A second virus-encoded proteinase involved in proteolytic processing of poliovirus polyprotein. Cell 45, 761–770.

Tsou, Y.L., Lin, Y.W., Chang, H.W., Lin, H.Y., Shao, H.Y., Yu, S.L., Liu, C.C., Chitra, E., Sia, C., Chow, Y.H., 2013. Heat shock protein 90: Role in *Enterovirus* 71 entry and assembly and potential target for therapy. PLoS One 8, e77133.

Turner, R.E., Hayden, F.G., 1992. Efficacy of oral WIN 54954 for the prophylaxis of experimental rhinovirus infections. Antivir. Res. 17 (Suppl. 1), 92.

Turner, P.C., Young, D.C., Flanegan, J.I., Moyer, R.W., 1989. Interference with vaccinia virus growth caused by insertion of the coding sequence for poliovirus protease 2A. Virology 173, 509–521.

Tyndall, J.D., Fairlie, D.P., 2001. Macrocycles mimic the extended peptide conformation recognized by aspartic, serine, cysteine and metallo proteases. Curr. Med. Chem. 8, 893–907.

Tyndall, J.D., Nall, T., Fairlie, D.P., 2005. Proteases universally recognize beta strands in their active sites. Chem. Rev. 105, 973–999.

Ulferts, R., van der Linden, L., Thibaut, H.J., Lanke, K.H., Leyssen, P., Coutard, B., De Palma, A.M., Canard, B., Neyts, J., van Kuppeveld, F.J., 2013. Selective serotonin reuptake inhibitor fluoxetine inhibits replication of human enteroviruses B and D by targeting viral protein 2C. Antimicrob. Agents Chemother. 57, 1952–1956.

van der Schaar, H.M., Leyssen, P., Thibaut, H.J., de Palma, A., van der Linden, L., Lanke, K.H., Lacroix, C., Verbeken, E., Conrath, K., Macleod, A.M., Mitchell, D.R., Palmer, N.J., van de Poel, H., Andrews, M., Neyts, J., van Kuppeveld, F.J., 2013. A novel, broad-spectrum inhibitor of *Enterovirus* replication that targets host cell factor phosphatidylinositol 4-kinase IIIβ. Antimicrob. Agents Chemother. 57, 4971–4981.

van der Schaar, H.M., van der Linden, L., Lanke, K.H., Strating, J.R., Pürstinger, G., de Vries, E., de Haan, C.A.M., Neyts, J., van Kuppeveld, F.J., 2012. Coxsackievirus mutants that can bypass host factor PI4KIIIβ and the need for high levels of PI4P lipids for replication. Cell Res. 22, 1576–1592.

van der Linden, L., Wolthers, K.C., van Kuppeveld, F.J.M., 2015. Replication and inhibitors of enteroviruses and parechoviruses. Viruses 7, 4529–4562.

Vance, L.M., Moscufo, N., Chow, M., Heinz, B.A., 1997. Poliovirus 2C region functions during encapsidation of viral RNA. J. Virol. 71, 8759–8765.

Veals, J., Nemiu, A.A., Kumari, A.A., O'Connell, J., Cox, S., Wright-Minogue, J., Buontempo, P., DeMartino, J., Lin, C.C., Cayen, M.N., 1996. Absorption, bioavailability, metabolism and dose proportionality of a new antipicornavirus agent, SCG 48973, in mice.Interscience Conference on Antimicrobial Agents and Chemotherapy, New Orleans, 16–19 September, 1996, (Abstract H97).

Veber, D.F., Johnson, S.R., Cheng, H.Y., Smith, B.R., Ward, K.W., Kopple, K.D., 2002. Molecular properties that influence the oral bioavailability of drug candidates. J. Med. Chem. 45, 2615–2623.

Velu, A.B., Chen, G.W., Hsieh, P.T., Horng, J.T., Hsu, J.T.A., Hsieh, H.P., Chen, T.C., Weng, K.F., Shih, S.R., 2014. BPR-3P0128 inhibits RNA-dependent RNA polymerase elongation and VPg uridylylation activities of *Enterovirus* 71. Antivir. Res. 112, 18–25.

Verlinden, Y., Cuconati, A., Wimmer, E., Rombaut, B., 2000. The antiviral compound 5-(3,4-dichlorophenyl) methylhydantoin inhibits the post-synthetic cleavages and the assembly of poliovirus in a cell-free system. Antivir. Res. 48, 61–69.

Walker, P.A., Leong, L.E.C., Porter, A.G., 1995. Sequence and structural determinants of the interaction between the 5′-noncoding region of picornavirus RNA and rhinovirus protease 3C. J. Biol. Chem. 270, 14510–14516.

Walker, E.J., Younessi, P., Fulcher, A.J., McCuaig, R., Thomas, B.J., Bardin, P.G., Jans, D.A., Ghildyal, R., 2013. Rhinovirus 3C protease facilitates specific nucleoporin cleavage and mislocalisation of nuclear proteins in infected host cells. PLoS One 8, e71316.

Wang, D., Fang, L., Li, K., Zhong, H., Fan, J., Ouyang, C., Zhang, H., Duan, E., Luo, R., Zhang, Z., Liu, X., Chen, H., Xiao, S., 2012. Foot-and-mouth disease virus 3C protease cleaves NEMO to impair innate immune signaling. J. Virol. 86, 9311–9322.

Wang, D., Fang, L., Wei, D., Zhang, H., Luo, R., Chen, H., Li, K., Xiao, S., 2014. Hepatitis A virus 3C protease cleaves NEMO to impair induction of β interferon. J. Virol. 88, 10252–10258.

Warner, S., Hartley, C.A., Stevenson, R.A., Ficorilli, N., Varrasso, A., Studdert, M.J., Crabb, B.S., 2001. Evidence that Equine rhinitis A virus VP1 is a target of neutralizing antibodies and participates directly in receptor binding. J. Virol. 75, 9274–9281.

Watters, K., Palmenberg, A.C., 2011. Differential processing of nuclear pore complex proteins by rhinovirus 2A proteases from different species and serotypes. J. Virol. 85, 10874–10883.

Weidman, M.K., Yalamanchili, P., Ng, B., Tsai, W., Dasgupta, A., 2001. Poliovirus 3C protease-mediated degradation of transcriptional activator p53 requires a cellular activity. Virology 291, 260–271.

Weng, K.F., Li, M.L., Hung, C.T., Shih, S.R., 2009. *Enterovirus* 71 3C protease cleaves a novel target CstF-64 and inhibits cellular polyadenylation. PLoS Pathog. 5, e1000593.

Wikel, J.H., Paget, C.J., DeLong, D.C., Nelson, J.D., Wu, C.Y., Paschal, J.W., Dinner, A., Templeton, R.J., Chaney, M.O., Jones, N.D., Chamberlin, J.W., 1980. Synthesis of syn and anti isomers of 6-[[(hydroxyimino)phenyl]methyl]-1-[(1-methylethyl)sulfonyl]-1H-benz imidazol-2-amine. Inhibitors of rhinovirus multiplication. J. Med. Chem. 23, 368–372.

Wolthers, K.C., Benschop, K.S., Schinkel, J., Molenkamp, R., Bergevoet, R.M., Spijkerman, I.J., Kraakman, H.C., Pajkrt, D., 2008. Human parechoviruses as an important viral cause of sepsis like illness and meningitis in young children. Clin. Infect. Dis. 47, 358–363.

Wong, S.S., Yip, C.C., Lau, S.K., Yuen, K.Y., 2010. Human enterovirus 71 and hand, foot and mouth disease. Epidemiol. Infect. 138, 1071–1089.

Wright-Minogue, J., Cox, S., Buontempo, P.J., DeMartino, J., Shelton, A.M., Ferrari, E., Albin, R., Girijavallabhan, V., O'Connell, J.F., 1996. SCH 48973: a potent orally active anti enterovirus compound. Interscience Conference on Antimicrobial Agents and Chemotherapy, New Orleans, 16–19 September, 1996, (Abstract H98).

Xing, L., Huhtala, M., Pietiainen, V., Kapyla, J., Vuorinen, K., Marjomaki, V., Heino, J., Johnson, M.S., Hyypia, T., Cheng, R.H., 2004. Structural and functional analysis of integrin alpha2I domain interaction with echovirus 1. J. Biol. Chem. 279, 11632–11638.

Xu, J., Qian, Y., Wang, S., Serrano, J.M., Li, W., Huang, Z., Lu, S., 2010. EV71: an emerging infectious disease vaccine target in the Far East? Vaccine 28, 3516–3521.

Yalamanchili, P., Datta, U., Dasgupta, A., 1997a. Inhibition of host cell transcription by poliovirus: Cleavage of transcription factor CREB by poliovirus-encoded protease 3C$^{pro}$. J. Virol. 71, 1220–1226.

Yalamanchili, P., Weidman, K., Dasgupta, A., 1997b. Cleavage of transcriptional activator Oct-1 by poliovirus encoded protease 3C$^{pro}$. Virology 239, 176–185.

Yamashita, T., Kobayashi, S., Sakae, K., Nakata, S., Chiba, S., Ishihara, Y., Isomura, S., 1991. Isolation of cytopathic small round viruses with BS-C-1 cells from patients with gastroenteritis. J. Infect. Dis. 164, 954–957.

Yamayoshi, S., Yamashita, Y., Li, J., Hanagata, N., Minowa, T., Takemura, T., Koike, S., 2009. Scavenger receptor B2 is a cellular receptor for *Enterovirus* 71. Nat. Med. 15, 798–801.

Yang, F., Ren, L., Xiong, Z., Li, J., Xiao, Y., Zhao, R., 2009. *Enterovirus* 71 outbreak in People's Republic of China in 2008. J. Clin. Microbiol. 47, 2351–2352.

Yin, J., Bergmann, E.M., Cherney, M.M., Lall, M.S., Jain, R.P., Vederas, J.C., James, M.N., 2005. Dual modes of modification of hepatitis a virus 3C protease by a serine-derived β-lactone: Selective crystallization and formation of a functional catalytic triad in the active site. J. Mol. Biol. 354, 854–871.

Yin, J., Paul, A.V., Wimmer, E., Rieder, E., 2003. Functional dissection of a poliovirus cis-acting replication element (PV-*cre*(2C)): Analysis of single- and dual-*cre* viral genomes and proteins that bind specifically to PV-*cre* RNA. J. Virol. 77, 5152–5166.

Ypma-Wong, M.F., Dewalt, P.G., Johnson, V.H., Lamb, J.G., Semler, B.L., 1988. Protein 3CD is the major poliovirus proteinase responsible for cleavage of the P1 capsid precursor. Virology 166, 265–270.

Zhang, B., Morace, G., Gauss-Müller, V., Kusov, Y., 2007a. Poly(A) binding protein, C-terminally truncated by the hepatitis A virus proteinase 3C, inhibits viral translation. Nucleic Acids Res. 35, 5975–5984.

Zhang, A., Nanni, R.G., Oren, D.A.A., Rozhon, E.J., Arnold, E., 1992. Three dimensional structure-relationship for antiviral agents that interact with picornavirus capsid. Semin. Virol. 3, 453–471.

Zhang, B., Seitz, S., Kusov, Y., Zell, R., Gauss-Müller, V., 2007b. RNA interaction and cleavage of poly (C)-binding protein 2 by hepatitis A virus protease. Biochem. Biophys. Res. Commun. 364, 725–730.

Zhang, Y., Tan, X.J., Wang, H.Y., Yan, D.M., Zhu, S.L., Wang, D.Y., Ji, F., Wang, X.J., Gao, Y.J., Chen, L., An, H.Q., Li, D.X., Wang, S.W., Xu, A.Q., Wang, Z.J., Xu, W.B., 2009. An outbreak of hand, foot, and mouth disease associated with subgenotype C4 of human enterovirus 71 in Shandong, China. J. Clin. Virol. 44, 262–267.

Zhao, Y.N., Jiang, Q.W., Jiang, R.J., Chen, L., Perlin, D.S., 2005. Echovirus 30, Jiangsu Province, China. Emerg. Infect. Dis. 11, 562–567.

Zhao, R., Pevear, D.C., Kremer, M., Giranda, V.L., Kofron, J.A., Kuhn, R.J., Rossmann, M.G., 1996. Human rhinovirus 3 at 3.0 Å resolution. Structure 4, 1205–1220.

Zhou, L., Luo, Y., Wu, Y., Tsao, J., Luo, M., 2000. Sialylation of the host receptor may modulate entry of demyelinating persistent Theiler's virus. J. Virol. 74, 1477–1485.

Zoll, J., Hulshof, S.E., Lanke, K., Lunel, F.V., Melchers, W.J., Schoondermark-van de Ven, E., Roivainen, M., Galama, J.M.D., van Kuppeveld, F.J.M., 2009. Saffold virus, a human Theiler's-like *Cardiovirus*, is ubiquitous and causes infection early in life. PLoS Pathog. 5, e1000416.

## FURTHER READING

Banerjee, R., Dasgupta, A., 2001. Interaction of picornavirus 2C polypeptide with the viral negative-strand RNA. J. Gen. Virol. 82, 2621–2627.

# Chapter 11

# Structural Insight Into the Viral 3C-Like Protease Inhibitors: Comparative SAR/QSAR Approaches

Nilanjan Adhikari*, Sandip K. Baidya*, Achintya Saha**, Tarun Jha*

*Jadavpur University, Kolkata, West Bengal, India; **University of Calcutta, Kolkata, West Bengal, India

## 1   INTRODUCTION

In early 2003, about 8500 people were diagnosed across the world with severe acute respiratory syndrome (SARS). Among them, almost 800 died due to its first outbreak. The disease was broken out and turned into an epidemic in Guangdong of South China. Two cases of SARS infections were noticed in Taiwan and Singapore due to improper handling of the samples in the research laboratory. During April 2004, a "mini outbreak" of infections took place in a research laboratory of Beijing that, in turn, led to a chain of infections across three generations. Fortunately, the total number of SARS-infected people was only nine that time. This incidence threatens us about the mini outbreaks of SARS globally at any time (Anand et al., 2005). The SARS is caused by the SARS-coronavirus (SARS-CoV) that belongs to the family of Coronaviridae. This family also includes viruses, such as feline infectious peritonitis virus, murine hepatitis virus, bovine coronavirus, transmissible gastroenteritis virus (TGEV), as well as human coronavirus 229E (Kim et al., 2015). The SARS is considered as a global threat to the health (Khan, 2013; Perlman and Netland, 2009). Moreover, water and food-borne viral gastroenteritis may be caused by the noroviruses that belong to the family Calciviridae (Atmar, 2010; Patel et al., 2009). Therefore, there is an urgent need to develop small molecule antiviral drugs to combat these viruses. The picornavirus belongs to the family of viruses namely Picornaviridae, Calciviridae, and Coronaviridae (Mandadapu et al., 2013a). A number of pathogeneses in human may occur due to these viruses leading to economic and medical burden. For example, human rhinovirus (HRV) is the major reason

Viral Proteases and Their Inhibitors. http://dx.doi.org/10.1016/B978-0-12-809712-0.00011-3

**317**

for upper respiratory tract infection (Ren et al., 2012; Turner and Couch, 2007; Winther, 2011) whereas nonpolio enteroviruses are responsible for symptomatic infections with 10–15 million cases per year in United States (McMinn, 2012; Solomon et al., 2010). Depending on the similarity of the polycistronic organization of the genome, common and posttranscriptional strategies along with the conserved region of domain homology in viral proteins, the virus families are related phylogenetically though these families are not related morphologically (Anand et al., 2005; Cavanagh, 1997; Cowley et al., 2000). Coronavirus is found to be responsible for causing a number of diseases not only in human but also in animals, though the human coronavirus has not been taken into account seriously before the SARS outbreak (Anand et al., 2005). Human coronavirus (HCoV) OC43 and 229E may be responsible for illness in the upper portion of the respiratory tract along with common cold-like conditions (Myint, 1995). The HCoV 229E is the only strain till date that can be cultured in cell culture technique efficiently. The symptoms of SARS include rigor, malaise, high degree of fever, cough, headache, and dyspnoea. The symptoms may also lead to produce interstitial infiltrates in lungs that may be treated through ventilation and intubation (Lee et al., 2003). Not only the lungs but also other organs may be affected by SARS infection (such as liver, kidney, and gastrointestinal tract). Therefore, the SARS infection may be treated as a cause of systemic infection. Face-to-face contacts may be supposed to be the reason for transmission of the pathogen though other routes are also possible. A number of inhibitors against SARS-CoV 3CL$^{pro}$ and HRV 3C$^{pro}$ were reported and the process of development of new antivirals against this class has been continued for a decade. In the present report, quantitative structure–activity relationships (QSARs) techniques have been explored to understand the relation between the SARS-CoV 3CL$^{pro}$ and HRV 3C$^{pro}$ enzyme inhibitory activity with the physicochemical and structural properties of these inhibitors developed till now. This approach may be a useful strategy to design and develop novel and potential SARS-CoV 3CL$^{pro}$ and HRV 3C$^{pro}$ inhibitors to combat the dreadful viral infections.

## 2   GENOME STRUCTURE OF SARS-CoV AND ITS REPLICATIONS

Among the known RNA viruses, the coronaviruses are enveloped, (+) stranded RNA viruses with the largest single-stranded RNA genome (27–31 kb approximately). The RNAs are polyadenylated and 5′capped. These are translated into large polyproteins following their entry into the host cell. The polyproteins are proteolytically cleaved by viral proteinases resulting in viral gene products. The RNA polymerase (pol) is found to be encoded by the genome of SARS-CoV and four structural proteins that commonly include the spike glycoprotein (S), envelope (E), membrane (M), and nucleocapsid (N) proteins in the order of Pol-S-E-M-N. The spike protein (S) is the antigenic determinant for coronavirus and is found to be involved in receptor binding. The E protein plays a significant role during viral assembly. The M glycoprotein transmembrane envelope is found abundantly and is responsible for budding of the virus, and the N protein is related to the viral RNA packaging (Holmes, 2003; Shigeta and Yamase, 2005; Zhai et al., 2007).

# 3  STRUCTURE AND FUNCTIONS OF CORONAVIRUS MAIN PROTEASES

Coronavirus contains a positive-stranded RNA with a single-stranded, large size (27–31 kb) RNA genome. Two overlapping polyproteins, that is, pp1a (450 kDa approx.) and pp1ab (750 kDa approx.) are encoded by the replicase gene with more than 20,000 nucleotides (Herold et al., 1993). These two polyproteins regulate the replication, as well as transcription processes in the virus (Thiel et al., 2001). Proteolysis helps to liberate nonstructural proteins (nsp) from these polyproteins. The proteolytic cleavage is mainly controlled by the viral protein-ase termed as $M^{pro}$ (Anand et al., 2005). The $M^{pro}$ is a cysteine proteinase which is synonymous or rather called as 3C-like protease ($3CL^{pro}$) as the substrate specificity resembles picornavirus 3C-protease ($3C^{pro}$), though both of these vi-ruses are structurally less similar (Anand et al., 2002, 2005). The $3CL^{pro}$ is found to cleave the polyprotein at 11 conserved region including Leu-Gln↓ sequences which are initiated by the autolytic cleavage of the enzyme from pp1a and pp1ab (Hegyi and Ziebuhr, 2002b; Ziebuhr et al., 2000). The pp1a and pp1ab poly-proteins help to release functional polypeptides by papain-like protease ($PL^{pro}$) and $3CL^{pro}$ is located in the nonstructural protein regions, namely nsp3 and nsp5 (Fig. 11.1) via proteolytic reaction mechanisms (Grum-Tokars et al., 2008).

These $PL^{pro}$ and $3CL^{pro}$ are processed by the replicase through autocatalytic mechanisms. The $PL^{pro}$ is found to be responsible for cleaving 3 sites, whereas $3CL^{pro}$ cleaves 11 sites in the viral genome (Grum-Tokars et al., 2008). The SARS-$3CL^{pro}$ cleavage sites are shown in Table 11.1. This type of cleavage is found to be conserved in the $3CL^{pro}$ as evidenced from the experimental data

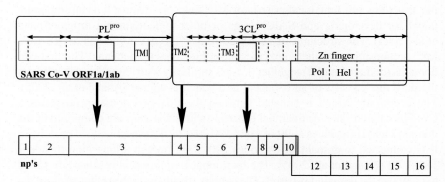

**FIGURE 11.1    SARS-CoV genomic RNA encoding viral replicase polyproteins pp1a and pp1ab (3 and 11 sites are recognized and processed by $PL^{pro}$ and $3CL^{pro}$, respectively).** Hel, Helicase coding regions; Pol, polymerase; TM, transmembrane. *(Adapted from Grum-Tokars, V., Ratia, K., Begaye, A., Baker, S.C., Mesecar, A.D., 2008. Evaluating the 3C-like protease activity of SARS-Coronavirus: recommendations for standardized assays for drug discovery, Virus Res. 133, 63–73.)*

**TABLE 11.1 SARS-CoV 3CL^pro Cleavage Sites and the Canonical Recognition Sequences (11 Recognition Sequences are Shown from P5′ to P6 Positions at the Respective Locations)**

| Sl. No. | 3CL^pro cleavage sites | P5′ | P4′ | P3′ | P2′ | P1′ | P1 | P2 | P3 | P4 | P5 | P6 |
|---|---|---|---|---|---|---|---|---|---|---|---|---|
| 01 | nsp4/5 | LYS | ARG | PHE | GLY | SER | GLN | LEU | VAL | ALA | SER | THR |
| 02 | nsp5/6 | LYS | LYS | PHE | LYS | GLY | GLN | PHE | THR | VAL | GLY | SER |
| 03 | nsp6/7 | ASP | SER | MET | LYS | SER | GLN | VAL | THR | ALA | VAL | LYS |
| 04 | nsp7/8 | GLU | SER | ALA | ILE | ALA | GLN | LEU | THR | ALA | ARG | ASN |
| 05 | nsp8/9 | SER | LEU | GLU | ASN | ASN | GLN | LEU | LYS | VAL | ALA | SER |
| 06 | nsp9/10 | THR | ALA | ASN | GLY | ALA | GLN | LEU | ARG | VAL | THR | ALA |
| 07 | nsp10–12 | SER | ALA | ASP | ALA | SER | GLN | MET | LEU | PRO | GLU | ARG |
| 08 | nsp12/13 | CYS | ALA | GLY | VAL | ALA | GLN | LEU | VAL | THR | HIS | PRO |
| 09 | nsp13/14 | THR | VAL | ASN | GLU | ALA | GLN | LEU | THR | ALA | VAL | ASN |
| 10 | nsp14/15 | VAL | ASN | GLU | LEU | SER | GLN | LEU | ARG | THR | PHE | THR |
| 11 | nsp15/16 | TRP | ALA | GLN | SER | ALA | GLN | LEU | LYS | PRO | TYR | PHE |

Source: Adapted from Grum-Tokars, V., Ratia, K., Begaye, A., Baker, S.C., Mesecar, A.D., 2008. Evaluating the 3C-like protease activity of SARS-Coronavirus: recommendations for standardized assays for drug discovery, Virus Res. 133, 63–73.

(Anand et al., 2003) and the sequence of genomic structure (Marra et al., 2003; Rota et al., 2003).

Three noncanonical M^pro cleavage sites are observed in SARS coronavirus polyproteins having Val, Met, or Phe amino acid residues at P2 position whereas the same cleavage site is found dissimilar in other coronaviruses. Therefore, the structural and functional criteria of M^pro helps to identify it as an important target for developing anti-SARS drugs or other anticoronaviral drugs (Anand et al., 2005). The structures of HCoV 229E M^pro, TGEV M^pro, and SARS-CoV M^pro demonstrate that these enzymes have three distinct domains. The first two domains (domain I and II) together possess similarity with chymotrypsin whereas the third one consists of an α-helical fold which is unique (Anand et al., 2005). The active site which is situated between the first two domains possesses a Cys-His catalytic site. Antiparallel β-barrels with six strands are composed of the domain I and II (residues 8–99 of I and 100–183 of II, respectively). The domain II is connected to domain III (residues 200–300) through a long loop (residues 184–199) (Anand et al., 2005). The hydrophobic amino acid residues are found to compose the domain I β-barrel. The α-helix (residues 53–58) helps to close the β-barrel like a lid. The domain I is bigger than domain II, as well as the homologous domain II of chymotrypsin and 3C^pro of HAV (Allaire et al., 1994; Bergmann et al., 1997; Tsukada and Blow, 1985). Moreover, a number of secondary structural elements are found to be missing in coronavirus $M^{pro}$ compared to HAV $3C^{pro}$ (such as strands b2II and cII along with the linking loop). The Gly135 to Ser146 form a portion of the barrel though domain II possesses maximum consecutive turns and loops. Moreover, the structural alignment of coronavirus $M^{pro}$ domain II with the picornavirus $3C^{pro}$ domain II is found to be different. Superimposition of domain I of TGEV $M^{pro}$, with HAV $3C^{pro}$ domain I results in a root mean square deviation (rmsd) of 1.85 Å whereas superimposition of domain II of both of these enzymes yields a rmsd of 3.25 Å. The overall rmsd for the $C_\alpha$ atoms between their structures is $>2$ Å for all 300 $C_\alpha$ positions and the three $M^{pro}$ structures possess similarity among themselves (Anand et al., 2005). The helical domain III is the most variable domain that exhibits a better overlapping between HCoV $M^{pro}$ and TGEV $M^{pro}$ compared to the SARS-CoV $M^{pro}$, with each other. Moreover, TGEV and HCoV 229E (belongs to group I coronavirus) show 61% sequence similarity whereas SARS-CoV (belongs to group II coronavirus) exhibits 40% and 44% sequence similarity with HCoV 229E and TGEV, respectively (Anand et al., 2003). A high degree of conserved region (42%–48%) between the domain I and II is observed while comparing group I coronavirus $M^{pro}$, and group II SARS-CoV $M^{pro}$. The domain III comparatively exhibits a lower degree of sequence similarity (36%–40%) between these two groups coronaviral enzymes (Anand et al., 2005). The X-ray crystallography structures of SARS-CoV $M^{pro}$, TGEV $M^{pro}$, and HCoV 229E $M^{pro}$ show that these form dimers (Anand et al., 2002, 2003; Yang et al., 2003). Moreover, it was also confirmed that the dimer form is enzymatically active but the monomeric form is not active (Anand et al., 2005; Fan et al., 2004). The dimerization process is found to be mandatory for enzyme activity and this process helps to discriminate the coronavirus $M^{pro}$, and the picornavirus $M^{pro}$ distinctly.

## 4 CATALYTIC SITE OF SARS-CoV $M^{PRO}$

A catalytic dyad is formed by Cys145 and His41 at the SARS-CoV active site, whereas other cysteine and serine protease are found to form a catalytic triad. A water molecule is found to have hydrogen bonding interaction with His41 and Asp187. Moreover, if the cysteine residue is replaced with serine at the enzyme active site, the enzymatic activity of SARS-CoV $M^{pro}$ is decreased. For coronaviral main protease, as well as the picornaviral protease, the cysteine residue is located at the same place of the active site of the His41 imidazole ring plane (distance 3.5–4 Å). For hydrogen bonding interaction between the side chains, the sulfur atom of cysteine residue should be along with the same plane of imidazole function (Anand et al., 2005).

## 5 SUBSTRATE BINDING SITES OF SARS-CoV $M^{PRO}$

The substrate binding sites are found to be conserved in all coronavirus main proteases as suggested by the experimental observations (Anand et al., 2003, 2005). The X-ray crystallographic study between the inhibitor-SARS CoV $M^{pro}$ suggests that the imidazole function of His163 is located at the bottom of the S1 site of M^pro to donate hydrogen bond to the backbone carbonyl function of glutamine. For interaction with glutamine at S1 site, the histidine amino acid residue has to be remained unaltered over a broad range of pH. This may be possible through two interactions involved in the imidazole ring. It may either stack to the phenyl ring of Phe140 or may accept a hydrogen bond from the hydroxyl function of Tyr161. Replacement of the His163 is found to abolish the proteolytic activity (Hegyi et al., 2002a; Ziebuhr et al., 2000). All these residues discussed are found to be conserved not only in SARS-CoV M^pro but also in all other coronavirus main proteases. Moreover, residues Ile51, Met151, Glu166, and His172 of the S1 pocket take part in the conformation of SARS-CoV M^pro (Anand et al., 2005). Regarding the S2 specificity site, all the coronaviruses M^pro consists of a leucine residue at the S2 cleavage site. This S2 site is hydrophobic in nature and is composed of side chain amino acid residues, such as His41, Thr47, Met49, Tyr53, and Met165. The longer methionine residue may restrict the S2 pocket and requires slight spatial orientation to accommodate the substrate leucine residue. Due to the presence of Ala46 residue and differences in amino acid sequences, the S2 pocket is bigger in SARS-CoV M^pro compared to HCoV 229E M^pro and TGEV M^pro. In SARS-CoV M^pro, a stretch of amino acid sequences is observed in 40–50 residues that help to enlarge the size by forming a helix which is not observed in other coronaviruses. This bigger size may be effective in the substrate binding (Anand et al., 2005). Apart from the S1 and S2 pockets, some other substrate binding pockets should be taken into consideration. At the P4 position, small amino acid residues may be preferable (such as Val, Thr, Ser, and Pro) whereas no specificity at the P3 position is observed for coronavirus M^pro. At the P4 position, some amino acid residues are found to be conserved in SARS-CoV M^pro (such as Met165 and Thr190). Moreover, P5 amino acid side chains are found to interact with the main chain at Pro168, Thr190, and Gln192 in SARS-CoV M^pro, and help like a linker between domain II and III.

## 6   RNA INTERFERENCE AND VACCINES OF SARS-CoV

Apart from antivirals to fight against these coronaviruses, RNA interference (RNAi) and vaccine development may be a useful strategy though it is a challenging task. The RNAi is an important tool for gene silencing. Apart from the use of RNAi in cancer and genetic disorder (Wang et al., 2004), development of siRNA inhibitors in SARS infection may be a boon for the treatment of the disease (Li et al., 2005). The replication of the SARS may be inhibited effectively through RNAi in vero cells. Therefore, siRNA therapy may be effective to combat SARS infection (Wang et al., 2004). Short hairpin RNA (shRNA) may be useful to target the N gene sequence of SARS coronavirus and to inhibit shRNA of SARS-CoV antigen expression (Tao et al., 2005; Zhai et al., 2007). These results suggest that gene silencing through RNAi may effectively inhibit the SARS-CoV antigen expression, and, therefore, RNAi approach may be effectively utilized as possible therapy for inhibiting SARS-CoV infection. Moreover, the RNAi is used to target the replicase enzyme of human SARS virus. It not only targets the hSARS gene but also produces inhibitory effects on the SARS RNA virus expression (Zhai et al., 2007). As far as the development of SARS vaccines is concerned, the inactivated SARS-CoV along with the full-length S protein and an attenuated weak virus, and recombinant SARS protein may be used (Jiang et al., 2005; Zhai et al., 2007). The S protein and the inactivated virus were reported to be used to neutralize antibodies. The attenuated or the weak form of the virus might be used to induce immunity, as well as to neutralize antibodies (Finlay et al., 2004). The development of recombinant vaccines may be a useful strategy to prevent SARS infections. It mainly depends on the best antigen identification, as well as the choice of expression system. The S glycoprotein of SARS-CoV along with its truncated form may be targeted for development of recombinant vaccine as the best candidate (Babcock et al., 2004; Bisht et al., 2004; Buchholz et al., 2004; Yang et al., 2004; Zhai et al., 2007). A number of reports were published regarding recombinant S protein vaccine against different SARS-CoV through aryl delivery system (Pogrebnyak et al., 2005; Tuboly et al., 2000).

## 7 DEVELOPMENT OF QSAR MODELS

QSAR is a useful tool to understand the relation between the structural and physicochemical properties of the drug molecules, and their biological activity which may be useful for predicting the activity or toxicity profile of drugs (Gupta, 2007; Verma and Hansch, 2009). The data required for developing the QSAR models are collected from the literature (see individual QSAR for corresponding references). The $IC_{50}$ (molar concentration required to produce 50% inhibition of the enzyme), $EC_{50}$ (effective concentration), or $K_i$ (binding affinity) data are obviously considered as the biological activity term or dependent variable. The independent variables include physicochemical parameters [such as hydrophobicity, molar refractivity, dipole moment along with different axes, molecular weight (MW), polar surface area (PSA), and polar volume, as well as surface area (SA) and volume] and many topological parameters, such as Kier's molecular connectivity indices, Balaban indices, etc. Regarding the statistics of QSAR models, $N$ is used to indicate the number of compounds in the set, $R$ to indicate the correlation coefficient of the QSAR model obtained, $R^2$ refers to the squared correlation coefficient exhibiting the goodness of fit, $q^2$ indicates square of the leave-one-out cross-validated correlation coefficient (represents the internal validation of the model), $R_A^2$ refers to the adjusted $R^2$, $F$ value represents the Fischer statistics (Fischer ratio) that actually means the ratio between the explained and unexplained variance for a particular degree of freedom, $P$ stands for the probability factor related to $F$-ratio, SEE means the standard error of estimate, $Q$ is the quality factor that can be a measure of chance correlation. A high $Q$ represents the high predictivity, as well as the lack of over-fitting of the model.

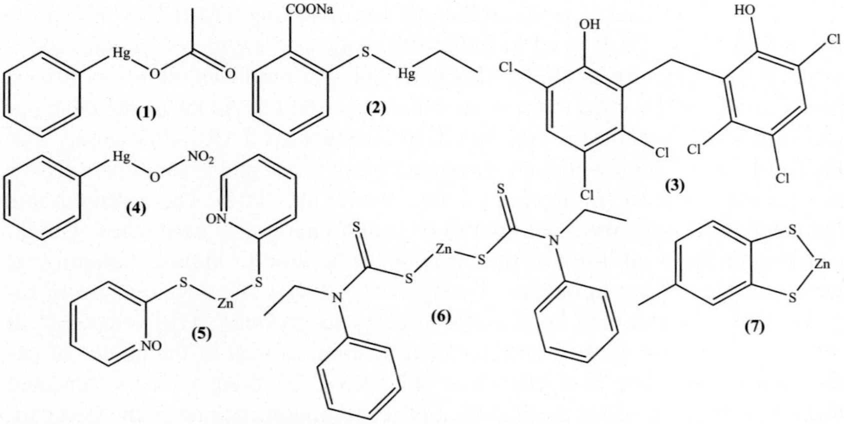

FIGURE 11.2  **Metal-conjugated SARS 3CL^pro inhibitors.**

Compounds that misfit in the correlation are considered as outliers and are usually removed from the regression. We discuss here the QSAR models obtained for different categories of SARS-CoV 3CL^pro and HRV 3C^pro inhibitors.

## 7.1  Coronaviral 3CL^pro Inhibitors

### 7.1.1  Metal-Conjugated SARS-CoV 3CL^pro Inhibitors

Hsu et al. (2004) reported some metal-conjugated compounds as promising SARS-CoV 3CL^pro inhibitors (Fig. 11.2; Table 11.2). The model obtained was as shown by Eq. (11.1):

$$pK_i = 4.806\,(\pm0.369) + 0.013\,(\pm0.003)\,\mathrm{PSA} \qquad (11.1)$$

$N = 5$, $R = 0.910$, $R^2 = 0.828$, $R_A^2 = 0.771$, $F\,(1, 3) = 14.459$, $P < 0.03195$, SEE $= 0.229$, $q^2 = 0.643$, $Q = 3.974$, Outlier = Compounds **1, 3**

This model suggested that the increasing value of the PSA may contribute positively to the binding enzyme. Compounds with a higher PSA (Compounds **4–6**, Table 11.2) have higher activity than compounds with a lower PSA (Compounds **2, 7**, Table 11.2). Compound **1** has a lower PSA but higher activity whereas compound **3** having the higher PSA has lower activity. These molecules are not explained properly by this model. Therefore, these molecules (Compounds **1, 3**, Table 11.2) were considered as outliers. They may have different mechanism(s) of action(s).

### 7.1.2  Some Small Molecule SARS-CoV 3CL^pro Inhibitors

Blanchard et al. (2004) reported some SARS-CoV 3CL^pro inhibitors (Fig. 11.3; Table 11.3). The QSAR model for these compounds was as shown by Eq. (11.2).

$$p\,\mathrm{IC}_{50} = 4.845\,(\pm0.064) + 0.002\,(\pm0.000)\,\mathrm{PSA} \qquad (11.2)$$

**TABLE 11.2** The Biological Activity and Physicochemical Parameters of Metal-Conjugated SARS-CoV 3CL$^{pro}$ Inhibitors (Fig. 11.2) for QSAR Model [Eq. (11.1)]

| Compound | Obsd$^b$ | Calcd$^c$ | Res$^d$ | Del res$^e$ | Pred$^f$ | PSA |
|---|---|---|---|---|---|---|
| 1$^a$ | 6.155 | 5.625 | 0.530 | 0.716 | 5.439 | 69.998 |
| 2 | 5.620 | 5.511 | 0.109 | 0.168 | 5.452 | 61.487 |
| 3$^a$ | 4.863 | 5.791 | −0.928 | −1.124 | 5.987 | 82.389 |
| 4 | 6.523 | 6.506 | 0.017 | 0.029 | 6.494 | 135.592 |
| 5 | 6.770 | 6.444 | 0.326 | 0.514 | 6.256 | 130.932 |
| 6 | 6.000 | 6.213 | −0.213 | −0.267 | 6.267 | 113.794 |
| 7 | 5.854 | 5.694 | 0.160 | 0.204 | 5.649 | 75.130 |

$^a$Considered as outliers.
$^b$Observed or experimental activity.
$^c$Calculated activity of compounds according to Eq. (11.1).
$^d$Difference between the observed and calculated activity.
$^e$Difference between the observed and leave-one-out cross-validated activity.
$^f$Leave-one-out cross-validated activity.

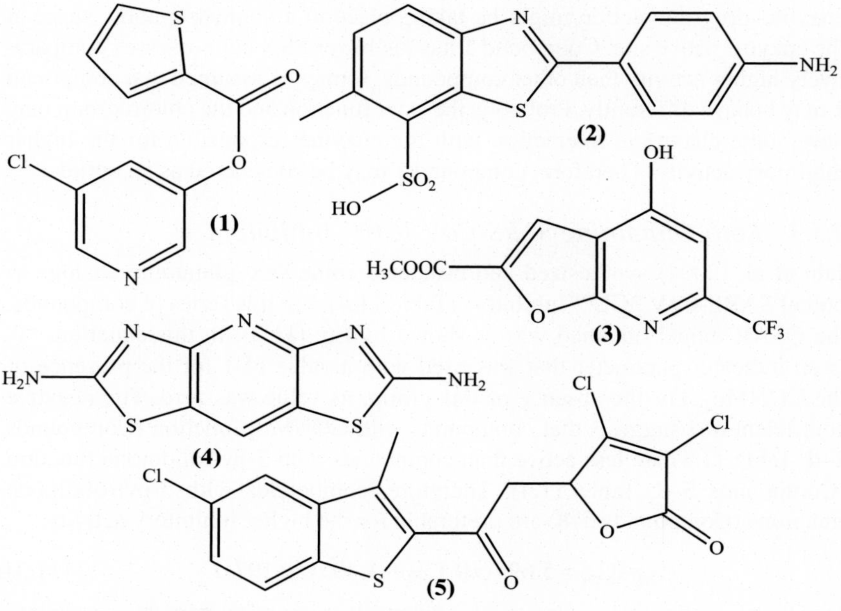

FIGURE 11.3    SARS-CoV 3CL protease inhibitors.

**TABLE 11.3 The Biological Activity and Physicochemical Parameters of Some Small Molecule SARS-CoV 3CL$^{pro}$ Inhibitors (Fig. 11.3) for QSAR Model [Eq. (11.2)]**

| Compound | Obsd | Calcd | Res | Del res | Pred | PSA |
|---|---|---|---|---|---|---|
| 1[a] | 6.301 | 5.520 | 0.781 | 1.152 | 5.149 | 132.553 |
| 2 | 5.367 | 5.510 | −0.144 | −0.182 | 5.549 | 209.011 |
| 3 | 5.155 | 5.517 | −0.362 | −0.482 | 5.637 | 153.950 |
| 4 | 5.585 | 5.494 | 0.091 | 0.772 | 4.814 | 327.913 |
| 5 | 5.155 | 5.520 | −0.366 | −0.549 | 5.704 | 129.743 |

[a]Considered as outliers.

$N = 4$, $R = 0.984$, $R^2 = 0.969$, $R_A^2 = 0.953$, $F(1, 2) = 62.388$, $P < 0.01565$, SEE $= 0.044$, $q^2 = 0.838$, $Q = 22.364$, Outlier = Compound **1**.

This model also exhibited that the PSA of the molecule might be conducive to the enzyme inhibitory activity of the compounds. As obvious from Table 11.3, compounds **2** and **4** with a higher PSA have higher activity than compounds with the lower PSA. The sulfone and amino functions of compound **2** and the disubstituted amino acid function of compound **4** may produce higher PSA as compared to the trichloro-substituted compound **5** and the monohydroxy trifluoro substituted ester analog (compound **2**). It also suggested that the enzyme–drug interaction might be taking place in a nonhydrophobic space at the enzyme active site. Compound **1** has the lower PSA but possesses comparatively higher activity than other compounds. It may be assumed that compound **1** may behave differently. Probably, the ester function and the chloro group may have some electronic interaction with the enzyme responsible for the higher inhibitory activity. Therefore, compound **1** may be considered as an outlier.

### 7.1.3 Keto-Glutamine SARS-CoV 3CL$^{pro}$ Inhibitors

Jain et al. (2004) synthesized and evaluated some keto-glutamine analogs as potent SARS-CoV 3CL$^{pro}$ inhibitors (Table 11.4). For this series of compounds, the QSAR model obtained was as shown by Eq. (11.3). In this equation, "I" is an indicator parameter that was used with a value of 1 for the presence of the CONMe$_2$. For the absence of this group, its value was zero. The negative coefficient of I suggests that compounds with CONMe$_2$ function (Compounds **1–4**, Table 11.4) are less active than compounds with 3-pyrrolidinone function (Compounds **5–8**, Table 11.4). Therefore, compounds with 3-pyrrolidinone functions (Compounds **5–8**) are preferable for the higher inhibitory activity.

$$p IC_{50} = 5.699(\pm 0.139) - 1.405(\pm 0.197)I \quad (11.3)$$

$N = 8$, $R = 0.946$, $R^2 = 0.894$, $R_A^2 = 0.877$, $F(1, 6) = 50.793$, $P < 0.00038$, SEE $= 0.279$, $q^2 = 0.812$, $Q = 3.391$

**TABLE 11.4 The Biological Activity and Physicochemical Parameters of Keto-Glutamine SARS-CoV 3CL$^{pro}$ Inhibitors for QSAR Model [Eq. (11.3)]**

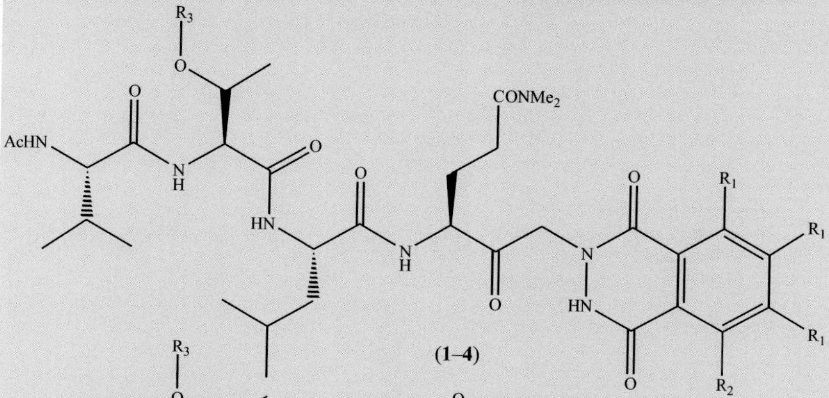

(1–4)

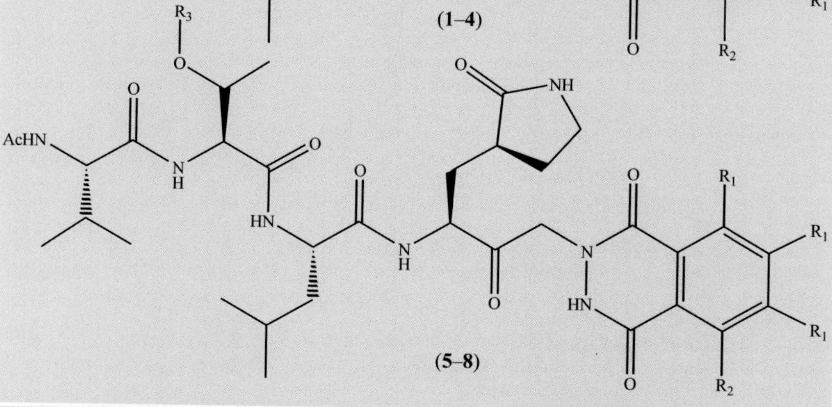

(5–8)

| Compound | R$_1$ | R$_2$ | R$_3$ | Obsd | Calcd | Res | Del res | Pred | I |
|---|---|---|---|---|---|---|---|---|---|
| 1 | H | H | Benzyl | 4.194 | 4.294 | −0.101 | −0.134 | 4.328 | 1 |
| 2 | H | H | H | 4.553 | 4.294 | 0.259 | 0.345 | 4.208 | 1 |
| 3 | H | NO$_2$ | Benzyl | 4.155 | 4.294 | −0.139 | −0.186 | 4.341 | 1 |
| 4 | H | NO$_2$ | H | 4.276 | 4.294 | −0.019 | −0.025 | 4.301 | 1 |
| 5 | H | H | Benzyl | 5.569 | 5.699 | −0.131 | −0.174 | 5.743 | 0 |
| 6 | H | H | H | 5.538 | 5.699 | −0.162 | −0.215 | 5.753 | 0 |
| 7 | H | NO$_2$ | Benzyl | 6.222 | 5.699 | 0.523 | 0.697 | 5.525 | 0 |
| 8 | H | NO$_2$ | H | 5.469 | 5.699 | −0.231 | −0.308 | 5.776 | 0 |

## 7.1.4 Lopinavir-Like SARS- CoV 3CL$^{pro}$ Inhibitors

Wu et al. (2004) reported some Lopinavir-like inhibitors of SARS-3CL$^{pro}$ (Table 11.5). The model obtained was as by Eq. (11.4), where SA refers to surface area. Eq. (11.4) suggested that the increase in value of the SA may be

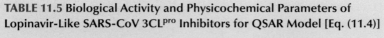

**TABLE 11.5** Biological Activity and Physicochemical Parameters of Lopinavir-Like SARS-CoV 3CL$^{pro}$ Inhibitors for QSAR Model [Eq. (11.4)]

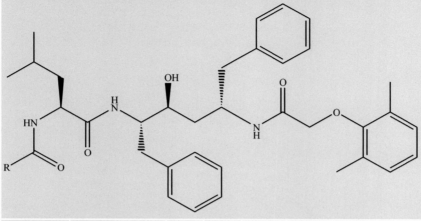

| Compound | R | Obsd | Calcd | Res | Del res | Pred | SA |
|---|---|---|---|---|---|---|---|
| 1 | 5-NO$_2$-2-furyl | 4.602 | 4.606 | −0.004 | −0.006 | 4.608 | 1018.970 |
| 2$^a$ | β-Napthylmethyl | 4.398 | 4.543 | −0.146 | −0.160 | 4.558 | 1057.910 |
| 3 | m-MePh | 4.602 | 4.589 | 0.013 | 0.016 | 4.587 | 1029.690 |
| 4 | Pyridyl-2-yl-sulfanylmethyl | 4.456 | 4.560 | −0.104 | −0.116 | 4.572 | 1047.550 |
| 5 | Indole-2-yl | 4.620 | 4.548 | 0.072 | 0.079 | 4.540 | 1055.240 |
| 6 | 2-Thioxo-4-thiazolidinone-3-ylmethyl | 4.638 | 4.621 | 0.017 | 0.025 | 4.613 | 1010.130 |
| 7 | p-CF$_3$Ph | 4.602 | 4.555 | 0.047 | 0.052 | 4.550 | 1050.650 |
| 8$^a$ | p-OMePhCH=CH | 4.620 | 4.489 | 0.131 | 0.159 | 4.461 | 1091.790 |
| 9 | (CH$_2$)$_2$NHBoc | 4.398 | 4.471 | −0.073 | −0.097 | 4.494 | 1102.490 |
| 10$^a$ | 6-ClChromene-4-yl | 4.602 | 4.506 | 0.096 | 0.110 | 4.492 | 1081.040 |
| 11 | Naphthalene-2-yloxymethyl | 4.398 | 4.447 | −0.049 | −0.077 | 4.475 | 1117.600 |

$^a$Considered as outliers.

detrimental to the activity. The compounds **2, 8,** and **10** (Table 11.5) may behave in a different fashion and, therefore, they were considered as outliers.

$$p \, IC_{50} = 6.968 \, (\pm 0.617) - 0.002 \, (\pm 0.001) \, SA \qquad (11.4)$$

$N = 8$, $R = 0.849$, $R^2 = 0.721$, $R_A^2 = 0.675$, $F \, (1, 6) = 15.531$, $P < 0.00762$, SEE $= 0.059$, $q^2 = 0.618$, $Q = 14.390$, Outlier = Compounds **2, 8, 10**

## 7.1.5 Anilide-Based SARS-CoV 3CL$^{pro}$ Inhibitors

Shie et al. (2005a) reported a series of potent anilide inhibitors against SARS-3CL$^{pro}$ (Fig. 11.4; Table 11.6), for which the QSAR model obtained was as shown by Eq. (11.5). This model showed the importance of dipole moment along the X-axis ($D_X$), MW, PSA, and volume (Vol) for controlling the enzyme inhibition. The positive coefficient of the dipole moment along X-axis suggested that the bulky substitutions along X-axis may favor the activity. Moreover, the MW was also shown to have a positive impact on the activity, whereas PSA was shown to have the negative effect. Therefore, it may be suggested that molecules with the bigger size along with bulky substituents may be conducive to the inhibition. Moreover, the volume is found to have a parabolic relation with the enzyme inhibition. The optimum value of the volume is 6500. Compounds **1, 7, 9,** and **14** were considered as outliers as these molecules may work with a different mechanism(s).

$$p\,IC_{50} = 13.858\,(\pm 2.004) + 0.066\,(\pm 0.022)\,D_X + 0.011(\pm 0.002)\,MW$$
$$-0.004\,(\pm 0.001)\,PSA - 0.013(\pm 0.002)\,Vol - 0.000001(\pm 0.000)\,Vol^2 \quad (11.5)$$

$N = 26$, $R = 0.904$, $R^2 = 0.817$, $R_A^2 = 0.771$, $F\,(5,\,20) = 17.862$, $P < 0.00000$, SEE $= 0.185$, $q^2 = 0.680$, $Q = 4.886$, $Vol_{Opt} = 6500$, Outlier $=$ Compounds **1, 7, 9, 14**

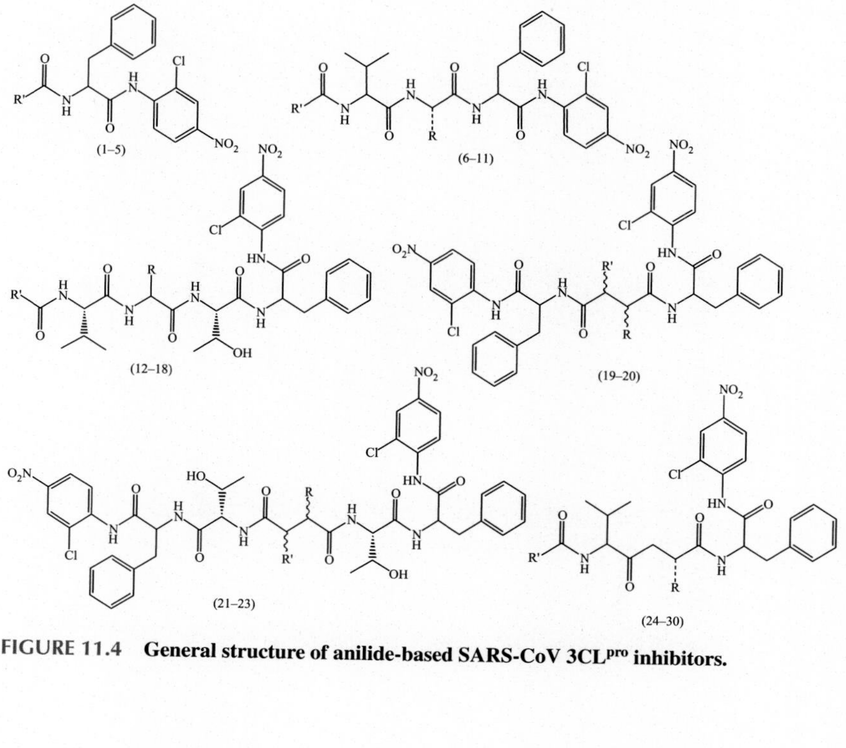

**FIGURE 11.4**  **General structure of anilide-based SARS-CoV 3CL$^{pro}$ inhibitors.**

**TABLE 11.6 Biological Activity and Physicochemical Parameters of Anilide-Based SARS-CoV 3CL$^{pro}$ Inhibitors (Fig. 11.4) for QSAR Model [Eq. (11.5)]**

| Compound | R | R' | Obsd | Calcd | Res | Del res | Pred | $D_X$ | MW | PSA | Vol |
|---|---|---|---|---|---|---|---|---|---|---|---|
| 1[a] | — | Me$_2$NC$_6$H$_4$ | 7.222 | 6.730 | 0.492 | 1.161 | 6.061 | −3.032 | 466.917 | 132.531 | 1184.630 |
| 2 | — | C$_{14}$H$_{29}$CH(Br) | 5.523 | 5.533 | −0.010 | −0.025 | 5.548 | −3.999 | 637.048 | 164.054 | 1505.220 |
| 3 | — | 3,4-(NH$_2$)$_2$C$_6$H$_3$ | 5.699 | 6.054 | −0.355 | −0.589 | 6.288 | −2.424 | 453.878 | 266.781 | 1142.890 |
| 4 | — | (Indol-3-yl)-CH=CH | 5.523 | 5.614 | −0.091 | −0.109 | 5.632 | −3.716 | 488.922 | 177.675 | 1290.600 |
| 5 | — | (2-NH$_2$-1,3-thiazol-4-yl)-C(NOCH$_3$) | 5.155 | 4.900 | 0.255 | 0.559 | 4.596 | −4.072 | 502.931 | 304.411 | 1273.020 |
| 6 | i-Bu | Et | 5.155 | 5.298 | −0.143 | −0.155 | 5.310 | 0.087 | 588.095 | 150.968 | 1561.070 |
| 7[a] | i-Bu | Ph | 5.398 | 4.970 | 0.428 | 0.492 | 4.906 | 1.527 | 636.138 | 151.962 | 1742.770 |
| 8 | i-Bu | Morpholino | 4.721 | 4.791 | −0.070 | −0.076 | 4.797 | −0.806 | 645.146 | 167.607 | 1715.620 |
| 9[a] | i-Bu | Thien-2-yl | 5.301 | 4.887 | 0.414 | 0.457 | 4.844 | 1.890 | 642.165 | 192.606 | 1716.130 |
| 10 | PhCH$_2$ | 5-Me-isoxazol-3-yl | 5.155 | 5.477 | −0.322 | −0.400 | 5.555 | 4.564 | 675.131 | 209.168 | 1697.490 |
| 11 | PhCH$_2$ | Thien-2-yl | 5.301 | 5.261 | 0.040 | 0.044 | 5.257 | 1.903 | 676.182 | 191.190 | 1708.000 |
| 12 | i-Bu | Et | 5.155 | 4.697 | 0.458 | 0.551 | 4.603 | −1.876 | 689.199 | 176.439 | 1789.010 |
| 13 | i-Bu | Morpholino | 4.796 | 4.899 | −0.103 | −0.118 | 4.914 | 2.387 | 746.250 | 203.573 | 1980.620 |

| | | | | | | | | | | |
|---|---|---|---|---|---|---|---|---|---|---|
| 14[a] | PhCH$_2$ | t-Bu | 5.699 | 5.095 | 0.604 | 0.704 | 4.995 | 1.521 | 751.268 | 163.488 | 2001.440 |
| 15 | PhCH$_2$ | 5-Me-isoxazol-3-yl | 5.301 | 5.347 | −0.046 | −0.050 | 5.351 | 4.917 | 776.235 | 216.166 | 1996.140 |
| 16 | PhCH$_2$ | PhCH$_2$O | 5.222 | 5.407 | −0.185 | −0.222 | 5.444 | 2.696 | 801.284 | 180.409 | 2087.700 |
| 17 | 4-FC$_6$H$_4$CH$_2$ | Et | 5.301 | 5.350 | −0.049 | −0.053 | 5.354 | 2.908 | 741.206 | 169.119 | 1916.260 |
| 18 | 4-FC$_6$H$_4$CH$_2$ | Ph | 5.699 | 5.519 | 0.180 | 0.208 | 5.491 | 2.716 | 789.248 | 173.610 | 2002.030 |
| 19 | (S)-OH | H | 5.398 | 5.175 | 0.223 | 0.304 | 5.094 | 7.882 | 737.543 | 291.835 | 1855.940 |
| 20 | (R)-OH | (R)-OH | 5.301 | 5.261 | 0.040 | 0.067 | 5.234 | 9.257 | 753.542 | 323.375 | 1853.190 |
| 21 | H | H | 5.699 | 5.737 | −0.038 | −0.050 | 5.749 | 6.265 | 923.751 | 325.291 | 2258.430 |
| 22 | (S)-OH | H | 5.699 | 5.764 | −0.065 | −0.090 | 5.789 | 7.011 | 939.751 | 353.359 | 2264.460 |
| 23 | (R)-OH | (R)-OH | 5.699 | 5.846 | −0.147 | −0.216 | 5.915 | 7.863 | 955.750 | 375.570 | 2288.530 |
| 24 | i-Bu | Et | 4.569 | 5.055 | −0.486 | −0.537 | 5.106 | −0.653 | 587.107 | 136.197 | 1605.950 |
| 25 | i-Bu | Ph | 4.678 | 5.003 | −0.325 | −0.355 | 5.033 | −1.172 | 635.150 | 126.423 | 1706.880 |
| 26 | i-Bu | t-BuO | 4.721 | 4.964 | −0.243 | −0.261 | 4.983 | −0.444 | 631.159 | 143.473 | 1698.180 |
| 27 | i-Bu | Morpholino | 4.538 | 4.844 | −0.306 | −0.338 | 4.875 | −0.935 | 644.158 | 141.182 | 1741.990 |
| 28 | i-Bu | Thien-2-yl | 4.658 | 4.853 | −0.196 | −0.210 | 4.868 | −0.247 | 641.177 | 170.511 | 1703.650 |
| 29 | PhCH$_2$ | 5-Me-isoxazol-3-yl | 5.222 | 5.105 | 0.117 | 0.126 | 5.096 | 1.855 | 674.143 | 197.782 | 1725.490 |
| 30 | PhCH$_2$ | Thien-2-yl | 4.796 | 4.864 | −0.069 | −0.082 | 4.878 | −1.901 | 675.194 | 171.248 | 1727.980 |

[a]Considered as outliers.

### 7.1.6   Peptidomimetic α,β Unsaturated Esters as SARS-CoV 3CL$^{pro}$ Inhibitors

Shie et al. (2005b) reported a series of peptidomimetic α,β unsaturated esters as promising SARS-3CL$^{pro}$ inhibitors (Table 11.7). The QSAR model obtained for them was as shown by Eq. (11.6).

$$pIC_{50} = 14.827\,(\pm1.418) - 0.016\,(\pm0.002)\,\text{MW} - 0.006\,(\pm0.001)\,\text{PSA} \quad (11.6)$$

$N = 15$, $R = 0.906$, $R^2 = 0.822$, $R_A^2 = 0.792$, $F\,(2,\,12) = 27.656$, $P < 0.00003$, SEE = 0.227, $q^2 = 0.710$, $Q = 3.991$, Outlier = Compound **8**

It was observed from Eq. (11.6) that increase in the value of both the MW, as well as the PSA may be detrimental to the activity. Thus, the model suggested that the smaller molecules with less steric bulk might favor the activity. Moreover, the enzyme-drug interaction would be more favored in non-hydrophobic space. It was observed that compounds having unsaturation at the R' position (Compounds **11–16**, Table 11.7) possess lower MW compared to the other molecules in the dataset and possess higher inhibitory activity. In compound **13**, both the phenyl rings might be accommodated in the S2 and S3 pockets. Moreover, the (dimethylamino) cinnamyl function adopts a coplanar rigid structure at the end terminal, which may help it in forming hydrogen bonding with amino acids residues Glu166, Glu189, and Glu192 at the enzyme active site. Compound **8** may behave differently and hence, this was considered as an outlier.

### 7.1.7   Benzotriazole Esters as SARS-CoV 3CL$^{pro}$ Inhibitors

Wu et al. (2006) reported some benzotriazole esters as promising SARS-3CL$^{pro}$ mechanism-based inhibitors (Table 11.8). The model obtained for this series is as shown by Eq. (11.7).

$$pK_i = 3.814\,(\pm1.495) - 1.405\,(\pm0.115)\,D_Y + 0.014\,(\pm0.005)\,\text{MW} \quad (11.7)$$

$N = 11$, $R = 0.974$, $R^2 = 0.949$, $R_A^2 = 0.937$, $F\,(2,\,8) = 75.154$, $P < 0.00001$, SEE = 0.282, $q^2 = 0.893$, $Q = 3.454$

It was observed from Eq. (11.7) that increasing value of the dipole moment along $Y$-axis ($D_Y$) may lead to a decrease in the activity, whereas increasing value of the MW may be conducive to the activity. It also suggests that compounds with the higher molecular bulk with lower steric effect may be favorable for the higher inhibitory activity. Compounds with ester functions (Compounds **1–8**, Table 11.8) is better active than compounds with acetyl function (Compounds **9–11**) as these molecules (Compounds **1–8**) possess higher bulkiness. Therefore, it may be assumed that the ester analogs impart less steric effect with the enzyme and hence, produce higher activity.

**TABLE 11.7** Biological Activity and Physicochemical Parameters of Peptidomimetic α,β Unsaturated Esters as SARS-CoV 3CL^pro Inhibitors for QSAR Model [Eq. (11.6)]

| Compound | R | R' | R" | X | Obsd | Calcd | Res | Del res | Pred | MW | PSA |
|---|---|---|---|---|---|---|---|---|---|---|---|
| 1 | Ph | 5-Me-isoxazolyl-3-yl | — | NH | 4.097 | 4.173 | −0.076 | −0.125 | 4.222 | 581.660 | 233.331 |
| 2 | Ph | PhCH$_2$O | — | NH | 4.071 | 4.195 | −0.124 | −0.157 | 4.228 | 606.709 | 175.330 |
| 3 | 4-FPh | 5-Me-isoxazolyl-3-yl | Ph | CH$_2$ | 4.409 | 4.563 | −0.154 | −0.169 | 4.578 | 591.670 | 150.782 |
| 4 | 4-FPh | PhCH$_2$O | Ph | CH$_2$ | 4.509 | 4.544 | −0.036 | −0.046 | 4.555 | 616.719 | 99.131 |
| 5 | Ph | 5-Me-isoxazolyl-3-yl | Ph | CH$_2$ | 4.886 | 4.815 | 0.072 | 0.077 | 4.809 | 573.679 | 150.790 |
| 6 | Ph | PhCH$_2$O | Ph | CH$_2$ | 4.420 | 4.796 | −0.376 | −0.441 | 4.861 | 598.729 | 99.149 |
| 7 | 4-FPh | 5-Me-isoxazolyl-3-yl | Ph | NH | 4.678 | 4.483 | 0.194 | 0.217 | 4.461 | 592.658 | 161.032 |

*(Continued)*

**TABLE 11.7 Biological Activity and Physicochemical Parameters of Peptidomimetic $\alpha,\beta$ Unsaturated Esters as SARS-CoV 3CL[pro]Inhibitors for QSAR Model [Eq. (11.6)] (cont.)**

| Compound | R | R' | R'' | X | Obsd | Calcd | Res | Del res | Pred | MW | PSA |
|---|---|---|---|---|---|---|---|---|---|---|---|
| 8[a] | 4-FPh | PhCH$_2$O | Ph | NH | 4.959 | 4.502 | 0.457 | 0.592 | 4.366 | 617.707 | 103.559 |
| 9 | Ph | 5-Me-isoxazolyl-3-yl | Ph | NH | 4.523 | 4.735 | −0.212 | −0.230 | 4.753 | 574.667 | 161.010 |
| 10 | Ph | PhCH$_2$O | Ph | NH | 4.959 | 4.754 | 0.205 | 0.239 | 4.720 | 599.716 | 103.573 |
| 11 | Ph | BiPhCH=CH | Ph | NH | 5.000 | 5.167 | −0.167 | −0.195 | 5.195 | 572.693 | 98.047 |
| 12 | Ph | 4-NO$_2$PhCH=CH | Ph | NH | 5.301 | 5.054 | 0.247 | 0.317 | 4.984 | 541.594 | 183.426 |
| 13 | Ph | 4-Me$_2$NPhCH=CH | Ph | NH | 6.000 | 5.662 | 0.338 | 0.515 | 5.485 | 539.665 | 92.953 |
| 14 | Ph | 2,4-diOMePhCH=CH | Ph | NH | 5.000 | 5.278 | −0.278 | −0.326 | 5.326 | 556.649 | 115.705 |
| 15 | Ph | 3-Benzo[1,3]dioxol-5-yl-CH=CH | Ph | NH | 5.155 | 5.398 | −0.243 | −0.308 | 5.463 | 540.606 | 131.977 |
| 16 | Ph | 3-Benzo[1,3]dioxol-5-yl-CH=CH | — | NH | 5.000 | 4.847 | 0.153 | 0.203 | 4.797 | 547.599 | 202.480 |

[a]*Considered as outliers.*

**TABLE 11.8** Biological Activity and Physicochemical Parameters of Benzotriazole Esters as SARS-CoV 3CL$^{pro}$ Inhibitors for QSAR Model [Eq. (11.7)]

| Compound | Ar | X | Obsd | Calcd | Res | Del res | Pred | $D_Y$ | MW |
|---|---|---|---|---|---|---|---|---|---|
| 1 | 2-NH$_2$Ph | O | 7.710 | 7.592 | 0.118 | 0.186 | 7.524 | −0.218 | 254.244 |
| 2 | 4-N(Me)$_2$Ph | O | 7.759 | 7.880 | −0.120 | −0.138 | 7.897 | −0.150 | 282.297 |
| 3 | 4-NHMePh | O | 7.917 | 8.083 | −0.165 | −0.211 | 8.129 | −0.431 | 268.271 |
| 4 | 4-N(Et)$_2$Ph | O | 7.955 | 8.311 | −0.356 | −0.662 | 8.617 | −0.185 | 310.350 |
| 5 | 5-Benzimidazolyl | O | 7.640 | 7.690 | −0.050 | −0.056 | 7.696 | −0.045 | 279.254 |
| 6 | 5-Indolyl | O | 8.125 | 7.611 | 0.514 | 0.577 | 7.548 | 0.002 | 278.265 |
| 7 | 2-Indolyl | O | 7.910 | 7.970 | −0.060 | −0.070 | 7.980 | −0.254 | 278.265 |
| 8 | 5-F-2-Indolyl | O | 7.860 | 7.595 | 0.265 | 0.320 | 7.540 | 0.188 | 296.256 |
| 9 | 4-N(Et)$_2$Ph | CH$_2$ | 6.000 | 5.782 | 0.218 | 0.437 | 5.563 | 1.597 | 308.378 |
| 10 | 4-NHMePh | CH$_2$ | 5.347 | 5.529 | −0.182 | −0.310 | 5.657 | 1.368 | 266.298 |
| 11 | 4-N(Me)$_2$Ph | CH$_2$ | 5.174 | 5.357 | −0.183 | −0.295 | 5.469 | 1.627 | 280.324 |

## 7.1.8 A Diverse Set of SARS-CoV 3CL$^{pro}$ Inhibitors

Chen et al. (2006a) reported some diverse chemical entities through virtual screening, surface plasmon resonance and fluorescence resonance energy transfer based assays as promising against SARS-CoV 3CL$^{pro}$ (Fig. 11.5; Table 11.9). The QSAR model obtained was as shown by Eq. (11.8):

$$pIC_{50} = 2.508\,(\pm0.417) - 0.012\,(\pm0.003)\,PSA \qquad (11.8)$$

$N = 6$, $R = 0.923$, $R^2 = 0.851$, $R_A^2 = 0.814$, $F\,(1, 4) = 22.908$, $P < 0.00874$, SEE = 0.167, $q^2 = 0.746$, $Q = 5.527$, Outlier = Compounds **5**, **7**

It is observed from Eq. (11.8) that increasing the value of the PSA may be detrimental to the activity. Thus it suggested that less polar molecules may have better inhibitory activity. Due to the presence of electronegative function (such as carboxyl, chloro, etc.), the molecule may have larger PSA. Compounds **5** and **7** (Table 11.9), though possess lower PSA, have higher activity and this could not be explained by this model. Thus these compounds might be supposed to involve the different mechanism of action for producing the higher activity. Therefore, these compounds are considered as outliers.

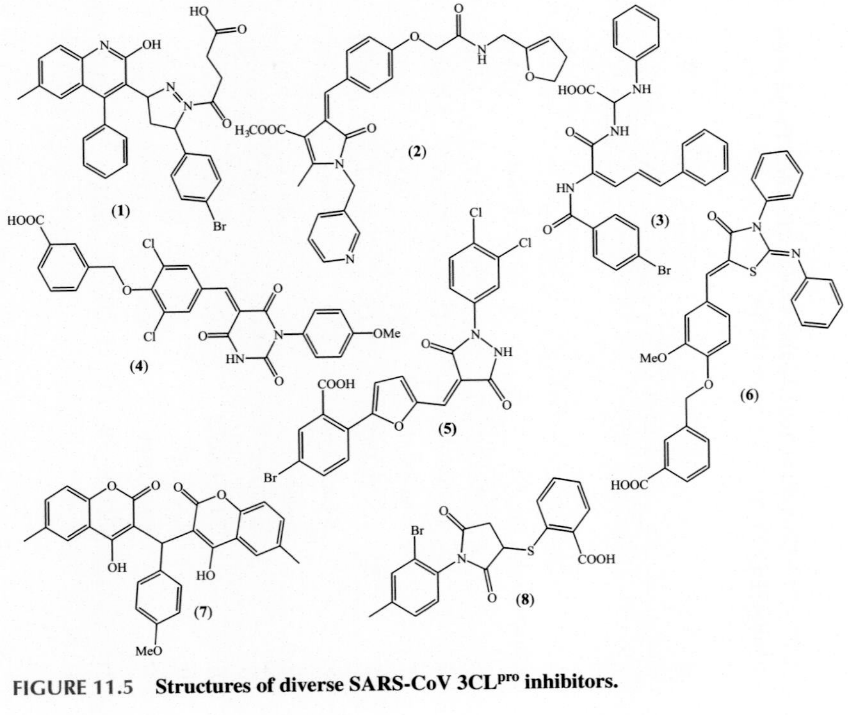

FIGURE 11.5 Structures of diverse SARS-CoV 3CL$^{pro}$ inhibitors.

**TABLE 11.9** Biological Activity and Physicochemical Parameters of a Diverse Set of SARS-CoV 3CL$^{pro}$ Inhibitors (Fig. 11.5) for QSAR Model [Eq. (11.8)]

| Compound | Obsd | Calcd | Res | Del res | Pred | PSA |
|---|---|---|---|---|---|---|
| 1 | 4.301 | 4.522 | −0.221 | −0.258 | 4.559 | 152.625 |
| 2 | 4.094 | 4.306 | −0.212 | −0.318 | 4.413 | 129.832 |
| 3 | 4.375 | 4.610 | −0.235 | −0.269 | 4.644 | 161.878 |
| 4 | 5.164 | 5.158 | 0.006 | 0.027 | 5.136 | 219.559 |
| 5[a] | 5.037 | 4.709 | 0.328 | 0.383 | 4.653 | 172.296 |
| 6 | 4.668 | 4.582 | 0.086 | 0.099 | 4.569 | 158.861 |
| 7[a] | 5.020 | 4.415 | 0.605 | 0.769 | 4.251 | 141.289 |
| 8 | 4.250 | 4.607 | −0.357 | −0.408 | 4.658 | 161.493 |

[a]Considered as outliers.

## 7.1.9 Isatin Analogs as SARS-CoV 3CL$^{pro}$ Inhibitors

Zhou et al. (2006) reported some isatin analogs as SARS-CoV 3CL$^{pro}$ inhibitors (Table 11.10), for which the correlation obtained was as in Eq. (11.9):

$$pIC_{50} = -5.298\,(\pm1.846) + 0.020\,(\pm0.004)\,SA \qquad (11.9)$$

$N = 7$, $R = 0.929$, $R^2 = 0.862$, $R_A^2 = 0.835$, $F\,(1, 5) = 31.349$, $P < 0.00251$, SEE = 0.347, $q^2 = 0.688$, $Q = 2.677$, Outlier = Compound **4**

It was observed from Eq. (11.9) that increasing the SA of these molecules may impart higher inhibitory activity. Bulky substitution at the $R_1$ position, such as β-napthylmethyl (compounds **3** and **8**, Table 11.10) may impart higher SA and hence, produce higher activity. Thus substitution with –CONH$_2$ function at the $R_2$ position in place of iodo function may have a better effect (compound **8** vs. **3**, compound **6** vs. **1**, and compound **7** vs. **2**, Table 11.10). Similarly, bulky aryl function may be more favorable than the alkyl function. The larger SA may help the molecule to occupy more space in the enzyme active site to have better binding interaction as evidenced by the molecular docking analysis (Zhou et al., 2006). Compound **8** having maximum SA exhibits hydrogen bonding with His41 and Cys145 through the keto functions of the isatin moiety. Moreover, the carboxamide function at the $R_2$ position makes hydrogen bonding with Phe140 and His163. The β-napthyl moiety (Compound **8**) fits well into the hydrophobic S2 pocket whereas smaller and less bulky substituents, such as methyl (Compound **4**), n-propyl (Compound **5**), n-butyl (Compound **6**), and benzyl (Compound **7**) do not accommodate well into the S2 pocket. It is not clear why the compound **4** behaves aberrantly though possessing a comparable good SA. Therefore, compound **4** may be considered as an outlier.

TABLE 11.10 Biological Activity and Physicochemical Parameters of Isatin Analogs as SARS-CoV 3CL$^{pro}$ Inhibitors for QSAR Model [Eq. (11.9)]

| Compound | R$_1$ | R$_2$ | Obsd | Calcd | Res | Del res | Pred | SA |
|---|---|---|---|---|---|---|---|---|
| 1 | n-Butyl | I | 4.180 | 4.665 | −0.485 | −0.564 | 4.745 | 469.003 |
| 2 | Benzyl | I | 4.301 | 4.902 | −0.601 | −0.686 | 4.987 | 485.295 |
| 3 | β-Napthylmethyl | I | 5.959 | 5.730 | 0.228 | 0.329 | 5.630 | 542.453 |
| 4$^a$ | Me | CONH$_2$ | 4.149 | 3.613 | 0.535 | 1.243 | 2.906 | 396.444 |
| 5 | n-Propyl | CONH$_2$ | 4.602 | 4.421 | 0.181 | 0.223 | 4.379 | 452.138 |
| 6 | n-Butyl | CONH$_2$ | 4.721 | 4.869 | −0.148 | −0.169 | 4.890 | 483.059 |
| 7 | Benzyl | CONH$_2$ | 4.903 | 5.103 | −0.199 | −0.231 | 5.134 | 499.156 |
| 8 | β-Napthylmethyl | CONH$_2$ | 6.432 | 5.944 | 0.488 | 0.829 | 5.603 | 557.185 |

$^a$Considered as outliers.

## 7.1.10    A Diverse Set of Potent SARS-CoV 3CL$^{pro}$ Inhibitors

Tsai et al. (2006) reported a series of SARS-CoV 3CL$^{pro}$ inhibitors through pharmacophore mapping and virtual screening approach (Fig. 11.6; Table 11.11). For this, the QSAR model obtained was as shown by Eq. (11.10).

$$pIC_{50} = 3.050\,(\pm 0.544) + 1.044\,(\pm 0.151)\,CMR - 0.204\,(\pm 0.033)\,D_Y \\ - 0.010\,(\pm 0.001)\,Vol \tag{11.10}$$

$N = 24$, $R = 0.915$, $R^2 = 0.836$, $R_A^2 = 0.812$, $F\,(3,\,20) = 34.080$, $P < 0.00000$, SEE $= 0.282$, $q^2 = 0.754$, $Q = 3.245$, Outlier $=$ Compounds **9, 13, 24**

It was observed from Eq. (11.10) that increasing the value of the molar refractivity (CMR) and decreasing the value of the dipole moment along $Y$-axis ($D_Y$), as well as the volume (Vol) may contribute positively to the enzyme inhibitory activity. It was, therefore, suggested that increasing the total molecular bulk may increase the activity whereas bulky substituent along $Y$-axis may be detrimental to the activity. Bulky substitution along $Y$-axis may produce some unfavorable steric interaction with the enzyme. Therefore, the bulky molecule with less steric effect may be favorable for the activity. Compounds **9, 13,** and **24** (Table 11.11) may act through different mechanism(s) of action and hence, they were considered as outliers.

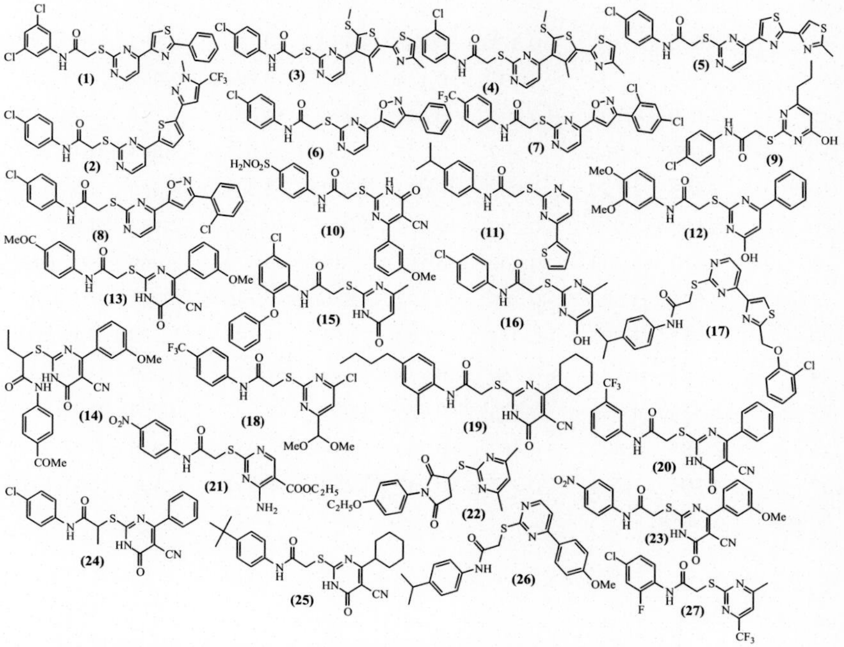

**FIGURE 11.6    Structures of some potent SARS-CoV 3CL$^{pro}$ inhibitors.**

TABLE 11.11 Biological Activity and Physicochemical Parameters of a Diverse Set of Potent SARS-CoV 3CL$^{pro}$ Inhibitors (Fig. 11.6) for QSAR Model [Eq. (11.10)]

| Compound | Obsd | Calcd | Res | Del res | Pred | CMR | $D_Y$ | Vol |
|---|---|---|---|---|---|---|---|---|
| 1 | 5.523 | 5.092 | 0.431 | 0.557 | 4.966 | 12.519 | −3.187 | 1169.180 |
| 2 | 5.000 | 4.615 | 0.385 | 0.484 | 4.516 | 12.431 | −5.072 | 1263.640 |
| 3 | 4.959 | 5.005 | −0.046 | −0.057 | 5.016 | 14.035 | −2.218 | 1326.330 |
| 4 | 4.921 | 4.902 | 0.019 | 0.022 | 4.899 | 14.035 | −2.750 | 1351.480 |
| 5 | 4.854 | 4.876 | −0.022 | −0.026 | 4.880 | 12.090 | −4.583 | 1180.750 |
| 6 | 4.824 | 4.491 | 0.333 | 0.355 | 4.469 | 11.433 | −3.308 | 1132.790 |
| 7 | 4.824 | 4.455 | 0.369 | 0.411 | 4.413 | 12.434 | −3.499 | 1251.610 |
| 8 | 4.824 | 4.812 | 0.012 | 0.014 | 4.810 | 11.924 | −3.530 | 1148.570 |
| 9[a] | 4.523 | 3.698 | 0.825 | 0.948 | 3.575 | 8.952 | −2.453 | 948.018 |
| 10 | 4.398 | 4.131 | 0.267 | 0.283 | 4.114 | 11.976 | −0.619 | 1182.810 |
| 11 | 4.398 | 4.313 | 0.085 | 0.091 | 4.307 | 10.628 | −3.433 | 1070.780 |
| 12 | 4.347 | 4.220 | 0.127 | 0.135 | 4.212 | 10.814 | −3.237 | 1099.490 |
| 13[a] | 4.222 | 3.362 | 0.860 | 1.193 | 3.029 | 11.698 | 2.871 | 1180.190 |
| 14 | 4.222 | 4.050 | 0.172 | 0.226 | 3.996 | 12.626 | 1.981 | 1209.220 |
| 15 | 4.000 | 4.371 | −0.371 | −0.469 | 4.469 | 10.749 | −0.795 | 1019.780 |
| 16 | 3.699 | 3.794 | −0.095 | −0.124 | 3.823 | 8.024 | −2.450 | 833.212 |
| 17 | 3.699 | 4.193 | −0.494 | −0.699 | 4.398 | 14.036 | −1.711 | 1424.160 |
| 18 | 3.699 | 3.986 | −0.287 | −0.331 | 4.030 | 9.615 | −3.899 | 1013.380 |
| 19 | 3.699 | 3.759 | −0.060 | −0.073 | 3.772 | 12.531 | −0.188 | 1284.230 |
| 20 | 3.699 | 3.998 | −0.299 | −0.315 | 4.014 | 10.628 | −2.143 | 1085.260 |
| 21 | 3.602 | 3.703 | −0.101 | −0.112 | 3.714 | 9.476 | −2.069 | 996.643 |
| 22 | 3.523 | 3.808 | −0.285 | −0.327 | 3.850 | 9.710 | −3.228 | 1033.240 |
| 23 | 3.523 | 3.557 | −0.034 | −0.040 | 3.563 | 11.346 | 1.299 | 1149.300 |
| 24[a] | 3.523 | 4.283 | −0.760 | −0.871 | 4.394 | 11.073 | −0.759 | 1066.380 |
| 25 | 3.456 | 4.031 | −0.575 | −0.625 | 4.081 | 12.067 | 0.024 | 1192.450 |
| 26 | 3.398 | 3.887 | −0.489 | −0.539 | 3.937 | 11.435 | −1.858 | 1182.620 |
| 27 | 3.301 | 3.264 | 0.037 | 0.047 | 3.255 | 8.397 | −0.405 | 901.161 |

[a]Considered as outliers.

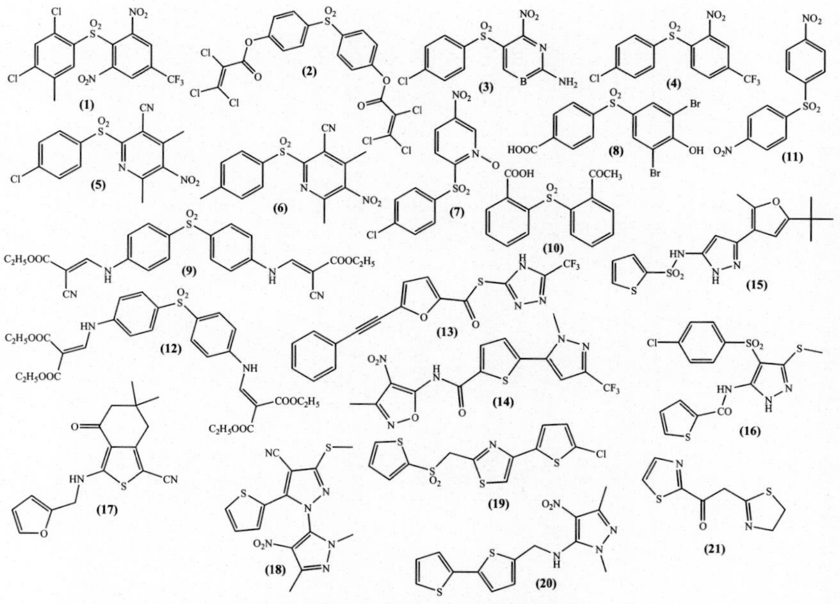

FIGURE 11.7 **Structures of nonpeptide SARS-CoVM$^{pro}$ inhibitors.**

## 7.1.11 A Series of Nonpeptide SARS-CoV 3CL$^{pro}$ Inhibitors

Lu et al. (2006) reported a series of nonpeptide SARS-CoV M$^{pro}$ inhibitors (Fig. 11.7; Table 11.12) through structure-based drug design approach, for which the QSAR model obtained was as shown by Eq. (11.11). It was observed from this equation that the increase in the value of the SA and the polar volume (Pol Vol) may be conducive to the activity, whereas the increasing in the value of volume and dipole moment along $X$-axis ($D_X$) might be detrimental to the activity. Thus it could be suggested that bulky substitutions along $X$-axis might produce unfavorable steric hindrance that may lower the activity. Moreover, this model also revealed that compounds having higher polar volume may favor the activity compared to compounds with lower polar volume. In compound **1** (Table 11.12), one of the nitro groups is closer to the imidazole function of His41 and thus there may be some electrostatic interaction between them leading to better activity. Moreover, the phenyl ring may form π–π interactions with His237 at the enzyme active site leading to potent activity. Compounds **1** and **8** (Table 11.12) might act in a different manner and hence, they were considered as outliers.

$$pIC_{50} = 2.664\,(\pm 0.339) + 0.020\,(\pm 0.003)\,SA - 0.010\,(\pm 0.002)\,Vol$$
$$- 0.075\,(\pm 0.015)\,D_X + 0.003\,(\pm 0.001)\,PolVol \qquad (11.11)$$

$N = 19$, $R = 0.911$, $R^2 = 0.830$, $R_A^2 = 0.781$, $F\,(1, 14) = 17.089$, $P < 0.00003$, SEE $= 0.178$, $q^2 = 0.605$, $Q = 5.118$, Outlier = Compounds **1, 8**

TABLE 11.12 Biological Activity and Physicochemical Parameters of a Series of Nonpeptide SARS-CoV 3CL$^{pro}$ Inhibitors (Fig. 11.7) for QSAR Model [Eq. (11.11)]

| Compound | Obsd | Calcd | Res | Del res | Pred | $D_X$ | Pol Vol | SA | Vol |
|---|---|---|---|---|---|---|---|---|---|
| 1[a] | 6.523 | 4.965 | 1.558 | 1.879 | 4.644 | −1.109 | 177.492 | 547.807 | 939.699 |
| 2 | 6.046 | 5.466 | 0.579 | 0.931 | 5.115 | −1.905 | 182.783 | 733.597 | 1184.370 |
| 3 | 5.222 | 5.057 | 0.165 | 0.212 | 5.010 | −0.568 | 251.310 | 497.859 | 803.261 |
| 4 | 4.921 | 5.232 | −0.311 | −0.389 | 5.310 | −4.454 | 158.771 | 509.445 | 843.770 |
| 5 | 4.886 | 4.955 | −0.069 | −0.073 | 4.960 | 0.390 | 216.773 | 526.936 | 873.061 |
| 6 | 4.886 | 5.053 | −0.167 | −0.180 | 5.066 | −0.772 | 211.376 | 534.514 | 881.205 |
| 7 | 4.824 | 4.559 | 0.265 | 0.366 | 4.458 | 7.202 | 245.339 | 483.718 | 773.796 |
| 8[a] | 4.796 | 5.293 | −0.497 | −0.851 | 5.647 | 0.780 | 299.398 | 829.728 | 1389.050 |
| 9 | 4.796 | 5.013 | −0.217 | −0.235 | 5.031 | −0.547 | 171.214 | 524.341 | 865.596 |
| 10 | 4.796 | 4.991 | −0.196 | −0.242 | 5.038 | −1.866 | 136.667 | 477.650 | 800.051 |
| 11 | 4.602 | 4.655 | −0.053 | −0.072 | 4.674 | 6.493 | 261.992 | 497.278 | 787.197 |
| 12 | 4.495 | 4.811 | −0.316 | −1.040 | 5.534 | 2.675 | 283.177 | 773.777 | 1382.880 |
| 13 | 5.523 | 5.228 | 0.295 | 0.340 | 5.183 | −0.933 | 175.089 | 612.441 | 988.794 |
| 14 | 5.301 | 5.179 | 0.122 | 0.141 | 5.160 | −1.175 | 255.605 | 614.602 | 1017.590 |
| 15 | 5.000 | 5.216 | −0.216 | −0.247 | 5.247 | −2.424 | 233.334 | 638.284 | 1077.980 |
| 16 | 4.824 | 5.028 | −0.204 | −0.294 | 5.118 | −2.348 | 289.728 | 543.817 | 943.419 |
| 17 | 4.796 | 4.945 | −0.149 | −0.176 | 4.972 | 0.299 | 141.590 | 526.835 | 873.189 |
| 18 | 4.745 | 4.559 | 0.186 | 0.334 | 4.410 | 5.607 | 371.972 | 528.119 | 903.617 |
| 19 | 4.745 | 4.905 | −0.161 | −0.207 | 4.952 | 3.748 | 158.165 | 534.331 | 834.408 |
| 20 | 4.699 | 4.788 | −0.089 | −0.111 | 4.810 | 4.862 | 190.284 | 564.125 | 911.548 |
| 21 | 4.398 | 4.921 | −0.523 | −0.711 | 5.109 | −0.151 | 98.696 | 388.069 | 601.894 |

[a]Considered as outliers.

### 7.1.12 Quercetin-3-β-Galactoside SARS-CoV 3CL^pro Inhibitors

Chen et al. (2006b) reported some quercetin-3-β-galactoside and its analogs as promising SARS-CoV 3CL^pro inhibitors (Table 11.13), for which the QSAR model obtained was as in Eq. (11.12):

$$pIC_{50} = 6.112\,(\pm 0.057) - 0.005\,(\pm 0.0002)\,PSA \tag{11.12}$$

$N = 4$, $R = 0.999$, $R^2 = 0.998$, $R_A^2 = 0.996$, $F\,(1, 2) = 842.36$, $P < 0.00119$, SEE $= 0.008$, $q^2 = 0.984$, $Q = 124.875$, Outlier $=$ Compound **4**

It was observed from Eq. (11.12) that decreasing the value of the PSA would have the positive effect on the biological activity. It meant that less polar molecules would be preferred to the high polar molecules. Due to the presence of a number of hydroxyl groups, these molecules may interact with the enzyme as hydrogen bond acceptors. The molecular modeling study revealed that the side chain of Gln189 forms four hydrogen bonds with compound **5** (Table 11.13), whereas two hydrogen bonding interactions are observed with the nitrogen atom of Glu166. It was, however, observed that compound **4** having the highest PSA value due to the presence of two galactose rings was less active. Probably, compound **4** might behave in an aberrant fashion and hence, it was considered as an outlier.

### 7.1.13 Phthalhydrazide Ketones as Potent SARS-CoV 3CL^pro Inhibitors

Zhang et al. (2007) synthesized and evaluated some phthalhydrazide ketones (Table 11.14) and heteroatomic ester as potential SARS-3CL^pro inhibitors. The QSAR model developed for this set of compounds was as shown by Eq. (11.13):

$$pIC_{50} = 8.108\,(\pm 0.131) - 0.009\,(\pm 0.001)\,PolVol \tag{11.13}$$

$N = 6$, $R = 0.970$, $R^2 = 0.942$, $R_A^2 = 0.927$, $F\,(1, 4) = 64.539$, $P < 0.00130$, SEE $= 0.072$, $q^2 = 0.739$, $Q = 13.472$, Outlier $=$ Compounds **2, 7**

Eq. (11.13) suggested that high polar volume of the compound would not favor the activity. A molecular modeling study had revealed that the halopyridine moiety of the compounds was well accommodated in the S1 binding pocket where it could have van der Waals interactions. Moreover, it was observed that the halogen atom does not interact with the enzyme and is directed toward the solvent exposed area. The furyl group of compound **3** is located near the catalytic Cys145 residue where it can have hydrophobic interaction. Compounds **2** and **7** being a misfit in the correlation were excluded.

### 7.1.14 Some Peptidomimetic SARS-CoV 3CL^pro Inhibitors

Ghosh et al. (2007) reported some peptidomimetic SARS-CoV 3CL^pro inhibitors (Table 11.15), for which a QSAR model obtained was as:

$$pIC_{50} = 5.009\,(\pm 0.174) - 0.504\,(\pm 0.067)\,D_Z \tag{11.14}$$

**TABLE 11.13 Biological Activity and Physicochemical Parameters of Quercetin-3-β-Galactoside SARS-CoV 3CL^pro Inhibitors for QSAR Model [Eq. (11.12)]**

| Compound | R₁ | R₂ | Obsd | Calcd | Res | Del res | Pred | PSA |
|---|---|---|---|---|---|---|---|---|
| 1 | H | L-Fucose | 4.617 | 4.568 | 0.049 | 0.099 | 4.518 | 306.747 |
| 2 | H | D-Arabinose | 4.500 | 4.486 | 0.014 | 0.019 | 4.481 | 333.869 |
| 3 | H | D-Glucose | 4.311 | 4.378 | −0.067 | −0.084 | 4.395 | 369.951 |
| 4[a] | D-Galactose | D-Galactose | 4.211 | 4.166 | 0.045 | 0.236 | 3.975 | 440.138 |
| 5 | H | D-Galactose | 4.369 | 4.410 | −0.041 | −0.052 | 4.420 | 359.227 |

*[a]Considered as outliers.*

**TABLE 11.14** Biological Activity and Physicochemical Parameters of Phthalhydrazide Ketones as Potent SARS-CoV 3CL$^{pro}$ Inhibitors for QSAR Model [Eq. (11.13)]

| Compound | Ar | X | Obsd | Calcd | Res | Del res | Pred | Pol Vol |
|---|---|---|---|---|---|---|---|---|
| 1 | 2-Furyl | Cl | 7.222 | 7.134 | 0.088 | 0.115 | 7.106 | 93.111 |
| 2[a] | Benzofuran-2-yl | Cl | 6.770 | 7.086 | −0.317 | −0.388 | 7.158 | 99.866 |
| 3 | 2-Furyl | Br | 7.301 | 7.144 | 0.157 | 0.211 | 7.090 | 91.748 |
| 4 | 2-Indolyl | Cl | 7.187 | 6.993 | 0.194 | 0.222 | 6.965 | 113.106 |
| 5 | 2-Benzothiophenyl | Cl | 7.022 | 6.889 | 0.133 | 0.157 | 6.866 | 127.948 |
| 6 | Thiazole-4-yl | Cl | 6.569 | 6.562 | 0.006 | 0.031 | 6.537 | 174.454 |
| 7[a] | 3-OMePh | Cl | 6.469 | 6.962 | −0.493 | −0.564 | 7.032 | 117.647 |
| 8 | 5-(4-ClPh)-Furan-2-yl | Cl | 7.201 | 6.969 | 0.231 | 0.264 | 6.936 | 116.549 |

[a]*Considered as outliers.*

**TABLE 11.15 Biological Activity and Physicochemical Parameters of Some Peptidomimetic SARS-CoV 3CL$^{pro}$ Inhibitors for QSAR Model [Eq. (11.14)]**

| Compound | R | $R_1$ | Obsd | Calcd | Res | Del res | Pred | $D_Z$ |
|---|---|---|---|---|---|---|---|---|
| 1 | 5-Me-isoxazolyl-3-yl | Benzyl | 3.060 | 3.334 | −0.273 | −0.359 | 3.419 | 3.323 |
| 2 | 5-Me-isoxazolyl-3-yl | $CH_2CH{=}C(Me)_2$ | 3.097 | 3.365 | −0.268 | −0.348 | 3.444 | 3.260 |
| 3 | $CH(CH_2OH)NHBoc$ | $CH_2CH{=}C(Me)_2$ | 4.097 | 4.062 | 0.035 | 0.041 | 4.055 | 1.879 |
| 4 | $CH(CH_2OH)NHBoc$ | $i$-Butyl | 5.000 | 5.049 | −0.049 | −0.106 | 5.106 | −0.080 |
| 5 | $CH(CH_2OH)NHBoc$ | Benzyl | 4.824 | 4.786 | 0.038 | 0.061 | 4.763 | 0.442 |
| 6 | 5-Me-isoxazolyl-3-yl | Benzyl | 3.523 | 3.334 | 0.189 | 0.249 | 3.274 | 3.323 |
| 7 | 5-Me-isoxazolyl-3-yl | $i$-Butyl | 3.699 | 3.371 | 0.328 | 0.424 | 3.275 | 3.249 |

$N = 7$, $R = 0.959$, $R^2 = 0.919$, $R_A^2 = 0.903$, $F$ (1, 5) = 56.603, $P < 0.00066$, SEE = 0.243, $q^2 = 0.860$, $Q = 3.947$

It was observed from Eq. (11.14) that increasing the value of the dipole moment along Z-axis $(D_Z)$ will lead to decrease the enzyme inhibitory activity. It thus suggested that the bulky substituent along Z-axis will not be conducive to the activity. The long chain linear aminobutoxy derivatives (Compounds **4** and **5**, Table 11.15) are better than the isoxazole analogs (Compounds **1**, **2**, **6**, and **7**, Table 11.15) as the isoxazole moiety may produce more bulkiness that may impart unfavorable steric effect with the enzyme.

### 7.1.15 Arylmethylene Ketones and Fluorinated Methylene Ketones as SARS-CoV 3CL$^{pro}$ Inhibitors

Zhang et al. (2008) reported some arylmethylene ketones and fluorinated methylene ketones as SARS-CoV 3CL$^{pro}$ inhibitors (Table 11.16). The QSAR model for them was as shown by Eq. (11.15), where the indicator variable "I" stands for a value of unity for the ester group. A positive coefficient of it suggested that the ester group may be favorable for imparting the higher inhibitory activity. Compounds bearing ester functions (Compounds **2–4**, Table 11.16) are highly active compared to the nonester derivatives (Compounds **5–8**, Table 11.16). Compounds **5–7** are found to be oriented from S1 to S4 pocket and the furan oxygen atom forms hydrogen bonds with the amino function of Glu166. Moreover, it is observed that compound **1** though having ester function, may behave in an aberrant fashion. Therefore, it was considered as an outlier.

$$pIC_{50} = 4.452\,(\pm 0.133) + 2.789\,(\pm 0.203)\,I \qquad (11.15)$$

$N = 7$, $R = 0.987$, $R^2 = 0.974$, $R_A^2 = 0.969$, $F$ (1, 5) = 188.18, $P < 0.00004$, SEE = 0.266, $q^2 = 0.954$, $Q = 3.711$, Outlier = Compound **1**

### 7.1.16 Chloropyridine Esters as Potent SARS-CoV 3CL$^{pro}$ Inhibitors

The QSAR model obtained for a series of chloropyridine esters reported by Niu et al. (2008) as potent SARS-CoV 3CL$^{pro}$ inhibitors (Table 11.17) was as shown by Eq. (11.16), which suggested that increasing the molar refractivity and decreasing the total dipole moment may favor 3CL protease inhibitory activity

$$pIC_{50} = 3.460\,(\pm 0.429) + 0.470\,(\pm 0.063)\,CMR - 0.067\,(\pm 0.019)\,D_{Tot} \qquad (11.16)$$

$N = 10$, $R = 0.945$, $R^2 = 0.892$, $R_A^2 = 0.861$, $F$ (2, 7) = 29.024, $P < 0.00041$, SEE = 0.148, $q^2 = 0.778$, $Q = 6.385$, Outlier = Compound **5, 6**.

This also suggested that the smaller molecules with less steric effect may be conducive to the inhibitory activity. The α-naphthyl (Compound **11**, Table 11.17) and the 2-oxochromene function (Compound **12**, Table 11.17) at Ar position yield less dipole moment and better molar refractivity compared to the nitrophenyl (Compounds **9** and **10**), the chlorophenyl (Compound **8**) or the

TABLE 11.16 Biological Activity and Physicochemical Parameters of Arylmethylene Ketones and Fluorinated Methylene Ketones as SARS-CoV 3CL$^{pro}$ Inhibitors for QSAR Model [Eq. (11.15)]

| Compound | R | X | Y | Z | W | Obsd | Calcd | Res | Del res | Pred | I |
|---|---|---|---|---|---|---|---|---|---|---|---|
| 1[a] | H | H | CH | O | — | 5.102 | 6.706 | −1.604 | −2.139 | 7.241 | 1 |
| 2 | H | Br | CH | O | — | 7.301 | 6.706 | 0.595 | 0.793 | 6.508 | 1 |
| 3 | H | Cl | CH | O | — | 7.222 | 6.706 | 0.515 | 0.687 | 6.535 | 1 |
| 4 | 4-ClPh | Cl | CH | O | — | 7.201 | 6.706 | 0.494 | 0.659 | 6.542 | 1 |
| 5 | 4-ClPh | Br | CH | CH$_2$ | — | 4.886 | 4.452 | 0.434 | 0.579 | 4.307 | 0 |
| 6 | 4-ClPh | Br | CH | CH | F | 4.553 | 4.452 | 0.101 | 0.134 | 4.418 | 0 |
| 7 | 4-ClPh | Br | CH | C | F,F | 4.244 | 4.452 | −0.208 | −0.277 | 4.521 | 0 |
| 8 | 4-ClPh | Br | N | CH$_2$ | — | 4.125 | 4.452 | −0.327 | −0.436 | 4.561 | 0 |

[a]Considered as outliers.

**TABLE 11.17** Biological Activity and Physicochemical Parameters of Chloropyridine Esters as Potent SARS-CoV 3CL$^{pro}$ Inhibitors for QSAR Model [Eq. (11.16)]

(1–5)                (6–12)

| Compound | R | Ar | Obsd | Calcd | Res | Del res | Pred | CMR | $D_{Tot}$ |
|---|---|---|---|---|---|---|---|---|---|
| 1 | 4-ClPh | — | 7.201 | 7.087 | 0.114 | 0.156 | 7.045 | 8.349 | 3.682 |
| 2 | 4-NO$_2$Ph | — | 7.222 | 7.022 | 0.200 | 0.252 | 6.970 | 8.469 | 5.001 |
| 3 | 2-NO$_2$,4-ClPh | — | 6.914 | 6.789 | 0.125 | 0.184 | 6.730 | 8.961 | 10.027 |
| 4 | 2-NO$_2$Ph | — | 6.682 | 6.594 | 0.088 | 0.130 | 6.552 | 8.469 | 10.059 |
| 5[a] | 3-NO$_2$Ph | — | 6.301 | 6.727 | -0.426 | -0.534 | 6.835 | 8.469 | 8.490 |
| 6[a] | — | 4-Pyr | 6.785 | 6.378 | 0.407 | 0.568 | 6.217 | 5.922 | 0.861 |
| 7 | — | 3-Pyr | 6.157 | 6.348 | -0.191 | -0.266 | 6.423 | 5.922 | 1.216 |
| 8 | — | 4-ClPh | 6.363 | 6.633 | -0.270 | -0.331 | 6.693 | 6.624 | 1.092 |
| 9 | — | 2-NO$_2$Ph | 6.478 | 6.260 | 0.217 | 0.305 | 6.172 | 6.744 | 6.047 |
| 10 | — | 3-NO$_2$Ph | 6.165 | 6.371 | -0.206 | -0.252 | 6.417 | 6.744 | 4.736 |
| 11 | — | α-Napthyl | 6.907 | 7.022 | -0.115 | -0.157 | 7.064 | 7.821 | 2.014 |
| 12 | — | 2-Oxo-chromene | 6.967 | 6.909 | 0.058 | 0.071 | 6.896 | 7.607 | 2.361 |

[a]Considered as outliers.

pyridyl analog (Compound **7**) and thus compounds **11** and **12** are more potent than compounds **7–10**. Furyl derivatives (Compounds **1–4**) are better inhibitors as compared to the other aryl ester analogs (Compounds **7–12**) as they have higher molar refractivity despite having comparatively moderate bulky *p*-chlorophenyl or the *p*-nitrophenyl groups at R position. A slight reduction in the activity is noticed for the disubstituted aryl function (Compound **3**) and the alteration of the nitro function at the 2nd position of the phenyl ring (Compound **4**) in contrast to the 4th position (Compound **2**), which increases the bulkiness or total bulk, and reduces the activity slightly. It is observed from the molecular modeling study that increasing the length of the side chain may increase the interaction between S2 and S4 pocket and the inhibitor that can be reflected by the QSAR model. Compound **5** possesses the higher molar refractivity while compound **6** possesses the lowest value of total dipole moment but it is not reflected in their activity. Probably, these compounds behave differently from other compounds in the dataset and hence were outliers.

### 7.1.17 Cinanserin Analogs as Promising SARS-CoV 3CL$^{pro}$ Inhibitors

Yang et al. (2008) reported some cinanserin analogs as SARS-CoV 3CL$^{pro}$ inhibitors (Table 11.18), for which the QSAR model obtained was as shown by Eq. (11.17). This equation clearly exhibited that high molecular volume will not be favorable to the activity. Thus compounds having aryl (Compound **4**), the heteroaryl (Compounds **6**, **7**), or the long chain amide function (Compounds **1**, **2**) at Y position have the lower activity than compounds having at this position the unsaturation (Compounds **8**, **9**) or the ester function (Compound **5**). Compound **8** enters into the deep S1 pocket and has hydrophobic interactions. However, compound **3** has the lowest volume but it does not show the highest activity. Probably, this compound (Compound **3**) may behave in a different manner with the enzyme and hence it is considered as an outlier.

$$pIC_{50} = 24.586\,(\pm 3.262) - 0.020\,(\pm 0.003)\,\text{Vol} \tag{11.17}$$

$N = 8$, $R = 0.931$, $R^2 = 0.867$, $R_A^2 = 0.845$, $F\,(1, 6) = 39.280$, $P < 0.00077$, SEE = 0.347, $q^2 = 0.686$, $Q = 2.683$, Outlier = Compound **3**

### 7.1.18 Trifluoromethyl, Benzothiazolyl, and Thiazolyl Ketone Compounds as Promising SARS-CoV 3CL$^{pro}$ Inhibitors

The QSAR model derived for some trifluoromethyl, benzothiazolyl, and thiazolyl ketone compounds with peptide side chain reported by Regnier et al. (2009) as promising SARS-CoV 3CL$^{pro}$ inhibitors (Table 11.19) was as shown by Eq. (11.18)

$$pK_i = 6.096\,(\pm 0.893) - 0.027\,(\pm 0.008)\,\text{Pol Vol} + 0.0001\,(\pm 0.000)\,\text{Pol Vol}^2 \tag{11.18}$$

$N = 15$, $R = 0.856$, $R^2 = 0.732$, $R_A^2 = 0.687$, $F\,(2, 12) = 16.394$, $P < 0.00037$, SEE = 0.375, $q^2 = 0.536$, $Q = 2.283$, Pol Vol$_{opt} = 135$.

**TABLE 11.18** Biological Activity and Physicochemical Parameters of Cinanserin Analogs as Promising SARS-CoV $3CL^{pro}$ Inhibitors for QSAR Model [Eq. (11.17)]

| Compound | X | Y | Obsd | Calcd | Res | Del res | Pred | Vol |
|---|---|---|---|---|---|---|---|---|
| 1 | H | $(CH_2)_3N(Me)_2$ | 3.491 | 3.964 | −0.473 | −0.542 | 4.033 | 1020.630 |
| 2 | CN | $(CH_2)_3N(Me)_2$ | 3.903 | 3.771 | 0.132 | 0.158 | 3.745 | 1034.690 |
| 3[a] | H | $CH_2C{\equiv}CH$ | 4.706 | 5.450 | −0.745 | −1.488 | 6.194 | 912.242 |
| 4 | H | Benzyl | 3.686 | 3.477 | 0.209 | 0.280 | 3.406 | 1056.120 |
| 5 | H | $(CH_2)_2COOMe$ | 4.870 | 4.741 | 0.128 | 0.157 | 4.713 | 963.919 |
| 6 | H | 2-Pyridylmethyl | 3.533 | 3.638 | −0.104 | −0.130 | 3.663 | 1044.400 |
| 7 | H | 3-Pyridylmethyl | 3.457 | 3.663 | −0.206 | −0.254 | 3.711 | 1042.560 |
| 8 | H | $COCH{=}CHPh$ | 5.975 | 5.019 | 0.955 | 1.318 | 4.657 | 943.645 |
| 9 | CN | $COC(CN){=}CHPh$ | 4.360 | 4.257 | 0.103 | 0.116 | 4.244 | 999.248 |

[a]Considered as outliers.

TABLE 11.19 Biological Activity and Physicochemical Parameters of Trifluoromethyl, Benzothiazolyl, and Thiazolyl Ketone Compounds as Promising SARS-CoV 3CL$^{pro}$ Inhibitors for QSAR Model [Eq. (11.18)]

| Compound | AA | X | Y | Obsd | Calcd | Res | Del res | Pred | Pol Vol |
|---|---|---|---|---|---|---|---|---|---|
| 1 | Cbz-Val-Leu-NH | OH | CF$_3$ | 3.936 | 3.959 | -0.023 | -0.026 | 3.961 | 416.522 |
| 2 | Cbz-Ala-Val-Leu-NH | NH$_2$ | CF$_3$ | 3.870 | 3.939 | -0.069 | -0.078 | 3.947 | 415.318 |
| 3 | Cbz-Val-Leu-NH | N(Et)$_2$ | CF$_3$ | 3.440 | 3.882 | -0.442 | -0.499 | 3.939 | 411.755 |
| 4 | Cbz-Val-Leu-NH | Morpholine | CF$_3$ | 4.678 | 4.267 | 0.411 | 0.455 | 4.223 | 434.513 |
| 5 | Cbz-Val-Leu-NH | N(Me)Benzyl | CF$_3$ | 4.467 | 3.779 | 0.688 | 0.795 | 3.672 | 405.044 |
| 6 | Cbz-Ala-Val-Leu-NH | N(Et)$_2$ | CF$_3$ | 3.527 | 3.426 | 0.101 | 0.135 | 3.393 | 379.365 |
| 7 | Cbz-Leu-NH | N(Et)$_2$ | CF$_3$ | 3.234 | 3.469 | -0.235 | -0.324 | 3.558 | 133.810 |
| 8 | Cbz-Val-Leu-NH | Morpholine | Thiazole-2-yl | 3.321 | 3.520 | -0.200 | -0.253 | 3.574 | 386.735 |
| 9 | Cbz-Val-Leu-NH | N(Et)$_2$ | Thiazole-2-yl | 3.951 | 4.131 | -0.180 | -0.199 | 4.150 | 426.813 |
| 10 | Cbz-Val-Leu-NH | 2-Oxo-pyrrolidin-3-yl | Thiazole-2-yl | 5.658 | 5.214 | 0.443 | 0.862 | 4.795 | 480.804 |
| 11 | Cbz-Val-Leu-NH | N(Et)$_2$ | Thiazole-2-yl | 4.345 | 4.787 | -0.442 | -0.576 | 4.921 | 461.239 |
| 12 | Cbz-Leu-NH | N(Et)$_2$ | Thiazole-2-yl | 3.335 | 3.316 | 0.020 | 0.026 | 3.310 | 146.432 |
| 13 | Cbz-Val-NH | N(Et)$_2$ | Thiazole-2-yl | 3.212 | 3.415 | -0.203 | -0.273 | 3.485 | 138.098 |
| 14 | Cbz-Val-Leu-NH | N(Et)$_2$ | Benzohiazole-2-yl | 4.307 | 4.604 | -0.297 | -0.355 | 4.662 | 452.238 |
| 15 | Cbz-Val-NH | N(Et)$_2$ | Benzohiazole-2-yl | 3.799 | 3.370 | 0.429 | 0.567 | 3.232 | 141.781 |

Eq. (11.18) showed that the enzyme inhibitory activity was correlated with the polar volume of the molecules through a parabolic relation. It, therefore, suggests that the activity would decrease upto an optimum value of polar volume (Pol Vol$_{opt}$ = 135) and beyond that will start increasing. Compound with the 2-oxo-pyrrolidin-3-yl function (Compound **10**, Table 11.19) possesses the higher polar volume and hence, possess the maximum inhibition. Comparing the activity of this compound with those of compounds **8, 10–13**, it may be suggested that the 2-oxo-pyrrolidin-3-yl function in compound **10** is favorable than the diethylamino function in compounds **11–13** and morpholino function in compound **8** at the X position. Moreover, the benzothiazole-2-yl function in compounds **14, 15** is favorable than the thiazole function in compounds **8, 12, 13**. The bulky group, such as the morpholino in compound **4** and the benzyl-methylamino function in compound **5** are favorable than the smaller substituents, such as the hydroxyl in compound **1**, the amino group in compound **2**, and the diethylamino group in compounds **3, 6**, and **7** at X position. Comparing compound **14** with **15**, it may be inferred that the bulky amino acid moiety (compound **14**) is favorable than smaller amino acid functions (Compound **15**), as bulky functions may produce the higher polar volume. The molecular modeling study revealed that the benzyloxycarbonyl moiety of compound **10** did not make any hydrophobic interaction rather had hydrogen bonding interactions with Glu166 through its adjacent amino function.

### 7.1.19    Pyrazolone Analogs as Promising SARS-CoV 3CL$^{pro}$ Inhibitors

Ramajayam et al. (2010) reported some pyrazolone analogs as promising SARS-CoV 3CL$^{pro}$ inhibitors (Table 11.20), for which the QSAR model obtained was as:

$$pIC_{50} = 9.292\,(\pm 0.347) - 0.281\,(\pm 0.074)\,C\log P \qquad (11.19)$$

$N = 7$, $R = 0.862$, $R^2 = 0.744$, $R_A^2 = 0.692$, $F\,(1, 5) = 14.512$, $P < 0.01251$, SEE = 0.119, $q^2 = 0.529$, $Q = 7.244$, Outlier = Compounds **1, 5**

Eq. (11.19) thus suggested that increasing value of the hydrophobicity of these molecules may be detrimental to the activity. Compounds with the smaller halogen substitution, such as fluorine at R position (Compound **8**, Table 11.20) are better than compounds with the bigger halogen substituents, such as the chloro (Compounds **2** and **6**, Table 11.20). Further, a dihalo substituted compound, such as compound **7** was shown to be less active as compared to mono-halo-substituted analogs (Compounds **2, 6**, and **8**). The cyano (Compound **4**) and the nitro (Compound **9**) substitutions also produced the higher activity as compared to the methoxy substitution (Compound **3**). The docking study suggested that the N1-phenyl group was located near to the S1' pocket. One of the oxygen atoms of the nitro group forms a hydrogen bond with Gly143. The keto function of the pyrazolone ring was also found to form another hydrogen bond

TABLE 11.20 Biological Activity and Physicochemical Parameters of Pyrazolone Analogs as Promising SARS-CoV 3CL$^{pro}$ Inhibitors for QSAR Model [Eq. (11.19)]

| Compound | R | Obsd | Calcd | Res | Del res | Pred | C Log P |
|---|---|---|---|---|---|---|---|
| 1[a] | H | 7.745 | 7.998 | −0.254 | −0.297 | 8.042 | 4.380 |
| 2 | 4-Cl | 7.857 | 7.760 | 0.097 | 0.115 | 7.742 | 5.093 |
| 3 | 4-OMe | 7.921 | 8.025 | −0.105 | −0.125 | 8.046 | 4.299 |
| 4 | 4-CN | 8.260 | 8.188 | 0.072 | 0.113 | 8.147 | 3.813 |
| 5[a] | 4-OCF$_3$ | 7.377 | 7.655 | −0.278 | −0.377 | 7.754 | 5.408 |
| 6 | 3-Cl | 7.967 | 7.760 | 0.207 | 0.245 | 7.722 | 5.093 |
| 7 | 3,4-diCl | 7.614 | 7.562 | 0.052 | 0.089 | 7.526 | 5.686 |
| 8 | 4-F | 8.167 | 7.951 | 0.217 | 0.247 | 7.920 | 4.523 |
| 9 | 3-NO$_2$ | 8.076 | 8.084 | −0.008 | −0.011 | 8.087 | 4.123 |

[a]Considered as outliers.

with Glu166. The C-3 phenyl ring was found to be well-accommodated in the S2 pocket. The benzylidene ring without any carboxyl functions lost the activity. Therefore, it may be assumed that hydrogen bonding interaction is more important than the hydrophobic interaction. The oxygen atom of the carboxyl group forms a hydrogen bond with Gln192. Therefore, apart from S2 pocket, none of the aryl functions has exhibited hydrophobic interactions, whereas three hydrogen bonding interactions were observed. Compounds **1** and **5** considered as outliers.

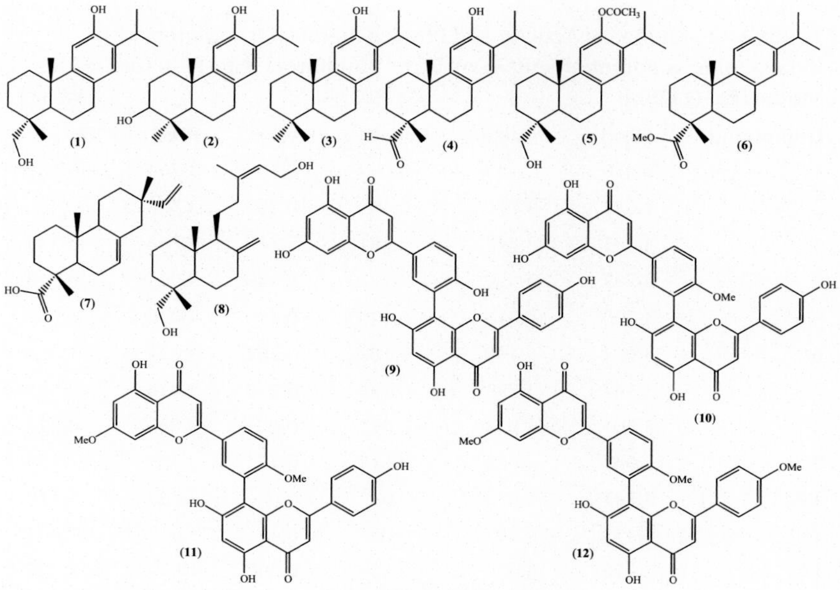

**FIGURE 11.8**  **General structure of some biflavonoid analogs.**

## 7.1.20 Biflavonoids as Potential SARS-CoV 3CL^pro Inhibitors

Ryu et al. (2010) reported a series of biflavonoids (Fig. 11.8; Table 11.21) from *Torreya nucifera* having potential SARS-CoV 3CL^pro inhibitory activity. The QSAR model obtained for them was as shown by Eq. (11.20). It was observed from Eq. (11.20) that the increasing value of the dipole moment along $X$-axis may be conducive to the activity. Thus, the bulky substitution at X-axis of these molecules may be favorable for activity. Compounds **10–12** (Table 11.21) possess higher dipole moment due to much bulky aryl groups as compared to the compounds **1–2**, **4–8** and, therefore, have higher activity. Compounds **3** and **9** exhibited the aberrant behavior and thus were considered as outliers.

$$p\text{IC}_{50} = 3.833(\pm0.037) + 0.212(\pm0.026)D_X \qquad (11.20)$$

$N = 10$, $R = 0.946$, $R^2 = 0.895$, $R_A^2 = 0.882$, $F\,(1,\,8) = 68.528$, $P < 0.00003$, SEE $= 0.114$, $q^2 = 0.805$, $Q = 8.298$, Outlier = Compounds **3, 9**

## 7.1.21 A Series of Some Promising SARS-CoV 3CL^pro Inhibitors

Nguyen et al. (2011) reported some promising SARS-CoV 3CL^pro inhibitors through virtual screening (Fig. 11.9; Table 11.22). The QSAR model obtained for these compounds was as shown by Eq. (11.21), which exhibited that the activity is well correlated with the hydrophobicity of the molecules. The docking

**TABLE 11.21 Biological Activity and Physicochemical Parameters of Biflavonoids as Potential SARS-CoV 3CL$^{pro}$ Inhibitors (Fig. 11.8) for QSAR Model [Eq. (11.20)]**

| Compound | Obsd | Calcd | Res | Del res | Pred | $D_X$ |
|---|---|---|---|---|---|---|
| 1 | 3.656 | 3.749 | −0.093 | −0.108 | 3.764 | −0.708 |
| 2 | 3.632 | 3.811 | −0.179 | −0.204 | 3.835 | −0.460 |
| 3$^a$ | 4.305 | 3.843 | 0.461 | 0.518 | 3.786 | −0.333 |
| 4 | 3.787 | 3.920 | −0.133 | −0.146 | 3.934 | −0.028 |
| 5 | 3.890 | 4.018 | −0.129 | −0.140 | 4.030 | 0.364 |
| 6 | 3.684 | 3.498 | 0.187 | 0.261 | 3.423 | −1.708 |
| 7 | 3.547 | 3.783 | −0.235 | −0.270 | 3.818 | −0.573 |
| 8 | 3.861 | 3.933 | −0.072 | −0.079 | 3.940 | 0.024 |
| 9$^a$ | 5.081 | 4.462 | 0.619 | 0.777 | 4.303 | 2.129 |
| 10 | 4.141 | 4.223 | −0.082 | −0.092 | 4.233 | 1.179 |
| 11 | 4.495 | 4.581 | −0.087 | −0.121 | 4.615 | 2.603 |
| 12 | 4.416 | 4.672 | −0.257 | −0.398 | 4.814 | 2.965 |

$^a$Considered as outliers.

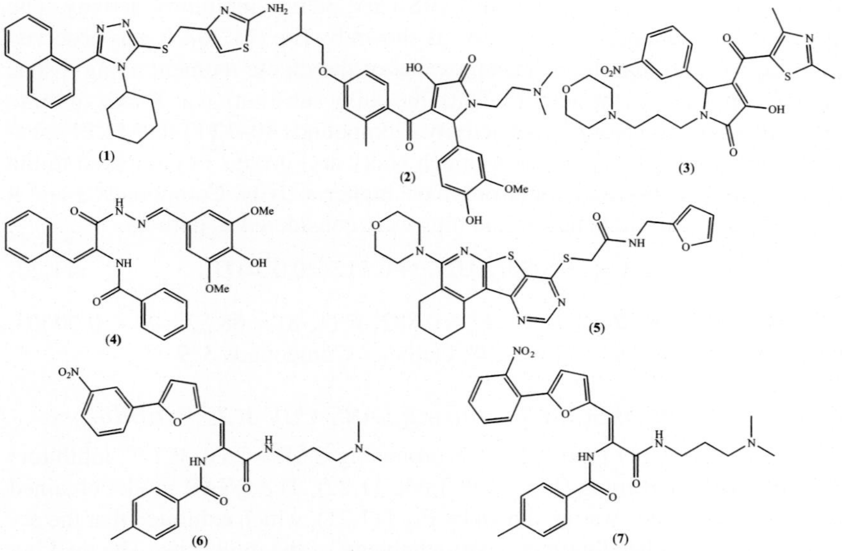

FIGURE 11.9   General structure of some potent SARS-CoV 3CL$^{pro}$ inhibitors.

**TABLE 11.22** Biological Activity and Physicochemical Parameters of a Series of Some Promising SARS-CoV 3CL^pro Inhibitors (Fig. 11.9) for QSAR Model [Eq. (11.21)]

| Compound | Obsd | Calcd | Res | Del res | Pred | C Log P |
|----------|-------|-------|--------|---------|-------|---------|
| 1 | 7.234 | 7.235 | −0.001 | −0.002 | 7.235 | 4.707 |
| 2[a] | 7.202 | 7.283 | −0.081 | −0.108 | 7.310 | 5.049 |
| 3 | 6.994 | 7.016 | −0.022 | −0.060 | 7.054 | 3.135 |
| 4[a] | 7.113 | 7.286 | −0.173 | −0.233 | 7.346 | 5.069 |
| 5 | 7.042 | 7.079 | −0.037 | −0.057 | 7.099 | 3.587 |
| 6 | 7.414 | 7.242 | 0.172 | 0.208 | 7.206 | 4.752 |
| 7 | 7.383 | 7.242 | 0.142 | 0.171 | 7.212 | 4.752 |

[a]Considered as outliers.

study had revealed that compound **7** (Table 11.22) had good hydrophobic interactions with His41, Phe140, Leu141, Cys145, His163, Glu166, Gly170, and His172 apart from a number of hydrogen bonding interactions (the nitro group with Gly143, methacrylamide group with Phe140, one of the oxygen atoms of the nitro group with Cys145). The nitrophenyl group was found to be the most crucial moiety to enter into the S1 pocket for imparting potent inhibition. Compounds **2** and **4** though possessed a higher value of hydrophobicity but less activity than expected, hence, they were considered as outliers.

$$pIC_{50} = 6.238\,(\pm0.213) + 0.233\,(\pm0.050)\,C\log P \tag{11.21}$$

$N = 5$, $R = 0.937$, $R^2 = 0.878$, $R_A^2 = 0.837$, $F\,(1, 3) = 21.594$, $P < 0.01879$, SEE = 0.077, $q^2 = 0.703$, $Q = 12.169$, Outlier = Compounds **2, 4**

### 7.1.22   Peptidomimetic SARS-CoV 3CL^pro Inhibitors

Some peptidomimetic SARS-CoV 3CL^pro inhibitors (Table 11.23) were synthesized and evaluated by Akaji et al. (2011) and the QSAR model obtained for this [Eq. (11.22)] indicated that the activity is controlled by a single indicator parameter "I" used for an imidazolyl-4-yl methyl substituent at the $R_1$ position. The positive coefficient of this indicated that such a substituent would conducive to the activity. The reason of this may be that this substituent might have better steric fitting in the S1 pocket of the enzyme formed by Phe140, Leu141, and Glu166. Compounds **4** and **8** (Table 11.23) were considered as outliers.

$$pIC_{50} = 4.319\,(\pm0.168) + 2.409\,(\pm0.213)\,I \tag{11.22}$$

$N = 8$, $R = 0.977$, $R^2 = 0.955$, $R_A^2 = 0.948$, $F\,(1, 6) = 128.20$, $P < 0.00003$, SEE = 0.291, $q^2 = 0.929$, $Q = 3.397$, Outlier = Compounds **4, 8**

**TABLE 11.23 Biological Activity and Physicochemical Parameters of Peptidomimetic SARS-CoV 3CL$^{pro}$ Inhibitors for QSAR Model [Eq. (11.22)]**

| Compound | R | R₁ | R₂ | Obsd | Calcd | Res | Del res | Pred | I |
|---|---|---|---|---|---|---|---|---|---|
| 1 | i-Butyl | (CH₂)₂CONMe₂ | Me | 4.432 | 4.319 | 0.112 | 0.169 | 4.263 | 0 |
| 2 | i-Butyl | c-Hexylmethyl | Me | 4.208 | 4.319 | −0.112 | −0.168 | 4.375 | 0 |
| 3 | i-Butyl | 2-Thiophenylmethyl | Me | 4.319 | 4.319 | −0.001 | −0.001 | 4.320 | 0 |
| 4ᵃ | i-Butyl | Imidazole-4-ylmethyl | Me | 5.244 | 6.127 | −0.882 | −1.030 | 6.274 | 1 |
| 5 | Benzyl | Imidazole-4-ylmethyl | Me | 6.409 | 6.127 | 0.282 | 0.329 | 6.080 | 1 |
| 6 | c-Hexylmethyl | Imidazole-4-ylmethyl | Me | 7.187 | 6.127 | 1.061 | 1.237 | 5.950 | 1 |
| 7 | c-Hexylmethyl | Imidazole-4-ylmethyl | Me | 6.569 | 6.127 | 0.442 | 0.516 | 6.053 | 1 |
| 8ᵃ | c-Hexylmethyl | Imidazole-4-ylmethyl | CH₂CONH₂ | 4.000 | 6.127 | −2.127 | −2.481 | 6.481 | 1 |
| 9 | c-Hexylmethyl | Imidazole-4-ylmethyl | CH₂OH | 6.469 | 6.127 | 0.342 | 0.399 | 6.070 | 1 |
| 10 | c-Hexylmethyl | Imidazole-4-ylmethyl | CH(OH)Me | 7.009 | 6.127 | 0.882 | 1.029 | 5.980 | 1 |

ᵃConsidered as outliers

## 7.1.23 Flavonoids as SARS-CoV 3CL$^{pro}$ Inhibitors

Nguyen et al. (2012) reported some flavonoids from *Pichia pastoris* (Fig. 11.10; Table 11.24) having SARS-CoV 3CL$^{pro}$ inhibitory activity. For these compounds, the inhibition activity was shown to be correlated with the PSA of the molecule [Eq. (11.23)], suggesting that highly polar molecules may have better activity. Substituents like hydroxy might give better PSA, leading to better

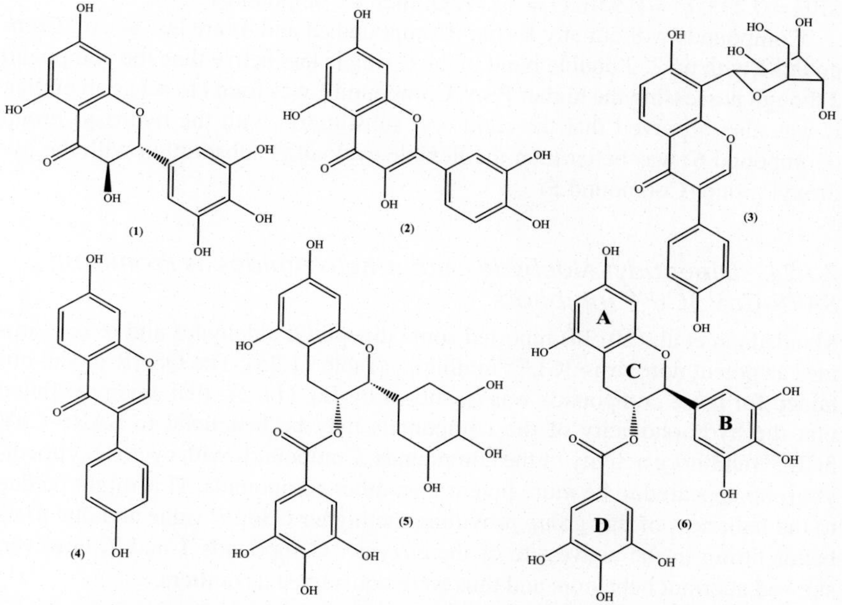

FIGURE 11.10 General structure of some flavonoids as SARS-CoV 3CL$^{pro}$ inhibitors.

TABLE 11.24 Biological Activity and Physicochemical Parameters of Flavonoids as SARS-CoV 3CL$^{pro}$ Inhibitors (Fig. 11.10) for QSAR Model [Eq. (11.23)]

| Compound | Obsd | Calcd | Res | Del res | Pred | PSA |
|---|---|---|---|---|---|---|
| 1[a] | 3.439 | 3.873 | −0.435 | −0.525 | 3.964 | 314.921 |
| 2 | 4.137 | 3.794 | 0.343 | 0.412 | 3.725 | 290.968 |
| 3 | 3.419 | 3.644 | −0.225 | −0.291 | 3.710 | 245.862 |
| 4 | 3.455 | 3.320 | 0.134 | 0.392 | 3.062 | 148.364 |
| 5 | 4.137 | 4.052 | 0.085 | 0.117 | 4.020 | 368.594 |
| 6 | 4.328 | 4.230 | 0.097 | 0.195 | 4.133 | 422.452 |

[a]Considered as outliers.

activity and also such substituents might form the hydrogen bonds. A molecular docking study showed that the galloyl group forms hydrogen bonds with Leu141, Gly143, Ser144, and His163 at the enzyme active site.

$$pIC_{50} = 2.859\,(\pm0.339) + 0.004\,(\pm0.001)\,PSA \tag{11.23}$$

$N = 5$, $R = 0.880$, $R^2 = 0.774$, $R_A^2 = 0.699$, $F\,(1, 3) = 10.292$, $P < 0.04903$, SEE $= 0.233$, $q^2 = 0.556$, $Q = 3.777$, Outlier = Compound **1**

Compounds without any B ring (Compounds **3** and **4**) are less active. Compound **1** with no 2, 3 double bond in the C ring is less active than the compound **2** though possessing the higher PSA. Compound **1** was found to act as an outlier. It was also observed that the rigid aryl substitution with the hydroxyl group (Compound **6**) was better than the flexible cycloalkyl substitution with the hydroxyl group (Compound **5**).

### 7.1.24 Dipeptidyl Aldehydes and α-Keto Amides as Promising SARS-CoV 3CL$^{pro}$ Inhibitors

Mandadapu et al. (2013b) reported some dipeptidyl aldehydes and α-keto amides as potent norovirus 3CL$^{pro}$ inhibitors (Table 11.25). The QSAR model obtained for these compounds was as shown by Eq. (11.24) that again exhibited that the hydrophobicity of the compounds may be beneficial to SARS-CoV 3CL$^{pro}$ inhibitory activity of the compounds. Compounds with cyclohexylmethyl group appeared to be more potent than other compounds. This might be due to the bulkiness of this group providing the higher C log $P$ value and due to its better fitting in the active site of the enzyme. Compounds **1** and **7,** however, showed aberrant behaviors and thus were considered as outliers.

$$pIC_{50} = 4.890\,(\pm0.129) + 0.553\,(\pm0.085)\,C \log P \tag{11.24}$$

$N = 8$, $R = 0.936$, $R^2 = 0.876$, $R_A^2 = 0.855$, $F\,(1, 6) = 42.329$, $P < 0.00063$, SEE $= 0.164$, $q^2 = 0.781$, $Q = 5.707$, Outlier = Compounds **1, 7**

### 7.1.25 A Series of Dipeptide-Type SARS CoV 3CL$^{pro}$ Inhibitors

Thanigaimalai et al. (2013a) reported a series of dipeptide-type SARS-CoV 3CL$^{pro}$ inhibitors (Table 11.26), for which the QSAR model obtained was as shown by Eq. (11.25).

$$pK_i = -67.682\,(\pm\,23.859) - 0.082\,(\pm0.014)\,SA + 0.145\,(\pm0.034)\,Vol \\ - 0.00001\,(\pm0.000)\,Vol^2 \tag{11.25}$$

$N = 17$, $R = 0.865$, $R^2 = 0.749$, $R_A^2 = 0.691$, $F\,(3, 13) = 12.899$, $P < 0.00034$, SEE $= 0.557$, $q^2 = 0.548$, $Q = 1.553$, $Vol_{opt} = 7250$, Outlier = Compound **14**

**TABLE 11.25** Biological Activity and Physicochemical Parameters of Dipeptidyl Aldehydes and $\alpha$-Keto Amides as Promising SARS-CoV 3CL$^{pro}$ Inhibitors for QSAR Model [Eq. (11.24)]

| Compound | R$_1$ | R$_2$ | R$_3$ | Obsd | Calcd | Res | Del res | Pred | C Log P |
|---|---|---|---|---|---|---|---|---|---|
| 1[a] | Benzyl | i-But | CHO | 6.222 | 5.632 | 0.590 | 0.666 | 5.556 | 1.075 |
| 2 | Benzyl | n-Pr | CHO | 5.215 | 5.435 | -0.221 | -0.278 | 5.493 | 0.676 |
| 3 | Benzyl | i-But | CHO | 5.347 | 5.696 | -0.349 | -0.389 | 5.736 | 1.205 |
| 4 | Benzyl | (c-Hex)methyl | CHO | 6.301 | 6.220 | 0.081 | 0.122 | 6.179 | 2.268 |
| 5 | Benzyl | Benzyl | CHO | 5.292 | 5.613 | -0.320 | -0.364 | 5.657 | 1.036 |
| 6 | 4-FBenzyl | i-But | CHO | 5.745 | 5.703 | 0.042 | 0.047 | 5.698 | 1.218 |
| 7[a] | m-FBenzyl | i-But | CHO | 6.155 | 5.703 | 0.452 | 0.504 | 5.651 | 1.218 |
| 8 | 2-Phenethyl | i-But | CHO | 5.721 | 5.826 | -0.105 | -0.117 | 5.838 | 1.468 |
| 9 | (2-c-Hex)ethyl | i-But | CHO | 6.222 | 6.359 | -0.137 | -0.274 | 6.496 | 2.550 |
| 10 | Benzyl | i-But | C(OH)(SO$_3$Na) CONHc-Pr | 5.276 | 5.309 | -0.033 | -0.047 | 5.323 | 0.419 |

[a]Considered as outliers.

TABLE 11.26 Biological Activity and Physicochemical Parameters of a Series of Dipeptide-Type SARS CoV 3CL[pro] Inhibitors for QSAR Model [Eq. (11.25)]

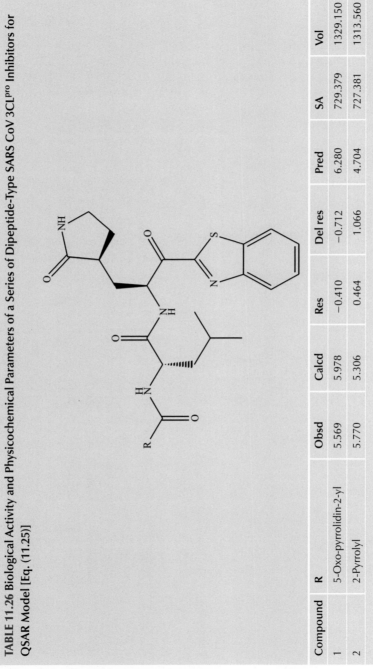

| Compound | R | Obsd | Calcd | Res | Del res | Pred | SA | Vol |
|---|---|---|---|---|---|---|---|---|
| 1 | 5-Oxo-pyrrolidin-2-yl | 5.569 | 5.978 | −0.410 | −0.712 | 6.280 | 729.379 | 1329.150 |
| 2 | 2-Pyrrolyl | 5.770 | 5.306 | 0.464 | 1.066 | 4.704 | 727.381 | 1313.560 |

| | | | | | | | |
|---|---|---|---|---|---|---|---|
| 3 | 2-Indolyl | 7.187 | 7.437 | −0.250 | −0.294 | 7.481 | 795.312 | 1463.480 |
| 4 | 5-OMe-Indole-2-yl | 7.174 | 6.483 | 0.691 | 0.761 | 6.413 | 840.460 | 1523.020 |
| 5 | 5-OH-Indole-2-yl | 6.796 | 7.262 | −0.466 | −0.524 | 7.320 | 810.363 | 1486.140 |
| 6 | 5-Cl-Indole-2-yl | 7.553 | 7.276 | 0.277 | 0.310 | 7.243 | 818.124 | 1500.870 |
| 7 | 6-OMe-Indole-2-yl | 6.481 | 6.843 | −0.362 | −0.392 | 6.874 | 836.137 | 1524.160 |
| 8 | 4-OMe-Indole-2-yl | 8.201 | 7.584 | 0.616 | 0.718 | 7.482 | 834.821 | 1541.680 |
| 9 | 4-O-i-Pr-Indole-2-yl | 7.319 | 7.720 | −0.402 | −0.535 | 7.853 | 865.057 | 1612.100 |
| 10 | 4-O-i-But-Indole-2-yl | 7.523 | 7.143 | 0.380 | 0.914 | 6.608 | 901.604 | 1687.260 |
| 11 | 4-OH-Indole-2-yl | 7.585 | 7.337 | 0.248 | 0.284 | 7.302 | 804.247 | 1476.920 |
| 12 | 3-Me,5-OMe- Indole-2-yl | 5.174 | 5.199 | −0.025 | −0.040 | 5.213 | 894.897 | 1601.640 |
| 13 | 3-Et,5-OMe- Indole-2-yl | 5.125 | 5.626 | −0.501 | −0.818 | 5.943 | 913.184 | 1661.880 |
| 14[a] | Benzimidazole-2-yl | 7.658 | 6.210 | 1.447 | 1.648 | 6.010 | 832.069 | 1499.850 |
| 15 | Benzthiazole-2-yl | 6.097 | 6.343 | −0.246 | −0.275 | 6.372 | 830.805 | 1500.860 |
| 16 | 2,3-dihydroindole-2-yl | 6.921 | 6.510 | 0.410 | 0.451 | 6.470 | 842.313 | 1527.410 |
| 17 | Benzofuran-2-yl | 4.854 | 5.812 | −0.958 | −1.187 | 6.041 | 834.923 | 1495.060 |
| 18 | Indole-3-yl | 6.167 | 7.082 | −0.915 | −1.012 | 7.180 | 811.144 | 1463.120 |

[a]Considered as outlier.

It was suggested from Eq. (11.25) that decreasing value of the SA may be conducive to the enzyme inhibition. However, the volume of the molecules was found to exhibit a parabolic relation with the enzyme inhibitory activity. It, therefore, suggested that increase in volume may be responsible for enhancing the activity only up to an optimum value of 7250. Beyond this value, the activity would decrease. Thus it indicated that molecules with limited bulk or with substituents with limited bulk might be favorable to the activity. Thus indole derivatives with less bulky substitution (Compounds **3–11**, Table 11.26) resulted in higher activity than those with a greater bulk (Compounds **12, 13**). Compared to the indole analogs, the oxopyrrolidine (Compound **1**), the pyrrole (Compound **2**), the benzothiazole (Compound **15**), and the benzofuran (Compound **17**) analogs were comparatively less active. However, it could not be explained by the model why benzimidazole analog (Compound **14**) had higher activity as compared to the benzothiazole (Compound **15**) and benzofuran analogs (Compound **17**). Probably, this compound may behave differently as compared to the other compounds in the dataset. Therefore, this compound is considered as an outlier.

### 7.1.26 A Series of Novel Dipeptide-Type SARS-CoV 3CL $^{pro}$ Inhibitors

In the subsequent study, Thanigaimalai et al. (2013b) reported a series of dipeptide-type SARS-CoV 3CL protease inhibitors (Table 11.27) whose activity was shown to be controlled by the molar refractivity (CMR) and the polar volume (Pol Vol) of the compounds [Eq. (11.26)]. Since the correlation was quadratic with respect to both CMR and Pol Vol, it suggested that compounds with limited bulk and polarity may have a better binding affinity. Several compounds, however, were treated as outliers.

$$pK_i = -177.032\,(\pm 33.763) + 18.238\,(\pm 3.699)\,CMR - 0.598\,(\pm 0.124)\,CMR^2$$
$$+0.334\,(\pm 0.077)\,Pol\,Vol - 0.001\,(\pm 0.000)\,Pol\,Vol^2 \qquad (11.26)$$

$N = 19$, $R = 0.874$, $R^2 = 0.764$, $R_A^2 = 0.697$, $F\,(4, 14) = 11.331$, $P < 0.00026$, SEE $= 0.275$, $q^2 = 0.577$, $Q = 3.178$, $CMR_{Opt} = 15.249$, Pol $Vol_{opt} = 167$, Outlier $=$ Compounds **1, 6, 7, 15, 21, 24, 25**

### 7.1.27 A Series of N-(Benzo [1,2,3]Triazol-1-yl)-N-(Benzyl) Acetamido) Phenyl) Carboxamides as SARS-CoV 3CL $^{pro}$ Inhibitors

Turlington et al. (2013) reported a series of *N*-(benzo [1,2,3]triazol-1-yl)-*N*-(benzyl)acetamido) phenyl) carboxamides as promising SARS-CoV 3CL $^{pro}$ inhibitors (Table 11.28). The QSAR model obtained for these compounds [Eq. (11.27)] suggested that highly hydrophobic ($C \log P > 4.1$) molecule with high molar refractivity but the less MW will be conducive to the activity. With the

TABLE 11.27 Biological Activity and Physicochemical Parameters of a Series of Novel Dipeptide-Type SARS-CoV 3CL $^{pro}$ Inhibitors for QSAR Model [Eq. (11.26)]

| Compound | R | $R_1$ | Obsd | Calcd | Res | Del res | Pred | CMR | Pol Vol |
|---|---|---|---|---|---|---|---|---|---|
| 1[a] | i-But | i-But | 5.229 | 4.658 | 0.571 | 1.189 | 4.040 | 13.395 | 246.218 |
| 2 | O-t-But | i-But | 4.638 | 4.998 | -0.359 | -0.645 | 5.284 | 13.548 | 277.670 |
| 3 | OBnz | i-But | 6.337 | 5.834 | 0.503 | 0.559 | 5.778 | 14.667 | 249.660 |
| 4 | OBnz | n-But | 5.796 | 5.921 | -0.125 | -0.147 | 5.943 | 14.667 | 283.458 |
| 5 | OBnz | i-Pr | 5.767 | 5.501 | 0.266 | 0.308 | 5.459 | 14.204 | 242.641 |
| 6[b] | OBnz | Sec-But | 4.538 | 5.851 | -1.313 | -1.454 | 5.992 | 14.667 | 251.687 |
| 7[a] | OBnz | $(CH_2)_2SMe$ | 5.027 | 5.035 | -0.008 | -0.058 | 5.085 | 15.010 | 348.627 |

(Continued)

TABLE 11.27 Biological Activity and Physicochemical Parameters of a Series of Novel Dipeptide-Type SARS-CoV 3CL$^{pro}$ Inhibitors for QSAR Model [Eq. (11.26)] (cont.)

| Compound | R | $R_1$ | Obsd | Calcd | Res | Del res | Pred | CMR | Pol Vol |
|---|---|---|---|---|---|---|---|---|---|
| 8 | OBnz | Bnz | 5.921 | 5.826 | 0.095 | 0.104 | 5.817 | 15.787 | 260.830 |
| 9 | Bnz | i-But | 5.495 | 5.688 | −0.193 | −0.223 | 5.717 | 14.514 | 241.735 |
| 10 | 4-OMe-Bnz | i-But | 6.377 | 6.019 | 0.358 | 0.397 | 5.980 | 15.131 | 266.950 |
| 11 | 4-OMe-Phenethyl | i-But | 6.215 | 5.928 | 0.287 | 0.312 | 5.903 | 15.595 | 282.725 |
| 12 | 3-Pyridylethyl | i-But | 5.131 | 5.657 | −0.527 | −0.726 | 5.857 | 14.767 | 231.621 |
| 13 | PhCH=CH | i-But | 6.161 | 5.743 | 0.418 | 0.575 | 5.587 | 15.257 | 234.084 |
| 14 | 4-OMe-PhCH=CH | i-But | 6.155 | 5.774 | 0.380 | 0.418 | 5.737 | 15.874 | 260.311 |
| 15[a] | 3,4-diOMe-PhCH=CH | i-But | 5.886 | 5.205 | 0.681 | 0.951 | 4.935 | 16.491 | 294.794 |
| 16 | Phenoxymethyl | i-But | 6.252 | 5.930 | 0.322 | 0.364 | 5.888 | 14.667 | 267.498 |
| 17 | 4-OMe-Phenoxymethyl | i-But | 5.807 | 5.881 | −0.074 | −0.084 | 5.891 | 15.284 | 301.759 |
| 18 | 4-OH-Phenoxymethyl | i-But | 5.076 | 5.790 | −0.715 | −0.853 | 5.929 | 14.821 | 307.175 |
| 19 | 3-NMe$_2$-Phenoxymethyl | i-But | 6.076 | 5.739 | 0.337 | 0.366 | 5.709 | 15.964 | 281.487 |
| 20 | 4-OMe-PhNHCH$_2$ | i-But | 5.495 | 5.822 | −0.327 | −0.365 | 5.859 | 15.500 | 303.153 |
| 21[a] | 3-OMe-PhNHCH$_2$ | i-But | 6.409 | 5.779 | 0.630 | 0.708 | 5.701 | 15.500 | 307.108 |
| 22 | 2-OMe-PhNHCH$_2$ | i-But | 6.481 | 5.918 | 0.564 | 0.621 | 5.860 | 15.500 | 291.600 |
| 23 | OBnz | i-But | 6.180 | 5.874 | 0.307 | 0.330 | 5.850 | 15.745 | 269.597 |
| 24[a] | OBnz | i-But | 4.432 | 5.560 | −1.128 | −1.299 | 5.731 | 16.209 | 275.135 |
| 25[a] | OBnz | i-But | 4.284 | 5.416 | −1.132 | −1.405 | 5.689 | 16.362 | 276.355 |
| 26 | OBnz | i-But | 5.602 | 5.417 | 0.185 | 0.230 | 5.372 | 16.362 | 274.643 |

[a]Considered as outlier.

TABLE 11.28 Biological Activity and Physicochemical Parameters of a Series of $N$-(benzo[1,2,3]triazol-1-yl)-$N$-(benzyl) acetamido) phenyl) carboxamides as SARS-CoV 3CL$^{pro}$ Inhibitors for QSAR Model [Eq. (11.27)]

| Compound | R | $R_1$ | Obsd | Calcd | Res | Del res | Pred | C Log P | CMR | MW |
|---|---|---|---|---|---|---|---|---|---|---|
| 1 | NHCOMe | CONH-t-But | 5.112 | 5.374 | −0.262 | −0.326 | 5.439 | 3.135 | 14.067 | 504.604 |
| 2 | NHSO$_2$Me | CONH-t-But | 4.597 | 4.665 | −0.068 | −0.117 | 4.714 | 2.925 | 14.440 | 540.658 |
| 3 | NHCOEt | CONH-t-But | 5.161 | 5.034 | 0.127 | 0.142 | 5.019 | 3.664 | 14.530 | 518.630 |
| 4 | NHCO-i-Pr | CONH-t-But | 5.387 | 5.030 | 0.357 | 0.421 | 4.966 | 3.973 | 14.994 | 532.657 |
| 5[a] | NHCO-t-But | CONH-t-But | 4.648 | 5.151 | −0.504 | −0.641 | 5.289 | 4.372 | 15.458 | 546.684 |
| 6 | NHCO-c-Pr | CONH-t-But | 5.041 | 4.993 | 0.048 | 0.055 | 4.986 | 3.719 | 14.857 | 530.641 |
| 7 | NHCO-c-But | CONH-t-But | 5.420 | 4.943 | 0.477 | 0.575 | 4.845 | 4.048 | 15.281 | 544.668 |
| 8 | NHCO-5-(Isoxazole) | CONH-t-But | 4.585 | 4.440 | 0.145 | 0.190 | 4.395 | 3.410 | 15.117 | 557.624 |
| 9 | NHCOEt | H | 5.538 | 5.400 | 0.138 | 0.163 | 5.374 | 3.267 | 11.807 | 419.499 |

*(Continued)*

TABLE 11.28 Biological Activity and Physicochemical Parameters of a Series of $N$-(benzo[1,2,3]triazol-1-yl)-$N$-(benzyl) acetamido) phenyl) carboxamides as SARS-CoV 3CL$^{pro}$ Inhibitors for QSAR Model [Eq. (11.27)] (cont.)

| Compound | R | R$_1$ | Obsd | Calcd | Res | Del res | Pred | C Log $P$ | CMR | MW |
|---|---|---|---|---|---|---|---|---|---|---|
| 10 | NHCO-$i$-Pr | H | 5.444 | 5.256 | 0.188 | 0.212 | 5.231 | 3.576 | 12.271 | 433.526 |
| 11 | NHCO-$t$-But | H | 4.876 | 5.198 | −0.322 | −0.364 | 5.240 | 3.975 | 12.735 | 447.553 |
| 12 | NHCO(CH$_2$)$_2$OMe | H | 5.469 | 5.563 | −0.094 | −0.133 | 5.602 | 2.862 | 12.424 | 449.525 |
| 13 | NHCO-$c$-Pr | H | 5.387 | 5.334 | 0.053 | 0.061 | 5.326 | 3.322 | 12.133 | 431.510 |
| 14 | NHCO-$c$-But | H | 5.092 | 5.136 | −0.044 | −0.050 | 5.141 | 3.651 | 12.557 | 445.537 |
| 15$^a$ | NHCO-$c$-Hex | H | 4.656 | 5.432 | −0.776 | −0.857 | 5.512 | 4.769 | 13.485 | 473.590 |
| 16 | NHCOO-$i$-but | H | 4.987 | 4.862 | 0.125 | 0.179 | 4.808 | 4.645 | 12.888 | 463.552 |
| 17 | NHMe | H | 5.678 | 5.770 | −0.093 | −0.131 | 5.809 | 3.218 | 10.844 | 377.463 |
| 18 | NHBnz | H | 5.824 | 6.029 | −0.205 | −0.277 | 6.101 | 4.666 | 13.355 | 453.559 |
| 19 | Ph | H | 7.292 | 7.037 | 0.255 | 0.595 | 6.697 | 5.607 | 12.522 | 424.517 |
| 20 | 3-Pyridyl | H | 6.013 | 5.549 | 0.465 | 0.527 | 5.486 | 4.672 | 12.680 | 440.520 |
| 21 | 2-OMe-3-Pyridyl | H | 6.155 | 6.115 | 0.040 | 0.066 | 6.089 | 5.520 | 13.297 | 470.546 |
| 22 | 4-Pyridyl | H | 5.699 | 5.549 | 0.150 | 0.171 | 5.528 | 4.672 | 12.680 | 440.520 |
| 23 | 2-OMe-Pyrimidin-5-yl | H | 4.903 | 5.104 | −0.201 | −0.296 | 5.199 | 5.011 | 13.086 | 471.534 |

$^a$Considered as outlier.

adjustment of such parameters, compounds **19–22** of Table 11.28 were found to have higher activity.

$$pIC_{50} = 12.976(\pm1.456) - 4.268(\pm0.773)CLogP + 0.520(\pm0.089)C\,Log\,P^2$$
$$+1.681(\pm0.279)CMR - 0.045(\pm0.007)MW \tag{11.27}$$

$N = 21$, $R = 0.941$, $R^2 = 0.885$, $R_A^2 = 0.857$, $F\,(4, 16) = 30.898$, $P < 0.00000$, SEE $= 0.228$, $q^2 = 0.799$, $Q = 4.127$, $C\log P_{Opt} = 4.104$, Outlier $=$ Compounds **5, 15**

### 7.1.28   Tripeptidyl Transition State Norwalk Virus 3C-Like Protease Inhibitors

A QSAR model obtained [Eq. (11.28)] for some tripeptidyl transition state Norwalk virus 3C-like protease inhibitors (Table 11.29) reported by Prior et al. (2013) suggested that the PSA of the molecules will control their activity and that a –CHO group at their X-position, for which an indicator parameter "I" was used, will have an added advantage. While the PSA may affect the activity due to a polar interaction of the molecule, the –CHO group might be involved in some hydrogen bond interactions with some residue of the active site.

$$pIC_{50} = 1.825(\pm0.582) + 0.023(\pm0.003)PSA + 1.614(\pm0.189)I \tag{11.28}$$

$N = 8$, $R = 0.972$, $R^2 = 0.944$, $R_A^2 = 0.921$, $F\,(2, 5) = 42.078$, $P < 0.00074$, SEE $= 0.236$, $q^2 = 0.742$, $Q = 4.119$

### 7.1.29   5-Sulfonyl Isatin Derivatives as Promising SARS-CoV 3CL^pro Inhibitors

Liu et al. (2014) reported a series of 5-sulfonyl isatin derivatives as potent SARS-CoV 3CL protease inhibitors (Table 11.30), for which the QSAR model was as shown by Eq. (11.29).

$$pIC_{50} = 9.333(\pm1.030) + 0.376(\pm0.079)C\,Log\,P - 3.002(\pm0.537)D_{Tot}$$
$$+0.379(\pm0.064)D_{Tot}^2 - 1.047(\pm0.244)I \tag{11.29}$$

$N = 21$, $R = 0.894$, $R^2 = 0.800$, $R_A^2 = 0.750$, $F\,(4, 16) = 15.999$, $P < 0.00002$, SEE $= 0.307$, $q^2 = 0.655$, $Q = 2.912$, $D_{Tot\,(opt)} = 3.960$, Outlier $=$ Compounds **5, 8, 16, 18**

It was observed from Eq. (11.29) that increasing value of hydrophobicity may be favorable for the activity and that the moderate dipole moment of the compound will also be conducive to the inhibition of the enzyme. However, a negative value of the indicator variable "I", which was used with a value of 1 for compounds having a β-napthylmethyl function at $R_1$-position, indicated

**TABLE 11.29 Biological Activity and Physicochemical Parameters of Tripeptidyl Transition State Norwalk Virus 3C-like Protease Inhibitors for QSAR Model [Eq. (11.28)]**

| Compound | R | X | Obsd | Calcd | Res | Del res | Pred | PSA | I |
|---|---|---|---|---|---|---|---|---|---|
| 1 | Ph | CHO | 6.678 | 6.898 | −0.220 | −0.284 | 6.962 | 153.457 | 1 |
| 2 | (S)-α-Napthyl | CHO | 6.854 | 6.789 | 0.065 | 0.082 | 6.772 | 148.629 | 1 |
| 3 | (R)-α-Napthyl | CHO | 6.155 | 5.822 | 0.333 | 0.647 | 5.508 | 105.728 | 1 |
| 4 | (S)-β-Napthyl | CHO | 6.824 | 6.898 | −0.074 | −0.096 | 6.920 | 153.477 | 1 |
| 5 | (S)-Biph | CHO | 6.745 | 6.848 | −0.103 | −0.131 | 6.876 | 151.233 | 1 |
| 6 | (S)-α-Napthyl | COONH-i-Pr | 5.585 | 5.656 | −0.071 | −0.107 | 5.692 | 169.965 | 0 |
| 7 | (S)-α-Napthyl | CH(OH)SO₃Na | 6.620 | 6.372 | 0.248 | 0.641 | 5.979 | 201.709 | 0 |
| 8 | (S)-α-Napthyl | CH(OH)P(O)(OEt)₂ | 4.450 | 4.627 | −0.177 | −0.569 | 5.019 | 124.302 | 0 |

**TABLE 11.30** Biological Activity and Physicochemical Parameters of 5-Sulfonyl Isatin Derivatives as Promising SARS-CoV 3CL$^{pro}$ Inhibitors for QSAR Model [Eq. (11.29)]

| Compound | R$_1$ | R$_2$ | Obsd | Calcd | Res | Del res | Pred | C Log P | D$_{Tot}$ | I |
|---|---|---|---|---|---|---|---|---|---|---|
| 1 | H | 4-Me-Piperazine | 4.115 | 4.836 | −0.721 | −0.829 | 4.944 | 0.530 | 5.641 | 0 |
| 2 | H | 3-ClBnz-Piperazine | 4.499 | 4.860 | −0.361 | −0.402 | 4.901 | 2.252 | 2.880 | 0 |
| 3 | H | 3,4,5-triOMeBnz-Piperazine | 4.494 | 4.836 | −0.342 | −0.380 | 4.874 | 1.552 | 5.150 | 0 |
| 4 | H | 4-Phenethyl-piperazine | 4.457 | 4.917 | −0.460 | −0.519 | 4.976 | 2.508 | 2.909 | 0 |
| 5$^a$ | H | (Furan-2-carbonyl)-piperazine | 4.997 | 5.301 | −0.304 | −0.982 | 5.979 | 0.311 | 1.277 | 0 |
| 6 | H | 2-Pyridyl-(4-piperazine) | 4.290 | 4.445 | −0.156 | −0.182 | 4.472 | 1.013 | 3.230 | 0 |
| 7 | H | Piperidine | 5.352 | 5.164 | 0.187 | 0.217 | 5.135 | 1.216 | 5.866 | 0 |
| 8$^a$ | H | Morpholine | 4.898 | 4.519 | 0.379 | 0.463 | 4.435 | −0.031 | 5.346 | 0 |
| 9 | H | 4-Me-Piperidine | 5.928 | 5.282 | 0.646 | 0.752 | 5.176 | 1.735 | 5.839 | 0 |
| 10 | H | 2-Me-Piperidine | 5.648 | 5.294 | 0.354 | 0.413 | 5.235 | 1.735 | 5.857 | 0 |

*(Continued)*

TABLE 11.30 Biological Activity and Physicochemical Parameters of 5-Sulfonyl Isatin Derivatives as Promising SARS-CoV 3CL$^{PRO}$ Inhibitors for QSAR Model [Eq. (11.29)] (cont.)

| Compound | $R_1$ | $R_2$ | Obsd | Calcd | Res | Del res | Pred | C Log P | $D_{tot}$ | I |
|---|---|---|---|---|---|---|---|---|---|---|
| 11 | H | 3,5-diMe-Piperidine | 5.367 | 5.409 | −0.042 | −0.051 | 5.417 | 2.254 | 5.827 | 0 |
| 12 | Me | 4-Me-Piperazine | 4.927 | 4.724 | .203 | 0.251 | 4.676 | 0.585 | 2.191 | 0 |
| 13 | Bnz | 4-Me-Piperazine | 4.173 | 4.768 | −0.595 | −0.690 | 4.862 | 2.353 | 3.395 | 0 |
| 14 | β-NapthylICH$_2$ | 4-Me-Piperazine | 4.081 | 4.758 | −0.677 | −0.834 | 4.915 | 3.527 | 5.326 | 1 |
| 15 | β-NapthylICH$_2$ | 4-Phenethyl-piperazine | 4.858 | 4.922 | −0.064 | −0.084 | 4.942 | 5.505 | 3.322 | 1 |
| 16$^a$ | β-NapthylICH$_2$ | 2-Pyridyl-(4-piperazine) | 5.258 | 4.520 | 0.738 | 0.933 | 4.325 | 4.010 | 3.398 | 1 |
| 17 | β-NapthylICH$_2$ | Piperidine | 4.854 | 4.929 | −0.075 | −0.095 | 4.949 | 4.213 | 2.309 | 1 |
| 18$^a$ | Me | Morpholine | 5.004 | 4.147 | 0.857 | 1.167 | 3.837 | 0.024 | 3.492 | 0 |
| 19 | Bnz | Morpholine | 4.858 | 4.989 | −0.131 | −0.145 | 5.004 | 1.792 | 5.334 | 0 |
| 20 | β-NapthylICH$_2$ | Morpholine | 4.399 | 4.617 | −0.218 | −0.282 | 4.681 | 2.966 | 5.337 | 1 |
| 21 | Bnz | 4-Me-Piperazine | 5.983 | 5.392 | 0.591 | 0.752 | 5.231 | 3.558 | 2.403 | 0 |
| 22 | β-NapthylICH$_2$ | 4-Me-Piperazine | 5.772 | 5.660 | 0.112 | 0.181 | 5.592 | 4.732 | 6.204 | 1 |
| 23 | Me | 4-Me-Piperazine | 4.749 | 4.900 | −0.151 | −0.167 | 4.916 | 1.790 | 2.466 | 0 |
| 24 | Bnz | 3,5-diMe-Piperidine | 5.550 | 5.504 | 0.046 | 0.064 | 5.485 | 4.077 | 2.452 | 0 |
| 25 | β-NapthylICH$_2$ | 3,5-diMe-Piperidine | 5.328 | 5.144 | 0.184 | 0.234 | 5.094 | 5.251 | 2.424 | 1 |

$^a$Considered as outlier.

that such a substituent would not be preferred, probably such a substituent might create steric hindrance in the interaction of the compounds with the receptor.

### 7.1.30  Some Substituted Pyrazoles and Substituted Pyrimidines as SARS-CoV 3CL$^{pro}$ Inhibitors

Mohamed et al. (2015) recently reported some substituted pyrazoles and substituted pyrimidines as promising SARS-CoV 3CL$^{pro}$ inhibitors (Fig. 11.11; Table 11.31). The QSAR model [Eq. (11.30)] obtained for these compounds simply suggested that highly polar fraction of the molecule with the high value of its $X$-component of dipole moment ($D_X$) will not be conducive to the activity.

$$pIC_{50} = 6.437\,(\pm 0.052) - 0.162\,(\pm 0.036)\,D_x \qquad (11.30)$$

$N = 8$, $R = 0.881$, $R^2 = 0.776$, $R_A^2 = 0.739$, $F\,(1,\,6) = 20.776$, $P < 0.00386$, SEE $= 0.139$, $q^2 = 0.657$, $Q = 6.338$, Outlier $=$ Compounds **1, 2, 5, 12**

### 7.1.31  Dipeptidyl Norovirus 3CL$^{pro}$ Inhibitors

Galasiti et al. (2015) recently reported a series of dipeptidyl norovirus 3CL$^{pro}$ inhibitors having potent inhibitory activity (Table 11.32). The QSAR model obtained for these inhibitors was as shown by Eq. (11.31). This equation simply suggested that while the $z$-component of the dipole moment will be favorable to the activity, its moderate PSA will have an adverse effect.

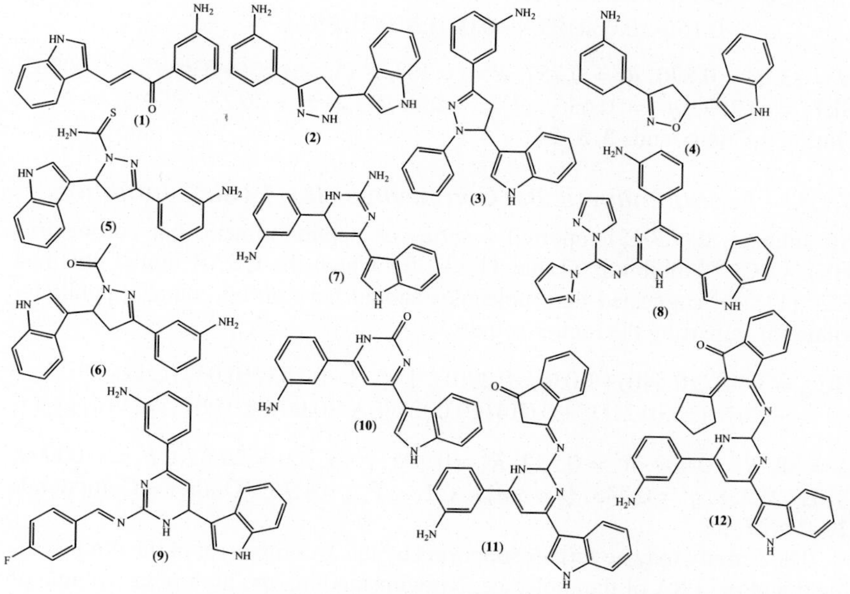

FIGURE 11.11    **General structure of substituted pyrazoles and substituted pyrimidines.**

**TABLE 11.31** Biological Activity and Physicochemical Parameters of Some Substituted Pyrazoles and Substituted Pyrimidines as SARS-CoV 3CL$^{pro}$ Inhibitors (Fig. 11.8) for QSAR Model [Eq. (11.30)]

| Compound | Obsd | Calcd | Res | Del res | Pred | $D_X$ |
|---|---|---|---|---|---|---|
| 1[a] | 6.009 | 6.357 | −0.348 | −0.382 | 6.391 | 0.582 |
| 2[a] | 6.745 | 6.320 | 0.425 | 0.475 | 6.269 | 0.928 |
| 3 | 6.229 | 6.232 | −0.003 | −0.004 | 6.233 | 1.757 |
| 4 | 6.155 | 6.344 | −0.189 | −0.208 | 6.363 | 0.704 |
| 5[a] | 6.161 | 6.628 | −0.467 | −0.672 | 6.833 | −1.977 |
| 6 | 6.244 | 6.365 | −0.121 | −0.132 | 6.376 | 0.504 |
| 7 | 6.420 | 6.468 | −0.048 | −0.054 | 6.474 | −0.470 |
| 8 | 6.347 | 6.383 | −0.036 | −0.040 | 6.386 | 0.333 |
| 9 | 6.018 | 6.120 | −0.103 | −0.169 | 6.187 | 2.811 |
| 10 | 6.854 | 6.650 | 0.203 | 0.313 | 6.541 | −2.190 |
| 11 | 6.638 | 6.398 | 0.240 | 0.262 | 6.376 | 0.189 |
| 12[a] | 6.921 | 6.475 | 0.446 | 0.501 | 6.420 | −0.533 |

[a]Considered as outliers.

$$pIC_{50} = 19.197\,(\pm2.508) + 0.105\,(\pm0.036)\,D_Z + 0.002\,(\pm0.001)\,D_Z^2 \\ -0.165\,(\pm0.032)\,PSA + 0.001\,(\pm0.000)\,PSA^2 \tag{11.31}$$

$N = 23$, $R = 0.870$, $R^2 = 0.757$, $R_A^2 = 0.703$, $F\,(4,\,18) = 14.036$, $P < 0.00002$, SEE $= 0.269$, $q^2 = 0.636$, $Q = 3.234$, $D_{Zopt} = -26.25$, $PSA_{opt} = 82.5$, Outlier $=$ Compounds **3, 5, 14**

### 7.1.32 Peptidomimetic Bat Coronavirus HKU4 3CL$^{pro}$ Inhibitors

St. John et al. (2015) reported a series of peptidomimetic bat coronavirus HKU4 3CL$^{pro}$ inhibitors (Table 11.33), for which the QSAR model obtained [Eq. (11.32)] suggested that molecules should have an optimum lipophilicity value for imparting the higher activity.

$$pIC_{50} = 3.401\,(\pm0.540) + 1.030\,(\pm0.290)\,C\,Log\,P - 0.108\,(\pm0.037)\,C\,Log\,P^2 \\ + 0.074\,(\pm0.012)\,D_X + 0.014\,(\pm0.002)\,PSA - 0.009\,(\pm0.002)\,Pol\,Vol \tag{11.32}$$

$N = 38$, $R = 0.907$, $R^2 = 0.822$, $R_A^2 = 0.795$, $F\,(5,\,32) = 29.620$, $P < 0.00000$, SEE $= 0.238$, $q^2 = 0.745$, $Q = 3.811$, $C\,Log\,P_{opt} = 4.769$, Outlier $=$ Compounds **20–22, 25, 35**.

The activity may be further supported by the $X$-component of the dipole moment and the PSA of the molecule. Notwithstanding, the high polar volume of the molecule will be delirious to the activity.

**TABLE 11.32** Biological Activity and Physicochemical Parameters of Dipeptidyl Norovirus 3CL$^{pro}$ Inhibitors for QSAR Model [Eq. (11.31)]

| Compound | R$_1$ | R$_2$ | X | Obsd | Calcd | Res | Del res | Pred | D$_Z$ | PSA |
|---|---|---|---|---|---|---|---|---|---|---|
| 1 | (c-Hex)CH$_2$ | o-Cl | CHO | 6.097 | 6.381 | −0.284 | −0.326 | 6.423 | −1.786 | 121.901 |
| 2 | (c-Hex)CH$_2$ | o-Cl | CH(OH)SO$_3$Na | 6.155 | 6.524 | −0.369 | −0.449 | 6.604 | −64.476 | 152.366 |
| 3$^a$ | (c-Hex)CH$_2$ | m-Cl | CHO | 7.000 | 6.260 | 0.740 | 0.834 | 6.166 | −2.721 | 121.894 |
| 4 | (c-Hex)CH$_2$ | m-Cl | CH(OH)SO$_3$Na | 7.000 | 6.678 | 0.322 | 0.401 | 6.599 | −65.412 | 152.335 |
| 5$^a$ | (c-Hex)CH$_2$ | m-Cl | COCONH-c-Pr | 6.602 | 5.767 | 0.835 | 1.068 | 5.534 | −1.938 | 184.498 |
| 6 | c-Hex | m-Cl | CHO | 5.292 | 5.547 | −0.255 | −0.289 | 5.582 | −2.687 | 144.957 |
| 7 | i-Butyl | m-Cl | CHO | 6.046 | 5.995 | 0.050 | 0.061 | 5.985 | 0.903 | 147.129 |
| 8 | i-Butyl | m-Cl | COCONH-c-Pr | 5.260 | 5.496 | −0.237 | −0.275 | 5.534 | −2.592 | 148.775 |
| 9 | (c-Hex)CH$_2$ | p-Cl | CHO | 6.149 | 6.122 | 0.027 | 0.030 | 6.119 | −3.833 | 121.895 |

*(Continued)*

TABLE 11.32 Biological Activity and Physicochemical Parameters of Dipeptidyl Norovirus 3CL$^{pro}$ Inhibitors for QSAR Model [Eq. (11.31)] (cont.)

| Compound | $R_1$ | $R_2$ | X | Obsd | Calcd | Res | Del res | Pred | $D_Z$ | PSA |
|---|---|---|---|---|---|---|---|---|---|---|
| 10 | (c-Hex)CH$_2$ | o-F | CHO | 6.046 | 6.474 | −0.428 | −0.509 | 6.554 | −1.085 | 121.894 |
| 11 | (c-Hex)CH$_2$ | m-F | CHO | 5.921 | 6.277 | −0.356 | −0.402 | 6.323 | −2.592 | 121.894 |
| 12 | (c-Hex)CH$_2$ | m-F | COCONH-c-Pr | 5.456 | 5.635 | −0.179 | −0.198 | 5.654 | −2.663 | 140.867 |
| 13 | (c-Hex)CH$_2$ | m-Br | CHO | 6.523 | 6.258 | 0.265 | 0.299 | 6.224 | −2.739 | 121.893 |
| 14[a] | (c-Hex)CH$_2$ | m-Br | CH(OH)P(O)(OEt)$_2$ | 5.187 | 5.648 | −0.461 | −0.556 | 5.743 | −5.135 | 130.026 |
| 15 | (c-Hex)CH$_2$ | m-Br | CH(OH)SO$_3$Na | 6.824 | 6.681 | 0.143 | 0.178 | 6.646 | −65.428 | 152.344 |
| 16 | (c-Hex)CH$_2$ | m-Br | COCONH-c-Pr | 5.824 | 5.616 | 0.208 | 0.231 | 5.592 | −2.809 | 140.890 |
| 17 | (c-Hex)CH$_2$ | m-I | CHO | 6.456 | 6.258 | 0.198 | 0.223 | 6.233 | −2.740 | 121.891 |
| 18 | (c-Hex)CH$_2$ | m-I | CH(OH)SO$_3$Na | 6.824 | 6.681 | 0.143 | 0.178 | 6.645 | −65.428 | 152.355 |
| 19 | (c-Hex)CH$_2$ | m-OMe | CHO | 5.854 | 5.560 | 0.294 | 0.343 | 5.511 | −4.235 | 136.191 |
| 20 | (c-Hex)CH$_2$ | m-NHBoc | CHO | 5.347 | 5.720 | −0.373 | −0.457 | 5.803 | −0.205 | 161.925 |
| 21 | (c-Hex)CH$_2$ | m-NO$_2$ | CHO | 6.222 | 6.569 | −0.347 | −1.243 | 7.465 | −4.328 | 212.100 |
| 22 | i-Butyl | H | CHO | 6.222 | 6.125 | 0.097 | 0.129 | 6.093 | 1.790 | 147.121 |
| 23 | i-Butyl | H | CH(OH)SO$_3$Na | 6.097 | 6.023 | 0.073 | 0.190 | 5.907 | −55.699 | 199.888 |
| 24 | i-Butyl | H | COCONH-c-Pr | 5.469 | 5.552 | −0.083 | −0.095 | 5.564 | −2.016 | 150.177 |
| 25 | (c-Hex)CH$_2$ | H | CHO | 6.523 | 6.245 | 0.278 | 0.314 | 6.209 | −2.848 | 121.888 |
| 26 | (c-Hex)CH$_2$ | H | CH(OH)SO$_3$Na | 6.398 | 6.698 | −0.300 | −0.376 | 6.774 | −65.535 | 152.353 |

[a]Considered as outliers.

**TABLE 11.33 Biological Activity and Physicochemical Parameters of Peptidomimetic Bat Coronavirus HKU4 3CL$^{pro}$ Inhibitors for QSAR Model [Eq. (11.32)]**

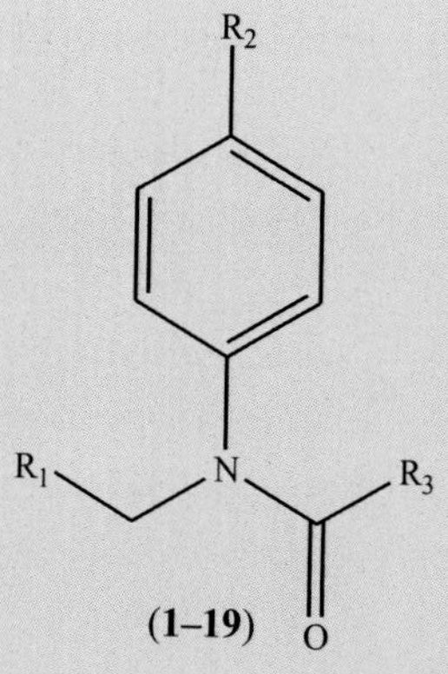

(1–19) 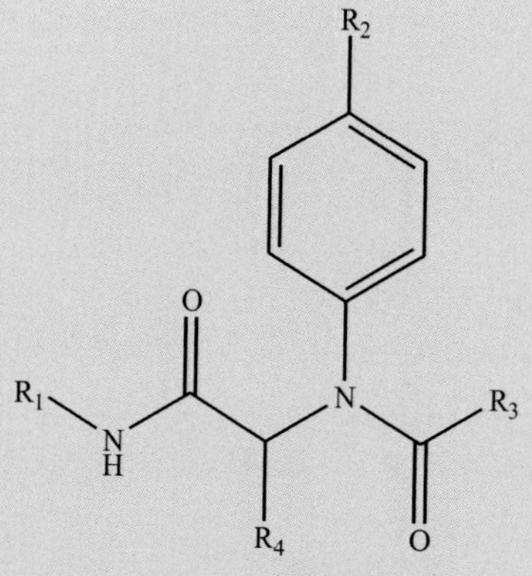  (20–43)

| Compound | R$_1$ | R$_2$ | R$_3$ | R$_4$ | Obsd | Calcd | Res | Del res | Pred | C Log P | $D_X$ | PSA | Pol Vol |
|---|---|---|---|---|---|---|---|---|---|---|---|---|---|
| 1 | 3-Thienyl | NHCO-(3-thienyl) | (2-Benzotriazolyl)methyl | — | 6.481 | 6.327 | 0.154 | 0.195 | 6.286 | 4.065 | 6.553 | 237.419 | 302.836 |
| 2 | 3-Thienyl | NHCOPh | (2-Benzotriazolyl)methyl | — | 6.387 | 5.987 | 0.400 | 0.451 | 5.936 | 4.227 | 6.409 | 194.671 | 291.571 |
| 3 | 3-Thienyl | NHCO-c-but | (2-Benzotriazolyl)methyl | — | 5.921 | 5.657 | 0.264 | 0.280 | 5.641 | 3.651 | 3.217 | 187.416 | 288.775 |
| 4 | 3-Thienyl | NHCO-c-Pent | (2-Benzotriazolyl)methyl | — | 5.921 | 6.005 | −0.084 | −0.097 | 6.018 | 4.769 | 6.487 | 179.306 | 266.098 |
| 5 | 3-Thienyl | 4-Pyr | (2-Benzotriazolyl)methyl | — | 5.824 | 5.854 | −0.030 | −0.034 | 5.857 | 3.474 | 4.957 | 220.355 | 317.406 |
| 6 | 3-Thienyl | NHCO-i-Pr | (2-Benzotriazolyl)methyl | — | 5.796 | 5.580 | 0.216 | 0.232 | 5.564 | 3.576 | −1.272 | 184.840 | 252.498 |

(Continued)

TABLE 11.33 Biological Activity and Physicochemical Parameters of Peptidomimetic Bat Coronavirus HKU4 3CL$^{pro}$ Inhibitors for QSAR Model [Eq. (11.32)] (cont.)

| Compound | R$_1$ | R$_2$ | R$_3$ | R$_4$ | Obsd | Calcd | Res | Del res | Pred | C Log P | $D_X$ | PSA | Pol Vol |
|---|---|---|---|---|---|---|---|---|---|---|---|---|---|
| 7 | 3-Thienyl | (2-OMe-Pyr)-3-yl | (2-Benzotriazolyl)methyl | — | 5.770 | 6.032 | −0.263 | −0.303 | 6.073 | 4.210 | 7.061 | 211.031 | 316.028 |
| 8 | 3-Thienyl | NHCO-c-Pr | (2-Benzotriazolyl)methyl | — | 5.721 | 5.432 | 0.289 | 0.304 | 5.417 | 3.322 | −1.281 | 191.518 | 274.817 |
| 9 | 3-Thienyl | (2-OMe-Pyr)-5-yl | (2-Benzotriazolyl)methyl | — | 5.699 | 5.698 | 0.001 | 0.001 | 5.698 | 3.624 | 4.380 | 233.228 | 364.577 |
| 10 | 3-Thienyl | NHCOCF$_3$ | (2-Benzotriazolyl)methyl | — | 5.658 | 5.718 | −0.060 | −0.077 | 5.734 | 3.847 | −3.437 | 195.593 | 237.299 |
| 11 | 3-Thienyl | 3-Pyr | (2-Benzotriazolyl)methyl | — | 5.620 | 5.885 | −0.265 | −0.298 | 5.918 | 3.474 | 5.735 | 220.889 | 320.479 |
| 12 | 3-Thienyl | NHCOO-i-butyl | (2-Benzotriazolyl)methyl | — | 5.553 | 5.581 | −0.028 | −0.032 | 5.585 | 4.645 | −2.550 | 186.127 | 263.229 |
| 13 | 3-Thienyl | NHCOEt | (2-Benzotriazolyl)methyl | — | 5.509 | 5.467 | 0.041 | 0.044 | 5.465 | 3.267 | −1.330 | 185.781 | 256.440 |
| 14 | 3-Thienyl | NHCOCH$_2$OMe | (2-Benzotriazolyl)methyl | — | 5.509 | 5.151 | 0.358 | 0.412 | 5.097 | 2.532 | −0.411 | 207.442 | 297.225 |
| 15 | 3-Thienyl | NHCO(CH$_2$)$_2$OMe | (2-Benzotriazolyl)methyl | — | 5.432 | 5.243 | 0.189 | 0.206 | 5.226 | 2.862 | −2.343 | 196.610 | 274.182 |
| 16 | 3-Thienyl | NHMe | (2-Benzotriazolyl)methyl | — | 5.319 | 5.329 | −0.011 | −0.011 | 5.330 | 3.218 | −2.694 | 156.296 | 216.523 |
| 17 | 3-Thienyl | NHCH$_2$Ph | (2-Benzotriazolyl)methyl | — | 5.276 | 5.440 | −0.165 | −0.181 | 5.457 | 4.666 | −2.698 | 155.649 | 236.128 |
| 18 | 3-Thienyl | NHCO-t-butyl | (2-Benzotriazolyl)methyl | — | 5.056 | 5.180 | −0.124 | −0.135 | 5.190 | 3.975 | −3.963 | 155.121 | 259.023 |
| 19 | 3-Thienyl | NHCO-i-Pr | (1,2,3-Triazole)1-yl | — | 4.796 | 4.709 | 0.087 | 0.130 | 4.666 | 1.982 | −1.272 | 165.375 | 238.400 |
| 20[a] | t-Butyl | t-Butyl | (1-Me-imidazole)-4-yl | 3-Pyr | 5.886 | 4.965 | 0.921 | 1.120 | 4.766 | 3.604 | −0.676 | 71.113 | 178.314 |
| 21[a] | Benzyl | NHCO-i-Pr | (2-Benzotriazolyl)methyl | 3-FPh | 5.824 | 5.112 | 0.712 | 0.818 | 5.006 | 5.114 | −3.858 | 123.131 | 220.187 |
| 22[a] | t-Butyl | NHCOMe | (2-Benzotriazolyl)methyl | 3-Thienyl | 5.745 | 4.926 | 0.819 | 0.913 | 4.832 | 3.135 | −4.596 | 175.901 | 290.678 |
| 23 | t-Butyl | 2-CNPh | 2-Furyl | 3-Pyr | 5.658 | 5.407 | 0.251 | 0.281 | 5.377 | 4.147 | −1.284 | 93.556 | 152.746 |

| No. | R1 | R2 | R3 | R4 | | | | | | | | | |
|---|---|---|---|---|---|---|---|---|---|---|---|---|---|
| 24 | t-Butyl | | (2-Indolyl)methyl | 3-Pyr | 5.658 | 5.256 | 0.402 | 0.488 | 5.169 | 5.466 | -0.015 | 67.832 | 132.826 |
| 25[a] | t-Butyl | NHCO-c-Pr | (2-Benzotriazolyl)methyl | 3-Thienyl | 5.569 | 5.142 | 0.427 | 0.486 | 5.082 | 3.719 | -1.920 | 189.368 | 330.556 |
| 26 | Benzyl | NHCO-c-butyl | (2-Benzotriazolyl)methyl | 3-FPh | 5.469 | 5.463 | 0.005 | 0.007 | 5.461 | 5.189 | 6.476 | 138.478 | 281.050 |
| 27 | t-Butyl | OCH$_2$F | (2-Indolyl)methyl | 3-Pyr | 5.409 | 5.452 | -0.043 | -0.050 | 5.459 | 3.732 | 0.712 | 84.008 | 137.546 |
| 28 | t-Butyl | NHCOPh | (2-Benzotriazolyl)methyl | 3-Thienyl | 5.377 | 5.268 | 0.108 | 0.138 | 5.239 | 4.624 | 2.529 | 170.484 | 336.095 |
| 29 | t-Butyl | i-Pr | (2-Benzotriazolyl)methyl | 3-Pyr | 5.161 | 5.406 | -0.245 | -0.286 | 5.447 | 5.067 | -0.043 | 67.838 | 120.946 |
| 30 | Benzyl | NHCO-c-Pr | (2-Benzotriazolyl)methyl | 3-FPh | 5.155 | 5.449 | -0.294 | -0.318 | 5.473 | 4.860 | -1.115 | 156.448 | 248.714 |
| 31 | t-Butyl | t-butyl | (2-Oxazo-5-yl)methyl | 3-Pyr | 5.066 | 5.048 | 0.018 | 0.020 | 5.045 | 3.155 | -0.616 | 83.224 | 161.469 |
| 32 | t-Butyl | t-butyl | (2-Imidazo-4-yl)methyl | 3-Pyr | 5.032 | 5.386 | -0.354 | -0.393 | 5.424 | 3.305 | 1.332 | 110.996 | 177.477 |
| 33 | t-Butyl | Me | 2-Furyl | 3-Pyr | 5.022 | 5.203 | -0.181 | -0.227 | 5.249 | 3.949 | 0.627 | 47.293 | 125.546 |
| 34 | t-Butyl | t-Butyl | (2-Imidazo-4-yl)methyl | 3-Pyr | 4.955 | 5.192 | -0.238 | -0.262 | 5.217 | 3.305 | -1.670 | 99.248 | 162.438 |
| 35[a] | t-Butyl | NHCO(2-benzotria) | (2-Benzotriazolyl)methyl | 3-Thienyl | 4.833 | 5.512 | -0.680 | -0.751 | 5.584 | 3.410 | 1.015 | 237.356 | 360.408 |
| 36 | t-Butyl | t-Butyl | (2-Imidazo-4-yl)methyl | 2-Pyrim-5-yl | 4.812 | 4.817 | -0.004 | -0.005 | 4.817 | 2.348 | -2.326 | 122.867 | 182.158 |
| 37 | t-Butyl | NHCO-i-Pr | (2-Benzotriazolyl)methyl | 3-Thienyl | 4.764 | 5.072 | -0.307 | -0.345 | 5.110 | 3.973 | -4.585 | 159.577 | 277.847 |
| 38 | t-Butyl | NHCOEt | (2-Benzotriazolyl)methyl | 3-Thienyl | 4.738 | 5.013 | -0.276 | -0.310 | 5.048 | 3.664 | -4.603 | 165.613 | 286.848 |
| 39 | t-Butyl | NHCOMe | (2-Benzotriazolyl)methyl | 3-FPh | 4.728 | 5.072 | -0.344 | -0.371 | 5.099 | 3.632 | -3.646 | 123.951 | 218.588 |
| 40 | t-Butyl | NH$_2$ | (2-Benzotriazolyl)methyl | 3-Thienyl | 4.658 | 5.036 | -0.379 | -0.417 | 5.075 | 2.889 | -3.545 | 194.693 | 295.576 |
| 41 | t-Butyl | t-butyl | 2-Furyl | 3-Pyr | 4.449 | 4.699 | -0.251 | -0.524 | 4.972 | 6.149 | -1.580 | 35.972 | 119.956 |
| 42 | t-Butyl | NHCO-i-Pr | (2-Benzimidazolyl)methyl | 3-Thienyl | 4.281 | 4.940 | -0.658 | -0.770 | 5.052 | 4.161 | -3.325 | 125.993 | 261.488 |
| 43 | t-Butyl | t-Butyl | (2-Oxazo-5-yl)methyl | 2-Pyrim-5-yl | 4.255 | 4.634 | -0.379 | -0.497 | 4.752 | 2.198 | -1.272 | 106.846 | 180.953 |

[a]Considered as outliers.

## 7.1.33 Substituted Furan Analogs as Promising SARS CoV 3C$^{pro}$ Inhibitors

Kumar et al. (2016) recently reported a series of substituted furan analogs as promising SARS-CoV 3C$^{pro}$ inhibitors (Table 11.34). The QSAR model obtained for these compounds was as shown by Eq. (11.33)

$$pIC_{50} = 4.555\,(\pm 0.067) + 0.522\,(\pm 0.141)\,I_1 + 0.597\,(\pm 0.141)\,I_2 \qquad (11.33)$$

$N = 11$, $R = 0.874$, $R^2 = 0.763$, $R_A^2 = 0.704$, $F\,(2,\,8) = 12.907$, $P < 0.00313$, SEE $= 0.176$, $q^2 = 0.536$, $Q = 4.966$, Outlier = Compound **6**.

**TABLE 11.34 Biological Activity and Physicochemical Parameters of Substituted Furan Analogs as Promising SARS CoV 3C$^{pro}$ Inhibitors for QSAR Model [Eq. (11.33)]**

| Compound | $R_1$ | $R_2$ | $R_3$ | Obsd | Calcd | Res | Del res | Pred | $I_1$ | $I_2$ |
|---|---|---|---|---|---|---|---|---|---|---|
| 1 | COOH | Cl | 3-COOH | 4.350 | 4.634 | −0.285 | −0.325 | 4.675 | 0 | 0 |
| 2 | COOH | Cl | H | 4.785 | 4.634 | 0.151 | 0.172 | 4.613 | 0 | 0 |
| 3 | COOH | Cl | 4-F | 4.695 | 4.634 | 0.060 | 0.069 | 4.626 | 0 | 0 |
| 4 | COOH | Cl | 4-i-Pr | 5.222 | 5.077 | 0.145 | 0.290 | 4.932 | 1 | 0 |
| 5 | COOH | Cl | 4-t-But | 5.237 | 5.151 | 0.086 | 0.171 | 5.066 | 0 | 1 |
| 6[a] | H | H | 3-COOH | 5.194 | 4.634 | 0.559 | 0.639 | 4.555 | 0 | 0 |
| 7 | COOH | H | H | 4.385 | 4.634 | −0.249 | −0.285 | 4.670 | 0 | 0 |
| 8 | COOH | H | 4-F | 4.426 | 4.634 | −0.208 | −0.238 | 4.664 | 0 | 0 |
| 9 | COOH | H | 4-i-Pr | 4.932 | 5.077 | −0.145 | −0.290 | 5.222 | 1 | 0 |
| 10 | COOH | H | 4-t-But | 5.066 | 5.151 | −0.086 | −0.171 | 5.237 | 0 | 1 |
| 11 | COOH | H | 4-CN | 4.728 | 4.634 | 0.094 | 0.107 | 4.621 | 0 | 0 |
| 12 | COOH | H | 4-OMe | 4.513 | 4.634 | −0.122 | −0.139 | 4.652 | 0 | 0 |

[a]Considered as outlier.

Where the activity is shown to be correlated with two indicator variables, $I_1$ and $I_2$. $I_1$ and $I_2$, with a value of 1 each, indicate the presence of the $i$-propyl and the $t$-butyl moiety at $R_3$-position, respectively. The positive coefficients of both these parameters suggested that the $i$-propyl and the $t$-butyl functions at $R_3$-position will be favorable for inhibitory activity. Compound **6** behaved aberrantly and therefore it was considered as an outlier. It may be observed that the $t$-butyl substitution at $R_3$-position in compounds **5** and **10** gave better activity than the $i$-butyl substitution at the same position in compounds **4** and **9**.

## 7.2  QSAR Model Development on Human Rhinovirus 3C$^{pro}$ Inhibitors

### 7.2.1  A Series of Michael Acceptor Type HRV 3C$^{pro}$ Inhibitors

Dragovich et al. (1998a) reported a series of Michael acceptor type potent HRV 3C$^{pro}$ inhibitors (Fig. 11.12; Table 11.35), for which the QSAR model [Eq. (11.34)] exhibited that the positive effect on the activity of the compounds will be produced by the Z-component of the dipole moment and the SA of the compounds till it attains an optimum value. These two properties indicate the same kind of electronic interactions of the molecule with the receptor.

$$pEC_{50} = -56.622(\pm 19.902) + 0.423(\pm 0.117)D_Z + 0.102(\pm 0.021)D_Z^2$$
$$+ 0.134(\pm 0.041)SA - 0.0001(\pm 0.00002)SA^2 - 0.018(\pm 0.004)PolVol$$
$$(11.34)$$

$N = 33$, $R = 0.841$, $R^2 = 0.708$, $R_A^2 = 0.654$, $F(5, 27) = 13.087$, $P < 0.00000$, SEE $= 0.411$, $q^2 = 0.584$, $Q = 2.046$, $D_{Zopt} = -2.074$, $SA_{opt} = 670$, Outlier = Compounds **4**, **9**, **11**, **15**, **16**, **18**, **21**, **23**, **30**, **34**

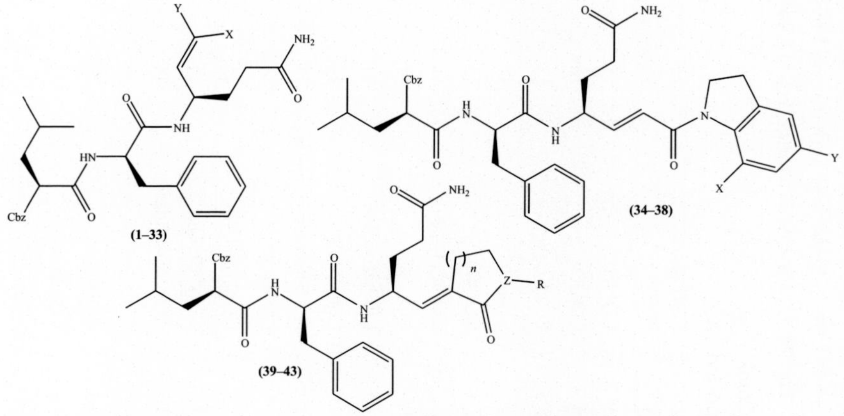

FIGURE 11.12    **General structure of Michael acceptor type HRV 3C$^{pro}$ inhibitors.**

TABLE 11.35 Biological Activity and Physicochemical Parameters of a Series of Michael Acceptor Type HRV 3C$^{pro}$ Inhibitors (Fig. 11.12) for QSAR Model [Eq. (11.34)]

| Com-pound | X | Y | Z | n | R | Obsd | Calcd | Res | Del res | Pred | $D_Z$ | SA | Pol Vol |
|---|---|---|---|---|---|---|---|---|---|---|---|---|---|
| 1 | COOMe | H | — | — | — | — | 5.886 | 5.268 | 0.618 | 0.699 | 5.187 | 0.447 | 892.098 | 240.514 |
| 2 | H | COOMe | — | — | — | — | 5.495 | 5.269 | 0.226 | 0.258 | 5.237 | 0.501 | 889.632 | 239.189 |
| 3 | COOEt | H | — | — | — | — | 6.268 | 5.819 | 0.449 | 0.599 | 5.668 | 1.733 | 919.558 | 252.873 |
| 4 | COOEt | Me | — | — | — | — | 5.523 | 5.922 | −0.399 | −0.523 | 6.046 | 1.697 | 925.084 | 241.339 |
| 5[a] | COO-c-Pent | H | — | — | — | — | 6.252 | 5.258 | 0.994 | 1.075 | 5.177 | −0.640 | 995.992 | 259.677 |
| 6 | COO-c-Hex | H | — | — | — | — | 5.444 | 5.322 | 0.122 | 0.136 | 5.308 | −0.610 | 1019.220 | 240.828 |
| 7 | COOBnz | H | — | — | — | — | 5.495 | 5.206 | 0.288 | 0.313 | 5.182 | −0.667 | 1004.590 | 263.005 |
| 8 | COOCH$_2$-t-But | H | — | — | — | — | 6.301 | 5.633 | 0.668 | 0.757 | 5.544 | 0.277 | 972.403 | 237.458 |
| 9[a] | CONMe$_2$ | H | — | — | — | — | 4.252 | 5.764 | −1.512 | −1.750 | 6.002 | 0.314 | 932.597 | 206.888 |
| 10 | COPyrrolidine | H | — | — | — | — | 4.658 | 5.090 | −0.432 | −0.457 | 5.114 | −2.650 | 983.638 | 266.905 |
| 11[a] | CON(Me)Ph | H | — | — | — | — | 3.801 | 4.981 | −1.180 | −1.258 | 5.059 | −2.564 | 1025.570 | 263.713 |
| 12 | COTetrahydroquinoline | H | — | — | — | — | 3.900 | 4.573 | −0.673 | −0.749 | 4.649 | −2.685 | 1058.120 | 288.062 |
| 13 | COIndoline | H | — | — | — | — | 4.796 | 4.564 | 0.232 | 0.255 | 4.541 | −2.671 | 1048.840 | 299.918 |
| 14 | CON(Me)OMe | H | — | — | — | — | 5.398 | 5.246 | 0.152 | 0.160 | 5.238 | −1.033 | 944.039 | 248.567 |
| 15 | CON(Me)OH | H | — | — | — | — | 4.377 | 5.268 | −0.892 | −0.980 | 5.357 | 0.311 | 915.220 | 260.579 |
| 16 | COIsoxazolidine | H | — | — | — | — | 4.347 | 5.246 | −0.899 | −1.000 | 5.346 | −4.621 | 971.546 | 280.447 |
| 17 | CO[1,2]Oxazinan | H | — | — | — | — | 4.796 | 4.997 | −0.201 | −0.213 | 5.009 | −4.087 | 989.697 | 298.674 |
| 18 | COPyrrole | H | — | — | — | — | 5.854 | 4.952 | 0.901 | 0.964 | 4.890 | −2.527 | 967.760 | 284.960 |
| 19 | COIndole | H | — | — | — | — | 5.745 | 4.837 | 0.907 | 0.959 | 4.786 | −2.454 | 1032.130 | 277.918 |

| No. | $R_1$ | $R_2$ | $R_3$ | X | n | $R_4$ | | | | | | | | |
|---|---|---|---|---|---|---|---|---|---|---|---|---|---|---|
| $20^a$ | COMe | H | — | — | — | — | 5.699 | 5.118 | 0.581 | 0.773 | 4.926 | 0.619 | 870.192 | 236.384 |
| 21 | CO-t-Butyl | H | — | — | — | — | 5.770 | 5.811 | −0.042 | −0.049 | 5.819 | 0.326 | 945.814 | 208.111 |
| 22 | COPh | H | — | — | — | — | 5.398 | 5.160 | 0.238 | 0.255 | 5.143 | −2.434 | 985.661 | 256.507 |
| 23 | 4-OMeCOPh | H | — | — | — | — | 4.658 | 5.147 | −0.489 | −0.582 | 5.240 | −2.980 | 1030.290 | 240.131 |
| 24 | 4-$NO_2$COPh | H | — | — | — | — | 4.495 | 4.660 | −0.165 | −0.183 | 4.678 | −4.312 | 1025.260 | 333.702 |
| $25^a$ | 4-CNCOPh | H | — | — | — | — | 4.301 | 4.714 | −0.413 | −0.457 | 4.758 | −2.116 | 1021.670 | 302.664 |
| $26^a$ | CO-2-(1,3-Benzodioxole) | H | — | — | — | — | 5.495 | 4.830 | 0.665 | 0.706 | 4.789 | −2.247 | 1034.870 | 276.486 |
| 27 | CO-2-Furyl | H | — | — | — | — | 5.620 | 5.317 | 0.303 | 0.320 | 5.300 | −0.875 | 962.609 | 248.127 |
| 28 | $SO_2Ph$ | H | — | — | — | — | 3.699 | 4.439 | −0.740 | −1.075 | 4.774 | −3.287 | 1005.240 | 355.178 |
| 29 | CN | H | — | — | — | — | 4.745 | 4.971 | −0.226 | −0.326 | 5.070 | 0.503 | 865.270 | 244.002 |
| 30 | C=NOMe | H | — | — | — | — | 4.000 | 5.161 | −1.161 | −1.226 | 5.226 | −0.997 | 932.010 | 253.879 |
| 31 | 2-Oxopyrrolidine | H | — | — | — | — | 6.051 | 5.354 | 0.697 | 0.799 | 5.252 | −5.150 | 982.770 | 283.742 |
| 32 | 2-Oxooxazolidine | H | — | — | — | — | 5.796 | 5.369 | 0.427 | 0.514 | 5.282 | −6.068 | 968.853 | 321.802 |
| 33 | 3-Me-2-oxo-imidazolidine | H | — | — | — | — | 5.301 | 4.939 | 0.362 | 0.410 | 4.891 | −5.335 | 1007.300 | 340.586 |
| 34 | — | H | $NO_2$ | — | — | — | 4.770 | 5.228 | −0.458 | −1.727 | 6.496 | −8.144 | 1088.210 | 380.953 |
| 35 | — | H | F | — | — | — | 4.959 | 4.645 | 0.314 | 0.347 | 4.612 | −3.694 | 1058.160 | 289.980 |
| 36 | — | Cl | H | — | — | — | 4.252 | 4.490 | −0.238 | −0.266 | 4.518 | −3.307 | 1060.860 | 301.647 |
| 37 | — | H | Cl | — | — | — | 5.252 | 4.531 | 0.721 | 0.859 | 4.393 | −3.370 | 1071.540 | 282.633 |
| 38 | — | H | Br | — | — | — | 4.328 | 4.455 | −0.127 | −0.159 | 4.486 | −3.179 | 1077.580 | 281.707 |
| 39 | — | — | — | O | 1 | — | 5.284 | 4.876 | 0.408 | 0.509 | 4.775 | −4.057 | 907.495 | 280.430 |
| $40^a$ | — | — | — | O | 2 | — | 4.796 | 5.025 | −0.229 | −0.260 | 5.056 | −3.775 | 930.567 | 273.891 |
| 41 | — | — | — | N | 1 | COMe | 6.149 | 5.364 | 0.785 | 0.912 | 5.236 | −5.619 | 969.035 | 301.375 |
| $42^a$ | — | — | — | N | 1 | COOMe | 4.553 | 4.685 | −0.132 | −0.156 | 4.709 | −3.719 | 990.921 | 333.132 |
| 43 | — | — | — | N | 1 | OMe | 4.745 | 5.195 | −0.450 | −0.506 | 5.250 | −5.047 | 954.961 | 298.399 |

*aConsidered as outlier.*

## 7.2.2 A Series of Peptide-Derived HRV 3C$^{pro}$ Inhibitors

Dragovich et al. (1998b) reported a series of peptide-derived potent HRV 3C$^{pro}$ inhibitors (Table 11.36), the QSAR model [Eq. (11.35)] suggested that the activity would be primarily controlled by the hydrophobicity of the molecule. The polar volume and the total dipole moment of the compounds would also help to increase the activity of the compounds. "I" is an indicator parameter indicating the presence of $i$-butyl moiety at R$_3$ position. Its negative coefficient suggested that $i$-butyl function is not favorable at R$_3$-position. This might be creating some steric problem.

$$pEC_{50} = 5.030(\pm 0.717) + 0.549(\pm 0.060)C \text{ Log } P - 0.076(\pm 0.032)D_Y$$
$$+ 0.160(\pm 0.042)D_{Tot} - 0.007(\pm 0.001)MW$$
$$+ 0.008(\pm 0.001)PolVol - 0.869(\pm 0.109)I \quad (11.35)$$

$N = 64$, $R = 0.858$, $R^2 = 0.736$, $R_A^2 = 0.708$, $F(6, 57) = 26.421$, $P < 0.00000$, SEE $= 0.361$, $q^2 = 0.648$, $Q = 2.377$, Outlier $=$ Compounds **5, 9, 11, 20, 25, 26, 40, 42, 51, 60, 69, 71, 73, 77**

## 7.2.3 Ketomethylene Containing Peptide-Based HRV 3C$^{pro}$ Inhibitors

Dragovich et al. (1999a) reported some ketomethylene group containing peptide-based HRV 3C$^{pro}$ inhibitors (Table 11.37). The QSAR model obtained for them was as shown by Eq. (11.36):

$$pEC_{50} = 6.795(\pm 0.154) - 0.154(\pm 0.036)D_Y - 0.434(\pm 0.190)I_1$$
$$+ 0.601(\pm 0.196)I_2 \quad (11.36)$$

$N = 14$, $R = 0.927$, $R^2 = 0.859$, $R_A^2 = 0.816$, $F(3, 10) = 20.273$, $P < 0.00014$, SEE $= 0.266$, $q^2 = 0.727$, $Q = 3.485$

It is observed from Eq. (11.36) that decreasing value of the dipole moment along $Y$-axis may increase the activity. This model also showed the importance of two indicator variables, $I_1$ and $I_2$, each of which was used with a value of unity $i$-butyl and the $i$-propyl substituent at R$_2$ position, respectively. The negative coefficient of $I_1$ suggested that $i$-butyl group will not be conducive to the activity, while the positive coefficient of $I_2$ suggested that $i$-propyl group would be favorable to the activity. These facts are observed from the $i$-butyl substituted compounds **1–3** and **5–7** and the $i$-propyl substituted compounds **4, 8–11**.

## 7.2.4 Peptide-Based HRV 3C$^{pro}$ Inhibitors

Dragovich et al. (1999b) reported some HRV 3C$^{pro}$ inhibitors (Table 11.38). The QSAR model is shown in Eq. (11.37) obtained for them suggested that the high polarity of the compound in the $X$-direction will be detrimental to the inhibition potency of the compound. The simple structure–activity relationship study of

TABLE 11.36 Biological Activity and Physicochemical Parameters of a Series of Peptide-Derived HRV 3C$^{pro}$ Inhibitors for QSAR Model [Eq. (11.35)]

| Compound | R$_1$ | R$_2$ | R$_3$ | R$_4$ | Obsd | Calcd | Res | Del res | Pred | C Log P | D$_Y$ | D$_{Tot}$ | MW | Pol Vol | I |
|---|---|---|---|---|---|---|---|---|---|---|---|---|---|---|---|
| 1 | (CH$_2$)$_2$CONH$_2$ | Bnz | i-But | Cbz | 6.268 | 5.643 | 0.625 | 0.644 | 5.623 | 4.402 | -0.526 | 2.710 | 594.698 | 413.706 | 1 |
| 2 | (CH$_2$)$_2$CONHTr | Bnz | i-But | Cbz | 4.000 | 4.542 | -0.542 | -1.136 | 5.136 | 6.581 | -1.113 | 1.932 | 825.002 | 330.906 | 1 |
| 3 | (CH$_2$)$_2$CONHMe | Bnz | i-But | Cbz | 5.252 | 5.273 | -0.021 | -0.023 | 5.275 | 4.582 | 2.135 | 2.543 | 608.725 | 398.617 | 1 |
| 4 | (CH$_2$)$_2$CONMe$_2$ | Bnz | i-But | Cbz | 5.398 | 5.323 | 0.075 | 0.083 | 5.315 | 4.954 | 2.531 | 3.027 | 622.752 | 389.226 | 1 |
| 5[a] | (CH$_2$)$_2$COOH | Bnz | i-But | Cbz | 4.854 | 5.724 | -0.870 | -0.933 | 5.787 | 5.138 | -2.225 | 2.775 | 595.683 | 337.877 | 1 |
| 6 | (CH$_2$)$_2$COMe | Bnz | i-But | Cbz | 5.796 | 5.550 | 0.246 | 0.266 | 5.530 | 5.278 | -0.507 | 2.894 | 593.710 | 318.394 | 1 |
| 7 | (CH$_2$)$_2$SOMe | Bnz | i-But | Cbz | 5.796 | 5.602 | 0.194 | 0.204 | 5.592 | 4.342 | -2.187 | 3.447 | 613.765 | 387.787 | 1 |
| 8 | CH$_2$NHCOMe | Bnz | i-But | Cbz | 5.658 | 5.155 | 0.502 | 0.564 | 5.094 | 4.379 | 5.417 | 7.240 | 594.698 | 313.448 | 1 |
| 9[a] | CH$_2$NHCONH$_2$ | Bnz | i-But | Cbz | 4.495 | 5.681 | -1.186 | -1.274 | 5.769 | 4.015 | -0.559 | 2.258 | 595.687 | 465.089 | 1 |
| 10 | CH$_2$OCONH$_2$ | Bnz | i-But | Cbz | 5.796 | 5.587 | 0.209 | 0.229 | 5.567 | 4.459 | -0.530 | 0.909 | 596.671 | 446.942 | 1 |
| 11[a] | (CH$_2$)$_2$CONH$_2$ | H | i-But | Cbz | 3.851 | 5.073 | -1.222 | -1.323 | 5.174 | 2.675 | 2.269 | 5.487 | 504.576 | 315.571 | 1 |
| 12 | (CH$_2$)$_2$CONH$_2$ | Me | i-But | Cbz | 4.699 | 5.020 | -0.321 | -0.357 | 5.056 | 2.984 | -0.004 | 4.193 | 518.602 | 291.439 | 1 |

(Continued)

TABLE 11.36 Biological Activity and Physicochemical Parameters of a Series of Peptide-Derived HRV 3C$^{pro}$ Inhibitors for QSAR Model [Eq. (11.35)] (cont.)

| Compound | R$_1$ | R$_2$ | R$_3$ | R$_4$ | Obsd | Calcd | Res | Del res | Pred | C Log P | D$_y$ | D$_{Tot}$ | MW | Pol Vol | I |
|---|---|---|---|---|---|---|---|---|---|---|---|---|---|---|---|
| 13 | (CH$_2$)$_2$CONH$_2$ | Et | i-But | Cbz | 5.222 | 5.181 | 0.041 | 0.044 | 5.178 | 3.513 | 0.010 | 4.173 | 532.629 | 295.871 | 1 |
| 14 | (CH$_2$)$_2$CONH$_2$ | n-Pr | i-But | Cbz | 5.301 | 5.517 | -0.216 | -0.233 | 5.534 | 4.042 | 2.612 | 3.991 | 546.656 | 381.286 | 1 |
| 15 | (CH$_2$)$_2$CONH$_2$ | i-But | i-But | Cbz | 5.268 | 5.452 | -0.184 | -0.199 | 5.467 | 4.441 | -0.030 | 4.239 | 560.682 | 301.966 | 1 |
| 16 | (CH$_2$)$_2$CONH$_2$ | CH$_2$SMe | i-But | Cbz | 5.051 | 5.422 | -0.371 | -0.399 | 5.449 | 3.513 | -0.768 | 5.152 | 564.694 | 342.413 | 1 |
| 17 | (CH$_2$)$_2$CONH$_2$ | CH$_2$SEt | i-But | Cbz | 5.000 | 5.564 | -0.564 | -0.606 | 5.606 | 4.042 | -0.744 | 5.171 | 578.721 | 342.413 | 1 |
| 18 | (CH$_2$)$_2$CONH$_2$ | CH$_2$C-Hex | i-But | Cbz | 5.721 | 5.933 | -0.212 | -0.230 | 5.951 | 5.634 | 0.021 | 2.242 | 600.746 | 394.126 | 1 |
| 19 | (CH$_2$)$_2$CONH$_2$ | 4-FBnz | i-But | Cbz | 5.745 | 5.543 | 0.201 | 0.209 | 5.536 | 4.545 | -2.108 | 2.393 | 612.689 | 386.902 | 1 |
| 20[a] | (CH$_2$)$_2$CONH$_2$ | 4-MeBnz | i-But | Cbz | 6.745 | 5.717 | 1.028 | 1.071 | 5.673 | 4.901 | -2.119 | 2.623 | 608.725 | 378.323 | 1 |
| 21 | (CH$_2$)$_2$CONH$_2$ | 4-OHBnz | i-But | Cbz | 5.276 | 5.308 | -0.032 | -0.033 | 5.309 | 3.735 | -1.100 | 2.275 | 610.698 | 428.516 | 1 |
| 22 | (CH$_2$)$_2$CONH$_2$ | 4-OAcBnz | i-But | Cbz | 4.959 | 4.819 | 0.139 | 0.153 | 4.806 | 3.751 | 4.402 | 6.233 | 652.735 | 384.882 | 1 |
| 23 | (CH$_2$)$_2$CONH$_2$ | 4-OMeBnz | i-But | Cbz | 5.770 | 5.303 | 0.467 | 0.545 | 5.224 | 4.321 | 6.526 | 8.296 | 624.725 | 375.690 | 1 |
| 24 | (CH$_2$)$_2$CONH$_2$ | 4-OPO$_3$H$_2$Bnz | i-But | Cbz | 4.854 | 4.636 | 0.218 | 0.322 | 4.532 | 2.397 | 6.797 | 9.078 | 691.686 | 481.642 | 1 |
| 25[a] | (CH$_2$)$_2$CONH$_2$ | 4-CH$_2$OHBnz | i-But | Cbz | 6.260 | 5.157 | 1.103 | 1.197 | 5.062 | 3.364 | -2.137 | 3.570 | 624.725 | 398.350 | 1 |
| 26[a] | (CH$_2$)$_2$CONH$_2$ | 4-CH$_2$OMeBnz | i-But | Cbz | 4.409 | 5.260 | -0.851 | -0.882 | 5.291 | 4.200 | 0.299 | 3.964 | 638.751 | 397.994 | 1 |
| 27 | (CH$_2$)$_2$CONH$_2$ | 4-(CH$_2$)$_2$OHBnz | i-But | Cbz | 5.456 | 5.026 | 0.430 | 0.463 | 4.993 | 3.593 | -2.114 | 2.115 | 638.751 | 410.287 | 1 |
| 28 | (CH$_2$)$_2$CONH$_2$ | 4-CNBnz | i-But | Cbz | 5.252 | 5.483 | -0.232 | -0.244 | 5.496 | 3.835 | -2.202 | 2.611 | 619.708 | 436.415 | 1 |
| 29 | (CH$_2$)$_2$CONH$_2$ | CH$_2$-2-Imidazol | i-But | Cbz | 4.569 | 4.524 | 0.045 | 0.049 | 4.519 | 2.001 | 5.302 | 7.567 | 584.664 | 371.088 | 1 |

| # | | | | | | | | | | | | | | |
|---|---|---|---|---|---|---|---|---|---|---|---|---|---|---|
| 30 | (CH$_2$)$_2$CONH$_2$ | CH$_2$-2-(N-Melmid) | i-But | Cbz | 5.000 | 4.958 | 0.042 | 0.047 | 4.953 | 2.017 | −0.320 | 3.569 | 598.691 | 471.960 | 1 |
| 31 | (CH$_2$)$_2$CONH$_2$ | CH$_2$-2-Thienyl | i-But | Cbz | 6.252 | 5.770 | 0.481 | 0.542 | 5.710 | 4.048 | −0.074 | 2.329 | 600.726 | 491.708 | 1 |
| 32 | (CH$_2$)$_2$CONH$_2$ | CH(R-OH)Me | i-But | Cbz | 4.252 | 4.698 | −0.446 | −0.499 | 4.750 | 2.437 | −0.602 | 3.575 | 548.629 | 320.804 | 1 |
| 33 | (CH$_2$)$_2$CONH$_2$ | Bnz | H | Cbz | 5.252 | 5.644 | −0.392 | −0.451 | 5.703 | 2.636 | 6.383 | 7.818 | 538.592 | 342.483 | 0 |
| 34 | (CH$_2$)$_2$CONH$_2$ | Bnz | Me | Cbz | 5.699 | 5.673 | 0.026 | 0.030 | 5.669 | 2.945 | 6.386 | 7.863 | 552.619 | 338.858 | 0 |
| 35 | (CH$_2$)$_2$CONH$_2$ | Bnz | i-Pr | Cbz | 6.420 | 6.292 | 0.128 | 0.137 | 6.283 | 3.873 | −0.517 | 2.730 | 580.672 | 422.592 | 0 |
| 36 | (CH$_2$)$_2$CONH$_2$ | Bnz | CH(S-Me)Et | Cbz | 6.102 | 6.405 | −0.303 | −0.326 | 6.429 | 4.402 | −0.561 | 2.688 | 594.698 | 417.655 | 0 |
| 37 | (CH$_2$)$_2$CONH$_2$ | Bnz | t-But | Cbz | 6.495 | 6.325 | 0.170 | 0.182 | 6.313 | 4.272 | −0.577 | 2.633 | 594.698 | 414.693 | 0 |
| 38 | (CH$_2$)$_2$CONH$_2$ | Bnz | (CH$_2$)$_2$SMe | Cbz | 5.854 | 6.212 | −0.358 | −0.401 | 6.255 | 3.093 | −0.565 | 3.842 | 612.737 | 483.809 | 0 |
| 39 | (CH$_2$)$_2$CONH$_2$ | Bnz | CH$_2$SMe | Cbz | 6.745 | 6.005 | 0.740 | 0.798 | 5.946 | 3.329 | −0.522 | 1.569 | 598.710 | 467.024 | 0 |
| 40[a] | (CH$_2$)$_2$CONH$_2$ | Bnz | CH(R-Me)S-i-Pr | Cbz | 5.000 | 5.942 | −0.942 | −1.017 | 6.017 | 4.476 | −1.336 | 1.448 | 640.790 | 402.437 | 0 |
| 41 | (CH$_2$)$_2$CONH$_2$ | Bnz | c-Hex | Cbz | 6.000 | 6.254 | −0.254 | −0.277 | 6.277 | 5.066 | 0.233 | 2.017 | 620.736 | 398.422 | 0 |
| 42 | (CH$_2$)$_2$CONH$_2$ | Bnz | CH$_2$c-Hex | Cbz | 5.921 | 6.531 | −0.610 | −0.681 | 6.602 | 5.595 | 0.242 | 2.071 | 634.762 | 422.208 | 0 |
| 43 | (CH$_2$)$_2$CONH$_2$ | Bnz | Bnz | Cbz | 6.252 | 6.009 | 0.243 | 0.256 | 5.996 | 4.363 | 0.637 | 2.656 | 628.715 | 412.373 | 0 |
| 44 | (CH$_2$)$_2$CONH$_2$ | Bnz | CH$_2$SPh | Cbz | 6.921 | 5.975 | 0.946 | 1.048 | 5.873 | 4.975 | 0.134 | 0.894 | 660.780 | 432.119 | 0 |
| 45 | (CH$_2$)$_2$CONH$_2$ | Bnz | CH$_2$SBnz | Cbz | 6.699 | 6.057 | 0.642 | 0.711 | 5.988 | 5.252 | 0.595 | 1.560 | 674.806 | 432.894 | 0 |
| 46 | (CH$_2$)$_2$CONH$_2$ | Bnz | Phenethyl | Cbz | 6.602 | 6.088 | 0.514 | 0.552 | 6.050 | 4.892 | 0.147 | 2.107 | 642.741 | 406.351 | 0 |
| 47 | (CH$_2$)$_2$CONH$_2$ | Bnz | CH$_2$OH | Cbz | 5.745 | 5.409 | 0.336 | 0.373 | 5.371 | 1.944 | 5.520 | 7.889 | 568.618 | 377.703 | 0 |
| 48 | (CH$_2$)$_2$CONH$_2$ | Bnz | CH(R-OH)Me | Cbz | 5.745 | 5.382 | 0.363 | 0.393 | 5.352 | 2.253 | 5.016 | 6.785 | 582.645 | 384.249 | 0 |
| 49 | (CH$_2$)$_2$CONH$_2$ | Bnz | CMe$_2$OH | Cbz | 6.180 | 5.584 | 0.596 | 0.631 | 5.550 | 2.652 | 0.924 | 3.276 | 596.671 | 424.567 | 0 |
| 50 | (CH$_2$)$_2$CONH$_2$ | Bnz | CMe$_2$CH$_2$OH | Cbz | 5.886 | 5.452 | 0.434 | 0.468 | 5.418 | 2.285 | 0.346 | 3.352 | 610.698 | 436.415 | 0 |

*(Continued)*

TABLE 11.36 Biological Activity and Physicochemical Parameters of a Series of Peptide-Derived HRV 3C$^{pro}$ Inhibitors for QSAR Model [Eq. (11.35)] (cont.)

| Compound | R$_1$ | R$_2$ | R$_3$ | R$_4$ | Obsd | Calcd | Res | Del res | Pred | C Log P | D$_\gamma$ | D$_{Tot}$ | MW | Pol Vol | I |
|---|---|---|---|---|---|---|---|---|---|---|---|---|---|---|---|
| 51[a] | (CH$_2$)$_2$CONH$_2$ | Bnz | (CH$_2$)$_4$NH$_2$ | Cbz | 3.688 | 5.803 | −2.115 | −2.293 | 5.981 | 2.645 | −0.448 | 2.678 | 609.713 | 473.406 | 0 |
| 52 | (CH$_2$)$_2$CONH$_2$ | Bnz | (CH$_2$)$_2$Morphol | Cbz | 5.292 | 5.349 | −0.056 | −0.063 | 5.356 | 2.418 | −0.519 | 2.657 | 651.750 | 460.861 | 0 |
| 53 | (CH$_2$)$_2$CONH$_2$ | Bnz | (CH$_2$)$_3$Morphol | Cbz | 5.149 | 5.564 | −0.415 | −0.465 | 5.613 | 2.947 | −1.018 | 2.732 | 665.776 | 463.920 | 0 |
| 54 | (CH$_2$)$_2$CONH$_2$ | Bnz | CH$_2$COOH | Cbz | 5.620 | 5.486 | 0.134 | 0.153 | 5.467 | 2.366 | 4.934 | 8.528 | 596.628 | 365.901 | 0 |
| 55 | (CH$_2$)$_2$CONH$_2$ | Bnz | (CH$_2$)$_2$COOH | Cbz | 5.260 | 5.101 | 0.158 | 0.199 | 5.061 | 2.057 | 8.211 | 10.633 | 610.655 | 341.532 | 0 |
| 56 | (CH$_2$)$_2$CONH$_2$ | Bnz | CH$_2$CONMe$_2$ | Cbz | 5.229 | 5.213 | 0.016 | 0.018 | 5.211 | 2.309 | −0.516 | 0.519 | 623.697 | 464.062 | 0 |
| 57 | (CH$_2$)$_2$CONH$_2$ | Bnz | i-But | 2-MeCbz | 6.000 | 5.732 | 0.268 | 0.279 | 5.721 | 5.049 | 0.227 | 3.329 | 608.725 | 391.567 | 1 |
| 58 | (CH$_2$)$_2$CONH$_2$ | Bnz | i-But | 2-ClCbz | 6.201 | 5.874 | 0.327 | 0.345 | 5.855 | 5.313 | −1.058 | 3.636 | 629.144 | 391.664 | 1 |
| 59 | (CH$_2$)$_2$CONH$_2$ | Bnz | i-But | COOCH$_2$(4-Pyr) | 4.252 | 4.818 | −0.566 | −0.614 | 4.866 | 3.103 | 0.604 | 2.039 | 595.687 | 407.187 | 1 |
| 60[b] | (CH$_2$)$_2$CONH$_2$ | Bnz | i-But | COOMe | 5.886 | 5.071 | 0.815 | 0.881 | 5.006 | 2.243 | −1.063 | 3.298 | 518.602 | 366.847 | 1 |
| 61 | (CH$_2$)$_2$CONH$_2$ | Bnz | i-But | COO-c-Hex | 5.119 | 5.702 | −0.583 | −0.601 | 5.720 | 4.274 | −0.617 | 3.216 | 586.720 | 410.790 | 1 |
| 62 | (CH$_2$)$_2$CONH$_2$ | Bnz | i-But | COO-t-But | 5.347 | 5.683 | −0.336 | −0.359 | 5.706 | 3.480 | −0.904 | 4.859 | 560.682 | 393.366 | 1 |
| 63 | (CH$_2$)$_2$CONH$_2$ | Bnz | i-But | COSMe | 5.959 | 5.787 | 0.172 | 0.188 | 5.770 | 3.236 | −1.830 | 4.270 | 534.668 | 398.801 | 1 |

| 64 | (CH$_2$)$_2$CONH$_2$ | Bnz | i-But | COSEt | 6.337 | 6.009 | 0.328 | 0.364 | 5.974 | 3.765 | −1.847 | 4.246 | 548.695 | 414.000 | 1 |
| 65 | (CH$_2$)$_2$CONH$_2$ | Bnz | i-But | COS-c-Pent | 6.745 | 6.060 | 0.685 | 0.737 | 6.007 | 4.708 | −1.713 | 3.770 | 588.759 | 410.214 | 1 |
| 66 | (CH$_2$)$_2$CONH$_2$ | Bnz | i-But | COSBnz | 6.569 | 5.908 | 0.661 | 0.693 | 5.876 | 4.940 | −1.782 | 2.960 | 610.764 | 410.261 | 1 |
| 67 | (CH$_2$)$_2$CONH$_2$ | Bnz | i-But | CO-2-Napthyl | 6.000 | 5.632 | 0.368 | 0.410 | 5.590 | 4.898 | −3.035 | 3.347 | 614.731 | 336.399 | 1 |
| 68 | (CH$_2$)$_2$CONH$_2$ | Bnz | i-But | COPh | 5.284 | 5.551 | −0.267 | −0.275 | 5.559 | 3.724 | −0.370 | 3.364 | 564.673 | 401.162 | 1 |
| 69[a] | (CH$_2$)$_2$CONH$_2$ | Bnz | i-But | COPhOPh | 5.284 | 6.351 | −1.067 | −1.295 | 6.579 | 5.822 | −0.349 | 4.902 | 656.768 | 452.361 | 1 |
| 70 | (CH$_2$)$_2$CONH$_2$ | Bnz | i-But | COMe | 4.854 | 4.855 | −0.001 | −0.001 | 4.855 | 2.073 | 0.390 | 3.421 | 502.603 | 342.891 | 1 |
| 71[a] | (CH$_2$)$_2$CONH$_2$ | Bnz | i-But | CO-i-Pr | 6.000 | 5.185 | 0.815 | 0.854 | 5.146 | 2.911 | 0.267 | 3.332 | 530.656 | 369.905 | 1 |
| 72 | (CH$_2$)$_2$CONH$_2$ | Bnz | i-But | CO-t-But | 5.745 | 5.312 | 0.433 | 0.448 | 5.296 | 3.310 | 0.354 | 3.399 | 544.683 | 377.655 | 1 |
| 73[a] | (CH$_2$)$_2$CONH$_2$ | Bnz | i-But | CO-c-Pent | 6.222 | 5.281 | 0.941 | 0.975 | 5.247 | 3.545 | 0.062 | 2.830 | 556.694 | 377.076 | 1 |
| 74 | (CH$_2$)$_2$CONH$_2$ | Bnz | i-But | COCH$_2$OH | 4.523 | 4.666 | −0.143 | −0.164 | 4.687 | 1.913 | 1.111 | 2.402 | 518.602 | 379.856 | 1 |
| 75 | (CH$_2$)$_2$CONH$_2$ | Bnz | i-But | CO(CH$_2$)$_2$OH | 4.721 | 4.816 | −0.095 | −0.102 | 4.824 | 1.746 | 1.759 | 4.449 | 532.629 | 398.315 | 1 |
| 76 | (CH$_2$)$_2$CONH$_2$ | Bnz | i-But | CON(Me)Bnz | 5.252 | 5.496 | −0.244 | −0.254 | 5.506 | 4.072 | 0.071 | 2.815 | 607.740 | 438.202 | 1 |
| 77[a] | (CH$_2$)$_2$CONH$_2$ | Bnz | i-But | Ac-L-Val | 4.201 | 5.106 | −0.906 | −0.998 | 5.198 | 3.802 | 4.372 | 4.664 | 574.709 | 374.435 | 1 |
| 78 | (CH$_2$)$_2$CONH$_2$ | Bnz | i-But | Ac-L-Ala | 4.699 | 5.077 | −0.378 | −0.397 | 5.096 | 2.874 | 2.249 | 4.207 | 546.656 | 385.237 | 1 |

[a]Considered as outlier.

**TABLE 11.37 Biological Activity and Physicochemical Parameters of Ktomethylene Containing Peptide-Based HRV 3C$^{pro}$ Inhibitors for QSAR Model [Eq. (11.36)]**

| Compound | Ar | Y | X | $R_1$ | $R_2$ | Obsd | Calcd | Res | Del res | Pred | $D_Y$ | $I_1$ | $I_2$ |
|---|---|---|---|---|---|---|---|---|---|---|---|---|---|
| 1 | Bnz | O | NH | Bnz | i-But | 6.268 | 6.245 | 0.023 | 0.029 | 6.238 | 0.755 | 1 | 0 |
| 2 | Bnz | O | CH$_2$ | Bnz | i-But | 6.444 | 6.207 | 0.237 | 0.307 | 6.137 | 1.002 | 1 | 0 |
| 3 | Bnz | S | CH$_2$ | Bnz | i-But | 6.167 | 6.284 | −0.117 | −0.146 | 6.313 | 0.498 | 1 | 0 |

| | | | | | | | | | | | | |
|---|---|---|---|---|---|---|---|---|---|---|---|---|
| 4 | Bnz | S | $CH_2$ | Bnz | i-Pr | 6.721 | 7.317 | −0.596 | −0.778 | 7.499 | 0.508 | 0 | 1 |
| 5 | c-Pent | S | $CH_2$ | Bnz | i-But | 6.721 | 6.827 | −0.106 | −0.141 | 6.863 | −3.025 | 1 | 0 |
| 6 | c-Pent | S | $CH_2$ | 4F-Bnz | i-But | 6.553 | 6.536 | 0.016 | 0.020 | 6.533 | −1.139 | 1 | 0 |
| 7 | c-Pent | S | $CH_2$ | 4-MeBnz | i-But | 6.796 | 6.850 | −0.054 | −0.073 | 6.869 | −3.172 | 1 | 0 |
| 8 | c-Pent | S | $CH_2$ | Bnz | i-Pr | 7.699 | 7.861 | −0.162 | −0.226 | 7.925 | −3.021 | 0 | 1 |
| 9 | c-Pent | S | $CH_2$ | 4F-Bnz | i-Pr | 7.699 | 7.570 | 0.129 | 0.162 | 7.537 | −1.130 | 0 | 1 |
| 10 | c-Pent | S | $CH_2$ | 4-MeBnz | i-Pr | 8.222 | 7.883 | 0.338 | 0.480 | 7.742 | −3.168 | 0 | 1 |
| 11 | c-Pent | S | $CH_2$ | 4-$CF_3$Bnz | i-Pr | 7.301 | 7.011 | 0.290 | 0.485 | 6.816 | 2.495 | 0 | 1 |
| 12 | c-Pent | S | $CH_2$ | 4F-Bnz | Bnz | 6.319 | 6.445 | −0.126 | −0.227 | 6.545 | 2.268 | 0 | 0 |
| 13 | c-Pent | S | $CH_2$ | 4-MeBnz | Bnz | 6.854 | 6.759 | 0.095 | 0.143 | 6.710 | 0.233 | 0 | 0 |
| 14 | c-Pent | S | $CH_2$ | Bnz | t-But | 7.301 | 7.270 | 0.031 | 0.060 | 7.241 | −3.084 | 0 | 0 |

**TABLE 11.38 Biological Activity and Physicochemical Parameters of Peptide-Based HRV 3C$^{pro}$ Inhibitors for QSAR Model [Eq. (11.37)]**

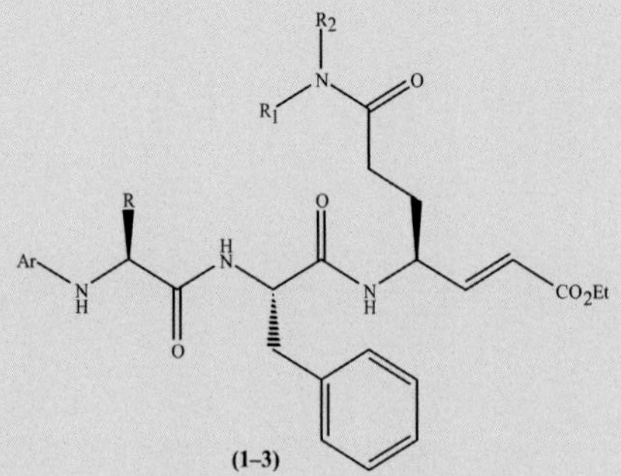

(1–3)

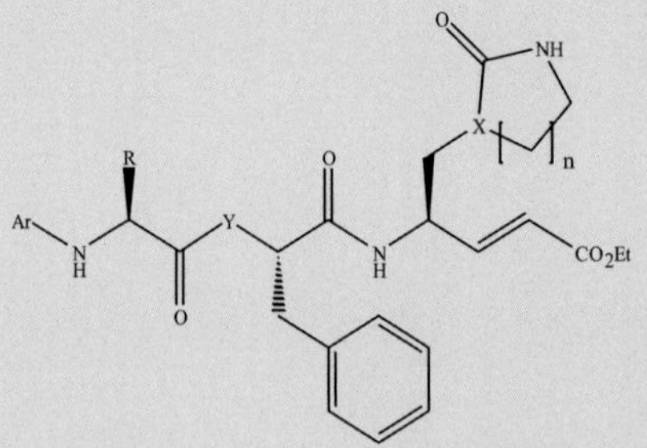

(4–13)

| Compound | Ar | Ar$_1$ | R | R$_1$ | R$_2$ | n | X | Y | Obsd | Calcd | Res | Del res | Pred | D$_X$ |
|---|---|---|---|---|---|---|---|---|---|---|---|---|---|---|
| 1[a] | Cbz | Bnz | i-But | H | H | — | — | — | 6.268 | 4.694 | 1.573 | 2.178 | 4.090 | 2.748 |
| 2 | Cbz | Bnz | i-But | Me | H | — | — | — | 3.125 | 4.926 | −1.801 | −2.359 | 5.484 | 2.545 |
| 3 | Cbz | Bnz | i-But | Me | Me | — | — | — | 4.222 | 4.419 | −0.197 | −0.296 | 4.517 | 2.988 |
| 4 | Cbz | Bnz | i-But | — | — | 1 | (S)-CH | NH | 7.000 | 7.112 | −0.112 | −0.122 | 7.122 | 0.640 |
| 5 | Cbz | Bnz | i-But | — | — | 2 | (S)-CH | NH | 7.523 | 8.286 | −0.763 | −0.922 | 8.445 | −0.383 |
| 6 | Cbz | Bnz | i-But | — | — | 1 | (R)-CH | NH | 5.796 | 7.112 | −1.316 | −1.432 | 7.228 | 0.640 |
| 7 | Cbz | Bnz | i-But | — | — | 1 | N | NH | 6.222 | 6.177 | 0.045 | 0.050 | 6.172 | 1.456 |
| 8 | Cbz | Bnz | i-Pr | — | — | 1 | (S)-CH | NH | 7.523 | 7.128 | 0.395 | 0.430 | 7.093 | 0.627 |
| 9 | 5-Me-isoxazole-3-CO | 4F-Bnz | i-Pr | — | — | 1 | (S)-CH | NH | 8.000 | 7.938 | 0.062 | 0.071 | 7.929 | −0.079 |
| 10 | 5-Me-isoxazole-3-CO | 4F-Bnz | i-Pr | — | — | 2 | (S)-CH | NH | 7.699 | 6.937 | 0.762 | 0.826 | 6.873 | 0.793 |
| 11 | 5-Me-isoxazole-3-CO | 4F-Bnz | i-Pr | — | — | 1 | (S)-CH | CH$_2$ | 8.301 | 8.522 | −0.221 | −0.278 | 8.579 | −0.589 |
| 12 | 5-Me-isoxazole-3-CO | 4F-Bnz | i-Pr | — | — | 2 | (S)-CH | CH$_2$ | 9.000 | 7.510 | 1.490 | 1.651 | 7.349 | 0.293 |
| 13 | 5-Me-isoxazole-3-CO | 4F-Bnz | t-But | — | — | 1 | (S)-CH | NH | 8.000 | 7.915 | 0.085 | 0.098 | 7.902 | −0.059 |

[a]Considered as outlier.

these compounds had suggested that (S)-conformation of the compound had better activity than (R)-conformation.

$$pEC_{50} = 7.890(\pm0.309) - 1.383(\pm0.239)D_X \quad (11.37)$$

$N = 12$, $R = 0.877$, $R^2 = 0.769$, $R_A^2 = 0.746$, $F(1, 10) = 33.377$, $P < 0.00018$, SEE $= 0.876$, $q^2 = 0.652$, $Q = 1.001$, Outlier = Compound **1**

Dragovich et al. (1999c) had also reported some tripeptidyl $N$-methyl amino acids as HRV 3C$^{pro}$ inhibitors (Table 11.39), for which the QSAR model obtained [Eq. (11.38)] had suggested that the activity will be totally governed by the total dipole moment of the compound. This indicated the strong electronic interaction between the molecule and the receptor.

$$pEC_{50} = 6.094(\pm0.122) + 0.075(\pm0.019)D_{Tot} \quad (11.38)$$

$N = 8$, $R = 0.852$, $R^2 = 0.726$, $R_A^2 = 0.680$, $F(1, 6) = 15.869$, $P < 0.00725$, SEE $= 0.173$, $q^2 = 0.537$, $Q = 4.925$, Outlier = Compounds **7, 10**

However, for some ketone containing tripeptidyl HRV 3C$^{pro}$ inhibitors (Table 11.40) reported by Dragovich et al. (2000), the QSAR model [Eq. (11.39)] exhibited the importance of only $Z$-component of the dipole moment of the compound. In the $Z$-direction, the polarity of the compound may be more favorable for electronic interaction with the receptor.

$$pK_i = 3.662(\pm0.583) + 2.862(\pm0.713)D_Z \quad (11.39)$$

$N = 6$, $R = 0.895$, $R^2 = 0.801$, $R_A^2 = 0.751$, $F(1, 4) = 16.096$, $P < 0.01597$, SEE $= 0.490$, $q^2 = 0.594$, $Q = 1.827$, Outlier = Compounds **5, 7**

### 7.2.5 Depsipeptidyl HRV 3C$^{pro}$ Inhibitors

Webber et al. (2001) reported some depsipeptidyl HRV 3C$^{pro}$ inhibitors (Table 11.41), for which the QSAR model obtained was as is shown by Eq. (11.40). This model suggested that in this case, the polar volume of the molecule will not be conducive to the activity.

$$pEC_{50} = 18.601(\pm2.905) - 0.029(\pm0.008)PolVol \quad (11.40)$$

$N = 7$, $R = 0.866$, $R^2 = 0.750$, $R_A^2 = 0.700$, $F(1, 5) = 15.030$, $P < 0.01168$, SEE $= 0.499$, $q^2 = 0.533$, $Q = 1.735$, Outlier = Compounds **2, 6**

### 7.2.6 2-Pyridone Containing Peptidomimetics as HRV 3C$^{pro}$ Inhibitors

For some 2-pyridone containing peptidomimetics as promising HRV 3C$^{pro}$ inhibitors (Table 11.42) reported by Dragovich et al. (2002), the QSAR model [Eq. (11.41)] had, however, suggested that the SA of the molecule will be favorable to the activity, probably because of dispersion interaction between the active surface of the molecule and that of the receptor. However, a very high

**TABLE 11.39 Biological Activity and Physicochemical Parameters of Some Tripeptidyl *N*-methyl Amino Acids for QSAR Model [Eq. (11.38)]**

| Compound | R | $R_1$ | $R_2$ | Ar | Obsd | Calcd | Res | Del res | Pred | $D_{Tot}$ |
|---|---|---|---|---|---|---|---|---|---|---|
| 1 | NHCBz | H | *i*-But | Bnz | 6.268 | 6.165 | 0.103 | 0.160 | 6.108 | 1.102 |
| 2 | NHCBz | Me | *i*-But | Bnz | 6.000 | 6.187 | −0.187 | −0.272 | 6.272 | 1.542 |
| 3 | CO-S-c-Pent | Me | *i*-But | Bnz | 6.495 | 6.351 | 0.144 | 0.163 | 6.331 | 4.724 |
| 4 | CO-S-c-Pent | Me | *i*-Pr | Bnz | 6.721 | 6.353 | 0.368 | 0.418 | 6.303 | 4.759 |
| 5 | CO-S-c-Pent | Me | $CH_2SPh$ | Bnz | 6.268 | 6.331 | −0.064 | −0.073 | 6.341 | 4.336 |
| 6 | 5-Methyl-isoxazole-3-carbonyl | Me | *i*-But | Bnz | 6.854 | 6.547 | 0.307 | 0.364 | 6.490 | 8.521 |
| 7[a] | 5-Methyl-isoxazole-3-carbonyl | Me | *i*-But | 4-FBnz | 6.398 | 6.594 | −0.197 | −0.249 | 6.647 | 9.449 |
| 8 | 5-Methyl-isoxazole-3-carbonyl | Me | $CH_2$(1-Napthyl) | 4-FBnz | 6.699 | 6.592 | 0.107 | 0.135 | 6.564 | 9.400 |
| 9 | 5-Methyl-isoxazole-3-carbonyl | Me | $CH_2$(2-Napthyl) | 4-FBnz | 6.796 | 6.641 | 0.155 | 0.214 | 6.581 | 10.349 |
| 10[a] | 5-Methyl-isoxazole-3-carbonyl | Me | $CH_2$(4-Imidazole) | 4-FBnz | 5.745 | 6.482 | −0.737 | −0.831 | 6.575 | 7.262 |

[a]Considered as outlier.

**TABLE 11.40** Biological Activity and Physicochemical Parameters of Some Ketone Containing Tripeptidyl HRV 3C$^{pro}$ Inhibitors for QSAR Model [Eq. (11.39)]

| Compound | R | R$_1$ | R$_2$ | R$_3$ | Obsd | Calcd | Res | Del res | Pred | D$_Z$ |
|---|---|---|---|---|---|---|---|---|---|---|
| 1 | Benzothiazol-2-yl | NHCOMe | =O | | 5.770 | 5.512 | 0.257 | 0.321 | 5.449 | 0.645 |
| 2 | Benzothiazol-2-yl | CH$_2$CONH$_2$ | =O | | 5.456 | 5.458 | −0.002 | −0.003 | 5.459 | 0.596 |
| 3 | Benzothiazol-2-yl | 2-Oxo-pyrrolidin-3-yl | =O | | 7.187 | 6.227 | 0.960 | 1.266 | 5.921 | 1.287 |
| 4 | Benzothiazol-2-yl | 2-Oxo-pyrrolidin-3-yl | | H | 4.469 | 5.225 | −0.756 | −1.245 | 5.714 | 0.386 |
| 5$^a$ | Thiazol-2-yl | 2-Oxo-pyrrolidin-3-yl | OH | | 6.155 | 6.350 | −0.195 | −0.290 | 6.445 | 1.399 |
| 6 | 2-Pyridyl | 2-Oxo-pyrrolidin-3-yl | =O | | 6.770 | 5.742 | 1.028 | 1.182 | 5.588 | 0.851 |
| 7$^a$ | Benzothiophen-2-yl | 2-Oxo-pyrrolidin-3-yl | =O | | 5.328 | 6.387 | −1.059 | −1.639 | 6.966 | 1.431 |
| 8 | Ph | 2-Oxo-pyrrolidin-3-yl | =O | | 5.495 | 5.727 | −0.232 | −0.268 | 5.763 | 0.838 |

$^a$Considered as outlier.

TABLE 11.41 Biological Activity and Physicochemical Parameters of Depsipeptidyl HRV 3C$^{pro}$ Inhibitors for QSAR Model [Eq. (11.40)]

| Compound | R | $R_1$ | X | Obsd | Calcd | Res | Del res | Pred | Pol Vol |
|---|---|---|---|---|---|---|---|---|---|
| 1 | S-c-pent | $(CH_2)_2CONH_2$ | O | 6.000 | 6.466 | −0.466 | −0.692 | 6.692 | 426.668 |
| 2[a] | S-c-pent | $(CH_2)_2CONH_2$ | NH | 6.721 | 6.244 | 0.477 | 0.894 | 5.827 | 438.416 |
| 3 | S-c-pent | $(CH_2)_2CONH_2$ | $CH_2$ | 7.699 | 7.329 | 0.370 | 0.418 | 7.281 | 380.864 |
| 4 | 5-Methyl-isoxazole | $(CH_2)_2CONH_2$ | O | 7.000 | 7.360 | −0.360 | −0.407 | 7.407 | 379.172 |
| 5 | 5-Methyl-isoxazole | $(CH_2)_2CONH_2$ | NH | 6.377 | 7.042 | −0.665 | −0.762 | 7.138 | 396.070 |
| 6[a] | 5-Methyl-isoxazole | $(CH_2)_2CONH_2$ | $CH_2$ | 7.000 | 7.773 | −0.773 | −0.974 | 7.974 | 357.284 |
| 7 | 5-Methyl-isoxazole | 5-(2-Oxo-pyrrolidin-3-yl) | O | 8.155 | 7.516 | 0.639 | 0.739 | 7.416 | 370.905 |
| 8 | 5-Methyl-isoxazole | 5-(2-Oxo-pyrrolidin-3-yl) | NH | 8.000 | 7.379 | 0.621 | 0.703 | 7.297 | 378.176 |
| 9 | 5-Methyl-isoxazole | 5-(2-Oxo-pyrrolidin-3-yl) | $CH_2$ | 8.301 | 8.144 | 0.157 | 0.258 | 8.043 | 337.594 |

[a]Considered as outlier.

**TABLE 11.42 Biological Activity and Physicochemical Parameters of 2-Pyridone Containing Peptidomimetics as HRV 3C$^{pro}$ Inhibitors for QSAR Model [Eq. (11.41)]**

| Compound | R$_1$ | R$_2$ | R$_3$ | Obsd | Calcd | Res | Del res | Pred | SA |
|---|---|---|---|---|---|---|---|---|---|
| 1 | (CH$_2$)$_2$CONH$_2$ | Bnz | OBnz | 7.481 | 7.493 | −0.012 | −0.013 | 7.495 | 882.377 |
| 2 | (CH$_2$)$_2$CONH$_2$ | 4-FBnz | OBnz | 7.854 | 7.492 | 0.362 | 0.410 | 7.444 | 883.062 |
| 3 | (CH$_2$)$_2$CONH$_2$ | 3,4-diFBnz | OBnz | 8.523 | 7.492 | 1.031 | 1.165 | 7.358 | 882.859 |
| 4 | (CH$_2$)$_2$CONH$_2$ | CH$_2$-c-Hex | OBnz | 6.750 | 7.368 | −0.618 | −0.850 | 7.600 | 906.641 |
| 5[a] | (CH$_2$)$_2$CONH$_2$ | Bnz | Me | 6.052 | 5.531 | 0.520 | 9.821 | −3.770 | 749.850 |
| 6 | (CH$_2$)$_2$CONH$_2$ | Bnz | c-Pent | 7.469 | 7.194 | 0.275 | 0.325 | 7.144 | 825.095 |
| 7 | (CH$_2$)$_2$CONH$_2$ | Bnz | [1,3]Dithiolane-2-yl | 6.983 | 7.190 | −0.207 | −0.245 | 7.228 | 824.774 |
| 8 | (CH$_2$)$_2$CONH$_2$ | Bnz | Tetrahydrofuran-2-yl | 5.750 | 7.049 | −1.299 | −1.550 | 7.299 | 814.614 |
| 9 | (CH$_2$)$_2$CONH$_2$ | Bnz | t-Butyl | 6.286 | 7.120 | −0.835 | −0.991 | 7.277 | 819.529 |
| 10 | (CH$_2$)$_2$CONH$_2$ | Bnz | 5-Me-Benzoxazol-3-yl | 7.796 | 7.394 | 0.402 | 0.466 | 7.330 | 845.229 |
| 11 | (CH$_2$)$_2$CONH$_2$ | Bnz | 5-Cl-Benzoxazol-3-yl | 7.620 | 7.315 | 0.305 | 0.357 | 7.263 | 836.004 |
| 12 | (R)-2-Oxopyrrodine-3-yl | Bnz | OBnz | 8.523 | 7.476 | 1.047 | 1.204 | 7.319 | 888.453 |
| 13 | (S)-2-Oxopyrrodine-3-yl | Bnz | OBnz | 6.329 | 7.300 | −0.971 | −1.592 | 7.921 | 914.072 |

[a]Considered as outlier.

SA was shown to be detrimental to the activity, probably because of the steric problem.

$$pEC_{50} = -670.463(\pm111.246)+1.568(\pm0.259)SA-0.001(\pm0.0001)SA^2 \quad (11.41)$$

$N$ = 12, $R$ = 0.905, $R^2$ = 0.819, $R_A^2$ = 0.779, $F$ (2, 9) = 20.403, $P$ < 0.00045, SEE = 0.413, $q^2$ = 0.715, $Q$ = 2.191, $SA_{opt}$ = 784, Outlier = Compound **5**

### 7.2.7 Michael Acceptor Containing Irreversible HRV 3C$^{pro}$ Inhibitors

Johnson et al. (2002) reported some Michael acceptor containing irreversible HRV 3C$^{pro}$ inhibitors (Table 11.43). For these inhibitors, the QSAR model [Eq. (11.42)] indicated that the activity of compounds will simply depend upon the presence or absence of β-naphthyl group or chromene ring at the R-position of the compound. Of the two indicator variables, $I_1$ and $I_2$, the former with a value of 1 indicated the presence of β-naphthyl group and the latter with a value of unity indicated the presence of chromene ring at the R-position. The positive coefficients of both the variables indicated the favorable contribution of either of the substituent. Both might have steric interactions with the active site of the enzyme.

$$pEC_{50} = 5.951(\pm0.086)+0.833(\pm0.172)I_1 +0.819(\pm0.172)I_2 \quad (11.42)$$

$N$ = 10, $R$ = 0.917, $R^2$ = 0.841, $R_A^2$ = 0.796, $F$ (2, 7) = 18.528, $P$ < 0.00160, SEE = 0.210, $q^2$ = 0.765, $Q$ = 4.367, Outlier = Compounds **7, 11**

### 7.2.8 2-Pyridone Containing Peptidomimetics as HRV 3C$^{pro}$ Inhibitors

Dragovich et al. (2003) reported a series of 2-pyridone containing peptidomimetics as HRV 3C$^{pro}$ inhibitors (Table 11.44). The QSAR model obtained for these compounds was as shown by Eq. (11.43). This model suggested that have less bulky molecules with small $X$-component of their dipole moment will be favored. Simultaneously, the high value of the PSA of these molecules may be conducive to activity. Further, the positive coefficient of the indicator variable "I" that was defined with a value of unity for a benzyl substituent at $R_1$-position indicated that such a substituent should be desired for the better activity of the compound. This benzyl group might have steric interaction with the receptor.

$$pEC_{50} = 7.013(\pm2.741)-0.471(\pm0.091)CMR - 0.329(\pm0.042)D_X \\ + 0.032(\pm0.009)PSA + 0.675(\pm0.165)I \quad (11.43)$$

$N$ = 18, $R$ = 0.947, $R^2$ = 0.897, $R_A^2$ = 0.865, $F$ (4, 13) = 28.323, $P$ < 0.00000, SEE = 0.175, $q^2$ = 0.737, $Q$ = 5.411

**TABLE 11.43 Biological Activity and Physicochemical Parameters of Michael Acceptor Containing Irreversible HRV 3C$^{pro}$ Inhibitors for QSAR Model [Eq. (11.42)]**

| Compound | R | Obsd | Calcd | Res | Del res | Pred | $I_1$ | $I_2$ |
|---|---|---|---|---|---|---|---|---|
| 1 | 3-BrPhCH=CH | 5.854 | 5.951 | -0.098 | -0.117 | 5.971 | 0 | 0 |
| 2 | 3-Br,4-MePhCH=CH | 5.889 | 5.951 | -0.062 | -0.074 | 5.964 | 0 | 0 |
| 3 | 3-Br,4-FPhCH=CH | 6.201 | 5.951 | 0.249 | 0.299 | 5.902 | 0 | 0 |
| 4 | Benzo[1,3]dioxole | 5.742 | 5.951 | 0.209 | 0.251 | 5.993 | 0 | 0 |
| 5 | 5-Bromo-benzo[1,3]dioxole | 6.310 | 5.951 | 0.358 | 0.430 | 5.880 | 0 | 0 |
| 6 | 2-Methyl-5-phenyl-furan | 5.712 | 5.951 | 0.239 | 0.287 | 5.999 | 0 | 0 |
| 7$^a$ | 2H-Chromene-3-yl | 5.924 | 6.488 | -0.564 | -0.846 | 6.770 | 0 | 1 |
| 8 | 6-Chloro-2H-chromene-3-yl | 6.796 | 6.488 | 0.308 | 0.461 | 6.335 | 0 | 1 |
| 9 | 6-Bromo-2H-chromene-3-yl | 6.745 | 6.488 | 0.256 | 0.385 | 6.360 | 0 | 1 |
| 10 | Napthaleny-2-yl | 6.824 | 6.541 | 0.283 | 0.424 | 6.400 | 1 | 0 |
| 11$^a$ | 6-Methyl-napthaleny-2-yl | 6.056 | 6.541 | -0.486 | -0.729 | 6.784 | 1 | 0 |
| 12 | 7-Bromo-napthaleny-2-yl | 6.745 | 6.541 | 0.203 | 0.305 | 6.440 | 1 | 0 |

$^a$Considered as outlier.

**TABLE 11.44 Biological Activity and Physicochemical Parameters of 2-Pyridone Containing Peptidomimetics as HRV 3C$^{pro}$ Inhibitors for QSAR Model [Eq. (11.43)]**

| Compound | R$_1$ | R$_2$ | Obsd | Calcd | Res | Del res | Pred | CMR | $D_X$ | PSA | I |
|---|---|---|---|---|---|---|---|---|---|---|---|
| 1 | Et | CH$_2$(3,4-F)Ph | 7.959 | 7.724 | 0.234 | 0.463 | 7.495 | 15.443 | −2.386 | 226.148 | 0 |
| 2 | i-Pr | CH$_2$(3,4-F)Ph | 7.108 | 7.394 | −0.286 | −0.507 | 7.615 | 15.907 | −2.400 | 222.485 | 0 |
| 3 | Et | CH$_2$CCH | 7.237 | 7.243 | −0.006 | −0.008 | 7.245 | 13.625 | 1.971 | 229.100 | 0 |
| 4 | i-Pr | CH$_2$CCH | 6.759 | 6.808 | −0.049 | −0.054 | 6.814 | 14.089 | 1.926 | 221.859 | 0 |

| 6 | CH$_2$-t-But | CH$_2$CCH | 6.559 | 6.345 | 0.214 | 0.312 | 6.247 | 15.017 | 2.128 | 223.123 | 0 |
|---|---|---|---|---|---|---|---|---|---|---|---|
| 7 | c-But | CH$_2$CCH | 7.046 | 6.959 | 0.087 | 0.110 | 6.936 | 14.376 | 0.459 | 215.681 | 0 |
| 8 | c-Pent | CH$_2$CCH | 6.759 | 6.756 | 0.003 | 0.003 | 6.756 | 14.839 | 0.494 | 216.549 | 0 |
| 9 | c-Hex | CH$_2$CCH | 6.391 | 6.438 | -0.047 | -0.056 | 6.447 | 15.303 | 0.536 | 213.857 | 0 |
| 10 | c-Hept | CH$_2$CCH | 6.090 | 6.234 | -0.143 | -0.200 | 6.291 | 15.767 | 0.581 | 214.763 | 0 |
| 11 | Bnz | CH$_2$CCH | 7.444 | 7.298 | 0.145 | 0.292 | 7.152 | 15.673 | 0.533 | 225.091 | 1 |
| 12 | Et | Et | 7.328 | 7.356 | -0.028 | -0.047 | 7.375 | 13.364 | 2.199 | 231.167 | 0 |
| 13 | i-Pr | Et | 6.492 | 6.808 | -0.316 | -0.371 | 6.863 | 13.828 | 2.155 | 220.347 | 0 |
| 14 | t-But | Et | 6.393 | 6.402 | -0.009 | -0.011 | 6.404 | 14.292 | 2.297 | 215.939 | 0 |
| 15 | CH$_2$-t-But | Et | 6.509 | 6.460 | 0.049 | 0.071 | 6.437 | 14.756 | 2.353 | 225.186 | 0 |
| 16 | c-But | Et | 7.119 | 6.943 | 0.176 | 0.260 | 6.859 | 14.115 | 0.733 | 214.152 | 0 |
| 17 | c-Hex | Et | 6.504 | 6.424 | 0.080 | 0.098 | 6.407 | 15.042 | 0.807 | 212.348 | 0 |
| 18 | Bnz | Et | 7.194 | 7.339 | -0.145 | -0.292 | 7.486 | 15.412 | 0.797 | 225.237 | 1 |

## 8 OVERVIEW AND CONCLUSIONS

A total of 43 QSAR models (33 for SARS-CoV 3CL$^{pro}$ inhibitors and 10 for HRV 3C$^{pro}$ inhibitors) have been reported here to get an insight into the relation between the enzyme inhibitory activities of the antiviral compounds and their physicochemical and structural properties. QSAR models exhibited that the physicochemical parameters, such as dipole moment, PSA, polar volume, hydrophobicity, molar refractivity, SA, and molecular volume of the compounds play a crucial role in controlling both SARS-CoV 3CL$^{pro}$ and HRV 3C$^{pro}$ inhibitory activities. Moreover, some structural indicator variables were found to play an important role for inhibition of these enzymes. In many cases, the dipole moment and the PSA were found to be dominant factors. The bulk of the inhibitors and their flexibility and polarity also appeared to play crucial roles in the inhibition of the enzyme. Most of the QSAR models exhibited a direct correlation of dipole moment with the 3CL$^{pro}$ or 3C$^{pro}$ inhibitory activity, where a majority of them showed the positive effect of dipole moment on activity but few showed the negative effect, too. These positive and negative effects may be attributed to the orientation of the inhibitor molecules in the active site of the enzyme.

The PSA and the polarity of the inhibitors were some other important factors that were found in some cases to influence the activity. With their positive coefficients in the correlation, they indicated the attractive electronic interactions of the molecules with the enzyme, and with negative coefficient, they indicated the repulsive interaction. In many cases, the polar volume was found to govern the activity. The polar volume also indicated the attractive or repulsive dispersion interaction between the molecule and the receptor.

Among all, the hydrophobicity of the molecules had its own role. In any drug-receptor interaction, hydrophobicity is found to play an important role because in most of the enzymes the active site has a hydrophobic pocket which plays an important role in the activity of the site. In most of the cases, the bulky portion of the molecule tries to occupy this hydrophobic pocket where it may have hydrophobic interaction. The molecular volume, MW, or molar refractivity all sometimes become synonymous to the hydrophobic property of the molecules. If they are not found related to hydrophobicity and are crucial for the activity, then it means that the inhibitor-enzyme interaction involves dispersion interaction. The QSAR models discussed here may be a great help to design some new, more potent compounds in any given category of SARS-CoV 3CL$^{pro}$ or HRV 3C$^{pro}$ inhibitors.

## REFERENCES

Akaji, K., Konno, H., Mitsui, H., Teruya, K., Shimamoto, Y., Hattori, Y., Ozaki, T., Kusunoki, M., Sanjoh, A., 2011. Structure-based design, synthesis, and evaluation of peptide-mimetic SARS 3CL protease inhibitors. J. Med. Chem. 54, 7962–7973.

Allaire, M., Chernaia, M.M., Malcolm, B.A., James, M.N., 1994. Picornaviral 3C cysteine proteinases have a fold similar to chymotrypsin-like serine proteinases. Nature 369, 72–76.

Anand, K., Palm, G.J., Mesters, J.R., Siddell, S.G., Ziebuhr, J., Hilgenfeld, R., 2002. Structure of coronavirus main proteinase reveals combination of a chymotrypsin fold with an extra alpha-helical domain. EMBO J. 21, 3213–3224.

Anand, K., Ziebuhr, J., Wadhwani, P., Mesters, J.R., Hilgenfeld, R., 2003. Coronavirus main proteinase (3CLpro) structure: basis for design of anti-SARS drugs. Science 300, 1763–1767.

Anand, K., Yang, H., Bartlam, M., Rao, Z., Hilgenfeld, R., 2005. Coronavirus main proteinase: target for antiviral drug therapy. In: Schmidt, A., Wolff, M.H., Weber, O. (Eds.), Coronaviruses with Special Emphasis on First Insights Concerning SARS. Birkhauser Verlag, Basel, Switzerland, pp. 173–199.

Atmar, R.L., 2010. Noroviruses—state of the art. Food Environ. Virol. 2, 117–126.

Babcock, G.J., Esshaki, D.J., Thomas, Jr., W.D., Ambrosino, D.M., 2004. Amino acids 270 to 510 of the severe acute respiratory syndrome coronavirus spike protein are required for interaction with receptor. J. Virol. 78, 4552–4560.

Bergmann, E.M., Mosimann, S.C., Chernaia, M.M., Malcolm, B.A., James, M.N., 1997. The refined crystal structure of the 3C gene product from hepatitis A virus: specific proteinase activity and RNA recognition. J. Virol. 71, 2436–2448.

Bisht, H., Roberts, A., Vogel, L., Bukreyev, A., Collins, P.L., Murphy, B.R., Subbarao, K., Moss, B., 2004. Severe acute respiratory syndrome coronavirus spike protein expressed by attenuated vaccinia virus protectively immunizes mice. Proc. Natl. Acad. Sci. USA 101, 6641–6646.

Blanchard, J.E., Elowe, N.H., Huitema, C., Fortin, P.D., Cechetto, J.D., Eltis, L.D., Brown, E.D., 2004. High-throughput screening identifies inhibitors of the SARS coronavirus main proteinase. Chem. Biol. 11, 1445–1453.

Buchholz, U.J., Bukreyev, A., Yang, L., Lamirande, E.W., Murphy, B.R., Subbarao, K., Collins, P.L., 2004. Contributions of the structural proteins of severe acute respiratory syndrome coronavirus to protective immunity. Proc. Natl. Acad. Sci. U.S.A. 101, 9804–9809.

Cavanagh, D., 1997. Nidovirales: a new order comprising Coronaviridae and Arteriviridae. Arch. Virol. 142, 629–633.

Chen, L., Chen, S., Gui, C., Shen, J., Shen, X., Jiang, H., 2006a. Discovering severe acute respiratory syndrome coronavirus 3CL protease inhibitors: virtual screening, surface plasmon resonance, and fluorescence resonance energy transfer assays. J. Biomol. Screen 11, 915–921.

Chen, L., Li, J., Luo, C., Liu, H., Xu, W., Chen, G., Liew, O.W., Zhu, W., Puah, C.M., Shen, X., Jiang, H., 2006b. Binding interaction of quercetin-3-beta-galactoside and its synthetic derivatives with SARS-CoV 3CL (pro): structure–activity relationship studies reveal salient pharmacophore features. Bioorg. Med. Chem. 14, 8295–8306.

Cowley, J.A., Dimmock, C.M., Spann, K.M., Walker, P.J., 2000. Gill-associated virus of *Penaeus monodon* prawns: an invertebrate virus with ORF1a and ORF1b genes related to arteri- and coronaviruses. J. Gen. Virol. 81, 1473–1484.

Dragovich, P.S., Webber, S.E., Babine, R.E., Fuhrman, S.A., Patick, A.K., Matthews, D.A., Lee, C.A., Reich, S.H., Prins, T.J., Marakovits, J.T., Littlefield, E.S., Zhou, R., Tikhe, J., Ford, C.E., Wallace, M.B., Meador, III, J.W., Ferre, R.A., Brown, E.L., Binford, S.L., Harr, J.E., DeLisle, D.M., Worland, S.T., 1998a. Structure-based design, synthesis, and biological evaluation of irreversible human rhinovirus 3C protease inhibitors. 1. Michael acceptor structure–activity studies. J. Med. Chem. 41, 2806–2818.

Dragovich, P.S., Webber, S.E., Babine, R.E., Fuhrman, S.A., Patick, A.K., Matthews, D.A., Reich, S.H., Marakovits, J.T., Prins, T.J., Zhou, R., Tikhe, J., Littlefield, E.S., Bleckman, T.M., Wallace, M.B., Little, T.L., Ford, C.E., Meador, III, J.W., Ferre, R.A., Brown, E.L., Binford, S.L., DeLisle, D.M., Worland, S.T., 1998b. Structure-based design, synthesis, and biological evaluation of irreversible human rhinovirus 3C protease inhibitors. 2. Peptide structure–activity studies. J. Med. Chem. 41, 2819–2834.

Dragovich, P.S., Prins, T.J., Zhou, R., Fuhrman, S.A., Patick, A.K., Matthews, D.A., Ford, C.E., Meador, III, J.W., Ferre, R.A., Worland, S.T., 1999a. Structure-based design, synthesis, and biological evaluation of irreversible human rhinovirus 3C protease inhibitors. 3. Structure–activity studies of ketomethylene-containing peptidomimetics. J. Med. Chem. 42, 1203–1212.

Dragovich, P.S., Prins, T.J., Zhou, R., Webber, S.E., Marakovits, J.T., Fuhrman, S.A., Patick, A.K., Matthews, D.A., Lee, C.A., Ford, C.E., Burke, B.J., Rejto, P.A., Hendrickson, T.F., Tuntland, T., Brown, E.L., Meador, III, J.W., Ferre, R.A., Harr, J.E., Kosa, M.B., Worland, S.T., 1999b. Structure-based design, synthesis, and biological evaluation of irreversible human rhinovirus 3C protease inhibitors. 4. Incorporation of P1 lactam moieties as L-glutamine replacements. J. Med. Chem. 42, 1213–1224.

Dragovich, P.S., Webber, S.E., Prins, T.J., Zhou, R., Marakovits, J.T., Tikhe, J.G., Fuhrman, S.A., Patick, A.K., Matthews, D.A., Ford, C.E., Brown, E.L., Binford, S.L., Meador, III, J.W., Ferre, R.A., Worland, S.T., 1999c. Structure-based design of irreversible, tripeptidyl human rhinovirus 3C protease inhibitors containing N-methyl amino acids. Bioorg. Med. Chem. Lett. 9, 2189–2194.

Dragovich, P.S., Zhou, R., Webber, S.E., Prins, T.J., Kwok, A.K., Okano, K., Fuhrman, S.A., Zalman, L.S., Maldonado, F.C., Brown, E.L., Meador, III, J.W., Patick, A.K., Ford, C.E., Brothers, M.A., Binford, S.L., Matthews, D.A., Ferre, R.A., Worland, S.T., 2000. Structure-based design of ketone-containing, tripeptidyl human rhinovirus 3C protease inhibitors. Bioorg. Med. Chem. Lett. 10, 45–48.

Dragovich, P.S., Prins, T.J., Zhou, R., Brown, E.L., Maldonado, F.C., Fuhrman, S.A., Zalman, L.S., Tuntland, T., Lee, C.A., Patick, A.K., Matthews, D.A., Hendrickson, T.F., Kosa, M.B., Liu, B., Batugo, M.R., Gleeson, J.P., Sakata, S.K., Chen, L., Guzman, M.C., Meador, III, J.W., Ferre, R.A., Worland, S.T., 2002. Structure-based design, synthesis, and biological evaluation of irreversible human rhinovirus 3C protease inhibitors. 6. Structure–activity studies of orally bioavailable, 2-pyridone-containing peptidomimetics. J. Med. Chem. 45, 1607–1623.

Dragovich, P.S., Prins, T.J., Zhou, R., Johnson, T.O., Hua, Y., Luu, H.T., Sakata, S.K., Brown, E.L., Maldonado, F.C., Tuntland, T., Lee, C.A., Fuhrman, S.A., Zalman, L.S., Patick, A.K., Matthews, D.A., Wu, E.Y., Guo, M., Borer, B.C., Nayyar, N.K., Moran, T., Chen, L., Rejto, P.A., Rose, P.W., Guzman, M.C., Dovalsantos, E.Z., Lee, S., McGee, K., Mohajeri, M., Liese, A., Tao, J., Kosa, M.B., Liu, B., Batugo, M.R., Gleeson, J.P., Wu, Z.P., Liu, J., Meador, III, J.W., Ferre, R.A., 2003. Structure-based design, synthesis, and biological evaluation of irreversible human rhinovirus 3C protease inhibitors. 8. Pharmacological optimization of orally bioavailable 2-pyridone-containing peptidomimetics. J. Med. Chem. 46, 4572–4585.

Fan, K., Wei, P., Feng, Q., Chen, S., Huang, C., Ma, L., Lai, B., Pei, J., Liu, Y., Chen, J., Lai, L., 2004. Biosynthesis, purification, and substrate specificity of severe acute respiratory syndrome coronavirus 3C-like proteinase. J. Biol. Chem. 279, 1637–1642.

Finlay, B.B., See, R.H., Brunham, R.C., 2004. Rapid response research to emerging infectious diseases: lessons from SARS. Nat. Rev. Microbiol. 2, 602–607.

Galasiti, Kankanamalage, A.C., Kim, Y., Weerawarna, P.M., Uy, R.A., Damalanka, V.C., Mandadapu, S.R., Alliston, K.R., Mehzabeen, N., Battaile, K.P., Lovell, S., Chang, K.O., Groutas, W.C., 2015. Structure-guided design and optimization of dipeptidyl inhibitors of norovirus 3CL protease.Structure–activity relationships and biochemical, X-ray crystallographic, cell-based, and in vivo studies. J. Med. Chem. 58, 3144–3155.

Ghosh, A.K., Xi, K., Grum-Tokars, V., Xu, X., Ratia, K., Fu, W., Houser, K.V., Baker, S.C., Johnson, M.E., Mesecar, A.D., 2007. Structure-based design, synthesis, and biological evaluation of peptidomimetic SARS-CoV 3CLpro inhibitors. Bioorg. Med. Chem. Lett. 17, 5876–5880.

Grum-Tokars, V., Ratia, K., Begaye, A., Baker, S.C., Mesecar, A.D., 2008. Evaluating the 3C-like protease activity of SARS-Coronavirus: recommendations for standardized assays for drug discovery. Virus Res. 133, 63–73.

Gupta, S.P., 2007. Quantitative structure–activity relationship studies on zinc-containing metalloproteinase inhibitors. Chem. Rev. 107, 3042–3087.

Hegyi, A., Friebe, A., Gorbalenya, A.E., Ziebuhr, J., 2002a. Mutational analysis of the active centre of coronavirus 3C-like proteases. J. Gen. Virol. 83, 581–593.

Hegyi, A., Ziebuhr, J., 2002b. Conservation of substrate specificities among coronavirus main proteases. J. Gen. Virol. 83, 595–599.

Herold, J., Raabe, T., Schelle-Prinz, B., Siddell, S.G., 1993. Nucleotide sequence of the human coronavirus 229E RNA polymerase locus. Virol. 195, 680–691.

Holmes, K.V., 2003. SARS coronavirus: a new challenge for prevention and therapy. J. Clin. Invest. 111, 1605–1609.

Hsu, J.T., Kuo, C.J., Hsieh, H.P., Wang, Y.C., Huang, K.K., Lin, C.P., Huang, P.F., Chen, X., Liang, P.H., 2004. Evaluation of metal-conjugated compounds as inhibitors of 3CL protease of SARS-CoV. FEBS. Lett. 574, 116–120.

Jain, R.P., Pettersson, H.I., Zhang, J., Aull, K.D., Fortin, P.D., Huitema, C., Eltis, L.D., Parrish, J.C., James, M.N., Wishart, D.S., Vederas, J.C., 2004. Synthesis and evaluation of keto-glutamine analogues as potent inhibitors of severe acute respiratory syndrome 3CLpro. J. Med. Chem. 47, 6113–6116.

Jiang, S., He, Y., Liu, S., 2005. SARS vaccine development. Emerg. Infect. Dis. 11, 1016–1020.

Johnson, T.O., Hua, Y., Luu, H.T., Brown, E.L., Chan, F., Chu, S.S., Dragovich, P.S., Eastman, B.W., Ferre, R.A., Fuhrman, S.A., Hendrickson, T.F., Maldonado, F.C., Matthews, D.A., Meador, III, J.W., Patick, A.K., Reich, S.H., Skalitzky, D.J., Worland, S.T., Yang, M., Zalman, L.S., 2002. Structure-based design of a parallel synthetic array directed toward the discovery of irreversible inhibitors of human rhinovirus 3C protease. J. Med. Chem. 45, 2016–2023.

Khan, G., 2013. A novel coronavirus capable of lethal human infections: an emerging picture. Virol. J. 10, 66.

Kim, Y., Shivanna, V., Narayanan, S., Prior, A.M., Weerasekara, S., Hua, D.H., Kankanamalage, A.C., Groutas, W.C., Chang, K.O., 2015. Broad-spectrum inhibitors against 3C-like proteases of feline coronaviruses and feline caliciviruses. J. Virol. 89, 4942–4950.

Lee, N., Hui, D., Wu, A., Chan, P., Cameron, P., Joynt, G.M., Ahuja, A., Yung, M.Y., Leung, C.B., To, K.F., Lui, S.F., Szeto, C.C., Chung, S., Sung, J.J., 2003. A major outbreak of severe acute respiratory syndrome in Hong Kong. N. Engl. J. Med. 348, 1986–1994.

Li, B.J., Tang, Q., Cheng, D., Qin, C., Xie, F.Y., Wei, Q., Xu, J., Liu, Y., Zheng, B.J., Woodle, M.C., Zhong, N., Lu, P.Y., 2005. Using siRNA in prophylactic and therapeutic regimens against SARS coronavirus in Rhesus macaque. Nat. Med. 11, 944–951.

Liu, W., Zhu, H.M., Niu, G.J., Shi, E.Z., Chen, J., Sun, B., Chen, W.Q., Zhou, H.G., Yang, C., 2014. Synthesis, modification and docking studies of 5-sulfonyl isatin derivatives as SARS-CoV 3C-like protease inhibitors. Bioorg. Med. Chem. 22, 292–302.

Lu, I.L., Mahindroo, N., Liang, P.H., Peng, Y.H., Kuo, C.J., Tsai, K.C., Hsieh, H.P., Chao, Y.S., Wu, S.Y., 2006. Structure-based drug design and structural biology study of novel nonpeptide inhibitors of severe acute respiratory syndrome coronavirus main protease. J. Med. Chem. 49, 5154–5161.

Mandadapu, S.R., Gunnam, M.R., Tiew, K.C., Uy, R.A., Prior, A.M., Alliston, K.R., Hua, D.H., Kim, Y., Chang, K.O., Groutas, W.C., 2013a. Inhibition of norovirus 3CL protease by bisulfite adducts of transition state inhibitors. Bioorg. Med. Chem. Lett. 23, 62–65.

Mandadapu, S.R., Weerawarna, P.M., Prior, A.M., Uy, R.A., Aravapalli, S., Alliston, K.R., Lushington, G.H., Kim, Y., Hua, D.H., Chang, K.O., Groutas, W.C., 2013b. Macrocyclic inhibitors of 3C and 3C-like proteases of picornavirus, norovirus, and coronavirus. Bioorg. Med. Chem. Lett. 23, 3709–3712.

Marra, M.A., Jones, S.J., Astell, C.R., Holt, R.A., Brooks-Wilson, A., Butterfield, Y.S., Khattra, J., Asano, J.K., Barber, S.A., Chan, S.Y., Cloutier, A., Coughlin, S.M., Freeman, D., Girn, N., Griffith, O.L., Leach, S.R., Mayo, M., McDonald, H., Montgomery, S.B., Pandoh, P.K., Petrescu, A.S., Robertson, A.G., Schein, J.E., Siddiqui, A., Smailus, D.E., Stott, J.M., Yang, G.S., Plummer, F., Andonov, A., Artsob, H., Bastien, N., Bernard, K., Booth, T.F., Bowness, D., Czub, M., Drebot, M., Fernando, L., Flick, R., Garbutt, M., Gray, M., Grolla, A., Jones, S., Feldmann, H., Meyers, A., Kabani, A., Li, Y., Normand, S., Stroher, U., Tipples, G.A., Tyler, S., Vogrig, R., Ward, D., Watson, B., Brunham, R.C., Krajden, M., Petric, M., Skowronski, D.M., Upton, C., Roper, R.L., 2003. The genome sequence of the SARS-associated coronavirus. Science 300, 1399–1404.

McMinn, P.C., 2012. Recent advances in the molecular epidemiology and control of human enterovirus 71 infection. Curr. Opin. Virol. 2, 199–205.

Mohamed, S.F., Ibrahiem, A.A., Amr Abd El-Galil, E., Abdalla, M.M., 2015. SARS-CoV 3C-like protease inhibitors of some newly synthesized substituted pyrazoles and substituted pyrimidines based on 1-(3-aminophenyl)-3-(1H-indol-3-yl)prop-2-en-1-one. Int. J. Pharmacol. 11, 749–756.

Myint, S.H., 1995. Human coronavirus infections. In: Siddell, S.G. (Ed.), The Coronaviridae. Plenum press, New York, pp. 389–401.

Nguyen, T.T., Ryu, H.J., Lee, S.H., Hwang, S., Cha, J., Breton, V., Kim, D., 2012. Flavonoid-mediated inhibition of SARS coronavirus 3C-like protease expressed in Pichia pastoris. Biotechnol. Lett. 34, 831–838.

Nguyen, T.T., Ryu, H.J., Lee, S.H., Hwang, S., Breton, V., Rhee, J.H., Kim, D., 2011. Virtual screening identification of novel severe acute respiratory syndrome 3C-like protease inhibitors and in vitro confirmation. Bioorg. Med. Chem. Lett. 21, 3088–3091.

Niu, C., Yin, J., Zhang, J., Vederas, J.C., James, M.N., 2008. Molecular docking identifies the binding of 3-chloropyridine moieties specifically to the S1 pocket of SARS-CoV $M^{pro}$. Bioorg. Med. Chem. 16, 293–302.

Patel, M.M., Hall, A.J., Vinjé, J., Parashar, U.D., 2009. Noroviruses: a comprehensive review. J. Clin. Virol. 44, 1–8.

Perlman, S., Netland, J., 2009. Coronaviruses post-SARS: update on replication and pathogenesis. Nat. Rev. Microbiol. 7, 439–450.

Pogrebnyak, N., Golovkin, M., Andrianov, V., Spitsin, S., Smirnov, Y., Egolf, R., Koprowski, H., 2005. Severe acute respiratory syndrome (SARS) S protein production in plants: development of recombinant vaccine. Proc. Natl. Acad. Sci. USA 102, 9062–9067.

Prior, A.M., Kim, Y., Weerasekara, S., Moroze, M., Alliston, K.R., Uy, R.A., Groutas, W.C., Chang, K.O., Hua, D.H., 2013. Design, synthesis, and bioevaluation of viral 3C and 3C-like protease inhibitors. Bioorg. Med. Chem. Lett. 23, 6317–6320.

Ramajayam, R., Tan, K.P., Liu, H.G., Liang, P.H., 2010. Synthesis and evaluation of pyrazolone compounds as SARS-coronavirus 3C-like protease inhibitors. Bioorg. Med. Chem. 18, 7849–7854.

Regnier, T., Sarma, D., Hidaka, K., Bacha, U., Freire, E., Hayashi, Y., Kiso, Y., 2009. New developments for the design, synthesis and biological evaluation of potent SARS-CoV 3CL (pro) inhibitors. Bioorg. Med. Chem. Lett. 19, 2722–2727.

Ren, L., Xiang, Z., Guo, L., Wang, J., 2012. Viral infections of the lower respiratory tract. Curr. Infect. Dis. Rep. 14, 284–291.

Rota, P.A., Oberste, M.S., Monroe, S.S., Nix, W.A., Campagnoli, R., Icenogle, J.P., Peñaranda, S., Bankamp, B., Maher, K., Chen, M.H., Tong, S., Tamin, A., Lowe, L., Frace, M., DeRisi, J.L., Chen, Q., Wang, D., Erdman, D.D., Peret, T.C., Burns, C., Ksiazek, T.G., Rollin, P.E., Sanchez, A., Liffick, S., Holloway, B., Limor, J., McCaustland, K., Olsen-Rasmussen, M., Fouchier, R., Günther, S., Osterhaus, A.D., Drosten, C., Pallansch, M.A., Anderson, L.J., Bellini, W.J., 2003. Characterization of a novel coronavirus associated with severe acute respiratory syndrome. Science 300, 1394–1399.

Ryu, Y.B., Jeong, H.J., Kim, J.H., Kim, Y.M., Park, J.Y., Kim, D., Nguyen, T.T., Park, S.J., Chang, J.S., Park, K.H., Rho, M.C., Lee, W.S., 2010. Biflavonoids from Torreya nucifera displaying SARS-CoV 3CL (pro) inhibition. Bioorg. Med. Chem. 18, 7940–7947.

Shie, J.J., Fang, J.M., Kuo, C.J., Kuo, T.H., Liang, P.H., Huang, H.J., Yang, W.B., Lin, C.H., Chen, J.L., Wu, Y.T., Wong, C.H., 2005a. Discovery of potent anilide inhibitors against the severe acute respiratory syndrome 3CL protease. J. Med. Chem. 48, 4469–4473.

Shie, J.J., Fang, J.M., Kuo, T.H., Kuo, C.J., Liang, P.H., Huang, H.J., Wu, Y.T., Jan, J.T., Cheng, Y.S., Wong, C.H., 2005b. Inhibition of the severe acute respiratory syndrome 3CL protease by peptidomimetic alpha, beta-unsaturated esters. Bioorg. Med. Chem. 13, 5240–5252.

Shigeta, S., Yamase, T., 2005. Current status of anti-SARS agents. Antivir. Chem. Chemother. 16, 23–31.

Solomon, T., Lewthwaite, P., Perera, D., Cardosa, M.J., McMinn, P., Ooi, M.H., 2010. Virology, epidemiology, pathogenesis, and control of enterovirus 71. Lancet. Infect. Dis. 10, 778–790.

St. John, S.E., Tomar, S., Stauffer, S.R., Mesecar, A.D., 2015. Targeting zoonotic viruses: structure-based inhibition of the 3C-like protease from bat coronavirus HKU4-The likely reservoir host to the human coronavirus that causes middle east respiratory syndrome (MERS). Bioorg. Med. Chem. 23, 6036–6048.

Tao, P., Zhang, J., Tang, N., Zhang, B.Q., He, T.C., Huang, A.L., 2005. Potent and specific inhibition of SARS-CoV antigen expression by RNA interference. Chin. Med. J. (Engl.) 118, 714–719.

Thanigaimalai, P., Konno, S., Yamamoto, T., Koiwai, Y., Taguchi, A., Takayama, K., Yakushiji, F., Akaji, K., Chen, S.E., Naser-Tavakolian, A., Schön, A., Freire, E., Hayashi, Y., 2013a. Development of potent dipeptide-type SARS-CoV 3CL protease inhibitors with novel P3 scaffolds: design, synthesis, biological evaluation, and docking studies. Eur. J. Med. Chem. 68, 372–384.

Thanigaimalai, P., Konno, S., Yamamoto, T., Koiwai, Y., Taguchi, A., Takayama, K., Yakushiji, F., Akaji, K., Kiso, Y., Kawasaki, Y., Chen, S.E., Naser-Tavakolian, A., Schön, A., Freire, E., Hayashi, Y., 2013b. Design, synthesis, and biological evaluation of novel dipeptide-type SARS-CoV 3CL protease inhibitors: structure–activity relationship study. Eur. J. Med. Chem. 65, 436–447.

Thiel, V., Herold, J., Schelle, B., Siddell, S.G., 2001. Viral replicase gene products suffice for coronavirus discontinuous transcription. J. Virol. 75, 6676–6681.

Tsai, K.C., Chen, S.Y., Liang, P.H., Lu, I.L., Mahindroo, N., Hsieh, H.P., Chao, Y.S., Liu, L., Liu, D., Lien, W., Lin, T.H., Wu, S.Y., 2006. Discovery of a novel family of SARS-CoV protease inhibitors by virtual screening and 3D-QSAR studies. J. Med. Chem. 49, 3485–3495.

Tsukada, H., Blow, D.M., 1985. Structure of alpha-chymotrypsin refined at 1.68 Å resolution. J. Mol. Biol. 184, 703–711.

Tuboly, T., Yu, W., Bailey, A., Degrandis, S., Du, S., Erickson, L., Nagy, E., 2000. Immunogenicity of porcine transmissible gastroenteritis virus spike protein expressed in plants. Vaccine 18, 2023–2028.

Turlington, M., Chun, A., Tomar, S., Eggler, A., Grum-Tokars, V., Jacobs, J., Daniels, J.S., Dawson, E., Saldanha, A., Chase, P., Baez-Santos, Y.M., Lindsley, C.W., Hodder, P., Mesecar, A.D., Stauffer, S.R., 2013. Discovery of N-(benzo [1,2,3]triazol-1-yl)-N-(benzyl)acetamido)phenyl) carboxamides as severe acute respiratory syndrome coronavirus (SARS-CoV) 3CLpro inhibitors: identification of ML300 and noncovalent nanomolar inhibitors with an induced-fit binding. Bioorg. Med. Chem. Lett. 23, 6172–6177.

Turner, R.B., Couch, R.B., 2007. Rhiniviruses. In: Knipe, D.M., Howley, P.M. (Eds.), Fields Virology. Lippincott Williams & Wilkins, Philadelphia, PA, pp. 895–909.

Verma, R.P., Hansch, C., 2009. Camptothecins: a SAR/QSAR study. Chem. Rev. 109, 213–235.

Wang, Z., Ren, L., Zhao, X., Hung, T., Meng, A., Wang, J., Chen, Y.G., 2004. Inhibition of severe acute respiratory syndrome virus replication by small interfering RNAs in mammalian cells. J. Virol. 78, 7523–7527.

Webber, S.E., Marakovits, J.T., Dragovich, P.S., Prins, T.J., Zhou, R., Fuhrman, S.A., Patick, A.K., Matthews, D.A., Lee, C.A., Srinivasan, B., Moran, T., Ford, C.E., Brothers, M.A., Harr, J.E., Meador, III, J.W., Ferre, R.A., Worland, S.T., 2001. Design and synthesis of irreversible depsipeptidyl human rhinovirus 3C protease inhibitors. Bioorg. Med. Chem. Lett. 11, 2683–2686.

Winther, B., 2011. Rhinovirus infections in the upper airway. Proc. Am. Thorac. Soc. 8, 79–89.

Wu, C.Y., Jan, J.T., Ma, S.H., Kuo, C.J., Juan, H.F., Cheng, Y.S., Hsu, H.H., Huang, H.C., Wu, D., Brik, A., Liang, F.S., Liu, R.S., Fang, J.M., Chen, S.T., Liang, P.H., Wong, C.H., 2004. Small molecules targeting severe acute respiratory syndrome human coronavirus. Proc. Natl. Acad. Sci. USA 101, 10012–10017.

Wu, C.Y., King, K.Y., Kuo, C.J., Fang, J.M., Wu, Y.T., Ho, M.Y., Liao, C.L., Shie, J.J., Liang, P.H., Wong, C.H., 2006. Stable benzotriazole esters as mechanism-based inactivators of the severe acute respiratory syndrome 3CL protease. Chem. Biol. 13, 261–268.

Yang, H., Yang, M., Ding, Y., Liu, Y., Lou, Z., Zhou, Z., Sun, L., Mo, L., Ye, S., Pang, H., Gao, G.F., Anand, K., Bartlam, M., Hilgenfeld, R., Rao, Z., 2003. The crystal structures of severe acute respiratory syndrome virus main protease and its complex with an inhibitor. Proc. Natl. Acad. Sci. USA 100, 13190–13195.

Yang, Q., Chen, L., He, X., Gao, Z., Shen, X., Bai, D., 2008. Design and synthesis of cinanserin analogs as severe acute respiratory syndrome coronavirus 3CL protease inhibitors. Chem. Pharm. Bull. 56, 1400–1405.

Yang, Z.Y., Kong, W.P., Huang, Y., Roberts, A., Murphy, B.R., Subbarao, K., Nabel, G.J., 2004. A DNA vaccine induces SARS coronavirus neutralization and protective immunity in mice. Nature 428, 561–564.

Zhai, S., Liu, W., Yan, B., 2007. Recent patents on treatment of severe acute respiratory syndrome (SARS). Recent Pat. Antiinfect. Drug Discov. 2, 1–10.

Zhang, J., Pettersson, H.I., Huitema, C., Niu, C., Yin, J., James, M.N., Eltis, L.D., Vederas, J.C., 2007. Design, synthesis, and evaluation of inhibitors for severe acute respiratory syndrome 3C-like protease based on phthalhydrazide ketones or heteroaromatic esters. J. Med. Chem. 50, 1850–1864.

Zhang, J., Huitema, C., Niu, C., Yin, J., James, M.N., Eltis, L.D., Vederas, J.C., 2008. Aryl methylene ketones and fluorinated methylene ketones as reversible inhibitors for severe acute respiratory syndrome (SARS) 3C-like proteinase. Bioorg. Chem. 36, 229–240.

Zhou, L., Liu, Y., Zhang, W., Wei, P., Huang, C., Pei, J., Yuan, Y., Lai, L., 2006. Isatin compounds as noncovalent SARS coronavirus 3C-like protease inhibitors. J. Med. Chem. 49, 3440–3443.

Ziebuhr, J., Snijder, E.J., Gorbalenya, A.E., 2000. Virus-encoded proteinases and proteolytic processing in the Nidovirales. J. Gen. Virol. 81, 853–879.

## FURTHER READING

Ghosh, A.K., Takayama, J., Aubin, Y., Ratia, K., Chaudhuri, R., Baez, Y., Sleeman, K., Coughlin, M., Nichols, D.B., Mulhearn, D.C., Prabhakar, B.S., Baker, S.C., Johnson, M.E., Mesecar, A.D., 2009. Structure-based design, synthesis, and biological evaluation of a series of novel and reversible inhibitors for the severe acute respiratory syndrome-coronavirus papain-like protease. J. Med. Chem. 52, 5228–5240.

Ghosh, A.K., Takayama, J., Rao, K.V., Ratia, K., Chaudhuri, R., Mulhearn, D.C., Lee, H., Nichols, D.B., Baliji, S., Baker, S.C., Johnson, M.E., Mesecar, A.D., 2010. Severe acute respiratory syndrome coronavirus papain-like novel protease inhibitors: design, synthesis, protein-ligand X-ray structure and biological evaluation. J. Med. Chem. 53, 4968–4979.

Song, Y.H., Kim, D.W., Curtis-Long, M.J., Yuk, H.J., Wang, Y., Zhuang, N., Lee, K.H., Jeon, K.S., Park, K.H., 2014. Papain-like protease (PLpro) inhibitory effects of cinnamic amides from *Tribulus terrestris* fruits. Biol. Pharm. Bull. 37, 1021–1028.

Chapter 12

# Herpesvirus Proteases: Structure, Function, and Inhibition

**Kriti Kashyap, Rita Kakkar**
*University of Delhi, Delhi, India*

## 1 INTRODUCTION

The members of the herpesviridae family are large, double-stranded DNA viruses (Roizman and Pellett, 2001) that afflict most species of the animal kingdom throughout the world. Of these, nine herpesviruses are known to affect humans and are classified into three subfamilies based on differences in cell tropism, immunological response, genome size, and genome content (Khayat et al., 2003). The α-subfamily consists of herpes simplex virus type 1 (HSV-1), HSV-2, and varicella-zoster virus (VZV). The human cytomegalovirus (HCMV), human herpesvirus 6A (HHV6A), HHV6B, and HHV7 comprise the β-subfamily, while the Epstein–Barr virus (EBV) and Kaposi's sarcoma related herpesvirus (KSHV, also known as HHV8) make up the γ-subfamily. Herpesviruses are responsible for a wide variety of maladies, ranging in severity from the harmless occasional cold sore (HSV-1), and childhood chickenpox (VZV) to more problematic illnesses like malignancies caused by EBV (Rickinson and Kieff, 1996) and KSHV (Levy, 1997), as well as infections caused by HCMV that can prove fatal to immunocompromised individuals (Britt and Alford, 1996). A recent case of an immunocompetent woman suffering from a sudden hepatic failure due to HHV-6 infection has also been reported (Charnot-Katsikas et al., 2016). An interesting feature of all herpesviruses is their long-term latency; that is, they tend to stay in the latent form without replicating for long periods of time and get reactivated, even after many years, to start viral replication again. For example, individuals affected by chickenpox may get shingles many years after being infected, with VZV due to activation of the latent virus (Arvin, 1996). This feature has made it difficult for scientists to develop antiviral drugs against this virus family.

Herpesviruses have circular, double-stranded DNA genomes. A number of proteins are encoded by these genomes. However, it has been reported that herpesvirus replication in cell culture does not require all of these enzymes

**Viral Proteases and Their Inhibitors.** http://dx.doi.org/10.1016/B978-0-12-809712-0.00012-5

**411**

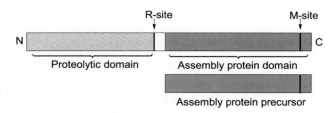

FIGURE 12.1 **Herpesvirus protease precursor, the translation product of the viral gene open reading frame, contains the proteolytic domain and the substrate assembly protein domain.** The C-terminal product of R-site cleavage contains a few amino acid residues in addition to the assembly protein. M-site cleavage of assembly protein precursor gives the assembly protein.

(Roizman and Sears, 1996). Therefore, care must be taken while deciding the target for antiviral drugs against this virus family.

Herpesvirus proteases are serine proteases that are encoded by all herpesvirus genomes and are responsible for the maturational processing of the viral assembly protein. These proteases are synthesized in fusion, with their assembly protein substrate in the form of a protease precursor (Fig. 12.1). This precursor undergoes autocleavage at the maturation site (M-site) and the release site (R-site). From the assembly protein, a scaffold is built which is required in the capsid formation. It is into this capsid that the viral genome is subsequently packed to produce mature infectious virions. Therefore, the protease plays an important role in the life cycle of the virus. Also, many mutagenesis studies on HSV-1 (Gao et al., 1994; Matusick-Kumar et al., 1995a; Preston et al., 1983) showed that, if the protease was somehow inactivated, either by mutating the protease coding region or its cleavage sites, the viral life cycle was hindered and infectious virions were not produced. These studies consolidated the position of the herpesvirus proteases as prime targets for antiviral drugs.

## 2 SYNTHESIS OF HERPESVIRUS PROTEASE AND ITS SUBSTRATE

The viral gene open reading frame (ORF) that encodes the precursor form of the protease also contains nested genes that encodes the assembly protein. The translation product of this gene, therefore, has the catalytic protease domain on one end (N-terminal) and the assembly protein domain on the other (C-terminal) (Fig. 12.1). The shorter nested genes are in-frame, with the large ORF encoding the protease precursor and are also 3′-coterminal with it. These nested genes have their own promoter TATA boxes and are capable of transcription independent of the protease precursor gene (Liu and Roizman, 1991a,b, 1992; Welch et al., 1991a). The HSV-1 precursor protease gene (UL26) has one smaller nested gene (UL26.5) which encodes the substrate assembly protein and is capable of independent transcription to a shorter mRNA and its subsequent translation to protein (Fig. 12.2A) (Liu and Roizman, 1991a,b, 1992). Similarly, CMV protease

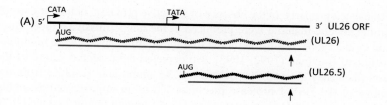

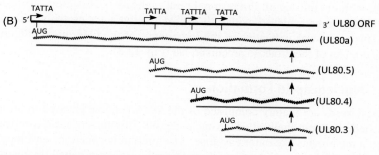

**FIGURE 12.2** Organization of the (A) UL26 ORF of HSV-1 protease and (B) UL80 ORF of HCMV protease is shown along with potential promoter boxes. Both transcriptional and translational expression of the respective nested genes are represented with respective mRNAs (*wavy lines*) and proteins (*red lines*) encoded by them. The nested genes from which the mRNAs are transcribed are indicated in the parentheses next to them.

precursor genes [APNG1 in case of simian cytomegalovirus (SCMV) and UL80 in case of HCMV] encode four nested, in-frame, 3′-coterminal genes, one of which encodes the substrate assembly protein (Fig. 12.2B). All four nested genes have their own promoter TATA elements, translational start codons and all use a common 3′ polyadenylation signal (Welch et al., 1991a). The protease precursor genes of other herpesviruses also have nested, inframe, 3′-coterminal genes, with EBV gene (BVRF2) having two such nested genes (Baer et al., 1984) and VZV gene (VZV33) having up to eight nested genes (Davison and Scott, 1986). Therefore, we can say that the protease precursor contains not only the catalytic protease but also its substrate. Hence, it carries out two types of cleavage: (1) autocleavage, and (2) cleavage of the assembly protein precursor (*trans* cleavage).

# 3   ROLE OF HERPESVIRUS PROTEASES IN VIRAL REPLICATION

## 3.1   Essential Nature of Protease

The essential nature of the protease was first established by Preston et al. (1983). They worked on a temperature sensitive HSV-1 mutant, 17*ts*VP1201, that had a mutated coding region for the protease, which led to the protease not being translated properly. In the absence of an active protease, defective capsids lacking the genetic material were produced at the nonpermissive temperatures (Preston et al., 1983). It must be noted that these viral capsids were noninfectious. Liu and Roizman (1992, 1993) also depicted the essential nature of the protease when they showed that deletion of 32 amino acid residues from the N-terminal of the HSV-1 protease precursor led to an unprocessed assembly protein. Gao et al. (1994) characterized an HSV-1 mutant that lacked the UL26 gene and, as a result, did not express the 635 amino acids protease precursor, Pra (though the assembly protein ICP35 was expressed due to the presence of UL26.5 gene) (Fig. 12.3). This mutant was unable to replicate, but complementing it, with the full-length Pra restored viral replication. However, this restoration could not be achieved, with only the 247 amino acids proteolytic domain, $N_o$. This result was indicative of the importance of the C-terminal 388 amino acids domain of the protease precursor, Na (Fig. 12.3).

## 3.2 Nucleocapsid Formation

Herpesviruses are mainly composed of the DNA genome encased in an inner icosahedral protein shell called the nucleocapsid. This nucleocapsid is surrounded by a membrane-like outer envelope. Also present between the envelope and the nucleocapsid is a thick layer of an amorphous protein called the tegument or matrix. Three types of capsids varying in protein and DNA compositions

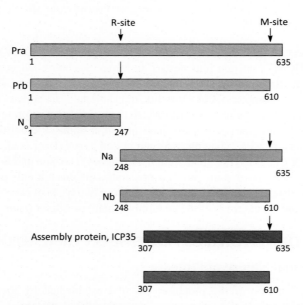

FIGURE 12.3 **HSV-1 full-length protease precursor (Pra) and assembly protein precursor ICP35, along with their cleavage products.**

are known to be isolated from the HSV-1 infected cells—capsids A, B, and C. Capsids A and C have similar protein content, but only C capsids contain the entire viral genome, and are capable of maturing into infectious virions. B capsids contain the assembly protein, which is not found in C capsids (Gibson and Roizman, 1972, 1974). The assembly protein (called infected cell polypeptide 35, ICP35 in case of HSV) found in B capsids is the most abundant substrate of the herpesevirus protease. This assembly protein assists in the formation of an inner scaffold that is subsequently required for proper nucleocapsid assembly. This scaffold results from the interactions of major capsid protein VP5 and other viral proteins—VP19C, VP23, and VP26, arranged in an icosahedral lattice (Desai et al., 1994). Another important interaction is the interaction of the assembly protein precursor, with the major capsid protein, VP5, through its last 25 amino acids, that is, the residues distal to the M-site (Desai and Person, 1996; Kennard et al., 1995; Thomsen et al., 1995; Wood et al., 1997). A minimal 12-amino acids hydrophobic domain in the C-terminus of HSV-1 assembly protein has been shown to interact directly with VP5 (Hong et al., 1996). Similarly, it was shown that a 16-amino acids motif near the C-terminus of the CMV assembly protein precursor is sufficient to interact with VP5 (Beaudet-Miller et al., 1996). Therefore, after the formation of B capsids, the viral protease cleaves the assembly protein at the M-site and releases this scaffold, which is followed by DNA packaging. It is interesting to note that these interaction motifs are conserved within the herpesvirus subfamilies (Loutsch et al., 1994). This conservation allows the assembly protein of one herpesvirus to support the capsid assembly of another herpesvirus of the same subfamily (Haanes et al., 1995). The significance of these distal M-site residues was also shown by Matusick-Kumar et al. (1994), who observed that replication of the HSV-1 mutant, ΔICP35, which was deficient of assembly protein, was restricted. Although the DNA replication of the mutant occurred normally, it was not encapsidated. Studies by the same group on another HSV-1 mutant which expressed only the protease precursor truncated at the M-site (Prb) showed that this mutant could not replicate on its own, even though it had a functional proteolytic domain. However, Prb could replicate when complemented, with either uncleaved ICP35 or with Na, the C-terminal product of R-site cleavage of the full-length protease precursor (Fig. 12.3) (Matusick-Kumar et al., 1995b; Robertson et al., 1997).

Matusick-Kumar et al. (1995a) claimed that R-site cleavage is necessary for viral growth, when they showed that an HSV-1 R-site modified mutant (A247S mutation) was incapable of replicating, even when complemented with unprocessed ICP35. This mutant, though capable of cleaving its M-site and that of ICP35, was incapable of releasing the catalytic domain $N_o$. However, when this mutant (A247S) was complemented, with an inactive protease precursor, S129C, with a normal R-site, viral replication was restored (Robertson et al., 1996). The S129C mutant was enzymatically inactive but acted as a substrate for the proteolytically active A247S mutant protease. Together they were shown to restore viral replication, which suggested that the S129C protease had some essential function. The $N_o$ domain was inactive in the S129C mutant; therefore, this led to the conclusion that this essential function belonged to the Nb domain. However, this theory was ruled out (Robertson et al., 1996), and it was suggested that $N_o$ has other functions besides the enzymatic function. This study was also able to establish that the proteolytic processing occurs inside the capsid because, when infected cell-lysates were sedimented, Prb (precursor of $N_o$ and Nb) was isolated from the fraction in which B capsids sediment.

## 3.3 Protease Forms and Their Activity

The amino acid residues extending from the N-terminal of the full-length protease (Pra) to the R-site define the catalytic domain of the protease, and this domain is termed as $N_o$ (Fig. 12.3). In HSV-1, the first 247 residues define this domain, and it has been shown that it is catalytically active not only in mammalian cells (Liu and Roizman, 1992) but also in *Escherichia coli* (Weinheimer et al., 1993). In case of HCMV protease, this domain is limited to the first 256 residues translated from the UL80 ORF and is also catalytically active when expressed in *E. coli* (Baum et al., 1993).

However, $N_o$ is not the only active form of the protease, as the full-length protease, Pra, has also been shown to cleave at the M-site of the assembly protein precursors (Liu and Roizman, 1991a). Restricting cleavage at the R-site or M-site of HSV-1 (Godefroy and Guenet, 1995; Liu and Roizman, 1993; Weinheimer et al., 1993), SCMV, and HCMV (Jones et al., 1994; Welch et al., 1993, 1995) proteases did not inhibit the catalytic activity of Pra to cleave the M-site of the assembly protein precursor. Jones et al. (1994) showed that both the full-length protease (Pra) and the minimal protease ($N_o$) of HCMV cleave the assembly protein precursor to the same extent.

Three additional cleavage sites are present in case of HCMV protease: (1) the internal site (I-site) present between Ala-143 and Ala-144, (2) the cryptic site (C-site) between Ala-209 and Ser-210 (Fig. 12.4) (Baum et al., 1993; Burck et al., 1994; Holwerda et al., 1994; Jones et al., 1994; O'Boyle et al., 1995), and (3) the recently proposed "tail" site (T-site) between Ala675 and Ala676 (Brignole and Gibson, 2007). These additional sites are not observed in other herpesvirus proteases. Earlier, it was proposed that cleavage at the I-site was responsible for inactivating the protease and, therefore, it was even called as the "inactivation site" by many (Baum et al., 1993; Jones et al., 1994; Welch et al., 1993), as both the N-terminal and the C-terminal products of this cleavage are proteolytically inactive alone (Baum et al., 1993; Burck et al., 1994; Jones et al., 1994; O'Boyle et al., 1995). However, it was shown that this cleavage results into a two-chain enzyme (Fig. 12.4) that is active (Holwerda et al., 1994; O'Boyle et al., 1995). Holwerda et al. (1994) showed that cleavage of the 30 kDa protease at the I-site results in a 16 kDa amino-terminal chain ($A_n$) and a 14 kDa carboxyl-terminal chain ($A_c$), and that these two chains remain associated in a catalytically active enzyme form. Mutation of the I-site resulted in isolation

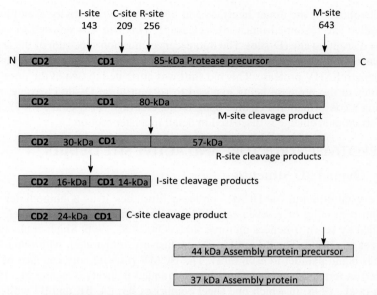

FIGURE 12.4   **HCMV full-length protease precursor and assembly protein precursor along with their cleavage products.** The herpesvirus protease conserved domains reported by Welch et al. (1991b) are also depicted.

of the one-chain protease, and it was revealed that both forms of protease cleave the assembly protein precursor. Also, the active site serine residue was identified on the 16 kDa $A_n$ chain by diisopropyl fluorophosphate labeling. O'Boyle et al. (1995) showed that, when the N-terminal chain and the C-terminal chain were translated separately from gene constructs of the two chains, and were mixed, an active enzyme was obtained, with an activity similar to the wild type protease. Characterization of a mutant HCMV protease, with a defective I-site revealed that I-site cleavage, by itself, is not essential for infectious HCMV virions (Chan et al., 2002). This mutation had little effect on virus infectivity, but an increase in the abundance of the amino-terminal C-site cleavage product, $C_n$, was observed in the I-site mutant. Therefore, C-site cleavage was suggested as an alternative processing pathway for HCMV assembling (Chan et al., 2002). When a mutant with a blocked C-site was analyzed, ~40% decrease in infectivity was observed, and a mutant, with both the I-site and C-site blocked showed a 90% decrease in infectivity. As a result, it was shown that, though cleavage at these sites is not essential, it enhances infectivity of HCMV (Loveland et al., 2005). T-site cleavage was observed when a urea-denatured HCMV protease precursor was renatured (Brignole and Gibson, 2007). More reports on this cleavage site are, however, not available.

   An additional cleavage site was also identified in KSHV protease at the Ala203-Ser204 position (Pray et al., 1999). Autoproteolysis at this site led to the disruption of the dimer interface and loss of catalytic activity even though the active site remained intact after cleavage. This site was referred to as the dimer disruption site (D-site). The fluorescence emission spectrum of the D-site cleavage product was 20% weaker, and had a 10 nm red shift compared to the full-length KSHV protease. This red shift was attributed to two of the three Trp residues in the protease being exposed to solvent upon D-site cleavage, suggesting a conformational change upon D-site cleavage (Pray et al., 1999). This conformational change is discussed in detail in a later section.

# 4  PROTEASE STRUCTURE AND ACTIVE SITE DETAILS

## 4.1  Overall 3D Structure

The crystal structure of HCMV protease illustrates it as a homodimer $\alpha/\beta$ protein consisting of a seven-stranded $\beta$-barrel core, mostly antiparallel, surrounded by seven $\alpha$-helices on three sides (Qiu et al., 1996; Shieh et al., 1996; Tong et al., 1996). The $\beta$-barrel is not exclusively antiparallel, with two of the strands being parallel to each other. The HCMV $\beta$-barrel core can also be described in terms of two orthogonal four-stranded $\beta$-sheets (Chen et al., 1996; Tong et al., 1996) in which one sheet comprises $\beta3$, $\beta4$, $\beta1$, and $\beta7$ while the other consists of $\beta3$, $\beta2$, $\beta6$, and $\beta5$ (Tong et al., 1996). Strand $\beta3$ runs on one side of the barrel and is common between the two sheets. The short three-residue $\beta8$ strand is not a part of the $\beta$-barrel, and is hydrogen bonded to $\beta3$ (Tong et al., 1996). The $\beta$-barrel core is a highly hydrophobic region, as the side chains of all amino acids pointing toward inside of the barrel are hydrophobic. One opening of this $\beta$-barrel is narrow, and two helices are located near it, while the other opening is wider. The middle of the barrel is surrounded by four helices, making this a hydrophobic region.

The overall crystal structures of VZV (Qiu et al., 1997), HSV-1 and HSV-2 (Hoog et al., 1997), KSHV (Reiling et al., 2000), and EBV (Buisson et al., 2002) proteases were reported to be similar to that of the HCMV protease, with the "$\beta$-barrel core surrounded by $\alpha$-helices" structure being conserved among them all. However, differences were reported in the dimer interfaces and loop regions of the proteases. The VZV protease has an additional loop containing a small $\alpha$-helix, called the AA loop (Qiu et al., 1997). The corresponding region in HCMV protease is partially disordered, and is nine residues longer (Qiu et al., 1996; Shieh et al., 1996; Tong et al., 1996). The dimer interface of HCMV protease has the A6 helices of the two monomers almost parallel to each other. However, the corresponding helices in HSV-2 (Hoog et al., 1997) and VZV (Qiu et al., 1997) have a different orientation, and shows a 30 degrees twist. The dimer interface of the KSHV protease resembles that of the HCMV protease, but the orientation of its $\beta$-barrel core resembles that of the $\alpha$-herpesvirus proteases (HSV-2 and VZV). This leads to the dimer interface $\alpha5$ helices rotating 30 degrees away from each other (Reiling et al., 2000).

## 4.2   Active Site Details

Herpesvirus proteases have conserved serine residues, as well as conserved cysteine residues, and also a unique amino acid sequence. Therefore, prior to biochemical studies, it could not be established by homology studies whether these proteases were serine proteases or cysteine proteases. The first evidence for these enzymes being serine proteases was demonstrated when Liu and Roizman (1992) showed that the HSV-1 protease was inhibited by well known serine protease inhibitors like diisopropyl fluorophosphates, N-tosyl-L-phenylalanine chloromethyl ketone, and phenylmethylsulfonyl fluoride. However, they were not inhibited by cysteine protease inhibitors, such as iodoacetamide and cystatin. Experiments that labeled Ser129 in case of the HSV-1 protease (DiIanni et al., 1994) and Ser132 in case of the HCMV protease (Holwerda et al., 1994; Stevens et al., 1994), with serine protease inhibitor [$^3$H]diisopropyl fluorophosphates showed irreversible inactivation of the proteases. Although the conserved Cys161 residue of HCMV is in the proximity of the active site Ser132, catalytic activity was not affected by its replacement, ruling out its involvement in catalysis (Baum et al., 1996a). The same conclusions could be drawn from similar substitution experiments carried out on HSV-1 Cys152 (Liu and Roizman, 1993). Mutagenesis studies carried out on SCMV (Welch et al., 1993) suggested His47 as a potential active site residue. When His61 and His148 of the HSV-1 protease (Deckman et al., 1992; DiIanni et al., 1993a; Liu and Roizman, 1992), His47 in the SCMV protease (Welch et al., 1993), and His63 in the HCMV protease (Cox et al., 1995) were replaced, catalytic activities of the respective enzymes were lost.

Crystallographic studies showed that, unlike classical serine proteases, no acidic residues are present in the vicinity of His63 and a catalytic triad typical of classical serine proteases (Ser-His-Asp/Glu) is not possible (Chen et al., 1996). However, a conserved His157 was observed just ideally placed to form a hydrogen bond, with the imidazole of His63 of the HCMV protease, suggesting a unique variation of the classical serine protease triad in HCMV: Ser132-His63-His157 (Chen et al., 1996). The SCMV protease counterpart of HCMV His157 is His142 (Welch et al., 1993). However, replacement of the second histidine residue in SCMV protease (Welch et al., 1993) and HCMV protease (Cox et al., 1995) did not abolish activity, though it was severely reduced. This suggests that, though the second His residue plays an important role, its presence is not essential for catalytic activity.

## 4.3   Binding Pockets

A significant change in surface topology was observed in the crystal structure of inhibitor bound-HCMV protease (Tong et al., 1998). This change resulted in the formation of S1 and S3 binding pockets, which form a large depression in the protease surface, and were not observed in the crystal structure of the free enzyme (Chen et al., 1996; Shieh et al., 1996; Tong et al., 1996). The S1 pocket is composed of Leu32 (loop L2), Ser 132, Leu133, Arg165, and Arg166 residues. The small size of this pocket governs the preference of a small residue like Ala in the P1 position. The salt bridges between Glu31 (loop L2), Arg137 (loop L9), Arg 165, Ser135, and Arg166 residues make up the S3 pocket. These residues render a hydrophilic nature to this pocket, with the side chain of Arg166 being present at the base of this pocket. This Arg residue is inaccessible to the solvent and is also highly conserved among herpesvirus proteases (Tong et al., 1998). Both these pockets are mostly connected to each other, and this explains why the P3 and P1 side chains of substrates are less solvent exposed, while P2 and P4 side chains are more solvent exposed.

## 4.4 Catalytic Mechanism

Being serine proteases, the catalytic mechanism of substrate hydrolysis by herpesvirus proteases involves initial attack of the active-site serine nucleophile on the carbonyl carbon of the scissile bond of the substrate. A covalent tetrahedral intermediate is formed in the process, which is followed by loss of water, resulting in an enzyme-serine acyl intermediate for further hydrolysis. All serine proteases are known to stabilize the distortion of their carbonyl carbon in the tetrahedral intermediate by means of an "oxyanion hole." In this, the backbone amides or other hydrogen bond donors form a pocket into which the negatively charged oxygen atom fits perfectly, thereby stabilizing the negative charge. The crystal structure of HCMV protease complexed, with peptidomimetic inhibitors revealed the presence of two water molecules that are part of the oxyanion hole. The structure showed that this oxyanion is stabilized by two hydrogen bonds: one with the main chain amide of Arg165, the other being an indirect hydrogen bond to the guanidinium side chain of Arg166 via these two water molecules (Khayat et al., 2003). It is interesting to note that, upon mutagenesis of Arg166, the catalytic activity was lost, while mutagenesis of Arg165 had little effect, which is suggestive of a lesser role of Arg165 (Liang et al., 1998). The Ser-His-His catalytic triad and the oxyanion loop (composed of Arg residues and water molecules) are conserved among all herpesvirus proteases, as can be observed in the crystal structures of HSV-1, HSV-2 (Hoog et al., 1997), VZV (Qiu et al., 1997), KSHV (Reiling et al., 2000), and EBV (Buisson et al., 2002) proteases.

## 4.5 Importance of Dimeric Form of Protease

Initial reports of substrate catalysis by HCMV and HSV proteases showed very low turnover numbers, along with the catalytic activities being orders of magnitude lesser than that of classical serine proteases. Hall and Darke (1995) noted that kosmotropes, or water structure-forming cosolvents like glycerol and citrate, activate the HSV-1 protease (Table 12.1). Also, high concentrations of certain salts like sodium sulfate and potassium phosphate were shown to confer conformational

**TABLE 12.1 Effect of Kosmotropes on $K_d$ of Dimerization**

| Protease (References) | Concentration of kosmotrope | Temperature, pH | $K_d$ (μM) |
|---|---|---|---|
| HCMV (Darke et al., 1996) | 10% glycerol<br>20% glycerol | 30°C, 7.5 | 6.6<br>0.55 |
| HCMV (Cole, 1996) | 0% glycerol<br>10% glycerol<br>20% glycerol | 20°C, 7.5 | 59<br>20<br>5.7 |
| HSV-1 (Schmidt and Darke, 1997) | No salt (20% glycerol)<br>0.2 M phosphate (20% glycerol)<br>0.4 M phosphate (20% glycerol)<br>0.2 M citrate (20% glycerol)<br>0.4 M citrate (20% glycerol) | 15°C, 7.5 | >1.6<br>1.04<br>0.704<br>0.964<br>0.24 |

HCMV, Human cytomegalovirus; HSV-1, herpes simplex virus type 1.

changes on both the HSV-1 protease and its substrate, as well as stimulated protease activity by a 100-fold (Yamanaka et al., 1995). Similar results were obtained when glycerol was used to activate HCMV protease (Burck et al., 1994). Later, Darke et al. (1996) recognized that HCMV protease exists in a monomer–dimer equilibrium, and that the dimer is the active species. All the aforementioned conditions were conducive to shift this equilibrium toward the active dimeric form and, therefore, were responsible for increased activity (Cole, 1996; Darke et al., 1996; Margosiak et al., 1996; Schmidt and Darke, 1997).

Dimeric protease structures were observed in X-ray crystallographic studies of HSV-1 and HSV-2 proteases (Hoog et al., 1997), HCMV protease (Chen et al., 1996; Qiu et al., 1996; Shieh et al., 1996; Tong et al., 1996), VZV protease (Qiu et al., 1997), and EBV protease (Buisson et al., 2002). Substantial experimental evidence exists that is consistent, with this active dimer hypothesis. Experiments at high protease concentrations have reported increased specific activity, with lower activities being observed upon dilution (Darke et al., 1996; Margosiak et al., 1996). This observation would not have been made if the monomer were the active species since, in that case, a constant specific activity, irrespective of protease concentration, would have been observed. It was also observed that, greater the enzyme volumes loaded onto gel filtration columns, greater was the amount of enzyme recovered as dimer (Darke et al., 1996). An inverse relation was noted between the protease activity and temperature, indicative of the dimer being destabilized at higher temperatures (Darke et al., 1996; Margosiak et al., 1996).

Various studies to measure the $K_d$ of dimerization (Cole, 1996; Darke et al., 1996; Margosiak et al., 1996) have established that glycerol increases the affinity between monomeric subunits, thereby facilitating dimerization (Table 12.1). Analysis of $K_m$ and $k_{cat}$ values (Hall and Darke, 1995; Yamanaka et al., 1995) has shown that the kinetic parameter most affected by the presence of kosmotropes is $K_m$, which is indicative of conformational changes at play. Yamanaka et al. (1995) made use of the chromophore reagent 1-anilino-8-naphthalene sulfonate (ANS) to determine whether protease simulation was a result of conformational change or not. The fluorescent properties of ANS are known to change on binding to hydrophobic regions of proteins. A novel peak was observed when it was made to bind to HSV-1 protease in the presence of sodium sulfate. However, this peak was lost upon substrate binding, suggesting that a conformational change had occurred that involved loss of the hydrophobic region in the previous case (Yamanaka et al., 1995). Tong et al. (1996) also observed a change in organization of the dimer when HCMV protease was exposed to 0.15 M sodium sulfate. Therefore, these conformational changes may be due to changes in the dimeric state. We discuss these conformational changes in detail in the subsequent section.

## 5 CONFORMATIONAL CHANGES AND ENZYMATIC REGULATION

The first experimental evidence of a conformational change in herpesvirus proteases upon substrate/inhibitor binding was given by Bonneau et al. (1997). In a previous study, they had reported a series of peptidyl-activated carbonyl inhibitors for HCMV protease (Ogilvie et al., 1997), which they further used to study the effect of inhibitor binding on the conformation of the protease. The fluorescence spectra of the enzyme, in presence of the previously mentioned inhibitors, showed a blue shift which was attributed to a change in environment of a Trp residue in proximity of the active site. It was suggested that this residue, which was identified as Trp42, finds itself in a more hydrophobic environment upon inhibitor binding. The X-ray structure of HCMV protease (Chen et al., 1996) clearly shows that the aromatic side chain of Trp42 is not in proximity of the region that these inhibitors occupy, thereby ruling out the possibility that the blue shift was due to the direct interaction of the inhibitors and the Trp residue (Bonneau et al., 1997). Therefore, a conformational change was said to have taken place upon inhibitor binding. The near-UV CD spectra results were consistent with a conformational change upon inhibitor binding and also the structural transition occurring around Trp42. The CD spectra of inhibitor-bound protease revealed that, after the conformational change, the indole ring of the Trp42 residue is buried in a hydrophobic area of the enzyme and becomes immobilized, though it was conformationally mobile in the free enzyme. From crystal structure analysis it was known that Trp42 is situated at the end of helix $\alpha$A, next to a missing surface loop (45–53). It also lies near loop L9 which is present between this residue and the region occupied by the active site machinery. Based on these observations, it was speculated that loop L9 plays an important role in substrate/inhibitor binding and contributes to the conformational change observed (Bonneau et al., 1997). It is worth noting that covalent adduct formation between the inhibitor and the protease is not a prerequisite for this conformational change to occur. A peptidyl-methyl ketone inhibitor, which does not form a covalent adduct with the protease, was observed to induce the same conformational change in the protease upon binding (Bonneau et al., 1997). Therefore, it was suspected that contacts between the peptide chain of inhibitor and the protease binding sites were responsible for this transition. The interactions of this methyl ketone inhibitor and its analogues, with the protease were further analyzed by a combination of NMR techniques to elucidate the induced-fit mechanism of the herpesviral proteases (LaPlante et al., 1999). Peptidyl-methyl and activated ketone inhibitors of HCMV protease were shown to exist in an extended, rigid structure in solution that was reportedly maintained by the bulky P3 *tert*-butyl side chain. This conformation of the peptide inihibitors was similar to the conformation in the enzyme-inhibitor complex. In other words, the peptide inhibitors exist in their bioactive conformation in solution and, upon binding to the protease, very little change in peptide conformation is observed. This signifies that very little adjustment occurs in the bound ligand during the course of conformational changes occurring in the protease. The following conformational changes were said to occur in the protease upon ligand binding: (1) movement of strand β5, containing the active site Ser132, which makes it more ordered; (2) loop L9 becomes more ordered, and these changes are responsible for the burying and immobilization of Trp-42; (3) a large change in the position of loop L2 results in the formation of a cleft that is occupied by the side chains of P1 and P3; and (4) repositioning of Arg165 and Arg166 alters the structure of the oxyanion loop (LaPlante et al., 1999). Therefore, a substantial conformational change was observed in HCMV protease upon ligand binding, solidifying its position as an induced-fit protein. These experimental observations of an induced-fit mechanism were corroborated by molecular dynamics simulations (de Oliveira et al., 2003) that were able to duplicate almost all experimental observations of this mechanism (Bonneau et al., 1997; LaPlante et al., 1999; Tong et al., 1998). These MD simulations involved four models: inhibitor-bound dimer, inhibitor-bound monomer, free solution dimer, and free solution monomer. Analysis of the inhibitor-bound monomer revealed a conformational change in αF which causes a change in the closely linked helix αG, which further induces repositioning of loop L10, the oxyanion loop containing Arg165 and Arg166. These changes were also observed upon mutagenesis of the dimer interface residue Ser225 to Tyr, which was accompanied by a drastic reduction in catalytic activity (Batra et al., 2001). Therefore, it was suggested that conformational changes that occur on perturbing the dimer interface disrupt the oxyanion stabilizing machinery.

A link between the active site of the protease and its dimer interface was established by Marnett et al. (2004). An optimized hexapeptide organophosphonate inhibitor, biotinyl-Pro-Val-Tyr-Tbg-Gln-Ala$^P$-(OPh)$_2$ (**1**) (Fig. 12.5), against KSHV protease, which phosphonylated the active site serine and locked it in a conformation most relevant to catalysis, was prepared,. It was observed

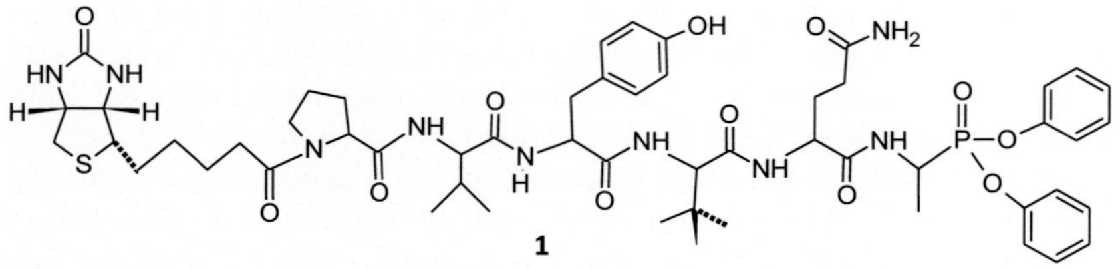

**FIGURE 12.5 Organophosphonate inhibitor of KSHV protease: biotin-Pro-Val-Tyr-Tbg-Gln-Ala^P-(OPh)_2.**

that, on inhibition, the dimer–monomer equilibrium drastically shifted toward the dimer form. This was of great interest because the dimer interface is spatially distal to the active site, but still a change in the dimer–monomer equilibrium occurred when the active site was modified upon inhibition. In this way a structural link was established between the active site and the dimer interface. Two helices, α5 and α6, were shown to play an important role in this structural pathway. The suggested mechanism involved phosphonylation of the active site serine by the inhibitor, followed by stabilization of the negative charge in the oxyanion hole by formation of hydrogen bonds to Arg143 and Arg142 located on the oxyanion loop. This loop had previously been shown to form hydrogen bonds and electrostatic interactions with residues in helix 6 (Reiling et al., 2000). Lastly, helix 6 is directly connected to helix 5, which is the main dimer interface contact. In this way, the active site of KSHV protease is structurally connected to the distant dimer interface. This vast network of hydrogen bonds and electrostatic interactions was suggested to confer the extra stability to the dimer upon inhibitor binding (Marnett et al., 2004).

This link between the active site and dimerization allowed Nomura et al. (2005) to study the mechanism of enzyme catalysis by monitoring the structural changes that accompany dimerization. They suggested that regulation of enzymatic activity was dependent on a local "helical switch", instead of a global unfolding of the protease. A carefully placed disulfide bond in KSHV protease, by mutations G145C-V219C, stabilized the interaction of helix 6 with the oxyanion hole of the active site. Under conditions conducive to disulfide bond formation (oxidizing conditions), both dimeric quaternary structure and increased enzymatic activity were attained, which highlighted that the activity was dependent on disulfide formation and, in turn, on the interaction of helix 6 with the active site. The importance of this interaction is evident from the startling observation that upon introducing such an intramolecular disulfide bond in an inactive monomer protease, the first active monomer was obtained (Nomura et al., 2005). NMR studies by the same group reported that dimerization is responsible for folding of helices 5 and 6 which, in turn, positions the oxyanion loop containing Arg142 and Arg143, thereby stabilizing the transition state. Therefore, dimerization was reported as the regulatory mechanism of these enzymes. Dimer formation is a concentration-dependent step, which makes enzyme activation also a function of concentration. The enzyme concentration in cytosol ($\sim 5$ nM) is four orders of magnitude lower than in immature capsids (100 µM). Therefore, the biological significance of this is that it prevents premature activation of the protease in the cytosol, which would result only into immature capsids and no mature capsids would be formed (Nomura et al., 2005).

The unique autolysis site of KSHV protease (D-site) removes 27 amino acids from the C-terminal of the protease, which includes the last turn of helix 5 and complete helix 6. This leads to irreversible disruption of the dimer interface and inactivation of the protease. Dissociation of the dimeric protease was accompanied by a 31% loss of α-helicity, which was attributed to the disordering of helices 5 and 6 upon dimer disruption (Nomura et al., 2006). Only these helices are affected by the dimerization, while the β-barrel and other areas distant from the dimer interface are not affected by dissociation. Therefore, it was shown that, though the catalytic machinery is perturbed by dissociation and activity is lost, the substrate binding, which is mediated through the β-barrel, and loops that are distant from the dimer interface, remains undisturbed. As a result, the truncated protease can bind substrates but cannot catalyze proteolysis (Nomura et al., 2006). These observations showed that the mechanism behind the irreversible inactivation of KSHV protease through D-site cleavage and the reversible regulation of herpesvirus proteases by dimerization is the same. Both structural changes involve the stabilization or destabilization of the oxyanion loop-helix 6 interaction.

## 6  INHIBITION OF HERPESVIRUS PROTEASES

When classic serine protease inhibitors were tested against herpesvirus proteases, only weak inhibition was observed (Burck et al., 1994; DiIanni et al., 1993b, 1994; Holwerda et al., 1994; Stevens et al., 1994). The reason for this poor inhibition was attributed to the presence of a second His residue in the catalytic triad instead of the negatively charged Asp residue, typical of catalytic triads of classic serine proteases. This, in turn, made the first His residue of the catalytic triad less reactive toward these inhibitors (Holwerda, 1997). Proteinaceous inhibitors against classic serine proteases also failed to inhibit herpesvirus proteases, probably because of the unique substrate sequence specificity of herpesvirus proteases (Waxman and Darke, 2000). Many inhibitors with modest potencies have been reported for herpesvirus proteases. These include: (1) peptidomimetic or substrate based competitive inhibitors that mimic the natural substrate sequence (Holskin et al., 1995; LaFemina et al., 1996; LaPlante et al., 1998); (2) oxazolone-based active-site inhibitors (Pinto et al., 1996); (3) peptidic inhibitors with C-terminal activated carbonyl, which showed submicromolar potencies but had high $EC_{50}$ values (Ogilvie et al., 1997); (4) monocyclic β-lactam based inhibitors (Borthwick et al., 1998; Déziel and Malenfant, 1998; Yoakim et al., 1998a,b); (5) oxazinone based active-site inhibitors (Jarvest et al., 1996, 1997, 1999; Pinto et al., 1999); and (6) cysteine-specific inhibitors that interacted not only with the active-site serine but also Cys161 of HCMV protease (Baum et al., 1996a,b; Flynn et al., 1997). These inhibitors have been reviewed in detail by Waxman and Darke (2000). In this section, we discuss new classes of inhibitors that have been reported in recent years, along with the advancements that have occurred in some of the previously reported inhibitor classes.

## 6.1    Inhibitors Based on β-Lactam and *trans*-Lactam

Based on monocyclic β-lactam (also called azetidinone) inhibitors, Gerona-Navarro et al. (2004, 2005) attempted to develop the first noncovalent inhibitors against HCMV protease. First, 1-acyl-β-lactam inhibitors (like **2**) (Fig. 12.6) were reported in this study that had an additional carboxyl group at C-4 position of β-lactam in order to have an extra interaction, with the guanidine group of Arg165 or Arg166 of the viral protease. These inhibitors showed activity against HCMV protease in the range 10–50 μM but had high toxicity. Then, in order to remove the covalent interaction of the inhibitors, with the protease, the carbonyl group of the β-lactam ring was removed to generate a series of 1-benzyloxycarbonylazetidine inhibitors (like **3**) (Fig. 12.6), which showed better activity but were still toxic. Recently, azetidine-containing dipeptide inhibitors of HCMV have been reported (Pérez-Faginas et al., 2011). A prototype azetidine (**4**) containing dipeptide was selected and three points of modification (indicated in Fig. 12.6) were chosen to enhance the activity of the inhibitors and reduce toxicity. It was reported that these inhibitors absolutely required an aliphatic or no substituent at the C-terminal carboxamide group (A), and an aliphatic side chain at the C-terminal (B). A benzyloxycarbonyl group was also absolutely necessary at the N-terminus (C).

Following β-lactams, the utility of pyrrolidine-5,5-*trans*-lactams as HCMV protease inhibitors was also investigated (Borthwick et al., 2000, 2002a,b). These compounds had the potential to access the specificity sites, S1′, S1, and S3, of

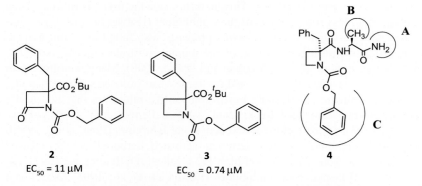

**2**
$EC_{50} = 11\,\mu M$

**3**
$EC_{50} = 0.74\,\mu M$

FIGURE 12.6   **β-Lactam and *trans*-lactam based inhibitors of herpesvirus proteases.**

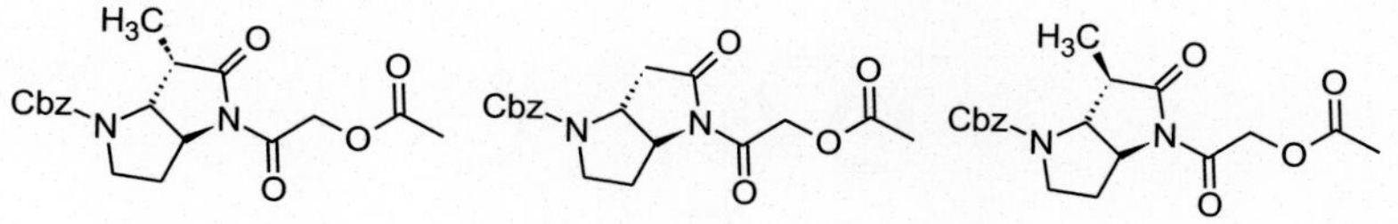

**5**
α-Methyl
$IC_{50}$ = 13 μM

**6**
Desmethyl
$IC_{50}$ = 102 μM

**7**
β-Methyl
$IC_{50}$ = 805 μM

FIGURE 12.7 **Inhibitors based on pyrrolidine-5,5-*trans*-lactam.**

HCMV protease. β-Methyl-, desmethyl-, and α-methyl-templates of this ring system were developed and their potency was tested against the protease. The activity against HCMV protease decreased in the order α-methyl > desmethyl > β-methyl series (Fig. 12.7). The acetoxyacetyl derivative **5** of the α-methyl series was the best candidate in the series. In terms of the substituent on the lactam N, the activity against HCMV protease decreased in the order acyl > ester > sulfonyl > carbamate (Borthwick et al., 2000).

However, due to their reactive functionalities, these compounds were metabolically vulnerable. Hence, efforts were made to develop an inhibitor that was stable enough to survive the harsh conditions of metabolism in the plasma, yet reach the target site, with appropriate potency to bind the active site of the protease (Borthwick et al., 2002b). As a result, a series of inhibitors based on the dansylproline α-methyl pyrrolidine-5,5-*trans*-lactam nucleus (Fig. 12.8) was developed. The mechanism of action of three of the most potent compounds in this series (**8**, **9**, and **10**) (Table 12.2) was studied, and thiazole **10** was reported as the most potent and stable acylating inhibitor of the HCMV protease (Borthwick et al., 2002b).

## 6.2 Benzoxazinone-Based Inhibitors

Recently, computational studies investigating the structure-activity relationship of benzoxazinone-based inhibitors reported for HSV-1 protease by Jarvest et al. (1996, 1997, 1999) were carried out by Mello et al. (2015). The most active inhibitor (**11**) (Fig. 12.9) reported by them was observed to form four hydrogen

**TABLE 12.2 Stability and Inhibitory Parameters of Dansylproline *trans*-Lactam Inhibitors of HCMV (8–10)**

| Compound | R | Human plasma stability, $t_{1/2}$ (h) | $IC_{50}$ (μM) | $K_i$ (μM) |
|---|---|---|---|---|
| **8** | CO-(Z)dimethylcyclopropyl | 20 | 5.1 | 1.1 |
| **9** | $CONHCH_2Ph$ | >33 | 6.6 | 2.6 |
| **10** | 2-Thiazole | >50 | 2.1 | 0.4 |

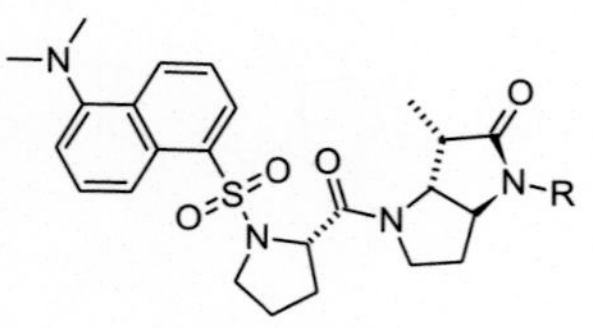

**8–10**

**FIGURE 12.8** Template of inhibitors based on dansylproline α-methyl pyrrolidine-5,5-*trans*-lactam.

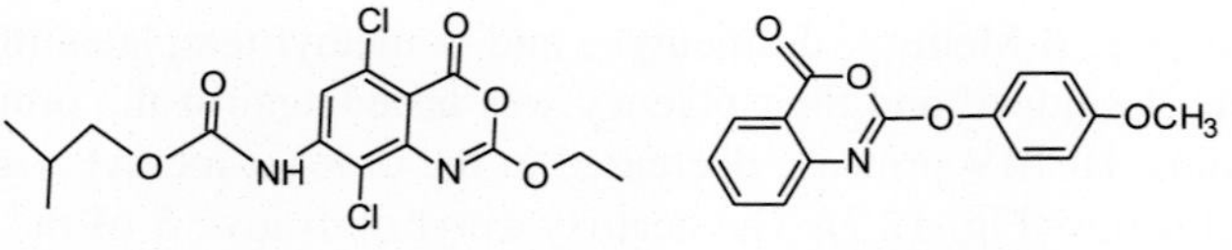

**11** IC$_{50}$ = 1.5 µM          **12** IC$_{50}$ = 2.5 µM

**FIGURE 12.9** Benzoxazinone based inhibitors.

bonds: one each with Thr132 and Ser129 (active-site nucleophile), and two hydrogen bonds with Arg157. The hydrogen bonds with Arg157 make it less available to stabilize the oxyanion formation, and inactivates the enzyme. This study also reported novel hydrophobic interactions of these inhibitors with Leu38, Leu130, Ala131, Cys152, Ala153, and Ile154 residues. The best overall profile was, however, of derivative **12** and not **11**, the most active derivative. This was due to the better druglikeness and drugscore of **12** over **11** (Mello et al., 2015).

## 6.3 Dimer Disrupting/Allosteric Inhibitors

The knowledge that dimerization regulates protease activity provided an allosteric site that could be targeted by inhibitors. Shimba et al. (2004) were the first to report dimer-disrupting inhibitors of herpesvirus proteases. Like other herpesvirus proteases, the dimer interface of the KSHV protease is primarily made up of helix α5 of one monomer interacting with α5 helix of the other monomer, and also helices α1 and α2 of the other monomer. The peptide inhibitor suggested by Shimba et al. (2004) was designed to mimic this interfacial α5 helix. In order to form a stable peptide inhibitor, the α5 helix sequence was "grafted" onto a stable α-helical scaffold of avian pancreatic polypeptide (aPP), resulting in a 30-amino-acid helical peptide (PSQPTYPGDDAPLEDLMA-FAIDLSFYLGVV), referred to as the KSDD (KSHV protease dimer disrupter) peptide (Shimba et al., 2004). This chimeric peptide was revealed to disrupt KSHV protease dimerization, simultaneously inhibiting KSHV enzyme activity by 50% when ~200-fold molar excess of inhibitor was present. Mutagenesis of interfacial residues was carried out to show that inhibition by this inhibitor involved peptide binding to the dimer interface (Shimba et al., 2004).

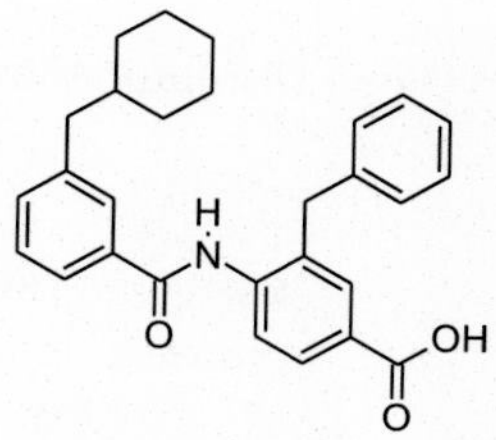

DD1

FIGURE 12.10    α-Helical mimetic dimer disruptor inhibitor of KSHV protease.

Shahian et al. (2009) then aimed at developing small molecule dimer disruptors and identified DD2 as the first ever small molecule inhibitor of a herpesvirus protease that acts by preferentially selecting a partially folded zymogen. A library of small α-helical mimetics was screened, and six molecules were identified that inhibited HCMV protease activity by at least 50%. DD1 (Fig. 12.10), the most potent of these six compounds with an $IC_{50} = 8.8$ μM, became the basis of a series of DD1 analogs being designed that had modifications in the 4-benzoylaminobenzoic acid scaffold. DD2, a benzyl-substituted 4-(pyridine-2-amido) benzoic acid, was so developed and had an $IC_{50} = 3.1$ μM (Table 12.3). Structure-activity relationship stressed on the correct positioning of the cyclohexylmethyl side chain, as activity was observed to decrease if its position was changed to *ortho* or *para*. DD2 was suggested to interact with the interface residues like Trp109. Trp109 is positioned behind the α5 helix, and interacts with the α5 helix of the other monomer through its indole ring, which is extended toward the hydrophobic side chains of Met197 and Ile201 of this helix. Met197 and Ile201 are positioned in a manner to satisfy i and i + 4 positions of this helix. It was suggested that the hydrophobic side chains of DD2 mimic this structural feature of Met197 and Ile 201, and interact with Trp 109 in a similar way. A "monomer trap" model was proposed for inhibition by DD2 (Shahian et al., 2009). DD2 was suggested to compete in the monomer–dimer equilibrium of the KSHV protease which has a $K_d = 1.2$ μM and bind to one of the zymogen conformations. Trapping the monomer in an inactive state interferes, with the interactions critical for dimer formation, thereby inhibiting the protease. DD2 binds to the protease at a binding pocket ~15 Å from the active site. This pocket is formed in the partially disordered monomer by a conformational change that occurs when the aromatic hot spot Trp109 changes the rotomeric state (Lee et al., 2011). Interestingly, DD2 was shown to be active against the HCMV protease also, with an $IC_{50} = 5.0$ μM. This led to the possibility of a broad-spectrum inhibitor targeting the dimer interface in all herpesvirus proteases. The observation that interface residues in other herpesvirus proteases are also involved in hydrophobic interactions similar to those found in the KSHV protease augmented interest in this possibility (Shahian et al., 2009).

In an attempt to develop broad-spectrum allosteric inhibitors of herpesvirus proteases, Gable et al. (2014) synthesized a series of DD2 analogs that had

**TABLE 12.3 Broad-Spectrum Dimer Disruptor Allosteric Inhibitors of Herpesvirus Proteases**

| Compound | IC$_{50}$ (μM) | | | |
| --- | --- | --- | --- | --- |
| | KSHV protease (γ) | EBV protease (γ) | CMV protease (β) | HSV-2 protease (α) |
| DD2 | 1.5 | 7.7 | 7.4 | 52.0 |
| 2-Tetrazole | 1.0 | 4.0 | 4.7 | 36.7 |
| 3-Sulfonamide | 3.6 | 19.0 | 38.8 | 94.0 |

CMV, Cytomegalovirus; EBV, Epstein–Barr virus; HSV-2, herpes simplex virus type 2; KSHV, Kaposi's sarcoma related herpesvirus.

polar anionic or nonionic functional groups instead of the carboxylate of DD2. The potency and mechanism of action of these inhibitors was evaluated against HSV-2 protease ($\alpha$-Subfamily), HCMV protease ($\beta$-Subfamily), EBV, and KSHV proteases ($\gamma$-Subfamily). DD2, along with the 2-tetrazole analog and the 3-sulfonamide analog, showed micromolar potencies against all the proteases (Table 12.3). The tetrazole analog is the most potent, followed by DD2 and lastly the sulfonamide analog. All three inhibitors were most active against the $\gamma$-subfamily, followed by $\beta$- and $\alpha$-subfamilies of herpesvirus proteases. Both DD2 analogs were shown to bind to the same binding pocket as DD2 and have an inhibition mechanism (Gable et al., 2014) similar to that of DD2 inhibiting KSHV protease (Shahian et al., 2009). Structure–activity relationships suggested an important role for the carboxylate or other functional groups having similar biological properties within these inhibitors. Further analysis by cocrystallization studies of the inhibitor-KSHV protease complex revealed a novel interaction involving a salt bridge formation between Arg82 and the anionic substituent on these inhibitors. The importance of this interaction was established by Arg82Gln mutation, which resulted in an $IC_{50}$ value three- to fourfold higher than that for the wild-type enzyme (Gable et al., 2014).

## 6.4  Miscellaneous

Metal-mediated inhibition has been reported in a class of bis-imidazolemethane (BIM) inhibitors against HCMV protease (Dhanak et al., 2000). These inhibitors were found to be active only in the presence of zinc ions and their mechanism of action was reported to involve formation of a ternary complex of zinc, inhibitor and protease, typical of zinc ion mediated inhibition of classical serine proteases like trypsin (Katz et al., 1998). Modification of substituents on the BIM template (Fig. 12.11A), however, did not enhance the activity of these inhibitors beyond the micromolar level (Dhanak et al., 2000).

Gopalsamy et al. (2004) reported naphthalene derivatives as non-covalent, competitive inhibitors of HCMV protease with potency of a few micromolar. Styryl naphthalene was identified as the template molecule. Optimization

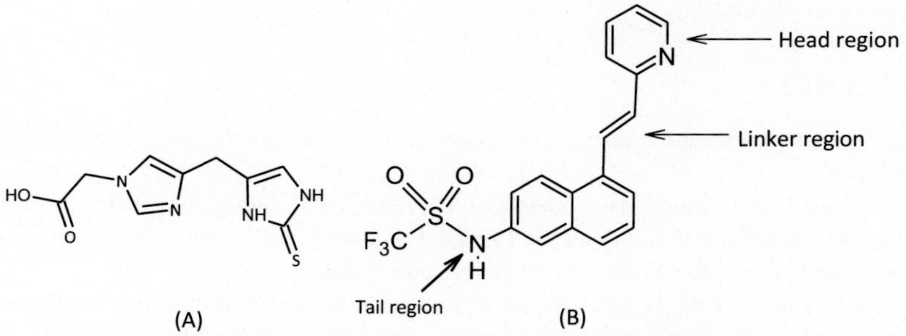

FIGURE 12.11    Templates of (A) BIM inhibitors and (B) Styryl naphthalene inhibitors.

of three regions of this template (Fig. 12.11B) led to the identification of 1,6-disubstituted naphthalenes as potent inhibitors. Replacement of the pyridyl head-group ($IC_{50}$ = 66 µM) by quinolinyl groups considerably enhanced potency, with $IC_{50}$ as low as 3.2 µM being observed for 6-chloro-2-quinolinyl. Modification of the tailpiece, however, did not enhance potency much. The best substitution was observed, with trifluoromethylsulfonamide ($IC_{50}$ = 2.7 µM). Substitutions in the linker region also did not improve potency of these inhibitors, although it was observed that naphthylmethyl ether and reverse naphthyl ether substituents gave active inhibitors of similar potencies.

## 7 CONCLUSIONS

Research for inhibitors against herpesvirus proteases has progressed considerably in the last two decades. New inhibitors based on dipeptide azetidines and *trans*-lactams, with micromolar potencies have been reported with the dansylproline *trans*-lactam inhibitors projecting high stabilities in the human plasma. Elucidation of the conformational changes that accompany substrate/inhibitor binding revealed dimerization as the process regulating enzymatic activity and also established herpesviral proteases as induced-fit enzymes. The structural link between the active site and dimer interface of herpesvirus proteases provided an insight into their induced-fit mechanism. The fact that dimerization regulates activity led to the development of a new class of inhibitors against KSHV proteases that target the dimer interface instead of the active site and inhibit herpesvirus proteases by interfering with dimer formation. The small molecule dimer disruptors act by selectively binding to a partially folded zymogen, which prevents proper dimer formation, with concomitant protease inhibition. Since residues at the dimer interface of all herpesvirus proteases are involved in similar interactions as the interface residues of KSHV protease, the DD inhibitors have been developed into broad-spectrum allosteric inhibitors that target dimer interfaces of all herpesvirus proteases. The information gained in the last two decades has, therefore, allowed development of stable inhibitors with even nanomolar potencies. However, further improvement of these inhibitors is required in order to develop them into antiherpetic drugs targeting the herpesvirus proteases.

## REFERENCES

Arvin, A.M., 1996. Varicella-zoster virus: overview and clinical manifestations. Semin. Dermatol. 15, 4–7.
Baer, R., Bankier, A.T., Biggin, M.D., Deininger, P.L., Farrell, P.J., Gibson, T.J., Hatfull, G., Hudson, G.S., Satchwell, S.C., Séguin, C., Tuffnell, P.S., Barrell, B.G., 1984. DNA sequence and expression of the B95-8 Epstein–Barr virus genome. Nature 310, 207–211.
Batra, R., Khayat, R., Tong, L., 2001. Molecular mechanism for dimerization to regulate the catalytic activity of human cytomegalovirus protease. Nat. Struct. Mol. Biol. 8, 810–817.

Baum, E.Z., Bebernitz, G.A., Hulmes, J.D., Muzithras, V.P., Jones, T.R., Gluzman, Y., 1993. Expression and analysis of the human cytomegalovirus UL80-encoded protease: identification of autoproteolytic sites. J. Virol. 67, 497–506.

Baum, E.Z., Ding, W.-D., Siegel, M.M., Hulmes, J., Bebernitz, G.A., Sridharan, L., Tabei, K., Krishnamurthy, G., Carofiglio, T., Groves, J.T., Bloom, J.D., DiGrandi, M., Bradley, M., Ellestad, G., Seddon, A.P., Gluzman, Y., 1996a. Flavins inhibit human cytomegalovirus UL80 protease via disulfide bond formation. Biochemistry 35, 5847–5855.

Baum, E.Z., Siegel, M.M., Bebernitz, G.A., Hulmes, J.D., Sridharan, L., Sun, L., Tabei, K., Johnston, S.H., Wildey, M.J., Nygaard, J., Jones, T.R., Gluzman, Y., 1996b. Inhibition of human cytomegalovirus UL80 protease by specific intramolecular disulfide bond formation. Biochemistry 35, 5838–5846.

Beaudet-Miller, M., Zhang, R., Durkin, J., Gibson, W., Kwong, A.D., Hong, Z., 1996. Virus-specific interaction between the human cytomegalovirus major capsid protein and the C terminus of the assembly protein precursor. J. Virol. 70, 8081–8088.

Bonneau, P.R., Grand-Maître, C., Greenwood, D.J., Lagacé, L., LaPlante, S.R., Massariol, M.-J., Ogilvie, W.W., O'Meara, J.A., Kawai, S.H., 1997. Evidence of a conformational change in the human cytomegalovirus protease upon binding of peptidyl-activated carbonyl inhibitors. Biochemistry 36, 12644–12652.

Borthwick, A.D., Weingarten, G., Haley, T.M., Tomaszewski, M., Wang, W., Hu, Z., Bedard, J., Jin, H., Yuen, L., Mansour, T.S., 1998. Design and synthesis of monocyclic β-lactams as mechanism-based inhibitors of human cytomegalovirus protease. Bioorg. Med. Chem. Lett. 8, 365–370.

Borthwick, A.D., Angier, S.J., Crame, A.J., Exall, A.M., Haley, T.M., Hart, G.J., Mason, A.M., Pennell, A.M., Weingarten, G.G., 2000. Design and synthesis of pyrrolidine-5,5-*trans*-lactams (5-Oxo-hexahydro-pyrrolo[3,2-*b*]pyrroles) as novel mechanism-based inhibitors of human cytomegalovirus protease. 1. The alpha-methyl-*trans*-lactam template. J. Med. Chem. 43, 4452–4464.

Borthwick, A.D., Crame, A.J., Ertl, P.F., Exall, A.M., Haley, T.M., Hart, G.J., Mason, A.M., Pennell, A.M., Singh, O.M., Weingarten, G.G., Woolven, J.M., 2002a. Design and synthesis of pyrrolidine-5,5-*trans*-lactams (5-Oxohexahydropyrrolo[3,2-*b*]pyrroles) as novel mechanism-based inhibitors of human cytomegalovirus protease. 2. Potency and chirality. J. Med. Chem. 45, 1–18.

Borthwick, A.D., Exall, A.M., Haley, T.M., Jackson, D.L., Mason, A.M., Weingarten, G.G., 2002b. Pyrrolidine-5,5-*trans*-lactams as novel mechanism-based inhibitors of human cytomegalovirus protease. Part 3: Potency and plasma stability. Bioorg. Med. Chem. Lett. 12, 1719–1722.

Brignole, E.J., Gibson, W., 2007. Enzymatic activities of human cytomegalovirus maturational protease assemblin and its precursor (pPR, pUL80a): Maximal activity of pPR requires self-interaction through its scaffolding domain. J. Virol. 81, 4091–4103.

Britt, W.J., Alford, C.A., 1996. Cytomegalovirus. In: Fields, B.N., Knipe, D.M., Howley, P.M. (Eds.), Fields Virol. Lippincott-Raven, Philadelphia, pp. 2493–2523.

Buisson, M., Hernandez, J.F., Lascoux, D., Schoehn, G., Forest, E., Arlaud, G., Seigneurin, J.M., Ruigrok, R.W., Burmeister, W.P., 2002. The crystal structure of the Epstein–Barr virus protease shows rearrangement of the processed C terminus. J. Mol. Biol. 324, 89–103.

Burck, P.J., Berg, D.H., Luk, T.P., Sassmannshausen, L.M., Wakulchik, M., Smith, D.P., Hsiung, H.M., Becker, G.W., Gibson, W., Villarreal, E.C., 1994. Human cytomegalovirus maturational proteinase: expression in *Escherichia coli*, purification, and enzymatic characterization by using peptide substrate mimics of natural cleavage sites. J. Virol. 68, 2937–2946.

Chan, C.-K., Brignole, E.J., Gibson, W., 2002. Cytomegalovirus assemblin (pUL80a): cleavage at internal site not essential for virus growth; proteinase absent from virions. J. Virol. 76, 8667–8674.

Charnot-Katsikas, A., Baewer, D., Cook, L., David, M.Z., 2016. Fulminant hepatic failure attributed to infection with human herpesvirus 6 (HHV-6) in an immunocompetent woman: a case report and review of the literature. J. Clin. Virol. 75, 27–32.

Chen, P., Tsuge, H., Almassy, R.J., Gribskov, C.L., Katoh, S., Vanderpool, D.L., Margosiak, S.A., Pinko, C., Matthews, D.A., Kan, C.C., 1996. Structure of the human cytomegalovirus protease catalytic domain reveals a novel serine protease fold and catalytic triad. Cell 86, 835–843.

Cole, J.L., 1996. Characterization of human cytomegalovirus protease dimerization by analytical centrifugation. Biochem. 35, 15601–15610.

Cox, G.A., Wakulchik, M., Sassmannshausen, L.M., Gibson, W., Villarreal, E.C., 1995. Human cytomegalovirus proteinase: candidate glutamic acid identified as third member of putative active-site triad. J. Virol. 69, 4524–4528.

Darke, P.L., Cole, J.L., Waxman, L., Hall, D.L., Sardana, M.K., Kuo, L.C., 1996. Active human cytomegalovirus protease is a dimer. J. Biol. Chem. 271, 7445–7449.

Davison, A.J., Scott, J.E., 1986. The complete DNA sequence of varicella-zoster virus. J. Gen. Virol. 67, 1759–1816.

de Oliveira, C.A., Guimarães, C.R., Barreiro, G., de Alencastro, R.B., 2003. Investigation of the induced-fit mechanism and catalytic activity of the human cytomegalovirus protease homodimers via molecular dynamics simulations. Proteins 52, 483–491.

Deckman, I.C., Hagen, M., McCann, III, P.J., 1992. Herpes simplex virus type 1 protease expressed in *Escherichia coli* exhibits autoprocessing and specific cleavage of the ICP35 assembly protein. J. Virol. 66, 7362–7367.

Desai, P., Person, S., 1996. Molecular interactions between the HSV-1 capsid proteins as measured by the yeast two-hybrid system. Virol. 220, 516–521.

Desai, P., Watkins, S.C., Person, S., 1994. The size and symmetry of B capsids of herpes simplex virus type 1 are determined by the gene products of the UL26 open reading frame. J. Virol. 68, 5365–5374.

Déziel, R., Malenfant, E., 1998. Inhibition of human cytomegalovirus protease $N_o$ with monocyclic β-lactams. Bioorg. Med. Chem. Lett. 8, 1437–1442.

Dhanak, D., Burton, G., Christmann, L.T., Darcy, M.G., Elrod, K.C., Kaura, A., Keenan, R.M., Link, J.O., Peishoff, C.E., Shah, D.H., 2000. Metal mediated protease inhibition: design and synthesis of inhibitors of the human cytomegalovirus (hCMV) protease. Bioorg. Med. Chem. Lett. 10, 2279–2282.

DiIanni, C.L., Drier, D.A., Deckman, I.C., McCann, III, P.J., Liu, F., Roizman, B., Colonno, R.J., Cordingley, M.G., 1993a. Identification of the herpes simplex virus-1 protease cleavage sites by direct sequence analysis of autoproteolytic cleavage products. J. Biol. Chem. 268, 2048–2051.

DiIanni, C.L., Mapelli, C., Drier, D.A., Tsao, J., Natarajan, S., Riexinger, D., Festin, S.M., Bolgar, M., Yamanaka, G., Weinheimer, S.P., Meyers, C.A., Colonno, R.J., Cordingley, M.G., 1993b. In vitro activity of the herpes simplex virus type 1 protease with peptide substrates. J. Biol. Chem. 268, 25449–25454.

DiIanni, C.L., Stevens, J.T., Bolgar, M., O'Boyle, II, D.R., Weinheimer, S.P., Colonno, R.J., 1994. Identification of the serine residue at the active site of the herpes simplex virus type 1 protease. J. Biol. Chem. 269, 12672–12676.

Flynn, D.L., Becker, D.P., Dilworth, V.M., Highkin, M.K., Hippenmeyer, P.J., Houseman, K.A., Levine, L.M., Li, M., Moormann, A.E., Rankin, A., Toth, M.V., Villamil, C.I., Wittwer, A.J., Holwerda, B.C., 1997. The herpesvirus protease: mechanistic studies and discovery of inhibitors of the human cytomegalovirus protease. Drug Des. Discov. 15, 3–15.

Gable, J.E., Lee, G.M., Jaishankar, P., Hearn, B.R., Waddling, C.A., Renslo, A.R., Craik, C.S., 2014. Broad-spectrum allosteric inhibition of herpesvirus proteases. Biochemistry 53, 4648–4660.

Gao, M., Matusick-Kumar, L., Hurlburt, W., DiTusa, S.F., Newcomb, W.W., Brown, J.C., McCann, III, P.J., Deckman, I., Colonno, R.J., 1994. The protease of herpes simplex virus type 1 is essential for functional capsid formation and viral growth. J. Virol. 68, 3702–3712.

Gerona-Navarro, G., Bonache, M.A., Alías, M., Pérez-de-Vega, M.J., García-López, M.T., López, P., Cativiela, C., Gonzalez-Muñiz, R., 2004. Simple access to novel azetidine-containing conformationally restricted amino acids by chemoselective reduction of β-lactams. Tetrahedron Lett. 45, 2193–2196.

Gerona-Navarro, G., Pérez de Vega, M.J., García-López, M.T., Andrei, G., Snoeck, R., De Clercq, E., Balzarini, J., González-Muñiz, R., 2005. From 1-acyl-β-lactam human cytomegalovirus protease inhibitors to 1-benzyloxycarbonylazetidines with improved antiviral activity. A straightforward approach to convert covalent to noncovalent inhibitors. J. Med. Chem. 48, 2612–2621.

Gibson, W., Roizman, B., 1972. Proteins specified by herpes simplex virus. VIII. Characterization and composition of multiple capsid forms of subtypes 1 and 2. J. Virol. 10, 1044–1052.

Gibson, W., Roizman, B., 1974. Proteins specified by herpes simplex virus. X. Staining and radiolabeling properties of B capsid and virion proteins in polyacrylamide gels. J. Virol. 13, 155–165.

Godefroy, S., Guenet, C., 1995. Autoprocessing of HSV-1 protease: effect of deletions on autoproteolysis. FEBS Lett. 357, 168–172.

Gopalsamy, A., Lim, K., Ellingboe, J.W., Mitsner, B., Nikitenko, A., Upeslacis, J., Mansour, T.S., Olson, M.W., Bebernitz, G.A., Grinberg, D., Feld, B., Moy, F.J., O'Connell, J., 2004. Design and syntheses of 1,6-naphthalene derivatives as selective HCMV protease inhibitors. J. Med. Chem. 47, 1893–1899.

Haanes, E.J., Thomsen, D.R., Martin, S., Homa, F.L., Lowery, D.E., 1995. The bovine herpesvirus 1 maturational proteinase and scaffold proteins can substitute for the homologous herpes simplex virus type 1 proteins in the formation of hybrid type B capsids. J. Virol. 69, 7375–7379.

Hall, D.L., Darke, P.L., 1995. Activation of the herpes simplex virus type 1 protease. J. Biol. Chem. 270, 22697–22700.

Holskin, B.P., Bukhtlyarova, M., Dunn, B.M., Baur, P., Dechastonay, J., Pennington, M.W., 1995. A continuous fluorescence-based assay of human cytomegalovirus protease using a peptide substrate. Anal. Biochem. 227, 148–155.

Holwerda, B.C., 1997. Herpesvirus proteases: targets for novel antiviral drugs. Antiviral Res. 35, 1–21.

Holwerda, B.C., Wittwer, A.J., Duffin, K.L., Smith, C., Toth, M.V., Carr, L.S., Wiegand, R.C., Bryant, M.L., 1994. Activity of two-chain recombinant human cytomegalovirus protease. J. Biol. Chem. 269, 25911–25915.

Hong, Z., Beaudet-Miller, M., Durkin, J., Zhang, R.M., Kwong, A.D., 1996. Identification of a minimal hydrophobic domain in the herpes simplex virus type 1 scaffolding protein which is required for interaction with the major capsid protein. J. Virol. 70, 533–540.

Hoog, S.S., Smith, W.W., Qiu, X., Janson, C.A., Hellmig, B., McQueney, M.S., O'Donnell, K., O'Shannessy, D., DiLella, A.G., Debouck, C., Abdel-Meguid, S.S., 1997. Active site cavity of herpesvirus proteases revealed by the crystal structure of herpes simplex virus protease/inhibitor complex. Biochemistry 36, 14023–14029.

Jarvest, R.L., Parratt, M.J., Debouck, C.M., Gorniak, J.G., Jennings, L.J., Serafinowska, H.T., Strickler, J.E., 1996. Inhibition of HSV-1 protease by benzoxazinones. Bioorg. Med. Chem. Lett. 6, 2463–2466.

Jarvest, R.L., Connor, S.C., Gorniak, J.G., Jennings, L.J., Serafinowska, H.T., West, A., 1997. Potent selective thienoxazinone inhibitors of herpes proteases. Bioorg. Med. Chem. Lett. 7, 1733–1738.

Jarvest, R.L., Pinto, I.L., Ashman, S.M., Dabrowski, C.E., Fernandez, A.V., Jennings, L.J., Lavery, P., Tew, D.G., 1999. Inhibition of herpes proteases and antiviral activity of 2-substituted thieno[2,3-*d*]oxazinones. Bioorg. Med. Chem. Lett. 9, 443–448.

Jones, T.R., Sun, L., Bebernitz, G.A., Muzithras, V.P., Kim, H.J., Johnston, S.H., Baum, E.Z., 1994. Proteolytic activity of human cytomegalovirus UL80 protease cleavage site mutants. J. Virol. 68, 3742–3752.

Katz, B.A., Clark, J.M., Finer-Moore, J.S., Jenkins, T.E., Johnson, C.R., Ross, M.J., Luong, C., Moore, W.R., Stroud, R.M., 1998. Design of potent selective zinc-mediated serine protease inhibitors. Nature 391, 608–612.

Kennard, J., Rixon, F.J., McDougall, I.M., Tatman, J.D., Preston, V.G., 1995. The 25 amino acid residues at the carboxy terminus of the herpes simplex virus type-1 UL26.5 protein are required for the formation of the capsid shell around the scaffold. J. Gen. Virol. 76, 1611–1621.

Khayat, R., Batra, R., Qian, C., Halmos, T., Bailey, M., Tong, L., 2003. Structural and biochemical studies of inhibitor binding to human cytomegalovirus protease. Biochem. 42, 885–891.

LaFemina, R.L., Bakshi, K., Long, W.L., Pramanik, B., Veloski, C.A., Wolanski, B.S., Marcy, A.I., Hazuda, D.J., 1996. Characterization of a soluble stable human cytomegalovirus protease and inhibition by M-site peptide mimics. J. Virol. 70, 4819–4824.

LaPlante, S.R., Aubry, N., Bonneau, P.R., Cameron, D.R., Lagacé, L., Massariol, M.-J., Montpetit, H., Plouffe, C., Kawai, S.H., Fulton, B.D., Chen, Z., Ni, F., 1998. Human cytomegalovirus protease complexes its substrate recognition sequences in an extended peptide conformation. Biochem. 37, 9793–9801.

LaPlante, S.R., Bonneau, P.R., Aubry, N., Cameron, D.R., Deziel, R., Grand-Maitre, C., Plouffe, C., Tong, L., Kawai, S.H., 1999. Characterization of the human cytomegalovirus protease as an induced-fit serine protease and the implications to the design of mechanism-based inhibitors. J. Am. Chem. Soc. 121, 2974–2986.

Lee, G.M., Shahian, T., Baharuddin, A., Gable, J.E., Craik, C.S., 2011. Enzyme inhibition by allosteric capture of an inactive conformation. J. Mol. Biol. 411, 999–1016.

Levy, J.A., 1997. Three new human herpesviruses (HHV6, 7, and 8). Lancet 349, 558–563.

Liang, P.-H., Brun, K.A., Field, J.A., O'Donnell, K., Doyle, M.L., Green, S.M., Baker, A.E., Blackburn, M.N., Abdel-Meguid, S.S., 1998. Site-directed mutagenesis probing the catalytic role of arginines 165 and 166 of human cytomegalovirus protease. Biochemistry 37, 5923–5929.

Liu, F., Roizman, B., 1991a. The herpes simplex virus 1 gene encoding a protease also contains within its coding domain the gene encoding the more abundant substrate. J. Virol. 65, 5149–5156.

Liu, F., Roizman, B., 1991b. The promoter, transcriptional unit, and coding sequence of herpes simplex virus 1 family 35 proteins are contained within and in frame with the UL26 open reading frame. J. Virol. 65, 206–212.

Liu, F., Roizman, B., 1992. Differentiation of multiple domains in the herpes simplex virus 1 protease encoded by the UL26 gene. Proc. Natl. Acad. Sci. USA 89, 2076–2080.

Liu, F., Roizman, B., 1993. Characterization of the protease and other products of amino-terminus-proximal cleavage of the herpes simplex virus 1 UL26 protein. J. Virol. 67, 1300–1309.

Loutsch, J.M., Galvin, N.J., Bryant, M.L., Holwerda, B.C., 1994. Cloning and sequence analysis of murine cytomegalovirus protease and capsid assembly protein genes. Biochem. Biophys.l Res. Commun. 203, 472–478.

Loveland, A.N., Chan, C.K., Brignole, E.J., Gibson, W., 2005. Cleavage of human cytomegalovirus protease pUL80a at internal and cryptic sites is not essential but enhances infectivity. J. Virol. 79, 12961–12968.

Margosiak, S.A., Vanderpool, D.L., Sisson, W., Pinko, C., Kan, C.-C., 1996. Dimerization of the human cytomegalovirus protease: kinetic and biochemical characterization of the catalytic homodimer. Biochemistry 35, 5300–5307.

Marnett, A.B., Nomura, A.M., Shimba, N., Ortiz de Montellano, P.R., Craik, C.S., 2004. Communication between the active sites and dimer interface of a herpesvirus protease revealed by a transition-state inhibitor. Proc. Natl. Acad. Sc. USA 101, 6870–6875.

Matusick-Kumar, L., Hurlburt, W., Weinheimer, S.P., Newcomb, W.W., Brown, J.C., Gao, M., 1994. Phenotype of the herpes simplex virus type 1 protease substrate ICP35 mutant virus. J. Virol. 68, 5384–5394.

Matusick-Kumar, L., McCann, III, P.J., Robertson, B.J., Newcomb, W.W., Brown, J.C., Gao, M., 1995a. Release of the catalytic domain $N_0$ from the herpes simplex virus type 1 protease is required for viral growth. J. Virol. 69, 7113–7121.

Matusick-Kumar, L., Newcomb, W.W., Brown, J.C., McCann, III, P.J., Hurlburt, W., Weinheimer, S.P., Gao, M., 1995b. The C-Terminal 25 amino acids of the protease and its substrate ICP35 of herpes simplex virus type 1 are involved in the formation of sealed capsids. J. Virol. 69, 4347–4356.

Mello, J.F., Botelho, N.C., Souza, A.M., Oliveira, R., Brito, M.A., Abrahim-Vieira Bde, A., Sodero, A.C., Castro, H.C., Cabral, L.M., Miceli, L.A., Rodrigues, C.R., 2015. Computational studies of benzoxazinone derivatives as antiviral agents against herpes virus type 1 protease. Molecules 20, 10689–10704.

Nomura, A.M., Marnett, A.B., Shimba, N., Dötsch, V., Craik, C.S., 2005. Induced structure of a helical switch as a mechanism to regulate enzymatic activity. Nat. Struct. Mol. Biol. 12, 1019–1020.

Nomura, A.M., Marnett, A.B., Shimba, N., Dötsch, V., Craik, C.S., 2006. One functional switch mediates reversible and irreversible inactivation of a herpesvirus protease. Biochemistry 45, 3572–3579.

O'Boyle, II, D.R., Wager-Smith, K., Stevens, III, J.T., Weinheimer, S.P., 1995. The effect of internal autocleavage on kinetic properties of the human cytomegalovirus protease catalytic domain. J. Biol. Chem. 270, 4753–4758.

Ogilvie, W., Bailey, M., Poupart, M.-A., Abraham, A., Bhavsar, A., Bonneau, P., Bordeleau, J., Bousquet, Y., Chabot, C., Duceppe, J.-S., Fazal, G., Goulet, S., Grand-Maitre, C., Guse, I., Halmos, T., Lavallée, P., Leach, M., Malenfant, E., O'Meara, J., Plante, R., Plouffe, C., Poirier, M., Soucy, F., Yoakim, C., Déziel, R., 1997. Peptidomimetic inhibitors of the human cytomegalovirus protease. J. Med. Chem. 40, 4113–4135.

Pérez-Faginas, P., Aranda, M.T., García-López, M.T., Snoeck, R., Andrei, G., Balzarini, J., González-Muñiz, R., 2011. Synthesis and SAR studies on azetidine-containing dipeptides as HCMV inhibitors. Bioorg. Med. Chem. 19, 1155–1161.

Pinto, I.L., West, A., Debouck, C.M., DiLella, A.G., Gorniak, J.G., O'Donnell, K.C., O'Shannessy, D.J., Patel, A., Jarvest, R.L., 1996. Novel, selective mechanism-based inhibitors of the herpes proteases. Bioorg. Med. Chem. Lett. 6, 2467–2472.

Pinto, I.L., Jarvest, R.L., Clarke, B., Dabrowski, C.E., Fenwick, A., Gorczyca, M.M., Jennings, L.J., Lavery, P., Sternberg, E.J., Tew, D.G., West, A., 1999. Inhibition of human cytomegalovirus protease by enedione derivatives of thieno[2,3-$d$]oxazinones through a novel dual acylation/ alkylation mechanism. Bioorg. Med. Chem. Lett. 9, 449–452.

Pray, T.R., Nomura, A.M., Pennington, M.W., Craik, C.S., 1999. Auto-inactivation by cleavage within the dimer interface of Kaposi's sarcoma-associated herpesvirus protease. J. Mol. Biol. 289, 197–203.

Preston, V.G., Coates, J.A.V., Rixon, F.J., 1983. Identification and characterization of a herpes simplex virus gene product required for encapsidation of virus DNA. J. Virol. 45, 1056–1064.

Qiu, X., Culp, J.S., DiLella, A.G., Hellmig, B., Hoog, S.S., Janson, C.A., Smith, W.W., Abdel-Meguid, S.S., 1996. Unique fold and active site in cytomegalovirus protease. Nature 383, 275–279.

Qiu, X., Janson, C.A., Culp, J.S., Richardson, S.B., Debouck, C., Smith, W.W., Abdel-Meguid, S.S., 1997. Crystal structure of varicella-zoster virus protease. Proc. Natl. Acad. Sci. USA 94, 2874–2879.

Reiling, K.K., Pray, T.R., Craik, C.S., Stroud, R.M., 2000. Functional consequences of the Kaposi's sarcoma-associated herpesvirus protease structure: regulation of activity and dimerization by conserved structural elements. Biochemistry 39, 12796–12803.

Rickinson, A.B., Kieff, E., 1996. Epstein–Barr virus. In: Fields, B.N., Knipe, D.M., Howley, P.M. (Eds.), Fields Virology. Lippincott-Raven, Philadelphia, pp. 2397–2446.

Robertson, B.J., McCann, III, P.J., Matusick-Kumar, L., Newcomb, W.W., Brown, J.C., Colonno, R.J., Gao, M., 1996. Separate functional domains of the herpes simplex virus type 1 protease: evidence for cleavage inside capsids. J. Virol. 70, 4317–4328.

Robertson, B.J., McCann, III, P.J., Matusick-Kumar, L., Preston, V.G., Gao, M., 1997. Na, an auto-proteolytic product of the herpes simplex virus type 1 protease, can functionally substitute for the assembly protein ICP35. J. Virol. 71, 1683–1687.

Roizman, B., Pellett, P.E., 2001. The family herpesviridae: a brief introduction. In: Knipe, D.M., Howley, P.M. (Eds.), Fields Virology. Lippincott Williams & Wilkins, Philadelphia, pp. 2479–2499.

Roizman, B., Sears, A.E., 1996. Herpes simplex viruses and their replication. In: Fields, B.N., Knipe, D.M., Howley, P.M. (Eds.), Fields Virology. Lippincott-Raven, Philadelphia, pp. 2231–2296.

Schmidt, U., Darke, P.L., 1997. Dimerization and activation of the herpes simplex virus type 1 protease. J. Biol. Chem. 272, 7732–7735.

Shahian, T., Lee, G.M., Lazic, A., Arnold, L.A., Velusamy, P., Roels, C.M., Guy, R.K., Craik, C.S., 2009. Inhibition of a viral enzyme by a small molecule dimer disruptor. Nat. Chem. Biol. 9, 640–646.

Shieh, H.-S., Kurumbail, R.G., Stevens, A.M., Stegeman, R.A., Sturman, E.J., Pak, J.Y., Wittwer, A.J., Palmier, M.O., Wiegand, R.C., Holwerda, B.C., Stallings, W.C., 1996. Three-dimensional structure of human cytomegalovirus protease. Nature 383, 279–282.

Shimba, N., Nomura, A.M., Marnett, A.B., Craik, C.S., 2004. Herpesvirus protease inhibition by dimer disruption. J. Virol. 78, 6657–6665.

Stevens, J.T., Mapelli, C., Tsao, J., Hail, M., O'Boyle, II, D., Weinheimer, S.P., Diianni, C.L., 1994. In vitro proteolytic activity and active-site identification of the human cytomegalovirus protease. Eur. J. Biochem. 226, 361–367.

Thomsen, D.R., Newcomb, W.W., Brown, J.C., Homa, F.L., 1995. Assembly of the herpes simplex virus capsid: requirement for the carboxyl-terminal twenty-five amino acids of the proteins encoded by the UL26 and UL26.5 genes. J. Virol. 69, 3690–3703.

Tong, L., Qian, C., Massariol, M.-J., Bonneau, P.R., Cordingley, M.G., Lagacé, L., 1996. A new serine-protease fold revealed by the crystal structure of human cytomegalovirus protease. Nature 383, 272–275.

Tong, L., Qian, C., Massariol, M.-J., Déziel, R., Yoakim, C., Lagacé, L., 1998. Conserved mode of peptidomimetic inhibition and substrate recognition of human cytomegalovirus protease. Nat. Struct. Mol. Biol. 5, 819–826.

Waxman, L., Darke, P.L., 2000. The herpesvirus proteases as targets for antiviral chemotherapy. Antivir. Chem. Chemother. 11, 1–22.

Weinheimer, S.P., McCann, III, P.J., O'Boyle, II, D.R., Stevens, J.T., Boyd, B.A., Drier, D.A., Yamanaka, G.A., Dilanni, C.L., Deckman, I.C., Cordingley, M.G., 1993. Autoproteolysis of herpes simplex virus type 1 protease releases an active catalytic domain found in intermediate capsid particles. J. Virol. 67, 5813–5822.

Welch, A.R., McNally, L.M., Gibson, W., 1991a. Cytomegalovirus assembly protein nested gene family: four 3′-coterminal transcripts encode four in-frame, overlapping proteins. J. Virol. 65, 4091–4100.

Welch, A.R., Woods, A.S., McNally, L.M., Cotter, R.J., Gibson, W., 1991b. A herpesvirus maturational proteinase, assemblin: identification of its gene, putative active site domain, and cleavage site. Proc. Natl. Acad. Sci. USA 88, 10792–10796.

Welch, A.R., McNally, L.M., Hall, M.R., Gibson, W., 1993. Herpesvirus proteinase: site-directed mutagenesis used to study maturational, release, and inactivation cleavage sites of precursor and to identify a possible catalytic site serine and histidine. J. Virol. 67, 7360–7372.

Welch, A.R., Villarreal, E.C., Gibson, W., 1995. Cytomegalovirus protein substrates are not cleaved by the herpes simplex virus type 1 proteinase. J. Virol. 69, 341–347.

Wood, L.J., Baxter, M.K., Plafker, S.M., Gibson, W., 1997. Human cytomegalovirus capsid assembly protein precursor (pUL80.5) interacts with itself and with the major capsid protein (pUL86) through two different domains. J. Virol. 71, 179–190.

Yamanaka, G., Dilanni, C.L., O'Boyle, II, D.R., Stevens, J., Weinheimer, S.P., Deckman, I.C., Matusick-Kumar, L., Colonno, R.J., 1995. Stimulation of the herpes simplex virus type I protease by antichaeotrophic salts. J. Biol. Chem. 270, 30168–30172.

Yoakim, C., Ogilvie, W.W., Cameron, D.R., Chabot, C., Guse, I., Haché, B., Naud, J., O'Meara, J.A., Plante, R., Déziel, R., 1998a. β-Lactam derivatives as inhibitors of human cytomegalovirus protease. J. Med. Chem. 41, 2882–2891.

Yoakim, C., Ogilvie, W.W., Cameron, D.R., Chabot, C., Grand-Maître, C., Guse, I., Haché, B., Kawai, S., Naud, J., O'Meara, J.A., Plante, R., Déziel, R., 1998b. Potent β-lactam inhibitors of human cytomegalovirus protease. Antivir. Chem. Chemother. 9, 379–387.

Chapter 13

# Design and Development of Inhibitors of Herpes Viral Proteases and Their SAR and QSAR

**Dimitra Hadjipavlou-Litina\*, Satya P. Gupta\*\***

*\*Aristotle University of Thessaloniki, Thessaloniki, Greece; \*\*National Institute of Technical Teachers' Training and Research, Bhopal, Madhya Pradesh, India*

## 1  INTRODUCTION

Herpes viruses are a large family of DNA viruses (enveloped dsDNA viruses) (Tong, 2002), which cause a number of diseases infecting vertebrates, as well as invertebrates and fungi. The genomes of herpes viruses are double-stranded DNA circles of approximately 150 kilobases, encoding at least a dozen of enzymes (Waxman and Darke, 2000). DNA genome is packaged within an inner "nucleocapsid" structure. The virus particle consists of four parts:

1. the membrane-like outer envelope;
2. an amorphous tegument between the envelope and nucleocapsid (Jarvest and Dabrowski, 2002);
3. a stable assembly of proteins forming the nucleocapsid;
4. the core contained within the nucleocapsid.

The icosahedral nucleocapsids of herpes viruses have diameters of about 1250 Å, enclosing DNA genomes of 130–250 kbp. The size of the enveloped virions is much larger, with diameter up to 3000 Å, depending on the thickness of an amorphous protein layer between the nucleocapsid and the envelope (Tong, 2002).

The herpes viruses family, also known as Herpesviridae, has been divided into eight distinct viruses and three subfamilies ($\alpha$, $\beta$, $\gamma$) known to cause diseases in humans (Baron, 1996; Ryan and Ray, 2004; Speck-Planche and Cordeiro, 2011). However, herpes simplex type 1 and 2 have been more extensively studied. Human herpes viruses (HHVs) are associated with several

**Viral Proteases and Their Inhibitors. http://dx.doi.org/10.1016/B978-0-12-809712-0.00013-7**

**TABLE 13.1 HHV Classification**

| Type | Subfamily | Primary target cell |
| --- | --- | --- |
| HHV-1, HSV-1 | α | Mucoepithelia |
| HHV-2, HSV-2 | α | Mucoepithelia |
| HHV-3, VZV | α | Mucoepithelia |
| HHV-4, EBV, lymphocryptovirus | γ | B cells and epithelial cells |
| HHV-5, HCMV | β | Monocyte, lymphocyte, and epithelial cells |
| HHV-6, roseolovirus, herpes lymphotropic virus | β | T cells |
| HHV-7, roseolovirus | β | T cells |
| HHV-8, KSHV | γ | Lymphocyte and other cells |

EBV, Epstein–Barr virus; HCMV, human cytomegalovirus; HSV, herpes simplex virus; HHV, human herpes virus; KSHV, Kaposi's sarcoma-associated herpes virus; VZV, varicella zoster virus.

malignancies, side-effects, or coconditions like Alzheimer's disease. They are notorious for their ability to form latent and recurrent infections. They can remain silent for years waiting for reactivation. HHVs classification is given in Table 13.1.

It has been observed that with time, HHVs become resistant to the available commercial drugs like acyclovir, whose final target is the DNA-polymerase. Taking this fact under consideration there is an increasing interest to find new antiherpes agents that may be able to inhibit not only the principal target, but also other biological targets which could be involved in molecular functions related to the virulence of herpes viruses. It must be considered that enzymes encoded by the virus are more appealing than host enzymes or receptors, because complete selective inhibition of the viral target is less likely to have side effects on the patient.

The finding that all herpes viruses encode a unique protease that is essential for viral replication has afforded a potential new therapeutic target (Liu and Roizman 1991; Welch et al., 1991). The herpes virus protease is the only protease that has been identified so far in the herpes virus genome. The proteases within the individual herpes virus subfamilies are highly conserved. Since the herpes protease is essential for the production of infectious virus, it represents a valid target.

## 2 HERPES VIRUS PROTEASE

Herpes virus protease was first identified in 1991 from studies on herpes simplex virus-1 (HSV-1) and human cytomegalovirus (HCMV) (Liu and Roizman, 1991; Welch et al., 1991). All HHVs express a dimeric serine protease that is essential for the viral life cycle. Herpes protease is used in construction of the capsid which occurs in the third phase of viral infection during viral assembly. The first phase is attachment, entry, and release of DNA. In the second phase, DNA replication occurs and the viral enzymes are produced along with the viral protease. In the third phase, the viral protease is used in the maturation of the capsid required for packaging DNA before release. The viral protease must localize to the capsid inside the nucleus before expressing its activity (Robertson et al., 1996). After completion of capsid assembly, autocleavage by the protease releases the scaffold, permitting DNA packaging (Gibson et al., 1995). This occurs in the nucleus of the cell, so any inhibitor has to penetrate two membranes, the cellular membrane and the nuclear membrane, to reach the target protease. Thus an inhibitor would have to be capable of penetrating not only in the cells, but also into the capsid structure within the nucleus of the infected cells. Poor antiviral activity might occur due to the nucleocapsid barrier. Genetic deletion of the viral protease in HSV-1 precludes capsid maturation, confirming that the protease is necessary for successful viral replication and thus validating the protease as a potential therapeutic target (Gao et al., 1994). Similarly, knockdown of the maturational protease in murine cytomegalovirus, a model of β herpesvirus infection, causes a significantly reduced viral load (Jiang et al., 2012).

Herpes virus proteases are not related to any other enzymes and, therefore, they could not be readily identified by homology searches in DNA or protein databases. Hence, these enzymes were identified by their function in processing capsid assembly proteins during viral replication. The active site of the herpesvirus protease is located on the surface of the β-barrel and contains a novel Ser132-His63-His157 catalytic triad within the less tractable active site. Herpes protease belongs to the serine protease family and shows a varying degree of sequence homology across the herpes virus family, and a highly conserved P4–P1′ cleavage motif in which proteolysis occurs between alanine and serine residues. The cleavage sites are unique and highly conserved across the herpes viruses. The protease is expressed as part of a polyprotein. As shown in Fig. 13.1, the catalytic domain is contained within the N-terminal third of the protein and the remainder comprises a structural "scaffold" protein, which is independently expressed in excess to the polyprotein from an internal initiation codon. The polyprotein is cleaved by protease at two sites: one at the C-terminus of the protease catalytic domain, the release or R-site, and the other close to C-terminus of the scaffold protein, the maturation or M-site (Ertl et al., 2000). The cleavage

**FIGURE 13.1 Diagrammatic representation of herpesvirus protease/scaffold polyprotein, showing the cleavage of the polyprotein by protease at two sites, the release site (R-site) and the maturation site (M-site).**

of the M-site follows assembly of the viral procapsids and precedes packaging of the viral DNA. The M-site has a consensus sequence of (V/L)-X-A-S with cleavage between A and S. This site is conserved among the herpes viruses and it has been shown that herpes virus proteases have a novel structure, and their essential role in capsid maturation makes them a potential target for the development of antiherpetic drugs.

As a serine protease, herpes virus protease catalysis proceeds in two chemical steps, with initial cleavage of the scissile amide bond by nucleophilic attack of serine upon the carbonyl of the amide, generating an acyl-enzyme ester intermediate and the C-terminal cleavage product. Subsequent water hydrolysis of the intermediate regenerates free enzyme and the N-terminal cleavage product. An "oxyanion hole" is found in serine proteases, wherein a developing negative charge on the scissile amide carbonyl is stabilized during nucleophilic attack (Waxman and Darke, 2000). While the classical serine proteases behave mostly as lock-and-key enzymes, structural and biochemical studies show that herpes virus proteases behave in an induced-fit manner (LaPlante et al., 1999), with large conformational changes upon the binding of the peptide inhibitors (Castro and Martinez, 2000; Tong, 2002). These changes help to define the S3 and S1 binding pockets of the protease, which are absent in the free enzyme structures. An attractive feature of the herpes virus protease is the well-defined substrate specificity. High substrate specificity suggests that it should be possible to eventually develop a peptidomimetic inhibitor.

Initial attempts to exploit HHV proteases as therapeutic targets were directed at the active site. These attempts were relied heavily on chemical "warheads" for covalent inhibition and/or peptidomimetic scaffolds (Marnett et al., 2004; Waxman and Darke, 2000). All HHV proteases have a relatively shallow substrate binding pocket with a strict preference for alanine at P1 and serine at P1'. In addition, substrate binding is reported to occur through an induced-fit mechanism (Batra et al., 2001; LaPlante et al., 1999; Lazic et al., 2007). Being both shallow and dynamic, this active site is particularly challenging to be inhibited.

X-ray crystallography has determined a single domain protein fold for proteases of at least three different herpesviruses. Dimerization is a unique feature for serine proteases and has been found to be essential for the activity of cytomegalovirus (CMV) and HSV proteases. It plays an essential role in inhibitor discovery. In general, the inhibitors of classical serine proteases that act at active site residues do not readily inactivate the herpesvirus proteases, but the recent crystal structures of the herpesvirus proteases now allow more direct interpretation of ligand structure–activity relationships (SARs). Inhibition of the viral encoded enzymes for the development of antiviral agents is more appealing than the inhibition of host enzymes, because the former has less possibility to have side effects. The detailed analysis of function of herpesvirus protease in replication of the virus was established by various authors (Gao et al., 1994; Matusick-Kumar et al., 1995). The herpes virus proteases as targets for antiviral chemotherapy have been discussed in detail by Waxman and Darke (2000). Prior to this also, there have been reviews on protease biochemistry (Gibson, 1996) and its potential as a target for antiviral chemotherapy (Flynn et al., 1997; Holwerda, 1997). In herpes virus replication, the protease plays the role in DNA packaging, the last step in viral replication. Initially, the enveloped virus binds to the cell surface and upon fusion with the cellular membrane releases the nucleocapsid to the cytoplasm. Without entry of the capsid, the viral genome is transferred to the nucleus, where transcription takes place. Genes transcribed late in the replication cycle code for the protein components that form new capsids. Protein components of new capsids are transported to the nucleus along with the assembly protein and the protease precursor. The assembly protein provides a scaffold for correct capsid construction. After the basic capsid structure is completed, the assembly protein is proteolytically cleaved by the viral protease and leaves the capsid prior to, or during, DNA packaging. If proteolysis is prevented, DNA packaging does not occur (Gao et al., 1994).

## 3  SEQUENCE SPECIFICITY IN HERPESVIRUS PROTEASES AND CATALYTIC MECHANISM

Sequence specificity in herpesvirus proteases is important for optimization of synthetic substrates for mechanistic studies. Characterization of sequence specificity is also important for the design of peptide-based protease inhibitors. Herpes virus proteases are synthesized as precursors that undergo autoproteolytic cleavage at the R- and M-sites (Gibson and Hall, 1997). For both R- and M-sites, the naturally occurring consensus cleavage sequence is (V, L, I)-X-A↓S, where X is a polar amino acid. The pattern of conservation among all of the herpes virus protease cleavage sites suggests essential recognition elements for cleavage susceptibility in the P3, P1, and P1′ residues, and less dependence upon amino acid side chain identity on the C-terminal side of P1′. The efficient peptide cleavage in HSV-1 protease, having naturally occurring amino acids, requires 5–8 residues on both sides of scissile bond (Darke et al., 1994; DiIanni et al., 1993), while for HCMV and HHV-6 only 4 residues are required on both sides (Sardana et al., 1994; Tigue and Kay, 1998). The catalytic efficiency of herpesvirus proteases with peptide substrates is low in comparison to other serine proteases. The low activity of these proteases is attributed to the dissociation of active dimer to inactive monomer in, in vitro assay.

In order to explore essential features of herpes virus protease substrate recognition, synthetic peptides have been of great importance. Pattern of conservation in the natural substrate sequences has indicated that residues at positions P3, P1, and P1′ may be more sensitive to substitution. In a structure of a substrate analog bound to HCMV protease P1 Ala side chain was shown to be directed into a small pocket of the enzyme (Tong et al., 1998). The P3 residue (Val, Leu, or Ile) required a branched aliphatic side chain for efficient cleavage and substitution with Gly residue was found to eliminate cleavage

(Sardana et al., 1994). Tong et al. (1998) had observed that in the inhibited structure, the P3 side chain had extended into a large cavity that had a hydrophilic component with the guanidine group of Arg166 near the back of the pocket. It was, however, hard to explain as to how a hydrophobic P3 residue in both substrate, as well as inhibitors could interact with hydrophilic S3 pocket. The alcohol functionality of P1′ Ser does not appear critical for substrate specificity. However, the coexpression of HSV-1 or HCMV proteases with their cognate substrate proteins demonstrates the necessity of the P3 to P1′ conserved sequence of (V/L)-X-A↓S (Godefroy and Guenet, 1995; Jones et al., 1994; Liu and Roizman, 1991; McCann et al., 1994; Welch et al., 1993).

It is quite likely that the binding of the natural protein substrates does not involve the regions adjacent to the cleavage site. An entropic advantage for protein substrate binding over the freely flexible peptides might be provided by conformational constraints upon the cleaved region. Factors contributing to better binding of substrate to protease might be provided by X-ray structure or kinetics studies.

Regarding the catalytic mechanism of herpes virus proteases, it is essential to restate that these proteases are serine proteases with an active site triad Ser-His-His in which serine was found to play an active role. In a complex of a ketoamide with the HCMV protease, the formation of a covalent bond from Ser132 to an active site trapping carbonyl was observed (Tong et al., 1998). The three dimensional structures (3Ds) of the herpes virus proteases showed that, in addition to the active site serine, conserved His63 and His148 (HCMV numbering) were in the active site, with His63 positioned for hydrogen bonding to the nucleophilic serine (Chen et al., 1996; Hoog et al., 1997; Qiu et al., 1996, 1997; Shieh et al., 1996; Tong et al., 1996). It was demonstrated that changes in the absolute conserved His148 in HSV-1 protease altered or abolished the ability of the virus to replicate in the culture (Register and Shafer, 1997). The third member of the catalytic site has not been found so far to be very important in catalytic role of the protease.

The catalytic mechanism of herpesvirus protease involves two steps. It initially cleaves the scissile amide bond by nucleophilic attack of serine on the carbonyl of the amide, producing an acyl-enzyme ester intermediate and C-terminal cleavage product. It is followed by water hydrolysis of the intermediate to regenerate free enzyme and the N-terminal cleavage product. There has been no report of ester hydrolysis, expected of serine protease, although it has been found that peptides of the appropriate sequence with a thioester replacement of the scissile amide are substrates for the HCMV protease. This facile esterolysis may allow definition of acyl-enzyme hydrolysis or product release rates. However, the herpes virus proteases are known to require dimerization for their significant activity. X-ray crystallographic studies have shown the existence of dimers in the structures of VZV, HSV-2 and HCMV proteases (Hoog et al., 1997; Qiu et al., 1996, 1997; Shieh et al., 1996; Tong et al., 1996). It may be that dimerization is a regulatory mechanism whereby activity is only realized when it is needed after the protease has been concentrated within the nascent capsid.

There can be an equilibrium between the dimer and monomer of the protease. If an enzyme monomer, in equilibrium with the dimer, is inactive and does not bind the inhibitor or substrate, and the dimer is active, then the monomer acts as a buffer to the loss of active dimers. For the herpes virus proteases, kinetic stabilization of dimers by use of low temperature, brief assays, or high concentrations of antichaeotropes are ways to avoid the complication (Darke and Kuo, 1998; Liang et al., 1998).

# 4  DESIGN AND DEVELOPMENT OF INHIBITORS OF HERPES VIRUS PROTEASES

Through studying the structure–function relationships of these enzymes, researchers built-up an understanding of their allosteric regulation (Hoog et al., 1997; Marnett et al., 2004; Qiu et al., 1997; Shimba et al., 2004; Tong et al. 1996). Each monomer has an independent active site (Tong, 2002). In the monomeric state, the enzyme is inactive and partially disordered. As a dimer, the enzyme is active, and the disordered C-terminal residues of the monomer form two helices, one that functions as a major contact surface at the dimer interface and the other that interacts with the catalytic site. This disorder-to-order transition links the dimer interface to the catalytic site (Lee et al., 2011; Nomura et al., 2005). Given the evidence supporting an allosteric link between protease dimerization and activation, Gable et al. (2014) focused their efforts on identifying molecules that target the dimer interface (Lazic et al., 2007; Marnett et al., 2004). In doing so, they previously identified a small molecule inhibitor (1) of Kaposi's sarcoma-associated herpes virus (KSHV) protease designated as DD2 (Shahian et al., 2009). Small molecules that allosterically regulate protein function have been increasingly developed as an alternative to classical active site inhibitors (Lee and Craik, 2009; Wells and McClendon, 2007) and this is the case for the homodimeric HHV proteases. DD2, a benzyl-substituted 4-(pyridine-2-amido)benzoic acid, is a helical peptide mimetic allosteric inhibitor that prevents the disorder-to-order transition that activates KSHV protease, trapping an inactive monomeric state (Lee et al., 2011; Shahian et al., 2009). The primary DD2 binding pocket, ~15 Å from the active site, is formed by conformational changes that occur only in the partially disordered monomer. The presence of a conserved aromatic hot spot in all HHV proteases suggests the potential for the development of broadly antiherpetic small molecules that allosterically inhibit HHV protease enzyme activity by disrupting protein–protein interactions. Gable et al. (2014) attempted to determine whether DD2 or its analogs could be allosteric inhibitors of herpesvirus proteases. To accomplish this, they generated a series of compounds in which the carboxylate of DD2 was replaced by polar anionic or nonionic functional groups, for example, 2–4, and assessed the inhibitory activity of these compounds.

1, (DD2)          2          3

4

Gable's group findings supported a model in which compounds **3** and **4**, trapping the inactive monomer and preventing homodimerization, allosterically inhibited HHV proteases in a manner analogous to DD2 inhibition of KSHV protease. In this model, they bind to a pocket at the dimer interface, $>10$ Å from the active site. Binding to this site prevents C-terminal helices 5 and 6 from folding against the hydrophobic dimer interface. While these two helices are in a disordered state, the oxyanion hole of the active site, normally formed in part by the two conserved arginines in contact with helix 6, cannot adopt the correct conformation that allows efficient proteolysis of peptide substrates to occur (Gable et al., 2014). Compound **3** was identified and demonstrated to have comparable or improved potency against all HHV proteases due to the larger van der Waals surface area of the tetrazole, compared to the carboxylate functional group of DD2, without additional rotatable bonds as found in compound **4**. SAR for this class of inhibitors suggested a role for carboxylate or carboxylate bioisosteres in inhibiting HHV proteases, namely formation of a salt-bridge between Arg82 and the anionic substituent of this class of inhibitors. Compounds **3** and **4** rely on the malleability of protein–protein interfaces, binding a cryptic pocket not apparent from apo structures (Sundberg and Mariuzza, 2000; Teague, 2003).

The application of X-ray crystallography offered useful structural information to the design and optimization of proteases inhibitors (Borthwick et al., 1998; Dunn et al., 1999; Holwerda, 1997; Waxman and Darke, 2000) and led to the development of several mechanism-based peptide and heterocyclic inhibitors, reversible or irreversible (Borthwick, 2005). The mechanism-based inhibition is an irreversible form of enzyme inhibition that occurs when an enzyme binds a substrate analog and forms an irreversible complex with it through a covalent bond during the "normal" catalysis reaction. This is also known as suicide inhibition or suicide inactivation. The basis for design of mechanism-based inhibitors includes the nucleophilic Ser hydroxyl attack upon a peptide carbonyl to form a tetrahedral intermediate, followed by the loss of water to generate an enzyme-serine acyl intermediate for subsequent hydrolysis. Thus serine protease inhibitors exploit nucleophilicity of the serine or histidine, as well as the oxyanion hole used in stabilization of the tetrahedral intermediate (Flynn et al., 1997; Waxman and Darke, 2000). These inhibitors can be roughly divided into three categories: active site inhibitors, cysteine modifiers, and natural products. The cysteine modifiers make covalent changes to Cys161 and Cys138 of HCMV protease (Baum et al., 1996a,b) and inhibit the protease possibly by the creation of steric hindrance in the active site, as Cys161 is likely to be in the S1' pocket. The natural products are identified from high-throughput screening, but their mechanism of inhibition is currently unknown (Chu et al., 1996). Chu et al. (1996) identified a novel fungal metabolite Sch65676 (**5**) with inhibitory activity against HCMV protease, but the mechanism of inhibition has not been delineated.

5 (Sch65676)

In general, it seemed that 3, 4-dichloroisocoumarin, peptide chloromethyl ketones, and sulfonyl fluorides were of limited activity. The active site inhibitors included peptidomimetic compounds (LaFemina et al., 1996; Ogilvie et al., 1997), lactams (Borthwick et al., 1998), oxazinones (Jarvest et al., 1996), and benzothiopyran-4-ones (Dhanak et al., 1998) (Fig. 13.2). A common feature among these inhibitors is the presence of an activated carbonyl, for example, R-ketoamide or trifluoromethyl ketone groups as in peptidomimetic compounds.

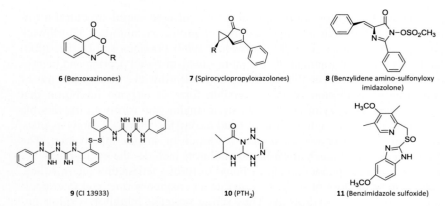

**6** (Benzoxazinones)    **7** (Spirocyclopropyloxazolones)    **8** (Benzylidene amino-sulfonyloxy imidazolone)

**9** (CI 13933)    **10** (PTH₂)    **11** (Benzimidazole sulfoxide)

FIGURE 13.2    **Structures of some herpesvirus protease inhibitors.**

The benzoxazinones (**6**, Fig. 13.2) were identified in 1960 as secondary metabolites of grasses and currently are known as serine protease inhibitors (Arroyo et al., 2010; Jarvest et al., 1996). A 1:1 adduct is formed with the protease, consistent with formation of an acyl-enzyme complex (Jarvest et al., 1996). SARs studies had pointed out that compounds with certain amino acids or alkyl groups in the 2-position (a possible P1 pocket) possessed increased potency. Further evidences also pointed the interactions with the residues His61, His148, Arg156, and Arg157 to be important for HSV-1 protease inhibition (Hemmateenejad and Elyasi, 2009; Hoog et al., 1997; Jarvest et al., 1996; Waxman and Darke, 2000). It has been further pointed out that the inhibitory potency of these compounds was inversely dependent on the size of the 5-substituent (Jarvest et al., 1996). Thus replacement of the benzene ring with a thiophene ring, reducing the steric bulk at this position, led to compounds with micromolar and submicromolar activity, for example, N-acyl analogs of 5-methylthieno[2,3-*d*]oxazinone.

Several mechanism based inhibitors were reported that were found to interact with the catalytic serine, for example,.a series of enedione-thieno[2,3-dloxazinones] derivatives as represented by **12** (Pinto et al., 1999) and some spirocyclopropyl oxazolones, and benzylidine *N*-sulphonyloxyimidazolones as represented by **13** and **14**, respectively (Pinto et al., 1996). They were reported as being potent and selective CMV protease inhibitors, acting by not only acylating the catalytic serine but also alkylating Cys161 of the protease. Spirocyclopropyl oxazolones (**13**) and the benzylidine *N*-sulphonyloxyimidazolones (**14**) were shown to be submicromolar inhibitors of HSV-2 proteases with some selectivity relative to the mammalian serine protease elastase, trypsin, and chymotrypsin (Martinez et al., 2001; Pinto et al., 1996).

**12** (Enedione-thieno[2,3-dloxazinones])

**13** (Spirocyclopropyl oxazolones)          **14** (Benzylidine *N*-sulphonyloxyimidazolones)

Cinnamyl derivatives of thieno[2,3-*d*]oxazinones (**15**) were also found to be mechanism-based inhibitors of the HSV-2, varicella-zoster virus and CMV herpesvirus proteases with nanomolar potency (Jarvest et al., 1999). A variety of derivatives of benzothiazole (**16**) were prepared in an attempt to find compounds with increased antiviral potency. In general, potency against the HCMV protease was found to be increased with electron withdrawing groups and decreased with electron donating groups. However, most of the tested derivatives were found to be less active than the unsubstituted parent compound (Borthwick, 2005). Substitution on the benzothiazole ring with electron donating or withdrawing groups had only a small effect on antiviral potency. Modeling studies suggested that the benzothiazole function makes a better hydrophobic interaction in the S1′ pocket than the smaller thiazole group. Hydrophobic interactions are favorable and account for greater potency.

**15** (Thieno[2,3-*d*]oxazinones)          **16** (Benzothiazole molecule)

The monocyclic β-lactam nucleus (**17**) has been used as a scaffold. Acylation of the active site serine takes place rapidly and the acylated serine is more slowly hydrolyzed to regenerate the active enzyme. Novel nonpeptidic β-lactam inhibitors, incorporating a benzyl or heterocyclic thiomethyl side-chain at C-4, and a substituted urea functionality at N-1, were synthesized and tested (Yoakim et al., 1998a,b). All were found to have $IC_{50}$ values in the low micromolar range.

17 (β-Lactam nucleus)

Mechanism-based inhibitors of HCMV protease were designed based on the pyrrolidine-5,5-*trans*-lactam ring system (Borthwick et al., 2000). SAR studies on this series of pyrrolidine-5,5-*trans*-lactams had defined the relative stereochemisty of the methyl substituent adjacent to the lactam carbonyl and the functionality on the lactam nitrogen and the mechanism of action against the HCMV δAla protease. It was found that activity decreased on moving from the α-methyl to the desmethyl to the β-methyl series. Higher inhibition activity was observed to be correlated with the Cbz-protected α-methyl-5,5-*trans*-lactam template which had low micromolar activity against the viral enzyme.

Through screening of a compound library, Matsumoto et al. (2001) discovered 4 effective inhibitors of HSV-1 protease which included a 1,4-dihydroxynaphthalene (**18**) and three naphthoquinones (**19–21**). They were found to effectively inhibit HSV-1 protease.

18          19          20          21

Consistent with their unique amino acid sequences, the crystal structures of herpes virus proteases show that they share a novel polypeptide backbone fold and do not show homology with other proteins, making them ideal therapeutic targets (Tong, 2002). Knowing the sequences of the two cleavage sites in the HCMV protease, Ogilvie et al. (1997) synthesized a series of peptidomimetic inhibitors that form disulfide bonds with cysteine residues near the active site and yield a cross-linked enzyme (Di Grandi et al., 2003). Flavin analogs and riboflavin inhibit HCMV protease in the same manner (Baum et al., 1996a).

Peptides with a C-terminal activated ketone (Bonneau et al., 1997; Ogilvie et al., 1997) render competitive inhibitors, with the formation of reversible hemiketal adducts with the hydroxyl group of the catalytic serine residue. The basis for this series of compounds is the N-terminal side of the M-site and the observation that the minimum length required for an inhibitor with low micromolar activity is the tetrapeptide. A shorter peptide region results in a loss in potency. It should be noted that the formation of a reversible, covalent bond from the active site Ser to the carbonyl group of an inhibitor would be expected to add significantly to the binding affinity of the supporting inhibitor scaffold. The activated carbonyl compounds are only slightly less active than a similar tetrapeptide carboxylate, which will not form such a linkage. The binding mode of the carboxylate is fundamentally different. The large increase in potency upon change from amide to carboxylate is accompanied by a shift in the binding mode of the P1 residue. A shift of the carboxylate toward the Arg166 in the oxyanion hole is conceivable. The shallowness of the active site cleft of HCMV protease indicates that it may be difficult to design a peptidomimetic molecule, with suitable affinity to effectively inhibit the enzyme. The major disadvantages of all peptide inhibitors are that they are generally poorly absorbed and often cleaved by other proteases. Substrate and product sequences are the basis of peptide inhibitors. The screening of a totally random combinatorial library of tetrapeptides: Ac-Tyr-Tbg-Asn-Gly-NH, Ac-Tyr-Tbg-Asn-Gly-OH, Ac-Val-Val-Asn-Gly-NH$_2$, and Ac-Val-Val-Asn-Gly-OH made from both natural and unnatural amino acids were found to be competitive inhibitors in the micromolar range of potency. Smaller peptides interacting with only three or four subsites on the enzyme may form the basis for further inhibitor development. The problems involved with the peptidomimetic inhibitors is that they display poor cell penetration and are easily subjected to degradation in cell culture (LaFemina et al., 1996; Ogilvie et al., 1997). These facts suggest that the low molecular weight inhibitors are more suitable than the peptidomimetic inhibitors. A peptidomimetic inhibitor is a small protein-like chain designed to mimic a peptide. It typically arises either from modification of an existing peptide, or by designing a similar system that mimics a peptide.

Since the determination of crystal structures of herpes virus proteases bound with substrate or inhibitors, the design of potent inhibitors has been tremendously facilitated. Advances in, in silico methodologies have further facilitated the design and development of drugs. In cases where the 3D structure of the target is not available, homology modeling could be used as a tool to predict the 3D structure of the protein/enzyme before conducting the docking evaluation. Other parameters, such as druglikeness and drugscore calculations, may also be used to rationalize how the physicochemical properties may influence in vivo studies (Zuegg and Cooper, 2012). In addition, 2D QSAR studies have been of great use for predictive and diagnostic aspects of drug design (Gupta, 2011).

Thus on a series of benzoxazinones (**22**) studied by Jarvest et al. (1996), the SAR, comparative modeling, molecular docking, and in silico pharmacokinetic studies were performed by Mello et al. (2015) in order to find the ways to have more potent derivatives.

A comparison of the two groups, oxybenzoxazinones (X = O) and amino-benzoxazinones (X = NH), had shown better antiviral activity for the former, since the biological activity could be related to electronegativity and hydrogen bond acceptor nature of the atom, and in this respect the oxygen is better than nitrogen. The SAR analysis of the series suggested that bulky, lipophilic and aliphatic substitutents at $R^1$-position would decrease or even abolish the activity, while the high electronic density would favor the activity. The hydrophobic and bulky substituents at $R^2$-position were found to be favorable to the activity. Hydrophobic and polarizable groups located at $R^3$-position, as well as $R^4$-position seemed to be good for the inhibitory activity, and thus the occurrence of hydrophobic interactions or dispersion interactions should be considered for these compounds.

**22** (Benzoxazinones)

In comparative molecular modeling study, Mello et al. (2015) used the crystal structure of HSV-2 with PDB Id as 1AT3 as potential template as a result of a BLAST search and constructed two 3D structures of HSV-1 using SWISS-MODEL and Modeller, where the best model was found to be the one obtained with the Modeller. Both models presented quite separated catalytic residues, allowing the entry and cleavage of a polypeptide substrate at the active site.

Docking study was performed for all the compounds in the active site of the HSV-1 protease and their interactions analyzed. The most active compound (**23**) of the series was found to be involved in four hydrogen bonds with Thr132, Ser129,and Arg157 as shown in Fig. 13.3. Residues Arg157 and Ser129 were found to play important role in the activity of HSV-1 protease (Hemmateenejad and Elyasi, 2009; Waxman and Darke, 2000). Ser129 is a member of catalytic triad and a nucleophile when the enzyme is activated. The formation of the hydrogen bond between **23** and Ser129 seemed to be one of the most important

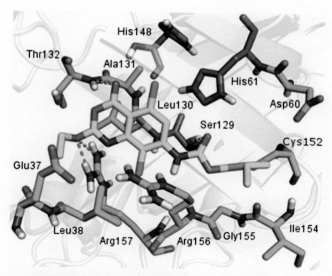

**FIGURE 13.3** **The binding of compound 23 with the active site of HSV-1 protease.** *(Source: Reproduced from Mello, J.F.R., Botelho, N.C., de Souz, A.M.T., de Oliveira, R., de Brito, M.A., de Abrahim-Vieira, B., Sodero, AC.R., Castro, H.C., Cabral, L.M., Miceli, L.A., Rodrigues. 2015. Computational studies of benzoxazinone derivatives as antiviral agents against herpes virus type 1 protease. Molecules 20, 10689–10704.)*

interactions to inhibit HSV-1 protease and this compound was the only member in the entire group that could interact with Ser129. Compound **23** also had hydrophobic interactions with Leu38, Leu130, Ala131, Cys152, Ala153, and Ile154 residues (Fig. 13.4) that further stabilized the inhibitor in the enzyme.

**23**

Docking of other compounds exhibited that bulky substituents at $R^2$-position or complete absence of any substituent at this position had detrimental effect on the activity, but the presence of chlorine atom at $R^3$-positions and specially at $R^4$-position increased the activity.

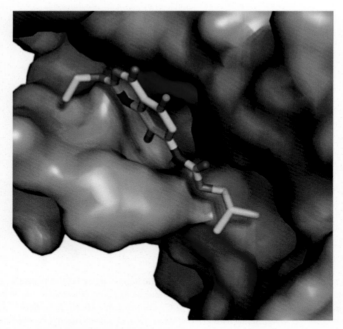

**FIGURE 13.4** **Hydrophobic interactions of compound 23 with Leu38, Leu130, Ala131, Cys152, Ala153, and Ile154 residues (*orange*).** *(Source: Reproduced from Mello, J.F.R., Botelho, N.C., de Souz, A.M.T., de Oliveira, R., de Brito, M.A., de Abrahim-Vieira, B., Sodero, AC.R., Castro, H.C., Cabral, L.M., Miceli, L.A., Rodrigues. 2015. Computational studies of benzoxazinone derivatives as antiviral agents against herpes virus type 1 protease. Molecules 20, 10689–10704.)*

The formation of a hydrogen bond of benzoxazinone **23** with Ser129 in the active site of HSV-1 was unique and this supported the hypothesis that this interaction could facilitate the approximation of the nucleophile Ser129 to the carbonyl. The proximity of the atoms consequently could lead to the reaction for covalent bond formation, which would be in accordance with the proposed mechanism of action (Jarvest et al., 1996). To evaluate this hypothesis, Mello et al. (2015) also performed the covalent docking between benzoxazinone **23** and HSV-1 protease. As expected, the results showed the presence of the covalent bond between the Ser129 side chain and the benzoxazinone (Fig. 13.5). Other interactions were also observed as a cation-π and a hydrogen bond with Arg156 with a distance of 3.1 Å (Fig. 13.5) (Hemmateenejad and Elyasi, 2009; Waxman and Darke, 2000). Although compound **23** was the most active compound, the best druglikeness and drugscore results were achieved for another compound, **24,** that was even better than the commercial drugs. The druglikeness evaluation revealed a positive value only for this compound in addition to the higher drugscore for it.

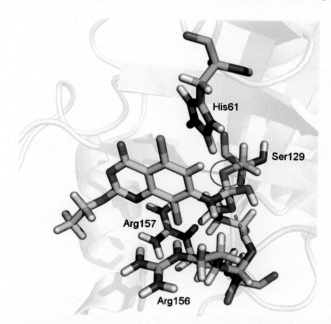

**24**

A symmetrically substituted disulfide compound, CL13933 (**25**), was identified as a potent inhibitor of HCMV UL80 protease by Baum et al. (1996b). At high concentration, this compound formed a covalent adduct with the protease on

**FIGURE 13.5** **Covalent docking between benzoxazinone 23 and HSV-1 protease, showing the presence of a covalent bond between Ser129 and the R²-substituent of the ligand.** *(Source: Reproduced from Mello, J.F.R., Botelho, N.C., de Souz, A.M.T., de Oliveira, R., de Brito, M.A., de Abrahim-Vieira, B., Sodero, AC.R., Castro, H.C., Cabral, L.M., Miceli, L.A., Rodrigues. 2015. Computational studies of benzoxazinone derivatives as antiviral agents against herpes virus type 1 protease. Molecules 20, 10689–10704.)*

Cys residues and at lower concentration it induced-specific intramolecular disulfide bond formation between Cys84 and Cys87, and between Cys138 and Cys161. It was suggested that disulfide bond formation between Cys138 and Cys161 was responsible for inhibition. In a subsequent communication, Baum et al. (1996a) showed that among the most potent inhibitors of HCMV protease identified by random screening of a chemical library was 1,4-dihydro-7,8-dimethyl 6H-pyrimido[1,2-*b*]-1,2,4,5-tetrazin-6-one (**26**) (PTH2). However, the actual inhibitory species was not PTH2, but its oxidized form PT (**27**), present in its solutions. PT inactivates the protease irreversibly and its activity was also ascribed to the formation of a disulfide bond between Cys138 and Cys161 that does not readily occur by air oxidation but is specifically induced by the interaction of the inhibitor with the protease.

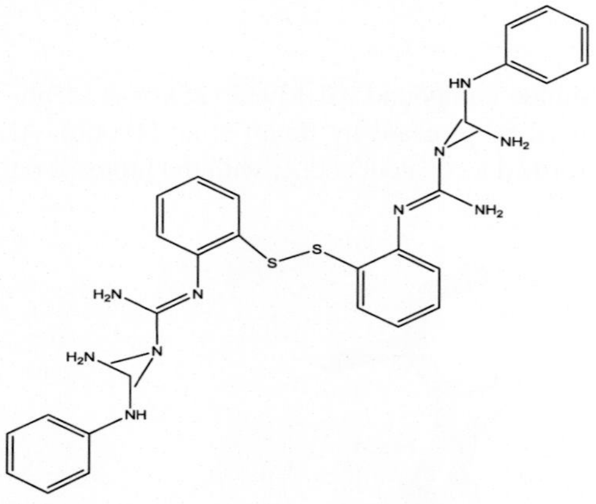

**25** (CL 13933)

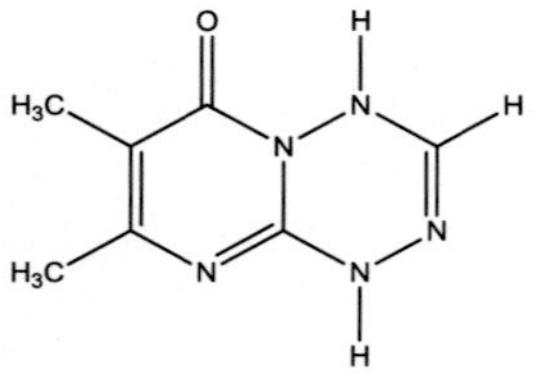

**26** (PTH$_2$)

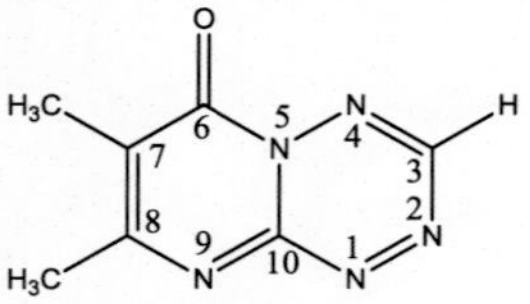

**27** (PT)

Borthwick et al. (2000) designed a novel class of mechanism-based inhibitors of HCMV protease based on the α-methylpyrrolidine-5,5-*trans*-lactam template (**28**) incorporating the natural substrate requirements of the consensus sequence of HCMV protease, where the SAR had defined the relative stereochemistry of the methyl substituent adjacent to the lactam carbonyl and the trend in the activity of the substituents on the lactam nitrogen as acyl > ester > sulfonyl > carbamate against HVMV δAla protease. It was finally concluded that α-methyl group at the 6-position and a carbonyl function on the nitrogen at the 2-position in the novel pyrrolidine-5,5-*trans*-lactam inhibitors, for example, **29** and **30,** were optimal for maximum potency against HCMV protease.

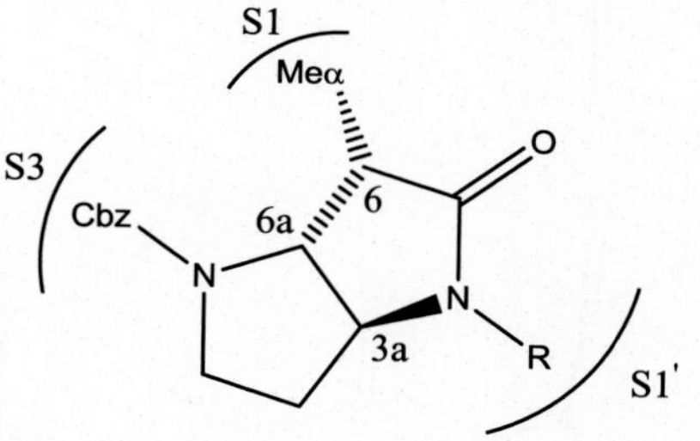

**28**

**29,** R = COMe; **30,** R = COCH₂OAc

In their next communication, Borthwick et al. (2002) had found that optimization of the substituent on the lactum nitrogen had indicated that acivity against HCMV δAla protease was in the order CO-cyclopropyl > COMe > CO₂Me > SO₂Me > CONHMe and the optimization of the functionality on the pyrrolidine nitrogen had given the highly potent dansyl-(S)-proline derivatives **31** and **32**, having $K_i$ values in the nanomolar range against HCMV Ala protease. These compounds were highly selective over the mammalian enzymes elastase, thrombin, and acetylcholine esterase.

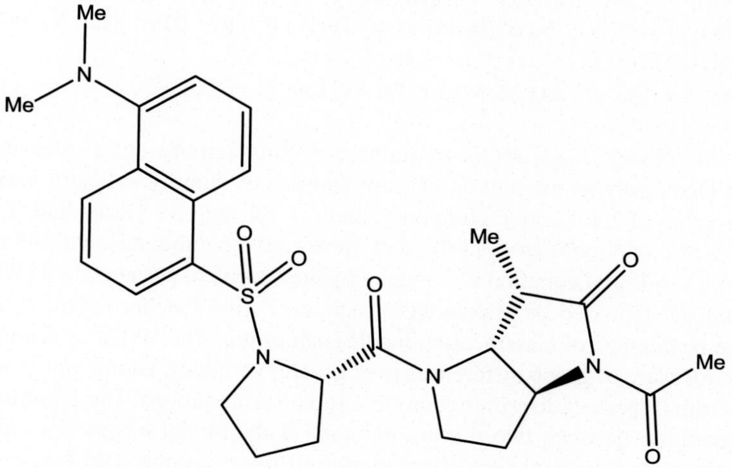

**31**

**32**

Speck-Planche and Cordeiro (2011) analyzed a very large database of herpes virus inhibitors acting against multiple targets. For a set of 1176 compounds, these authors obtained a discriminant model as

$$A_{HHV} = 1.11 \times 10^{-11}\ {}^{(Dip)}_{15} - 4.87 \times 10^{-10}\ {}^{(Ab-R2)}_{15} + 2.62 \times 10^{-3} PSseq3 - 4.59 \times 10^{-3} (Vseq2)$$
$$+ 0.34(OHp) + 2.34(ArCOOR) - 0.73(RCOOH) + 0.52(ArCONHR) - 0.62(RNR_2)$$
$$+ 2.37(SO_2) + 1.82(PO_4) + 1.37(Pyrld) + 1.40(TPH) + 0.26(C - 025) - 0.61(N - 067)$$
$$+ 1.10(N - 071) + 0.31$$
$$n = 1176,\ \lambda = 0.39,\ D^2 = 11.31,\ F(16,1159) = 111.64,\ P < 0.001 \tag{13.1}$$

In this model, $A_{HHV}$ refers to antiherpes virus activity due to the inhibition of DNA-polymerase and/or protease and acts as dummy indicator variable with a value of 1 for active compound and $-1$ for inactive compound. In the total set of compounds ($n = 1176$), 191 were active compounds and 985 inactive. The validity of equation was judged by the statistical parameters $\lambda$ (Wilks' lambda), $D^2$ (squared of Mahanobis distance), $F$ (the Fischer ratio), $P$ level, and the percentage of classification inside each group. The Wilks' $\lambda$ is a multivariate measure of group differences over several variables, taking values in the range from 0 (perfect discrimination) to 1 (no discrimination). The $D^2$ indicates the separation between two groups, showing if the model has an appropriate discriminatory power for the differentiation of those groups. The $F$ is $F$-ratio between the variances of calculated and observed activities. From all the parameters, Eq. (13.1) was found to represent a significant model. The sensitivity of the model in the prediction series was found to be 85.94% and the specificity 93.90%, for an accuracy of 92.60%.

Among the structural parameters used, $\mu_{15}^{(Dip)}$ was the spectral moments of order 15 weighted by the dipole moment and $\mu_{15}^{(Ab\text{-}R2)}$ was the spectral moments of order 15 weighted by the Abraham's molar refractivity. PSseq3 represented the polar surface of the sequence 3 (amino acids 121–180) of the proteins. The descriptor Vseq2 expressed the information related to the volume of the sequence 2 (amino acids 61–120) of the proteins. OHp represented the numbers of fragments containing primary alcohol OH groups. ArCOOR meant the number of aromatic esters. RCOOH represented the number of aliphatic carboxylic acids. ArCONHR was the number of N-substituted aromatic amides. $RNR_2$ referred to the number of fragments containing tertiary aliphatic amines and $SO_2$ represented the number of fragments containing sulfone groups. $PO_4$ stood for the number of phosphate groups and Pyrld represented the number of pyrrolidine fragments. TPH referred to the number of thiophene fragments and C-025 meant the number of fragments where an aromatic carbon atom was attached to an aliphatic group. N-067 was the number of fragments containing secondary aliphatic amines and N-071 represented the number of tertiary aromatic amines, in which nitrogen atom was attached to one aromatic ring and with two aliphatic groups. In this model, the variables with positive coefficient had the positive contribution to the activity and those with negative coefficient the negative contribution. Thus Speck-Planche and Cordeiro (2011) suggested that to increase the activity of the compounds the polarity of the bonds in the molecules should be increased and the size of the molecules reduced. The reduction of the size of the molecules will permit a better fit inside the cavity of the proteins. Further, the increase in the polar surface area in the sequence 3 of the protein was also suggested to be beneficial to the activity because then the molecule will be able to interact with amino acids of this sequence by polar interactions. Eq. (13.1) also suggested that the presence of aromatic esters, N-substituted aromatic amides, fragments containing sulfone groups, phosphate groups, pyrrolidine fragments, thiophene fragments, fragments where an aromatic carbon atom was attached to an aliphatic group, and tertiary aromatic amines in which nitrogen atom was attached to one aromatic ring and with two aliphatic groups will be beneficial to the activity, and the increase in the number of such functionalities will further increase the activity.

From this model, Speck-Planche and Cordeiro (2011) could assess the quantitative contributions of any fragment to the antiherpetic activity through the inhibition of DNA-polymerase and/or protease proteins. They selected 12 fragments for the calculation of the quantitative contributions (Fig. 13.6) which provided useful information about the molecular patterns that can be decisive for the development of the antiherpes agents, DNA-polymerase or protease inhibitors. The quantitative contributions of 12 fragments in inhibitors acting against three out of five targets studied were reported by these authors (Table 13.2).

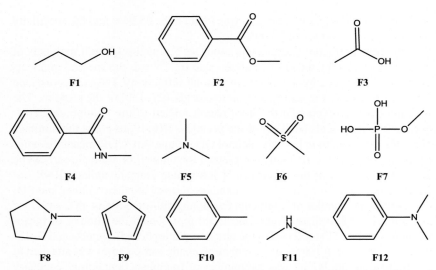

FIGURE 13.6 **The fragments important for the activity of herpes virus inhibitors selected by Speck-Planche and Cordeiro (2011).**

TABLE 13.2 Activity Contribution of Various Fragments in the Inhibitors Acting Against Various Herpesvirus Targets (Speck-Planche and Cordeiro, 2011)

| Fragment ID | HSV-1/DNA polymerase | CMV/CP P40[a] | EBV/PCS protein[b] |
|---|---|---|---|
| F1 | −0.82 | −0.14 | −0.49 |
| F2 | 1.44 | 2.13 | 1.77 |
| F3 | −1.89 | −1.20 | −1.56 |
| F4 | −0.38 | 0.31 | −0.05 |
| F5 | −1.78 | −1.10 | −1.45 |
| F6 | 1.82 | 2.51 | 2.16 |
| F7 | 0.67 | 1.35 | 1.00 |
| F8 | −0.41 | 0.27 | −0.08 |
| F9 | 0.24 | 0.92 | 0.57 |
| F10 | −0.90 | −0.22 | −0.57 |
| F11 | −1.77 | 1.09 | −1.44 |
| F12 | −0.07 | 0.62 | 0.27 |

CMV, Cytomegalovirus; EBV, Epstein–Barr virus; HSV, herpes simplex virus; PCS, protease capsid scaffold.
[a]*CP P40.*
[b]*Protease capsid scaffold protein.*

# 5 CONCLUSIONS

Herpes viruses represent one of the most prevalent viral families worldwide. There are eight known types of HHVs (HHV-1–HHV-8), which lead to variety of diseases ranging from relatively harmless simple ones to life threatening ones. The finding that all herpes viruses encode a unique protease that is essential for viral replication has afforded a potential new therapeutic target. Herpes virus proteases are not related to any other enzymes and, therefore, they could not be readily identified by homology searches in DNA or protein databases. Herpes virus proteases are serine proteases with an active site triad Ser-His-His in which serine was found to play an active role. An attractive feature of the herpes virus protease is the well-defined substrate specificity. High substrate specificity suggests that it should be possible to eventually develop a peptidomimetic inhibitor. Dimerization is a unique feature of herpes virus proteases, which plays an essential role in inhibitor discovery. In herpes virus replication, the protease plays the role in DNA packaging, the last step in viral replication. Characterization of sequence specificity is also important for the design of peptide-based protease inhibitors. The application of X-ray crystallography offered useful structural information to the design and optimization of proteases inhibitors and led to the development of several mechanism-based peptide and heterocyclic inhibitors, reversible or irreversible. The problems involved with the peptidomimetic inhibitors is that they display poor cell penetration and are easily subjected to degradation in cell culture. These facts suggested that the low molecular weight inhibitors are more suitable than the peptidomimetic inhibitors.

# REFERENCES

Arroyo, E., Chinchilla, N., Molinillo, J.M., Macias, F.A., Astola, A., Ortiz, M., Valdivia, M.M., 2010. Aneugenic effects of benzoxazinones in cultured human cells. Mutagen. Res. 695, 81–86, 2010.

Baron, S., 1996. Medical Microbiology, fourth ed. University of Texas Medical Branch, Galveston, TXPN.

Batra, R., Khayat, R., Tong, L., 2001. Molecular mechanism for dimerization to regulate the catalytic activity of human cytomegalovirus protease. Nat. Struct. Biol. 8, 810–817.

Baum, E.Z., Ding, W.-D., Siegel, M.M., Hulmes, J., Bebernitz, G.A., Sridharan, L., Tabei, K., Krishnamurthy, G., Carofiglio, T., Groves, J.T., Bloom, J.D., DiGrandi, M., Bradley, M., Ellestad, G., Seddon, A.P., Gluzman, Y., 1996a. Flavins inhibit cytomegalovirus protease. UL80 protease via disulfide bond formation. Biochemistry 35, 5847–5855.

Baum, E.Z., Siegel, M.M., Bebernitz, G.A., Hulmes, J.D., Sridharan, L., Sun, L., Tabei, K., Johnston, S.H., Wildey, M.J., Nygaard, J., Jones, T.R., Gluzman, Y., 1996b. Inhibition of human cytomegalovirus UL80 protease by specific intramolecular disulfide bond formation. Biochemistry 35, 5838–5846.

Bonneau, P.R., Grand-Maitre, C., Greenwood, D.J., Lagace, L., La Plante, S.R., Massariol, M.-J., Ogilvie, W.W., O'Meara, J.A., Kawai, S.H., 1997. Evidence of a conformational change in the human cytomegalovirus protease upon binding of peptidyl-activated carbonyl inhibitors. Biochemistry 36, 12644–12652.

Borthwick, A.D., 2005. Design of translactam HCMV protease inhibitors as potent antivirals. Med. Res. Rev. 25 (4), 427–452.

Borthwick, A.D., Weingarten, G., Haley, T.M., Tomaszewski, M., Wang, W., Hu, Z., Bedard, J., Jin, H., Youen, L., Mansour, T.S., 1998. Design and synthesis of monocyclic β-lactams as mechanism-based inhibitors of human cytomegalovirus protease. Bioorg. Med. Chem. Lett. 8, 365–370.

Borthwick, A.D., Angier, S.J., Crame, A.J., Exall, A.M., Halley, T.M., Hart, G.J., Mason, A.M., Pennell, A.M.K., Weingarten, G.G., 2000. Design and synthesis of pyrrolidine 5,5-*trans*-lactam (5-oxo-hexahydro-pyrrolo[3,2*b*]pyrroles) as novel mechanism-based inhibitors of human cytomegalovirus protease. 1. The α-methyl-*trans*-lactam template. J. Med. Chemi. 43, 4452–4464.

Borthwick, A.D., Crame, A.J., Ertl, P.F., 2002. Design and synthesis of pyrrolidine-5,5-*trans*-lactams (5-oxohexahydropyrrolo[3,2-*b*]pyrroles) as novel mechanism-based inhibitors of human cytomegalovirus protease. 2. Potency and chirality. J. Med. Chem. 45, 1–18.

Castro, A., Martinez, A., 2000. Novel agents for the treatment of human cytomegalovirus infection. Expert. Opin.Ther. Pat. 10, 165–177.

Chen, P., Tsuge, H., Almassy, R.J., Gribskov, C.L., Katoh, S., Vanderpooi, D.L., Margosiak, S.A., Pinko, C., Matthews, D.A., Kan, C.-C., 1996. Structure of the human cytomegalovirus protease catalytic domain reveals a novel serine protease fold and catalytic triad. Cell 86, 835–843.

Chu, M., Mierzwa, R., Truumees, I., King, A., Patel, M., Pichardo, J., Hart, A., Dasmahapatra, B., Das, P.R., Puar, M.S., 1996. Sch 65676: A novel fungal metabolite with inhibitory activity against the cytomegalovirus protease. Tetrahedron Lett. 37, 3943–3946.

Darke, P.L., Kuo, L.C. 1998. Low temperature assay for active HCMV protease in dimeric form. US Patent No. 5,891,661.

Darke, P.L., Chen, E., Hall, D.L., Sardana, M.K., Veloski, C.A., Lafemina, R.L., Shafer, J.A., Kuo, L.C., 1994. Purification of active herpes simplex virus 1 protease expressed in *Escherichia coli*. J. Biol. Chem. 269, 18708–18711.

Dhanak, D., Keenan, R.M., Burton, G., Kaura, A., Darcy, M.G., Shah, D.H., Ridgers, L.H., Breen, A., Lavery, P., Tew, D.G., West, A., 1998. Benzothiopyran-4-one based reversible inhibitors of the human cytomegalovirus (HCMV) protease. Bioorg. Med. Chem. Lett. 8, 3677–3682.

Di Grandi, M.J., Curran, K.J., Baum, E.Z., Bebernitz, G., Ellestad, G.A., Ding, W.D., Lang, S.A., Rossi, M., Bloom, J.D., 2003. Pyrimido[1,2-*b*]-1,2,4,5-tetrazin-6-ones as HCMV protease inhibitors: a new class of heterocycles with flavin-like redox properties. Bioorg. Med. Chem. Lett. 13, 3483–3486.

DiIanni, C.L., Mapelli, C., Drier, D.A., Tsao, J., Natarajan, S., Riexinger, D., Festin, S.M., Bolgar, M., Yamanaka, G., Weinheimer, S.P., Meyers, C.A., Colonno, R.J., Cordingley, M.G., 1993. In vitro activity of the herpes simplex virus type 1 protease with peptide substrates. J. Biol. Chem. 268, 25449–25454.

Dunn, B.M., Qiu, X.Y., Abdelmeguid, S.S., 1999. Human herpes proteases. In: Dunn, B.M. (Ed.), Proteases of Infectious Agents. Academic Press, San Diego, pp. 93–115.

Ertl, P., Russel, L., Angier, J., 2000. Herpesvirus protease assays. In: Kinchington, D. (Ed.), Antiviral Methods and Protocols. Springer.

Flynn, D.L., Becker, D.P., Dilworth, V.M., Highkin, M.K., Hippenmeyer, P.J., Houseman, K.A., Levine, L.M., Li, M., Moormann, A., Rankin, E., Toth, A., Villamil, M.V., Wittwer, A.J., Holwerda, B.C., 1997. The herpesvirus protease: mechanistic studies and discovery of inhibitors of the human cytomegalovirus protease. Drug Des. Discov. 15, 3–15.

Gable, J.E., Lee, G.M., Jaishankar, P., Hearn, B.R., Waddling, C.A., Renslo, A.R., Craik, C.S., 2014. Broad-spectrum allosteric inhibition of herpesvirus proteases. Biochemistry 53, 4648–4660.

Gao, M., Matusick-Kumar, L., Hurlburt, W., DiTusa, S.F., Newcomb, W.W., Brown, J.C., McCann, III, P.J., Deckman, I., Colonno, R.J., 1994. The protease of herpes simplex virus type 1 is essential for functional capsid formation and viral growth. J. Virol. 68, 3702–3712.

Gibson, W., 1996. Structure and assembly of the virion. Intervirology 39, 389–400.

Gibson, W., Hall, M.R.T., 1997. Assemblin, an essential herpesvirus proteinase. Drug Des. Discov. 15, 39–47.

Gibson, W., Welch, A.R., Hall, W.R.T., 1995. Assembling a herpesvirus serine maturational proteinase and a new molecular target for antivirals. Perspect. Drug Discov. Des. 2, 413–416.

Godefroy, S., Guenet, C., 1995. Autoprocessing of HSV-1 protease: effect of deletions on autoproteolysis. FEBS Lett. 357, 168–172.

Gupta, S.P., 2011. QSAR and Molecular Modeling. Springer, Basel/Anamaya.

Hemmateenejad, B., Elyasi, M.A., 2009. Segmented principal component analysis-regression approach to quantitative structure–activity relationship modeling. Anal. Chim. Acta 646, 30–38.

Holwerda, B.C., 1997. Herpesvirus proteases: targets for novel antiviral drugs. Antivir. Res. 35, 1–21.

Hoog, S.S., Smith, W.W., Qiu, X., Janson, C.A., Hellmig, B., McQueney, M.S., O'Donnell, K., O'Shannessy, D., DiLella, A.G., Debouck, C., Abdel-Meguid, S.S., 1997. Active site cavity of herpesvirus proteases revealed by the crystal structure of herpes simplex virus protease/inhibitor complex. Biochemistry 36, 14023–14029.

Jarvest, R.L., Dabrowski, C.E., 2002. Herpes virus and cytomegalovirus proteinase. In: Smith, H.J., Simons, C. (Eds.), Proteinase and Peptidase Inhibition—Recent Potential Targets for Drug Development. Taylor & Francis, London, New York, pp. 264–281.

Jarvest, R.L., Parratt, M.J., Debouck, C.M., Gorniak, J.G., Jennings, L.J., Serafinowska, H.T., Strickler, J.E., 1996. Inhibition of HSV-1 protease by benzoxazinones. Bioorg. Med. Chem. Lett. 6, 2463–2466.

Jarvest, R.L., Pinto, I.L., Ashman, S.M., Dabrowski, C.E., Fernandez, A.V., Jennings, L.J., Lavery, P., Tew, D.G., 1999. Iinhibition of herpes proteases and antiviral activity of 2-substituted thieno[2,3-d]oxazinones. Bioorg. Med. Chem. Lett. 9, 443–448.

Jiang, X.H., Gong, H., Chen, Y.C., Vu, G.P., Trang, P., Zhang, C.Y., Lu, S.W., Liu, F.Y., 2012. Effective inhibition of cytomegalovirus infection by external guide sequences in mice. Proc. Natl. Acad. Sci. USA 109, 13070–13075.

Jones, T.R., Sun, L., Bebernitz, G.A., Muzithras, V.P., Kim, H.J., Johnston, S.H., Baum, E.Z., 1994. Proteolytic activity of human cytomegalovirus UL80 protease cleavage site mutants. J. Virol. 68, 3742–3752.

LaFemina, R.L., Bakshi, K., Long, W.J., Pramanik, B., Veloski, C.A., Wolanski, B.S., Marcy, A.I., Hazuda, D.J., 1996. Characterization of a soluble stable human cytomegalovirus protease and inhibition by M-site peptide mimics. J. Virol. 70, 4819–4824.

LaPlante, S.R., Bonneau, P.R., Aubry, N., Cameron, D.R., Deziel, R., Grand-Maitre, E., Plouffe, C., Tong, L., Kawai, S.H., 1999. Characterization of the human cytomegalovirus protease as an induced-fit serine protease and the implications to the design of mechanism-based inhibitors. J. Am. Chem. Soc. 121, 2974–2986.

Lazic, A., Goetz, D.H., Nomura, A.M., Marnett, A.B., Craik, C.S., 2007. Substrate modulation of enzyme activity in the herpesvirus protease family. J. Mol. Biol. 373, 913–923.

Lee, G.M., Craik, C.S., 2009. Trapping moving targets with small molecules. Science 324, 213–215.

Lee, G.M., Shahian, T., Baharuddin, A., Gable, J.E., Craik, C.S., 2011. Enzyme inhibition by allosteric capture of an inactive conformation. J. Mol. Biol. 411, 999–1016.

Liang, P.-H., Brun, K.A., Field, J.A., O'Donnell, K., Doyle, M.L., Green, S.M., Baker, A.E., Blackburn, M.N., Abdel-Meguid, S.S., 1998. Site-directed mutagenesis probing the catalytic role of arginines 165 and 166 of human cytomegalovirus protease. Biochemistry 37, 5923–5929.

Liu, F., Roizman, B., 1991. The herpes simplex virus 1 gene encoding a protease also contains within its coding domain the gene encoding the more abundant substrate. J. Virol. 65, 5149–5156.

Marnett, A.B., Nomura, A.M., Shimba, N., Ortiz de Montellano, P.R., Craik, C.S., 2004. Communication between the active sites and dimer interface of a herpesvirus protease revealed by a transition state inhibitor. Proc. Natl. Acad. Sci. USA 101, 6870–6875.

Martinez, A., Castro, A., Gil, C., Perez, C., 2001. Recent strategies in the development of new human cytomegalovirus inhibitors. Med. Res. Rev. 21, 227–244.

Matsumoto, M., Misawa, S., Chiba, N., Takaku, H., Hayashi, H., 2001. Selective nonpeptidic inhibitors of herpes simplex virus type 1 and human cytomegalovirus proteases. Biol. Pharm. Bull. 24, 236–241.

Matusick-Kumar, III, L., Robertson, P.J.M., Newcomb, B.J., Brown, J.C., Gao, M., 1995. Release of the catalytic domain NO from the herpes simplex virus type 1 protease is required for viral growth. J. Virol. 69, 7113–7121.

McCann, P.J., O'Boyle, II, D.R., Deckman, I.C., 1994. Investigation of the specificity of the herpes simplex virus 1 protease by point mutagenesis of the autoproteolysis sites. J. Virol. 68, 526–529.

Mello, J.F.R., Botelho, N.C., de Souz, A.M.T., de Oliveira, R., de Brito, M.A., de Abrahim-Vieira, B., Sodero, A.C.R., Castro, H.C., Cabral, L.M., Miceli, L.A., Rodrigues, C.R., 2015. Computational studies of benzoxazinone derivatives as antiviral agents against herpes virus type 1 protease. Molecules 20, 10689–10704.

Nomura, A.M., Marnett, A.B., Shimba, N., Dotsch, V., Craik, C.S., 2005. Induced structure of a helical switch as a mechanism to regulate enzymatic activity. Nat. Struct. Mol. Biol. 12, 1019–1020.

Ogilvie, W., Bailey, M., Poupart, M.-A., Abraham, A., Bhavsar, A., Bonneau, P., Bordeleau, J., Bousquet, Y., Chabot, C., Duceppe, J.-S., Fazal, G., Goulet, S., Grand-Maitre, C., Guse, I., Halmos, T., Lavallee, P., Leach, M., Malenfant, E., O'Meara, J., Plante, R., Plouffe, C., Poirier, M., Soucy, F., Yoakim, C., Deziel, R., 1997. Peptidomimetic inhibitors of the human cytomegalovirus protease. J. Med. Chem. 40, 4113–4135.

Pinto, I.L., West, A., Debouck, C.M., DiLella, A.G., Gorniak, J.G., O'Donnell, K.C., O'Shannessy, D.J., Patel, A., Jarvest, R.L., 1996. Novel, selective mechanism-based inhibitors of the herpes proteases. Bioorg. Med. Chem. Lett. 6, 2467–2472.

Pinto, I.L., Jarvest FR.L., Clarke, B., Dabrowskib, C.E., Fenwick, A., Gorczyca, M.M., Jennings, U.J., Lavery, P., Sternberg, E.J., Tew, D.G., West, A., 1999. Inhibition of human cytomegalovirus protease by enedione derivatives of thieno[2,3-d]oxazinones through a novel dual acylation/alkylation mechanism. Bioorg. Med. Chem. Lett. 9, 449–452.

Qiu, X., Culp, J.S., DiLella, A.G., Hellmig, B., Hoog, S.S., Janson, C.A., Smith, W.W., Abdel-Meguid, S.S., 1996. Unique fold and active site in cytomegalovirus protease. Nature 383, 275–279.

Qiu, X., Janson, C.A., Culp, J.S., Richardson, S.B., Debouck, C., Smith, W.W., Abdel-Meguid, S.S., 1997. Crystal structure of varicella-zoster virus protease. Proc. Natl. Acad. Sci. USA 94, 2874–2879.

Register, R.B., Shafer, J.A., 1997. Alteration in catalytic activity and virus maturation produced by mutation of the conserved histidine residues of herpes simplex virus type 1 protease. J. Virol. 71, 8572–8581.

Robertson, B.J., McCann, III, P.J., Matusick-Kumar, L., Newcomb, W.W., Brown, J.C., Colonno, R.J., Gao, M., 1996. Separate functional domains of the herpes simplex virus type 1 protease; evidence for cleavage inside capsids. J. Virol. 70, 4317–4328.

Ryan, K.J, Ray, C.G, 2004. Sherris Medical Microbiology, fourth ed. McGraw Hill, Arizona.

Sardana, V.V., Wolfgang, J.A., Veloski, C.A., Long, W.J., LeGrow, K., Wolanski, B., Emini, E.A., LaFemina, R.L., 1994. Peptide substrate cleavage specificity of the human cytomegalovirus protease. J. Biol. Chem. 269, 14337–14340.

Shahian, T., Lee, G., Lazic, A., Arnold, A., Velusamy, P., Roels, C.M., Guy, R.K., Craik, C.S., 2009. Inhibition of a viral enzyme by a small molecule dimer disruptor. Nat. Chem. Biol. 9, 640–646.

Shieh, H.S., Kurumbail, R.G., Stevens, A.M., 1996. Three dimensional structure of human cytomegalovirus protease. Nature 383, 279–282.

Shimba, N., Nomura, A.M., Marnett, A.B., Craik, C.S., 2004. Herpesvirus protease inhibition by dimer disruption. J. Virol. 78, 6657–6665.

Speck-Planche, A., Cordeiro, M.N.D.S., 2011. Application of bioinformatics for the search of novel antiviral therapies: rational design of antiherpes agents. Curr. Bioinform. 6, 81–93.

Sundberg, E.J., Mariuzza, R.A., 2000. Luxury accommodations: the expanding role of structural plasticity in protein-protein interactions. Structure 8, R137–R142.

Teague, S.J., 2003. Implications of protein flexibility for drug discovery. Nat. Rev. Drug Discov. 2 (7), 527–541.

Tigue, N.J., Kay, J., 1998. Autoprocessing and peptide substrates for human herpesvirus 6 proteinase. J. Biol. Chem. 273, 26441–26446.

Tong, L., 2002. Viral proteases. Chem. Rev. 102, 4609–4626.

Tong, L., Qian, C., Massariol, M.J., Bonneau, P.R., Cordingley, M.G., Lagace, L., 1996. A new serine-protease fold revealed by the crystal structure of human cytomegalovirus protease. Nature 383, 272–275.

Tong, L., Qian, C., Massariol, M.-J., Deziel, R., Yoakim, C., Lagace, L., 1998. Conserved mode of peptidomimetic inhibition and substrate recognition of human cytomegalovirus protease. Nat. Struct. Biol. 5, 819–826.

Waxman, L., Darke, P.L., 2000. The herpesvirus proteases as targets for antiviral chemotherapy. Antivir. Chem. Chemother. 11, 1–22.

Welch, A.R., Woods, A.S., McNally, L.M., Cotter, R.J., Gibson, W., 1991. A herpesvirus maturational proteinase, assemblin: identification of its gene, putative active site domain, and cleavage site. Proc. Natl. Acad. Sci. US.A 88, 10792–10796.

Welch, A.R., McNally, L.M., Hall, M.R.T., Gibson, W., 1993. Herpesvirus proteinase: site-directed mutagenesis used to study maturational, release, and inactivation cleavage sites of precursor and to identify a possible catalytic site serine and histidine. J. Virol. 67, 7360–7372.

Wells, J.A., McClendon, C.L., 2007. Reaching for high hanging fruit in drug discovery at protein–protein interfaces. Nature 450, 1001–1009.

Yoakim, C., Ogilvie, W.W., Cameron, D.R., Chabot, C., Guse, I., Hache, B., Naud, J., O'Meara, J.A., Plante, R., Deziel, R., 1998a. β-Lactam derivatives as inhibitors of human cytomegalovirus protease. J. Med. Chem. 41, 2882–2891.

Yoakim, C., Ogilvie, W.W., Cameron, D.R., Chabot, C., Grand-Maitre, C., Guse, I., Hache, B., Kawai, S., Naud, J., O'Meara, J.A., Plante, R., Deziel, R., 1998b. Potent β-lactam inhibitors of human cytomegalovirus protease. Antivir. Chem. Chemother. 9, 379–387.

Zuegg, J., Cooper, M.A., 2012. Drug-likeness and increased hydrophobicity of commercially available compound libraries for drug screening. Curr. Top. Med. Chemi. 12, 1500–1513.

# Chapter 14

# Delicate Dance in Treating Hepatitis C Infections and Overcoming Resistance

**Dolly A. Parasrampuria**
*Global Clinical Pharmacology, Quantitative Sciences, Janssen Research & Development, Philadelphia, PA, United States*

## 1 INTRODUCTION

Hepatitis C virus (HCV) is a blood borne virus, which was discovered quite recently in 1989 (Choo et al., 1989). It is transmitted through blood transfusions of unscreened blood products, sharing of needles, inadequately sterilized medical equipment, sexual intercourse, or during birth to an infected mother. The incubation period following infections ranges from 2 weeks to 6 months. Most infected patients remain asymptomatic through the acute or chronic phases, or have symptoms of fatigue, abdominal pain, poor appetite, and jaundice. Approximately 15%–20% infected patients will naturally clear the virus from the body without treatment (CDC, 2015). Globally, about 130–150 million people live with chronic hepatitis. Annually, approximately 700,000 deaths are attributed to HCV (WHO, 2016). While a substantial proportion of infections (in 15%–45% of patients) resolve within 6 months without treatment; others develop chronic hepatitis. Out of these, 60%–70% develop chronic liver disease, 5%–20% developing cirrhosis over a 20–30 year period, and 1%–5% develop liver cancer and cirrhosis requiring transplantation (CDC, 2015). In the United States, approximately 2.7–3.9 million suffer from chronic hepatitis C (CDC, 2015).

HCV is genetically diverse with seven genotypes (GT) and greater than 50 subtypes (Fig. 14.1), making therapeutic interventions challenging as each genotype has different susceptibility to each drug (Bukh, 2016; CDC, 2015). The vast majority of infections comprises of GTs-1, -2, and -3. GT-1 with subtypes 1a and 1b is the most common HCV causing infections in the United States. Chronic HCV infection has been attributed to other diseases, such as non-Hodgkin lymphoma, glomerulonephritis, and even diabetes mellitus, which occur more frequently in HCV-infected patients (CDC, 2015).

**Viral Proteases and Their Inhibitors.** http://dx.doi.org/10.1016/B978-0-12-809712-0.00014-9

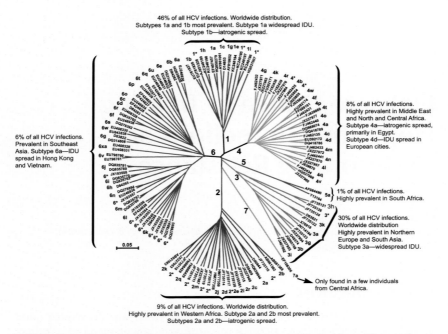

**FIGURE 14.1** **Phylogenetic tree of HCV types and subtypes with prevalence.** *(Source: Reproduced with permission from Bukh, J., 2016. The history of hepatitis C virus (HCV): basic research reveals unique features in phylogeny, evolution and the viral life cycle with new perspectives for epidemic control. J. Hepatol. 65 (1 Suppl), S2–S21. Copyright 2016 Elsevier.)*

Prior to 1998, HCV infections were only treatable with subcutaneous injections of interferon-α (IFN-α), with its associated risks and low cure rate. Later, long acting pegylated versions of interferon became available for use, which improved the cure rate. Subsequently the nonspecific nucleoside analog, ribavirin was added to improve sustained virologic response (SVR). These treatments were associated with significant side effects, inconvenience as IFN-α needs to be injected weekly for 48 weeks, and a poor response rate of only about 40%–50%. More recently direct-acting antivirals (DAA) have been developed, which when dosed alone or in combination are remarkably effective, with greater than 90% SVR (Jazwinski and Muir, 2011; Stättermayer et al., 2014). These DAA drugs are protease inhibitors that interfere with key steps in the viral replication cycle by inhibiting critical enzymes. Simeprevir, boceprevir, and telaprevir are NS3/NS4A serine protease inhibitors that target the proteolytic processing of HCV polyprotein. These target GT-1a and GT-1b. Daclatasvir is a NS5A inhibitor that disrupts formation of the replication complex and possibly the virion assembly. It has activity against all HCV genotypes. NS5B polymerase inhibitor, sofosbuvir, is a nucleotide analog that interferes with viral RNA replication and has activity against all genotypes. Increasingly,

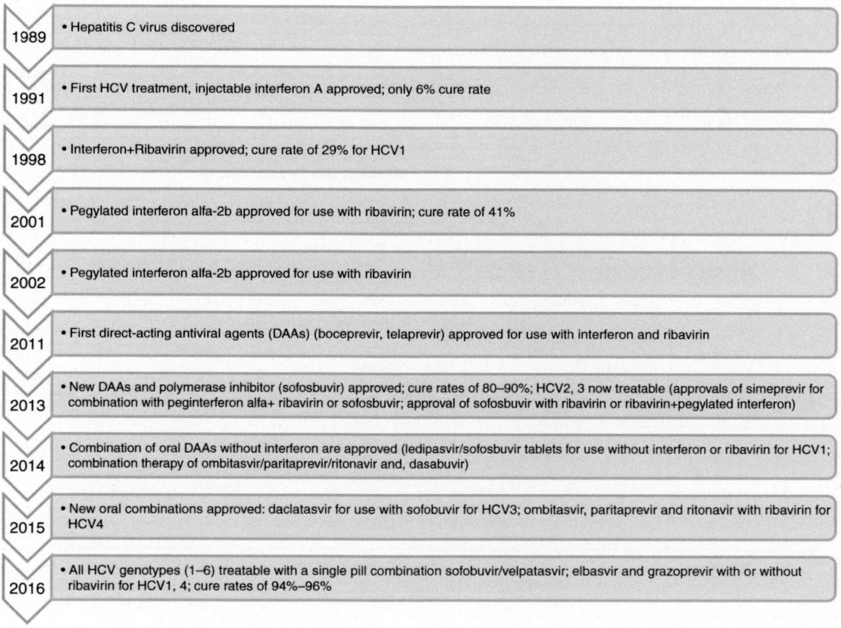

| 1989 | • Hepatitis C virus discovered |
|---|---|
| 1991 | • First HCV treatment, injectable interferon A approved; only 6% cure rate |
| 1998 | • Interferon+Ribavirin approved; cure rate of 29% for HCV1 |
| 2001 | • Pegylated interferon alfa-2b approved for use with ribavirin; cure rate of 41% |
| 2002 | • Pegylated interferon alfa-2b approved for use with ribavirin |
| 2011 | • First direct-acting antiviral agents (DAAs) (boceprevir, telaprevir) approved for use with interferon and ribavirin |
| 2013 | • New DAAs and polymerase inhibitor (sofosbuvir) approved; cure rates of 80–90%; HCV2, 3 now treatable (approvals of simeprevir for combination with peginterferon alfa+ ribavirin or sofosbuvir; approval of sofosbuvir with ribavirin or ribavirin+pegylated interferon) |
| 2014 | • Combination of oral DAAs without interferon are approved (ledipasvir/sofosbuvir tablets for use without interferon or ribavirin for HCV1; combination therapy of ombitasvir/paritaprevir/ritonavir and, dasabuvir) |
| 2015 | • New oral combinations approved: daclatasvir for use with sofosbuvir for HCV3; ombitasvir, paritaprevir and ritonavir with ribavirin for HCV4 |
| 2016 | • All HCV genotypes (1–6) treatable with a single pill combination sofosbuvir/velpatasvir; elbasvir and grazoprevir with or without ribavirin for HCV1, 4; cure rates of 94%–96% |

FIGURE 14.2    Key milestones for HCV drug approvals by the US FDA (Hepatitis Central, 2016; PhRMA, 2014).

HCV inhibitors are used in combination: sofosbuvir/velpatasvir indicated for all genotypes; ledipasvir/sofosbuvir indicated for GT-1; ombitasvir, paritaprevir, and ritonavir for GT-4; ombitasvir/paritaprevir/ritonavir/dasabuvir for GT-1; elbasvir/grazoprevir for GT-1 or 4. A timeline of the discovery of the HCV virus to development of therapies is presented in (Fig. 14.2) (Hepatitis Central, 2016; PhRMA, 2014).

## 2    STRUCTURE AND LIFE CYCLE OF HCV

The HCV virus has proven difficult to culture in vitro and the fact that the virus undergoes continuous evolution has made it difficult to fully understand the replication cycle and develop a vaccine. However, crystal structures of the translated and cleaved proteins have been analyzed in the context of substrate or inhibitor binding.

### 2.1    Structure

The HCV is a positive, single-stranded, linear RNA virus with a diameter of 45–86 nm (de Clercq and Li, 2016). It belongs to the Flaviviridae family, and

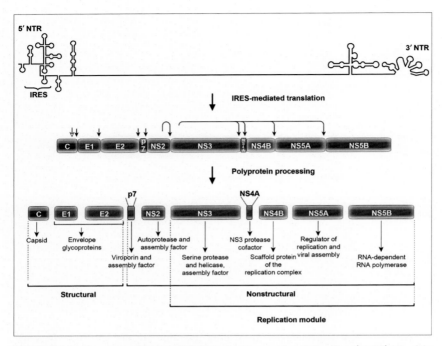

**FIGURE 14.3 Structure of the hepatitis C viral genome and proteins.** The 5′ and 3′ends of the HCV genome consists of loop like structures that form the nontranslated regions (NTR). Following internalization, which is mediated by internal ribosome entry sit (IRES), the hepatitis C virus (HCV) polyprotein is cleaved by viral and host proteases. The *vertical* and *curved arrows* show the cellular signal peptidase mediated cleavage, while the *joined arrows* show the viral protease mediated cleavage. The function of each protein is listed at the bottom of each protein. *(Source: Reproduced with permission from Dubuisson, J., Cosset, F.L., 2014. Virology and cell biology of the hepatitis C virus life cycle: an update. J. Hepatol. 61(1 Suppl), S3-S13. Copyright 2014 Elsevier.)*

shares structural and functional characteristics with other members of this family, such as dengue, chikungunya, and West Nile viruses. The virus has a lipid bilayer envelope surrounding a nucleocapsid that contains the RNA genome of 9.6–12.3 thousand nucleotides (Chevaliez and Pawlotsky, 2006). The HCV genome (Fig. 14.3) is about 9000 nucleotide in size and consists of a single open reading frame (ORF), which encodes a polyprotein of about 3011 residues (Grakoui et al., 1993). The polyprotein is cleaved into three structural and six nonstructural proteins. The nonstructural proteins are generated via viral protease activity, whereas host peptidases are involved in cleaving the nonstructural proteins (Neddermann et al., 1997).

The ORF is flanked by nontranslated regions. The 5′ nontranslated region contains four highly structured loop and knot like domains collectively forming the internal ribosome entry site (IRES). Many domains within IRES region are highly conserved in structure and primary sequence across RNA viruses (Lukavsky, 2009). The IRES commandeers the small ribosomal subunit 40 and the eukaryotic initiation factor 3 (eIF3) to initiate translation. Collier et al. (2002), have shown that the three-dimensional structure in the internal loop is conserved across HCV genotypes and is crucial to the IRES function of initiation of translation, and thus represents a target for antiviral drugs.

HCV forms a membranous web like structure in the cytoplasm of the host, where replication occurs. The membranous web is predominantly derived from host's rough endoplasmic reticulum, but also includes markers of other cellular components, such as early and later endosomes, coat protein vesicles, mitochondria, and liquid droplets (Romero-Brey et al., 2012). The main components of this web are double membrane vesicles, formed primarily through the NS5A viral protein activity. They exist as protrusions from the endoplasmic reticulum and remain attached to it with neck-like structures.

The nonstructural proteins include the envelope glycoproteins E1 (31 kDa) and E2 (70 kDa), and the core capsid protein C (21 kDa). The structural proteins are located near the N-terminal part of the ORF, while the remainder of the ORF encodes nonstructural proteins: NS2 (23 kDa), NS3 (70 kDa), NS4A (8 kDa), NS4B (27 kDa), NS5A (58 kDa), and NS5B (68 kDa). The NS2/NS3 junction is cleaved by a metal-dependent autoprotease through *cis* processing, while the other junctions are cleaved by the N-terminal region of the 631-residue NS3 with both helicase and serine protease activity, through *trans* processing (Bartenschlager et al., 1993). NS4A acts as a cofactor for NS3 and forms a heterodimeric complex with the protease (Tomei et al., 1993). The NS5A is a cytoplasmic phosphoprotein with a role in viral replication, assembly, as well as resistance to interferon, while the NS5B is a membrane associated viral RNA-dependent RNA polymerase, with a role in viral–host interactions (Dustin et al., 2016).

## 2.2 Life Cycle

The HCV virus has developed an elegant solution to its limited repertoire of genes for evading the immune system, replicating and propagating progeny. It does so by appropriating the host cell signaling and lipid pathways to replicate and survive and reengineers the cellular membrane structure to create a membranous birthing platform, as well as supplies the material for formation of the budding virion envelope (Paul et al., 2014).

There are seven main steps in the HCV viral replication cycle (Fig. 14.4). The virus travels to the liver through the blood stream. In the liver, it crosses the fenestrated endothelium of the liver sinusoids and attaches to the basolateral side of the hepatocytes. The attachment is thought to occur by the HCV glycoproteins in the context of host apolipoprotein E (Dubuisson and Cosset, 2014). The initiation of the viral life cycle is triggered by its binding to the glycosoaminoglycans and scavenger receptor SR-B1. Internalization of the virion-lipoprotein

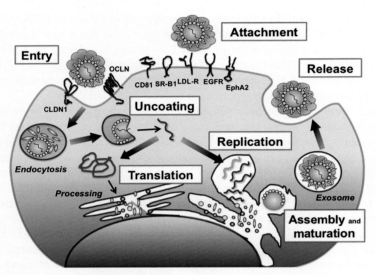

**FIGURE 14.4   HCV replication cycle.** *(Source: Reproduced with permission from Dustin, L.B., Bartolini, B., Capobianchi, M.R. et al., 2016. Hepatitis C virus: life cycle in cells, infection and host response, and analysis of molecular markers influencing the outcome of infection and response to therapy. Clin Microbiol Inf. 22, 826–832. Copyright 2016 Elsevier.)*

occurs by clathrin-mediated endocytosis following a series of complex interactions, with various cellular entry factors, tight-junction proteins, and signaling pathways. The virion then translocate to endosomes with Ras-related GTP binding protein 5 located on the actin stress fibers, where fusion with target membranes takes place. The host Ras-related GTP binding protein 5a is a crucial factor in HCV replication cycle. Subsequently, the viral RNA is released into the cytosol, where cap-independent translation is initiated by the viral IRES (Lukavsky, 2009; Otto and Puglisi, 2004). The single polyprotein is processed and then cleaved into 10 mature proteins by the viral, and host proteases. The nonstructural proteins NS3/4A, NS4B, NS5A/5B form the replication module (Fig. 14.3).

Replication of the virus involves translating the positive stranded RNA through a negative intermediate strand. The positive stranded RNA may be subsequently packaged into virions, with envelope proteins acquired from the endoplasmic reticulum of the host cells (Dustin et al., 2016; Lohmann, 2013). As the virion is assembled, it binds to host lipoproteins to form an envelope such that it is not recognized by the immune system. During the replication process, it also interferes with the host cell's ability to mount an interferon-based defense. Newly assembled virions are released by exocytosis through a secretory pathway. The newly minted lipid coated virions have low buoyant density and range in size from 40 to 80 mm (Niepmann, 2013; Paul et al., 2014).

## 3 STRUCTURE OF DRUG TARGET PROTEINS IN HCV

### 3.1 NS3/4A

Yao et al. (1999) reported the crystal structure of NS3. It is a compact, globular molecule with approximate dimensions of 65 × 65 × 75 Å (Fig. 14.4). The NS3/4A heterodimer consists of six subdomains. The protease domain has a dual β-barrel fold similar to other chymotrypsin-like serine proteases (Yan et al., 1998), while the helicase domain has two structurally similar β-α-β subdomains and a helicase α-helical subdomain consisting of seven helices, and three short β strands. The two protease catalytic sites are covalently linked/separated by a single strand that is exposed to solvent and is flexible to allow for conformational changes of the NS3/4A complex. The helicase active site and RNA interaction site are exposed to solvent, while the NS3 protease active site faces the interior. The substrate binding pocket is shallow, with an amino acid triad of serine-1165, histidine-1083, aspartate-1107 that are critical to protease inhibitor binding (Pizzi et al., 1994). Viral substrates compensate for this by allowing for an extended substrate which can engage in many hydrophobic and electrostatic interactions, which modulate the conformation of both the proximal and distal amino acids in the helicase region. Mutations in the viral genome can confer resistance to protease inhibitors by affecting the binding and conformational changes needed for formation of a favorable induced-fit (Xue et al., 2012).

NS3/4A has low substrate specificity, but membrane-association prior to substrate binding appears to compensate for the low specificity by orienting the substrate for an optimum fit within the binding pocket (Bartenschlager et al., 2013).

### 3.2 HCV N5A

NS5A is a 447-amino acid zinc-metalloprotein with 3 structural domains and an N-terminal amphipathic alpha helix (amino acids 5–25), found in phosphorylated (56 kDa) and hyperphosphorylated forms (58 kDa) (Penin et al., 2004; Tellinghuisen et al., 2004). The α-helix is contained in the N-terminal 30 amino acid sequence and serves to anchor the NS5A protein to the endoplasmic reticulum by intercalating the hydrophobic residues (Trp, Val, Leu, Ile, and Phe) of the helix with the membrane (Tellinghuisen et al., 2005). The hydrophilic residues remain facing the cytoplasm. This monotropic topology resulting in the amphipathic helix to be parallel to the lipid layer is found in other amphipathic α-helices. The binding of the NS5A to the ER occurs through a posttranslational mechanism (Brass et al., 2002). Mutations in the conserved residues at positions 6 and 7, where charged residues were replaced by Ala, or replacement of conserved Trp9 and Cys13 with Ala or addition by insertion of Ala at positions 8 or 11 resulted in conformational changes in the helix, but did not affect membrane anchoring. However, the effect on replication was dramatic, suggesting that the conserved regions may have other critical functions for HCV replication in addition anchoring to the ER membranes (Penin et al., 2004).

Domain I (28–213) is an N-terminus domain separated by a low complexity sequence from domain II (250–342), which is involved in RNA replication. Domain III (amino acids 356–447, located near the C-terminus domain is involved in virus assembly (Tellinghuisen et al., 2005). Both domains II and III appear to be unfolded (Appel et al., 2008). An interferon sensitivity determining region is located at amino acid positions 237–276 in the low complexity sequence and domain 2 regions, whereas an interferon/ribavirin resistance-determining region is located in domain 3 (amino acid positions 362–407) (Nakamoto et al., 2014). Domain I consists of two monomeric units 1a and 1b. Subunit 1a binds zinc through its four cysteine residues while subunit 1b forms a disulfide bond with the C-terminal. Subunits 1a and 1b dimerize and in this form, they facilitate replication. Domain II is involved in replication but appears to have significant sections of nonessential amino-acids. Domain III mediates colocalization of the core and NS5A to lipid droplets, which serve as a template for the formation of virions. Mutations in domain III that interfere with colocalization disrupt virus particulation (Appel et al., 2008).

## 3.3  HCV NS5B

NS5B (66 kDa), a membrane anchored RNA-dependent RNA polymerase, is crucial for HCV RNA synthesis. For enzymatic activity, the active site consisting of a Gly-Asp-Asp sequence needs to bind to magnesium ions (Love et al., 2003). NS5B shares structural similarity with other polymerases, with three distinct domains forming the "fingers", "palm," and "thumb", and a highly hydrophobic membrane anchoring C-terminus (Bressanelli et al., 1999; Ivashkina et al., 2002). Additionally, there is a unique loop like structure (amino acids 12–46) extending between the fingers and thumb that potentially holds the protein in the closed conformation of the active site (Labonté et al., 2002). This closed conformation binds the single-stranded RNA template for initiation of replication. Subsequently, the site expands to allow for binding the double-stranded RNA (Bartenschlager et al., 2013).

A five basic amino acid tunnel of 19 Å appears to be involved in NTP trafficking. One of the inhibitor binding sites has a cleft that is $30 \times 10 \times 10$ Å. The amino-acid positions that constitute the cleft are 374–378, 416–427, 469–490, 493–501, and 524–536. The interface with the inhibitor allows for hydrogen bonding. The binding site has a high number of conserved amino acid residues. NS5B-inhibitor binding appears to mediate HCV inhibition through multiple pathways: altering protein conformation and interference with RNA binding (Labonté et al., 2002).

## 3.4  Internal Ribosomal Entry Site (IRES)

Current therapeutic options derive heavily from the experience of treating HIV infections, thus, most drugs are protease inhibitors and target the replication steps mediated by NS3A/4, NS5A, and NS5B. When individual drugs fail to clear the virus, either due to polymorphisms, low potency, or low concentrations or mutations, a combination of drugs has been employed quite successfully and

SAR rates have increased to be greater than 90%. However, as discussed previously, these targets are hypermutable and hence, resistance develops rapidly. The highly conserved region of the IRES is being evaluated as a druggable target. In the IRES domain, 83 folded consensus secondary structures have been reported (Zhao and Wimmer, 2001).

The IRES consists of four stem-loops and a pseudoknot structure at the 5′ untranslated region of the HCV genome. Loops I and II are involved in replication whereas loops II–IV are involved in RNA translation in combination with some core coding nucleotides. Loop III is the largest loop and has been further subclassified into loops IIIa–IIIf. The apical part of loop II and loop IV may potentially interact with their complementary nucleotide sequences on either loop. The hairpin shaped apical section of loop IIIb binds eIF3, while the remainder of loop III,loops II, and IV constitute a double pseudoknot structure that binds the ribosomal 40S subunit in a stretched conformation (Spahn et al., 2001). The binding of HCV RNA to ribosomal subunits appears to be dependent on free cytosolic $Mg^{2+}$ at average concentration of 0.65 mM (Niepmann, 2013). To initiate translation, the HCV IRES requires 3 initiation factors, eIF1-3. Yamamoto et al. (2015) have described the structure of bound IRES and complex conformations during initiation of translations in great detail. Overall, they report a modular structure that adopts various different conformations to facilitate binding to the flat surface of the 40s ribosomal subunit, as well as binding the tRNA to the P-site. Benzimidazole works by preventing productive interaction of IRES to the 40S subunit as it freezes the domain II in the stretched position, thereby preventing the conformational changes needed for initiation of translation. Thus maintaining a flexible domain that allows optimal docking to the ribosomal subunits and induce conformational changes to bind tRNA are critical to maintaining HCV viability. Alternatively, drugs that interfere with various conformational changes that are essential to allow binding and initiation of translation represent a new approach for the treatment of hepatitis C. Several antisense oligonucleotides and aptamers targeting different sections of the IRES domains have shown activity in vitro and in animal studies (Davis and Seth, 2011).

# 4    DRUGS AGAINST HCV INFECTION

## 4.1    Interferon and Ribavirin

IFN-α was the first approved treatment for hepatitis C, but had low response rates of only about 6%–12%, for 6-month treatment duration (Hoofnagle et al., 1986). Pegylated interferons were later approved and allowed for less frequent dosing (Glue et al., 2000). The addition of the broad-spectrum antiviral agent, ribavirin, to the treatment regimen improved the response rate to about 35%–40% (McHutchison and Poynard, 1999). Exogenously administered interferons mimic the action of endogenous interferons in exerting antiviral activity through their effects on signaling and intracellular gene activation, as well as direct immunomodulatory effect (Feld and Hoofnagle, 2005).

By binding to specific cell surface receptors, they activate the Janus-activated kinase 1 (Jak1) and tyrosine kinase 2 (Tyk2) proteins, which phosphorylate the signal transducer and activator of transcription proteins 1 and 2 (STAT1 and STAT2). These translocate to the nucleus where the STAT1/2 complex combines with the interferon-regulatory factor 9 leading to the expression of multiple interferon-stimulated genes, many of which have antiviral activity (Zhu et al., 2003). INF-α also appears to affect viral RNA synthesis and assembly of replication complexes, as well as IRES-induced translation (Guo et al., 2004). IFN-α also has a more direct effect on the host immune system by activating natural killer cells and triggering the innate immune response to efficiently recognize and destroy the HCV-infected cells. In KIR2DL3+ patients who are homozygous for the *HLA-C1* alleles, INF-α induced HCV clearance is particularly effective, presumably because these patients have a lower activation threshold for natural killer cells (Rehermann, 2015). However, there is a delicate balance between the antiviral effect and the HCV effects on host immune function, and gene expression. Not only does the virus evade the immune system, possibly through creating a coat using host lipoproteins, but it also subverts the host immune system, protein synthesis, and signaling pathways to survive (Feld and Hoofnagle, 2005).

IFN-α sensitivity is NS5A region based, conferred by unique amino acid difference at codon 2218 and clusters of differences in codons 2154–2383 of the NS5A C-terminal region. Codons 2209–2248 appear to form the "sensitivity determining" region, with missense mutations occurring in this region in individuals that are resistant to IFN-α (Enomoto et al., 1995). Complete clinical responses (absence of HCV RNA) were observed in patients with 1–3 (intermediate type) or 4–11 (mutant type) amino acid substitutions in NS5A2209-2248, but no complete response was observed in patients with wild-type sequence (Enomoto et al., 1996). The response rate was higher in patients with more mutations in this region. The same region was also sensitive to IFN-β (Kurosaki et al., 1997). Thus, IFN-α may exert its antiviral effects through its effect on the host, as well as the virus.

Ribavirin, is a nucleoside analog, first discovered in 1970. In vitro, it showed broad spectrum activity as an antiviral. When dosed alone, although it showed improvement in liver function of patients, it failed to clear the virus (Di Bisceglie et al., 1995). In combination with IFN-α, it was effective in clearing the virus and improving response rates. Ribavirin does not appear to directly inhibit the HCV replication, but appears to have pleiotropic effects on infectability and clearance of HCV. Ribavarin appears to tilt the balance toward an "error catastrophe" by inducing mutations in all regions of the HCV genome, with significant mutations in the E1/E2 region affecting infectability and mutations in the NS5B region affecting viability (Contreras et al., 2002). Additionally, ribavirin appears to selectively boost $T_H1$ mediated clearance of HCV from the cell (Tam et al., 1999). Thus a combination of ribavirin and IFN-α appears to act synergistically to increase the error rate during replication, while also enhancing immune recognition and clearance of the HCV (Lanford et al., 2003).

## 4.2   Direct Acting Antivirals

The NS3 protein along with the NS4A peptide cofactor is a domain-specific helicase and serine protease that serves, with critical role in the cytosolic replication of the HCV virus (Yao et al., 1999). NS3/4A protease is also involved in evasion of the host innate response, and impairs host cell signaling through Toll-like receptor and mitochondrial antiviral signaling, both of which are involved in endogenous interferon production.

Thus NS3/4A inhibitors have been developed to use either alone or in combination with other drugs to overcome resistance and provide a wider genotype coverage. The first generation NS3/4A protease inhibitor drugs were boceprevir and telaprevir approved for treating HCV GT-1 infections. These drugs significantly improved upon the response rates of interferon/ribavirin therapy. However, they have a low genetic barrier for resistant variants. Second generation NS3/4A inhibitors include asunaprevir (approved in Japan), paritaprevir, simeprevir, vaniprevir and grazoprevir. These have improved potency but continue to suffer from a low genetic barrier to resistance (Ahmed and Felmlee, 2015).

Inhibitors of NS34A include linear peptidomimetics with a ketoamide moiety that interacts with the catalytic serine to form a reversible enzyme–inhibitor adduct, which decouples at a slow rate. Examples include telaprevir and boceprevir. Other drugs include noncovalent binding linear peptidomimetics and macrocyclic compounds, such as simeprevir and grazoprevir, which have improved resistance barrier and pan-genotypic activity (Bartenschlager et al., 2013; Cummings et al., 2010).

Approved HCV NS5A inhibitors include daclatasvir, ledipasvir, ombitasvir, and elbasvir (de Clercq and Li, 2016). Daclatasvir is a highly effective drug that works by blocking the dimerization of NS5A (Bartenschlager et al., 2013). Although the exact function or mechanism of action of NS5A remains unknown, inhibitors of NS5A block HCV replication by disrupting the formation of a membranous web in the cytoplasm, where HCV replication occurs (Romero-Brey et al., 2012). Approved nonnucleoside NS5B inhibitors include sofosbuvir and dasabuvir.

More recently, combination therapy has improved cure rates of HCV dramatically for all genotypes. A combination of sofosbuvir and velpatasvir was approved by the Food and Drug Administration (FDA) in June 2016 (Epclusa Approval Letter, 2016). This is the first therapy that treats all HCV genotypes and represents the first drug to be approved for GTs 2 and 3, without combination with ribavirin. After 12 months of therapy, pooled analyses showed 98% of patients achieved a SVR (not detectable HCV in blood), with 99% each for GT-1 and GT-2, 95% for GT-3, 100% each for GT-4 and GT-6, and 97% for GT-5 (Epclusa Cross Discipline Review, 2016).

Additional drug targets have been reported (Bartenschlager et al., 2013) but do not currently form part of routine HCV treatment modality.

## 5 TREATMENT FAILURE; CAUSES AND SOLUTIONS

Treatment failure for HCV can result from a multitude of reasons. The major reasons include drug-resistant viral variants, impact of quasispecies and genotype subtypes, noncompliance/interrupted treatment, intrinsic patient characteristics, such as cirrhosis, and extrinsic factors, such as prior therapy. With the availability of various direct acting agents and with combination therapy, many of these reasons are mitigated. Drug resistance to HCV drugs is a major issue in trying to eliminate the virus from the patient. The mechanisms involved in drug resistance can be manifold and be based on patient, as well as viral characteristics. Resistance may result in decreased intracellular drug concentrations, changes in binding pocket structure due to mutations, changes in gene expression that alter the susceptibility of the virus, etc.(Hughes and Andersson, 2015).

## 6 WHAT MAKES TREATING HCV UNIQUELY CHALLENGING: A BAG OF TRICKS

Viruses and bacteria have been around for a very long time. Despite a small genome, they demonstrate a remarkable ability to evade death. In part, their evolutionary survival tactic has been sloppy replication of the genome. The HCV virus has very few conserved structures critical to the survival and propagation of the virus. It has also developed tricks to hijack the host's lipoproteins to form a cloak and evade the immune system, as well as disrupt signaling pathways within the host cell that would recognize and clear the virus.

The HCV has a high replication rate ($10^{12}$ new virions per day) along with an error prone RNA-dependent essential RNA polymerase NS3/4A (error rate of $10^{-5}$–$10^{-4}$ per nucleotide). This results in the generation of heterogeneous populations of wild-type and mutant HCV (about 14.6% within genotype) in infected patients, resulting in quasispecies of HCV (de Clercq and Li, 2016). Mutations occur frequently and under selection pressure in the presence of protease inhibitors, resistance develops rapidly to antiviral agents (Rong et al., 2010). The presence of a heterogeneous population (quasispecies) offers a distinct evolutionary advantage to the virus as at any given time, there is a large number of variants pools available for replication and propagation in the presence of unfavorable host conditions that exist during antiviral therapy.

## 7 STRUCTURAL TRICKS TO RESIST HCV INHIBITORS

HCV NS3/4A has been an important drug target due to its critical role in viral replication. The NS3/4A binding pocket is shallow with two structural domains (Howe and Venkatraman, 2013) (Fig.14.5). The crystal structure of protease-substrate complexes indicates a conserved structure for the volume and "substrate envelope" (Ozen et al., 2013). Molecules that bind in the space between the helicase and serine protease domains can cause allosteric conformational change, and freeze the protein in the inactive closed conformation. The key amino-acids involved in inhibitor–protease/helicase interactions have been identified to be His57, Asp79, Asp81, Asp168, Met485, Cys525, and Asp527 (Xue et al., 2014).

Mutations in this region would be lethal to the virus (Romano et al., 2010). Thus the virus has developed an elegant solution by allowing mutations in the area outside the binding pocket, which weaken the inhibitor binding but spare the substrate binding capability of the domain. Inhibitors that fit completely within the substrate envelope would not develop resistance, whereas inhibitors that protrude from the substrate pocket will have a higher susceptibility to resistance. Natural substrates of NS3/4A appear to interact strongly with strands 135–139 and154–160 (Guan et al., 2014). Thus inhibitors that mimic natural substrates are more likely to resist loss of activity. Both linear and macrocyclic compounds have activity as NS3/4A protease inhibitors, although the macrocyclic compounds appear to be more potent. The location of the macrocycle was determined to be critical and determined if a drug molecule could avoid HCV resistance. P1–P3 macrocycles with a flexible P2 moiety formed optimal conformation to allow contact with the invariable triad. These are most likely to retain potency (Ali et al., 2013).

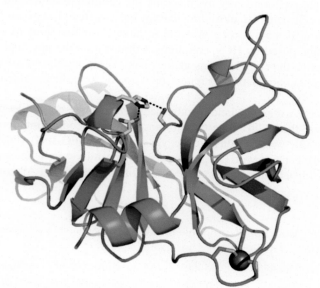

**FIGURE 14.5** **HCV NS3/4A crystal structure.** Ribbon diagram—two barrel subdomains; ball-and-stick: His57, Asp81, Ser139 forming the catalytic triad; blue-sphere: protease structural zinc ion; light orange—NS4A peptide. *(Source: Reproduced with permission from Howe, A.Y., Venkatraman, S., 2013. The discovery and development of boceprevir: a novel, first generation inhibitor of the hepatitis C virus NS3/4A serine protease. J. Clin.Transl. Hepatol. 1, 22–32. Copyright 2013 The second affiliated Hospital of Chongqing Medical University.)*

## 8 MUTATIONS AND POLYMORPHISMS TO TRICK HCV INHIBITORS

The HCV has a high replication rate coupled with a lack of proofreading capability that introduces mutations and results in a heterogeneous HCV population. Under selection pressure from inhibitor drugs, resistant HCV populations emerge. Many of these have high replication rates and they can crowd out the wild type of HCV rapidly. Some of these may have lower level of replicative fitness, but the lower sensitivity to drugs keeps them viable, until the selection pressure is removed and the wild type of HCV can recolonize. The hypermutability is one of the major tools that the HCV has in its armament. Most mutations are located in the drug target regions of the virus (Table 14.1).

Mutations confer resistance through a number of different mechanisms: (1) inhibitor-binding site interactions are attenuated while substrate binding is retained, (2) compensatory conformational changes in the binding pocket following inhibitor binding, (3) inhibitor fit in the active site is mutated, and (4) mutations may occur at distal locations to compensate for conformational distortion by an inhibitor (Kurt Yilmaz et al., 2016; Stross et al., 2016).

Designing drug molecules that can be effective with an ever-changing mix of mutations becomes particularly challenging as the HCV virus can function and replicate predominantly by hijacking the cellular structure and function of the host cell. Thus far, most drug therapy has targeted viral genes, however, in the future, it may also be possible to target host functions to overcome the challenge posed by HCV mutations. Combination therapy appears to be more effective than single drugs as they can exert a multipronged attack on the HCV replication pathways rendering it unable to replicate, assemble, and infect.

A genome-wide association study identified a single polynucleotide polymorphism in the IL28B gene on chromosome 19, encoding interferon-lambda-3 (INF-λ3), with response to pegylated-IFN-α (PEG-IFN) (Stättermayer et al., 2014). Patients with CC genotype at this position were associated with a twofold or higher improvement in SVR in Europeans, African-Americans, and

**TABLE 14.1 Mutations Responsible for Resistance Against DAAs (Dustin et al., 2016)**

| Class of HCV inhibitors | Mutations (AA positions related to resistance) |
| --- | --- |
| NS3 | 36, 41, 43, 54, 55, 56, 80, 122, 155, 156, 168, 170, 175 |
| NS5A | 159, 282, 316, 320, 321, 368, 414, 419, 421, 446, 448, 495, 553, 554, 556, 558, 559, 561 |
| NS5B | 24, 28, 30, 31, 32, 38, 58, 93 |

Hispanics. This polymorphism is associated with spontaneous HCV clearance, as well as exogenous interferon-dependent HCV clearance. The response rate was even better in East Asians. The SVR is highly correlated with the frequency of rs12979860 C-allele frequency in these groups, with higher frequency resulting in much higher SVRs (Ge et al., 2009).

Breakthrough mutations have been reported in patients receiving DAAs. These mutations occur more frequently in the presence of a selection pressure. Due to errors during replication, breakthrough mutations are often lost and replaced by the original primary variants. This property can be exploited to treat patients with breakthrough variants by switching HCV inhibitor drugs or using a combination of drugs. In patients who were treated with telaprevir but failed to achieve SVR, wild-type variants replaced breakthrough variants within three weeks for GT-1b and within 10 months for GT-1a (Ahmed and Felmlee, 2015). A few examples of breakthrough mutations associated with DAAs are listed in Table 14.2.

Genetic analysis has been helpful in identifying many mutations that may make a particular drug ineffective in certain populations. Concordance with clinical data has also been demonstrated. Many of these mutations are more prevalent in a particular geographical area and also in specific genotypes. Vidal et al. (2016) have identified a number of mutations against NS3 protease inhibitor drugs and also geographical prevalence of certain resistance associated variants. This information can be very helpful in selecting appropriate drugs to treat HCV in patients within a region or with certain genotypes, thus, rapidly clearing the HCV and also limiting the emergence of new mutations. For example, approximately 46% of HCV subtype 1a sequences in North America harbor the Q80K mutation that confers resistance to simeprevir, the prevalence of this mutation is very low in countries, such as Brazil and China. Thus in countries with a low rate of the drug resistant strain, use of simeprevir may result in superior treatment response than in North America (Vidal et al., 2016).

**TABLE 14.2** Examples of DAA HCV Inhibitors and Associated Breakthrough Mutations That Confer Resistance to the Corresponding Class of Drugs (Ahmed and Felmlee, 2015)

| HCV target protein | HCV inhibitors | Breakthrough mutations |
|---|---|---|
| NS3/4A | Telaprevir, boceprevir, simeprevir, faldaprevir, asunaprevir | Q80K, V36AM, T54A/S, R155K/T, A156S/T, D168N |
| NS5A | Daclatasvir, ledipasvir | M28, Q30, L31V, Y93H |
| NS5B | Sofosbuvir, dasabuvir | S282T, V321I, L159F |

## 9 CONCLUSIONS

HCV is a chronic disease with significant clinical burden, if left untreated. The virus represents a therapeutic challenge due to the presence of multiple genotypes and subtypes. Additionally, the high replication rate and errors in replication result in quasispecies, that can develop multidrug resistance rapidly through mutations. Developing a vaccine for HCV is challenging due to the genetic diversity and emergence of drug-resistance mutations. Despite the recent identification of HCV, it has been extensively studied. Within a short period of time, we have gone from an untreatable disease to almost 100% cure rates, with combination therapy with direct antiviral drugs.

## REFERENCES

Ahmed, A., Felmlee, D.J., 2015. Mechanisms of hepatitis C viral resistance to direct acting antivirals. Viruses 7, 6716–6729.

Ali, A., Aydin, C., Gildemeister, R., Romano, K.P., Cao, H., Ozen, A., Soumana, D., Newton, A., Petropoulos, C.J., Huang, W., Schiffer, C.A., 2013. Evaluating the role of macrocycles in the susceptibility of hepatitis C virus NS3/4A protease inhibitors to drug resistance. ACS Chem. Biol. 8, 1469–1478.

Appel, N., Zayas, M., Miller, S., Krijnse-Locker, J., Schaller, T., Friebe, P., Kallis, S., Engel, U., Bartenschlager, R., 2008. Essential role of domain III of nonstructural protein 5A for hepatitis C virus infectious particle assembly. PLoS Pathog. 4, e1000035.

Bartenschlager, R., Ahlborn-Laake, L., Mous, J., Jacobsen, H., 1993. Nonstructural protein 3 of the hepatitis C virus encodes a serine-type proteinase required for cleavage at the NS3/4 and NS4/5 junctions. J. Virol. 67, 3835–3844.

Bartenschlager, R., Lohmann, V., Penin, F., 2013. The molecular and structural basis of advanced antiviral therapy for hepatitis C virus infection. Nat. Rev. Microbiol. 11, 482–496.

Brass, V., Bieck, E., Montserret, R., Wölk, B., Hellings, J.A., Blum, H.E., Penin, F., Moradpour, D., 2002. An amino-terminal amphipathic alpha-helix mediates membrane association of the hepatitis C virus nonstructural protein 5A. J. Biol. Chem. 277, 8130–8139.

Bressanelli, S., Tomei, L., Roussel, A., Incitti, I., Vitale, R.L., Mathieu, M., De Francesco, R., Rey, F.A., 1999. Crystal structure of the RNA-dependent RNA polymerase of hepatitis C virus. Proc. Natl. Acad. Sci. USA 96, 13034–13039.

Bukh, J., 2016. The history of hepatitis C virus (HCV): asic research reveals unique features in phylogeny, evolution and the viral life cycle with new perspectives for epidemic control. J. Hepatol. 65 (1 Suppl), S2–S21.

Centers for Disease Control and Prevention. Viral hepatitis-hepatitis C information, 2015. Available from: http://www.cdc.gov/hepatitis/hcv/hcvfaq.htm#section1

Chevaliez, S., Pawlotsky, J.M., 2006. HCV Genome and Life Cycle. In: Tan, S.L. (Ed.), Hepatitis C Viruses: Genomes and Molecular Biology. Horizon Bioscience, Norfolk (UK), Chapter 1. Available from: https://www.ncbi.nlm.nih.gov/books/NBK1630/.

Choo, Q.L., Kuo, G., Weiner, A.J., Overby, L.R., Bradley, D.W., Houghton, M., 1989. Isolation of a cDNA clone derived from a blood-borne nonA, nonB viral hepatitis genome. Science 244, 359–362.

Collier, A.J., Gallego, J., Klinck, R., Cole, P.T., Harris, S.J., Harrison, G.P., Aboul-Ela, F., Varani, G., Walker, S., 2002. A conserved RNA structure within the HCV IRES eIF3-binding site. Nat. Struct. Biol. 9, 375–380.

Contreras, A.M., Hiasa, Y., He, W., Terella, A., Schmidt, E.V., Chung, R.T., 2002. Viral RNA mutations are region specific and increased by ribavirin in a full-length hepatitis C virus replication system. J. Virol. 76, 8505–8517.

Cummings, M.D., Lindberg, J., Lin, T.I., de Kock, H., Lenz, O., Lilja, E., Felländer, S., Baraznenok, V., Nyström, S., Nilsson, M., Vrang, L., Edlund, M., Rosenquist, A., Samuelsson, B., Raboisson, P., Simmen, K., 2010. Induced-fit binding of the macrocyclic noncovalent inhibitor TMC435 to its HCV NS3/NS4A protease target. Angew. Chem. (Int. Ed. English) 49, 1652–1655.

Davis, D.R., Seth, P.P., 2011. Therapeutic targeting of HCV internal ribosomal entry site RNA. Antivir. Chem. Chemother. 21, 117–128.

De Clercq, E., Li, G., 2016. Approved Antiviral Drugs over the Past 50 Years. Clin. Microbiol. Rev. 29, 695–747.

Di Bisceglie, A.M., Conjeevaram, H.S., Fried, M.W., Sallie, R., Park, Y., Yurdaydin, C., Swain, M., Kleiner, D.E., Mahaney, K., Hoofnagle, J.H., 1995. Ribavirin as therapy for chronic hepatitis C. A randomized, double-blind, placebo-controlled trial. Ann. Intern. Med. 123, 897–903.

Dubuisson, J., Cosset, F.L., 2014. Virology and cell biology of the hepatitis C virus life cycle: an update. J. Hepatol. 61 (1 Suppl), S3–S13.

Dustin, L.B., Bartolini, B., Capobianchi, M.R., Pistello, M., 2016. Hepatitis C virus: life cycle in cells, infection and host response, and analysis of molecular markers influencing the outcome of infection and response to therapy. Clin. Microbiol. Inf. 22 (10), 826–832.

Enomoto, N., Sakuma, I., Asahina, Y., Kurosaki, M., Murakami, T., Yamamoto, C., Izumi, N., Marumo, F., Sato, C., 1995. Comparison of full-length sequences of interferon-sensitive and resistant hepatitis C virus 1b. Sensitivity to interferon is conferred by amino acid substitutions in the NS5A region. J. Clin. Invest. 96, 224–230.

Enomoto, N., Sakuma, I., Asahina, Y., Kurosaki, M., Murakami, T., Yamamoto, C., Ogura, Y., Izumi, N., Marumo, F., Sato, C., 1996. Mutations in the nonstructural protein 5A gene and response to interferon in patients with chronic hepatitis C virus 1b infection. N. Engl. J. Med. 334, 77–81.

Epclusa Approval Letter, 2016. US Food and Drug Administration. Available from: http://www.accessdata.fda.gov/drugsatfda_docs/appletter/2016/208341Orig1s000ltr.pdf

Epclusa Cross Discipline Team Leader Review, 2016. US Food and Drug Administration. Available from: http://www.accessdata.fda.gov/drugsatfda_docs/nda/2016/208341Orig1s000CrossR.pdf

Feld, J.J., Hoofnagle, J.H., 2005. Mechanism of action of interferon and ribavirin in treatment of hepatitis C. Nature 436, 967–972.

Ge, D., Fellay, J., Thompson, A.J., Simon, J.S., Shianna, K.V., Urban, T.J., Heinzen, E.L., Qiu, P., Bertelsen, A.H., Muir, A.J., Sulkowski, M., McHutchison, J.G., Goldstein, D.B., 2009. Genetic variation in IL28B predicts hepatitis C treatment-induced viral clearance. Nature 461, 399–401.

Glue, P., Fang, J.W., Rouzier-Panis, R., Raffanel, C., Sabo, R., Gupta, S.K., Salfi, M., Jacobs, S., 2000. Pegylated interferon-alpha2b: pharmacokinetics, pharmacodynamics, safety, and preliminary efficacy data. Hepatitis C intervention therapy group. Clin. Pharmacol. Ther. 68, 556–567.

Grakoui, A., Wychowski, C., Lin, C., Feinstone, S.M., Rice, C.M., 1993. Expression and identification of hepatitis C virus polyprotein cleavage products. J. Virol. 67, 1385–1395.

Guan, Y., Sun, H., Li, Y., Pan, P., Li, D., Hou, T., 2014. The competitive binding between inhibitors and substrates of HCV NS3/4A protease: a general mechanism of drug resistance. Antivir. Res. 103, 60–70.

Guo, J.T., Sohn, J.A., Zhu, Q., Seeger, C., 2004. Mechanism of the interferon alpha response against hepatitis C virus replicons. Virol. 325, 71–81.

Hepatitis Central. Medications to treat hepatits C—a timeline, 2016. Available from: http://www.hepatitiscentral.com/medications-to-treat-hepatitis-c-a-timeline/

Hoofnagle, J.H., Mullen, K.D., Jones, D.B., Rustgi, V., Di Bisceglie, A., Peters, M., Waggoner, J.G., Park, Y., Jones, E.A., 1986. Treatment of chronic non-A,non-B hepatitis with recombinant human alpha interferon. A preliminary report. N. Engl. J. Med. 315, 1575–1578.

Howe, A.Y., Venkatraman, S., 2013. The discovery and development of boceprevir: a novel, first generation inhibitor of the hepatitis C virus NS3/4A serine protease. J. Clin.Transl. Hepatol. 1, 22–32.

Hughes, D., Andersson, D.I., 2015. Evolutionary consequences of drug resistance: shared principles across diverse targets and organisms. Nat. Rev. Genet. 16, 459–471.

Ivashkina, N., Wölk, B., Lohmann, V., Bartenschlager, R., Blum, H.E., Penin, F., Moradpour, D., 2002. The hepatitis C virus RNA-dependent RNA polymerase membrane insertion sequence is a transmembrane segment. J. Virol. 76, 13088–13093.

Jazwinski, A.B., Muir, A.J., 2011. Direct-acting antiviral medications for chronic hepatitis C virus infection. Gastroenterol. Hepatol. 7, 154–162.

Kurosaki, M., Enomoto, N., Murakami, T., Sakuma, I., Asahina, Y., Yamamoto, C., Ikeda, T., Tozuka, S., Izumi, N., Marumo, F., Sato, C., 1997. Analysis of genotypes and amino acid residues 2209 to 2248 of the NS5A region of hepatitis C virus in relation to the response to interferon-beta therapy. Hepatol. 25, 750–753.

Kurt Yilmaz, N., Swanstrom, R., Schiffer, C.A., 2016. Improving viral protease inhibitors to counter drug resistance. Trends. Microbiol. 24, 547–557.

Labonté, P., Axelrod, V., Agarwal, A., Aulabaugh, A., Amin, A., Mak, P., 2002. Modulation of hepatitis C virus RNA-dependent RNA polymerase activity by structure-based site-directed mutagenesis. J. Biolo. Chem. 277, 38838–38846.

Lanford, R.E., Guerra, B., Lee, H., Averett, D.R., Pfeiffer, B., Chavez, D., Notvall, L., Bigger, C., 2003. Antiviral effect and virus-host interactions in response to alpha interferon, gamma interferon, poly(i)-poly(c), tumor necrosis factor alpha, and ribavirin in hepatitis C virus subgenomic replicons. J. Virol. 77, 1092–1104.

Lohmann, V., 2013. Hepatitis C virus RNA replication. Curr. Top. Microbiol. Immunol. 369, 167–198.

Love, R.A., Parge, H.E., Yu, X., Hickey, M.J., Diehl, W., Gao, J., Wriggers, H., Ekker, A., Wang, L., Thomson, J.A., Dragovich, P.S., Fuhrman, S.A., 2003. Crystallographic identification of a noncompetitive inhibitor binding site on the hepatitis C virus NS5B RNA polymerase enzyme. J. Virol. 77, 7575–7581.

Lukavsky, P.J., 2009. Structure and function of HCV IRES domains. Virus Res. 139, 166–171.

McHutchison, J.G., Poynard, T., 1999. Combination therapy with interferon plus ribavirin for the initial treatment of chronic hepatitis C. Semin. Liver Dis. 19 (Suppl. 1), 57–65.

Nakamoto, S., Kanda, T., Wu, S., Shirasawa, H., Yokosuka, O., 2014. Hepatitis C virus NS5A inhibitors and drug resistance mutations. World J. Gastroenterol. 20, 2902–2912.

Neddermann, P., Tomei, L., Steinkuhler, C., Gallinari, P., Tramontano, A., De Francesco, R., 1997. The nonstructural proteins of the hepatitis C virus: structure and functions. J. Biol. Chem. 378, 469–476.

Niepmann, M., 2013. Hepatitis C virus RNA translation. Curr. Top. Microbiol. Immunol. 369, 143–166.

Otto, G.A., Puglisi, J.D., 2004. The pathway of HCV IRES-mediated translation initiation. Cell 119, 369–380.

Ozen, A., Sherman, W., Schiffer, C.A., 2013. Improving the resistance profile of hepatitis C NS3/4A inhibitors: dynamic substrate envelope guided design. J. Chem. Theory. Comput. 9, 5693–5705.

Paul, D., Madan, V., Bartenschlager, R., 2014. Hepatitis C virus RNA replication and assembly: living on the fat of the land. Cell Host Microbe 16, 569–579.

Penin, F., Brass, V., Appel, N., Ramboarina, S., Montserret, R., Ficheux, D., Blum, H.E., Bartenschlager, R., Moradpour, D., 2004. Structure and function of the membrane anchor domain of hepatitis C virus nonstructural protein 5A. J. Biol. Chem. 279, 40835–40843.

Pharmaceutical Research and Manufacturers of America (PhRMA). Twenty-five years of progress against hepatitis C: setbacks and stepping stones, 2014. Available from: http://www.phrma.org/sites/default/files/pdf/Hep-C-Report-2014-Stepping-Stones.pdf

Pizzi, E., Tramontano, A., Tomei, L., La Monica, N., Failla, C., Sardana, M., Wood, T., De Francesco, R., 1994. Molecular model fo the specificity pocket of the hepatitis C virus protease: Implications for substrate recognition. Proc. Natl. Acad. Sci. USA 91, 888–892.

Rehermann, B., 2015. Natural killer cells in viral hepatitis. Cell. Mol. Gastroenterol. Hepatol. 1, 578–588.

Romano, K.P., Ali, A., Royer, W.E., Schiffer, C.A., 2010. Drug resistance against HCV NS3/4A inhibitors is defined by the balance of substrate recognition versus inhibitor binding. Proc. Natl. Acad. Sci. USA 107, 20986–20991.

Romero-Brey, I., Merz, A., Chiramel, A., Lee, J.Y., Chlanda, P., Haselman, U., Santarella-Mellwig, R., Habermann, A., Hoppe, S., Kallis, S., Walther, P., Antony, C., Krijnse-Locker, J., Bartenschlager, R., 2012. Three-dimensional architecture and biogenesis of membrane structures associated with hepatitis C virus replication. PLoS Pathog. 8, e1003056.

Rong, L., Dahari, H., Ribeiro, R.M., Perelson, A.S., 2010. Rapid emergence of protease inhibitor resistance in hepatitis C virus. Sci. Transl. Med. 2, 30ra32.

Spahn, C.M., Kieft, J.S., Grassucci, R.A., Penczek, P.A., Zhou, K., Doudna, J.A., Frank, J., 2001. Hepatitis C virus IRES RNA-induced changes in the conformation of the 40s ribosomal subunit. Science 291, 1959–1962.

Stättermayer, A.F., Scherzer, T., Beinhardt, S., Rutter, K., Hofer, H., Ferenci, P., 2014. Review article: genetic factors that modify the outcome of viral hepatitis. Aliment. Pharmacol. Ther. 39, 1059–1070.

Stross, C., Shimakami, T., Haselow, K., Ahmad, M.Q., Zeuzem, S., Lange, C.M., Welsch, C., 2016. Natural HCV variants with increased replicative fitness due to NS3 helicase mutations in the C-terminal helix α18. Sci. Rep. 6, 19526.

Tam, R.C., Pai, B., Bard, J., Lim, C., Averett, D.R., Phan, U.T., Milovanovic, T., 1999. Ribavirin polarizes human T cell responses towards a Type 1 cytokine profile. J. Hepatol. 30, 376–382.

Tellinghuisen, T.L., Marcotrigiano, J., Gorbalenya, A.E., Rice, C.M., 2004. The NS5A protein of hepatitis C virus is a zinc metalloprotein. J. Biol. Chem. 279, 48576–48587.

Tellinghuisen, T.L., Marcotrigiano, J., Rice, C.M., 2005. Structure of the zinc-binding domain of an essential component of the hepatitis C virus replicase. Nature 435, 374–379.

Tomei, L., Failla, C., Santolini, E., De Francesco, R., La Monica, N., 1993. NS3 is a serine protease required for processing of hepatitis C virus polyprotein. J. Virol. 67, 4017–4026.

Vidal, L.L., Soares, M.A., Santos, A.F., 2016. NS3 protease polymorphisms and genetic barrier to drug resistance of distinct hepatitis C virus genotypes from worldwide treatment-naïve subjects. J Viral Hepatol. 23, 840–849.

World Health Organization (WHO) Hepatitis C Factsheet, 2016. Available from: http://www.who.int/mediacentre/factsheets/fs164/en/

Xue, W., Pan, D., Yang, Y., Liu, H., Yao, X., 2012. Molecular modeling study on the resistance mechanism of HCV NS3/4A serine protease mutants R155K, A156V and D168A to TMC435. Antivir. Res. 93, 126–137.

Xue, W., Yang, Y., Wang, X., Liu, H., Yao, X., 2014. Computational study on the inhibitor binding mode and allosteric regulation mechanism in hepatitis C virus NS3/4A protein. PLoS One 9, e87077.

Yamamoto, H., Collier, M., Loerke, J., Ismer, J., Schmidt, A., Hilal, T., Sprink, T., Yamamoto, K., Mielke, T., Bürger, J., Shaikh, T.R., Dabrowski, M., Hildebrand, P.W., Scheerer, P., Spahn, C.M., 2015. Molecular architecture of the ribosome-bound Hepatitis C Virus internal ribosomal entry site RNA. The EMBO J. 34, 3042–3058.

Yan, Y., Li, Y., Munshi, S., Sardana, V., Cole, J.L., Sardana, M., 1998. Complex of NS3 protease and NS4A peptide of BK strain hepatitis C virus: a 2.2 A resolution structure in a hexagonal crystal form. Protein Sci. 7, 837–847.

Yao, N., Reichert, P., Taremi, S.S., Prosise, W.W., Weber, P.C., 1999. Molecular views of viral polyprotein processing revealed by the crystal structure of the hepatitis C virus bifunctional protease-helicase. Structure 7, 1353–1363.

Zhao, W.D., Wimmer, E., 2001. Genetic analysis of a poliovirus/hepatitis C virus chimera: new structure for domain II of the internal ribosomal entry site of hepatitis C virus. J. Virol. 75, 3719–3730.

Zhu, H., Zhao, H., Collins, C.D., Eckenrode, S.E., Run, Q., McIndoe, R.A., Crawford, J.M., Nelson, D.R., She, J.X., Liu, C., 2003. Gene expression associated with interferon alfa antiviral activity in an HCV replicon cell line. Hepatol. 37, 1180–1188.

# Index

## A

Abraham's molar refractivity, 461
Acquired immunodeficiency syndrome (AIDS), 221
Acute flaccid paralysis (AFP), 265
Acyl dihydrouracils, palm site I inhibitors, 204
Acyl-enzyme hydrolysis, 446
Acyl enzyme intermediate, 111
Adenoviridae, 60
  *Atadenovirus*, 60
  *Aviadenovirus*, 60
  *Ichtadenovirus*, 60
  *Mastadenovirus*, 60
  *Siadenovirus*, 60
Adenovirus protease (AVP), 59
  activation pathway, 64
  AVP–pVIc complex, 63
  crystal structure of, 65, 67
  inhibitors, 64–72
  life cycle, 60
  maturation protease, 61–63
  structure of, 63–64
  substrates of, 62
  L1 52/55k substrate, cleavage of, 61
AFP. *See* Acute flaccid paralysis (AFP)
AIDS. *See* Acquired immunodeficiency syndrome (AIDS)
Ala131 residues, 427, 454
Ala153 residues, 427, 454
Alphavirus proteases, 8, 78–79, 107–108
  capsid domains, 109
  cleavage sites of nsP2 from, 126
  CP structural data of genus, 112
  genome, 107
  infection, 124
  life cycle of, 78
  new world alphaviruses, 8
    Eastern equine encephalitis virus (EEEV), 8
    Venezuelan equine encephalitis virus (VEEV), 8
    Western equine encephalitis virus (WEEV), 8
  nsP2 crystal structure of genus, 120

nsP2 domains, 108
old world, 8
  O'Nyong–Nyong virus (ONNV), 8
  Ross River virus (RRV), 8
  Semliki Forest virus (SFV), 8
  polyprotein processing in, 108
  virion structure of, 79
Amino acid residues, 416
Aminobenzoxazinones, 454
Amprenavir (APV), 231
Anilide-based SARS-CoV 3CL$^{pro}$ inhibitors
  biological activity and physicochemical parameters of, 325
  structure of, 329
1-Anilino-8-naphthalene sulfonate (ANS), 421
Anti-*Flavivirus* therapy, 174
Antivirals, broad-acting, 70
  cidofovir, 70
  ganciclovir, 70
  ribavirin, 70
Arg157, 427, 454
Arg166
  mutagenesis of, 420
  in oxyanion hole, 453
Aryl dihydrouracils, palm site I inhibitors, 204
Arylthiomethane, 234
Atazanavir (ATV), 236
ATV. *See* Atazanavir (ATV)
Aura virus capsid protein (AVCP), 105
  active and inactive, comparison of, 113
  monomer, structure of, 113
AVCP. *See* Aura virus capsid protein (AVCP)
AVP. *See* Adenovirus protease (AVP)
Azetidinone inhibitors, 426
Azidothymidine (AZT), 40
  chemical geometry of, 41

## B

Baicalin, chemical structures of, 96
Barmah Forest virus, 77
Benserazide, 48
Benzofuran derivatives, palm site II inhibitors, 205
Benzo-1,2,4-thiadiazine scaffold, HCV-RNA polymerase inhibitors, 200